Paul Griffin

QUANTITATIVE MODELS FOR SUPPLY CHAIN MANAGEMENT

INTERNATIONAL SERIES IN
OPERATIONS RESEARCH & MANAGEMENT SCIENCE

Frederick S. Hillier, Series Editor
Stanford University

Saigal, R. / *LINEAR PROGRAMMING: A Modern Integrated Analysis*

Nagurney, A. & Zhang, D. / *PROJECTED DYNAMICAL SYSTEMS AND VARIATIONAL INEQUALITIES WITH APPLICATIONS*

Padberg, M. & Rijal, M. / *LOCATION, SCHEDULING, DESIGN AND INTEGER PROGRAMMING*

Vanderbei, R. / *LINEAR PROGRAMMING: Foundations and Extensions*

Jaiswal, N.K. / *MILITARY OPERATIONS RESEARCH: Quantitative Decision Making*

Gal, T. & Greenberg, H. / *ADVANCES IN SENSITIVITY ANALYSIS AND PARAMETRIC PROGRAMMING*

Prabhu, N.U. / *FOUNDATIONS OF QUEUEING THEORY*

Fang, S.-C., Rajasekera, J.R. & Tsao, H.-S.J. / *ENTROPY OPTIMIZATION AND MATHEMATICAL PROGRAMMING*

Yu, G. / *OPERATIONS RESEARCH IN THE AIRLINE INDUSTRY*

Ho, T.-H. & Tang, C. S. / *PRODUCT VARIETY MANAGEMENT*

El-Taha, M. & Stidham , S. / *SAMPLE-PATH ANALYSIS OF QUEUEING SYSTEMS*

Miettinen, K. M. / *NONLINEAR MULTIOBJECTIVE OPTIMIZATION*

Chao, H. & Huntington, H. G. / *DESIGNING COMPETITIVE ELECTRICITY MARKETS*

Weglarz, J. / *PROJECT SCHEDULING: Recent Models, Algorithms & Applications*

Sahin, I. & Polatoglu, H. / *Quality, Warranty and Preventive Maintenance*

Tavares, L. V. / *Advanced Models for Project Management*

QUANTITATIVE MODELS FOR SUPPLY CHAIN MANAGEMENT

edited by

Sridhar Tayur
Carnegie Mellon University

Ram Ganeshan and Michael Magazine
University of Cincinnati

Kluwer Academic Publishers
Boston/Dordrecht/London

Distributors for North, Central and South America:
Kluwer Academic Publishers
101 Philip Drive
Assinippi Park
Norwell, Massachusetts 02061 USA
Telephone (781) 871-6600
Fax (781) 871-6528
E-Mail <kluwer@wkap.com>

Distributors for all other countries:
Kluwer Academic Publishers Group
Distribution Centre
Post Office Box 322
3300 AH Dordrecht, THE NETHERLANDS
Telephone 31 78 6392 392
Fax 31 78 6546 474
E-Mail <orderdept@wkap.nl>

 Electronic Services <http://www.wkap.nl>

Library of Congress Cataloging-in-Publication Data

Quantitative models for supply chain management / edited by Sridhar
    Tayur, Ram Ganeshan, and Michael Magazine.
       p.   cm. -- (International series in operations research &
  management science ; 17)
    Includes bibliographical references and index.
    ISBN 0-7923-8344-3
    1. Business logistics--Mathematical models. 2. Business
logistics--Data processing. 3. Inventory control--Mathematical
models. 4. Inventory control--Data processing. I. Tayur, Sridhar
R. II. Ganeshan, Ram. III. Magazine, Michael J. IV. Series.
HD38.5.Q83   1998
658.7--dc21

                                        98-37476
                                            CIP

Printed on acid-free paper.

CONTENTS

1.   **Introduction**   1
Sridhar R. Tayur, Ram Ganeshan
and Michael Magazine

2.   **Optimal Policies and Simulation Based**   7
Optimization for Capacitated Production
Inventory Systems
Roman Kapuscinski and Sridhar R. Tayur

3.   **Service Levels and Tail Probabilities in**   41
Multistage Capacitated Production-Inventory
Systems
Paul Glasserman

4.   **On (R, NQ) Policies Serial Inventory Systems**   71
Fangruo Chen

5.   **Competitive Supply Chain Inventory**   111
Management
Gèrard P. Cachon

6.   **Vehicle Routing and the Supply Chain**   147
Shoshana Anily and Julien Bramel

7.   **Supply Contracts with Quantity**   197
Commitments and Stochastic Demand
Ravi Anupindi and Yehuda Bassok

8.   **Supply Chain Contracting and Coordination**   233
with Stochastic Demand
Martin A. Lariviere

9.   **Designing Supply Contracts: Contract**   269
Type and Information Asymmetry
Charles J. Corbett and Christopher S. Tang

vi

10. **Modeling Supply Chain Contracts: A Review**   299
Andy A. Tsay, Steven Nahmias,
and Narendra Agrawal

11. **Modeling the Impact of Information on Inventories**   337
Ananth V. Iyer

12. **Modeling Impacts of Electronic Data Interchange Technology**   359
Sunder Kekre, Tridas Mukhopadhyay
and Kannan Srinivasan

13. **Business Cycles and Productivity in Capital Equipment Supply Chains**   381
Edward G. Anderson, Jr. and Charles H. Fine

14. **The Bullwhip Effect: Managerial Insights on the Impact of Forecasting and Information on Variability in a Supply Chain**   417
Frank Chen, Zvi Drezner, Jennifer K. Ryan
and David Simchi-Levi

15. **Value of Information Sharing and Comparison with Delayed Differentiation**   441
Srinagesh Gavirneni and Sridhar Tayur

16. **Managing Product Variety: An Operations Perspective**   467
Amit Garg and Hau L. Lee

17. **Retail Inventories and Consumer Choice**   491
Siddharth Mahajan and Garrett J. van Ryzin

18. **The Benefits of Design for Postponement**   553
Yossi Aviv and Awi Federgruen

19. **Stochastic Programming Models for Managing Product Variety**
Jayashankar M. Swaminathan
and Sridhar R. Tayur
585

20. **Global Sourcing Strategies Under Exchange Rate Uncertainty**
Panos Kouvelis
625

21. **Global Supply Chain Management: A Survey of Research and Applications**
Morris A. Cohen and Arnd Huchzermeir
669

22. **Managing Supply Chains in Emerging Markets**
Alan Scheller-Wolf and Sridhar R. Tayur
703

23. **Bottom-Up vs. Top-Down Approaches to Supply Chain Modeling**
Jeremy F. Shapiro
737

24. **Inventory Planning in Large Assembly Supply Chains**
Gerald E. Feigin
761

25. **Managing Inventory for Fashion Products**
Ananth Raman
789

26. **Inventory Control for Joint Manufacturing and Remanufacturing**
E.A. Van der Laan, M. Fleischman,
R. Dekker and L.N. Van Wassenhove
807

27. **A Taxonomic Review of Supply Chain Management Research**
Ram Ganeshan, Eric Jack, Michael J. Magazine
and Paul Stephens
839

Index

1 INTRODUCTION

Sridhar Tayur

Graduate School of Industrial Administration
Carnegie Mellon University

Ram Ganeshan and Michael Magazine

Department of Quantitative Analysis and Operations Management
University of Cincinnati

There is no doubt of the importance of quantitative models and computer based tools in decision making in today's business environment. This is especially true in the rapidly growing area of *supply chain management*. Anticipating this, a small group of researchers began working on some fundamental models in the early part of this decade, and have now been joined by a larger group of academics and industry researchers thus creating, in a very small time, a large body of research motivated by industry practices, current challenges and future expected needs. We thought this may be an appropriate time to edit a book that would, in an unified manner, provide a systematic summary of the large variety of new issues being considered, the new set of models being developed, the new techniques for analysis, and the computational methods that have become available recently. Every field has a "Golden Age" – a period of exciting developments, providing both a highly intellectual environment as well as making a strong impact on industrial practice; the field of Operations Management appears to be enjoying one right now.

This book contains invited chapters with the goal of providing a self-contained, sophisticated research summary – a snapshot at this point of time – in the area of *Quantitative Models for Supply Chain Management*. The broad category of topics (within this restricted focus), and the set of invited authors were selected by us, and the specific chapter topics were selected by the individual authors. We recognize that supply chain management is influenced by advances in other disciplines, such as Marketing, Information Systems, Economics and Logistics, and that many advances are occurring in non-quantitative aspects as well. Thus, we are aware that there will be aspects of supply chain management not covered here, but we believe we have captured many important developments in this exploding field.

The 26 chapters may be divided into 6 categories.

1. *Basic Concepts and Technical Material.*

   The chapters in this category focus on introducing basic concepts, providing mathematical background and validating algorithmic tools to solve operational problems in supply chains.

   Chapter 2, by Roman Kapuscinski and Sridhar Tayur, provides the framework – discrete-time, multi-stage, stochastic, capacitated production-inventory model with Markovian transitions – that pervades through much of the research in this area. After providing the mathematical tools to show the structure of optimal policies, the chapter describes and validates a computational procedure – a simulation-based optimization procedure – that can be used in a wide generality. Chapter 3, by Paul Glasserman, provides quick approximations for costs and service levels using recent results from large deviation theory, specifically based on exponential approximations to tail probabilities of inventory shortfalls. Both chapters 2 and 3 concentrate on base stock policies, primarily from a centralized perspective in the multi-stage setting.

In chapter 4, Fangruo Chen summarizes recent developments on re-order point, order quantity policies in serial systems, both from a centralized (by using echelon stock) and from a de-centralized (by using installation stocks only) perspectives. Key results include an efficient procedure for performance evaluation within these classes, simple algorithms for optimization, numerical findings on the benefits of centralization (a topic further explored in a later category) and the possible optimality of these policies in certain settings.

Chapter 5, by Gerard Cachon, moves away from the "central agent assumption" and explicitly considers de-centralized, multi-agent situations. The primary focus is to investigate whether independent agents, acting in their self interest, will choose policies that are optimal for the whole supply chain. If they do not, mechanisms are proposed that motivate agents to co-ordinate on the supply chain optimal behavior.

Shoshana Anily and Julien Bramel, in chapter 6, provide results from vehicle routing literature where inventory and transportation are integrated. Results include analysis of algorithms as well as practical algorithms usable for industrial size problems.

2. *Supply Contracts.*

In this category, the primary focus is in design and evaluation of supply contracts between independent agents in the supply chain.

Chapter 7, by Ravi Anupindi and Yehuda Bassok, presents a class of contracts based on quantity commitments. They study these contracts in a multi-period, multi-product settings by providing optimal policy structure and experimental studies based an numerical procedures. Some issues of channel co-ordination are explored within this class of contracts. Martin Lariviere continues to investigate channel co-ordination in decentralized supply chains in chapter 8 and focuses on price-only contracts and returns policies.

In chapter 9, Charles Corbett and Christopher Tang study supply contracts where there is an asymmetry in information between independent agents. They study three types of contracts – linear cost with or without side payments and non-linear cost with side payments – and provide insights into optimal behavior under different scenarios within a deterministic demand setting. Chapter 10, by Andy Tsay, Steven Nahmias, and Narendra Agrawal, provides a comprehensive literature review of supply contracts.

3. *Value of Information.*

The chapters in this category explicitly model the effect of information on decision making and on supply chain performance.

In chapter 11, Ananth Iyer studies the impact of information on demand uncertainty on inventories in the supply chain. The main focus is to evaluate the contractual effects of information sharing between the agents on supply chain performance.

Sunder Kekre, Tridas Mukhopadyay, and Kannan Srinivasan study the impact of Electronic Data Interchange (EDI) in chapter 12 by providing a modeling framework as well as through empirical studies using real data from the field.

The next two chapters investigate the Bullwhip effect. In chapter 13, Edward Anderson and Charles Fine examine its impact on the long-term productivity of capital suppliers. In chapter 14, Frank Chen, Zwi Drezner, Jennifer Ryan, and David Simchi-Levi quantify its effects on inventory, propose methods to reduce its negative effects, and demonstrate the value of centralized demand information.

Chapter 15, by Srinagesh Gavirneni and Sridhar Tayur, studies the relative benefits of sharing information and delaying differentiation (a topic studied in detail in the next category) and how the two strategies complement each other if done simultaneously. Experimental studies provide insights as to when information is most valuable, when one strategy is preferred over the other, and the value of combining the two strategies.

4. *Managing Product Variety.*

The chapters in this category analyze the effects of product variety and the different strategies to manage it.

In chapter 16, Amit Garg and Hau Lee review research focusing on lead time reduction strategies (such as production line structuring and quick response) and non-lead time reduction strategies (such as part commonality, postponement and operations re-sequencing).

Siddharth Mahajan and Garrett van Ryzin, in chapter 17, study the effects of consumer choice on operational features (such as inventory, assortment) using various choice models available from the economics and marketing literature.

Chapter 18, by Yossi Aviv and Awi Federgruen, provides an overview of analytical models that provide insights into the benefits of delayed differentiation. This chapter includes a systematic treatment on various types of complicating factors such as capacity limits, inability to hold inventory at some points, correlated demands, and multiple points of differentiation.

In chapter 19, Jayashankar Swaminathan and Sridhar Tayur provide stochastic programming models and effective computational procedures to study inventories of common components, the use of vanilla boxes for postponement, and the effect of assembly task sequencing on operational performance.

5. *International Operations.*

The three chapters in this category provide an overview of research in the emerging area of International Operations.

Panos Kouvelis, in chapter 20, presents an in-depth analysis of global sourcing strategies as operational hedging mechanisms for responding to fluctuating exchange rates. In chapter 21, Morris Cohen and Arnd Huchzermeier review the state-of-the-art of modeling approaches to support the design and management of global supply networks under multiple sources of uncertainty and different types of risk. The focus is on the integration of supply chain network optimization with real options pricing methods. In chapter 22, Alan Scheller-Wolf and Sridhar Tayur explore a model that captures some of the new challenges associated with emerging markets.

6. *Conceptual Issues and New Challenges.*

Chapter 23, by Jeremy Shapiro, provides a framework using mathematical programming techniques to analyze supply chains. Gerald Feigin, in chapter 24, describes a model for analyzing safety stock investment in large assembly supply chains typically found in computer, consumer electronics and automobile industries. Ananth Raman, in chapter 25, reviews papers dealing with issues related to fashion products – products with short life cycles and uncertain demand – and identifies opportunities for further research. In chapter 26, Erwin van der Laan, Moritz Fleishmann, Rommert Dekker and Luk van Wassenhove provide a framework and a quantitative model to study joint manufacturing and re-manufacturing. We close this edited volume with a taxonomic review of supply chain management research in chapter 27, by Ram Ganeshan, Eric Jack, Michael Magazine, and Paul Stephens.

We hope that this edited volume can serve as a graduate text, as a reference for researchers and as a guide for further development of this field.

2

OPTIMAL POLICIES AND SIMULATION-BASED OPTIMIZATION FOR CAPACITATED PRODUCTION INVENTORY SYSTEMS

Roman Kapuscinski

Michigan Business School
University of Michigan
Ann Arbor, MI 48104

Sridhar Tayur

Graduate School of Industrial Administration
Carnegie Mellon University
Pittsburgh, PA 15213

2.1 INTRODUCTION

The motivation for this stream of research has come from problems faced by diverse set of companies, such as IBM, AMD, Allegheny Ludlum, GE, Proctor and Gamble, Westinghouse, Intel, American Standard, McDonald's, and Caterpillar. Smaller local (to Pittsburgh) companies such as Sintermet, Blazer Diamond, ASKO and Northside Packing have also provided several interesting issues to pursue. At the heart of many of the problems is the interaction between demand variability and non-stationarity, available production capacity, holding costs of inventory (at different locations), lead times and desired service levels. The central goal of this research stream is to understand the interactions in simple single and multiple stage settings and to provide insights and implementable solutions for managing inventories in a cost-effective manner for complex systems. The goal of this chapter is to introduce in a systematic manner some recent advances in 'Discrete-time, Capacitated Production-Inventory Systems facing Stochastic Demands' and we limit ourselves to single product setting. The material here is collected from papers that have appeared in the literature: [31, 22, 23, 48, 49].

2.1.1 Quantitative Models for Supply Chain Management

A modern manufacturing network, consisting of multiple manufacturing facilities and several external vendors, can be modeled as a multi-stage, capacitated, assembly system; see Figure 2.1 for a representation of an IBM supply chain. Until 1991, the only major result that was available in *capacitated systems* was the structure of the optimal policy for a single product, single stage system facing a stationary demand process [16, 17]. Even for this case ('a simple model'), no computational method was available to compute the optimal parameters for a given instance. Since then, significant progress has been made in this area. In some sense, these models form the backbone of quantitative modeling for *Supply Chain Management.*

Among the many papers that are now available (since 1991), five papers on this topic – single product, capacitated systems in discrete time – make the following contributions: (1) Develop a method to compute the parameters for this simple model; (2) Find the optimal policy and provide a computational procedure for the case when this system faces a non-stationary (periodic) demand process (this paper generalizes [29, 30, 16, 17, 54]); (3) Study the stability of a multi-stage capacitated system operated by a base-stock policy; (4) Develop a computational method to compute best parameters for a multi-stage system operated by a base-stock policy; and (5) Develop a very quick and accurate approximation method for the same problem as above. Re-entrant Flows hops, Multi-product systems, Component commonality (and delayed differentiation) and other topics have been studied in greater detail since that time. Similarly, several papers and research themes in continuous time models are available as well.

10

No attempt has been made to provide a comprehensive literature review; however, most references may be obtained from the papers mentioned here. For a thorough survey of results (mainly in uncapacitated systems), see [53]. Other useful surveys and books include [27] and [6].

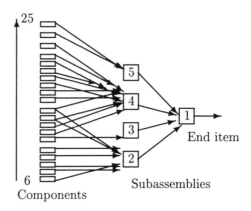

Figure 2.1 A typical supply chain: IBM assembly.

2.1.2 Basics

In presented here models, we will assume *time buckets*. Each bucket can be a day or a week long (or a month long) depending on the situation. These models will be called *discrete* time models. Each unit of time will be called a *period*. This approach is appropriate at the plant and system levels, while *continuous* time models are more appropriate at shop-floor and machine levels.

A base stock policy with order up to level z means that we produce in any period just enough to reach this target. If we cannot reach it, due to capacity limitations or lack of raw material, we do the best we can. An *echelon* base stock policy is exactly the same, except all quantities (mainly inventories) considered are cumulative to include this stage and all stages downstream (near the end product, close to the customer).

We will study these systems via a combination of analytical methods and simulation. Models are good for simple situations and to grasp concepts. To compute numbers for real world situations, simulation is preferred. The assumptions made are more realistic and the solutions obtained are more believable. Furthermore, certain flexibility that decision makers would prefer to have is better handled by simulation. In terms of acceptance by end users, a validated simulation has had better luck than complicated mathematical models. A shortcoming of simulation as compared to mathematical models is that it takes a much longer time to find answers. What we do then in reality is use models to get rough estimates and to provide intuition to fellow team members; then choose an alternative that shows most promise; finally, simulate to get accurate solutions.

2.1.3 Literature Survey: Papers before 1991

Clark and Scarf ([8]) developed a periodic review inventory control model for a serial uncapacitated system without setup costs. By using a discounted cost framework, they established that an *order up to* policy at each node is indeed optimal. Federgruen and Zipkin ([14, 15]) extended these results further. Muckstadt and others ([38]) conducted a computational study using the Clark and Scarf model ([8]). A continuous review version of the Clark and Scarf model is studied by [12]. An in depth analysis of an assembly structure with only two inputs, again by using the discounted cost framework is presented in [45]. Rosling ([43]) showed that under some initial conditions, an assembly system can be reduced to a serial system with modified lead times so that the results of [8] may be applied to this equivalent serial system.

A model of a production and distribution network in which manufacturing is modeled by a single node is described in [10]. Their work uses the earlier results of [29]. It also differs from much of the earlier work in an important way in that decentralized control is assumed and the model itself is a framework that combines separate models of production and distribution. A supply chain planning model that can be used to study production scale and scope economics is presented in [11].

A model for supply chain management which assumes decentralized control at each node in the manufacturing network, controlled by periodic review order up to inventory policies is presented in [35]. Once the service levels are set for each node, the overall relationships between cost and service can be obtained by applying this model. Although capacity considerations are not addressed, they allow for uncertainty in the supplier lead times.

The literature on inventory control systems and production-distribution systems is extensive and hence we limit our review to the work that is closely related to the theme of this paper. Similarly, there is a vast body of literature on single location production-inventory systems (addressing many aspects of interest) that is not reviewed here.

The Clark and Scarf model and many of its extensions, including [43], analyze the model within a discounted cost framework. These results are fairly involved and further, the computational procedures are not easy to describe or program. The assembly system inventory control problem in an average cost framework is studied in [34]. This analysis leads to an exact decomposition of the assembly system into several single location problems. Even this decomposition is not easy to handle, but [51] describes computational approximations that lead to a simplified computational procedure.

Except for [16, 17], not much work was done in capacitated systems in discrete time. In this survey paper, we will begin with their model and then

12

describe recent developments of several extensions of this basic model. The
notation we use may change between sections to remain consistent with the
papers that are being summarized.

2.2 SINGLE STAGE, SINGLE PRODUCT MODELS

The first progress since 1991 was the introduction of *shortfall* [1], and the connection that was made between the capacitated inventory model operated under a
base stock policy and a dam model that has been studied extensively by applied
probabilists.

Natural performance measures for production - inventory systems are costs
of inventories and backlogs. Even these costs may be considered in finite and
infinite horizons, average or discounted. Alternative performance measures are
based on service levels. We consider here cost-based approach. In section 2.3
we will also refer to service-based measures.

2.2.1 Basic Model

Tayur ([48]) provides a method to compute the optimal policy for a basic inventory problem addressed previously by [16]. We are to determine the base
stock level of a single item at a single location under periodic review, when

- the unit variable cost of purchase is c per item and there are no fixed
  costs;

- the holding cost (h) and stockout cost (p$>$ c) are per period and per item;

- demands in successive periods are non-negative i.i.d. random variables
  with known distributions (labeled by d);

- there is an infinite horizon and the costs are not discounted;

- all demands that are not satisfied by stock on hand are backordered;

- there is a finite production capacity, C, in every period;

- the cost is computed on the amount of inventory or backorder at the end
  of each period;

- we are to minimize the expected cost of holding plus penalty per period.

Recall that *inventory position* is defined as (stock on hand) + (stock on
order) - (backorders). An *order up-to* (or a base stock) policy with a critical
number z is one in which the inventory position (x) is raised to z if $x < z$, and
no production is done if $x \geq z$.

Unlike previous approaches, we provide a different construction of the sequence of problems that converge to the problem of interest. In particular,

---

[1]This shortfall type connection was known to queueing theorists before 1991.

we do not consider finite horizon problems of the capacitated problem and then take the limit as the number of periods go to infinity. Rather, we have a sequence of uncapacitated, multi-stage, serial infinite horizon versions, that converge to the desired system. We use results from uncapacitated multi-stage serial systems coupled with results in storage stochastic processes. Specifically, our steps are the following.

First, we show that the inventory model of interest is equivalent to a problem in dams. This suggests an analysis based not on the evolution of (inventory on hand minus backorder) process, but rather, by considering a shortfall process. The shortfall is defined as the *amount on order that has not yet been produced* because of the capacity constraint. If X_n is the shortfall at the end of period n, then $X_n = \max(0, X_{n-1} + d_n - C)$, where d_n is the demand in period n.

It is important to differentiate between backorders and shortfalls: the former represents what the customer did not obtain, while the latter represents what the manufacturer could not produce because of the capacity constraint. Thus, the backorder at the end of period n equals $\max(0, X_n - z)$ where z is the order up to level. The penalty cost p is on the backorder; there is no direct penalty on shortfall. Similarly, the amount of inventory at the end of period n is $\max(0, z - X_n)$. The cost in period n, therefore, equals $p\max(0, X_n - z) + h\max(0, z - X_n)$.

Second, we show that we can replace the single-stage capacitated inventory model by constructing a specially structured uncapacitated infinite-stage inventory model: This is simply a mathematical artifact. The sequence of multi-stage problems alluded to above will converge to this infinite stage system.

2.2.2   Connection with a Dam Model

Figure 2.2 shows the sample path of a typical single stage capacitated inventory system under periodic review that is operated by a base stock policy where excess demand is backlogged. The capacity (C) is 30, the order-upto level (z) is 45, and the inventory at time 0 (I_0) is 10. Let $d_1 = 15, d_2 = 9, d_3 = 37, d_4 = 21$ be the demands in the first four periods. Figure 2.3 shows the sample path of a dam (see [41, 42]) that has an infinite height, a release capability of at most C, an initial water level of 35. Let the rainfall in the first four periods be 15, 9, 37, and 21. The dam releases as much water as it can, and if the water level is less than C, the dam goes empty. The equivalence of the two sample path is obvious. Let $(a)^+$ stand for $\max(a,0)$. If Z_n is the content of the dam in period n just after release, then it satisfies $Z_n = (Z_{n-1} + d_{n-1} - C)^+$ and if X_n is the amount on order in period n that has not yet been produced, it satisfies $X_n = (X_{n-1} + d_{n-1} - C)^+$ (a similar recursion arises in the study of a D/G/1 queue also). Note that $\{X_n, n = 1, 2, \ldots\}$ is a Markov chain. This motivates us to study the capacitated inventory system in terms of the process X_n, and provide results in terms of the steady state distribution of $X = \lim_{n \to \infty} X_n$. Table 2.1 summarizes the equivalence between the capacitated inventory model and the dam model.

14

Figure 2.2 Sample Path of Inventory Model

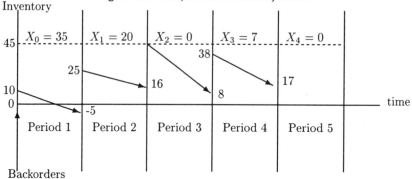

The order-upto level is 45
The Capacity is 30
Negative Inventories represent BackOrders
X_t is the amount in period t that is on order but not yet produced
Downward Sloping arrows represent demands; upward arrows denote production

Figure 2.3 Sample Path of Dam Model

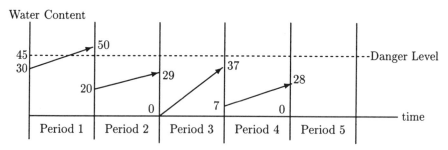

Upward arrows represent rainfall
Downward arrows represent release
One cannot release more than the water level
The danger level is simply shown to connect with the inventory model

Table 2.1 Comparison of dam and inventory models.

Dam Model	Inventory Model
Maximum release	Capacity
water content	amount not yet produced (shortfall)
empty dam	order upto level achieved
rainfall	demand
danger level crossed	backorders

2.2.3 Computing the Optimal Base-stock Level

Let $K(x)$ be the distribution of the input to the dam in any period. If $F(x)$ is the steady state distribution of the water content in the dam just after release. Then, the optimal value of the order upto level, z, in the capacitated inventory system satisfies

$$(F * K)(z) \quad = \quad \frac{p}{p+h}, \tag{2.1}$$

where $*$ represents convolution. $F(x)$ is known for all discrete distributions of the demand, and for Erlang distributions. Intuitively, we are adding the following two independent random variables: (1) demand in a period and (2) the amount on order at the beginning of the period that has not yet been produced. The necessary and sufficient condition for the distribution $F(x)$ to exist is that the expected demand (input to the dam) in a period be less than the capacity, C. Intuitively, the result states that it is the sum of two independent random variables (demand in a period and the amount on order not yet produced at the beginning of that period) that adds up to z. Penalty p is incurred if this sum crosses (at the end of the period) z and is proportional to the excess, and a holding cost (h) is imposed if the sum is less than z and is proportional to the amount on hand at the end of the period.

Example 1 *If the demand is exponentially distributed with mean rate λ ($K(u) = 1 - e^{-\lambda u}$, $u \geq 0$; $K(u) = 0$, otherwise) and the capacity is C, then the steady state distribution of the water content in the dam just after release is given by*

$$F(x) \quad = \quad 1 - e^{-\mu(x+C)} \quad (x \geq 0, \ \lambda C > 1)$$

where μ is the largest positive root of the equation

$$\mu \quad = \quad \lambda - \lambda e^{-(\mu+C)}.$$

Thus, the optimal order upto policy has a critical number z^*, and is obtained by solving $(F * K)(z^*) = \frac{p}{p+h}$.

Independently, the similarity of the basic inventory model operated via a base stock policy to a D/G/1 queue is recognized in [52].

2.2.4 Optimal Policy for an Extended Model

In the previous subsection we computed the optimal base stock level. In many situations it is very desirable to show that among all policies, a base stock policy is optimal. Federgruen and Zipkin ([16]) show this for the stationary case above. Kapuscinski and Tayur ([31]) provide a proof for an extension that allows for periodic demand as well as capacities by considering the following variant of the basic single-stage, single-item, discrete-time production - inventory model. Their proof applies for models where demand is Markov-modulated, lost demand is backlogged, and set of feasible points (initial inventory, ending inventory) is convex. That includes such cases as cyclical demand, capacity and minimum production constraints. Cost function needs to be linear in purchase (or production) quantity plus convex function of resulting inventory. All cost coefficients and constraints may be period-dependent. Here we present a simplest case that satisfies these conditions.

Specifically, assume that the demands (stochastic) follow a cyclical pattern with a cycle K. As before, there is a maximum production capacity (C) in any period; demands not satisfied in a period are backlogged to the next; there is holding cost (h) per unit of inventory per period and a cost of penalty (p) per unit of backlog per period. We want to find policies that minimize the finite-horizon costs, the infinite-horizon discounted cost and the infinite-horizon average cost (respectively) of operating this stage.

For all the three cases - finite-horizon cost, discounted infinite-horizon cost and infinite-horizon average cost - we show that an order up-to (or base-stock, critical-number) policy is optimal. This extends the results of [30] and [54] for uncapacitated, non-stationary model and [16, 17] for capacitated, stationary model. Our proof for the finite-horizon case follows standard steps. The proof of optimality for the infinite-horizon discounted case is simpler than that provided in [17] because we are able to use more recent results from [4]. To provide the optimality proof in the average cost case, we use the framework of [13], but our approach is different from that used by [16] for the stationary case. All proofs can be found in [31].

2.2.5 Sequence of Results Leading to Optimal Policy Structure

The following is the sequence of events at the beginning of a period: (1) some inventory or backlog exists; (2) a decision to increase the inventory is taken (limited by the production capacity); and (3) demand arrives. Holding or penalty costs are charged on the inventory after demand arrives. The notation is mostly standard. We have suppressed the time subscript in x, y and d below and these are assumed to be reals unless mentioned explicitly as integers. We will write them when necessary. We define

- x: inventory at the beginning of a period;

- y: inventory after ordering, but before demand arrives.

We assume that $\mathsf{E}d_i < \infty$ for all period types $i = 1, \ldots, K$.

Finite Horizon. As is standard practice, we begin with the finite-horizon case, which is not only the simplest situation to consider but also sets the framework for the infinite horizon cases.

Let $0 < \beta \leq 1$ be the discount factor. Define (recursively) $v_n(x)$ = minimum total expected discounted cost with a time horizon of n periods. Note (in this subsection only) that *we start in period n and count downward towards period 1*, the end of the horizon. The demand in period n is one of the K period types $(i = 1, \ldots K)$. We may assume that period 1 has type 1 demand and period $n = K + 1$ has type 1 demand again and so on. Thus:

$$v_0(.) = 0,$$
$$v_n(x) = \inf_{(x,y)\in A} \{c(y - x) + L_n(y) + \beta \mathsf{E}_{d_n} v_{n-1}(y - d_n)\}$$
$$\text{for } x \in R \text{ and } n \geq 1,$$

where

$$A = \{(x, y) \in R^2 | x \leq y \leq x + C\} = \bigcup_{x \in R} Y(x)$$

is the feasible set. For $x \in R$,

$$Y(x) = [x, x + C],$$

and for $y \in R$,

$$L_n(y) = \mathsf{E}_{d_n}(h(y - d_n)^+ + p(y - d_n)^-).$$

We can express v_n using additional functions J_n and I_n as follows.

$$v_n(x) = -cx + I_n(x), \ x \in R, \ n \geq 1;$$
$$I_n(x) = \min\{J_n(y) : y \in Y(x)\}, \ x \in R;$$
$$J_n(y) = cy + L_n(y) + \beta \mathsf{E} v_{n-1}(y - d_n), \ y \in R$$

with $v_0(.) = J_0(.) = I_0(.) = 0$.

Our first lemma generalizes Theorem 2 of [17] to the cyclic case. The proof is a direct adoption of theirs.

Lemma 1 *The set of feasible pairs (x, y), A (as defined above), is convex. For all $n \in N$:*
(a) The expected sum of holding and penalty costs, $L_n(y)$, is convex;
(b) J_n, v_n, and I_n are convex;
(c) $v_n \geq 0$; and
(d) For $n \geq 1$: $v_n(x_n) \to \infty$ when $|x_n| \to \infty$, and if $p > c$ then $J_n(y_n) \to \infty$ when $|y_n| \to \infty$.

Theorem 1 *Let y_n^* be the smallest value minimizing J_n. The optimal policy in period n is order up-to y_n^*. As $\lim_{|y_n| \to \infty} J_n(y_n) = \infty$, we have $y_n^* < \infty$.*

As a first property, we have the following.

Property 1 *For any $x \in R$, $v_{nK+i}(x)$ is increasing in n.*

Proof: We show by induction that $v_{m+K}(x) \geq v_m(x)$ for all x and for all m. First, $v_K(x) \geq 0 = v_0(x)$. Let $v_{K+m}(x) \geq v_m(x)$ for a certain m and all x. Then,

$$
\begin{aligned}
v_{K+m+1}(x) &= \min_{y \in Y(x)} \{c(y-x) + L_{K+m+1}(y) + \beta \mathsf{E} v_{K+m}(y - d_{K+m+1})\} \\
&\geq \min_{y \in Y(x)} \{c(y-x) + L_{m+1}(y) + \beta \mathsf{E} v_m(y - d_{m+1})\} \\
&= v_{m+1}(x). \ \square
\end{aligned}
$$

Note that the convexity of functions J_n, I_n, v_n implies continuity of these functions. One-sided derivatives exist at all points. Also, two-sided derivatives exist with the exception of a denumerable set of points. Points where two-sided derivatives do not exist are generated by mass points of demands. Although derivatives do not have to be continuous, they are monotonic and bounded on any compact set. Therefore, in this paper we will define them as right-hand-side limits. We will use $'$ to denote these derivatives.

We define the *myopic* solution to period i, $\bar{y}_i$, as the one that satisfies $c + L_i'(\bar{y}_i) = 0$. Property 2 provides a simple lower bound as in the uncapacitated case; see [54].

Property 2 *Assume that period 1 has the minimum myopic solution; $\bar{y}_1 = \min\{\bar{y}_i : i = 1, \ldots, K\}$ where $\bar{y}_i$ is the myopic solution for period i. Then: $y_n^* \geq \bar{y}_1$, $\forall n$.*

Strictly speaking, our next result is not a finite-horizon result. However, this appears to be the most appropriate point to state it. Part (c) of the technical lemma (Lemma A) is used for the next result. We will use this property in proving Property 6 in section 2.2.5 and when analyzing the infinite horizon average cost case in section 2.2.5.

Property 3 *For a given $0 < \beta \leq 1$, the sequence y_n^* is bounded. That is, $\limsup\{y_n^* : n \in N\} < \infty$.*

Property 4 *For any finite-horizon (n periods) problem consider a policy that produces up-to z_i in period $i = 1, \ldots, n$. For all i, the cost of such a policy is convex in z_i.*

Infinite Horizon: Discounted Model. We now move to the infinite-horizon discounted case. It was natural in the finite-horizon case to label periods as time to the end of the horizon. In the infinite-horizon case, we typically start the process at some point and continue indefinitely. To make the notation more intuitive, starting from this subsection, *we will number periods in increasing order*: following period n, we have period $n + 1$.

Federgruen and Zipkin ([17]) showed the next result for the stationary, capacitated case. We provide a simpler proof as we are able to use results from [4].

Theorem 2 *Let* $0 < \beta < 1$ *(discounted case). The optimal policy for the infinite-horizon is cyclic up-to level policy.*

Since $\mathsf{E}(d_i) < \infty$ for $i = 1, \ldots, K$, we have:

Property 5 *Let* $w_n(x, i) := v_{nK+i}(x)$ *for* $i = 1, \ldots, K$, $n \in N_0 := N \cup \{0\}$, *and* $x \in R$. *The limit,* $\lim_{n \to \infty} w_n(x, i) = w(x, i)$ *(where values of* w *are in* $R \cup \{\infty\}$*), exists and* $w < \infty$.

$w < \infty$ does not imply that stationary up-to levels are finite. However, it is possible to prove that (since $E(d_l) < \infty$ for $l = 1, \ldots, K$):

Property 6 *The up-to levels are finite (i.e.* $z_l < \infty$*).*

Average Cost Criterion. This case is the most difficult one to analyze. We need some technical results before the optimality of base-stock policies can be proved.

Property 7 *(Convexity) Consider the infinite-horizon case (with cyclic demands) and the class of up-to level policies, where levels* z_i *for* $= 1, \ldots, K$ *are period-type specific. For both the discounted cost and average cost criteria, the average cost is finite and convex in each of* z_i *'s.*

Let z_{max} and z_{min} be the maximum and minimum respectively among the levels for a given vector $(z_1, \ldots, z_K)$. The following two properties are used both in proving optimality of base stock policy, as well as in justifying use of IPA.

Property 8 *(Coupling)*
If z_{max} *is strongly regenerative, then for any two processes with starting points* $x_{i_0}^{(1)}$ *and* $x_{i_0}^{(2)}$ *respectively in period* i_0, *there exists (with probability 1) a period* n *such that starting from this period the two processes coincide.*

Property 9 *(Shortfall stability)*
Consider an up-to policy with a vector $z_1, \ldots, z_K$. *If* $\sum_{i=1}^{K} \mathsf{E}d_i < KC$ *then* $\mathsf{E}S_i < \infty$ *for all* i *'s, where* $S_i = z_i - y_i$ *is the shortfall in period type* i.

Average cost criterion is easy to analyze when either number of states process can take is finite or the one-period cost function is bounded (see [4]). Conditions when an optimal policy exists for semi-Markovian process with average cost objective function with denumerable state space and unbounded one-period cost function are derived in [13]. These conditions were used in [16] to derive optimality of up-to policy for capacitated stationary model. We extend it to a cyclical model as follows.

We first show that any policy can be dominated by a policy that requires reducing (any) backlog and not exceeding some stationary level A^*. Then we show that among such policies, the up-to policy is optimal. The main structure of our proof is based on results of [13], but the proof that the required conditions are satisfied is shown by a different method as compared to [16].

Fact 1 *If an optimal policy for the problem exists (including possibility of randomized policies), then it has the following form:*
a) for $x \leq -C$, $y = x + C$.
b) there exists $A^* < \infty$ such that for $y \geq A^*$ for all y's and for all period types the policy it is better to produce nothing rather than take any other action.

Theorem 3 *Consider a capacitated system with cyclic discrete demands and linear ordering, penalty, and holding costs. For the average cost criterion, the cyclic up-to policy is optimal.*

The proof of the following lemma can be found in [31].
Lemma A. *Consider the policy $\delta[0]$ (produce up-to 0). Let $\sum_{i=1}^{K} \mathsf{E} d_i < KC$ and $\mathsf{E}(d_i)^{2k+2} < \infty$ for a certain $k \geq 1$ and for all $i = 1, \ldots, K$. Consider a point process defined by points i, for which $y_i = 0$ (no backlog of previous demands). Let N be a random variable equal to time between two consecutive points of this point process. Then for any starting period-type:*
(a) $\mathsf{E}(N^k) < \infty$
(b) $\mathsf{E}(|y_i|^k) < \infty$ and $\mathsf{E}(|x_i|^k) < \infty$
(c) If $0 < \beta < 1$ then $\mathsf{E}(\sum_{n=0}^{\infty} |y_n|^k \beta^n), \mathsf{E}(\sum_{n=0}^{\infty} |x_n|^k \beta^n) < \infty$ for any x_0.
(d) For $0 < \beta \leq 1$, there exists $A \in R$, such that $J_n(0) \leq An$.

2.2.6 Basic Properties

We show several properties of the optimal policy including the following: (1) capacity smoothes the base-stock levels in a manner that is different from that due to holding costs; (2) the limit of finite horizon order up-to levels are bounded; (3) the optimal levels are higher than the minimum of the K myopic levels; (4) in an infinite-horizon average cost case, the optimal levels are lower than the maximum of the K stationary optimal levels and higher than the minimum of the K stationary optimal levels; (5) if demands are stochastically larger or the capacity is lower, the base-stock levels are higher; and (6) for $K = 2$, as the penalty cost is increased, the difference between the maximum and minimum levels is bounded by C under fairly general assumptions on demand distributions.

2.2.7 Computational Technique

Exact computation of optimal levels by analytical formulas appears difficult. We provide a simulation based method using infinitesimal perturbation analysis (IPA) to find these levels. The basic idea is simple: instead of using the derivative of the expected cost in a gradient search method, we use the expected value of the sample path derivative (obtained via simulation). To validate this approach and prove the optimality result for average cost case, we derive several technical properties of base-stock policies - convexity, regeneration, coupling, and stability.

The steps involved in derivation of these properties are similar for the model described here (non-stationary 1-stage model) to those described in the next section for stationary multi-stage model (one simple condition needs to be added). Therefore, we do not detail them here. See [20] for an excellent reference about IPA. A numerical study indicates that our IPA method is robust and finds solutions within a few minutes on a workstation. See [31].

2.3 SINGLE PRODUCT, SERIAL SYSTEM

It has been shown that for a serial capacitated system, the optimal policy (under the cost criteria discussed above) is not base-stock in general; see [47].

Parker and Kapuscinski [40] show, however, that if the last echelon's capacity does not exceed all other capacities, then it is possible to describe the structure of the optimal policy. In fairly general settings (stationary and non-stationary case, fixed delivery lead times, finite and infinite horizon) a "censored" modified base-stock policy is optimal.

2.3.1 A Simulation Based Optimization Procedure: Single Stage Simulation

Let s be any base stock level; the optimal value for s is what we eventually need. The notation used is consistent with the papers referenced on this topic where detailed proofs are available.

We need notation.

$$
\begin{aligned}
I_n &= \text{inventory} - \text{backlog in period } n; \\
R_n &= \text{production in period } n; \\
\ell &= \text{leadtime from production to inventory;} \\
T_n &= R_{n-1} + \cdots + R_{n-\ell} \\
&= \text{pipeline inventory;} \\
D_n &= \text{demand in period } n; \\
s &= \text{base-stock level;} \\
c &= \text{production capacity.}
\end{aligned}
$$

We always assume that $c > 0$ and $D_n \geq 0$ for all n. Under a (modified) base-stock policy, the production level in each period is set to try to restore the inventory position $I_n + T_n - D_n$ to s. If production were uncapacitated, this would be achieved by setting $R_n = D_n$. Since, however, R_n cannot exceed c, it may take multiple periods of production to offset demand in a single period. The system evolves as follows:

$$
I_{n+1} = I_n - D_n + R_{n-\ell} \tag{2.2}
$$

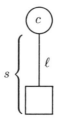

Figure 2.4   Illustration of a single stage

$$R_n \quad = \quad c \wedge [s + D_n - (I_n + T_n)]^+ \tag{2.3}$$
$$T_{n+1} \quad = \quad T_n + R_n - R_{n-\ell}. \tag{2.4}$$

x^+ denotes $\max(0, x)$, x^- denotes $\max(0, -x)$ and $a \wedge b$ denotes $\min(a, b)$.

What about costs? In period n the cost will be $C_n = hI_n^+ + pI^-$. The average cost over the long term will be $\frac{\sum_{n=1}^{N} C_n}{N}$ for a large N. N is the number of periods simulated.

2.3.2 A Simulation Based Optimization Procedure: Single Stage Gradient

We can certainly find the gradient of expected cost by simulating one more time and changing (say, increasing) only the order up to level by 1: the difference in cost between the two simulations is the gradient. This would be too much work especially if we had several stages whose order up to levels we want to optimize. Can we get the gradient by not doing any more simulations? Yes; and here goes the basic trick.

1. If we had inventory in period n, having started at a higher base stock level means an *increase* in holding by h.

2. If we had backlog in period n, we will *decrease* penalty by p.

Thus, the changes are either h or $-p$. We will just average these changes. The point is we are are claiming that average of changes is the change in the average. In this situation, and in several others, this claim is good.

How do we make the computer do this? In principle, we are just differentiating the recursions. The easiest way is to write the following two lines in the code to be executed in every period:

$$dCost \quad = \quad dCost + h \text{ if } I_n > 0$$
$$dCost \quad = \quad dCost - p \text{ if } I_n < 0$$

Then dividing dCost by N provides the derivative. Note that at the beginning of the simulation dCost is set to zero.

2.3.3 A Simulation Based Optimization Procedure: Serial System

Operation. We now link multiple stages in series. There are m stages: stage 1 supplies external demands and stage i supplies components for stage $i - 1$, $i = 2, \ldots, m$. Stage m draws raw materials from an unlimited supply. To specify an echelon-inventory base-stock policy for the system, we let

$$s^i = \text{echelon base-stock level for stage } i.$$

Naturally, we require $s^1 \leq s^2 \leq \cdots \leq s^m$. Let the variables R_n^i, T_n^i, ℓ^i, and c^i have the same meaning as before, applied to stage i. For $i = 2, \ldots, m$, let I_n^i be the installation inventory at stage i, and let $I_n^1 = I_n$, with I_n as in Section 2.1. In period n stage i sets production to try to restore the echelon inventory position

$$\sum_{j=1}^{i} (I_n^j + T_n^j) - D_n$$

to its base-stock level s^i.

Two features distinguish the multi-echelon system from a single stage: production at stage i, $i < m$, is constrained by available component inventory I_n^{i+1}, as well as by the capacity limit c^i; and for $i > 1$ the amount removed in period n from the store at stage i is the downstream production level R_n^{i-1}, rather than the external demand. Thus, for stage $i = 2, \ldots, m - 1$ we have

$$I_{n+1}^i = I_n^i - R_n^{i-1} + R_{n-\ell^i}^i \tag{2.5}$$

$$R_n^i = c^i \wedge \left[s^i + D_n - \sum_{j=1}^{i} (I_n^j + T_n^j) \right]^+ \wedge I_n^{i+1} \tag{2.6}$$

$$T_{n+1}^i = T_n^i + R_n^i - R_{n-\ell^i}^i. \tag{2.7}$$

At stage m, raw materials are unlimited so the last term in (2.6) is absent. To subsume these special cases in (2.5-2.7), we take $R_n^0 \equiv D_n$ and $I_n^{m+1} = \infty$ for all n. To complete our specification of the model, we need initial conditions; for simplicity, we take $I_1^1 = s^1$, $I_1^i = s^i - s^{i-1}$, $i = 2, \ldots, m$, and all other variables zero. In other words, the system starts with full inventory. For details see [22].

Similar to the single stage case, the derivatives with respect to the base stocks can be computed.

Performance Measures. We assign costs to inventories and backorders, and also measure performance through service levels. Thus, let

$$h^i = \text{holding cost rate for stage-} i \text{ inventory};$$
$$b = \text{backorder penalty rate at stage 1.}$$

Then the total cost incurred in period n is

$$C_n = [I_n^1]^- b + ([I_n^1]^+ + T_n^1) h^1 + \sum_{i=2}^{m} (I_n^i + T_n^i) h^i, \qquad (2.8)$$

Under appropriate conditions, we consider the finite-horizon average cost

$$c_N = \frac{1}{N} \sum_{n=1}^{N} \mathsf{E}[C_n], \qquad (2.9)$$

the infinite-horizon, α-discounted cost

$$c_{\alpha,\infty} = \mathsf{E}\left[\sum_{n=1}^{\infty} \alpha^n C_n\right], \qquad (2.10)$$

and the infinite-horizon average cost

$$c_\infty = \lim_{N \to \infty} \frac{1}{N} \sum_{n=1}^{N} C_n. \qquad (2.11)$$

In addition to finite-horizon average cost, infinite-horizon, α-discounted cost, and infinite-horizon average cost, we may consider 'service level'-based objectives.

We allow the possibility that $D_n = 0$; i.e., that in some periods there is no demand. By convention, if $D_n = 0$ then no stockouts occur in period n, regardless of the inventory levels in that period. Thus, writing $\mathbf{1}\{\cdot\}$ for the indicator of the event $\{\cdot\}$,

$$V_N = \frac{1}{N} \sum_{n=1}^{N} \mathbf{1}\{I_n^1 \geq D_n \text{ or } D_n = 0\} \qquad (2.12)$$

is the fraction of periods in which demands are filled immediately. *Type-1 service* is measured by $v_N = \mathsf{E}[V_N]$, the expected fraction of periods in which no backorders occur, or its steady-state counterpart

$$v_\infty = \lim_{N \to \infty} V_N. \qquad (2.13)$$

A discounted service criterion could be accommodated but has no clear interpretation.

This result can be easily modified to account for an alternative measure of service: the proportion of customer orders that are filled without backordering.

A final measure of service is the *fill rate* (Type-2 service), given by

$$v_\infty^{(f)} = 1 - \lim_{N \to \infty} \frac{\sum_{n=1}^{N} (D_n - I_n^1)^+}{\sum_{n=1}^{N} D_n}.$$

For independent, identically distributed demands, the denominator can be replaced with $N E[D_1]$ without changing the value of the limit. Thus, in this case, derivative estimation for $v_\infty^{(f)}$ reduces to derivative estimation for

$$\lim_{N \to \infty} \frac{1}{N} \sum_{n=1}^{N} (D_n - I_n^1)^+.$$

This case works just like (2.11), so we do not consider it separately.

Derivatives of State Variables: Recursions. We obtain derivatives of inventory and production levels by differentiating the recursions (2.5-2.7). Handling non-differentiable functions in these recursions requires some care, but the formal procedure is simple – it is usually possible to define one-sided derivatives.

We differentiate with respect to $s^* = s^{i^*}$ for some $i^* = 1, \ldots, m$. Since the base-stock levels must remain ordered, even under arbitrarily small perturbations, let us henceforth assume that

$$0 < s^1 < s^2 < \cdots < s^m.$$

Evidently, differentiating (2.5) and (2.7) yields:

$$\frac{dI_{n+1}^i}{ds^*} = \frac{dI_n^i}{ds^*} - \frac{dR_n^{i-1}}{ds^*} + \frac{dR_{n-\ell^i}^i}{ds^*}. \tag{2.14}$$

$$\frac{dT_{n+1}^i}{ds^*} = \frac{dT_n^i}{ds^*} + \frac{dR_n^i}{ds^*} - \frac{dR_{n-\ell^i}^i}{ds^*}. \tag{2.15}$$

For (2.6), we have several cases to consider, corresponding to which of the terms there attains the minimum. Let us say that stage i is *capacity bound* if the minimum in (2.6) is c^i; *demand bound* if the minimum is attained uniquely by $(s^i + D_n - (\sum I_n^j + T_n^j)) > 0$; and *supply bound* if the minimum is attained uniquely by $I_n^{i+1} > 0$. Then we have

$$\frac{dR_n^i}{ds^*} = \begin{cases} 0, & \text{if } i \text{ is capacity bound;} \\ \mathbf{1}\{i = i^*\} - \sum_{j=1}^{i}(I_n^j)' + (T_n^j)', & \text{if } i \text{ is demand bound;} \\ (I_n^{i+1})', & \text{if } i \text{ is supply bound;} \\ 0, & \text{if } R_n^i = 0. \end{cases} \tag{2.16}$$

The four cases included here are not exhaustive because we have not resolved all ties, but they are almost surely the only ones we need to consider. To complete the recursions, we need initial conditions. Differentiating the initial conditions for the state variables we get $(I_1^i)' = \mathbf{1}\{i^* = i\}$, $(I_1^i)' = \mathbf{1}\{i^* = i\} - \mathbf{1}\{i^* = i-1\}$, and zero for all other derivatives.

We briefly outline implementation details for simulating the model and evaluating derivatives in the course of the simulation. The basic state recursions (2.5-2.7) are easily implemented: in each period, one simply loops through the nodes from $i = 1$ to m, updating inventories according to (2.5), setting production levels according to (2.6), and advancing pipeline inventories according to (2.7). For the derivatives, one introduces algorithmic variables dI^i and dT^i recording derivatives of I^i and T^i, and variables dR_k^i, $k = 0, \ldots, \ell^i$, recording derivatives of production levels k periods earlier. These are updated in each period.

2.3.4 Validation of Technique

Validation of finite horizon derivatives – inventory and costs – is quite straightforward. We show that (right-side) derivatives exist with probability one at a given value of s, then appeal to Lipschitz continuity and finish by applying the dominated convergence theorem.

Justification of the Derivative Recursions. We now state results showing that (2.14-2.16) are valid and that the sample-path derivatives they generate are unbiased estimators of derivatives of expectations. For all n, we assume that the period-n demand D_n has a density on $(0, \infty)$, by which we mean that the function $x \mapsto P(D_n \leq x)$ is absolutely continuous for all $x > 0$; we do not exclude the possibility that $P(D_n = 0) > 0$.

Proposition 1 *If $\{D_n, n = 1, 2, \ldots\}$ are independent and if each D_n has a density on $(0, \infty)$, then the following hold:*

> *(i) For $i = 1, \ldots, m$ and $n = 1, 2, \ldots$, each I_n^i, R_n^i, and T_n^i is, with probability one, differentiable at $(s^1, \ldots, s^m)$ with respect to each s^j, $j = 1, \ldots, m$. Moreover, these derivatives satisfy (2.14-2.16).*

> *(ii) If in addition $\mathsf{E}[D_n] < \infty$ for all n, then $\mathsf{E}[I_n^i]'$, $\mathsf{E}[R_n^i]'$, and $\mathsf{E}[T_n^i]'$ exist and equal $\mathsf{E}[(I_n^i)']$, $\mathsf{E}[(R_n^i)']$, and $\mathsf{E}[(T_n^i)']$.*

The proof of part (ii) rests on a technical lemma that also underlies results in later sections. Essential to this lemma is the notion of a *Lipschitz* function, which is central to the method developed in this paper. A function ϕ mapping $S \subseteq \mathbf{R}$ into $\mathbf{R}$ is *Lipschitz* if there exists a constant k_ϕ, called the modulus, for which

$$|\phi(x) - \phi(y)| \leq k_\phi |x - y|. \tag{2.17}$$

The composition of Lipschitz functions with moduli k_1 and k_2 is Lipschitz with modulus $k_1 k_2$. A random function is Lipschitz with probability one if there exists a random variable K that serves as a path-wise modulus. The following result paraphrases Theorem 1.2 of Glasserman ([20]).

Lemma 2 *Let $\{X(s), s \in S\}$ be a random function with S an open subset of $\mathbf{R}$. Suppose that $\mathsf{E}[X(s)] < \infty$ for all $s \in S$. Suppose, further, that X is differentiable at $s_0 \in S$ with probability one, and that X is almost surely Lipschitz with modulus K_X satisfying $\mathsf{E}[K_X] < \infty$. Then $\mathsf{E}[X(s_0)]'$ exists and equals $\mathsf{E}[X'(s_0)]$.*

In light of this lemma, the key step in verifying that our derivative estimates are unbiased is showing that inventories and costs are, with probability one, Lipschitz functions of the base-stock levels having integrable moduli. The Lipschitz property is preserved by min, max and addition, so verifying it is straightforward in our application. We bound the modulus by bounding the derivatives.

Derivatives of Performance Measures. We now extend the analysis of derivatives of state variables to derivatives of performance measures.

Finite-Horizon Costs
We begin by considering finite-horizon costs. To differentiate $[I_n^1]^+$, notice that if $P(I_n^1 = 0) = 0$, then $([I_n^1]^+)' = \mathbf{1}\{I_n^1 > 0\}(I_n^1)'$; see the proof of Proposition 3.1(i). Thus, corresponding to (2.8) we have the path-wise derivative

$$C_n' = [(T_n^1)' + \mathbf{1}\{I_n^1 > 0\}(I_n^1)']h^1 - \mathbf{1}\{I_n^1 < 0\}(I_n^1)'b + \sum_{i=2}^{m}[(T_n^i)' + (I_n^i)']h^i. \quad (2.18)$$

This expression is easily evaluated in a simulation. The functions on the right side of (2.8) preserve the Lipschitz property, so we have

Theorem 4 *If, for $n = 1, 2, \ldots$, $\mathsf{E}[D_n] < \infty$, the D_n's are independent, and each D_n has a density on $(0, \infty)$, then C_n' exists with probability one, $n = 1, \ldots, N$, and*

$$\mathsf{E}\left[\frac{1}{N}\sum_{n=1}^{N} C_n'\right] = c_N'.$$

We turn next to the finite-horizon service level v_N. As defined in (2.12), V_N cannot be differentiated because it is not even continuous. To circumvent this difficulty, we replace the indicator in (2.12) with a conditional expectation. Let f_n be the density of D_n on $(0, \infty)$, keeping in mind that the integral of f_n may be less than one. Let

$$F_n(x) = \int_0^x f_n(t)\, dt. \quad (2.19)$$

Then,

$$v_n = \mathsf{E}[V_N] = \mathsf{E}[N^{-1}\sum_{n=1}^{N} \mathbf{1}\{D_n = 0 \text{ or } D_n \le I_n^1\}]$$

$$= N^{-1}\sum_{n=1}^{N} P(D_n = 0) + \mathsf{E}[N^{-1}\sum_{n=1}^{N} F_n(I_n^1)].$$

Since $P(D_n = 0)$ does not depend on s^*, we may work with

$$\tilde{V}_N = N^{-1}\sum_{n=1}^{N} F_n(I_n^1).$$

As a function of s^*, $\tilde{V}_N$ is differentiable except possibly on the zero-probability event that some I_n^1 equals zero, $n = 1, \ldots, N$. Off this event, we have

$$\tilde{V}_N' = N^{-1} \sum_{n=1}^{N} \mathbf{1}\{I_n^1 > 0\} f_n(I_n^1)(I_n^1)'.$$

By arguments similar to the one used for cost derivatives, we have

Theorem 5 *If, in addition to the conditions of Theorem 4, f_n is bounded for all n, then $\mathsf{E}[\tilde{V}_N'] = v_N'$.*

Discounted Costs

Since the inventory cannot exceed the echelon base-stock level and the backlog cannot exceed the cumulative demand, it follows that

$$\mathsf{E}[C_n] \leq \left(b + \sum_{j=1}^{m} h^j \right) \left(\sum_{j=1}^{m} s^j + m \sum_{k=1}^{n} \mathsf{E}[D_k] \right).$$

Consequently, with

$$\sup_{k \geq 1} \mathsf{E}[D_k] < \infty \qquad (2.20)$$

we ensure that $c_{\alpha,\infty}$ is finite. Condition (2.20) obviously holds if the demands have a common finite mean, but weaker conditions are clearly possible.

Computing a derivative estimate for $c_{\alpha,\infty}$ from its infinite-series representation is impractical. Instead, we use a method of Fox and Glynn ([18]) that replaces the infinite horizon with a random finite horizon. The combination of this method with IPA is new, but see Glynn, L'Ecuyer, and Adès ([26]) for a different derivative-estimation application. We need the following preliminary (known) result:

Lemma 3 *Suppose $c_{\alpha,\infty} < \infty$. Let L be a geometric random variable with $P(L = n) = \alpha^n(1 - \alpha)$, independent of the demands $\{D_1, D_2, \ldots\}$. Define $\tilde{C}_{\alpha,L} = \sum_{n=1}^{L} C_n$. Then $\mathsf{E}[\tilde{C}_{\alpha,L}] = c_{\alpha,\infty}$.*

This result provides an estimator of the infinite-horizon discounted cost from a finite number of transitions. The same idea leads to an unbiased estimator of $c_{\alpha,\infty}'$ from L transitions. With probability one, $\tilde{C}_{\alpha,L}$ is differentiable at any s^* and

$$\tilde{C}_{\alpha,L}' = \sum_{n=1}^{L} C_n',$$

with C_n' as given in (2.18). Moreover, $\tilde{C}_{\alpha,L}$ is Lipschitz and

$$L\left(b + \sum_{j=1}^{m} h^j \right)$$

is an integrable modulus. Combining these observations with Lemma 2 we obtain

Theorem 6 *In addition to the conditions of Theorem 4, suppose the mean demands satisfy (2.20). Then* $c'_{\alpha,\infty} = \mathsf{E}[\tilde{C}'_{\alpha,L}]$.

Average Costs

We now turn to derivative estimates for the infinite-horizon average costs c_∞ and v_∞ defined in (2.11) and (2.13) respectively. Not surprisingly, average costs require somewhat stronger assumptions than finite-horizon or discounted costs. Throughout this section, we assume that

$$\{D_n, n = 1, 2, \ldots\} \text{ are i.i.d. with finite expectation.} \qquad (2.21)$$

We examine the sequences $X_n = (I_n^i, R_{n-1}^i, \ldots, R_{n-\ell^i-1}^i)_{i=1}^m$, $n = 1, 2, \ldots$ and $\{Y_n, n = 1, 2, \ldots\}$, where the components of Y_n are the derivatives of the components of X_n. From the recursions (2.14 - 2.16), we obtain

Lemma 4 *If the D_n's are i.i.d., then $\{X_n, n \geq 1\}$ and $\{(X_n, Y_n), n \geq 1\}$ are Markov chains.*

To consider c_∞ and v_∞ (and their derivatives) we need to some general stability results. It is shown in Glasserman and Tayur ([23]) that if

$$\mathsf{E}[D_1] < \min\{c^1, \ldots, c^m\} \qquad (2.22)$$

then $\{X_n, n \geq 1\}$ is, in a precise sense, stable. This sense is discussed in Appendix B of [22], along with a proof of the following result:

Theorem 7 *Suppose (2.21) and (2.22) hold. Then $N^{-1} \sum_{n=1}^N C'_n \to c'_\infty$ with probability one, at almost every s^*. If, in addition, $\sup_x f(x) < \infty$, then $N^{-1} \sum_{n=1}^N \tilde{V}'_n \to v'_\infty$ with probability one, at almost every s^*.*

If $\lim_{N \to \infty} N^{-1} \sum_{n=1}^N \mathsf{E}[C'_n(s)]$ is continuous in s (as one would ordinarily expect), then the restriction to "almost every" s^* can be omitted from the statement of the theorem.

2.3.5 The Stationary Regime

Physical inventory levels are arguably the most natural descriptors of the state of the system. But, as is often the case in these types of systems, it turns out to be mathematically more convenient to work with echelon quantities. For $i = 1, \ldots, m$ define the period-n *shortfall* for echelon i by

$$Y_n^i = s^i - \sum_{j=1}^i I_n^j. \qquad (2.23)$$

Notice that:

$$\begin{aligned} Y^m_{n+1} &= Y^m_n + D_n - \min\{Y^m_n + D_n, c^m\} \\ &= \max\{0, Y^m_n + D_n - c^m\}. \end{aligned} \qquad (2.24)$$

For $i = 1, \ldots, m - 1$, the available inventory is limited to I^{i+1}_n, so we have

$$Y^i_{n+1} = \max\{0, Y^i_n + D_n - c^i, Y^{i+1}_n + D_n - (s^{i+1} - s^i)\}. \qquad (2.25)$$

Lemma 5 The echelon shortfalls satisfy $Y_{n+1} = \phi(Y_n, D_n)$ where $\phi : \mathbf{R}^m_+ \times \mathbf{R} \mapsto \mathbf{R}^m_+$ is defined by (2.24-2.25). In particular, ϕ is increasing and continuous.

Suppose that the demands form a stationary process. Without loss of generality, we may assume that D_n is defined for all integer n with $\{D_n, -\infty < n < \infty\}$ stationary. In this setting, through the method of Loynes, the conclusion of Lemma 5 is sufficiently strong to imply the existence of a stationary version of the echelon shortfalls. (In fact, it would suffice for ϕ to be increasing and continuous in its first argument for all values of its second argument.) Here and throughout, $\Rightarrow$ denotes convergence in distribution.

Lemma 6 Let $\{D_n, -\infty < n < \infty\}$ be stationary. There exists a (possibly infinite) stationary process $\{\tilde{Y}_n, -\infty < n < \infty\}$ satisfying $\tilde{Y}_{n+1} = \phi(\tilde{Y}_n, D_n)$ for all n, such that if $Y_0 = 0$, a.s., then $Y_n \Rightarrow \tilde{Y}_0$.

The natural stability condition rules out the possibility that $\tilde{Y}_0$ is defective:

Theorem 8 Suppose the demands $\{D_n, -\infty < n < \infty\}$ are ergodic as well as stationary. If

$$\mathsf{E}[D_0] < \min\{c^i : i = 1, \ldots, m\}, \qquad (2.26)$$

then $\tilde{Y}_0$ is almost surely finite. If for some i, $\mathsf{E}[D_0] > c^i$, then $\tilde{Y}^j_0 = \infty$, a.s., for all $j = 1, \ldots, i$.

Thus far, we have shown convergence of $\{Y_n, n \geq 0\}$ to the stationary distribution $\tilde{Y}_0$ only when $Y_0 = 0$. It remains to show that there is just one (finite) stationary distribution and that $\{Y_n, n \geq 0\}$ converges to it for all Y_0.

In general, a process $X = \{X_n, n \geq 0\}$ is said to *admit coupling* if for all states x_1 and x_2 it is possible to construct copies X^{x_i}, $i = 1, 2$, of X with $X^{x_i}_0 = x_i$, such that $X^{x_1}_n = X^{x_2}_n$ for all $n \geq N$, for some almost-surely finite N. The random time N is called a *coupling time* for the two processes. A process that admits coupling can have at most one stationary distribution, leading to the following result:

Theorem 9 Under the stability condition $\mathsf{E}[D_0] < \min_i c^i$, the echelon shortfall process admits coupling. Consequently, its stationary distribution is unique, and $Y_n \Rightarrow \tilde{Y}_0$ for all Y_0.

2.3.6   Regeneration

The previous section established conditions for the stability of the shortfall echelon process when demands are stationary and ergodic. We now examine the regenerative structure of $\{Y_n, n \geq 0\}$ when $\{D_n, n \geq 0\}$ is an i.i.d. sequence. (In Section 6 we relax the i.i.d. assumption.) Regenerative properties are valuable in establishing convergence of costs and also simulation estimators. Indeed, it was the simulation-based application in Glasserman and Tayur [22] that initially motivated this investigation.

We show that the stability condition of Section 2.3.5 suffices to ensure that $\{Y_n, n \geq 0\}$ possesses the regenerative structure of a *Harris ergodic* Markov chain. Under a stronger condition, we show that the vector of shortfalls returns to the origin infinitely often, with probability one.

Harris Recurrence.   Many of the attractive properties of classical regenerative processes have been shown to hold for the somewhat weaker regenerative structure of Harris recurrent Markov chains. We briefly review key definitions and results of this framework to apply them to our model. More extensive coverage can be found in Asmussen ([1]) and Nummelin ([39]); the treatment in Sigman ([46]) is particularly relevant to our application.

The general setting for Harris recurrence is a Markov chain $X = \{X_n, n \geq 0\}$ on a state space $\mathbf{S}$ with Borel sets $\mathcal{B}$. Let P_x denote the law of X when $X_0 = x$. Then X is Harris recurrent if there there exists a σ-finite measure π on $(\mathbf{S}, \mathcal{B})$, not identically zero, such that, for all $A \in \mathcal{B}$,

$$\pi(A) > 0 \Rightarrow P_x \left(\sum_{n=0}^{\infty} \mathbf{1}\{X_n \in A\} = \infty \right) = 1, \quad \text{for all } x \in \mathbf{S}. \qquad (2.27)$$

If π is finite (hence a probability, without loss of generality), then X is called *positive* Harris recurrent. If, in addition, X is aperiodic, then it is Harris *ergodic*.

However, if X is positive Harris recurrent and if $f : \mathbf{S} \to \mathbf{R}$ is π-integrable, then the regenerative ratio formula

$$\mathsf{E}_\pi[f(X_0)] = \frac{\mathsf{E}[\sum_{n=\tau_{k-1}}^{\tau_k - 1} f(X_n)]}{\mathsf{E}[\tau_k - \tau_{k-1}]} \qquad (2.28)$$

remains valid, as does the associated central limit theorem (under second-moment assumptions). Moreover, if X is Harris ergodic then for all initial conditions the distribution of X_n converges to π in *total variation;* that is,

$$\sup_{A \in \mathcal{B}} |P_x(X_n \in A) - \pi(A)| \to 0$$

as $n \to \infty$, for all $x \in \mathbf{S}$. Indeed, this total variation convergence to a probability measure completely characterizes Harris ergodicity.

A powerful tool in the analysis of Harris ergodic Markov chains is a connection with coupling; see for example Thorisson ([50]) and Sigman ([46]) for background. The main result is this: a Markov chain with an invariant probability measure admits coupling if and only if it is Harris ergodic. Since we already used a coupling argument for Y in Section 2.3.5, it is now easy to prove this:

Theorem 10 Let demands $\{D_n, n \geq 0\}$ be i.i.d. with $\mathsf{E}[D_0] < \min_i c^i$. Then $\{Y_n, n \geq 0\}$ is a Harris ergodic Markov chain.

Proof. Since $Y_{n+1} = \phi(Y_n, D_n)$, $n \geq 0$, Y is a Markov chain when D is i.i.d. We established in Theorem 8 that Y has an invariant (i.e., stationary) distribution and in Theorem 9 that Y admits coupling. Thus, Y is Harris ergodic. □

As a result of Theorem 10, Y inherits the regenerative structure of Harris ergodic Markov chains and the attendant ratio formula and convergence results. The same holds for the inventory levels:

Corollary 1 Under the conditions of Theorem 10, the inventory process $\{(I_n^1, \ldots, I_n^m), n \geq 0\}$ is a Harris ergodic Markov chain.

Proof. Equations (2.23) put Y_n and $I_n = (I_n^1, \ldots, I_n^m)$ in one-to-one correspondence for all n. Consequently, $I = \{I_n, n \geq 0\}$ is Markov if Y is, and I is Harris ergodic if Y is. □

2.3.7 Cost Implications

The stability results of the previous section make it possible to give a partial characterization of infinite-horizon costs, and this may be useful in optimization.

To each echelon i we assign an inventory holding cost h^i, $i = 1, \ldots, m$. Backorders at stage 1 are penalized at rate p. There is no fixed cost for production in a period; if there were, a base-stock policy would be unattractive. Costs are incurred at the end of each period, so the cost in period n is

$$f(Y_n) \overset{\triangle}{=} p(Y_n^1 - s^1)^+ + \sum_{i=1}^{m} h^i(s^i - Y_n^i)^+. \tag{2.29}$$

From the stability results of Sections 2.3.5 and 2.3.6, we obtain a partial characterization of the infinite-horizon average cost for any choice of parameters. Let $\tilde{Y}_0$ be as in Section 2.3.6, and define

$$F^i(y) = P(\tilde{Y}_0^i \leq y), \ y \in \mathbf{R},$$

for $i = 1, \ldots, m$. As a consequence of Theorem 8, we have

Corollary 2 Under the conditions of Theorem 8,

$$n^{-1} \sum_{i=0}^{n-1} f(Y_i) \to \mathsf{E}[f(\tilde{Y}_0)] = p \int_0^{s^1} (y - s^1) \, dF^1(y) + \sum_{i=1}^{m} h^i \int_0^{s^i} (s^i - y) \, dF^i(y),$$

with probability one, for all Y_0. The case $\mathsf{E}[f(\tilde{Y}_0)] = \infty$ is not excluded.

This result is a direct consequence of the strong law of large numbers for ergodic stationary sequences and the fact that f is non-negative. The form of $\mathsf{E}[f(\tilde{Y}_0)]$ is precisely what one would expect; our results guarantee that the limit holds, and may therefore be useful in finding optimal base-stock levels. In particular, this result can be used in the computation of optimal levels in the two cases where base stock policies are known to be optimal: a multi-stage uncapacitated system and a single-stage capacitated system.

Superficially, the expression in Corollary 2 is the type required for the optimization algorithm of van Houtum and Zijm ([51]) for multi-stage uncapacitated serial systems. In the uncapacitated case, F^i can be expressed in terms of the demand distribution K and s^j, $j \geq i$, and these s^j appear only as location parameters. The shortfall distributions are *nested* because the system decomposes by stages. However, in the presence of capacities, each F^i depends on $s^{i+1}, \ldots, s^m$, and $c^i, \ldots c^m$ in a more intricate way, and so the method of van Houtum and Zijm is not applicable.

2.3.8 Extensions

Let us discuss two obvious extensions.

Assembly System. In an assembly system, each node i requires components from a set of predecessor nodes. These are assembled into stage-i finished goods. By changing units if necessary, we may assume that components from predecessor stages are assembled in equal quantities.

To keep the notation simple, we consider a representative example, rather than the general case. Figure below depicts a three-node system in which node 1 assembles components supplied by nodes 2 and 3. Node 1 feeds external demands; the other nodes draw raw materials from infinite sources. The evolution of inventory at node 1 is characterized by

$$
\begin{aligned}
I^1_{n+1} &= I^1_n - D_n + R^1_{n-\ell^1} \\
R^1_n &= c^1 \wedge \left[s^1 + D_n - (I^1_n + T^1_n)\right]^+ \wedge I^2_n \wedge I^3_n \qquad (2.30) \\
T^1_{n+1} &= T^1_n + R^1_n - R^1_{n-\ell^1}.
\end{aligned}
$$

The assembly feature is reflected in the dependence of R^1_n on I^2_n and I^3_n in (2.30). Nodes 2 and 3 are characterized by the basic recursions (2.5-2.6), with obvious modifications to the indexing. The only notable difference is that now I^2_n and I^3_n are decreased by the same production level R^1_n each period.

Distribution System. Another variant of the serial system allows intermediate stages to supply multiple lower-echelon stages, typically in a tree topology. Our results extend without difficulties to such models. For ease of exposition we describe a less general setting—a serial system in which each stage faces external demands for components in addition to internal demands from the

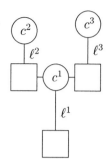

Figure 2.5   An assembly system

downstream stage. A manufacturer of electronic equipment, for example, may face demands for integrated circuits, and for circuits assembled into circuit packs, along with demands for finished goods.

To characterize the operation of such a system, we need to specify how each stage allocates inventory to internal and external demand. Rather than restrict ourselves to any one policy, we describe a class of policies consistent with our results. Let I_{0n}^i be the stage-i inventory reserved for external demands at stage i in period n, and let $I_{1n}^i = I_n^i - I_{0n}^i$ be the inventory available for downstream production. Suppose stage i has base-stock levels s_0^i and s_1^i to supply external and internal demands, and let $s^i = s_0^i + s_1^i$. Denote by D_n^i the external demand at stage i in period n.

The operation of stage 1 is unchanged. At stage i, production is now set according to

$$R_n^i = c^i \wedge \left[s^i + \sum_{j=1}^{i} (D_n^j - I_n^j - T_n^j) \right]^+ \wedge I_{1n}^{i+1}.$$

Also,

$$I_{n+1}^i = I_n^i - R_n^{i-1} - [D_n^i \wedge I_{0n}^i]^+ + R_{n-\ell^i}^i.$$

Pipeline inventory follows (2.7), just as before. It only remains to specify how I_{0n}^i and I_{1n}^i are determined.

A broad class of policies sets

$$I_{1n}^i = g^i (I_n^i, s_1^i + D_n^1 - I_{1n}^i - (\sum_{j=1}^{i-1} I_n^j + T_n^j), s_0^i + D_n^i - I_{0n}^i)$$

and $I_{0n}^i = I_n^i - I_{1n}^i$, for some function g^i; in other words, the inventory reserved for internal production is a function of the available inventory and of the short-falls in meeting internal and external demands. The choice of g^i determines the particular policy.

We briefly describe two types of policies. The first type gives strict priority to internal or external demands, allocating as much inventory as needed to the high-priority demand and using any excess for the other. The second type attempts to balance shortfalls, allocating inventory to minimize the resulting difference between inventories and base-stock levels. This type of policy implements relative priorities through the choice of s_0^i and s_1^i. The assumption that the system can, in fact, be balanced in each period is essential to many analytical approaches; our setting, of course, does not require this. For both types of policies just described, the particular functions g^i are tedious to write out but they are sufficiently smooth to allow derivative calculations.

2.3.9 *An Approximation using Tail Probabilities*

The IPA method could be computationally prohibitive for large problems, or for problems where the starting solution for the simulation is far way from its eventual optimal. So a quick (and accurate!) approximation is presented in [24]: Recall that our objective is to find base-stock levels that approximately minimize holding and backorder costs. The key step in our procedure approximates the distribution of echelon inventory by a sum of exponentials; the parameters of the exponentials are chosen to match asymptotically exact expressions. The computational requirements of the method are minimal. In a test bed of 72 problems, each with five production stages, the average relative error for our approximate optimization procedure is 1.9%.

Much of the technical development of this approximation is based on [21], which uses techniques from [1]. The next chapter discusses this in detail.

2.4   BRIEF SUMMARY OF RELATED TOPICS

1. RE-ENTRANT FLOW SHOPS. How do we handle multi-product *re-entrant* flow lines, a very typical topology in semi-conductor fabrication facilities? (Re-entrant flow lines have attracted significant interest from the research community in recent years because of their direct applicability to semi-conductor fabrication.) This question was motivated by a fab visit at AMD and is studied in [5]. Along the way we are able to study serial multi-product capacitated systems. The framework is similar to that described in section 2 above, but requires a far more detailed analysis.

2. CO-OPERATION IN SUPPLY CHAINS.

   The industrial supplier-customer relations have undergone radical changes in recent years as the philosophy behind managing manufacturing systems is influenced by several Japanese manufacturing practices. As more organizations realize that successful in-house implementation of Just-In-Time alone will have limited effect, they are seeking other members of their supply chain to change their operations. This has resulted in a certain level of co-operation, mainly in the areas of *supply contracts* and *information sharing*, that was lacking before. This is especially true when dealing with customized products, and is most commonly seen between suppliers and their larger customers.

In [19], we incorporate information flow between a supplier (or producer) and a customer in a capacitated setting of a simple supply chain. The customer faces i.i.d. end-product demand, and the supplier has a finite capacity in each period. We consider three situations: (1) a traditional model where there is no information to the supplier prior to a demand to him except from past data; (2) the supplier has the information of the (s,S) policy used by the customer as well as the end-product demand distribution; and (3) full information about the state of the customer. Each of these lead to a different non-stationary demand processes as seen by the supplier. Study of these three models enables us to understand the relationships between capacity, inventory and information at the supplier level. We show that order up-to policies continue to be optimal for models with information flow for the finite horizon, the infinite horizon discounted and the infinite horizon average cost cases. We develop a quick recursive solution procedure to compute the optimal parameters when capacity is infinite. For the finite capacity case, we develop and validate Infinitesimal Perturbation Analysis (IPA), as well as show how the solution for the uncapacitated system can be easily modified to obtain approximate values. Using these solution procedures we estimate the savings at the supplier due to information flow and study when information is most beneficial by varying capacity, holding costs, demand distributions and $S - s$ values.

See a later chapter in this volume for details.

2.5 SUMMARY

Several significant advances have occurred in the study of capacitated systems since 1991. This chapter informally provides an introduction to topics, and some techniques. The reference list below is not exhaustive at all, but should provide a good starting point. At least two very interesting developments (within a single product setting) are not considered in this review: (1) Supply Contracts and (2) International Supply Chains. Later chapters in this edited volume provide details on these topics.

References

[1] Asmussen, S., *Applied Probability and Queues*, Wiley, New York, 1987.

[2] Baccelli, F., W.A. Massey, and D. Towsley, "Acyclic Fork-Join Queueing Networks," *Journal of the Association for Computing Machinery*, 36 (1989), 615-642.

[3] Baccelli, F. and Z. Liu, "On a Class of Stochastic Recursive Sequences Arising in Queueing Theory," *The Annals of Probability*, 20 (1992), 350-374.

[4] Bertsekas, D.P., *Dynamic Programming: Deterministic and Stochastic Models*, Prentice-Hall, Englewood Cliffs, 1987.

[5] Bispo, C. and S. Tayur, "Re-entrant Flow Lines," *GSIA Working Paper*, CMU, Pittsburgh, 1996.

[6] Buzacott, J. and J. G. Shanthikumar, *Stochastic Models of Manufacturing Systems*, Prentice-Hall, New York, 1993.

[7] Chung, K.L., *A Course in Probability Theory*, Academic Press, New York, 1974.

[8] Clark A. and H. Scarf, "Optimal policies for a Multi-Echelon Inventory Problem," *Management Science*, 6 (1960), 474-490.

[9] Ciarallo, F., R. Akella, and T. E. Morton, "A Periodic Review, Production-Planning Model with Uncertain Capacity," *Management Science*, 40 (1994), 320-332.

[10] Cohen M.A. and H.L.Lee, "Scale Economics, Manufacturing Complexity, and Transportation Costs on Supply Chain Facility Networks," *Journal of Manufacturing and Operations Management*, 3 (1990), 269-292.

[11] Cohen M.A. and S. Moon, "Impact of Production Scale Economics, Manufacturing Complexity, and Transportation Costs on Supply Chain Facility Networks," *Journal of Manufacturing and Operations Management*, 3 (1990), 269-292.

[12] Debodt M. and S.C. Graves, "Continuous review policies for a Multi-Echelon Inventory Problem with Stochastic Demand," *Management Science*, 31 (1985), 1286-1299.

[13] Federgruen, A., P. Schweitzer, and H. Tijms, "Denumerable Undiscounted Decision Processes with Unbounded Rewards," *Mathematics of Operations Research*, 8 (1983), 298-314.

[14] Federgruen, A. and P. Zipkin, "Approximations of Dynamic Multilocation Production and Inventory Problems," *Management Science*, 30 (1984), 69-84.

[15] Federgruen, A. and P. Zipkin, "Computational Issues in an Infinite Horizon Multi-Echelon Inventory Model," *Operations Research*, 32 (1984), 218-236.

[16] Federgruen, A. and P. Zipkin, "An Inventory Model with Limited Production Capacity and Uncertain Demands I: The Average-cost Criterion," *Mathematics of Operations Research*, 11 (1986), 193-207.

[17] Federgruen, A. and P. Zipkin, "An Inventory Model with Limited Production Capacity and Uncertain Demands II: The Discounted-cost Criterion," *Mathematics of Operations Research*, 11 (1986), 208-215.

[18] Fox, B.L. and P.W. Glynn, "Simulating Discounted Costs," *Management Science*, 35 (1989), 1297-1315.

[19] Gavirneni, S., R. Kapuscinski, and S. Tayur, "Value of Information in Capacitated Supply Chains," *GSIA Working Paper*, CMU, Pittsburgh, October 1996.

[20] Glasserman, P., *Gradient Estimation via Perturbation Analysis*, Kluwer, Norwell, Massachusetts, 1991.

[21] Glasserman, P., "Bounds and Asymptotics for Planning Critical Safety Stocks," *Columbia University Working Paper*, 1994, to appear in *Operations Research.*

[22] Glasserman, P. and S. Tayur, "Sensitivity analysis for base stock levels in multi-echelon production-inventory system," *Management Science*, 41 (1995), 263-281.

[23] Glasserman, P. and S. Tayur, "The Stability of a Capacitated, Multi-Echelon Production-Inventory System under a Base-Stock Policy," *Operations Research*, 42 (1994), 913-925.

[24] Glasserman, P. and S. Tayur, "A Simple Approximation for a Multistage Capacitated Production-Inventory System," *Naval Research Logistics*, 43 (1996), 41-58.

[25] Glasserman, P. and D.D. Yao, "The Stochastic Vector Equation $Y_{n+1} = A_n \otimes Y_n \oplus B_n$ with Stationary Coefficients," *Columbia University Working Paper*, 1992, to appear in *Advances in Applied Probability.*

[26] Glynn, P.W., P. L'Ecuyer, and M. Adès, "Gradient Estimation for Ratios," *Proceedings of the Winter Simulation Conference,* The Society for Computer Simulation, San Diego, California, 1991, 986-993.

[27] Graves, S. C., A. Rinnoy Kan, and P. Zipkin (Eds.), *Logistics of Production and Inventory*, Handbooks in Operations Research and Management Science, Vol. 4, Elsevier North Holland, Amsterdam, 1992.

[28] Kamesam, P. V. and S. Tayur, "Algorithms for Multi-Stage, Capacitated Assembly Systems with Stochastic Demand," *IBM Report*, 1993.

[29] Karlin, S., "Dynamic Inventory Policy with Varying Stochastic Demands," *Management Science*, 6 (1960), 231-258.

[30] Karlin, S., "Optimal Policy for Dynamic Inventory Process with Stochastic Demands Subject to Seasonal Variations," *Journal of the Society for Industrial and Applied Mathematics*, 8 (1960), 611-629.

[31] Kapuscinski, R. and S. Tayur, "A Capacitated Production-Inventory Model with Periodic Demand," *GSIA Working Paper*, CMU, Pittsburgh, January 1996, to appear in *Operations Research.*

[32] Kapuscinski, R. and S. Tayur, "100% Reliable Quoted Lead Times," *GSIA Working Paper*, CMU, Pittsburgh, 1996.

[33] Karmarkar U., "Lot Sizes, Lead times and In-Process Inventories," *Management Science*, 33 (1987), 409-418.

[34] Langenhoff, L.J.G. and W.H.M. Zijm, "An Analytical Theory of Multi-Echelon Production/Distribution Systems," *Working Paper*, Eindhoven University of Technology, Department of Mathematics and Computer Sciences, 1989, to appear in *Statistica Neerlandica.*

[35] Lee H.L., C. Billington, and B. Carter, "Gaining Control of Inventory and Service through Design for Localization," *Working paper*, 1991.

[36] Morton, T., "The Non-Stationary Infinite Horizon Inventory Problem," *Management Science*, 24 (1978), 1474-1482.

[37] Morton, T. and D. Pentico, "The Finite Horizon Non-Stationary Stochastic Inventory Problem," *Management Science*, 41 (1995), 334-343.

[38] Muckstadt, J., M. Lambrecht and R. Luyten, "Protective Stocks in Multi-stage Production Systems," *International Journal of Production Research*, 6 (1984), 1001-1025.

[39] Nummelin, E., *General Irreducible Markov Chains and Non-Negative Operators*, Cambridge University Press, London, 1984.

[40] Parker, R. and R. Kapuscinski, "Is Base-Stock Policy Optimal for Capacitated Multi-Echelon Systems?," *Working Paper*, University of Michigan, Ann Arbor, 1998.

[41] Prabhu, U., *Queues and Inventories*, Wiley, New York, 1965.

[42] Prabhu, U., *Stochastic Storage Processes*, Springer-Verlag, New York, 1980.

[43] Rosling, K., "Optimal Inventory Policies for Assembly Systems under Random Demands," *Operations Research*, 37 (1989), 565-579.

[44] Scheller-Wolf, A. and S. Tayur, "Reducing International Risk through Quantity Contracts," *GSIA Working Paper*, CMU, Pittsburgh, February 1997.

[45] Schmidt, C. and S. Nahmias, "Optimal Policy for a Two-Stage Assembly System Under Random Demand," *Operations Research*, 33 (1985), 1130-1145.

[46] Sigman, K., "Queues as Harris Recurrent Markov Chains," *QUESTA*, 3 (1988), 179-198.

[47] Speck, C.J. and J. Van der Wal, "The capacitated multi-echelon inventory system with serial structure: the average cost criterion," *COSOR 91-39*, Dept. of Math. and Comp. Sci., Eindhoven Institute of Technology, Eindhoven, The Netherlands, 1991.

[48] Tayur, S., "Computing the Optimal Policy in Capacitated Inventory Models," *Stochastic Models*, 9 (1993), 585-598.

[49] Tayur, S., "Recent Developments in single product, discrete time, capacitated production-inventory systems," *Sadhana*, 22 (1997), 45-67.

[50] Thorisson, H., "The Coupling of Regenerative Processes," *Advances in Applied Probability*, 15 (1983), 531-561.

[51] Van Houtum, G.J. and W.H.M. Zijm, *Computational Procedures for Stochastic Multi-Echelon Production Systems*, Centre for Quantitative Methods, Nederlandse Philips Bedrijven, B.V. 1990.

[52] Van Houtum, G.J. and W.H.M. Zijm, "On multi-stage production/inventory systems under stochastic demand," *International Journal of Production Econonomics*, 35 (1994), 391-400.

40

[53] Van Houtum, G.J., K. Inderfuth, and W.H.M. Zijm, *Materials Coordination in Stochastic Multi-Echelon Systems*, University of Twente, December 1995.

[54] Zipkin, P., "Critical Number Policies for Inventory Models with Periodic Data," *Management Science*, 35 (1989), 71-80.

3 SERVICE LEVELS AND TAIL PROBABILITIES IN MULTISTAGE CAPACITATED PRODUCTION-INVENTORY SYSTEMS

Paul Glasserman

Columbia Business School
New York, NY 10027

3.1   INTRODUCTION

In most models of inventory systems with uncertain demands, inventory serves as a hedge against demand variability during a re-order leadtime — if replenishment were instantaneous, there would be no need to hold inventory. But the leadtime itself is generally not modeled in detail; rather, it is commonly represented by a fixed interval or, when it is taken to be stochastic, independent of the rest of the model. In some settings — for example when the leadtime is primarily due to transportation delays — a simple such model may be a fair representation of reality. But leadtimes can arise not only from external factors like transportation, but also from congestion effects internal to the operation of a system. In particular, when limits on production capacity are significant, the primary delay in replenishing stock may be due to backlogs in production created by the replenishment orders themselves, rather than to any external mechanism. In the terminology of Svoronos and Zipkin [33], leadtimes are *endogenous* to the model in such settings.

This chapter is concerned with a class of inventory systems in which limits on production capacity are modeled explicitly, and inventory is therefore held at least in part to compensate for the fact that the capacity required to meet demands may occasionally exceed the capacity available. Such systems have features in common with queueing systems and have in some cases been studied through models that draw on queueing theory (cf. Buzacott and Shanthikumar [4, 5], Lee and Zipkin [20] Veatch and Wein [35], Schraner [27], Song [30], Zipkin [37].) The models we treat here are extensions of classical inventory models — in particular, they are capacitated versions of the Clark-Scarf [6] model and multistage or multi-item extensions of the single-stage, single-item capacity constrained model of Federgruen and Zipkin [10, 11]. (See also Aviv and Federgruen [2], Federgruen [9], Kapuscinski and Tayur [19], Rosling [25], Speck and van der Wal [32], Tayur [34].)

Our objective is by no means to give a comprehensive treatment of inventory systems with limited production capacity. On the contrary, we intend to cover the use of just one specific tool for characterizing and approximating the performance of such systems. The principle underlying this tool is the following: if the tail of the distribution of demands is exponentially bounded, then the tail of the distribution of the *inventory shortfall* — the amount by which available inventory falls short of a target level — is approximately exponential; moreover, the exponent in this approximation is easily calculated. This exponent and some associated approximations are our primary object of study.

This may seem an unconventional focus of attention in an inventory setting where optimal policies, average or discounted costs, and expected backlogs and inventory levels tend to be emphasized. Our premise is that service levels are mainly determined by tail probabilities so when service levels are of primary interest (as they often are in practice) a precise descriptor of the tail of the shortfall can be an informative gauge of system performance. We will illustrate this principle through a variety of specific results. Moreover, we will see that the tail exponent captures quite a bit of model detail — in particular, it reflects both the average load on a production facility and the variability in demands and production.

We will consider a variety of models: single-stage and multistage, discrete review and continuous review, single-item and multi-item. An important assumption in all of these settings is that the system in question operates under a *base-stock* or *one-for-one* replenishment policy. In a single-stage stage, Federgruen and Zipkin [10, 11] showed that such a policy is optimal, as it is in the multistage uncapacitated Clark-Scarf [6] model. The Federgruen-Zipkin result has been extended to models with serially correlated demands by Aviv and Federgruen [2] and Kapucsinski and Tayur [19]. We do not suggest that base-stock policies remain optimal in the models treated here — only that they remain reasonable and convenient. Our intention is therefore to analyze the performance of capicitated models under a sensible policy rather than to try to identify the optimal policy.

The rest of this chapter is organized as follows. Section 3.2 introduces the basic model and approach through the example of a single-stage, single-item system, presenting both bounds and approximations to tail probabilities. Section 3.3 then sketches the underlying theoretical tools used for these results and those of later sections. Multistage versions of the simple model in Section 3.2 are treated in Section 3.4 and then again in Section 3.5 from the perspective of heavy-traffic. Section 3.6 illustrates the use of the techniques in previous sections with a continuous-review, assemble-to-order, multi-item system. Section 3.7 concludes the chapter.

3.2 A SINGLE-STAGE SYSTEM

Although the single-stage system is too simple to be of much intrinsic interest, it provides a useful setting in which to introduce the methods used throughout this chapter. The development in this section follows Glasserman [13].

3.2.1 Shortfall Formulation

We consider a storage facility supplying external demands and receiving stock from a production facility. Time is divided into periods of fixed length. In each period, demands arrive and are either filled or backordered. The system operates under a base-stock policy in which production is set in each period to restore inventory to a target level s while not exceeding the per-period capacity c of the production facility. Thus, if I_n denotes the net inventory (on-hand inventory minus backorders) at the start of period n, and if D_n is the demand in period n, then production in period n is $\min\{c, s - I_n + D_n\}$. The net inventory at the start of the next period is

$$
\begin{aligned}
I_{n+1} &= I_n - D_n + \min\{c, s - I_n + D_n\} \\
&= \min\{c + I_n - D_n, s\};
\end{aligned}
\tag{3.1}
$$

in particular, on-hand inventory never exceeds the target level s.

For the approach used throughout this chapter, it turns out to be convenient to replace the net inventory I_n with the *shortfall* $Y_n = s - I_n$, the amount by which the target inventory exceeds the net inventory. Clearly, Y_n and I_n provide equivalent information about the state of the system. In light of (3.1), Y_n is non-negative and satisfies

$$
\begin{aligned}
Y_{n+1} &= s - \min\{c + I_n - D_n, s\} \\
&= \max\{Y_n + D_n - c, 0\}.
\end{aligned}
\tag{3.2}
$$

This is a Lindley recursion and shows that the shortfall sequence coincides with the waiting-time sequence in a single-server queue with service times $\{D_n, n \geq 0\}$ and fixed interarrival time c. This correspondence is used in Tayur [34] and is part of the general treatment of queueing and inventory models in Prabhu [23]. It follows from (3.2) that if we assume demands are independent and identically distributed with

$$
E[D_1] < c,
\tag{3.3}
$$

then Y_n converges in distribution to a random variable Y satisfying

$$
Y =_d \max\{Y + D - c, 0\},
\tag{3.4}
$$

where $=_d$ denotes equality in distribution and D is a random variable independent of Y having the distribution of demands.

Various measures of performance are easily expressed in terms of Y. The long-run average proportion of periods in which no stockout occurs, the *stock availability*, is

$$
\alpha(s) = P(Y \leq s).
\tag{3.5}
$$

The *fill rate*, which is the long-run average proportion of demands met from stock, is given by

$$
\beta(s) = 1 - \frac{E[\max\{0, \min\{Y + D - c - s, D\}\}]}{E[D]}.
\tag{3.6}
$$

To see this, observe that when the shortfall is Y and the demand is D, the demand not met from stock is all of D if $Y > s$ (no on-hand inventory) and is the amount by which $Y + D - c$ exceeds s if $Y \leq s$. Thus, $E[\max\{0, \min\{Y + D - c - s, D\}\}]$ is the expected demand not met from stock in each period. Writing x^+ for $\max\{0, x\}$, the long-run average expected backlog is

$$
b(s) = E[(Y - s)^+].
\tag{3.7}
$$

If backorders are penalized at rate $p > 0$ and holding costs charged at rate $h > 0$, then the long-run average cost per period is

$$
\begin{aligned}
v(s) &= hE[(s - Y)^+] + pE[(Y - s)^+] \\
&= h(s - E[Y]) + (p + h)E[(Y - s)^+].
\end{aligned}
\tag{3.8}
$$

We develop approximations for these performance measures and for base-stock levels that achieve specified values of these measures.

Equation (3.2) and definitions (3.5)-(3.8) presuppose that production decisions are made after the total demand in a period is revealed. A model in which production must be set before the demand is known is equivalent to one with a leadtime equal to a single period. We return to this case in Section 3.4.3.

3.2.2 Approximations

To motivate our approximations, consider demands that are exponentially distributed with mean $1/\mu$. In this setting, as noted by Tayur [34], the stationary shortfall distribution coincides with the stationary waiting time in a D/M/1 queue. It follows that

$$P(Y > s) = (1 - \frac{\gamma}{\mu})e^{-\gamma s}, \tag{3.9}$$

where $\gamma > 0$ solves $\mu e^{-\gamma c}/(\mu - \gamma) = 1$. Indeed, the equality $P(Y > s) = Ce^{-\gamma s}$ holds for all G/M/1 queues; see Prabhu [23, p.109] or Asmussen [1, p.204]. Thus, when demands are exactly exponential, the tail of the shortfall distribution is exactly exponential.

This has several convenient and immediate consequences, some of which we record in the following result:

Proposition 1 *Suppose the demand distribution is $F_D(x) = 1 - \exp[-\mu x]$. Then, for all $s > 0$,*

(i) $1 - \alpha(s) = (1 - \gamma/\mu)e^{-\gamma s}$;

(ii) $b(s) = (1 - \gamma/\mu)\gamma^{-1}e^{-\gamma s}$;

(iii) $1 - \beta(s) = e^{-\gamma(s+c)}$.

Each of these expressions can be inverted to obtain the base-stock level required to meet a desired service level. For example, to achieve a stock availability of $1 - \delta$, for some $0 < \delta < 1$ (e.g., $\delta = .01$ for a 99% service level) we would set $\alpha(s_\delta) = \delta$ and solve to get

$$s_\delta = -\frac{1}{\gamma} \log \left(\frac{\delta \mu}{\mu - \gamma} \right).$$

The other two cases work similarly.

Once we move away from exponentially distributed demands, we cannot expect the shortfall distribution to be exactly exponential, nor can we expect to obtain exact results of the type in (1). But we can get surprisingly close.

We impose some mild assumptions on demands. We continue to assume that $\{D_n, n \geq 0\}$ are (non-negative and) i.i.d., and we denote their distribution by F_D. For simplicity, we assume F_D is continuous, though our results have predictable counterparts for integer-valued demands. The stability condition (3.3) is in force throughout. In addition, we assume that

$$P(D_1 > c) > 0;$$

otherwise, demands can always be met from the current period's production. Our most important assumption involves the moment generating function of $D_1 - c$, given by

$$\phi(\theta) = e^{-\theta c} \int_0^\infty e^{\theta x} \, dF_D(x).$$

We assume that there exists a $\theta_0 > 0$ at which

$$1 < \phi(\theta_0) < \infty. \tag{3.10}$$

This condition, together with the convexity of ϕ, and the fact that $\phi(0) = 1$ and $\phi'(0) = E[D_1 - c] < 0$, implies the existence of just one $\gamma > 0$ with $\phi(\gamma) = 1$, which then has $\phi'(\gamma) < \infty$; that is,

$$E[e^{\gamma(D_1 - c)}] = 1 \tag{3.11}$$

and $E[(D_1 - c)e^{\gamma(D_1 - c)}] < \infty$. The solution γ to (3.11) is called the *conjugate point* for the distribution of $D_1 - c$. It plays a central role in our approximations, beginning with Theorem 1, below. In the statement of the theorem, the notation $f(x) \sim g(x)$ means that $f(x)/g(x)$ converges to unity as $x \to \infty$.

Theorem 1 *There is a constant* C, $0 < C \le 1$, *depending on the demand distribution and the capacity, such that*

(i) *the stock availability satisfies* $1 - \alpha(s) \sim Ce^{-\gamma s}$;

(ii) *the average backlog satisfies* $b(s) \sim (C/\gamma)e^{-\gamma s}$;

(iii) *the fill rate satisfies* $1 - \beta(s) \sim (C/\gamma E[D])(1 - e^{-\gamma c})e^{-\gamma s}$.

This result follows from the general approach reviewed in Section 3.3. The conjugate point γ appearing in these approximations exists for all commonly used demand distributions; Table 3.1 gives examples. Generally, γ is not available explicitly but numerical solution requires little effort because γ is the unique strictly positive root of a convex function. A rough approximation to γ can be obtained by expanding $\log \phi$ to second-order,

$$\log \phi(\theta) \approx E[D - c]\theta + \frac{1}{2} \text{Var}[D - c]\theta^2,$$

and finding the $\theta > 0$ at which this quadratic equals 0, namely

$$\gamma \approx -\frac{2E[D - c]}{\text{Var}[D - c]}. \tag{3.12}$$

This two-moment approximation is exact for normally distributed $D - c$.

In practice, the constant C may be difficult to evaluate so further approximation is warranted. A simple upper bound is obtained by replacing C with 1, as will be clear from the derivation in Section 3.3 and the discussion of heavy

Table 3.1 The parameter γ for some demand distributions. The third column gives defining equations for γ along with the range in which it must lie.

Name	Density or Mass Function	Gamma
Exponential	$\mu e^{-\mu x}$	$(\frac{\mu}{\mu-\gamma})e^{-\gamma c} = 1$ in $(0, \mu)$
Normal	$(\sigma\sqrt{2\pi})^{-1}\exp[-\frac{(x-\mu)^2}{2\sigma^2}]$	$\gamma = 2(c - \mu)/\sigma^2$
Gamma	$\frac{\mu^m x^{m-1}}{\Gamma(m)}e^{-\mu x}$	$(\frac{\mu}{\mu-\gamma})^m e^{-\gamma c} = 1$ in $(0, \mu)$
Hyperexpnl.	$p\mu_1 e^{-\mu_1 x} + (1-p)\mu_2 e^{-\mu_2 x}$	$[p(\frac{\mu_1}{\mu_1-\gamma}) + (1-p)(\frac{\mu_2}{\mu_2-\gamma})]e^{-\gamma c} = 1$ in $(0, \min\{\mu_1, \mu_2\})$
Poisson	$e^{-\lambda}\lambda^k/k!$	$\lambda(e^\gamma - 1) - \gamma c = 0,\ \gamma > 0$
Neg. Binom.	$\binom{k-1}{m-1}p^m(1-p)^{k-m}$	$m\log[1 + p^{-1}(e^{-\gamma} - 1)] + \gamma c = 0,$ in $(0, -\log(1-p))$

traffic in Section 3.5. Somewhat better bounds are obtained through a method of Ross [26]. Define

$$C_- = \inf_{r \geq c} \left(E[\exp\{\gamma(D_1 - r)\}|D_1 > r]\right)^{-1} \qquad (3.13)$$

and

$$C_+ = \sup_{r \geq c} \left(E[\exp\{\gamma(D_1 - r)\}|D_1 > r]\right)^{-1}. \qquad (3.14)$$

Then we have

Theorem 2 *For all $s > 0$,*

(i) $C_- e^{-\gamma s} \leq 1 - \alpha(s) \leq C_+ e^{-\gamma s}$;

(ii) $(C_-/\gamma)e^{-\gamma s} \leq b(s) \leq (C_+/\gamma)e^{-\gamma s}$;

(iii) $(C_-/\gamma E[D])[1 - e^{-\gamma c}]e^{-\gamma s} \leq 1 - \beta(s) \leq (C_+/\gamma E[D])[1 - e^{-\gamma c}]e^{-\gamma s}$.

The use of these bounds is illustrated in Glasserman [13]. When demands are exponential, C_- and C_+ coincide and we recover (3.9).

As an application of Theorems 1 and 2, we consider the problem of setting the base-stock level to ensure that, over an infinite horizon, stockouts occur in at most a fraction $\delta > 0$ of periods. This is the problem of setting s so that $\alpha(s) \geq 1 - \delta$, with δ equal to, say, .01.

Corollary 1 *Let s_δ be the minimal base-stock level for which a stock availability of at least $1 - \delta$ is guaranteed; i.e., the minimal s satisfying $\alpha(s) \geq 1 - \delta$. Then*

$$\gamma^{-1}\log(C_-/\delta) \leq s_\delta \leq \gamma^{-1}\log(C_+/\delta) \quad \text{for all sufficiently small } \delta > 0, \quad (3.15)$$

and $s_\delta \leq -\gamma^{-1} \log \delta$ for all $\delta > 0$. If the demand distribution is continuous,

$$|s_\delta - \gamma^{-1} \log(C/\delta)| \to 0 \quad as \quad \delta \to 0. \tag{3.16}$$

Let s'_δ be the smallest s for which $\beta(s) \geq 1 - \delta$. Then

$$\gamma^{-1} \log \left(C_-[1 - e^{-\gamma c}]/\gamma \delta E[D_1]\right) \leq s'_\delta \leq \gamma^{-1} \log \left(C_+[1 - e^{-\gamma c}]/\gamma \delta E[D_1]\right) \tag{3.17}$$

for all sufficiently small $\delta > 0$. If demands are continuous, then

$$\left|s'_\delta - \gamma^{-1} \log \left(C[1 - e^{-\gamma c}]/\gamma \delta E[D_1]\right)\right| \to 0 \quad as \; \delta \to 0. \tag{3.18}$$

Proof. The bounds on s_δ follow from Theorem 2(i) by inverting the bounds on α and the fact that $C \leq 1$. The bounds in Theorem 2 are valid only for $s > 0$, so to invert them we need $\delta < C_+$ in (3.15) and $\delta < C_+[1 - e^{-\gamma c}]/(\gamma E[D_1])$ in (3.17), which is not a restriction since we are primarily interested in small δ.

For (3.16), notice from (3.15) that $s_\delta \to \infty$ as $\delta \to 0$, so from Theorem 1(i), $C^{-1} e^{\gamma s_\delta} P(Y > s_\delta) \to 1$; i.e., $C^{-1} e^{\gamma s_\delta} \delta \to 1$, implying that

$$|\log C^{-1} + \gamma s_\delta + \log \delta| \to 0,$$

which is equivalent to (3.16). The assertions regarding s'_δ are proved in the same way. □

A variant of Corollary 1 holds for the base-stock level minimizing the long-run average cost $v(s)$ defined in (3.8). We examine the optimal base-stock level as the backorder penalty p becomes large.

Corollary 2 *Suppose F_D is continuous; then v is convex. Suppose s_p minimizes v; then*

$$\gamma^{-1} \log((p+h)C_-/h) \leq s_p \leq \gamma^{-1} \log((p+h)C_+/h) \quad \text{for all sufficiently large } p, \tag{3.19}$$

and $s_p \leq \gamma^{-1} \log((p+h)/h)$ for all $p > 0$. Moreover,

$$|s_p - \gamma^{-1} \log((p+h)C/h)| \to 0 \quad as \; p \to \infty. \tag{3.20}$$

Proof. Differentiation of (3.8) yields $v'(s) = h - (p+h)P(Y > s)$, an increasing function of s, making v convex. The minimum of v is achieved at the point s_p satisfying $v'(s_p) = 0$; i.e., satisfying $P(Y > s_p) = h/(p+h)$. The bounds and limiting behavior of s_p thus follow from the bounds and asymptotics of the tail distribution of Y, just as in Corollary 1. □

The case in which production decisions are made before demands are revealed is equivalent to a model with a leadtime of 1 between production and availability of finished goods. Our approximations are easily adapted to this case; in particular, for $s \geq c$,

$$P(Y + D > s) = P(\max\{Y + D - c, 0\} > s - c) = P(Y > s - c) \sim Ce^{-\gamma(s-c)},$$

so the effect on the asymptotic behavior is to change the constant C to $Ce^{\gamma c}$. It follows that in (3.15) and (3.19), c should be added to both the lower and upper bounds (with the lower bound now valid for $s > c$, rather than $s > 0$), and in (3.16) and (3.20), c should be added to the limiting approximations.

Roundy and Muckstadt [24] have carried out a numerical study of some of the approximations from Glasserman [13] reviewed in this section and have proposed refined approximations. The refined approximations exploit (3.4) but also rely on the exponent γ to capture the tail of the shortfall distribution. Numerical experiments in Roundy and Muckstadt [24] indicate that the refinements can significantly improve the approximations, particularly in systems that are not very heavily utilized.

3.3   TAIL PROBABILITIES AND APPROXIMATIONS

The purpose of this section is to give some insight into the exponential form of the approximations to tail probabilities in the previous and subsequent sections. An excellent source for the tools used here is Chapter 12 of Asmussen [1], where many additional references to the related literature on ruin probabilities can also be found. Section 12.4 of Feller [12] treats similar results as an application of renewal theory and Laplace transforms. These ideas may also be viewed as special cases of the far more general techniques of large deviations, as in Dembo and Zeitouni [7].

3.3.1   Exponential Upper Bounds

Consider, first, a random variable X with a moment generating function that is finite in a neighborhood of the origin. This includes most of the distributions commonly used in inventory models (normal, gamma, negative binomial, and all bounded random variables) but excludes "heavy-tailed" distributions like the lognormal and Pareto distributions. Denote by ψ the *cumulant* generating function (cgf) of X, which is simply the logarithm of the moment generating function:

$$\psi(\theta) = \log E[e^{\theta X}]. \tag{3.21}$$

The following properties are easily verified: ψ is convex, $\psi(0) = 0$, $\psi'(0) = E[X]$, and $\psi''(0) = Var[X]$; see Kendall [18] for background.

The cgf arises naturally in bounding the tail of X. Define

$$\mathbf{1}_{(x,\infty)}(y) = \begin{cases} 0, & y \leq x; \\ 1, & y > x. \end{cases}$$

Observe that

$$\mathbf{1}_{(x,\infty)}(y) \leq e^{\theta(y-x)},$$

for any $\theta \geq 0$. Evaluating both sides at $y = X$ and taking expectations, we get

$$P(X > x) \leq E[e^{\theta(X-x)}] = e^{\psi(\theta)-\theta x} \equiv C_\theta e^{-\theta x} \tag{3.22}$$

for any $\theta \geq 0$ at which $\psi(\theta) < \infty$. This indicates that at least exponential upper bounds are often available for tail probabilities.

Now let X_i be i.i.d. with cgf ψ and set $S_n = X_1 + \cdots + X_n$ with $S_0 = 0$. Then

$$E[\exp(\theta S_n)] = E\left[\prod_{i=1}^{n} e^{\theta X_i}\right] = e^{n\psi(\theta)}.$$

Thus, arguing as in (3.22) we get

$$P(S_n > x) \leq e^{-\theta x + n\psi(\theta)}, \tag{3.23}$$

provided only that $\psi(\theta) < \infty$ and $\theta \geq 0$.

In Section 3.2.1 we characterized the stationary shortfall Y in (3.4) through the Lindley recursion (3.2). A standard property of the Lindley recursion (e.g., Section 3.7 of Asmussen [1]) implies that

$$Y =_d \max_{n \geq 0} S_n, \quad \text{with } X_i = D_i - c, \quad i = 1, 2, \ldots.$$

Thus, to bound the tail of the shortfall we need to bound the maximum of the random walk S_n. For the maximum to exceed x, at least one S_n must exceed x so we have the crude bound

$$P(\max_{n \geq 0} S_n > x) \leq \sum_{n \geq 0} P(S_n > x). \tag{3.24}$$

Applying (3.23), we find that

$$P(\max_{n \geq 0} S_n > x) \leq \sum_{n \geq 0} e^{-\theta x + n\psi(\theta)} \equiv C_\theta e^{-\theta x},$$

if

$$C_\theta = \sum_{n=0}^{\infty} e^{-n\psi(\theta)} < \infty.$$

(We use the notation C_θ to denote a generic constant depending on θ; its definition depends on the context.) This is a geometric series and is therefore convergent if and only if $\psi(\theta) < 0$. We thus find that the tail of Y decays at least as quickly as any exponential with decay rate less than

$$\sup\{\theta > 0 : \psi(\theta) < 0\}.$$

It further suggests that the actual decay rate for Y may be the smallest positive root of the equation $\psi(\theta) = 0$. Exponentiating both sides of the equation $\psi(\theta) = 0$, we find that this is the same as the smallest positive root of

$$1 = e^{\psi(\theta)} = E[e^{\theta X}] = E[e^{\theta(D-c)}],$$

and this is precisely the conjugate point γ defined in (3.11).

The development thus far should at least make it plausible that γ correctly measures the exponential decay of the tail of the stationary shortfall. However, as defined above, C_θ diverges as $\theta \uparrow \infty$; the bound in (3.24) suffices for identifying the exponent but is too crude to lead to a constant multiplying the exponential decay rate. For this, we need the sharper treatment of the next subsection.

3.3.2 Tail Asymptotics

We have already identified the stationary shortfall distribution with the maximum of a random walk having negative drift. That the tail of such a maximum is asymptotically exponential is a classical result; it can be established as a purely analytic result through Laplace transforms (as in Feller [12]) or through a change of measure argument (as in Asmussen [1] or Siegmund [29]). We present the change of measure argument because it is the one that most easily extends to multistage systems and to the other results discussed in this chapter.

We assume the conditions of Section 3.2.1 are in force; in particular, $E[D] < c$ and the conjugate point $\gamma > 0$ solving $\phi(\gamma) = 1$ (i.e., $\psi(\gamma) = 0$) exists. From the demand distribution F_D define a new distribution $\tilde{F}$ on $(-c, \infty)$ by

$$\tilde{F}(x) = \int_0^{x+c} e^{\gamma(t-c)} \, dF_D(t), \quad x > -c. \tag{3.25}$$

The condition $\phi(\gamma) = 1$ ensures that $\tilde{F}$ is indeed a probability distribution. Let $\{\tilde{X}_n, n \geq 0\}$ be i.i.d. with distribution $\tilde{F}$ and set

$$\tilde{S}_n = \tilde{X}_1 + \cdots + \tilde{X}_n, \ n \geq 1; \quad \tilde{S}_0 = 0. \tag{3.26}$$

This is the *conjugate* random walk associated with $\{S_n, n \geq 0\}$. A simple calculation shows that

$$E[\tilde{X}_1] = \phi'(\gamma) > 0, \tag{3.27}$$

so the new random walk $\tilde{S}_n$ has positive drift, whereas the original random walk S_n has drift $E[D_1 - c] < 0$.

We now use the argument in Asmussen [1, §12.5] to analyze $P(Y > x)$. For $x > 0$, let

$$T_x = \inf\{n \geq 1 : S_n > x\},$$

the first time the original random walk crosses level x, with the convention that $T_x = \infty$ if x is never crossed. Define $\tilde{T}_x$ from $\tilde{S}_n$ analogously. Because $\tilde{S}_n$ has positive drift, $\tilde{T}_x$ is finite with probability one. Clearly, the events $\{\max_{n \geq 0} S_n > x\}$ and $\{T_x < \infty\}$ are the same. Hence,

$$P(Y > x) = P(\max_{n \geq 0} S_n > x) = P(T_x < \infty).$$

The relation between F_D and $\tilde{F}$ and an application of Wald's identity give

$$P(T_x < \infty) = E[\exp(-\gamma \tilde{S}_{\tilde{T}_x})].$$

Rather than deriving this here, we refer the interested reader to Asmussen [1, §12.5]. Combining the previous two displays, we arrive at

$$P(Y > x) = e^{-\gamma x} E[\exp\{-\gamma(\tilde{S}_{\tilde{T}_x} - x)\}].$$

Observe that $\tilde{S}_{\tilde{T}_x} - x$ is the amount by which the random walk $\tilde{S}_n$ overshoots x when it crosses x for the first time. For large x one would expect that the magnitude of this overshoot would be insensitive to x; indeed, an application of the renewal theorem shows that the overshoot converges in distribution as $x \to \infty$. Because the function $u \mapsto \exp(-\gamma u)$ is bounded and continuous, it follows that

$$C \triangleq \lim_{x \to \infty} E[\exp\{-\gamma(\tilde{S}_{\tilde{T}_x} - x)\}] \tag{3.28}$$

exists. We have thus shown that

$$e^{\gamma x} P(Y > x) \to C,$$

which is to say that $P(Y > x) \sim C \exp(-\gamma x)$. The expression given for C shows that $C \leq 1$.

3.4 MULTISTAGE SYSTEMS

We now return to the types of models considered in Section 3.2 but generalize to multistage systems. Each system now consists of d nodes in series. Node 1 supplies external demands, node i draws material from node $i+1$, $i = 1, \ldots, d-1$, and node d draws from an unlimited supply of raw material. Node i has capacity c^i, $i = 1, \ldots, d$, in the sense that this is the maximum amount of material that can be processed at node i in each period.

3.4.1 Shortfall Formulation

Each node of the multistage system follows a base-stock policy for *echelon* inventory, operating as follows. Let I_n^1 be the net inventory (stock on hand minus backorders) at stage 1 and let I_n^i be the inventory available at stage i, $i = 2, \ldots, d$, all in period n. The echelon-i inventory at the start of period n is $I_n^1 + \cdots + I_n^i$, $i = 1, \ldots, d$; this drops by D_n upon the arrival of demands in that period. Subsequently, stage i sets production to restore the echelon-i inventory to a base-stock level s^i, while not exceeding its capacity c^i or the available supply I_n^{i+1} of predecessor inventory. Thus, production at node i in period n is given by

$$\min\{s^i + D_n - \sum_{j=1}^{i} I_n^j, c^i, I_n^{i+1}\}, \tag{3.29}$$

with $I_n^{d+1} \equiv \infty$ since node d is not constrained by upstream supply. Because the s^i's are target levels for *cumulative* stock, we always assume that $s^1 \leq s^2 \leq \cdots \leq s^d$.

As shown in Glasserman and Tayur [15], the dynamics of this system are conveniently represented through *echelon shortfalls*. The shortfall for echelon i at the start of period n is

$$Y_n^i = s^i - \sum_{j=1}^{i} I_n^j,$$

the difference between the target and actual echelon inventories. Using the expression in (3.29) for the production at stage i, Glasserman and Tayur [15] show that the shortfalls satisfy

$$
\begin{align}
Y_{n+1}^d &= \max\{0, Y_n^d + D_n - c^d\}; & (3.30) \\
Y_{n+1}^i &= \max\{0, Y_n^i + D_n - c^i, Y_n^{i+1} + D_n - (s^{i+1} - s^i)\}, & (3.31) \\
&\quad i = 1, \ldots, d-1. & (3.32)
\end{align}
$$

We show how these recursions lead to approximations.

3.4.2 Approximations

We continue to assume that demands are i.i.d. with distribution F_D. With $Y_n = (Y_n^1, \ldots, Y_n^d)$, Glasserman and Tayur [15] show that the process $\{Y_n, n \geq 1\}$ admits a finite stationary distribution to which it converges from all initial distributions, provided

$$E[D_1] < c^* \equiv \min\{c^1, \ldots, c^d\}. \tag{3.33}$$

For our approximations, we require that $P(D_1 > c^*) > 0$ and that there exist a $\theta_0 > 0$ at which

$$1 < E[e^{\theta_0(D_1 - c^*)}] < \infty.$$

We denote by γ the unique non-zero solution to

$$E[e^{\gamma(D_1 - c^*)}] = 1.$$

To simplify the exposition, we will initially impose the additional assumption that there is just one stage i^* with capacity c^* and thus that all other stages have strictly greater capacity. Later, we remove this condition. With i^* as just defined, let

$$\eta = \min_{j \geq i^*}\{(s^j - s^1) - (j-1)c^*\}. \tag{3.34}$$

This quantity provides the link between the analysis of single-stage and multistage systems. In the important special case that $i^* = d$ (meaning that the bottleneck stage is highest in the hierarchy), we have simply

$$\eta = (s^d - s^1) - (d-1)c^*. \tag{3.35}$$

We now have the following multistage version of Theorem 1(i):

Theorem 3 *Let $Y = (Y^1, \ldots, Y^d)$ have the stationary distribution of the shortfall process. Then*

$$P(Y^1 > x) \sim Ce^{-\gamma(x+\eta)} \qquad as \ x \to \infty, \tag{3.36}$$

where C is the constant for a single-stage system with capacity c^* and the same demand distribution as the multistage system.

Thus, the tail distribution of stock to serve external demands in a multistage system corresponds to that in a single-stage system with the minimal capacity, except that the constant C is replaced by $C \exp(-\gamma\eta)$. In particular, when $i^* = d$, tail probabilities ultimately depend only on c^d and $s^d - s^1$. Since Theorem 3 then suggests that the probability of a stockout admits the approximation $P(Y^1 > s^1) \approx C \exp[-\gamma(s^d - (d-1)c^*)]$, it futher suggests that the dominant features deterimining the ability to meet demands are the minimal capacity and the system-wide base-stock level s^d.

For any $s = (s^1, \ldots, s^d)$, the stock availability, the fill rate and the average backorders are defined for the multistage system just as for the single stage system but replacing Y with Y^1 in (3.5), (3.6) and (3.7). From Theorem 3 we get

Corollary 3 *Parts (i)-(iv) of Theorem 1 hold for multistage systems, with C replaced by $C \exp(-\gamma\eta)$ and c replaced by c^*.*

These approximations can also be formulated with respect to a single-stage system. For example, if $\alpha^*(s)$ is the stock availability for a single-stage system with capacity c^*, then

$$1 - \alpha(s) \sim e^{-\gamma\eta}(1 - \alpha^*(s)).$$

The interpretation of these results is in some cases clearer if we express the approximations in terms of the *incremental* base-stock levels $\Delta^i = s^i - s^{i-1}$, $i = 2, \ldots, d$. With this notation, we have

$$\eta = \min_{j \geq i^*}\{(\sum_{i=2}^{j} \Delta^i) - (j-1)c^*)\},$$

an expression not depending explicitly on s^1.

From Section 3.3, we know that the exponential approximation to the tail of the shortfall distribution in the single-stage system results from a link between the stationary shortfall and the maximum of a random walk. In the multistage setting, the key step is representing the stationary shortfall as the maximum of a process that is not itself a random walk but that is closely related to one. We detail the case in which $c^d < \min_{i \neq d} c^i$ (i.e., the uppermost node is the unique bottleneck) then discuss the general case.

Consider the graph in Figure 3.1. As indicated, the vertical arcs in column i all have length c^i, $i = 1, \ldots, d$, and diagonal transitions from column i to

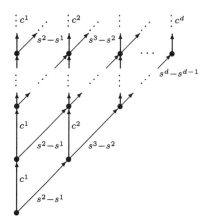

Figure 3.1 Each vertical arc in column i has length c^i; each diagonal arc from column i to column $i+1$ has length $s^{i+1} - s^i$.

column $i+1$ have length $s^{i+1} - s^i$. Let r_n be the length of the shortest n-step path through this graph, starting from the lowest node in column 1; specifically, $r_1 = \min(c^1, s^2 - s^1)$, $r_2 = \min(2c^1, c^1 + s^2 - s^1, s^2 - s^1 + c^2, s^3 - s^1)$, etc. We now have

Lemma 1 *Suppose $Y_0^i = 0$, for all $i = 1, \ldots, d$. Then*

$$Y_n^1 =_d \max_{1 \le j \le n} \left[\sum_{i=1}^{j} D_i - r_j \right]^{+} ;$$

consequently,

$$Y^1 =_d \max_{j \ge 1} \left[\sum_{i=1}^{j} D_i - r_j \right]^{+} .$$

Thus, although the distribution of Y^1 is not quite that of the maximum of a random walk, it is close to being one. In particular, as detailed in Glasserman [13] (and as is not hard to see directly) $r_n - nc^d = \eta$ for all sufficiently large n. So, the sums in the lemma look approximately like $S_j + \eta$.

Now fix $\Delta^2, \ldots, \Delta^d$ and let s_δ^1 be the corresponding minimal stage-1 base-stock level required to guarantee a stock availability of at least $1 - \delta$ for any $0 < \delta < 1$. Then we have

Corollary 4 *For any $\Delta^2, \ldots, \Delta^d$ and any $0 < \delta < 1$, $s_\delta^1 \le -\gamma^{-1} \log \delta - \eta$. If the demand distribution is continuous, then*

$$|s_\delta^1 - [\gamma^{-1} \log(C/\delta) - \eta]| \to 0$$

as $\delta \to 0$.

Corollary 4 can be supplemented with bounds, much as in Section 3.2.2. Approximating the base-stock level s_p^1 that minimizes system cost is more difficult, because holding costs are typically charged on inventory at all stages, not just the lowest stage. Hence, the optimal s_p^1 is no longer characterized by a simple condition on the tail distribution of Y^1. One might, however, choose to set base-stock levels by ensuring a certain stock availability at upper echelons and minimizing holding and penalty costs at stage 1 subject to that constraint on $s^2, \ldots, s^d$. For that approach, tail probabilities for $Y^2, \ldots, Y^d$ are relevant; we approximate these next.

For $k = 1, \ldots, d$, let i_k^* be the index of the stage with the smallest capacity among those in $\{k, k+1, \ldots, k\}$, which we assume is unique. (This holds if, e.g., no two capacities are equal.) Let c_k^* be the capacity of stage i_k^*. Define

$$\eta_k = \min_{j \geq i_k^*}\{ \sum_{i=k+1}^{j} \Delta_i - (j-k)c_k^*\}$$

and notice that η_k coincides with the η in (3.34) for the subsystem consisting of stages $k, k+1, \ldots, d$. Since the evolution of Y_n^k is unaffected by that of Y_n^i, $i = 1, \ldots, k-1$, a consequence of Theorem 3 is this:

Theorem 4 *For all $k = 1, \ldots, d$,*

$$P(Y^k > x) \sim C_k \exp[-\gamma_k(x + \eta_k)],$$

where γ_k solves $E[\exp\{\gamma_k(D_1 - c_k^*)\}] = 1$ and C_k is the constant C for a single-stage system with capacity c_k^*.

We can remove the requirement that there be just one stage with the minimal capacity c^* (or c_k^*) by making a simple modification. If, say, stages $i_1, \ldots, i_m$ all have capacity c^*, then η is determined by the lowest stage; that is, we have

$$\eta = \min_{j \geq i_1}\{(s^j - s^1) - (j-1)c^*\}. \tag{3.37}$$

The definition of η_k is modified analogously.

Because the distribution of each Y^k potentially depends on all of $\Delta^{k+1}, \ldots, \Delta^d$, Theorem 4 does not provide a direct solution to the problem of setting base-stock levels. However, it does provide a basis for approximating the distribution of shortfalls, and thus also for approximating costs. The simplest approximation sets

$$P(Y^1 > x) \approx Ce^{-\gamma(x+\eta)}, \quad E[Y^1] \approx Ce^{-\gamma\eta}/\gamma, \quad E[(Y^1-s^1)^+] \approx Ce^{-\gamma(s^1+\eta)}/\gamma, \tag{3.38}$$

and similarly for $Y^2, \ldots, Y^d$. If echelon-i inventory is charged a holding cost at rate h_i and if backorders are penalized at rate p, the long-run average cost per period becomes

$$\sum_{i=1}^{d} h_i(s^i - E[Y^i]) + (p + h_1 + \cdots + h_d)E[(Y^1 - s^1)^+],$$

which generalizes (3.8). Substituting for the expectations according to (3.38) results in a cost approximation.

A shortcoming of this simple approximation is that it is insensitive to all capacities except the smallest. A modification developed in Glasserman and Tayur [16] uses

$$P(Y^1 > x) \approx (1 - \exp(-\gamma[(s^2 - s^1) - c_1]^+))C'e^{-\gamma'x} + Ce^{-\gamma(x+\eta)},$$

where C', γ' are the constants for stage 1 viewed as a single-stage system in isolation. This approximation is also consistent with Theorem 3. A numerical study of this approximation, generalized to 5-node systems, is presented in Glasserman and Tayur [16]. Results there indicate that when used to select cost-minimizing base-stock levels the approximation produces values that are often close to optimal. The quality of the approximations appears to increase with the load on the system.

3.4.3 Systems with Leadtimes

Thus far, we have assumed that period-n production at stage $i + 1$ becomes available input to stage i in period $n + 1$, for all n and all $i = 1, \ldots, d - 1$. We now consider a modification in which there are fixed, exogenous leadtimes for each stage; these could model transportation times between stages, for example. Cumulative leadtimes are specified by positive integers $\ell^1 < \ell^2 < \cdots < \ell^d$ as follows. Stage-1 production becomes available to meet external demands after ℓ^1 periods; stage-$(i + 1)$ production becomes available input to stage-i after $\ell^{i+1} - \ell^i$ periods, $i = 1, \ldots, d - 1$. Thus, ℓ^i is the total leadtime from stage i to external demands. Our previous model had $\ell^i = i$, $i = 1, \ldots, d$. The natural counterpart to (3.37) is

$$\eta = \min_{j \geq i_1}\{(s^j - s^1) - (\ell^j - 1)c^*\}. \tag{3.39}$$

With this definition, we have

Theorem 5 *In a multistage system with cumulative leadtimes $\ell^1, \ldots, \ell^d$, the stage-1 shortfall satisfies (3.36) with γ the solution to $E[\exp\{\gamma(D_1 - c^*)\}] = 1$, C the constant for a single-stage system with capacity c^*, and η as in (3.39). In particular, if the capacity at stage d is strictly less than that at any other stage, then*

$$P(Y^1 > x) \sim C \exp[-\gamma\{(s^d - s^1) - (\ell^d - 1)c^d\}]e^{-\gamma x}.$$

See [13] for details and further discussion.

3.5 APPROXIMATING THE CONSTANT IN HEAVY TRAFFIC

All of the approximations developed thus far have been based on an exponent, generically denoted γ, and a constant multiplying the exponential, generically denoted C. In all cases, γ is easily computed, requiring at most finding the root of a convex function. The constant C is not, in general, easily computed. For the purpose of getting a rough indication of the shortfall tail, γ may suffice, but a more precise assessment requires a numerical value for C.

Building on fundamental work of Siegmund [28], Glasserman and Liu [14] show that in approximations for the shortfall and various performance measures the constant C can be replaced by a simpler expression $\exp(-\gamma\beta)$, with β reasonably easy to evaluate, under conditions of heavy traffic. We discuss these approximations next.

The term "heavy traffic" requires some elaboration in the present context. If the mean demand $E[D]$ is close to the bottleneck capacity c^*, then γ will be close to zero — intuitively, when the production facility is heavily loaded, we would expect the shortfall to have a heavier tail, and this is reflected in a smaller γ. We will in fact take $\gamma \downarrow 0$ as the defining property of heavy traffic. To make this precise we will embed the original demand distribution in a family of distributions in such a way that each value of γ uniquely determines a demand distribution. Moreover, as γ decreases to zero, the mean demand increases to the bottleneck capacity. One might say that we are taking limits through one particular path to heavy traffic.

A naive heavy traffic approximation notes that as $\gamma \downarrow 0$, $C \uparrow 1$ (as suggested by (3.28), for example). Indeed, the standard diffusion limit for the Lindley recursion (3.2) is reflected Brownian motion with negative drift on the positive real line; the stationary distribution of this process is a "pure" exponential — one with no mass at the origin. This might suggest an approximation of the form $P(Y > x) \approx \exp(-\gamma x)$, replacing C with 1 everywhere. Siegmund's method makes it possible to refine this approximation and replace C with an expression that approaches 1 as $\gamma \downarrow 0$ but is strictly less than 1 for positive γ.

Let F denote the common distribution of the random variables $X_n \overset{\triangle}{=} D_n - c^*$, $n \geq 1$, and let ψ denote the corresponding cgf. With each θ at which $\psi(\theta)$ is finite we associate the distribution

$$F_\theta(x) = \int_{-\infty}^{x} e^{\theta x - \psi(\theta)} \, dF(x);$$

this defines an *exponential family* of distributions. A simple calculation shows that the mean of F_θ is $\psi'(\theta)$. Suppose θ_0 satisfies $\psi'(\theta_0) = 0$; if such a θ_0 exists it is unique because ψ is convex. The distribution F_{θ_0} corresponds to the model in heavy traffic because a mean of zero for $D_n - c^*$ implies that the mean demand equals the bottleneck capacity.

For each $0 \leq \theta < \theta_0$ the conjugate point associated with F_θ is given by $\gamma(\theta) = \theta' - \theta$ where $\theta' > \theta_0$ is the point at which $\psi(\theta') = \psi(\theta)$. Conversely, every $\gamma \in (0, \gamma(0))$ determines a $\theta(\gamma) \in (0, \theta_0)$ at which $\psi(\theta(\gamma)) = \psi(\theta(\gamma) + \gamma)$. So, we can parameterize the distributions $\{F_\theta, \theta \in (0, \theta_0)\}$ by γ ranging from $\gamma(0)$ to 0. Heavy traffic corresponds to $\gamma \downarrow 0$.

Our approximations are sharpest when the distribution F is *strongly non-lattice*, a technical condition meaning that the characteristic function $g(\lambda) =$

$E[\exp(i\lambda X)]$ satisfies $\inf_{|\lambda|>\delta}|1 - g(\lambda)| > 0$ for each $\delta > 0$. This is equivalent to assuming that the demand distribution itself is strongly nonlattice. From a practical perspective this is essentially equivalent to requiring that the demand distribution be continuous.

We need some additional notation to state our main result. Let $S_n = \sum_{i=1}^{n} X_i$ and let

$$\tau_+ = \inf\{n \geq 1 : S_n > 0\}$$

be the first strong ascending ladder epoch for this random walk. Let $\beta = E_{\theta_0}[S_{\tau_+}^2]/(2E_{\theta_0}[S_{\tau_+}])$, the subscripts on the expectations indicating that they are taken with respect to the distribution F_{θ_0} of $D_n - c^*$ in heavy traffic. Set $\sigma^2 = \psi''(\theta_0)$, the heavy traffic variance.

Finally, let $j^* = \min\{1 \leq i \leq d : c^i = c^*\}$ be the index of the lowest bottleneck and define

$$\xi = \max_{i \geq j^*}\{(i - 1)c^* - \sum_{k=1}^{i-1} \Delta^k\}.$$

We now have

Theorem 6 *Suppose that F_0 is strongly nonlattice and that $\gamma \downarrow 0$, $b \to \infty$ in such a way that $\gamma b \to$ constant. Then*

(i) *the mean shortfall at node 1 satisfies $E[Y^1] = \gamma^{-1}e^{-\gamma(\beta\sigma-\xi)} + O(\gamma)$;*

(ii) *the stockout probability satisfies $P(Y^1 > b) = e^{-\gamma(b+\beta\sigma-\xi)} + o(\gamma^2)$;*

(iii) *the average backlog satisfies $E(Y^1 - b)^+ = \gamma^{-1}e^{-\gamma(b+\beta\sigma-\xi)} + o(\gamma)$;*

(iv) *the unfilled demand satisfies $u(b) = \gamma^{-1}e^{-\gamma(b+\beta-\xi)}(e^{\gamma c^*} - 1) + o(\gamma^{2-\epsilon})$, for all $\epsilon > 0$.*

This result is proved in Glasserman and Liu [14]. Although the statement and proof of this result are quite technical, the interpretation is ultimately straightforward: everywhere we previously had a constant C in an exponential approximation, we now have $\exp(-\gamma\beta\sigma)$. The parameters γ and σ are straightforward to evaluate. The constant β is more involved but Siegmund [28, p.716] gives an integral representation of β suitable for numerical evaluation. In the special case of the normal distribution, $\beta = 0.5826\ldots$.

The unfilled demand in (iv) is just the numerator of the ratio in (3.6), with Y replaced by Y^1. The fill rate — which is 1 minus the ratio of the unfilled demand to the mean demand — can be approximated using part (iv) of the theorem; a detailed analysis is given in Glasserman and Liu [14]. The case of discrete demand distributions is considered in Liu [21].

3.6 INVENTORY-LEADTIME TRADEOFFS

We now turn to a closely related but distinct set of models and issues that further illustrate the use of tail approximations to gain insights into service levels in production-inventory systems. This section is based on Glasserman and Wang [17] and Wang [36], which should be consulted for all proofs. Other references dealing with different aspects of broadly related problems include Baker, Magazine, and Nuttle [3], Buzacott and Shanthikumar [4], Ettl, Feigin, Lin, and Yao [8], Nemec and Nguyen [22], Schraner [27], Song [30], and Song, Xu, and Liu [31].

We consider a setting in which components are assembled-to-order into finished goods. Each component is produced by a dedicated facility or piece of equipment; production capacity for each component is limited but the assembly operation is uncapacitated. Production of each component follows a base-stock policy. A fixed delivery leadtime is advertised and service is measured by the fraction of orders that are filled within this leadtime. We refer to this measure as the fill rate, though it differs from the sense of this term in previous sections. When the promised leadtime is zero, this measure of service is the "off-the-shelf" fill rate.

For the mathematical model we formulate, an important difference between this setting and those of previous sections is that we now assume a continuous-review rather than discrete-review policy. We also explicitly model variability in the time between orders, in the sizes of orders, and in the time required to produce components. Specifically, we use the following notation, often modified by subscripts and superscripts:

$$
\begin{aligned}
A &= \text{order interarrival time;} \\
B &= \text{unit production interval;} \\
D &= \text{batch order size;} \\
R &= \text{response time;} \\
s &= \text{base-stock level;} \\
x &= \text{delivery leadtime.}
\end{aligned}
$$

This notation is perhaps best explained through reference to Figure 3.6. The circles are production facilities in which components are produced and the triangles are inventories of individual components. Generically, B^i is the random time required to produce a single unit of component i, and s^i is the base-stock level for this component. The components are potentially assembled into multiple products and A_j denotes a generic interarrival time for the jth product. A product of type j requires D_j^i units of component i, and this too is a random variable. All these random variables are assumed independent across orders, except that the components of the vector $(D_j^1, \ldots, D_j^d)$ of requirements for a finished product may be correlated. For the most part, we consider the case of a single type of finished good, in which case subscripts on A and D refer to the index of the order (e.g., A_n is the nth interarrival time). We denote by R the time elapsed between the arrival and filling of an order in stationarity. With x the promised delivery leadtime, the fill rate is given by $P(R \leq x)$ and the *unfill rate* by $P(R > x)$. The fill rate can be increased by reducing x or increasing

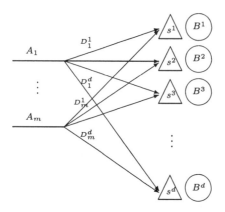

Figure 3.2 Multiple items assembled to order into multiple products. The demand process for product j is independent of those for other products and has generic interarrival time A_j. An order for product j requires a generic item portfolio: D_j^1 units of item $1, \ldots, D_j^d$ units of item d. Item i has generic production time B^i.

the s^i; our objective is to quantify the tradeoff between higher inventories and longer leadtimes.

As an indication of the sort of tradeoff we have in mind, we begin by describing the simplest possible setting. Orders for individual units of a single item arrive in a Poisson stream with rate λ; i.e., the A_n are exponentially distributed with mean $1/\lambda$. The time to produce an individual unit is exponentially distributed with mean $1/\mu$, with $\lambda < \mu$. The base-stock level for the sole item produced and stocked is s. In this case, the distribution of R is easily found in closed form and the unfill rate is

$$P(R > x) = \left(\frac{\lambda}{\mu}\right)^s \exp\{-(\mu - \lambda)x\}. \tag{3.40}$$

Consider the set of points in (x, s) space for which the expression on the right equals δ, thus resulting in a fill rate of $1 - \delta$, with $0 < \delta < 1$. So long as x and s are both positive, these points are characterized by the relation

$$s = \frac{\log \delta}{\log(\lambda/\mu)} - \frac{\mu - \lambda}{\log(\mu/\lambda)} x.$$

In other words, *the level curves of constant service are straight lines*, with slope $-b = -(\mu - \lambda)/\log(\mu/\lambda)$. Decreasing the leadtime by Δx without reducing the fill rate entails increasing s by $b\Delta x$; increasing s to $s + 1$ buys a reduction of $1/b$ in x.

Once we move beyond this Poisson-exponential setting, the simplicity of (3.40) is lost and the level curves are no longer exactly straight lines. Nevertheless, we show that the tradeoff between x and s is approximately linear at high fill rates. The Poisson-exponential setting does not provide much insight into the appropriate slope b in the general case, in part because it contains just two parameters λ and μ. Indeed, this setting could even be considered misleading because substituting the mean interarrival time and mean service time for $1/\lambda$ and $1/\mu$ in the general case is far from correct. We will see, however, that the appropriate generalization of b is quite easily characterized using tools similar to those developed in previous sections of this chapter.

3.6.1 Single-Item System

We begin with a result for the case of a single item. In this setting, we may omit the superscript i as we did above. We require that $E[D]E[B] < E[A]$ so that a steady-state response time exists. For any random variable Y, we let ψ_Y denote the cumulant generating function (3.21) of Y,

$$\psi_Y(\theta) = E[e^{\theta Y}].$$

Defining $X = \sum_{j=1}^{D} B_j - A$, we have

$$\psi_X(\theta) = E[e^{\theta(\sum_{j=1}^{D} B_j - A)}] = E[e^{\psi_B(\theta)D}]E[e^{-\theta A}] = \psi_D(\psi_B(\theta)) + \psi_A(-\theta),$$
(3.41)

in light of the assumed independence of the B_j, D, and A.

In the following result, the notation $\lim_{s+x\to\infty}$ refers to the limit as either $s \to \infty$ (through integer values), or $x \to \infty$, or both.

Theorem 7 If there is a $\gamma > 0$ at which $\psi_X(\gamma) = 0$, then with $\beta = \psi_B(\gamma)$,

$$\lim_{s+x\to\infty} e^{\gamma x + \beta s} P(R(s) > x) = C$$
(3.42)

for some constant $C > 0$.

Based on this result, we interpret $-\gamma/\beta$ as the approximate slope of the tradeoff between s and x at high fill rates. To see why, notice that (3.42) suggests the approximation $P(R(s) > x) \approx C \exp(-\gamma x - \beta s)$. A level curve of constant service is a set of (x, s) points for which $P(R(s) > x) = \delta$, for some $0 < \delta < 1$, and this is given approximately by the set of solutions to $C \exp(-\gamma x - \beta s) = \delta$. In the positive (x, s) orthant, these solutions form the line

$$s = -\frac{\gamma}{\beta}x + \frac{1}{\beta}\log(C/\delta).$$

Thus, a leadtime reduction of Δx entails an increase in s of $(\gamma/\beta)\Delta x$, and a 1-unit increase in s buys a reduction of β/γ in x, if the fill rate is to remain unchanged. It is not hard to verify that the ratio γ/β specializes to the ratios given in the previous section in the Poisson-exponential example.

The tradeoff parameters γ and β are not very easy to interpret intuitively and their calculation requires complete knowledge of the distributions of A, B and D. If we only have partial knowledge of the distributions — specifically, the means and variances — or if we want an expression that is simpler to interpret, we may approximate $\gamma > 0$ through a two-moment approximation for the ψ_X in (3.41), i.e., we set $\psi_X(\theta) \approx E[X]\theta + (1/2)\mathrm{Var}[X]\theta^2 = 0$ and solve to get

$$\gamma \approx -\frac{2E[X]}{\mathrm{Var}[X]}, \tag{3.43}$$

where $E[X] = E[B]E[D] - E[A]$ and $\mathrm{Var}[X] = E[D]\mathrm{Var}[B] + \mathrm{Var}[D](E[B])^2$, much as in (3.12). Similarly by the two-moment approximation for ψ_B we get

$$\beta = \psi_B(\gamma) \approx E[B]\gamma + \frac{1}{2}\mathrm{Var}[B]\gamma^2. \tag{3.44}$$

We can extend the setting in Theorem 7 to include an explicit assembly time U_n for the nth order after all the components it requires are available, with U_n i.i.d. and bounded. We assume no congestion and no finished goods inventory at the assembly stage, so U_n acts as an extra delay in the response time. This changes only the constant C in (3.42), as explained in Glasserman and Wang [17] or Wang [36].

As in previous sections, only a partial characterization of the constant C in (3.42) is generally available. However, in the important special case of Poisson arrivals, we obtain an explicit formula:

Proposition 2 *In the setting of Theorem 7, if arrivals are Poisson with rate λ, then*

$$C = \lambda^{-1}(\lambda + \gamma)(1 - \lambda E[D]E[B])(\psi_D'(\beta)\psi_B'(\gamma)(\lambda + \gamma) - 1)^{-1}.$$

A heavy-traffic approximation along the lines of that in Section 3.5 is also possible, though this has yet to be made rigorous.

3.6.2 Multiple Items, Single Product

Consider, next, a system with d items but a single product and thus just one arrival stream. Each item i has a base-stock level s^i, so we need to make an assumption about how these scale to get a limiting result. Let $s = s^1 + \cdots + s^d$ and $k_i = s^i/s$, $i = 1, \ldots, d$; we hold these ratios constant as s increases. This assumes that the proportion of total inventory held in each item remains constant, though we could just as easily assume that, e.g., the proportion of work content or holding cost for each item remains constant; this would merely change the constants k_i in the subsequent analysis. For each item i define X^i from A, B^i, and D^i paralleling the definition of X just before (3.41) and set

$$\psi_i(\theta) = \psi_{D^i}(\psi_{B^i}(\theta)) + \psi_A(-\theta), \tag{3.45}$$

in analogy with (3.41). Clearly, Theorem 7 applies to each item separately. Suppose $\gamma_i > 0$ solves $\psi_i(\gamma_i) = 0$ and set $\beta_i = \psi_{B^i}(\gamma_i)$, $\alpha_i = k_i\beta_i$. Then Theorem 7 implies

$$\lim_{s+x \to \infty} e^{\gamma_i x + \alpha_i s} P(R^i(s) > x) = C_i$$

for some $C_i > 0$. The response time for the full order is the maximum of the response times for the individual items required. Its behavior is a bit more subtle, because of the interactions among the multiple items.

Let $\gamma = \min_i \gamma_i$ and $\mathcal{I}_x = \{i : \gamma_i = \gamma\}$; these are the set of *leadtime-critical* items in the sense that their individual fill rates increase most slowly as x increases to ∞. Let $\alpha = \min_i \alpha_i$ and $\mathcal{I}_s = \{i : \alpha_i = \alpha\}$; these are similarly the set of *inventory-critical* items because their fill rates increase most slowly as s increases to ∞. These sets of items determine the product fill rate when x or s becomes large. To exclude trivial cases, we assume that for any two items i and j in $\mathcal{I}_x$ or $\mathcal{I}_s$, we have $P(X^i \neq X^j) > 0$. Indeed, the only case in which this fails is if the two items are always ordered in the same quantity and take the same time to produce, in which case they should be modeled as a single item.

Theorem 8 *Suppose the solutions* $\gamma_1, \ldots, \gamma_d$ *all exist. Then*

$$\lim_{x \to \infty} e^{\gamma x} P(R(s) > x) = \sum_{i \in \mathcal{I}_x} C_i e^{-\alpha_i s} \qquad (3.46)$$

and if either $|\mathcal{I}_s| = 1$ *or* $\{D^i, i \in \mathcal{I}_s\}$ *are independent, then*

$$\lim_{s \to \infty} e^{\alpha s} P(R(s) > x) = \sum_{i \in \mathcal{I}_s} C_i e^{-\gamma_i x}. \qquad (3.47)$$

The condition that the D^i be independent is far from necessary for (3.47), even if $|\mathcal{I}_s| > 1$. This issue is investigated in greater detail in Glasserman and Wang [17] and Wang [36].

If $\mathcal{I}_x = \mathcal{I}_s = \mathcal{I}$, Theorem 8 yields

$$\lim_{s+x \to \infty} e^{\gamma x + \alpha s} P(R(s) > x) = \sum_{i \in \mathcal{I}} C_i \equiv C_{\mathcal{I}}$$

which provides a simpler counterpart to Theorem 7 and suggests the approximation

$$P(R(s) > x) \approx C_{\mathcal{I}} e^{-\gamma x - \alpha s}.$$

It is only in this case that we can give $-\gamma/\alpha$ the simplest interpretation of the slope of the tradeoff between x and s. In the general case, $\mathcal{I}_x$ and $\mathcal{I}_s$ represent the sets of items that constrain the fill rate at long delivery intervals and high base-stock levels, respectively. When these are not the same, different tradeoffs apply in different regions. This is further explored through numerical examples in Glasserman and Wang [17] and Wang [36].

3.6.3 Multiple Items, Multiple Products

The final variant we consider allows multiple sets of items to be combined into m distinct products. In this setting, we require that arrivals of orders for the various products follow independent (compound) Poisson processes. (This assumption is necessary for the superposition of the arrival processes to be renewal.) We also need to vary our notation slightly to distinguish products from items: we use subscripts for products and continue to use superscripts for items. Orders for product j arrive at rate λ_j, and each order of product j requires D_j^i units of item i.

Let $\mathcal{I}_j = \{i : P(D_j^i > 0) > 0\}$ be the set of items required by product j; let $\mathcal{P}^i = \{j : P(D_j^i > 0) > 0\}$ be the set of products requiring item i. For each item i, the demand is the superposition of independent (compound) Poisson processes with $\lambda^i = \sum_{j \in \mathcal{P}^i} \lambda_j$; the batch size D^i is distributed as a mixture of $\{D_j^i\}$; i.e., with probability λ_j/λ^i, D^i is distributed as D_j^i, for $j \in \mathcal{P}^i$. With γ^i and α^i calculated just as before, Theorem 7 applies to R^i, the steady-state item-i response time. Let R_j be the steady-state response time for product j; then $R_j = \max_{i \in \mathcal{I}_j} R^i$. Define

$$\bar{\gamma}_j = \min_{i \in \mathcal{I}_j}\{\gamma^i\} \quad \text{and} \quad \mathcal{I}_j^x = \{i \in \mathcal{I}_j : \gamma^i = \bar{\gamma}_j\};$$

$\mathcal{I}_j^x$ is the set of leadtime-critical items for product j. Also define

$$\bar{\alpha}_j = \min_{i \in \mathcal{I}_j}\{\alpha^i\} \quad \text{and} \quad \mathcal{I}_j^s = \{i \in \mathcal{I}_j : \alpha^i = \bar{\alpha}_j\};$$

$\mathcal{I}_j^s$ is the set of inventory-critical items for product j. We now have

Theorem 9 *Suppose the solutions $\gamma^1, \ldots, \gamma^d$ all exist. Then*

$$\lim_{x \to \infty} e^{\bar{\gamma}_j x} P(R_j(s) > x) = \sum_{i \in \mathcal{I}_j^x} C_i e^{-\alpha^i s}, \tag{3.48}$$

and if $|\mathcal{I}_j^s| = 1$ or $\{D_j^i, i \in \mathcal{I}_j^s\}$ are independent, then

$$\lim_{s \to \infty} e^{\bar{\alpha}_j s} P(R_j(s) > x) = \sum_{i \in \mathcal{I}_j^s} C_i e^{-\gamma^i x}. \tag{3.49}$$

The same comments on the independence condition for the D_j^i after Theorem 8 also apply here. As with Theorem 8 the cleanest version of this result applies when $\mathcal{I}_j^x = \mathcal{I}_j^s$; i.e., the leadtime-critical and inventory-critical items coincide. Because the result has been specialized to the case of Poisson arrivals, Proposition 2 applied to each item i yields an expression for each C_i.

All of the tradeoffs discussed here are investigated numerically in Glasserman and Wang [17] and Wang [36]. Results there show that the asymptotics are remarkably accurate in tracing the level curves of constant service: as the delivery leadtime x and the base-stock levels s^i are varied according to the limiting tradeoffs, we find that the fill rate remaines nearly constant, as predicted. Also, Wang [36] investigates the use of these approximations in setting base-stock levels to minimize holding costs subject to a service-level constraint and finds that simple approximations give close-to-optimal base-stock levels.

3.7 CONCLUDING REMARKS

The purpose of this chapter has been to present a set of techniques for approximating tail distributions in production-inventory systems and to illustrate the use of these techniques in measuring system performance or gaining qualitative insight into the determinants of performance. A common feature of all the models considered in this chapter is a limit on production capacity. This sets up a conflict between product demand, which drives shortfalls up, and production, which drives shortfalls down. At the core of all these models is a process that looks roughly like a random walk reflected at the origin. So long as the production capacity outstrips demands on average, the random walk will tend to be driven down to the origin; but in the presence of production or demand variability, there may be periods in which production cannot keep up with demand. This is the source of the distributional tails and imperfect service we have sought to quantify.

Perhaps the most important conclusion of the modeling efforts reviewed here is that even a rough characterization of the tails of shortfall distributions or response-time distributions may lend useful insight. The theoretical underpinnings are rather technical, but in the end the results obtained are fairly easy to state. When combined with two-moment approximations to the key quantities, they become simple enough to have the potential for practical application.

Acknowledgements. Much of the work discussed here is the result of collaborations with Tai-Wen Liu, Sridhar Tayur, and Yashan Wang. It is a pleasure to acknowledge the countless discussions I have had with them in the past few years on these and other topics. This work has been supported by the National Science Foundation through grant DMI-9457189.

68

References

[1] ASMUSSEN, S. (1987). *Applied Probability and Queues.* Wiley, Chichester, England.

[2] AVIV, Y., AND A. FEDERGRUEN. (1997). Stochastic Inventory Models with Limited Production Capacities and Varying Parameters, *Probability in the Engineering and Information Sciences* **11**, 107–135.

[3] BAKER, K.R., MAGAZINE, M.J., AND NUTTLE, H. (1986). The Effect of Commonality on Safety Stock in a Simple Inventory Model, *Mgmt. Sci.*, **32**, 982-988.

[4] BUZACOTT, J.A., AND J.G. SHANTHIKUMAR. (1994). Safety Stock versus Safety Time in MRP Controlled Production Systems, *Mgmt. Sci.*, **40**, 1678-1689.

[5] BUZACOTT, J.A., S.M PRICE, AND J.G. SHANTHIKUMAR. (1992). Service Level in Multistage MRP and Base Stock Controlled Productions Systems, in *New Directions for Operations Research in Manufacturing*, G. Fendel, T. Gulledge, and A. James, eds., 445-463, Springer-Verlag, New York.

[6] CLARK, A.J., AND H. SCARF. (1960). Optimal Policies for a Multi-Echelon Inventory Problem, *Mgmt. Sci.*, **6**, 475-490.

[7] DEMBO, A., AND O. ZEITOUNI. (1993). *Large Deviations Techniques and Applications*, Jones and Bartlett, Boston.

[8] ETTL, M., FEIGIN, G.E., LIN, G.Y., AND YAO, D.D. (1995). A Supply-Chain Model with Base-Stock Control and Service Level Requirements, IBM Research Report, IBM T.J. Watson Research Center, Yorktown Heights, New York.

[9] FEDERGRUEN, A. (1993). Centralized Planning Models for Multi-Echelon Inventory Systems under Uncertainty, in *Logistics of Production and Inventory*, S.C. Graves, A.H.G. Rinnooy Kan, and P.H. Zipkin (eds.), North-Holland, Amsterdam, 133-173.

[10] FEDERGRUEN, A., AND P. ZIPKIN. (1986a). An Inventory Model with Limited Production Capacity and Uncertain Demands, I: The Average Cost Criterion. *Math. Oper. Res.* **11**, 193-207.

[11] FEDERGRUEN, A., AND P. ZIPKIN. (1986b). An Inventory Model with Limited Production Capacity and Uncertain Demands, II: The Discounted Cost Criterion. *Math. Oper. Res.* **11**, 208-215.

[12] FELLER, W. (1971). *An Introduction to Probability Theory and its Applications, Vol.2.* Second Edition, Wiley, New York.

[13] GLASSERMAN, P. (1997). Bounds and Asymptotics for Planning Critical Safety Stocks, *Oper. Res.* **45**, 244-257.

[14] GLASSERMAN, P., AND T.W. LIU. (1997) Corrected Diffusion Approximations for a Multistage Production-Inventory System, *Math. Oper. Res.* **22**, 186-201.

[15] GLASSERMAN, P., AND S. TAYUR. (1994). The Stability of a Capacitated, Multi-Echelon Production-Inventory System Under a Base-Stock Policy. *Oper. Res.* **42**, 913-925.

[16] GLASSERMAN, P., AND S. TAYUR. (1996). A Simple Approximation for a Multistage Capacitated Production-Inventory System. *Naval Research Logistics* **43**, 41-58.

[17] GLASSERMAN, P., AND Y. WANG. (1996). Leadtime-Inventory Trade-Offs in Assemble-to-Order Systems, *Oper. Res.*, to appear.

[18] KENDALL, M. (1987). *Advanced Theory of Statistics, Vol. II*, 5th Edition, Oxford Pub., New York.

[19] KAPUCSINZKI, R., AND S, TAYUR. (1995). A Capacitated Production-Inventory Model with Periodic Demand, *Oper. Res.*, to appear.

[20] LEE, Y., AND P. ZIPKIN. (1992). Tandem Queues with Planned Inventories. *Oper. Res.* **40**, 936-947.

[21] LIU, T.-W. (1995). Analysis and Simulation of a Multistage Production-Inventory System, Ph.D. thesis, Graduate School of Business, Columbia University, New York, NY.

[22] NEMEC, J.E., AND NGUYEN, V. (1997). Diffusion Approximations of Make-to-Stock, Produce-to-Order Systems with Backlog and Finite Backlog, Working Paper, MIT Operations Center, Cambridge, MA.

[23] PRABHU, U. (1965). *Queues and Inventories,* Wiley, New York.

[24] ROUNDY, R., AND MUCKSTADT, J.A. (1996). Heuristic Computation of Periodic-Review Base Stock Inventory Policies, Technical Report, School

of Operations Research and Industrial Engineering, Cornell University, Ithaca, New York.

[25] ROSLING, K. (1989). Optimal Inventory Policies for Assembly Systems under Random Demands. *Oper. Res.* **37**, 565-579.

[26] ROSS, S.M. (1974). Bounds on the Delay Distribution in GI/G/1 Queues. *J. Appl. Prob.* **11**, 417-421.

[27] SCHRANER, E. (1996). Capacity/Inventory Trade-Offs in Assemble-to-Order Systems, Working Paper, IBM T.J. Watson Research Center, Yorktown Heights, New York.

[28] SIEGMUND, D. (1979). Corrected Diffusion Approximations in Certain Random Walk Problems. *Adv. Appl. Prob.* **11**, 701-719.

[29] SIEGMUND, D. (1985). *Sequential Analysis: Tests and Confidence Intervals*, Springer, New York.

[30] SONG, J.S. (1997). On the Order Fill Rate in a Multi-Item Inventory System, Working Paper, Department of Industrial Engineering and Operations Research, Columbia University, New York. To appear in *Oper. Res.*

[31] SONG, J.S., XU, S.H., AND LIU, B. (1997). Order-Fulfillment Performance Measures in an Assemble-to-Order System with Stochastic Leadtime, Working Paper, Department of Industrial Engineering and Operations Research, Columbia University, New York. To appear in *Oper. Res.*

[32] SPECK, C.J., AND J. VAN DER WAL. (1991). The Capacitated Multi-Echelon Inventory System with Serial Structure, Working Paper, Department of Mathematics and Computer Science, Eindhoven University of Technology, Eindhoven, The Netherlands.

[33] SVORONOS, A., AND P. ZIPKIN. (1991). Evaluation of One-for-One Replenishment Policies for Multiechelon Inventory Systems, *Mgmt. Sci.* **37**, 68-83.

[34] TAYUR, S. (1992). Computing the Optimal Policy for Capacitated Inventory Models. *Comm. Statist — Stochastic Models* **9**, 585-598.

[35] VEATCH, M.H., AND L.M. WEIN. (1994). Optimal Control of a Two-Station Tandem Production/Inventory System, *Oper. Res.* **42**, 337-350.

[36] WANG, Y. (1998). *Service Levels in Production-Inventory Networks: Bottlenecks, Trade-Offs, and Optimization,* Doctoral dissertation, Columbia University Graduate School of Business, New York, NY.

[37] ZIPKIN, P. (1986). Models for Design and Control of Stochastic Multi-Item Batch Production Systems, *Oper. Res.*, **34**, 91-104.

4 ON (R,NQ) POLICIES IN SERIAL INVENTORY SYSTEMS

Fangruo Chen

Graduate School of Business
Columbia University
New York, NY 10027

4.1 INTRODUCTION

Supply chain management, which is about the management of the material and information flows in multi-stage production-distribution networks, has recently attracted much attention from academics and practitioners alike. Driven by fierce global competition and enabled by advanced information technology, many companies have taken initiatives to revamp their supply chains in order to reduce costs and at the same time increase responsiveness to changes in the marketplace. A remarkable success story is Wal-Mart, whose replenishment practices are hailed as the center piece of its competitive strategy.

Multi-echelon inventory models are central to supply chain management. The multi-echelon inventory theory began when Clark and Scarf (1960) published their seminal paper. The theory is now voluminous with categories by, e.g., demand characteristics (deterministic, stochastic) and network structures (series, assembly, distribution). This chapter focuses on a multi-stage, serial inventory system with stochastic demand.

Consider a production-distribution system where material is processed sequentially before being used to satisfy uncertain customer demand. The system consists of multiple stages representing the different stocking points in the production-distribution process. Material flow from one stage to the next requires a leadtime and incurs a fixed cost (in addition to a variable cost proportional to the flow quantity). Due to the value added, inventory becomes more expensive to carry as it moves downstream (closer to the customers). Customer demand unsatisfied from on-hand inventory is backlogged, incurring penalty costs. We assume that the entire supply chain is controlled by a central planner whose goal is to satisfy customer demand with minimum long-run average system-wide costs. (When the different stages are run by independent managers, the centralized solution can be used as a benchmark.)

The above model was originally proposed by Clark and Scarf (1962) as a generalization of the now classic Clark-Scarf (1960) model which does not allow setup costs at any stages except the most upstream stage. They introduced the important concept of echelon stock: a stage's echelon stock is the inventory position of the subsystem consisting of the stage itself and all its downstream stages.

The serial model can also be viewed as a generalization of the deterministic models studied by Roundy (1985,86), Maxwell and Muckstadt (1985), and Atkins and Sun (1995). They show that the so-called power-of-two policies are close to the optimal solution. Under the power-of-two structure, the reorder intervals (or order quantities) at all stages are restricted to be power-of-two multiples of a base time (or quantity) unit. This facilitates time/quantity coordination among the different stages.

For our serial model with random demand and setup costs, Clark and Scarf (1962) have pointed out correctly that the optimal policy does *not* have a simple structure. Thus, an optimal policy, even if it exists and is identified, would not be easy to implement. In other words, the "optimal" policy is no longer optimal or even attractive once the managerial effort of implementation is taken into account. Therefore, we turn to simple, cost-effective heuristic policies. Specifically, we consider (R, nQ) policies. An (R, nQ) policy operates as follows: whenever the inventory position (to be defined) at a stage is at or below R, order nQ units where n is the minimum integer required to increase the inventory position to above R. We call R the *reorder point* and Q the *base order quantity*. As with power-of-two policies, the base order quantities at the different stages satisfy an integer-ratio constraint: the base order quantity at an upstream stage is an integer multiple of the base order quantity at a downstream stage. If each stage uses an (R, nQ) policy to control its echelon stock, then we have an echelon-stock (R, nQ) policy. Otherwise, if each stage controls its installation stock (i.e., its local inventory position), then we have an installation-stock (R, nQ) policy.

Both types of policies are easy to implement. Installation-stock policies only require local inventory information, while echelon-stock policies require centralized demand information. Note that although the initial measurement of a stage's echelon stock requires the inventory information at every downstream stage, its update only requires the demand information at the point of sales, which is readily available for most companies with advanced communication networks such as EDI (Electronic Data Interchange).

This chapter summarizes the key results for both types of (R, nQ) policies in serial systems.[1] Since installation-stock (R, nQ) policies are special cases of echelon-stock (R, nQ) policies (as we will see later), most of the results are presented for the echelon-stock policies only. The goal is to provide the reader with a clean picture of the available results; so the proofs are either omitted or simply sketched. Section 4.2 deals with (R, nQ) policies in single-location systems since the results here are useful for serial systems. Section 4.3 formally introduces echelon-stock (R, nQ) policies and installation-stock (R, nQ) policies, defines notation, and makes several basic assumptions. Section 4.4 develops an efficient algorithm for computing the long-run average costs of an echelon-stock (R, nQ) policy. Section 4.5 focuses on optimization: how to determine the optimal reorder points and base order quantities for both types of policies. Section 4.6 discusses the impact of information technology on supply chain performance. Several hypotheses are set forth with some preliminary numerical results. Section 4.7 demonstrates that echelon-stock (R, nQ) policies are actually optimal (among all feasible policies) in a particular setting where the inventory transfers from one stage to the next must be integer multiples of a fixed (but stage-specific) lot size.

The following notation will be used throughout this chapter:

$E[X] = $ expectation of random variable X

$V[X] = $ variance of X

$(x)^+ = \max\{x, 0\}$

---

[1] For (R, nQ) policies in assembly systems, see Chen (1996b). For (R, nQ) policies in one-warehouse multi-retailer systems, see Axsater (1993b), Cachon (1995), and Chen and Zheng (1997).

$(x)^- = \max\{-x, 0\}$.

4.2 (R,NQ) POLICIES IN SINGLE-LOCATION SYSTEMS

Consider the following single-location inventory system. Customer demand for a single item arises periodically. The demands in different periods are independent and identically distributed random variables. These random variables are nonnegative and integer-valued. If demand exceeds the on-hand inventory, the excess is backlogged. Linear holding and backorder costs are assessed with h and p being the holding and backorder cost per unit per period respectively. Inventory is replenished from an outside source; each order arrives after a constant leadtime of L periods. Replenishment is controlled by an (R, nQ) policy: at the beginning of each period, if the inventory position y (on-hand inventory + outstanding orders − backorders) is at or below R, an order of size nQ is placed where n is the minimum integer such that $y + nQ > R$; otherwise, no order is placed. A fixed setup cost K is incurred for each Q units of order. The decision variables are the reorder point R and the base order quantity Q. The planning horizon is infinite, and the objective is to minimize the long-run average total cost in the system.

For clarity, we assume that for each period, the ordering decision is made at the beginning of the period, customer demand arrives during the period, and the holding and backorder costs are assessed at the end of the period.

Let $IP(t)$ be the inventory position at the beginning of period t after order placement (if any) and before demand occurrence. Thus $R+1 \leq IP(t) \leq R+Q$. Let $IL(t)$ be the inventory level (on-hand inventory minus backorders) at the end of period t. Write $D[t_1, t_2]$ for the total demand in periods $t_1, t_1 + 1, \cdots, t_2$. The following inventory balance equation is well known

$$IL(t + L) = IP(t) - D[t, t + L].$$

Notice that $IP(t)$ and $D[t, t+L]$ are independent. Therefore, given $IP(t) = y$, the expected holding and backorder costs in period $t + L$ can be expressed as

$$G(y) \overset{def}{=} E[h(y - D[t, t + L])^+ + p(D[t, t + L] - y)^+].$$

Clearly, $G(\cdot)$ is convex. If the Markov chain $\{IP(t)\}$ is irreducible then its steady state distribution is uniform over $R + 1, \cdots, R + Q$ (see Hadley and Whitin 1961). This condition is satisfied when, for example, the one-period demand equals 1 with a positive probability. This is a mild condition satisfied by most demand distributions. We make this assumption in this chapter, for the sake of brevity.[2] Therefore, the long-run average holding and backorder cost

---

[2]If the Markov chain is reducible then its steady state distribution is uniform over $r + \Delta, r + 2\Delta, \cdots, r + q\Delta$ where r, Δ, and q are integers with $\Delta > 1$, $R - \Delta < r \leq R$, and q being the largest integer so that $r + q\Delta \leq R + Q$. The value of r is determined by the initial inventory position. The results presented in this chapter can be easily extended to this general case. See Chen (1996b) for details.

is $\frac{1}{Q} \sum_{y=r+1}^{r+Q} G(y)$. Since the long-run average setup cost is $\mu K/Q$ where μ is the mean of the one-period demand, the long-run average total cost associated with the (R, nQ) policy is

$$C(R, Q) \stackrel{def}{=} \frac{\mu K + \sum_{y=R+1}^{R+Q} G(y)}{Q}. \tag{4.1}$$

This cost function has the same form as the cost function of (R,Q) policies, see Federgruen and Zheng (1992) and Zheng (1992). The following results follow directly from this observation. [An (R,Q) policy works in continuous-review systems where demand is for one unit at a time. Here the size of each order is exactly Q units.]

4.2.1 Optimization

Since $C(R, Q)$ is jointly convex in the decision variables, it can be easily minimized. Here is one way to do it. Fix Q and minimize $C(R, Q)$ over R. Since $G(\cdot)$ is convex, the Q smallest values of $G(\cdot)$ are achieved at Q contiguous points, which can be easily determined for each Q. The optimal R is the one so that the sum in (4.1) is over those Q points. The optimal Q is obtained by minimizing $C(Q) \stackrel{def}{=} \min_R C(R, Q)$, a unimodal function. Below is the algorithm.

Step 0.
Let $G(\cdot)$ be minimized at y^*.
$R^* := y^* - 1$, $Q^* := 1$, $C^* := \mu K + G(y^*)$.
$a := y^*$, $b := y^*$, $Q := 1$.
Step 1.
If $G(a - 1) < G(b + 1)$ then $C := (QC^* + G(a - 1))/(Q + 1)$, $a := a - 1$;
otherwise, $C := (QC^* + G(b + 1))/(Q + 1)$, $b := b + 1$.
If $C \geq C^*$, stop.
Otherwise, $Q := Q + 1$, $C^* := C$, $R^* := a - 1$, $Q^* = Q$, go to Step 1.

4.2.2 Sensitivity Analysis

The cost performance of (R, nQ) policies is insensitive to the choice of Q. Under the continuous approximation (i.e., both the customer demand and the inventory are represented by continuous variables), one can re-write the cost function as

$$C(R, Q) = \frac{\mu K + \int_R^{R+Q} G(y) dy}{Q}.$$

Following Zheng (1992), we have

$$\frac{C(\alpha Q^*)}{C(Q^*)} \leq \frac{1}{2}(\alpha + \frac{1}{\alpha}) \quad \text{for all } \alpha > 0.$$

Therefore, for example, if we use $Q = \sqrt{2}Q^*$ instead of the optimal Q^* then the relative cost increase is no more than 6%. Deviations from the optimal base order quantity may be necessary due to physical constraints (e.g., a full truckload) or coordination among different locations. Notice that the above inequality becomes an equality for the EOQ model (see, e.g., Hadley and Whitin 1963), suggesting that the (R, nQ) model is even more robust than the EOQ model with respect to the lot size. This result is very useful for multi-echelon, multi-location systems where quantity coordination is beneficial.

Remarks:
1. It was assumed above that a fixed setup cost is incurred for each Q units ordered. Therefore, for example, the setup cost for an order of $2Q$ units is $2K$. An alternative assumption is that a fixed cost is incurred for each order, independent of its size, as in Zheng and Chen (1992). The resulting cost function is no longer jointly convex in the decision variables. Consequently, a more complex algorithm is needed to determine the optimal control parameters. The sensitivity analysis, although more complex, still suggests that the cost performance is insensitive to the choice of Q. Moreover, under the new setup cost, it is known that (s,S) policies are optimal. In general, however, the relative cost difference between these two types of policies is quite small. In practice, both types of setup costs are likely to be present simultaneously. For example, an order for n truckloads may incur a fixed cost for arranging the entire order and a fixed cost for each individual truckload. Which model is the most appropriate depends on the relative magnitudes of the two types of setup costs.
2. The single-location (R, nQ) model originated in Morse (1959). Hadley and Whitin (1961) showed that the steady state distribution of $IP(t)$ is uniform over $R + 1, \cdots, R + Q$ under some minor conditions. Since the long-run average cost was written in a rather detailed and thus complicated form, the simplicity of (R, nQ) models was obscured. As a result, no simple algorithm to compute the optimal R and Q was deemed possible. Veinott (1965) demonstrated the optimality of (R, nQ) policies in inventory systems where each order is restricted to be integer multiples of Q. Simon (1968) and Richards (1975) studied continuous-review (R, nQ) models. There have been several attempts to compute optimal (R, nQ) policies, see Karlin and Fabens (1963), Wagner, O'Hagan and Lundh (1965) and Naddor (1975).

4.3 (R,NQ) POLICIES IN SERIAL SYSTEMS

This section defines (R, nQ) policies in serial systems. Unlike the single-location systems, (R, nQ) policies in serial systems have two variations with different informational requirements. Basic assumptions and preliminary notation are also introduced.

Consider a serial inventory system with N stages. Stage 1 orders from stage 2, 2 from 3, $\cdots$, and stage N orders from an outside supplier with unlimited stock. For convenience, the outside supplier is also referred to as stage $N + 1$. Each stage represents a stocking point in a production-distribution system.

The customer demands in different periods are independent and identically distributed random variables, which are nonnegative and integer-valued. When stage 1 runs out of stock, the excess demand is backlogged.

The replenishment policy is of the following type. Each stage controls a stage-specific inventory position according to a stage-specific (R, nQ) policy: when the inventory position falls to or below a *reorder point* R, the stage orders a *minimum* integer multiple of Q (base quantity) from its upstream stage to increase the inventory position to above R. In case the upstream stage does not have sufficient on-hand inventory to satisfy this order, a partial shipment is sent with the remainder backlogged at the upstream stage. The production/transportation leadtimes from one stage to the next are constant.

Let Q_i be the base quantity at stage i, $i = 1, \cdots, N$. We assume

$$Q_{i+1} = n_i Q_i, \quad i = 1, \cdots, N - 1 \tag{4.2}$$

where n_i is a positive integer. This integer-ratio constraint simplifies analysis significantly. It also simplifies material handling (e.g., packaging and bulk breaking) by restricting the shipments to each stage to multiples of a fixed quantity, which may represent a truckload or the size of a standard container. Moreover, the cost increase due to the constraint is likely to be insignificant because inventory costs tend to be insensitive to the choice of order quantities (see section 4.2).

We consider two variations of the above (R, nQ) policy. One is based on *echelon stock*: each stage replenishes its echelon stock with an *echelon reorder point*. The echelon stock at a stage is the inventory position of the subsystem consisting of the stage and all the downstream stages. Let R_i be the echelon reorder point at stage i, $i = 1, \cdots, N$. Therefore, under an echelon-stock (R, nQ) policy, stage i orders a multiple of Q_i from stage $i + 1$ every time its echelon stock falls to or below R_i. The other variation is based on *installation stock*: each stage controls its installation stock with an *installation reorder point*. The installation stock at a stage is just the inventory position of the stage itself (outstanding orders plus on-hand inventory minus backorders). Let r_i be the installation reorder point at stage i, $i = 1, \cdots, N$. Therefore, under an installation-stock (R, nQ) policy, stage i orders a multiple of Q_i from stage $i + 1$ every time its installation stock falls to or below r_i. Note that echelon-stock (R, nQ) policies require centralized demand information, while installation-stock (R, nQ) policies only require local 'demand' information, i.e., orders from the immediate downstream stage.

A fixed cost is incurred for each base quantity ordered at every stage. Linear holding costs are incurred at every stage, and linear backorder costs are incurred at stage 1 only. The decision variables are the reorder points and the base quantities. The objective is to minimize the long-run average total cost in the system.

The system has the following parameters:

μ = mean of the demand in one period
L_i = leadtime from stage $i + 1$ to stage i, a given nonnegative integer

K_i = fixed cost for each Q_i ordered at stage i,

H_i = installation holding cost per unit per period at stage i

h_i = echelon holding cost per unit per period at stage $i = H_i - H_{i+1} > 0$
   with $H_{N+1} = 0$

p = backorder cost per unit per period (at stage 1).

For any $t_1 < t_2$, let $D[t_1, t_2]$ be the total demand in periods $t_1, \cdots, t_2$, $D[t_1, t_2)$ the total demand in periods $t_1, \cdots, t_2 - 1$, $D(t_1, t_2)$ the total demand in periods $t_1 + 1, \cdots, t_2 - 1$.

For clarity, we assume that replenishment activities — ordering, shipping, and receiving — in a period occur at the beginning of the period. Demand occurs during the period. Holding and backorder costs are assessed at the end of the period. The following notation will be used for the rest of the chapter. Let t^- be an epoch after the replenishment activities but before demand in period t, and let t^+ be an epoch after demand in period t. Define

$I_i(t)$ = echelon inventory at stage i at t^+
   = on-hand inventory at stage i plus inventories on hand at,
      and in transit to, stages $1, \cdots, i - 1$

$B(t)$ = backorder level at stage 1 at t^+

$IL_i(t)$ = echelon inventory level at stage i at $t^+ = I_i(t) - B(t)$

$IL_i^-(t)$ = echelon inventory level at stage i at t^-
   = on-hand inventory at stage i plus inventories in transit to or
      on hand at stages $1, \cdots, i - 1$ minus backorders at stage 1 at t^-

$IP_i(t)$ = echelon inventory position at stage i at t^-
   = $IL_i^-(t)$ plus orders in transit to stage i

$ES_i(t)$ = echelon stock at stage i at t^-
   = $IP_i(t)$ plus outstanding orders of stage i that are backlogged
      at stage $i + 1$

$IS_i(t)$ = installation stock at stage i at t^-
   = outstanding orders at stage i (in transit to stage i or backlogged
      at stage $i + 1$) plus on-hand inventory at stage i minus
      backlogged orders from stage $i - 1$
   = $ES_i(t) - ES_{i-1}(t)$, $i \geq 2$; and $IS_1(t) = ES_1(t)$.

Since the outside supplier has unlimited supply, $ES_N(t) = IP_N(t)$ for all t. Note also that the echelon stock at a stage decreases as customer demands arrive at stage 1. To continuously monitor its echelon-stock level, a stage must have access to the demand information at stage 1 on a real-time basis. On the other hand, the installation stock at a stage is local information. When the time index t is suppressed, the notation represents the corresponding steady state variables.

We assume that the initial on-hand inventory at stage i is an integer multiple of Q_{i-1}, $i = 2, \cdots, N$. This is reasonable because each order placed by stage $i - 1$ is an integer multiple of Q_{i-1} and thus there is no incentive for stage i to keep a fraction of Q_{i-1} on hand. This assumption and the integer-ratio constraint (see (4.2)) imply that the installation stock at stage i is always an integer multiple of Q_{i-1}. This observation is true whether an echelon-stock (R, nQ) policy or an installation-stock (R, nQ) policy is in place. Consequently, without any loss of generality, we restrict r_i to be an integer multiple of Q_{i-1}, $i \geq 2$. Of course, r_1 can be any integer. No such restrictions are placed on R_i, $i = 1, \cdots, N$.

It can be shown that any installation-stock (R, nQ) policy, $(r_i, Q_i)_{i=1}^N$, is equivalent to an echelon-stock (R, nQ) policy, $(R_i, Q_i)_{i=1}^N$, as long as its echelon reorder points satisfy

$$R_1 = r_1, \quad \text{and} \quad R_i = R_{i-1} + Q_{i-1} + r_i, \quad i = 2, \cdots, N. \tag{4.3}$$

In other words, installation-stock (R, nQ) policies are special cases of echelon-stock (R, nQ) policies. See Axsater and Rosling (1993).

4.3.1 Related Literature

Section 4.2 describes the literature on single-location (R, nQ) models. To our knowledge, Chen and Zheng (1994a) are the first to adapt the single-location (R, nQ) policy to a multi-echelon system. As section 4.7 of this chapter will show, (R, nQ) policies are optimal (among all feasible policy) for serial systems where each stage can only order in fixed batches. Additional papers on (R, nQ) polices in serial systems will be mentioned in the later sections. As mentioned in introduction, this chapter does not consider studies of (R, nQ) policies in other multi-echelon systems (e.g., assembly and distribution systems). For those, see Axsater (1993b), Cachon (1995), Chen (1996b), and Chen and Zheng (1997).

There is, however, a large body of research on (R, Q) policies in multi-echelon systems. (An (R, nQ) policy reduces to an (R, Q) policy when demand is for a single unit at a time and a continuous-review system is in place.) De Bodt and Graves (1985) study echelon-stock (R, Q) policies in serial systems. They develop approximate cost expressions under the nestedness assumption: whenever a stage receives a shipment, a batch must be immediately sent to its downstream stage. Badinelli (1992) considers installation-stock (R, Q) policies in serial systems. He provides exact long-run average holding and backorder costs under the assumption that the installation stock at each stage is nonnegative. Installation-stock (R, Q) policies have also been studied in the context of one-warehouse multi-retailer systems, see, e.g., Deuermeyer and Schwarz (1981), Moinzadeh and Lee (1986), Lee and Moinzadeh (1987a,b), Svoronos and Zipkin (1988), and Axsater (1991,93a). The major reason why installation stock policies have received much attention is perhaps their modest informational requirement: installation stock is local inventory information, whereas echelon stock requires a certain degree of information centralization. But this advantage is quickly disappearing as more and more companies are equipped with advanced information technologies such as EDI.

Compared with serial systems, assembly systems with stochastic demand have attracted relatively less attention in the literature. Schmidt and Nahmias (1985) characterize an optimal policy for a system where two components are assembled into one end item. Rosling (1989) shows that a general assembly system (with any number of items) is equivalent to a serial system and thus can be solved by using the Clark-Scarf (1960) result. Both papers assume zero setup costs. Chen (1996b) shows that (R, nQ) policies are optimal for assembly systems where inventory transfers from one stage to another must be in integer multiples of a fixed base quantity (due to, e.g., fixed setup costs). He demonstrates that such an assembly system is still equivalent to a serial system with batch ordering, extending Rosling's result to assembly systems with batch transfers.

4.4 PERFORMANCE EVALUATION

This section shows how to determine the long-run average system-wide cost of an (R, nQ) policy in the serial system. Since installation-stock (R, nQ) policies are special cases of echelon-stock (R, nQ) policies, it suffices to consider the latter with control parameters $(R_i, Q_i)_{i=1}^{N}$.

First, recall that the on-hand inventory at stage $i+1$ is always a nonnegative integer multiple of Q_i, $i = 1, \cdots, N - 1$. By definition, $IL_{i+1}^{-}(t) - IP_i(t)$ is the on-hand inventory at stage $i + 1$. Therefore,

$$IL_{i+1}^{-}(t) - IP_i(t) = mQ_i, \quad m \geq 0 \quad \text{integer.} \tag{4.4}$$

On the other hand, the echelon-stock policy implies that $IP_i(t)$ is above R_i whenever stage $i + 1$ has positive on-hand inventory. This has the following implications. Suppose $IL_{i+1}^{-}(t) \leq R_i$. From (4.4), $IP_i(t) \leq R_i$ since $m \geq 0$. Thus the on-hand stock at stage $i + 1$ cannot be positive, i.e., $m = 0$ or $IP_i(t) = IL_{i+1}^{-}(t)$. Now suppose $IL_{i+1}^{-}(t) > R_i$. In this case, if $IP_i(t) \leq R_i$ then $m > 0$ (see (4.4)) implying that stage $i+1$ has positive on-hand inventory, which in turn implies that $IP_i(t) > R_i$, a contradiction. Thus $IP_i(t)$ must belong to the set $\{R_i + 1, \cdots, R_i + Q_i\}$. Note that there is a unique point in the set that satisfies (4.4). In short, $IL_{i+1}^{-}(t)$ uniquely determines $IP_i(t)$. Formally,

$$IP_i(t) = O_i[IL_{i+1}^{-}(t)] \quad \text{for } i = 1, \cdots, N - 1 \tag{4.5}$$

where

$$O_i[x] = \begin{cases} x & \text{if } x \leq R_i \\ x - nQ_i & \text{otherwise} \end{cases}$$

with n being the largest integer so that $x - nQ_i > R_i$.

Notice that

$$IL_i(t + L_i) = IP_i(t) - D[t, t + L_i] \tag{4.6}$$

and

$$IL_i^{-}(t + L_i) = IP_i(t) - D[t, t + L_i) \tag{4.7}$$

since $IL_i^{-}(t + L_i)$ is assessed before the demand in period $t + L_i$.

We now have a recursive procedure for determining the steady-state distributions of the key inventory variables in the serial system. First, consider stage N. Since the outside supplier is perfectly reliable, the echelon N inventory position behaves as if the whole serial system were a single location following the

(R_N, nQ_N) policy. Thus, IP_N, the steady-state echelon N inventory position, is uniformly distributed over $\{R_N + 1, \cdots, R_N + Q_N\}$ (see section 4.2). From (4.7), we have the distribution of IL_N^-. Now proceed to stage $N - 1$. Use (4.5) to obtain the distribution of IP_{N-1}, and then (4.7) to obtain the distribution of IL_{N-1}^-. Continuing in this fashion, we obtain the steady-state distributions of the echelon inventory positions at all stages. The steady-state distributions of the echelon inventory levels are obtained from (4.6).

We close this section with an expression for the long-run average cost of the echelon-stock (R, nQ) policy. First, note that the system-wide holding and backorder costs incurred in period t can be expressed as

$$\sum_{i=1}^{N} h_i I_i(t) + pB(t).$$

Since $I_i(t) = IL_i(t) + B(t)$ by definition, the above expression can be re-written as

$$\sum_{i=1}^{N} h_i IL_i(t) + (p + H_1)B(t).$$

Let $l_i = \sum_{j=i}^{N} L_j$ for $i = 1, \cdots, N$ with $l_{N+1} \equiv 0$, i.e., l_i is the total leadtime from the outside supplier to stage i. In period t, charge

$$\sum_{i=1}^{N} h_i IL_i(t + l_i) + (p + H_1)B(t + l_1). \tag{4.8}$$

Clearly, this cost-accounting system merely shifts costs in time and thus does not affect the long-run average cost.

Take any period t in steady state. For $i = 1, \cdots, N$, define

$$
\begin{aligned}
IP_i &= IP_i(t + l_{i+1}) \\
IL_i^- &= IL_i^-(t + l_i) \\
IL_i &= IL_i(t + l_i) \\
B &= B(t + l_1) = (IL_1)^- \\
D_i &= D_i[t + l_{i+1}, t + l_i] \\
D_i^- &= D_i[t + l_{i+1}, t + l_i).
\end{aligned}
$$

Notice that $D_N^-, D_{N-1}^-, \cdots, D_2^-, D_1$ are independent. From (4.6) and (4.7),

$$IL_i = IP_i - D_i, \quad i = 1, \cdots, N \tag{4.9}$$

and

$$IL_i^- = IP_i - D_i^-, \quad i = 1, \cdots, N. \tag{4.10}$$

Note that IP_i is independent of D_i and D_i^- for $i = 1, \cdots, N$. Figure 4.1 illustrates the top-down recursive procedure for determining the distributions

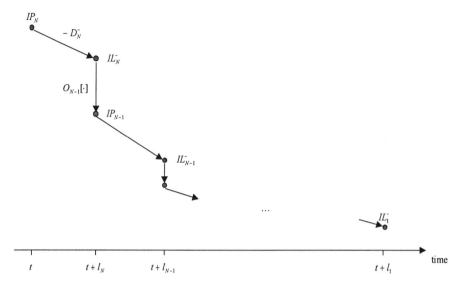

Figure 4.1 The Top-Down Recursive Procedure

of IP_i, which are then used to determine the distributions of IL_i via (4.9). The long-run average value of (4.8) is thus equal to

$$E[\sum_{i=1}^{N} h_i IL_i + (p + H_1)B].\qquad(4.11)$$

Let $\mathbf{R} = (R_1, \cdots, R_N)$ and $\mathbf{Q} = (Q_1, \cdots, Q_N)$. Combining the long-run average fixed costs with (4.11), we have the following long-run average system-wide cost

$$
\begin{aligned}
C(\mathbf{R}, \mathbf{Q}) \;\overset{def}{=}\; & \sum_{i=1}^{N} \frac{\mu K_i}{Q_i} + E[\sum_{i=1}^{N} h_i IL_i + (p + H_1)B] \\
= \; & \sum_{i=1}^{N} \frac{\mu K_i}{Q_i} + \sum_{i=2}^{N} h_i E[IL_i] + EG_1(IP_1) \qquad(4.12)
\end{aligned}
$$

where

$$
\begin{aligned}
G_1(y) \;=\; & E[h_1 IL_1 + (p + H_1)B \,|\, IP_1 = y] \\
= \; & E[h_1(y - D_1) + (p + H_1)(y - D_1)^-].
\end{aligned}
$$

4.5 OPTIMIZATION

This section has three subsections. The first and second subsections each present an algorithm for finding the optimal echelon and installation reorder points respectively. Both algorithms assume that the base order quantities are given. The third subsection describes several approaches to determining the optimal base order quantities for both echelon-stock and installation-stock policies.

4.5.1 Echelon Reorder Points

Consider the echelon-stock (R, nQ) policy, $(R_i, Q_i)_{i=1}^N$. Let its base order quantities be fixed. Thus the long-run average fixed cost is independent of the decision variables R_i. Let $C(\mathbf{R})$ be the long-run average system-wide holding and backorder cost, i.e., it is expression (4.12) without the fixed costs. We present an efficient procedure to identify the optimal echelon reorder points $\mathbf{R}^* = (R_1^*, \cdots, R_N^*)$.

The following random variables are useful. Define

$$Pr(U_i = u) = \frac{1}{Q_i}, \quad u = 1, \cdots, Q_i, \quad i = 1, \cdots, N$$

and

$$Pr(Z_i = z) = \frac{1}{n_i}, \quad z = 0, \cdots, n_i - 1, \quad i = 1, \cdots, N - 1.$$

(Recall that $Q_{i+1} = n_i Q_i$ where n_i is a positive integer, see (4.2).) We assume that these uniform random variables are independent, and they are all independent of the demand process. Let $V_N \equiv R_N$. Define recursively

$$V_i = \min\{R_i, V_{i+1} + Z_i Q_i - D_{i+1}^-\}, \quad i = 1, \cdots, N - 1. \tag{4.13}$$

Since the U's are independent of the demand process as well as the Z's, the V's are independent of the U's. From the above recursive definition, it is also clear that V_i is independent of D_i^- for $i = 1, \cdots, N$ and that V_{i+1} is independent of Z_i for $i = 1, \cdots, N - 1$.

Write $X_1 \overset{d}{=} X_2$ if the two random variables X_1 and X_2 have the same distribution. From Chen (1995),

$$IP_i \overset{d}{=} V_i + U_i, \quad i = 1, \cdots, N. \tag{4.14}$$

Therefore, IP_i is precisely the steady-state inventory position of the standard single-location (R, nQ) model with $R = V_i$ and $Q = Q_i$. A key feature here is that the reorder point of the single-location model is *random*, and it is jointly determined by the control parameters (echelon reorder points and base quantities) at stages $i, i+1, \cdots, N$ and the leadtime demands at stages $i+1, \cdots, N$. We will refer to V_i as the *effective reorder point* at stage i. Consequently, the N-stage model can be *decomposed* into N single-stage (R, nQ) models with reorder point V_i and base quantity Q_i, $i = 1, \cdots, N$. The linkage between these single-stage models is captured by (4.13).

We now present an algorithm for identifying $\mathbf{R}^*$. Note that $G_1(\cdot)$ is convex and has a finite minimum point. Define

$$\bar{G}_1(y) = EG_1(y + U_1).$$

Thus $\bar{G}_1(\cdot)$ is convex. Let $\bar{Y}_1$ be its minimum point. Note that

$$C(\mathbf{R}) \quad = \quad \sum_{j=2}^{N} h_j E[IL_j] + EG_1(IP_1)$$

$$\overset{(4.14)}{=} \sum_{j=2}^{N} h_j E[IL_j] + E\bar{G}_1(V_1)$$

$$\overset{(4.13)}{=} \sum_{j=2}^{N} h_j E[IL_j] + E\bar{G}_1(\min\{R_1, V_2 + Z_1 Q_1 - D_2^-\}). \quad (4.15)$$

From section 4.4, the distribution of IL_j, $j \geq 2$, is independent of R_1. As a result, the first term in (4.15), $\sum_{j=2}^{N} h_j E[IL_j]$, is independent of R_1. Therefore, the optimal R_1 minimizes

$$E\bar{G}_1(\min\{R_1, V_2 + Z_1 Q_1 - D_2^-\}).$$

Since $\bar{G}_1(\min\{R_1, W\}) \geq \bar{G}_1(\min\{\bar{Y}_1, W\})$ for any value of W, $R_1^* = \bar{Y}_1$, which is independent of the reorder points at the upstream stages! Now set $R_1 = \bar{Y}_1$.

To continue the optimization process, we define a sequence of functions recursively. Suppose $\bar{G}_i(\cdot)$ is defined, $i = 1, \cdots, N$. Let $\bar{Y}_i$ be a finite minimum point of $\bar{G}_i(\cdot)$. Define for $i = 1, \cdots, N-1$,

$$\bar{G}_{i+1}(y) = h_{i+1} E(y + U_{i+1} - D_{i+1}) + E\bar{G}_i(\min\{\bar{Y}_i, y + Z_i Q_i - D_{i+1}^-\}). \quad (4.16)$$

It is easy to see that $\bar{G}_i(\cdot)$ are all convex functions. Note that $IL_2 = IP_2 - D_2 \overset{d}{=} U_2 + V_2 - D_2$ and that V_2 is independent of U_2, Z_1, D_2, and D_2^-. Thus, from (4.15) and (4.16)

$$C(\mathbf{R}) \quad = \quad \sum_{j=3}^{N} h_j E[IL_j] + E\bar{G}_2(V_2)$$

$$\overset{(4.13)}{=} \sum_{j=3}^{N} h_j E[IL_j] + E\bar{G}_2(\min\{R_2, V_3 + Z_2 Q_2 - D_3^-\}).$$

Similarly, $R_2^* = \bar{Y}_2$. Continuing in this fashion, we have $R_i^* = \bar{Y}_i$ for $i = 1, \cdots, N$. Note that we obtain the optimal echelon reorder points by sequentially minimizing N convex functions. The minimum holding and backorder cost is $\bar{G}_N(\bar{Y}_N)$.

4.5.2 Installation Reorder Points

Consider the installation-stock (R, nQ) policy, $(r_i, Q_i)_{i=1}^{N}$. Let the base order quantities be fixed. The goal is to determine the optimal installation reorder points, denoted by r_i^*, $i = 1, \cdots, N$.

As mentioned in section 4.3, installation-stock (R, nQ) policies are special cases of echelon-stock (R, nQ) policies. From (4.3) and the fact that r_i is an integer multiple of Q_{i-1}, for $i \geq 2$, one can formulate the problem of finding r_i^* as

$$\mathcal{P} \quad \min \quad C(\mathbf{R})$$
$$\text{s.t.} \quad R_{i+1} - R_i = m_i Q_i, \quad m_i \text{ integer}$$
$$i = 1, \cdots, N - 1.$$

Let $\mathbf{R}^0 = (R_1^0, \cdots, R_N^0)$ be an optimal solution to $\mathcal{P}$. Thus $r_1^* = R_1^0$ and $r_i^* = R_i^0 - R_{i-1}^0 - Q_{i-1}$ for $i \geq 2$.

Problem $\mathcal{P}$ is, however, difficult to solve. One can develop bounds and search for the optimal solution, see Chen (1995). This approach becomes computationally infeasible for large problem instances (e.g., large N). Below, we describe a heuristic algorithm that finds near-optimal installation reorder points.

The heuristic algorithm is based on the following empirical observation. In many numerical examples, we found that $\mathbf{R}^*$, the optimal echelon reorder points, contains useful information about $\mathbf{r}^*$, the optimal installation reorder points. Obviously, if $\mathbf{R}^*$ is a feasible solution to $\mathcal{P}$, then the problem is solved: $r_1^* = R_1^*$ and $r_i^* = R_i^* - R_{i-1}^* - Q_{i-1}$ for $i \geq 2$. Interestingly, even when $\mathbf{R}^*$ is infeasible, one can still obtain r_i^*, $i \geq 2$, by rounding $R_i^* - R_{i-1}^* - Q_{i-1}$ to an integer multiple of Q_{i-1}. Formally, take any $i \geq 2$. Let r_i^- (resp., r_i^+) be the maximum (resp., minimum) integer multiple of Q_{i-1} that is less (resp., larger) than or equal to $R_i^* - R_{i-1}^* - Q_{i-1}$. Then, r_i^* is either r_i^- or r_i^+.

Based on this empirical observation, we restrict installation reorder points at stages $2, \cdots, N$ to the following set

$$(r_2, \cdots, r_N) \in \mathcal{R}_{-1} \overset{def}{=} \{r_2^-, r_2^+\} \times \cdots \times \{r_N^-, r_N^+\}.$$

Note that if $R_i^* - R_{i-1}^* - Q_{i-1}$ is an integer multiple of Q_{i-1}, then $r_i^- = r_i^+$; otherwise, $r_i^- + Q_{i-1} = r_i^+$. Thus, the set $\{r_i^-, r_i^+\}$ contains at most two points, which implies that there are at most 2^{N-1} combinations of installation reorder points in $\mathcal{R}_{-1}$.

Here is the algorithm. First, determine r_i^- and r_i^+, $i = 2, \cdots, N$, by rounding $R_i^* - R_{i-1}^* - Q_{i-1}$ down and up to an integer multiple of Q_{i-1}. Then, for each point in $\mathcal{R}_{-1}$, find the optimal corresponding r_1. This is easy to do since for any fixed installation reorder points at stages $2, \cdots, N$, the cost function is convex in r_1 (see Chen 1995). The heuristic solution is the best one obtained in this way. The computational complexity of this algorithm is about 2^{N-1} times the effort of evaluating a single policy.

4.5.3 Base Order Quantities

We proceed to determine the optimal base order quantities (or just base quantities) for echelon-stock (R, nQ) policies. Here is how. First, bound the cost function from both above and below by simple functions of the control parameters. These bounds are obtained by over- and under-charging a penalty cost to each upstream stage for holding less-than-adequate stock. Each bound is the sum of N single-stage cost functions. Substituting these bounds for the exact cost function, we effectively decouple the N-stage system into N single-stage systems. Then, solve these single-stage problems to obtain near-optimal base quantities. Since inventory costs tend to be insensitive to the choice of the base quantities (see section 4.2), we believe that the base quantities obtained here can also be used for installation-stock (R, nQ) policies. The proofs for most of the results here will be omitted; they can be found in Chen and Zheng (1998).

4.5.3.1 Upper-Bound Function. Define recursively a sequence of functions $G^i(\cdot)$ for $i = 1, \cdots, N$. Let $G^1(y) = G_1(y)$ for any integer y. Let Y_i be the minimum point of $G^i(\cdot)$. For $i = 1, \cdots, N-1$ and any integer y, define

$$G^{i,i+1}(y) = \begin{cases} G^i(y) - G^i(Y_i) & y \leq Y_i \\ 0 & \text{otherwise} \end{cases}$$

and

$$G^{i+1}(y) = E[h_{i+1}(y - D_{i+1}) + G^{i,i+1}(y - D_{i+1}^-)].$$

Since $G^1(\cdot)$ is convex, $G^{1,2}(\cdot)$ is convex and decreasing. Thus $G^2(\cdot)$ is convex. Similarly, $G^i(\cdot)$ are all convex. Note that $G^{i,i+1}(IL_{i+1}^-)$ is the induced-penalty cost charged to stage $i + 1$ in the Clark-Scarf model (see Chen and Zheng 1994b). Figure 4.2 illustrates the induced-penalty cost function.

One reason why the exact cost function $C(\mathbf{R}, \mathbf{Q})$ is difficult to minimize is that the stages are "coupled" in the sense that IP_i depends not only on the control parameters at stage i but also on those at the upstream stages. It can be shown that

$$G^i(IP_i) \leq G^{i,i+1}(IL_{i+1}^-) + G^i(ES_i), \quad i = 1, \cdots, N-1.$$

This inequality provides a way to decouple the system since IL_{i+1}^- is independent of, and ES_i is completely determined by, the control policy at stage i. Since ES_i is uniformly distributed from $R_i + 1$ to $R_i + Q_i$,

$$EG^i(IP_i) \leq EG^{i,i+1}(IL_{i+1}^-) + \sum_{y=R_i+1}^{R_i+Q_i} G^i(y)/Q_i, \quad i = 1, \cdots, N-1. \quad (4.17)$$

Define

$$C^i(R, Q) = \frac{\mu K_i + \sum_{y=R+1}^{R+Q} G^i(y)}{Q}, \quad i = 1, 2, \cdots, N.$$

Therefore

$$
\begin{aligned}
C(\mathbf{R}, \mathbf{Q}) \quad &= \quad \sum_{i=1}^{N} \frac{\mu K_i}{Q_i} + \sum_{i=2}^{N} h_i E[IL_i] + EG^1(IP_1) \\
&\overset{(4.17)}{\leq} \quad \sum_{i=2}^{N} \frac{\mu K_i}{Q_i} + \sum_{i=2}^{N} h_i E[IL_i] + EG^{1,2}(IL_2^-)
\end{aligned}
$$

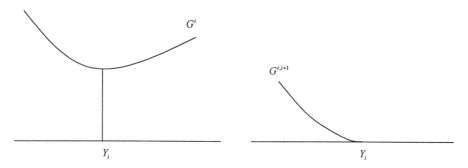

Figure 4.2   Induced-Penalty Cost for the Upper-Bound Function

$$+C^1(R_1, Q_1)$$

$$\overset{(4.9)(4.10)}{=} \sum_{i=2}^{N} \frac{\mu K_i}{Q_i} + \sum_{i=3}^{N} h_i E[IL_i] + E[h_2(IP_2 - D_2) +$$

$$G^{1,2}(IP_2 - D_2^-)] + C^1(R_1, Q_1)$$

$$= \sum_{i=2}^{N} \frac{\mu K_i}{Q_i} + \sum_{i=3}^{N} h_i E[IL_i] + EG^2(IP_2) + C^1(R_1, Q_1)$$

$$= \cdots$$

$$\leq \sum_{i=1}^{N} C^i(R_i, Q_i). \tag{4.18}$$

4.5.3.2 Lower-Bound Function. Define recursively a sequence of functions $\underline{G}_i(\cdot)$, $i = 1, \cdots, N$. Let $\underline{G}_1(y) = G_1(y)$ for all y. Suppose we have $\underline{G}_i(\cdot)$. For any integer R and any positive integer Q, define

$$C_i(R, Q) = \frac{\mu K_i + \sum_{y=R+1}^{R+Q} \underline{G}_i(y)}{Q}.$$

For fixed Q, let $C_i(R, Q)$ be minimized at $R = R_i(Q)$. Let $C_i(R_i(Q), Q)$ be minimized at Q_i^0. Set $R_i^0 = R_i(Q_i^0)$ and $C_i^0 = C_i(R_i^0, Q_i^0)$. Then, for any integer y, define

$$G_{i,i}(y) = \begin{cases} \underline{G}_i(y) & R_i^0 + 1 \leq y \leq R_i^0 + Q_i^0 \\ C_i^0 & \text{otherwise,} \end{cases}$$

$$G_{i,i+1}(y) = \begin{cases} \underline{G}_i(y) - C_i^0 & y \leq R_i^0 \\ 0 & \text{otherwise} \end{cases}$$

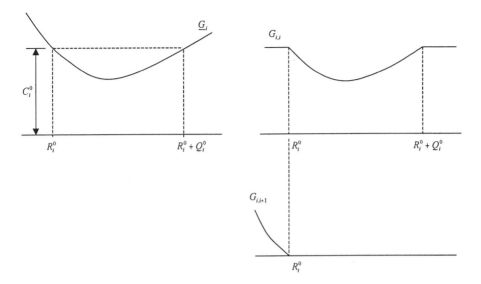

Figure 4.3 Induced-Penalty Cost for the Lower-Bound Function

and

$$\underline{G}_{i+1}(y) = E[h_{i+1}(y - D_{i+1}) + G_{i,i+1}(y - D_{i+1}^-)].$$

Note that $G_{i,i+1}(\cdot)$ is the induced-penalty cost used by Chen and Zheng (1994b) to construct lower bounds for several production-distribution networks. See Figure 4.3 for an illustration.

Since $\underline{G}_1(\cdot)$ is convex, $C_1(R, Q)$ has the form of the cost function of a single-stage (R, Q) policy and thus can be easily minimized (see Federgruen and Zheng 1992). Based on the single-stage results, one can recursively show that $\underline{G}_i(\cdot)$ are all convex. Therefore, $C_i(R, Q)$ all have the form of a single-stage cost function. Let $C_i(Q) = C_i(R_i(Q), Q)$. It can be shown that

$$C(\mathbf{R}, \mathbf{Q}) \geq \sum_{i=1}^{N} C_i(Q_i). \tag{4.19}$$

4.5.3.3 Alternative Lower-Bound Functions. By reallocating the setup costs among the stages, we can obtain better lower-bound functions. To see the intuition, consider a two-stage system where K_1 is much larger than K_2 so that $Q_1^0 > Q_2^0$. In this case, it is conceivable that the optimal base quantities at the two stages must be the same due to the constraint $Q_1 \leq Q_2$. Now allocate part of K_1 to K_2. This reduces Q_1^0 and increases the induced-penalty cost charged to stage 2. But for any feasible policy with $Q_1 = Q_2$, the allocation does not change the total setup cost. The result is a tighter lower-bound function.

In general, let $\tilde{K}_i$ be the new setup cost at stage i, $i = 1, \cdots, N$. Suppose

$$K_1 + \cdots + K_i \geq \tilde{K}_1 + \cdots + \tilde{K}_i, \quad i = 1, \cdots, N. \qquad (4.20)$$

Take any echelon-stock (R, nQ) policy. Since $Q_1 \leq \cdots \leq Q_N$, we have

$$
\begin{aligned}
\sum_{i=1}^{N} \frac{K_i}{Q_i} &= \sum_{i=1}^{N-1} \left(\frac{1}{Q_i} - \frac{1}{Q_{i+1}} \right) \sum_{j=1}^{i} K_j + \frac{1}{Q_N} \sum_{j=1}^{N} K_j \\
&\geq \sum_{i=1}^{N-1} \left(\frac{1}{Q_i} - \frac{1}{Q_{i+1}} \right) \sum_{j=1}^{i} \tilde{K}_j + \frac{1}{Q_N} \sum_{j=1}^{N} \tilde{K}_j \\
&= \sum_{i=1}^{N} \frac{\tilde{K}_i}{Q_i}.
\end{aligned}
$$

Therefore

$$C(\mathbf{R}, \mathbf{Q}) \geq \sum_{i=1}^{N} \frac{\mu \tilde{K}_i}{Q_i} + \sum_{i=2}^{N} h_i E(IL_i) + EG_1(IP_1).$$

Now treat the right side of the above inequality as a new cost function and follow section 4.5.3.2 to obtain a new lower-bound function.[3]

An allocation satisfying (4.20) can be obtained by solving a deterministic counterpart of the multi-stage problem. Consider

$$
P_d : \quad \min \quad \sum_{i=1}^{N} (K_i/T_i + \hbar_i T_i)
$$

$$
\text{s.t.} \quad T_{i+1} \geq T_i, \quad i = 1, \cdots, N - 1
$$

where

$$\hbar_i = \frac{\mu p^2 h_i}{2(p + H_i)(p + H_{i+1})}, \quad i = 1, \cdots, N.$$

This problem can be easily solved, see, e.g., Muckstadt and Roundy (1993). Let $(T_1^*, \cdots, T_N^*)$ be the optimal solution. Let

$$\tilde{K}_i = \hbar_i (T_i^*)^2, \quad i = 1, \cdots, N. \qquad (4.21)$$

These new setup costs satisfy (4.20). [4]

---

[3]Many lower bounds have been derived by reallocating setup costs, see, e.g., Atkins and Iyogun (1987), Atkins (1990), and Rosling (1993).

[4]The solution to P_d provides a lower bound on the long-run average cost of any feasible policy in the deterministic counterpart of the serial system where demand arrives at a constant rate μ. See Atkins and Sun (1995) and Chen (1998).

4.5.3.4 Algorithms. The exact cost function 4.12 is difficult to minimize over the base quantities. However, one can obtain near-optimal base quantities by minimizing the upper- and lower-bound functions developed above. For an illustration, consider the upper-bound function in (4.18). A similar algorithm can be developed based on the lower-bound functions (see Chen and Zheng 1998).

For fixed Q_i, let $R^i(Q_i)$ be the optimal R_i that minimizes $C^i(R_i, Q_i)$ or equivalently $\sum_{y=R_i+1}^{R_i+Q_i} G^i(y)$. This minimization is easy to do, see Federgruen and Zheng (1992). Define $C^i(Q_i) = C^i(R^i(Q_i), Q_i)$. Now consider

$$P_u : \quad \min \quad \sum_{i=1}^{N} C^i(Q_i)$$

$$s.t. \quad Q_{i+1} = n_i Q_i$$

$$n_i \geq 1, \quad \text{integer}, \quad i = 1, \cdots, N-1.$$

This problem can be solved in two steps.

First, consider the following relaxation of P_u:

$$P_u^- : \quad \min \quad \sum_{i=1}^{N} C^i(Q_i)$$

$$s.t. \quad Q_{i+1} \geq Q_i, \quad i = 1, \cdots, N-1.$$

A simple clustering technique solves this problem. Let $S = \{1, 2, \cdots, N\}$. For any $i, j \in S$ with $i \leq j$, the set $\{i, i+1, \cdots, j\}$ is called a *cluster*. For any cluster c, define

$$Q_c = \text{argmin}_Q \sum_{i \in c} C^i(Q).$$

(The minimization is over all positive integers. Thus Q_c is a positive integer.) A *partition* of S is a set of disjoint clusters whose union is S. A partition, $\{c(1), \cdots, c(n)\}$, is optimal if and only if

- $Q_{c(1)} \leq Q_{c(2)} \leq \cdots \leq Q_{c(n)}$, and

- for each cluster $c(k) = \{l_1, \cdots, l_2\}$, there does not exist an l with $l_1 \leq l < l_2$ so that $Q_{c^-(k)} < Q_{c^+(k)}$ where $c^-(k) = \{l_1, \cdots, l\}$ and $c^+(k) = \{l+1, \cdots, l_2\}$.

Muckstadt and Roundy (1993) provides an algorithm for finding an optimal partition. Let $\{c(1), \cdots, c(n)\}$ be an optimal partition. Let $\overline{Q}_i = Q_{c(k)}$ for $i \in c(k)$, $k = 1, 2, \cdots, n$. Then $(\overline{Q}_1, \overline{Q}_2, \cdots, \overline{Q}_N)$ is the optimal solution to P_u^-.

The second step is to round the above solution to a set of power-of-two integers. Let $Q_i^u = 2^{m_i}$, $i = 1, \cdots, N$, where m_i is the unique integer with $2^{m_i}/\sqrt{2} \leq \overline{Q}_i < 2^{m_i}\sqrt{2}$. (Since $\overline{Q}_i$ is a positive integer, $m_i \geq 0$ or $Q_i^u \geq 1$.) For example, if $N = 3$ and $(\overline{Q}_1, \overline{Q}_2, \overline{Q}_3) = (1, 3, 9)$ then $(Q_1^u, Q_2^u, Q_3^u) = (1, 4, 8)$. These power-of-two integers are the base quantities: Q_i^u is the base quantity at stage i, $i = 1, \cdots, N$.

The heuristic algorithms (based on the upper- and lower-bound functions) have been tested in numerical examples. First, we determined three sets of base quantities: one based on the upper-bound function in section 4.5.3.1, one based on the lower-bound function in section 4.5.3.2, and the third based on the alternative lower-bound function using the new setup costs in (4.21). For each set of base quantities, we determined the optimal echelon reorder points by using the algorithm described in section 4.5.1. This led to three heuristic solutions. They were all very easy to compute, and the computational time grew linearly in N. There was no clear dominance among these different solutions, however. The best heuristic among the three was always very close to the optimal solution. As N increased, the performance gap did not deteriorate. (This observation was based on a comparison between the long-run average cost of the heuristic and a lower bound on the long-run average costs of all feasible echelon-stock (R, nQ) policies. Such a lower bound was obtained by minimizing the lower-bound function with the integer-ration constraint $Q_{i+1} = n_i Q_i$ replaced by the precedence constraint $Q_{i+1} \geq Q_i$.)

We also compared the heuristic solutions with the EOQ solutions. For multi-echelon stochastic inventory systems, it is widely suggested that the order quantities can be obtained by solving the deterministic counterpart of the problem. Call the order quantities obtained in this way the EOQ's. In all the numerical examples considered, our heuristic base order quantities dominated the EOQ solution. Moreover, the EOQ's were always smaller than the heuristic base quantities.

4.6 IMPACT OF INFORMATION TECHNOLOGY

This section presents some preliminary findings about the impact of advanced information technology on supply chain performance.

There has been an unprecedented proliferation of information technology in supply chains. Well known examples include the checkout scanners in super-markets and the hand-held computers used by the delivery and repair person-nel (Fuchsberg 1990). Wal-Mart has its own satellite communication system to transmit sales data to its warehouses and suppliers (Stalk, Evans and Shulman 1992). Some suppliers in the auto industry are even required to have the ability to transmit and receive messages electronically (Wessell 1987).

This trend has a profound impact on how the supply chain is managed. Con-sider McKesson, for example. The company provides its customers (i.e., drug stores) with hand-held computers. A clerk in the drug stores walks the aisles every week. When a particular drug is running short, he/she uses a scanner to read a McKesson-supplied label, and the computer records the information and transmits an order to McKesson. The computer system has streamlined and drastically improved the ordering process, saving time and cost (Wessell 1987). Lower ordering costs lead to smaller order quantities and more frequent deliveries. One estimate indicates that manufacturers using EDI have experi-enced a decrease in retail order cycle times, sometimes by as much as 10-15 days (Tyndall 1988). Moreover, replenishment decisions can now be based on readily available, centralized demand/inventory information instead of local inventory information. In sum, information technology has led to *shorter lead-times, smaller batch sizes, and centralized information.* We will consider the impact of the changes in these dimensions on the supply-chain bottom line.

As mentioned earlier, echelon-stock policies require centralized demand information (i.e., each stage must have access to the demand data at the point of sale), whereas installation-stock policies use local information. Recall that installation-stock policies are special cases of echelon-stock policies in serial systems, reflecting the value of centralized demand information. Our task is to compare these two types of policies and to study how their performance changes as a result of shorter leadtimes and smaller batch sizes.

The numerical study assumed that the base order quantities were given. The only decision variables were the reorder points, echelon or installation, which we determined by using the algorithms developed in sections 4.5.1 and 4.5.2. Therefore, the focus was on the long-run average holding and backorder costs. (The costs reported below refer to these costs only.) The demand process was compound Poisson.[5] The distribution of the demand size D was geometric, i.e.,

$$Pr(D = x) = (1 - \alpha)^{x-1}\alpha, \quad x = 1, 2, \cdots$$

where $0 < \alpha < 1$.[6] Thus $E[D] = 1/\alpha$ and $V[D] = (1 - \alpha)/\alpha^2$. Let $D(1)$ be the total demand in one unit of time. It can be easily verified that $E[D(1)] = \lambda/\alpha$ and $V[D(1)] = \lambda(2 - \alpha)/\alpha^2$ with a coefficient of variation

$$cv = \frac{\sqrt{V[D(1)]}}{E[D(1)]} = \sqrt{\frac{2 - \alpha}{\lambda}}.$$

A wide range of cv can be obtained by using different values of α and λ.

The numerical study has two parts. The first part was designed to study the impact of leadtimes, batch sizes and the number of stages on the supply chain costs, and the second part focused on the value of centralized demand information.

4.6.1 Leadtimes, Batch Sizes and the Number of Stages

Let $h_i = h$, $L_i = L/N$, and $Q_i = Q$ for $i = 1, \cdots, N$. Thus L is the total lead-time from the outside supplier to the point of sales. Consequently, a numerical example is specified by the vector $(N, \alpha, \lambda, h, p, L, Q)$. There are three sets of numerical examples.

The first set illustrates the relationship between the minimum long-run average cost (or the total cost) and the total leadtime L. We considered examples with $N = 4, 8$; $\alpha = 0.4$; $\lambda = 0.21$; $h = 0.1$; $p = 10$; and $Q = 32, 64$. (Thus $cv = 2.76$.) Figures 4.4 and 4.5 indicate that for echelon-stock policies, the

---

[5] Although this chapter has so far dealt exclusively with i.i.d. demands, it is not difficult to extend the results to the compound Poisson case.

[6] Such a demand process is often referred to as a stuttering Poisson process.

total cost is *concave and increasing* in L. The same holds approximately for installation-stock policies. This is reminiscent of an observation from single-stage systems that the minimum cost is, approximately, proportional to the standard deviation of the leadtime demand (see, e.g., Scarf 1958, Hadley and Whitin 1963, and Gallego and Moon 1993). For our system, the standard deviation of the demand during the total leadtime is

$$\sqrt{\frac{\lambda(2-\alpha)}{\alpha^2}} L$$

which is concave in L. The concave relationship seems to have been preserved in multi-stage systems.

The second set of numerical examples illustrates the relationship between the total cost and the batch size Q. Here $N = 4, 8$; $\alpha = 0.4$; $\lambda = 0.21$; $h = 0.1$; $p = 10$; and $L = 30, 60$. (Thus $cv = 2.76$.) Figures 4.6 and 4.7 suggest that for echelon-stock policies, the total cost is *convex and increasing* in Q. The same holds approximately for installation-stock policies. Interesting, in the standard single-stage (R, Q) model, when the reorder point is chosen optimally for each Q, the holding and backorder cost is convex and increasing in Q and the function becomes linear asymptotically (Zheng 1992).

The third set of numerical examples illustrates the relationship between the total cost and the number of stages N. Here $\alpha = 0.4$; $\lambda = 0.21$; $h = 1/N$; $p = 10$; $L = 30$; and $Q = 32$. (Thus $H_1 = 1$ for all the examples.) Figure 4.8 shows that for echelon-stock policies, the total cost is *concave and increasing* in N. The same holds approximately for installation-stock policies. In other words, reducing the number of supply-chain intermediaries has increasing marginal returns.

4.6.2 Value of Centralized Demand Information

Let $h_i = 1/N$ and $L_i = L$ for $i = 1, \cdots, N$. Thus $H_1 \equiv 1$. Let $(Q_1, \cdots, Q_N) = (mQ_1^0, \cdots, mQ_N^0)$ where m is a positive integer and $(Q_1^0, \cdots, Q_N^0)$ is fixed for each value of N. Table 4.1 summarizes the values of key parameters. The four values of cv correspond to four different combinations of α and λ: $(\alpha, \lambda) = (1, 4)$, $(1, 1)$, $(0.4, 0.4)$, $(0.4, 0.1)$. Table 4.2 provides the base quantities for each value of N. There are a total of 1,536 examples.

To assess the value of centralized demand information (or value of information), we compared the long-run average cost of the optimal echelon-stock policy with the long-run average cost of the near-optimal installation-stock policy (since the installation reorder points were obtained by a heuristic algorithm) for each of the 1,536 examples. Let the relative difference between the two costs, i.e., (installation cost - echelon cost)/echelon cost, be the value of information. Here are the findings.

1. The value of information has a fairly wide range. The highest value observed is about 9%, achieved in an example with $N = 6$, $cv = 1/2$, $L = 4$, $m = 4$, and $p = 20$. The mean is about 1.75%.

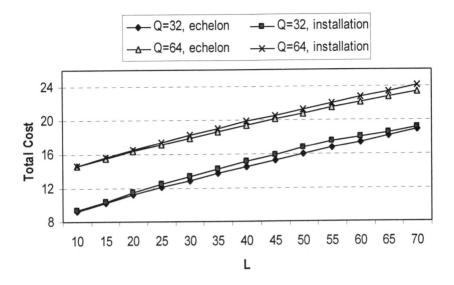

Figure 4.4 Total Cost for Examples with N=4

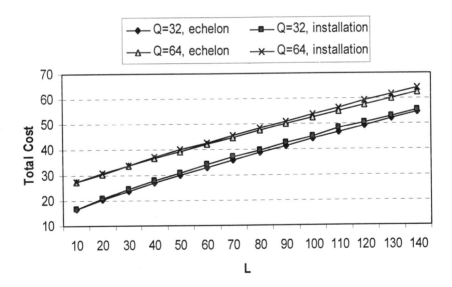

Figure 4.5 Total Cost for Examples with N=8

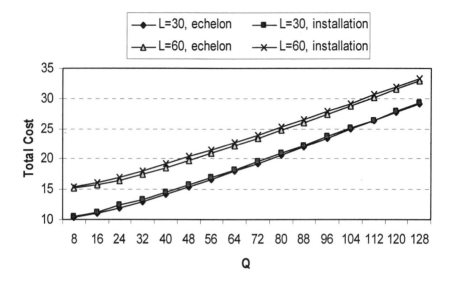

Figure 4.6 Total Cost for Examples with N=4

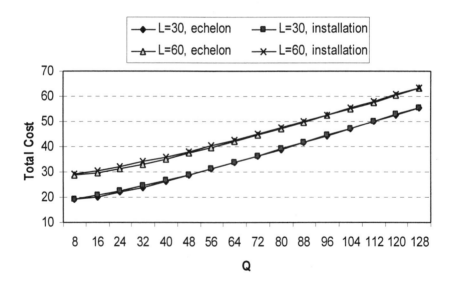

Figure 4.7 Total Cost for Examples with N=8

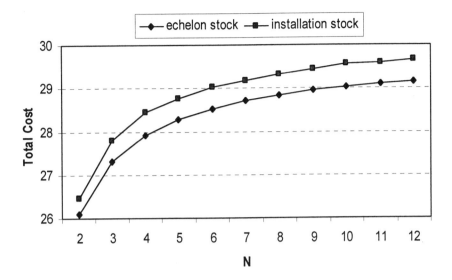

Figure 4.8   Total Cost and the Number of Stages

Table 4.1   System Parameters

Parameter	Values
N	2, 3, 4, 6, 8, 10
cv	1/2, 1, 2, 4
p	5, 10, 15, 20
L	1, 2, 3, 4
m	1, 2, 3, 4

Table 4.2   Base Quantities

N	$(Q_1^0, \cdots, Q_N^0)$
2	(8, 32)
3	(8, 16, 32)
4	(8, 8, 16, 32)
6	(8, 8, 16, 16, 32, 32)
8	(8, 8, 8, 16, 16, 16, 32, 32)
10	(8, 8, 8, 8, 16, 16, 16, 32, 32, 32)

2. Figures 4.9 to 4.13 show how the value of information depends on several key parameters of the model, i.e., N, L, m, cv, and p. The value of information seems to exhibit an upward trend as N, or L, or m increases. Increases in cv tend to decrease the value of information. The value of information is highest for the extreme (high or low) values of p.

Discussion: Under installation-stock (R, nQ) policies, the orders placed by a stage become the 'demand' at the upstream stage. Since every stage orders in batches, the demand process at an upstream stage is different from the actual customer demand at stage 1. (Lee, Padmanabhan and Whang 1997 call this the order-batching effect.) As the batch sizes increase, this distortion increases. It is thus not surprising that as m increases, the value of information increases. As N or L increases, the length of the serial channel increases. The numerical examples thus suggest that centralized demand information is more valuable in longer channels. This is intuitive. The relationship between the value of information and demand variability is surprising at first glance. Intuitively speaking, as demand becomes more unpredictable, it should be more beneficial to share the demand information at stage 1 with the upstream stages. The numerical examples suggest the opposite. One possible explanation is that as demand variability increases, the total cost increases so much that the (relative) value of information actually decreases. Finally, we can interpret p as a measure of the desired level of customer service: a larger value of p signals a higher level of service. As p increases, it becomes more important to replenish inventories in a timely fashion and thus the value of information should increase. Then, why is the value of information so high when $p = 5$? The reason might be, again, that a small value of p leads to a lower total cost, resulting in an increase in the (relative) value of information.

4.6.3 Related Literature

It has long been recognized that demand information and inventory are substitutes for one another. For example, advanced warnings from customers of their orders reduce inventory (Hariharan and Zipkin 1995). So does the additional demand information acquired through marketing means (Milgrom and Roberts 1988). The tradeoff between make-to-order and make-to-stock systems also reflects the value of demand information (see Federgruen and Katalan 1994 and Nguyen 1995 and the references therein). These examples focus on the value of better communication between the customer and the supplier. As this section demonstrates, there is also value in improving the communication between the supply-chain members. For further discussions on this topic, see, e.g., Lovejoy and Whang (1995), Cachon and Fisher (1996), Chen (1996a), Gavirneni, Kapuscinski and Tayur (1996), Lee, Padmanabhan and Whang (1997), and Lee, So and Tang (1997). This section draws heavily from Chen (1995).

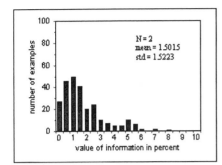

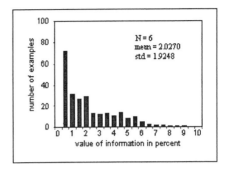

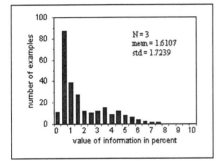

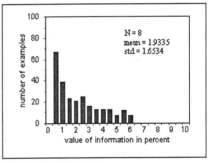

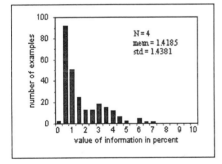

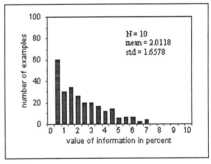

Figure 4.9   Value of Information for Different N's

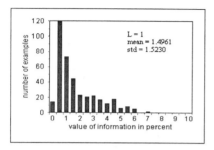
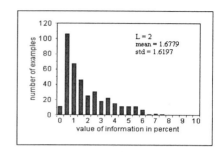
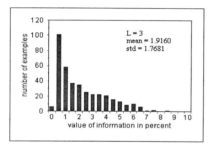
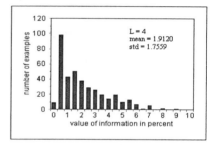

Figure 4.10   Value of Information for Different L's

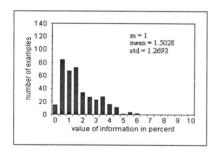
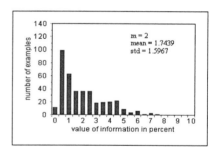
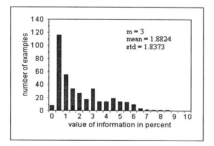
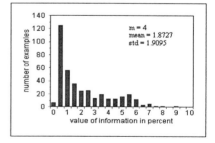

Figure 4.11   Value of Information for Different m's

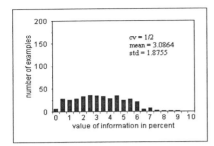
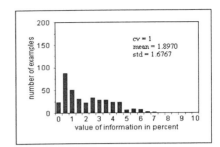
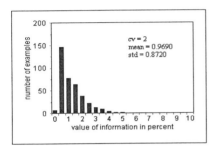
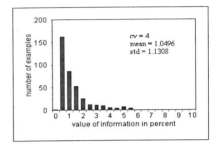

Figure 4.12 Value of Information for Different cv's

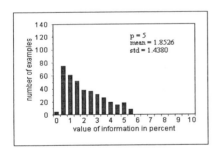
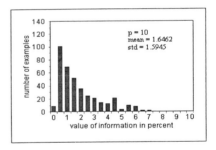
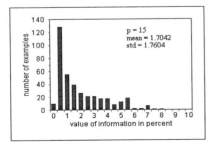
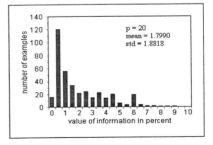

Figure 4.13 Value of Information for Different p's

4.7 OPTIMALITY RESULTS

This section demonstrates that (R, nQ) policies are actually optimal (among all feasible policies) for a serial system where the material flow is regulated in the sense that stage i can only order integer multiples of Q_i, $i = 1, \cdots, N$. These base order quantities may represent full truckloads in a distribution setting or full pallets in a production environment. The objective is to minimize the long-run average holding and backorder costs in the system. (Since the long-run average setup costs are fixed, they will be omitted in the following analysis.) Below, we will highlight the key ideas of the optimality proof. The details can be found in Chen (1996b).

We begin by considering the single-location inventory problem described in section 4.2. The replenishment policy can now take any form except that each order must be a positive integer multiple of Q, a given positive integer. Recall that $G(y)$ is the (conditional) expected holding and backorder costs charged to a period given that the inventory position at the beginning of the period after ordering is y. Let $\hat{G}(y) \stackrel{def}{=} \sum_{x=1}^{Q} G(y + x)$, a convex function of y, be minimized at $y = R^*$. (Note that $\hat{G}$ here corresponds to $\bar{G}$ in Chen 1996b.)

To determine an optimal policy for the above inventory problem, we first establish a lower bound on the long-run average cost of any feasible policy. We then show that the (R^*, nQ) policy actually achieves the lower bound. It is thus optimal.

Consider any feasible policy. Take any period t, and let y_t be the inventory position at the beginning of period t after ordering. This period's expected cost is $G(y_t)$. It follows from the convexity of $G(\cdot)$ that $G(y_t) \geq G(y_t')$ where $y_t' \stackrel{def}{=} y_t + nQ$ where n is the unique integer (positive or otherwise) so that $y_t' \in \{R^*+1, \cdots, R^*+Q\}$. The long-run average cost of the feasible policy must be greater than or equal to the long-run average value of $G(y_t')$. To determine the latter, consider the stochastic process $\{y_t'\}$. Let D_t be the demand in period t. Note that

$$y_{t+1} = y_t - D_t + mQ$$

for some nonnegative integer m. Since $y_t - y_t'$ and $y_{t+1} - y_{t+1}'$ are both integer multiples of Q,

$$y_{t+1}' = y_t' - D_t + m'Q$$

where m' is an integer. Moreover, given y_t' and D_t, the value of m' is unique since $R^* + 1 \leq y_{t+1}' \leq R^* + Q$. Since the demands in different periods are independent, $\{y_t'\}$ is a Markov chain with a finite state space $\{R^* + 1, \cdots, R^* + Q\}$. Since the steady state distribution of the Markov chain is uniform (see section 4.2), the long-run average value of $G(y_t')$ is

$$\frac{1}{Q} \sum_{y=R^*+1}^{R^*+Q} G(y) \qquad (4.22)$$

which is a lower bound on the long-run average cost of any feasible policy. Notice that this lower bound is precisely the long-run average holding and backorder costs of the (R^*, nQ) (see (4.1)).

Now consider the serial system described in section 4.3 under a constraint that each stage i can only order an integer multiple of Q_i, $i = 1, \cdots, N$. The approach is essentially the same as the one used above for single-location systems.

Let $g_i(y)$ be the expected one-period cost incurred at stage i in period t given $IP_i(t) = y$. The expected total cost in the system in period t, given the echelon inventory positions at the stages, is thus

$$\sum_{i=1}^{N} g_i(IP_i(t)). \tag{4.23}$$

Under the linear holding-backorder cost structure defined in section 4.3, we have

$$g_1(y) = E[h_1(y - D[t, t + L_1]) + (p + H_1)(y - D[t, t + L_1])^-]$$

and

$$g_i(y) = E[h_i(y - D[t, t + L_i])], \quad i = 2, \cdots, N.$$

Note that the optimality proof holds for more general cost structures.

We next establish a lower bound on the long-run average value of (4.23). First, note that the long-run average value of (4.23) is the same as the long-run average value of

$$\sum_{i=1}^{N} g_i(IP_i) \tag{4.24}$$

where IP_i are defined in section 4.4. Second, define a sequence of functions recursively. Recall that $G_1(y) = g_1(y)$ for all y. For $i = 1, \cdots, N$, assume that $G_i(\cdot)$ is defined and that $\hat{G}_i(y) \stackrel{def}{=} \sum_{x=1}^{Q_i} G_i(y + x)$ is quasiconvex and minimized at $y = R_i^*$, a finite integer. Define $O_i[\cdot]$ as in section 4.3 with R_i there replaced by R_i^*. Define

$$G_{i+1}(y) = g_{i+1}(y) + EG_i(O_i[y - D_{i+1}^-]), \quad i = 1, \cdots, N - 1. \tag{4.25}$$

The above assumptions can all be verified under the linear holding-backorder cost structure.[7] Third, one can show that under any feasible policy

$$G_i(IP_i) \geq G_i(O_i[IL_{i+1}^-]), \quad i = 1, \cdots, N - 1. \tag{4.26}$$

---

[7]Note that $\hat{G}_i$ here corresponds to $\bar{G}_i$ in Chen (1996b). Also, $\hat{G}_i$ is intimately related to $\bar{G}_i$ defined in section 4.5.1. In particular, $\hat{G}_i(y) = Q_i \bar{G}_i(y)$ for all y and $R_i^* = \bar{Y}_i$, $i = 1, \cdots, N$. By definition, $\hat{G}_1(y) = Q_1 \bar{G}_1(y)$ for all y and thus $R_1^* = \bar{Y}_1$. To see the intuition behind the

Finally, note that

$$
\begin{aligned}
C &\overset{def}{=} E[\sum_{i=1}^{N} g_i(IP_i)|IP_N] \\
&= E[\sum_{i=2}^{N} g_i(IP_i) + G_1(IP_1)|IP_N] \\
&\geq E[\sum_{i=2}^{N} g_i(IP_i) + G_1(O_1[IL_2^-])|IP_N] \\
&= E[\sum_{i=3}^{N} g_i(IP_i) + g_2(IP_2) + G_1(O_1[IP_2 - D_2^-])|IP_N] \\
&= E[\sum_{i=3}^{N} g_i(IP_i) + G_2(IP_2)|IP_N]
\end{aligned}
\tag{4.27}
$$

where the inequality follows from (4.26), the first equality from the definition of $G_1(\cdot)$, the second equality from (4.10), and the third equality from (4.25) and the fact that D_2^- is independent of IP_2 and IP_N. Repeating the above procedure, we have

$$
C \geq G_N(IP_N).
$$

Think of IP_N as the inventory position of a single-location inventory system. As shown in (4.22), the long-run average value of $G_N(IP_N)$ is bounded below by

$$
C^* \overset{def}{=} \frac{1}{Q_N} \sum_{y=R_N^*+1}^{R_N^*+Q_N} G_N(y) = \frac{1}{Q_N} \hat{G}_N(R_N^*).
$$

---

other relationships, notice that

$$
\begin{aligned}
\hat{G}_2(y) &= \sum_{x=1}^{Q_2} G_2(y+x) \\
&= \sum_{x=1}^{Q_2} g_2(y+x) + \sum_{x=1}^{Q_2} EG_1(O_1[y+x-D_2^-]) \\
&= Q_2 Eg_2(y+U_2) + Q_2 EG_1(O_1[y+U_2-D_2^-]) \\
&\overset{(a)}{=} Q_2 Eg_2(y+U_2) + Q_2 EG_1(O_1[y+U_1+Z_1 Q_1 - D_2^-]) \\
&\overset{(b)}{=} Q_2 Eg_2(y+U_2) + Q_2 EG_1(\min\{R_1^*, y+Z_1 Q_1 - D_2^-\} + U_1) \\
&= Q_2 Eg_2(y+U_2) + Q_2 E\bar{G}_1(\min\{R_1^*, y+Z_1 Q_1 - D_2^-\}) \\
&= Q_2 \bar{G}_2(y)
\end{aligned}
$$

where (a) follows since $U_2 \overset{d}{=} U_1 + Z_1 Q_1$ and (b) follows since $O_1[X+U_1] \overset{d}{=} \min\{R_1^*, X\} + U_1$ for any random variable X independent of U_1 (these results have been established in Chen 1995). Thus $R_2^* = \bar{Y}_2$, so on and so forth.

Therefore, C^* is a lower bound on the long-run average costs of all feasible policies in the serial system.

The lower bound C^* is achieved by the echelon-stock (R, nQ) policy with R_i^* as the reorder point at stage i, $i = 1, \cdots, N$. This is clear since under the policy, $IP_i = O_i[IL_{i+1}^-]$ (see (4.5)) and thus the inequality in (4.26) becomes an equality.

Remarks:

1. The above serial model with batch ordering is a generalization of the Clark-Scarf (1960) model which effectively assumes a base quantity of one for every stage. It is well known that base-stock policies are optimal for the Clark-Scarf model (Clark and Scarf 1960, Federgruen and Zipkin 1984, and Chen and Zheng 1994b). On the other hand, one can think of (R, nQ) policies as generalized base-stock policies, i.e., stage i orders every period to keep its echelon stock within an interval of length Q_i. Therefore, base-stock policies, when properly modified to accommodate the base order quantities, are still optimal when every stage orders in batches.

2. Veinott (1965) establishes the optimality of (R, nQ) policies in single-location inventory problems with batch ordering. We generalize this result to a multi-echelon system. Moreover, (R, nQ) policies are widely studied in the literature. The above optimality proof is a further testament of its attractiveness.

3. It is well known that it is extremely difficult to characterize an optimal policy for a multi-echelon, stochastic inventory system with setup costs at all echelons (see Clark and Scarf 1962 for an example). The existence of setup costs implies that it is economical to replenish inventories in batches. Substituting fixed base order quantities for the setup costs leads to a new formulation of the problem. This re-formulation makes the problem of finding an optimal policy tractable. Moreover, fixed base order quantities provide benefits such as standardized shipments.[8]

References

[1] Atkins, D., "A Survey of Lower Bounding Methodologies for Production/Inventory Models," *Annals of Operations Research*, 26 (1990), 9-28.

[2] Atkins, D. and P. Iyogun, "A Lower Bound on a Class of Coordinated Inventory/Production Problems," *Operations Research Letters*, 6 (1987), 63-67.

[3] Atkins, D. and D. Sun, "98%-Effective Lot-Sizing for Series Inventory System with Backlogging," *Operations Research*, 43 (1995), 335-345.

[4] Axsater, S. "Evaluation of (R,Q) Policies for Two-Level Inventory Systems with Poisson Demand," Working Paper, Lund University, Sweden, 1991.

[8]Financial support from the Faculty Research Fund, Columbia Business School, and from the National Science Foundation under grant SBR-9702461 is gratefully acknowledged.

[5] Axsater, S., "Exact and Approximate Evaluation of Batch Ordering Policies for Two-Level Inventory Systems," *Operations Research*, 41 (1993a), 777-785.

[6] Axsater, S., "Simple Evaluation of Echelon Stock (R,Q) Policies for Two-Level Inventory Systems," Working Paper, Lund University, Sweden, 1993b.

[7] Axsater, S. and K. Rosling, "Installation vs. Echelon Stock Policies for Multi-Level Inventory Control," *Management Science*, 39 (1993), 1274-1280.

[8] Badinelli, R., "A Model for Continuous-Review Pull Policies in Serial Inventory Systems," *Operations Research*, 40 (1992), 142-156.

[9] Cachon, G., "Exact Evaluation of Batch-Ordering Inventory Policies in Two-Echelon Supply Chains with Periodic Review," Working Paper, Duke University, 1995.

[10] Cachon, G. and M. Fisher, "Supply Chain Inventory Management and the Value of Shared Information," Working Paper, Duke University, 1996.

[11] Chen, F., "Echelon Reorder Points, Installation Reorder Points, and the Value of Centralized Demand Information," To appear in *Management Science*, 1995.

[12] Chen, F., "Decentralized Supply Chains Subject to Information Delays," Working Paper, Graduate School of Business, Columbia University, 1996a.

[13] Chen, F. "Optimal Policies for Multi-Echelon Inventory Problems with Batch Ordering," To appear in *Operations Research*, 1996b.

[14] Chen, F., "Stationary Policies in Multi-Echelon Inventory Systems with Deterministic Demand and Backlogging," To appear in *Operations Research* (May-June), 1998.

[15] Chen, F. and Y.-S. Zheng, "Evaluating Echelon Stock (R,nQ) Policies in Serial Production/Inventory Systems with Stochastic Demand," *Management Science* 40 (1994a), 1262-1275.

[16] Chen, F. and Y.-S. Zheng, "Lower Bounds for Multi-Echelon Stochastic Inventory Systems," *Management Science*, 40 (1994b), 1426-1443.

[17] Chen, F. and Y.-S. Zheng, "One-Warehouse Multi-Retailer Systems with Centralized Stock Information," *Operations Research* 45 (1997), 275-287.

[18] Chen, F. and Y.-S. Zheng, "Near-Optimal Echelon-Stock (R,nQ) Policies in Multi-Stage Serial Systems," To appear in *Operations Research* (July-August), 1998.

[19] Clark, A. and H. Scarf, "Optimal Policies for a Multi-Echelon Inventory Problem," *Management Science*, 6 (1960), 475-490.

[20] Clark, A. and H. Scarf, "Approximate Solutions to a Simple Multi-Echelon Inventory Problem," in *Studies in Applied Probability and Management Science*, 88-110, K. Arrow, S. Karlin and H. Scarf (eds.), Stanford University Press, Stanford, CA., 1962.

[21] De Bodt, M. and S. Graves, "Continuous Review Policies for a Multi-Echelon Inventory Problem with Stochastic Demand," *Management Science*, 31 (1985), 1286-1295.

[22] Deuermeyer, B. and L. Schwarz, "A Model for the Analysis of System Service Level in Warehouse/Retailer Distribution Systems: The Identical Retailer Case," in *Studies in the Management Sciences: The Multi-Level Production/Inventory Control Systems*, pp. 163-193, Vol. 16, L. Schwarz (ed.), North-Holland, Amsterdam, 1981.

[23] Federgruen, A. and Z. Katalan, "Make-to-Stock or Make-to-Order: that is the question; novel answers to an ancient debate," Graduate School of Business, Columbia University, 1994.

[24] Federgruen, A. and Y.-S. Zheng, "An Efficient Algorithm for Computing an Optimal (r,Q) Policy in Continuous Review Stochastic Inventory Systems," *Operations Research*, 40 (1992), 808-813.

[25] Federgruen, A. and P. Zipkin, "Computational Issues in an Infinite-Horizon, Multi-Echelon Inventory Model," *Operations Research*, 32 (1984), 818-836.

[26] Fuchsberg, G., "Hand-Held Computers Help Field Staff Cut Paper Work and Harvest More Data," *The Wall Street Journal*, January 30 (1990), p. B1.

[27] Gallego, G. and I. Moon, "The Distribution-Free Newsboy Problem: Review and Extensions," *Journal of the Operational Research Society*, 44 (1993), 825-834.

[28] Gavirneni, S., R. Kapuscinski, and S. Tayur, "Value of Information in Capacitated Supply Chains," Working Paper, Carnegie Mellon University.

[29] Hadley, G. and T. Whitin, "A Family of Inventory Models," *Management Science*, 7 (1961), 351-371.

[30] Hadley, G., and Whitin, T. M., *Analysis of Inventory Systems*, Prentice-Hall Inc., Englewood Cliffs, N.J., 1963.

[31] Hariharan, R. and P. Zipkin, "Customer-Order Information, Leadtimes, and Inventories," *Management Science*, 41 (1995), 1599-1607.

[32] Karlin, S., and Fabens, G., "A Stationary Inventory Model with Markovian Demand," Mathematical Methods in the Social Sciences, K. Arrow, S. Karlin and P. Suppes, Stanford University Press, Stanford, CA., 1963.

[33] Lee, H. L. and K. Moinzadeh, "Two-Parameter Approximations for Multi-Echelon Repairable Inventory Models with Batch Ordering Policy," *IIE Transactions* 19 (1987a), 140-149.

[34] Lee, H. L. and K. Moinzadeh, "Operating Characteristics of a Two-Echelon Inventory System for Repairable and Consumable Items Under Batch Ordering and Shipment Policy," *Naval Research Logistics Quarterly* 34 (1987b), 365-380.

108

[35] Lee, H. L., P. Padmanabhan and S. Whang, "Information Distortion in a Supply Chain: The Bullwhip Effect," *Management Science*, 43 (1997), 546-558.

[36] Lee, H. L., K. C. So, and C. S. Tang, "The Value of Information Sharing in a Two-Level Supply Chain," Working Paper, UCLA, 1997.

[37] Lovejoy, W. and S. Whang, "Response Time Design in Integrated Order Processing/Production Systems," *Operations Research* 43 (1995), 851-861.

[38] Maxwell, W. and J. Muckstadt, "Establishing Consistent and Realistic Reorder Intervals in Production-Distribution Systems," *Operations Research*, 33 (1985), 1316-1341.

[39] Milgrom, P. and J. Roberts, "Communication and Inventory as Substitutes in Organizing Production," *Scandinavian Journal of Economics*, 90 (1988), 275-289.

[40] Moinzadeh, K. and H. Lee, "Batch Size and Stocking Levels in Multi-Echelon Repairable Systems," *Management Science*, 32 (1986), 1567-1581.

[41] Morse, P. M., " Solutions of a Class of Discrete Time Tnventory Problems," *Operations Research*, 7 (1959), 67-78.

[42] Muckstadt, J. and R. Roundy, "Analysis of Multi-Stage Production Systems," in *Handbook in Operations Research and Management Science, Vol. 4, Logistics of Production and Inventory,* Ed. by S. Graves, A. Rinnooy Kan and P. Zipkin, North Holland, 1993.

[43] Naddor, E., "Optimal and Heuristic Decisions in Single and Multi-item Inventory Systems," *Management Science*, 21 (1975), 1234-1249.

[44] Nguyen, V., "On Base-Stock Policies for Make-to-Order/Make-to-Stock Production," Working Paper, MIT, 1995.

[45] Richards, F., "Comments on the Distribution of Inventory Position in a Continuous Review (s, S) Inventory System," *Operations Research*, 23(1975), 366-371.

[46] Rosling, K., "Optimal Inventory Policies for Assembly Systems Under Random Demand," *Operations Research* 37 (1989), 565-579.

[47] Rosling, K., "94%-Effective Lotsizing: User-Oriented Algorithms and Proofs," Working Paper, Linkoping Institute of Technology, Sweden, 1993.

[48] Roundy, R. O., "98% Effective Integer-Ratio Lot-Sizing for One-Warehouse Multi-Retailer Systems," *Management Science*, 31 (1985), 1416-1430.

[49] Roundy, R. O., "A 98% Effective Lot-Sizing Rule for a Multi-Product, Multi-Stage Production Inventory System," *Mathematics of Operations Research*, 11 (1986), 699-727.

[50] Scarf, H., "A Min-Max Solution of an Inventory Problem," in *Studies in the Mathematical Theory of Inventory and Production*, Stanford University Press, Stanford, CA, 1958.

[51] Schmidt, C. and S. Nahmias, "Optimal Policy for a Two-Stage Assembly System Under Random Demand," *Operations Research*, 33 (1985), 1130-1145.

[52] Simon, R. M., " The Uniform Distribution of Inventory Position for Continuous Review (s, Q) Policies," P-3938, The Rand Corp., Santa Monica, California, 1968.

[53] Stalk, G., P. Evans and L. Shulman, "Competing on Capabilities: The New Rules of Corporate Strategy," *Harvard Business Review*, 70(1992), 57-69.

[54] Svoronos, A. and P. Zipkin, "Estimating the Performance of Multi-Level Inventory Systems," *Operations Research*, 36 (1988), 57-72.

[55] Tyndall, G., "Supply Chain Management Innovations Spur Long-Term Strategic Retail Alliances," *Marketing News*, December 19 (1988), p. 10.

[56] Veinott, A., "The Optimal Inventory Policy for Batch Ordering," *Operations Research*, 13 (1965), 424-432.

[57] Wagner, H., O'Hagan, M., and Lundh, B., "An Empirical Study of Exact and Approximately Optimal Inventory Policies," *Management Science*, 11(1965), 690-723.

[58] Wessell, D., "Computer Finds a Role in Buying and Selling, Reshaping Businesses," *The Wall Street Journal*, March 18 (1987), p. 1.

[59] Zheng, Y.-S., "On Properties of Stochastic Inventory Systems," *Management Science*, 38 (1992), 87-103.

[60] Zheng, Y.-S. and F. Chen, "Inventory Policies with Quantized Ordering," *Naval Research Logistics*, 39 (1992), 285-305.

5 COMPETITIVE SUPPLY CHAIN INVENTORY MANAGEMENT

Gérard P. Cachon

Fuqua School of Business
Duke University

5.1 INTRODUCTION

Most supply chains are composed of independent agents with individual preferences. These agents could be distinct firms or they could even be managers within a single firm. In either case, it is expected that no single agent has control over the entire supply chain, and hence no agent has the power to optimize the supply chain. It is also reasonable to assume that each agent will attempt to optimize his own preference, knowing that all of the other agents will do the same. Will this competitive behavior lead the agents to choose policies that optimize overall supply chain performance? The answer is usually "no", due to supply chain externalities. An externality occurs whenever the action of one agent impacts another agent. For example, suppose agent i's action benefits agent j. Agent i will tend to do too little of that action because he does not consider the full benefit of the action on the supply chain (assuming increasing that action is costly to agent i). Similarly, suppose agent i's action confers an additional cost on agent j. In this case agent i will tend to do too much of that action. As will be discussed, many externalities exist in supply chain operations.

When competition degrades supply chain performance the agents can benefit from coordination. But how should they coordinate? They could each agree to work towards the general welfare of the supply chain, but this agreement still leaves each agent with an incentive to serve his own preference. The typical solution is for the agents to agree to a set of transfer payments that modifies their incentives, and hence modifies their behavior. Many types of transfer payments are possible.

This chapter reviews competitive supply chain inventory management. The topic is broadly interpreted, so important research from economics and marketing is included in addition to recent operations management research. Game theory is the primary methodology.

§5.2 begins with a model of a supplier that sells a product to a retailer that faces a downward sloping demand curve. In this setting Spengler (1950) obtains the classic double marginalization result: the retailer does not consider the supplier's profit margin when choosing his order quantity, so the retailer orders too little product. Techniques are discussed to encourage the retailer to choose the correct order quantity. §5.3 details a similar model, except the retailer faces stochastic demand at a fixed retail price. Again, the retailer purchases too little inventory because he does not consider the supplier's profit margin. Pasternack (1985) demonstrates that the optimal supply chain profit can be achieved when the supplier offers the retailer a buy-back contract (the supplier purchases unsold goods at a specified buy-back price). §5.4 discusses the use of quantity discounts to raise order quantities. There is a well developed literature on quantity discounts, so this section concentrates on describing the externalities that create the need for quantity discounts.

The primary model in this chapter is discussed in §5.5. This is a two echelon supply chain with a single supplier and a single retailer that faces stochastic demand. The firms incur holding and backorder costs. There are positive lead times between stages and the firms implement bases stock policies. Three externalities are identified in this setting. It is shown that there is a unique Nash equilibrium pair of base stock policies, i.e., there is only one pair of base stock policies such that no firm has a unilateral incentive to deviate. The Nash equilibrium is presumed to be the competitive solution. The optimal policies are never a Nash equilibrium, hence competition always deteriorates supply chain performance. A numerical study indicates that there is substantial variance in the magnitude of the competition penalty: In some cases it is modest (less than 5%), but in other cases it is enormous (over 100%).

Four coordination techniques are applied to this model in §5.6 . Cachon and Zipkin (1997) propose linear transfer payments based on actual inventory and backorder levels. Chen (1997) suggests linear transfer payments based on a special accounting of the inventory and backorder levels. Lee and Whang (1996) detail a non-linear transfer payment scheme that is related to Clark and Scarf's (1960) decomposition technique. Porteus (1997) suggests related transfer payments, but implements these payments with responsibility tokens.

Additional operations management research on competitive supply chain inventory management is discussed in §5.7. This section highlights research that identifies additional supply chain inventory management externalities and potential coordination solutions. The final section concludes.

5.2 DOUBLE MARGINALIZATION

Double marginalization is present in most supply chain models because it occurs whenever the supply chain's profits are divided among two or more players and at least one of the players influences demand. To explain in greater detail, consider a supply chain with a supplier and a retailer that sells a product in one period. The retailer can sell $q \geq 0$ units at price $p(q) \geq 0$. There exists a maximum possible sales quantity $\widehat{q}$, i.e., $p(\widehat{q}) = 0$. Over the interval $[0, \widehat{q}]$ assume that $p(q)$ is decreasing, continuous, twice differentiable and concave, i.e., $p'(q) < 0$ and $p''(q) \leq 0$. The supplier produces each unit for a cost $c \leq p(0)$, and sells each unit to the retailer for a wholesale price w.

The supplier first announces her wholesale price w and then the retailer chooses an order quantity q. The supplier produces and delivers this order to the retailer. Finally, the retailer sells the q units at price $p(q)$.

To analyze this game, begin with the *centralized solution*, which assumes a single agent controls the entire supply chain to maximize supply chain profits. Next, evaluate the *decentralized solution* which assumes the players make choices with the objective of maximizing their own profits. If the decentralized and centralized solutions differ, investigate how to modify the player's payoffs so that the new decentralized solution corresponds to the centralized solution. This three step "recipe" is commonplace in competitive supply chain inventory management research.

The centrally controlled supply chain's profits are

$$\Pi(q) = q(p(q) - c),$$

which only depends on the retailer's sales quantity. (The supplier's wholesale price decision merely creates a transfer payment between the two firms, so it does not influence supply chain profits.) Since profits are strictly concave in quantity over the interval $[0, \hat{q}]$, the optimal quantity q^o satisfies $\Pi'(q^o) = 0$,

$$p(q^o) - c + q^o p'(q^o) = 0. \tag{5.1}$$

(The interval boundaries are never optimal.)

Now assume the retailer must choose a quantity after observing the supplier's wholesale price. The retailer's profits are

$$\pi_r(q) = q(p(q) - w).$$

Since the retailer's profits are strictly concave, his optimal quantity q^* satisfies $\pi'_r(q^*) = 0$,

$$p(q^*) - w + q^* p'(q^*) = 0.$$

The supplier will always choose $w > c$ (otherwise she earns no profit), so it follows from (5.1) that

$$p(q^o) - w + q^o p'(q^o) < 0.$$

Hence, $q^o > q^*$. In words, the retailer orders less than the supply chain optimal quantity whenever the supplier earns positive profits. Spengler (1950) called this problem double marginalization because each firm only considers its own profit margin in making its decision, and does not consider the supply chain's profit margin.

Note that $q^* = q^o$ only when the supplier prices at marginal cost, $w = c$. Hence, *marginal cost pricing* is one solution to double marginalization. Of course, this is not a very good solution for the supplier, since $w = c$ implies she earns a zero profit. In effect, the supplier sells to the retailer her portion of the supply chain for free. A two-part tariff is a better strategy for the supplier. In particular, the supplier could choose to price at marginal cost $w = c$ (part one of the tariff) and also charge a fixed fee $\Pi(q^o)$ (part two of the tariff). The retailer will choose q^o to maximize his gross profits $\Pi(q^o)$, but of course the fixed fee eliminates those profits. Hence, total supply chain profits are optimal and awarded exclusively to the supplier. While marginal cost pricing combined with a fixed fee is a plum strategy for the supplier in this model, Saggi and Vettas (1998) demonstrate that marginal cost pricing is not an effective strategy for the supplier when there are multiple competing retailers.

Without the use of fixed payments, Jeuland and Shugan (1983) suggest the firms coordinate with a *profit sharing contract*. This contract specifies that the supplier earns $f\Pi(q)$, for $0 \leq f \leq 1$, and the retailer earns $(1 - f)\Pi(q)$. The wholesale price is now irrelevant to each firm's profits, so the retailer chooses q^o to maximize his profits. The supply chain earns the optimal profits.

5.3 BUY-BACK CONTRACTS

A buy-back contract specifies a price b at which the supplier will purchase unsold goods from a retailer. These contracts are common in many industries, e.g., publishing and personal computers (see Padmanabhan and Png, 1995).

To understand why a supplier might want to offer a buy-back contract, consider a supply chain with a single supplier and a single retailer. The retailer charges a fixed retail price $p > 0$ and faces stochastic demand. Let $\Phi(x)$ be the cumulative distribution function of demand and let $\phi(x)$ be the density function. Assume $\Phi(x)$ is continuous and differentiable. The sequence of events follows: (1) the supplier announces a wholesale price w and a buy-back price b; (2) the retailer chooses an order quantity q; (3) the supplier produces q units at marginal cost c and delivers these units to the retailer ($c < p$); (4) demand is realized and unsold goods are returned to the supplier. Assume the supplier earns nothing from the disposal of the returned units. Pasternack (1985) studies a generalized version of this model.

A central planner chooses only a production quantity to ship to the retailer, since both the wholesale price and the buy-back rate are mere transfer payments. The supply chain's profits are

$$\Pi(q) = -cq + p\left[(1 - \Phi(q))q + \int_0^q x\phi(x)dx\right].$$

The first term is the production cost and the second term is the expected sales revenue. This is a newsvendor problem, and it is well known that the optimal order quantity q^o satisfies

$$\Phi(q^o) = \frac{p - c}{p}. \tag{5.2}$$

The retailer's profits are

$$\pi_r(q) = -wq + p\left[(1 - \Phi(q))q + \int_0^q x\phi(x)dx\right] + b\int_0^q (q - x)\phi(x)dx.$$

The first term is the purchase costs, the second term is expected sales revenue and the third term is expected revenue from returns. Assuming $p > w > b$ (the retailers earns a profit on each unit sold and a loss on each unit returned), the retailer's profits are strictly concave and the optimal order quantity q^* satisfies

$$\Phi(q^*) = \frac{p - w}{p - b}. \tag{5.3}$$

When there is no buy-back, i.e., $w > c$ and $b = 0$, comparison of (5.2) and (5.3) reveals that $q^* < q^o$. In words, if the supplier prices above marginal cost and doesn't offer to purchase unsold goods, double marginalization causes the retailer to order less than the supply chain's optimal quantity. Since supply chain profits depend on q, the sum of the firms' profits will be less than maximum supply chain profits.

The supply chain could achieve its best performance if the supplier were willing to price at marginal cost, but as already mentioned, this is not an attractive solution to the supplier. Instead of lowering her wholesale price to marginal cost, (5.3) indicates that increasing b will raise the retailer's order quantity. In fact, the retailer chooses q^o whenever

$$\frac{p-w}{p-b} = \frac{p-c}{p}. \tag{5.4}$$

Let $\widehat{b}(w)$ be the buy-back that satisfies (5.4),

$$\widehat{b}(w) = p\left(\frac{w-c}{p-c}\right) = \frac{w-c}{\Phi(q^o)}.$$

So supply chain profits are maximized even if $w > c$ as long as $b = \widehat{b}(w)$. Note that $\widehat{b}(w)$ is independent of the demand distribution, so it could apply across multiple retailers facing heterogenous demand distributions.

The supplier's main concern is with her own profits, $\pi_s(w, b, q)$, and not with the supply chain's profits,

$$\pi_s(w, b, q) = q(w - c) - b\int_0^q x\phi(x)dx.$$

The first term is the supplier's sales revenue and the second term is the expected cost of purchasing unsold goods. Assuming the supplier chooses $b = \widehat{b}(w)$, the retailer will choose q^o, so the supplier's profits are

$$\pi_s\left(w, \widehat{b}(w), q^o\right) = q^o(w - c) - \widehat{b}(w)\int_0^{q^o} x\phi(x)dx.$$

Differentiate with respect to w

$$\frac{\partial \pi_s\left(w, \widehat{b}(w), q^o\right)}{\partial w} = q^o - \frac{p}{p-c}\int_0^{q^o} x\phi(x)dx$$

$$= \frac{1}{\Phi(q^o)}\int_0^{q^o} \Phi(x)dx$$

(The last step is done with integration by parts.) So the supplier's profits are increasing in her wholesale price. In fact, the supplier earns essentially all of the supply chain's profits when $w = p - \varepsilon$, for $\varepsilon \approx 0$. (The retailer's margin approaches zero, but the buy-back rate approaches w, thereby ensuring that the retailer still orders q^o.)

5.3.1 Related models

Several important extensions to Pasternack's model have been considered. Kandel (1996) and Emmons and Gilbert (1998) relax the assumption of a fixed retail price. Lariviere (1998) provides a more detailed analysis of price-only and buy-back contracts. For instance, he investigates the supplier's optimal wholesale price in a price-only contract.

Padmanabhan and Png (1997) demonstrate that buy-back policies can increase retail competition, thereby benefitting the supplier (even without stochastic demand). Butz (1997), Deneckere, Marvel and Peck (1997), and Deneckere, Marvel and Peck (1996) also study multiple retailer models. They suppose that the retail price depends on the total quantity retailers attempt to sell. While a retailer considers the impact of falling prices on his own inventories, he does not consider how falling prices reduces the value of inventory held by the other retailers. Therefore, after the retailers purchase their inventory they tend to sell this inventory too aggressively, depressing the market price. The retailers can anticipate this behavior when ordering, so they reduce their orders as their expectation for the market price decreases. In other words, retail competition decreases the incentive to hold inventory because it reduces the retailers' profit margin. The supplier can increase the retailers' orders if she mitigates retail competition. This can be achieved with *resale price maintenance contracts* that set a minimum price that retailers can charge.

Anupindi and Bassok (1998) investigate a model with one supplier and two retailers. Their model's main twist is to incorporate consumer search between the two retailers. They investigate competitive behavior when the retailers make independent decisions as well as when they jointly pool their inventories. In either case, they study how a supplier can benefit from offering a contract with a holding cost subsidy, which is essentially a buy-back contract. (The retailers incur holding costs on end of period inventory, so a holding cost subsidy lets the retailers virtually sell inventory back to the supplier.)

5.4 QUANTITY DISCOUNTS

There is a large literature on quantity discounts so this section concentrates on the reasons why they are used.

Jeuland and Shugan (1983) suggest that quantity discounts can mitigate double marginalization. Suppose the retailer pays $\omega(q)$ for q units, where for quantities less than q^o the marginal price paid is greater than the production cost, i.e., $\omega'(q|q < q^o) > c$, but $\omega'(q^o) = c$. It can be shown that the retailer will choose q^o since his marginal cost at q^o equals the supply chain's marginal cost, c. Further, the supplier earns positive profits since the average wholesale price per unit is greater than c. (See Moorthy, 1987, for additional technical requirements.)

Quantity discounts can also help manage operating costs. Suppose the supplier incurs a fixed order processing cost K_o for each retailer order and let q be a retailer's average order. Thus, the average order processing cost per unit is K_o/q, which is decreasing in q. A retailer does not incur this cost, so a retailer will order a smaller quantity than optimal for the supply chain. Quantity discounts will encourage the retailer to order more.

A retailer's order quantity also can influence the supplier's holding cost. Suppose the supplier incurs a production setup cost, so she will produce in batches. For simplicity, say mq is the supplier's batch size, where m is a positive even integer and q is the retailer's order quantity. Assuming constant retailer demand and a zero lead time between the supplier and the retailer, the supplier's average inventory equals $(m-1)q/2$. Now suppose the retailer doubles his order quantity to $\widehat{q} = 2q$. If the supplier doesn't change her batch size, her average inventory is now $(m/2-1)\widehat{q}/2$. Since

$$(m/2 - 1)\widehat{q}/2 < (m-1)q/2,$$

the supplier's average inventory has declined. This result holds in more general models. Hence, a retailer tends to order too little because he does not account for the holding cost savings the supplier earns from a larger order quantity. For more extensive treatment of quantity discounts see Lal and Staelin (1984), Lee and Rosenblatt (1986), Weng (1995), Boyaci and Gallego (1997), and Corbett and de Groote (1997).

5.5 COMPETITION IN THE SUPPLY CHAIN INVENTORY GAME

This section considers an infinite horizon, stochastic demand inventory game between one supplier and one retailer. The rules of the game are detailed and then the game is compared to other research. The optimal solution is described and the competitive solution is characterized. Finally, several coordination techniques are presented.

5.5.1 Model details

The supplier is stage 2 and the retailer is stage 1. Time is divided into an infinite number of discrete periods. Consumer demand at the retailer is stochastic, independent across periods and stationary. The following is the sequence of events during a period: (1) shipments arrive at each stage; (2) orders are submitted and shipments are released; (3) consumer demand occurs; (4) holding and backorder penalty costs are charged.

There is a lead time for shipments from the source to the supplier, L_2, and from the supplier to the retailer, L_1. Each firm may order any non-negative amount in each period. There is no fixed cost for placing or processing an order. Each firm pays a constant price per unit ordered.

The supplier is charged holding cost h_2 per period for each unit in her stock or on-route to the retailer. The retailer's holding cost is $h_2 + h_1$ per period for each unit in his stock. Assume $h_2 > 0$ and $h_1 \geq 0$.

Unmet demands are backlogged, and all backorders are ultimately filled. Both the retailer and the supplier may incur costs when demand is backordered. The retailer is charged $\alpha_1 p$ for each backorder, and the supplier $\alpha_2 p$, where $\alpha_1 + \alpha_2 = 1$ and $0 \leq \alpha_i \leq 1$. The parameter p is the total system backorder cost, and (α_1, α_2) specifies how this cost is divided between the firms. The parameters (α_1, α_2) are exogenous.

These backorder costs have several interpretations. They may represent the costs of financing receivables, if customers pay only upon the fulfillment of demands. (This requires a discounted-cost model to represent exactly, but the approximation here is standard in the average-cost context, analogous to the treatment of inventory financing costs.) Alternatively, they may be proxies for losses in customer good-will, which in turn lead to long-run declines in demand. Such costs need not affect the firms equally, which is why flexibility is allowed in the choice of $\alpha_i \in [0, 1]$. Finally, they provide an approximation to lost sales.

In period t just before demand define the following for stage i: *in-transit inventory*, IT_{it}, is all inventory in-transit between stages $i + 1$ and stage i; *inventory level*, IL_{it}, is inventory at stage i minus backorders at stage i (the supplier's backorders are unfilled retailer orders); and *inventory position*, IP_{it}, $IP_{it} = IL_{it} + IT_{it}$. Note that these are local inventory variables and not echelon inventory variables.

Each firm uses a base stock policy: Each period a firm orders a sufficient amount to raise its inventory position plus outstanding orders to that level. Define s_i as stage i's base stock level. In the inventory game's only move, the players simultaneously choose their strategies, $s_i \in \sigma = [0, S]$, where s_i equals player i's base stock level, σ is player i's strategy space and S is a very large constant. (S is sufficiently large that it never constrains the players.) A joint strategy s is a pair (s_1, s_2). After their choices, the players implement their policies over an infinite horizon. All model parameters are common knowledge (all information is known and verifiable to all players).

Let D^τ denote random total demand over τ periods, and μ^τ denote mean total demand over τ periods. Let ϕ^τ and Φ^τ be the density and distribution functions of demand over τ periods respectively. Assume $\Phi^1(x)$ is continuous, increasing and differentiable for $x \geq 0$, so the same is true of Φ^τ, $\tau > 0$. Furthermore, $\Phi^1(0) = 0$, so positive demand occurs in each period.

This model is identical to the Local Inventory game studied in Cachon and Zipkin (1997). They also consider a model in which firms track echelon inventories and find that the tracking method does influence strategic behavior. The notation in this model is generally consistent with Cachon and Zipkin (1997), but there are some differences. (In Cachon and Zipkin, 1997, an overbar on the local variables distinguishes them from the echelon variables. Echelon variables are not considered, so to avoid notational clutter overbars are not used.)

There are three externalities in this game:

1. The retailer ignores the supplier's backorder costs, so he tends to carry too little inventory;

2. The supplier ignores the retailer's backorder costs, so she tends to carry too little inventory;

3. The supplier ignores the retailer's holding costs, so she tends to carry too much inventory. An increase in the supplier's inventory leads to an increase in the retailer's inventory. (The supplier's average delivery time decreases, thereby raising the retailer's average inventory for a fixed

retailer base stock policy.) Higher retail inventory benefits the supplier, through lower backorder costs, but the supplier doesn't pay the retailer's holding costs, so she tends to raise the retailer's inventory more than she should.

Clearly, the second and third externalities conflict so *a priori* it is uncertain whether the supplier will carry too much or too little inventory. While a numerical study confirms that either outcome is possible, it is generally observed that supply chain inventory is too low in the competitive solution.

Chen (1997), Lee and Whang (1996), and Porteus (1997) study similar models. All assume stationary demand, serial supply chains, holding and backorder costs, fixed lead times and common knowledge. Chen (1997) assumes the players attempt to minimize total supply chain costs, so they don't have conflicting incentives. In his model there is a delay between when a stage submits an order and when its upstream supplier receives the order. In this model orders are transmitted instantly. Further, he considers a supply chain with four stages. Both Lee and Whang (1996) and Porteus (1997) study two stage supply chains, but they assume the upstream stage only cares about her local inventory, i.e., $\alpha_2 = 0$. Lee and Whang (1996) assume firms minimize discounted costs instead of average costs.

The supply chain inventory game differs from the models of the previous three sections. Unlike the double marginalization model in §5.2, demand is independent of the players' actions. Unlike the buy-back and quantity discount models, the retailers' orders are not always filled immediately because the supplier may have insufficient inventory.

5.5.2 Optimal Solution

The system optimal solution minimizes the total average cost per period. Clark and Scarf (1960), Federgruen and Zipkin (1984) and Chen and Zheng (1994) demonstrate that a base stock policy is optimal in this setting. The traditional method to find the optimal solution allocates costs to the firms in a particular way. Then, each firm's new cost function is minimized. This section briefly outlines this method.

Let $\widehat{G}_1^o(IL_{1t} - D^1)$ equal the retailer's charge in period t, where

$$\widehat{G}_1^o(x) = h_1[x]^+ + (h_2 + p)[x]^-.$$

(Relative to the retailer's actual costs, the retailer's holding cost is reduced by h_2 per unit, but his backorder penalty cost is increased by h_2 per unit.) Also in period t, define $G_1^o(IP_{1t})$ as the retailer's expected charge in period $t + L_1$, where

$$G_1^o(y) = E\left[\widehat{G}_1^o(y - D^{L_1+1})\right].$$

Define s_1^o as the value of y that minimizes $G_1^o(y)$:

$$\Phi^{L_1+1}(s_1^o) = \frac{h_2 + p}{h_1 + h_2 + p}. \tag{5.5}$$

This is the retailer's optimal base stock level. Define the induced penalty function,

$$\underline{G}_1^o(y) = G_1^o(\min\{s_1^o, y\}) - G_1^o(s_1^o),$$

and define

$$\widehat{G}_2^o(y) = h_2(y - \mu^1) + \underline{G}_1^o(y).$$

Note that $\underline{G}_1^o(y)$ is non-linear in y.

In period t charge the supplier $G_2^o(IP_{2t})$, where

$$G_2^o(y) = E\left[\widehat{G}_2^o(y + s_1^o - D^{L_2})\right].$$

The supplier's optimal base stock level, s_2^o, minimizes $G_2^o(\cdot)$.

5.5.3 Game analysis

Define $H_i(s_1, s_2)$ as player i's expected per period cost when players use base stock levels (s_1, s_2). The best reply mapping for firm i is a set-valued relationship associating each strategy s_j, $j \neq i$, with a subset of σ according to the following rules:

$$r_1(s_2) = \left\{s_1 \in \sigma \mid H_1(s_1, s_2) = \min_{x \in \sigma} H_1(x, s_2)\right\}$$

$$r_2(s_1) = \left\{s_2 \in \sigma \mid H_2(s_1, s_2) = \min_{x \in \sigma} H_2(s_1, x)\right\}.$$

A pure strategy Nash equilibrium is a pair of base stock levels (s_1^*, s_2^*), such that each player chooses a best reply to the other player's equilibrium base stock level:

$$s_2^* \in r_2(s_1^*) \quad s_1^* \in r_1(s_2^*).$$

Retailer's cost function. In each period, the retailer is charged $h_1 + h_2$ per unit held in inventory and $\alpha_1 p$ per unit backordered. Define $\widehat{G}_1(IL_{1t} - D^1)$ as the sum of these costs in period t,

$$\widehat{G}_1(y) = (h_1 + h_2)[y]^+ + \alpha_1 p[y]^-.$$

Define $G_1(IP_{1t})$ as the retailer's expected cost in period $t + L_1$,

$$
\begin{aligned}
G_1(y) &= E\left[\widehat{G}_1(y - D^{L_1+1})\right] \\
&= (h_1 + h_2)\left(y - \mu^{L_1+1}\right) + (h_1 + h_2 + \alpha p)\int_y^\infty (x - y)\phi^{L_1+1}(x)dx.
\end{aligned}
$$

The retailer's true expected cost depends on both his own base stock as well as the supplier's base stock. After the firms place their orders in period $t - L_2$, the supplier's inventory position equals s_2. After inventory arrives in period t, the supplier's inventory level equals $s_2 - D^{L_2}$. (The retailer orders D^{L_2} over periods $[t - L_2 + 1, t]$.) When $s_2 - D^{L_2} \geq 0$, the supplier completely fills the retailer's period t order, so $IP_{1t} = s_1$. When $s_2 - D^{L_2} < 0$, the supplier cannot fill all of the retailer's order, and $IP_{1t} = s_1 + s_2 - D^{L_2}$. Hence,

$$
\begin{aligned}
H_1(s_1, s_2) &= E\left[G_1\left(\min\left\{s_1 + s_2 - D^{L_2}, s_1\right\}\right)\right] \\
&= \Phi^{L_2}(s_2)G_1(s_1) + \int_{s_2}^{\infty} \phi^{L_2}(x)G_1(s_1 + s_2 - x)dx.
\end{aligned}
$$

Supplier's cost function. Define $\widehat{G}_2(IL_{1t} - D^1)$ as the supplier's actual period t backorder cost,

$$
\widehat{G}_2(y) = \alpha_2 p[y]^-,
$$

and $G_2(IP_{1t})$ as the supplier's expected period $t + L_1$ backorder cost,

$$
G_2(y) = E\left[\widehat{G}_2(y - D^{L_1+1})\right].
$$

Define

$$
\widehat{H}_2(s_1, x) = h_2\mu^{L_1} + h_2[x]^+ + G_2\left(s_1 + \min\left\{x, 0\right\}\right),
$$

so

$$
\begin{aligned}
H_2(s_1, s_2) &= E\left[\widehat{H}_2(s_1, s_2 - D^{L_2})\right] \\
&= h_2\mu^{L_1} + h_2 \int_0^{s_2} (s_2 - x)\phi^{L_2}(x)dx \\
&\quad + \Phi^{L_2}(s_2)G_2(s_1) + \int_{s_2}^{\infty} \phi^{L_2}(x)G_2(s_1 + s_2 - x)dx.
\end{aligned}
$$

The first term above is the expected holding cost for the units in-transit to the retailer (from Little's Law), the second term is the expected cost for inventory held at the supplier and the final two terms are the supplier's expected backorder cost.

Equilibrium analysis. The analysis of the game begins by characterizing the cost functions and the best reply mappings.

Theorem 1 $H_2(s_1, s_2)$ *is strictly convex in* s_2 *and* $H_1(s_1, s_2)$ *is strictly convex in* s_1.

Proof. It is sufficient to demonstrate that the second derivatives are positive. Differentiate $H_2(s_1, s_2)$:

$$
\begin{aligned}
\frac{\partial H_2(s_1, s_2)}{\partial s_2} &= h_2\Phi^{L_2}(s_2) + \int_{s_2}^{\infty} \phi^{L_2}(x)G_2'(s_1 + s_2 - x)dx \\
\frac{\partial^2 H_2(s_1, s_2)}{\partial s_2^2} &= h_2\phi^{L_2}(s_2) - \phi^{L_2}(s_2)G_2'(s_1) + \int_{s_2}^{\infty} \phi^{L_2}(x)G_2''(s_1 + s_2 - x)dx
\end{aligned}
$$

The second derivative is positive since $G_2'(y) \leq 0$, $G_2''(y) \geq 0$ and $h_2 \phi^{L_2}(s_2) \geq 0$. Differentiate $H_1(s_1, s_2)$,

$$\frac{\partial H_1(s_1, s_2)}{\partial s_1} = \Phi^{L_2}(s_2) G_1'(s_1) + \int_{s_2}^{\infty} \phi^{L_2}(x) G_1'(s_1 + s_2 - x) dx; \quad (5.6)$$

$$\frac{\partial^2 H_1(s_1, s_2)}{\partial s_1^2} = \Phi^{L_2}(s_2) G_1''(s_1) + \int_{s_2}^{\infty} \phi^{L_2}(x) G_1''(s_1 + s_2 - x) dx.$$

Since $G_1(\cdot)$ is strictly convex, $H_1(s_1, s_2)$ is strictly convex in s_1. $\square$

Since the cost functions are strictly convex, each player has a unique best reply to the other player's strategy, a useful result to demonstrate existence of an equilibrium. The next two theorems further characterize the best reply mappings. When both players care about backorder costs ($\alpha_1 > 0$, $\alpha_2 > 0$), each player will select a positive base stock. Further, as one player reduces its base stock, the other player will increase its base stock by a lesser amount. This result is used to demonstrate that there exists a unique Nash equilibrium.

Theorem 2 $r_2(s_1)$ is a function; when $\alpha_2 = 0$, $r_2(s_1) = 0$; and when $\alpha_2 > 0$, $r_2(s_1) > 0$ and $-1 < r_2'(s_1) < 0$.

Proof. From Theorem 1 $H_2(s_1, s_2)$ is strictly convex in s_2, so there a unique base stock that minimizes the supplier's cost. When $\alpha_2 = 0$ the supplier incurs no backorder costs, so she chooses $s_2 = 0$ to incur no holding costs. When $\alpha_2 > 0$, the supplier's first order condition determines her optimal s_2, but the first order condition is never satisfied at $s_2 = 0$, hence $r_2(s_1) = s_2 > 0$. Assume $s_2 > 0$. From the implicit function theorem,

$$r_2'(s_1) = -\left(\frac{\partial^2 H_2}{\partial s_2 \partial s_1} / \frac{\partial^2 H_2}{\partial s_2^2} \right) \quad (5.7)$$

$$= \frac{-\frac{\partial^2 H_2}{\partial s_2 \partial s_1}}{(h_2 - G_2'(s_1)) \phi^{L_2}(s_2) + \frac{\partial^2 H_2}{\partial s_2 \partial s_1}},$$

where

$$\frac{\partial^2 H_2}{\partial s_2^2} = (h_2 - G_2'(s_1)) \phi^{L_2}(s_2) + \int_{s_2}^{\infty} \phi^{L_2}(x) G_2''(s_1 + s_2 - x) dx,$$

$$\frac{\partial^2 H_2}{\partial s_2 \partial s_1} = \int_{s_2}^{\infty} \phi^{L_2}(x) G_2''(s_1 + s_2 - x) dx.$$

The cross partial of H_2 is positive because $G_2'' > 0$. The denominator of (5.7) is positive because $G_2'(s_1) \leq 0$ and $\phi^{L_2}(s_2) > 0$. So $-1 < r_2'(s_1) < 0$. $\square$

Theorem 3 $r_1(s_2)$ is a function; when $\alpha_1 = 0$, $r_1(s_2) = 0$; when $1 > \alpha_1 > 0$, $r_1(s_2) > 0$ and $-1 < r_1'(s_2) < 0$.

Proof. $r_1(s_2)$ is a function since H_1 is strictly convex in s_1. When $\alpha_1 = 0$ the retailer incurs no backorder costs, so he chooses $s_1 = 0$ to incur no holding costs.

When $\alpha_1 > 0$, the retailer's first order condition determines his optimal s_1, but the first order condition is never satisfied at $s_1 = 0$, hence $r_1(s_2) = s_1 > 0$. Assume $1 > \alpha_1 > 0$. From the implicit function theorem

$$
\begin{aligned}
r_1'(s_2) &= -\frac{\partial^2 H_1(s_1, s_2)}{\partial s_1 \partial s_2} \Big/ \frac{\partial^2 H_1(s_1, s_2)}{\partial s_1^2} \\
&= -\frac{\int_{s_2}^{\infty} \phi^{L_2}(x) G_1'''(s_1 + s_2 - x) dx}{\Phi^{L_2}(s_2) G_1''(s_1) + \int_{s_2}^{\infty} \phi^{L_2}(x) G_1'''(s_1 + s_2 - x) dx}.
\end{aligned}
$$

From Theorem 2, $s_2 > 0$ (because $\alpha_2 > 0$). Therefore, $\Phi^{L_2}(s_2) G_1''(s_1) > 0$. This implies $-1 < r_1'(s_2) < 0$. $\square$

Theorem 4 (s_1^*, s_2^*) *is the unique Nash equilibrium.*

Proof. From Theorem 1.2 in Fudenberg and Tirole (1991), a pure strategy Nash equilibrium exists if (1) each player's strategy space is a nonempty, compact convex subset of a Euclidean space, and (2) player i's cost function is continuous in s and quasi-convex in s_i. By the assumptions and Theorem 1, these conditions are met, so there is at least one equilibrium. It remains to show that there is a unique equilibrium. Assume $\alpha_1 = 0$; $r_1(s_2) = 0$, and since $r_2(\cdot)$ is a function, $r_2(0)$ is unique. Assume $\alpha_1 = 1$; $r_2(s_1) = 0$, and since $r_1(\cdot)$ is a function, $r_1(0)$ is unique. Now assume $0 < \alpha_1 < 1$. From Theorem 2, $r_2(s_1^*) = s_2^* > 0$. Suppose there are two equilibria, (s_1^*, s_2^*) and $(\hat{s}_1, \hat{s}_2)$. Assume $s_2^* < \hat{s}_2$. From Theorem 3, this implies that $\hat{s}_1 < s_1^*$. From the same theorem, $r_1'(s_2) > -1$, so $\hat{s}_2 - s_2^* > s_1^* - \hat{s}_1$. But from Theorem 2, $r_2'(s_1) > -1$, which implies $\hat{s}_2 - s_2^* < s_1^* - \hat{s}_1$, a contradiction. The analogous contradiction is obtained if $s_2^* > \hat{s}_2$ is assumed, so the equilibrium is unique. $\square$

Figures 1 and 2 display the best reply functions in this game as well as the Nash equilibrium and the optimal solution. In neither case does the Nash equilibrium coincide with the optimal solution.

Theorem 5 *Assuming* $\alpha_1 < 1$, $s_1^* + s_2^* < s_1^o + s_2^o$.

Proof. See Cachon and Zipkin (1997), Theorem 15 for proof. $\square$

It follows immediately from Theorem 5 that the system optimal solution is not a Nash equilibrium whenever $\alpha_1 < 1$. When $\alpha_1 = 1$ the system optimal solution can be a Nash equilibrium under a *very* special condition, see Cachon and Zipkin (1997). Therefore, competitive selection of inventory policies (virtually) always deteriorates supply chain performance (i.e., leads to higher than optimal cost).

A numerical study assesses the magnitude of the competitive penalty (the percentage increase of the Nash equilibrium cost over the optimal cost). The Nash equilibrium and the optimal policies are found for each of the 2625 scenarios constructed from the following parameters:

$$
\begin{aligned}
&\alpha_1 \in \{0, 0.1, 0.3, 0.5, 0.7, 0.9, 1\} \quad \alpha_2 = 1 - \alpha_1 \\
&L_1 \in \{1, 2, 4, 8, 16\} \quad\quad\quad\quad\quad L_2 \in \{1, 2, 4, 8, 16\} \\
&h_1 \in \{0.1, 0..3, 0.5, 0.7, 0.9\} \quad\quad h_2 = 1 - h_1 \\
&p \in \{1, 5, 25\}
\end{aligned}
$$

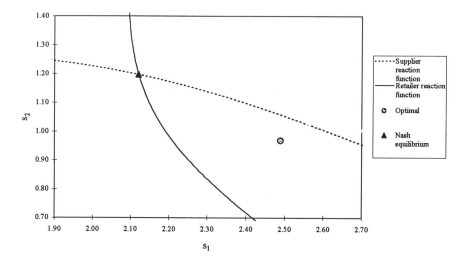

Figure 1: Reaction functions, $\alpha = 0.3$, $p = 5$, $h_1 = h_2 = 0.5$, $L_1 = L_2 = 1$

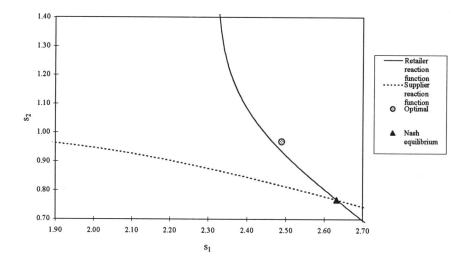

Figure 2: Reaction functions, $\alpha = 0.9$, $p = 5$, $h_1 = h_2 = 0.5$, $L_1 = L_2 = 1$

In each scenario demand is normally distributed with mean 1 and standard deviation 1/4. Table 1 summarizes the data.

Table 1: The distribution of competition penalties

α_1	Minimum	5th percentile	Median	95th percentile	Maximum
0	107%	117%	804%	5,930%	10,939%
0.1	5%	8%	37%	96%	119%
0.3	2%	3%	9%	19%	26%
0.5	1%	1%	3%	6%	8%
0.7	0%	0%	1%	4%	9%
0.9	0%	0%	1%	17%	45%
1	0%	0%	1%	34%	116%

When each firm cares equally about backorder costs, $\alpha_1 = \alpha_2$, the median competition penalty across scenarios is 3% and the maximum is 8%. When the backorder preferences are asymmetric, $\alpha_1 = 1$ or $\alpha_2 = 1$, then the competition penalty can be quite large. When the retailer cares little about backorders even the median competition penalty is large. This occurs because the retailer chooses to carry little inventory, hence the supplier can do little to prevent consumer backorders. On the other hand, when the supplier cares little about backorders the median competition penalty is low while the maximum penalty is large. This occurs because the supplier chooses to carry little inventory but in some cases the supplier carries little inventory in the optimal solution. That happens when the supplier's holding cost is large (when $h_1 = 0$ there is no advantage to keeping inventory at the supplier) or when L_2 is small relative to L_1. Overall, while competition does deteriorate supply chain performance, the magnitude of this problem is clearly context specific.

5.6 COORDINATION IN THE SUPPLY CHAIN INVENTORY GAME

According to Theorem 5, the optimal solution is virtually never a Nash equilibrium. Hence, the firms can lower total costs by cooperatively choosing the optimal base stock levels, (s_1^o, s_2^o). But any agreement to choose the optimal policy must eliminate each player's incentive to deviate. This is done by appropriately modifying each player's incentives with transfer payments.

5.6.1 Linear contracts

Cachon and Zipkin (1997) propose that the firms adopt a transfer payment contract with constant parameters $(\iota_1, \beta_2, \beta_1)$. This contract specifies that the period t transfer payment from the supplier to the retailer is

$$\iota_1 I_{1t} + \beta_2 B_{2t} + \beta_1 B_{1t},$$

where I_{1t} is the retailer's on-hand inventory, and B_{it} is stage i's backorders, all measured at the end of the period. There are no *a priori* sign restrictions on these parameters, e.g., $\iota_1 > 0$ represents a holding cost subsidy to the retailer and $\iota_1 < 0$ represents a holding fee. Note that $\iota_1 > 0$ is related to a buy-back contract: having the supplier compensate the retailer for his holding cost is like having the supplier (virtually) purchase some of the inventory back from the retailer.

Define $T_1(IP_{1t})$ as the expected transfer payment in period $t + L_1$ due to retailer inventory and backorders, where

$$T_1(y) = E[\iota_1[y - D^{L_1+1}]^+ + \beta_1[y - D^{L_1+1}]^-]$$

$$= \iota_1(y - \mu^{L_1+1}) + (\iota_1 + \beta_1) \int_y^\infty (x - y)\phi^{L_1+1}(x)dx.$$

Define $T(s_1, s_2)$ as the expected per period transfer payment from the supplier to the retailer,

$$T(s_1, s_2) = E[\beta_2 [s_2 - D^{L_2}]^- + T_1(s_1 + \min\{0, s_2 - D^{L_2}\})]$$

$$= \beta_2 \int_{s_2}^\infty \phi^{L_2}(x)(x - s_2)dx + \Phi^{L_2}(s_2)T_1(s_1)$$

$$+ \int_{s_2}^\infty \phi^{L_2}(x)T_1(s_1 + s_2 - x)dx.$$

Note that s_1 influences the retailer inventory and backorders, but not the supplier's backorders. Let $H_i^l(s_1, s_2)$ be player i's costs after accounting for the transfer payment,

$$H_1^l(s_1, s_2) = H_1(s_1, s_2) - T(s_1, s_2),$$
$$H_2^l(s_1, s_2) = H_2(s_1, s_2) + T(s_1, s_2).$$

The objective is to determine the set of contracts, $(\iota_1, \beta_2, \beta_1)$, such that (s_1^o, s_2^o) is a Nash equilibrium for the cost functions $H_i^l(s_1, s_2)$. With these contracts the firms can choose (s_1^o, s_2^o), thereby minimizing total costs, and also be assured that no player has an incentive to deviate.

To find the desired set of contracts, first assume that H_i^l is strictly convex in s_i, given that player j chooses s_j^o, $j \neq i$. Then determine the contracts in which s_i^o satisfies player i's first order condition, thereby minimizing player i's cost. Finally, determine the subset of these contracts that also satisfy the original strict convexity assumption.

The following are the first order conditions:

$$\frac{\partial H_1^l}{\partial s_1} = 0 = \Phi^{L_2}(s_2)(G_1'(s_1) - T_1'(s_1)) \tag{5.8}$$

$$+ \int_{s_2}^\infty \phi^{L_2}(x)[G_1'(s_1 + s_2 - x) - T_1'(s_1 + s_2 - x)]\, dx;$$

$$\frac{\partial H_2^l}{\partial s_2} = 0 = -\beta_2 + (h_2 + \beta_2)\Phi^{L_2}(s_2) \tag{5.9}$$

$$+ \int_{s_2}^{\infty} \phi^{L_2}(x)\left[G_2'(s_1 + s_2 - x) + T_1'(s_1 + s_2 - x)\right] dx.$$

Define $\gamma_2 = \Phi^{L_2}(s_2^o)$. (This is the supplier's in-stock probability, essentially her fill rate.) Furthermore, the supplier's first order condition in the optimal solution is

$$0 = -p + (p + h_2)\Phi^{L_2}(s_2^o) + (h_1 + h_2 + p)\int_{s_2^o}^{\infty} \phi^{L_2}(x)\Phi^{L_1+1}(s_1^o + s_2^o - x)dx,$$

or,

$$\int_{s_2^o}^{\infty} \phi^{L_2}(x)\Phi^{L_1+1}(s_1^o + s_2^o - x)dx = \frac{p - (p + h_2)\gamma_2}{h_1 + h_2 + p}. \tag{5.10}$$

Using (5.8), (5.9) and (5.10) yields the following two equations in three unknowns,

$$\alpha_2 p = \left(\frac{p}{h_1 + h_2}\right)\iota_1 - \beta_1, \tag{5.11}$$

$$h_2 = \left(\frac{h_2}{h_1 + h_2}\right)\iota_1 + \left(\frac{1 - \gamma_2}{\gamma_2}\right)\beta_2. \tag{5.12}$$

It remains to ensure that the costs functions are indeed strictly convex.

Theorem 6 *When the firms choose $(\iota_1, \beta_2, \beta_1)$ to satisfy (5.11) and (5.12), and*

$$\begin{array}{ll}
(i) & h_1 + h_2 > \iota_1 \geq 0 \\
(ii) & \beta_2 > 0 \\
(iii) & \alpha_1 p > \beta_1 \geq -\alpha_2 p,
\end{array}$$

then the optimal policy (s_1^o, s_2^o) is a Nash equilibrium.

Proof. When the following second order conditions are satisfied, H_i^l is strictly convex in s_i, assuming $s_j = s_j^o$, $j \neq i$:

$$\frac{\partial^2 H_1^l}{\partial s_1^2} = \Phi^{L_2}(s_2)\left(G_1''(s_1) - T_1''(s_1)\right)$$

$$+ \int_{s_2}^{\infty} \phi^{L_2}(x)\left(G_1''(s_1 + s_2 - x) - T_1''(s_1 + s_2 - x)\right) dx > 0;$$

$$\frac{\partial^2 H_2^l}{\partial s_2^2} = (h_2 + \beta_2)\phi^{L_2}(s_2)$$

$$+ \int_{s_2}^{\infty} \phi^{L_2}(x)\left(G_2''(s_1 + s_2 - x) + T_1''(s_1 + s_2 - x)\right) dx > 0.$$

The first inequality reduces to

$$h_1 + h_2 + \alpha_1 p - \iota_1 - \beta_1 > 0.$$

Substituting (5.11) yields $\iota_1 < h_1 + h_2$ and $\beta_1 < \alpha_1 p$. For the supplier sufficient conditions are

$$\alpha_2 p + \iota_1 + \beta_1 \;\geq\; 0;$$
$$h_2 + \beta_2 + \alpha_2 p + \beta_1 - (\alpha_2 p + \iota_1 + \beta_1)\,\Phi^{L_1+1}(s_1^o) \;>\; 0.$$

Combining the first inequality with (5.11) yields $\iota_1 \geq 0$ and $\beta_1 \geq -\alpha_2 p$. The second inequality, along with (5.11) and (5.12), yields $\beta_2 > 0$. $\square$

These are reasonable conditions: The first requires that the retailer's inventory subsidy not eliminate retailer holding costs; the second stipulates that the supplier be penalized for her backorders; and the third states that the supplier should not fully reimburse the retailer's backorder costs, and the retailer should not overcompensate the supplier's backorder costs.

To help interpret these results, consider the three extreme contracts where one of the parameters is set to zero:

$$
\begin{array}{llll}
i) & \iota_1 = 0 & \beta_2 = \frac{\gamma_2}{1-\gamma_2} h_2 & \beta_1 = -\alpha_2 p \\
ii) & \iota_1 = h_1 + h_2 & \beta_2 = 0 & \beta_1 = \alpha_1 p \\
iii) & \iota_1 = \alpha_2(h_1 + h_2) & \beta_2 = \frac{\gamma_2}{1-\gamma_2}\alpha_1 h_2 & \beta_1 = 0
\end{array}
$$

(Of these three contracts, the second does not meet the conditions in Theorem 6, because the supplier fully compensates the retailer for all of his costs. The retailer's incentive to choose the optimal policy is weak: s_1^o is a Nash equilibrium strategy, but any s_1 is too.)

With the first contract the retailer fully reimburses the supplier for the supplier's consumer backorder penalty. However, the supplier still carries inventory because she pays a penalty for her local backorders. With the third contract the supplier subsidizes the retailer's holding costs, but not fully (provided $\alpha_2 < 1$). In addition, the supplier is penalized for her backorders, but less than in the first contract. When the retailer incurs all backorder costs (i.e., $\alpha_1 = 1, \alpha_2 = 0$), only a supplier backorder penalty is required, $\beta_2 = \gamma_2 h_2/(1-\gamma_2)$. Hence, the traditional cost allocation scheme used to obtain the optimal solution is needlessly complex. (Recall that scheme requires a retailer holding cost subsidy, a retailer backorder penalty surcharge and a non-linear supplier backorder penalty.)

5.6.2 Accounting inventory

Chen (1997) proposes a linear incentive alignment scheme, but his scheme is not based on actual inventory levels. Instead, it is based on *accounting inventory*, where stage i's accounting inventory is the actual inventory level it would have if stage $i+1$ always fills its orders immediately. Since the supplier's lead time is assumed to be perfectly reliable, her accounting inventory always equals her actual inventory. On the other hand, the retailer's accounting inventory can be greater than his actual inventory level since the supplier will sometimes only partially fill an order.

To implement accounting inventory to align incentives one of the firms is assigned responsibility for all actual supply chain costs. (In Chen, 1997, this role is played by a single owner who is independent of the managers that actually operate each location.) While either firm can be given this honor, for ease of exposition, assume the supplier bears all costs. This means that each period the supplier pays the retailer's actual holding and backorder costs,

$$(h_1 + h_2)\left[IL_{1t} - D^1\right]^+ + \alpha_1 p\left[IL_{1t} - D^1\right]^-.$$

This leaves the retailer with zero costs, and hence little incentive to choose the optimal policy. To provide an incentive, each period the retailer pays the supplier h_1^a per accounting inventory, and p^a per accounting backorder. These costs are chosen so that s_1^o minimizes the retailer's payment to the supplier, and hence the retailer will choose s_1^o.

Define

$$\widehat{G}_1^a(x) = h_1^a[x]^+ + p^a[x]^-,$$

which is the retailer's payment to the supplier in a period that ends with accounting inventory x. Define

$$G_1^a(y) = E\left[\widehat{G}_1^a(y - D^{L_1+1})\right],$$

which is the retailer's expected payment in period $t + L_1$ when he begins period t with accounting inventory y. Since a retailer's accounting inventory always equals s_1 just before demand (because accounting inventory assumes the supplier always fills the retailer's orders), $G_1^a(s_1)$ is the retailer's expected cost per period. It is easy to confirm that $G_1^a(s_1)$ is strictly convex in s_1 (assuming $h_1^a > 0$ or $p^a > 0$), and the optimal s_1 is

$$\Phi^{L_1+1}(s_1) = \frac{p^a}{h_1^a + p^a}.$$

From (5.5), s_1^o is optimal for the retailer whenever

$$\frac{p^a}{h_1^a + p^a} = \frac{h_2 + p}{h_1 + h_2 + p}. \tag{5.13}$$

The supplier can choose from an infinite number of $(\widehat{p}^a, \widehat{h}_1^a)$ pairs that satisfy (5.13).

Note the similarity between (5.13) and (5.4). In each case the retailer faces a newsvendor problem so the supplier need only modify the retailer's costs in a manner that sets the retailer's critical fractile equal to the supply chain's optimal critical fractile. In each case the retailer is assured to receive his full order, so the retailer's decision is independent of the supplier's decision, and in turn independent of the demand distribution. In Cachon and Zipkin (1997) the retailer is not assured to receive his full order. Hence, the coordinating parameters depend on the supplier's base stock level and the demand distribution, $\gamma = \Phi^{L_2}(s_2^o)$.

Will the supplier choose s_2^o? The retailer's payment to the supplier is independent of s_2, so the supplier does not consider this payment when choosing

s_2. The supplier's actual cost equals the supply chain's cost (because she pays the retailer's actual cost in addition to her own), so (s_1^o, s_2^o) minimizes her actual cost. Hence, the supplier will choose s_2^o and a $(\widehat{p}^a, \widehat{h}_1^a)$ pair to induce the retailer to choose s_1^o.

While there are many $(\widehat{p}^a, \widehat{h}_1^a)$ pairs to choose from, the supplier's expected payment from the retailer depends on which pair is chosen. To see this, when the supplier chooses a $(\widehat{p}^a, \widehat{h}_1^a)$ pair,

$$G_1^a(s_1^o) = \widehat{p}^a \left(\frac{1 - \Phi^{L_1+1}(s_1^o)}{\Phi^{L_1+1}(s_1^o)} (s_1^o - \mu^{L_1+1}) \right.$$
$$\left. + \frac{1}{\Phi^{L_1+1}(s_1^o)} \int_{s_1^o}^{\infty} \phi^{L_1+1}(x)(x - s_1^o) dx \right).$$

The term in the above parentheses is independent of $(\widehat{p}^a, \widehat{h}_1^a)$, so the supplier can choose to make $G_1^a(s_1^o)$ arbitrarily small or large (by adjusting $\widehat{p}^a$). Hence, the supplier can offer the retailer a contract that leaves the retailer no worse off than in the Nash equilibrium, yet the retailer chooses the optimal policy. This is a great deal for the supplier because the supplier captures all of the benefits of coordination. (Since the retailer's cost is unchanged, the supplier's cost is reduced by the difference between the Nash equilibrium cost and the optimal cost.)

Notice that the accounting inventory incentive scheme essentially sells the supply chain to the supplier, thereby creating a single agent whose objective is to minimize total supply chain costs. The supplier then modifies the retailer's costs to make the retailer behave as the supplier wishes. Since the supplier buys the supply chain, the supplier can capture all of the benefits of coordination.

5.6.3 Non-linear payments

Lee and Whang (1996) propose an incentive alignment scheme that implements a combination of linear and non-linear transfer payments. Clark and Scarf (1960) also propose these payments to prove that (s_1^o, s_2^o) is optimal, but Lee and Whang demonstrate that with these payments (s_1^o, s_2^o) is a Nash equilibrium.

In the linear portion of this scheme the supplier pays the retailer h_2 for each unit in the retailer's inventory but the retailer pays the supplier $\alpha_2 p + h_2$ per backorder. Let $T_1^n(IL_{1t})$ be the retailer's expected transfer payment to the supplier in period $t + L_1$,

$$T_1^n(y) = -h_2 E[(y - D^{L_1+1})^+] + (\alpha_2 p + h_2) E[(y - D^{L_1+1})^-].$$

Define

$$G_1^n(y) = G_1(y) + T_1^n(y)$$
$$= h_1(y - \mu^{L_1+1}) + (h_1 + h_2 + p) \int_y^{\infty} (x - y)\phi^{L_1+1}(x) dx$$

so $G_1^n(IL_{1t})$ is the retailer's expected cost in period $t + L_1$.

The non-linear portion of this scheme is a payment from the supplier to the retailer whenever the supplier is unable to fill the retailer's order. Specifically, in period t the supplier pays the retailer $\underline{T}_1^n(s_1, IL_{1t})$,

$$\underline{T}_1^n(s_1, y) = G_1^n(\min\{y, s_1\}) - G_1^n(s_1).$$

When $IL_{1t} = s_1$, the supplier fills the retailer's period t order, so $\underline{T}_1^n(s_1, IL_{1t}) = 0$. When $IL_{1t} < s_1$, the supplier is unable to fill completely the retailer's period t order, so $\underline{T}_1^n(s_1, IL_{1t}) > 0$.

Taking all of the transfer payments into consideration, the retailer's expected per period cost is $H_1^n(s_1, s_2)$,

$$
\begin{aligned}
H_1^n(s_1, s_2) &= E[G_1^n(\min\{s_1 + s_2 - D^{L_2}, s_1\})] - E[\underline{T}_1^n(s_1, s_1 + s_2 - D^{L_2})] \\
&= G_1^n(s_1)
\end{aligned}
$$

So the retailer's expected cost is independent of s_2. Since $G_1^n(y) = G_1^o(y)$, s_1^o minimizes the retailer's cost.

Now consider the supplier's cost, assuming the retailer chooses s_1^o. (This assumption is non-trivial. If $s_1 \neq s_1^o$, the supplier's cost function is not necessarily convex.) Define

$$
\begin{aligned}
G_2^n(y) &= E\left[\hat{G}_2(y - D^{L_1+1})\right] - T_1^n(y) \\
&= h_2 E[(y - D^{L_1+1})^+] - h_2 E[(y - D^{L_1+1})^-] \\
&= h_2(y - \mu^{L_1+1})
\end{aligned}
$$

so $G_2^n(IL_{1t})$ is the supplier's period $t + L_1$ backorder cost minus the retailer's transfer payment. Define

$$\hat{H}_2^n(s_1, x) = h_2 \mu^{L_1} + h_2[x]^+ + G_2^n(s_1 + \min\{x, 0\}) + \underline{T}_1^n(s_1, s_1 + x)$$

and the supplier's cost including all transfers, $H_2^n(s_1, s_2)$,

$$
\begin{aligned}
H_2^n(s_1, s_2) &= E\left[\hat{H}_2^n(s_1, s_2 - D^{L_2})\right] \\
&= h_2 \mu^{L_1} + h_2 \int_0^{s_2} (s_2 - x)\phi^{L_2}(x)dx + \Phi^{L_2}(s_2)G_2^n(s_1) \\
&\quad + \int_{s_2}^{\infty} \phi^{L_2}(x)G_2^n(s_1 + s_2 - x)dx \\
&\quad + \int_{s_2}^{\infty} \phi^{L_2}(x)G_1^n(s_1 + s_2 - x)dx - \left(1 - \Phi^{L_2}(s_2)\right)G_1^n(s_1)
\end{aligned}
$$

Simplify the above,

$$
\begin{aligned}
H_2^n(s_1, s_2) &= h_2(s_2 + s_1 - \mu^{L_2+1}) + \int_{s_2}^{\infty} \phi^{L_2}(x)G_1^n(s_1 + s_2 - x)dx \\
&\quad - \left(1 - \Phi^{L_2}(s_2)\right)G_1^n(s_1).
\end{aligned}
$$

It is easy to show that $H_2^n(s_1^o, s_2) = G_2^o(s_2)$, so $H_2^n(s_1^o, s_2)$ is minimized with $s_2 = s_2^o$. Therefore, (s_1^o, s_2^o) is a Nash equilibrium since each player minimizes its cost given the strategy choice of the other player.

5.6.4 Responsibility tokens

Porteus (1997) proposes an incentive scheme, called responsibility tokens, that blends some of the features of the non-linear scheme in Lee and Whang (1996) with some of the features of accounting inventory. As with accounting inventory, the retailer receives perfectly reliable deliveries from the supplier, but the implementation of this assumption is different. Whenever the supplier is unable to fill a retailer order she issues responsibility tokens to cover the portion of the order she cannot fill. From the retailer's perspective these responsibility tokens are equivalent to real inventory. Tokens issued in period t are received in period $t + L_1$. The retailer incurs holding costs on these tokens and can use them to prevent backorder costs. To explain, suppose τ tokens are issued in period t and y is the retailer's actual inventory level in period $t + L_1$ when costs are measured. The retailer's actual holding and backorder costs are

$$(h_1 + h_2)[y]^+ + \alpha_1 p[y]^-,$$

but the supplier pays the retailer $\alpha_1 p \min\{[y]^-, \tau\}$ to compensate the retailer for those backorder charges that the retailer could have avoided if his tokens were real inventory. However, the retailer pays the supplier $(h_1 + h_2)[\tau - [y]^-]^+$ to reward the supplier for saving the retailer actual holding costs. Once the retailer's actual costs are combined with these transfer payments, the retailer's costs are

$$(h_1 + h_2)[y + \tau]^+ + \alpha_1 p[y + \tau]^-. \tag{5.14}$$

There are some additional transfer payments. As in Lee and Whang (1996), the supplier pays the retailer h_2 per unit of inventory and the retailer pays the supplier $\alpha_2 p + h_2$ per backorder, where inventory and backorders include tokens. Following the above example, the retailer receives $h_2[y + \tau]^+$ from the supplier but pays $(\alpha_2 p + h_2)[y + \tau]^-$ to the supplier. Combining these transfers with (5.14), the retailer's costs in period $t + L_1$ are

$$h_1[y + \tau]^+ + (p + h_2)[y + \tau]^-$$

Since $y + \tau = s_1 - D^{L_1+1}$, $G_1^n(s_1)$ is the retailer's expected cost per period, which is minimized by s_1^o.

Since s_1^o minimizes the retailer's actual costs plus his transfer payments for any s_2, the retailer will choose s_1^o. Given that $G_1^n(s_1^o)$ will be the retailer's cost, the supplier's costs must equal all remaining costs in the system, which equals $G_2^o(s_2)$. Hence, the supplier also will choose the optimal policy, s_2^o, and (s_1^o, s_2^o) is a Nash equilibrium.

Although there is a strong resemblance between responsibility tokens and the non-linear scheme in Lee and Whang (1996), there is a subtle difference.

With responsibility tokens the retailer receives perfectly reliable supply, so the retailer's costs are independent of the supplier's base stock. (This is also the case with accounting inventory.) Hence, it is the supplier that bears the actual cost consequence of her late deliveries. With the non-linear scheme the retailer doesn't receive perfectly reliable supply, but the supplier does pay the retailer an amount that exactly compensated the retailer for the expected cost consequence of any late delivery. Hence, in this case the retailer pays the actual cost consequence of late deliveries. This distinction is immaterial when all of the players are risk neutral (as is assumed in each case). However, it becomes relevant if one of the players were risk averse. For example, if the retailer were risk averse then the supply chain may be better off using responsibility tokens to let the risk neutral supplier bear the consequence of late deliveries. On the other hand, if the supplier were risk averse, the supply chain may be better off using the non-linear scheme since then the supplier only pays the expected consequence of her action. (In general, the most risk neutral player should bear the supply chain's risk. See Tirole, 1990, for a discussion of risk sharing among players with heterogenous tastes for risk.)

5.7 OTHER RESEARCH

There are many other supply chain inventory management models that differ significantly from the ones already described. The section reviews these models and highlights additional supply chain externalities and coordination techniques.

5.7.1 Multiple Retailers with Stochastic Demand

Multiple retailers substantially complicates supply chain inventory analysis. Optimal policies are not known and it is difficult even to find the best policy within a reasonable class of policies, e.g., reorder point policies. Nevertheless, some results have been obtained.

Cachon (1997) considers a two echelon supply chain with a single supplier and N retailers. Each retailer faces identically distributed Poisson demand. Inventory is review continuously and each location implements an (R, Q) policy: when a location's inventory position equals R it orders Q units from its supplier. There is a fixed lead time to replenish the supplier and a fixed lead time between the supplier and each retailer. There are holding costs and both the supplier and the retailers care about consumer backorder costs, as in §5.5.

In this model Axsater (1993) demonstrates how to find optimal (R, Q) policies assuming centralized control. Cachon (1997) assumes each location selects its reorder point to minimize its own costs given that each other player will do the same. The search for Nash equilibria in reorder point policies is complex because discrete demand implies that each player's strategy space in discrete (i.e., non-convex), which precludes the implementation of standard existence theorems. (It is common that there does not exist an equilibrium in games with non-convex strategy spaces.) However, it is shown that this is a supermodular game. Roughly speaking, in a supermodular game each player's strategy space is ordered and as player j chooses a higher strategy, player i will also wish to choose a higher strategy. In the inventory game the supplier's strategy space (reorder points) are given the natural ordering, i.e., a higher reorder point is a higher strategy, but the retailers are given the opposite ordering, i.e., a higher reorder point is a lower strategy. Hence, as the supplier increases her reorder point, a retailer will decrease his reorder point. See Milgrom and Roberts (1990) for a review of supermodular games. The supermodular property implies that there exists a pure strategy Nash equilibrium, and further, it provides an algorithm to find all Nash equilibria.

In a numerical study it is found that there can be multiple equilibria and the optimal reorder point policies can be a Nash equilibrium. These findings contrast with the results from the single retailer model. However, as in the single retailer model, the competition penalty is relatively moderate when the firms incur equal backorder costs but often quite large when the firms incur asymmetric backorder costs. Additional research is needed to better understand why these results occur. There are several candidate explanations: the multi-retailer model has a discrete strategy space, whereas the single retailer model has a continuous strategy space; and in the multi-retailer model the allocation of inventory among the retailers is an issue, whereas it is not in the single retailer model.

Anderson, Axsäter and Marklund (1996) study almost the same model, but they do not investigate competitive decision making. Instead, they provide a mechanism to decouple the supply chain: each location solves a problem that is independent of the reorder points chosen by the other locations. Specifically, the mechanism imposes on the supplier a linear backorder penalty cost. The supplier's problem is to choose a reorder point that minimizes her actual holding costs plus this backorder penalty cost. This problem depends on the retailers' ordering process, but the retailers' ordering process is independent of their reorder points. So the supplier can solve her problem independent of the retailers. The retailers minimize their actual holding and backorder costs assuming a deterministic lead time which equals the retailers' expected actual lead time. Anderson, et al. (1996) are unable to explicitly determine the correct backorder penalty, so they search for a good penalty via an iterative procedure. The procedure is not guaranteed to converge to the optimal reorder points, but they found that the procedure worked quite well in a numerical study.

In a significantly different model, Hausman and Erkip (1994) also study a decoupling technique to coordinate a supply chain model originally studied by Muckstadt and Thomas (1980). This model has multiple retail locations, multiple products and emergency shipments. Instead of transfer payments, they impose fill rate constraints on each location and show how to choose these fill rates so that each location selects near optimal policies.

5.7.2 Multiple Retailers with Deterministic Demand

Chen, Federgruen and Zheng (1997) consider a multi-echelon supply chain with multiple retailers. Each retailer faces deterministic demand that is decreasing in the retailer's price. Demand curves can differ across retailers. Firms incur holding and ordering costs. To manage each retailer's account, the supplier incurs a fee that is increasing and concave in the retailer's purchase quantity. Hence, the account management fee per unit sold is decreasing in sales. Retailers choose their retail prices and their order intervals (time between orders, which equals the order quantity divided by the demand rate). The supplier chooses her wholesale price and her order interval. They assume all of the firms choose power-of-two policies: a base order interval is chosen and then all firms choose their order interval to be a power of two multiple of this base interval. For fixed prices (and hence demand rates), Roundy (1985) demonstrates that power-of-two policies are guaranteed to be within 2% of the optimal cost.

Several incentive conflicts occur in this model. The retailers tend to price too high because they don't consider the supplier's profit when choosing their sales rate, i.e., double marginalization. They are also biased towards lower than optimal sales because they don't incur the account management fee. This externality is somewhat different than the order processing externality described in §5.4 because it is based on average sales and not the order frequency. Finally, the supplier's holding cost is non-increasing in the retailer's order interval, so the retailers tend to choose order intervals that are shorter than optimal.

Unlike in the stochastic demand models, in the deterministic model it is implicitly assumed that the supplier never makes a late delivery. Hence, the supplier's order interval has no impact on the retailers' costs. Only the supplier's wholesale price affects the retailers' costs.

Chen, et al. (1997) demonstrate that when the supplier chooses the optimal order interval, the supplier can announce a wholesale price schedule that makes the retailers choose the optimal retail prices and order intervals. This wholesale price per unit equals the supplier's marginal cost plus a sales volume fee plus an order interval fee. The sales volume fee equals the supplier's account management cost per unit given a retailer's sale volume, so the retailers exactly compensate the supplier for her account management costs. The order interval fee is non-increasing in a retailer's order interval length, and exactly compensates the supplier for her holding costs. With this scheme the supplier incurs her ordering costs, but the retailers incur all other costs, either directly or indirectly through the wholesale price. Further, the retailers receive all sales revenues, so they earn more than the supply chain's optimal profits. To gain the supplier's participation, the retailers clearly must share some of the profits via lump sum payments.

5.7.3 Capacity Allocation

Consider a supply chain with a single supplier and multiple retailers. Demand occurs in a single period. The supplier has limited capacity, so whenever the retailers order more than capacity, the capacity must be allocated among the retailers. How much will each retailer order and how much capacity will the supplier build?

Lee, Padmanabhan and Whang (1997) call this the *shortage game*. In their version of the game the supplier has uncertain yield, so even though the retailers know how much they all want, they don't know how much will be available. If they order more than the supplier produces, the supplier allocates the production proportional to their order quantities. Lee, et al. (1997) demonstrate that each retailer will order more than their desired quantity.

Cachon and Lariviere (1996) study a different shortage game. The supplier's capacity is known but each retailer has private information about his own demand: while a retailer knows his own expectation of demand, he does not know the demand expectations of the other retailers. Hence, each retailer is uncertain about how much the other retailers will order. There are several incentive problems in this setting. The retailers tend to purchase too little relative to the supply chain optimal, because the supplier charges above marginal cost, i.e., double marginalization. But when capacity is scarce, a retailer tends to order too much because he does not consider the value of scarce inventory to other retailers: the supply chain should allocate inventory to the retailer with the highest marginal value for inventory, but a retailer with a low marginal value may nevertheless receive a larger allocation by increasing his order. There are also incentive problems with the supplier. Since the supplier faces random demand (she doesn't know the retailers' expectations), and since she does not consider the retailers' profit margin on each unit, the supplier tends to build less than the supply chain optimal capacity.

Cachon and Lariviere (1996) identify a broad class of allocation schemes that induces truth telling behavior among the retailers, i.e., the retailers order their desired quantities. Another class of allocation schemes induces the retailers to inflate their orders. In a numerical study it is found that the order inflation schemes generally outperform the truth-inducing schemes. There are several reasons for this result. Double marginalization makes the retailers order too little, but an order inflation allocation scheme mitigates this problem. The problem with order inflation is that it might lead to a poor allocation of inventory among the retailers. The worst case scenario is that a retailer with low demand ends the period with excess inventory while a retailer with high demand ends the period with substantial shortages. However, this generally doesn't happen in equilibrium, because equilibrium orders correspond with needs, i.e., the retailer with the highest need inflates his order the most, thereby receiving the greatest allocation. So even though retailers inflate their orders, inventory allocations tend to be reasonable. Finally, retailer order inflation also induces the supplier to build more capacity, which can benefit everyone in the supply chain.

Several other shortage gaming models have been studied. In a multi-period model, Cachon and Lariviere (1997) study turn-and-earn allocation of scarce capacity, a scheme commonly used in the automobile industry. Mallik and

Harker (1997) study capacity allocation in an internal supply chain with multiple production managers. See Ha (1996) and Corbett (1998) for other supply chain inventory models with asymmetric information.

5.7.4 Information and Production Timing

Predicting demand for seasonal products is notoriously difficult. However, it has been frequently observed that early season sales provide excellent indications of total season sales (see Fisher and Raman, 1996). So retailers would like to be able to replenish inventory after observing early season sales. But delayed production is usually more expensive than early production, so suppliers prefer to have full production commitments well before the season begins. Hence, firms face a conflict between early cheap production with unpredictable demand and late expensive production with reliable demand.

Donohue (1996) studies this problem in a supply chain with one manufacturer and one buyer. The buyer places an initial order and then the manufacturer decides her initial production quantity. The buyer then observes a signal that improves the buyer's demand forecast. The buyer places a second order and the manufacturer decides her second production quantity. This second production is more expensive than the first. Finally, actual demand is realized. She demonstrates that a buy-back contract makes the firms choose the centrally optimal decisions. Further, the buy-back price is independent of the demand distributions, as in §5.3.

Barnes-Shuster, Bassok and Anupindi (1998) expand upon Donohue's model. In their model there are two demand periods. The sequence of events follows: the buyer makes firm orders for delivery in each period and also purchases options for additional units to be delivered after the first demand period; the supplier purchases raw material and makes an initial production; the buyer receives his first period order and stochastic demand occurs; the buyer receives his period two firm order plus he can exercise his options for a per unit exercise fee; the supplier delivers the buyer's period two firm order plus exercised options; and stochastic demand occurs. The buyer incurs a per unit holding cost for inventory carried from period one to two. The supplier's total production over the two periods is limited by the quantity of raw material purchased at the start of the game. Hence, the supplier is required to purchase sufficient raw material to cover the buyer's firm orders and potentially all of his options. In addition to the per unit cost of raw material, there is a per unit production cost which is higher in period two than in period one.

They find that marginal cost pricing of these options contracts doesn't coordinate the channel. The issue is the period two option exercise price. Since period one production is cheaper than period two production, and since there is an inventory holding cost at the retailer, the supplier may have more inventory at the start of period two than is needed to fill the buyer's period two firm order. From the supply chain's perspective, these excess units should certainly be moved to the retailer because they cannot fill period two demand while at the supplier. Only with a zero exercise price will the buyer necessarily order these additional units. However, a zero exercise price also means that the buyer will necessarily exercise all of his options. This may require the supplier to conduct some period two production, which has a greater than zero marginal cost. Hence, a zero exercise price may cause the buyer to exercise too many options from the supply chain's perspective, but a non-zero option may cause the buyer to exercise too few options. To summarize, marginal cost pricing doesn't work in this setting because the supply chain's marginal cost of moving additional units to the buyer in period two is not independent of the quantity moved.

Eppen and Iyer (1997) also study option-like contracts, which they call backup agreements. However, they do not consider the issue of channel coordination.

Tsay (1997) studies a setting in which the supplier makes one production, which occurs after the retailer's initial order, but before the retailer observes a signal of demand. The retailer places his firm order after observing this signal. The firm order may or may not be constrained by the initial order. If there is no link between them, the retailer is likely to provide a meaninglessly high initial order. Tsay (1997) proposes to link them via a *quantity flexibility contract*. If q is the initial order, these contracts specify that the firm order can be no greater than $(1 + \alpha)q$ and no less than $(1 - \omega)q$, for $\omega \in [0, 1]$ and $\alpha \geq -\omega$. Since the supplier allows the retailer to request up to $(1 + \alpha)q$, Tsay assumes the supplier produces $(1 + \alpha)q$, thereby guaranteeing that the retailer could indeed receive his maximum order. (See Cachon and Lariviere, 1997, for a discussion of a model in which it is not assumed that the supplier will necessarily live up to her commitments.) It is shown that these contracts can guarantee that each firm will choose supply chain optimal actions. Hence, $(1 + \alpha)q$ must equal the supply chain optimal production quantity.

Quantity flexibility contracts are more complex than buy-back contracts. A buy-back contract allows a retailer to return any portion of his initial order at a buy-back price which is less than the wholesale price. A quantity flexibility contract allows the firm to return a limited portion of his initial order for a full refund of the wholesale price. Further, a buy-back contract delivers to the retailer his initial order without an opportunity to increase the order. This makes sense when the retailer doesn't receive additional information between the initial order and the delivery (as is assumed in §5.3). However, because in Tsay's model the retailer does receive a demand signal, the quantity flexibility contract allows the retailer to increase his order somewhat. See Tsay and Lovejoy (1998) for additional discussion on the implementation of quantity flexibility contracts.

5.7.5 Internal Markets

Porteus and Whang (1991) consider a "supply chain" model with one owner, one manufacturing manager and several product managers. The manufacturing manager's costly effort affects the realization of capacity, which is subject to a random shock. The product managers' costly effort affects the realization of sales, which is also random. The owner wishes to maximize her profits, subject to the conditions that she cannot choose the effort levels of her managers. Further, she must offer them a sufficiently attractive contract to prevent them from seeking employment at another firm. Porteus and Whang (1991) suggest that the owner can achieve her objective through an internal market. The product managers receive all revenues from the sale of their product and pay the manufacturing manager the realized marginal value of capacity. (The marginal value of capacity is decreasing in the realization of capacity.) In addition, the product managers pay a fixed fee that equals their expected profits, so they are indifferent between working for the owner and their best outside opportunity. The manufacturing manager receives the expected marginal value of each unit of capacity she delivers, but also pays a fixed franchise fee to the owner that equals the manufacturing manager's expected profits. Interestingly, the owner loses money on average when operating this market: The average price the marketing managers pay for capacity is less than the price the manufacturing manager receives. Hence, the owner earns her profits through the franchise fees, and operates the internal market merely to induce optimal behavior. Kouvelis and Lariviere (1997) generalize the concept of an internal market, and apply it to several other settings.

5.7.6   The Beer Game

The Beer Game, Sterman (1989), is probably the most famous demonstration that decentralized decision making can lead to poor supply chain performance. There are several explanations for this result. Of the four players only the retailer observes demand, so it is difficult for upstream locations to know how much they should order. Indeed, optimal policies for this setting are not known, even if all of the players were to observe demand. Furthermore, poor judgment may also be a culprit: there is evidence that players forget to consider previously ordered inventory when choosing order quantities, leading them to order too much. Incentive conflicts are notably absent from this list of explanations: the players are told to minimize total supply chain costs.

5.7.7   Vendor Managed Inventory

With Vendor Managed Inventory (VMI) a supplier assumes responsibility for choosing a retailer's stocking level. In exchange for control of supply chain inventory, the supplier agrees to charge the retailer a constant wholesale price. Since the supplier determines shipments to the retailers, VMI generally requires electronic transmission of inventory and demand data from the retailer to the supplier, e.g., Electronic Data Interchange (EDI). The grocery industry provides several examples of successful VMI implementation. (These systems are also called Continuous Replenishment Programs, CRP, or Continuous Product Replenishment programs, CPR.)

Clark and Hammond (1997) studied retailers that adopted VMI with the Campbell Soup Company. The performance of these retailers was compared with retailers that implemented EDI but did not transfer control of inventory to Campbell Soup, i.e., they merely used EDI to submit orders electronically. They found that the VMI retailers experienced substantially better performance gains over the latter group, suggesting that transferring control provided significant benefits. Cachon and Fisher (1997) also studied Campbell Soup's implementation of VMI. However, they found that operating benefits could have been achieved even if the retailers had maintained control of their own inventories.

Several models have been developed to study VMI. Cachon (1997) allows a single supplier to choose operating policies for herself as well as all her retailers. It is assumed that the supplier chooses policies to optimize her own preferences, while leaving the retailers no worse off than they would be in the competitive solution. It is found that the supplier does not always choose the optimal policies. Nevertheless, the supplier frequently chooses better policies for the supply chain than the competitive solution. Hence, shifting control from one player to another does not eliminate all incentive conflicts, but often can mitigate them.

Narayanan and Raman (1997) study VMI between a supplier and a retailer. In addition to the supplier's product, the retailer sells a close substitute from another supplier. When a customer's favorite product is unavailable, the customer may switch to the other product which is always in stock. With VMI the retailer allows the supplier to choose his stock level of the supplier's product in exchange for a fixed transfer payment. They demonstrate that VMI does not achieve the supply chain optimal profit. The problem is that the supplier stocks too much of her product because she does not consider the revenue the retailer earns when customers switch to the other product.

5.8 CONCLUSION

The literature on competitive supply chain inventory management recognizes that supply chains are usually operated by independent agents with individual preferences. Game theory is the methodological tool to determine how the players will behave when they each seek to maximize their own welfare. The key issues include whether a Nash equilibrium exists, whether there is a unique Nash equilibrium and whether the optimal policies ever belong to the set of Nash equilibria. It is frequently found that competitive and optimal behavior do not coincide, in which case it is worthwhile to investigate coordination techniques.

Some coordination techniques are designed to manipulate the behavior of one firm to the advantage of another. For example, buy-back and quantity discount contracts can be used by a supplier to increase her profits at the expense of the retailer's profits. Other techniques are designed to make the optimal policies incentive compatible, regardless of whether all of the players actually prefer the optimal solution over the competitive solution. Most of the techniques implement transfer payments between the players, but there is considerable variation in the form of these payments (e.g., linear fees and subsidies, two-part tariffs). It may also be possible to coordinate a supply chain by imposing service constraints on the parties, by shifting control among the players (vendor managed inventories), or by operating internal markets.

References

Anderson, J., S. Axsater and J. Marklund, "Decentralized Multi-Echelon Inventory Control," Department of Industrial Engineering, Lund University, (1996).

Anupindi, R. and Y. Bassok, "Centralization of Stocks: Retailers vs. Manufacturer," Northwestern University working paper, (1998).

Axsater, S., "Exact and approximate evaluation of batch-ordering policies for two-level inventory systems," *Operations Research*, 41, 4 (1993), 777-785.

Barnes-Schuster, D., Y. Bassok and R. Anupindi, "Coordination and Flexibility in Supply Contracts with Options," University of Chicago working paper, (1998).

Boyaci, T. and G. Gallego, "Coordination Issues in Simple Supply Chains," Columbia University working paper, (1997).

Butz, D., "Vertical Price Controls with Uncertain Demand," *Journal of Law and Economics*, 40, (1997), 433-459.

Cachon, G., "Stock Wars: Inventory Competition in a Two Echelon Supply Chain," Fuqua School of Business working paper, (1997).

Cachon, G. and M. Fisher, "Campbell Soup's Continuous Product Replenishment Program: Evaluation and Enhanced Decision Rules," *Production and Operations Management*, 6, 3 (1997), 266-276.

Cachon, G. and M. Lariviere, "Capacity Choice and Allocation: Strategic Behavior and Supply Chain Performance," Fuqua School of Business working paper, Duke University, (1996).

Cachon, G. and M. Lariviere, "Capacity Allocation with Past Sales: When to Turn-and-Earn," Duke University working paper, (1997a)

Cachon, G. and M. Lariviere, "Contracting to Assure Supply, or What did the Supplier Know and When did He Know It?," Duke University working paper, (1997b).

Cachon, G. and P. Zipkin, "Competitive and Cooperative Inventory Policies in a Two-Stage Supply Chain," Duke University working paper, (1997).

Chen, F., "Decentralized Supply Chains Subject to Information Delays," Columbia University working paper, (1997).

144

Chen, F., A. Federgruen and Y. Zheng, "Coordination Mechanisms for Decentralized Distribution Systems," Columbia School of Business working paper, (1997).

Chen, F. and Y.-S. Zheng, "Lower Bounds for Multi-Echelon Stochastic Inventory Systems," *Management Science*, 40, 11 (1994), 1426-1443.

Clark, A.J. and H.E. Scarf, "Optimal Policies for a Multi-Echelon Inventory Problem," *Management Science*, 6, (1960), 475-490.

Clark, T. and J. Hammond, "Reengineering Channel Reordering Processes to Improve Total Supply Chain Performance," *Production and Operations Management*, 6, 3 (1997), 248-265.

Corbett, C. and X. de Groote, "Integrate Supply-Chain Lot Sizing under Asymmetric Information," University of California at Los Angeles working paper, (1997).

Corbett, C., "Stochastic Inventory Systems in a Supply Chain with Asymmetric Information," University of California at Los Angeles working paper, (1998).

Deneckere, R., H. Marvel and J. Peck, "Demand Uncertainty, Inventories, and Resale Price Maintenance," *Quarterly Journal of Economics*, 111, (1996), 885-913.

Deneckere, R., H. Marvel and J. Peck, "Demand Uncertainty and Price Maintenance: Markdowns as Destructive Competition," *American Economic Review*, 87, 4 (1997), 619-641.

Donohue, K., "Supply Contracts for Fashion Goods: Optimizing Channel Profits," Wharton School working paper, (1996).

Emmons, H. and S. Gilbert, "Returns Policies in Pricing and Inventory Decisions for Catalogue Goods," *Management Science*, 44, 2 (1998), 276-283.

Eppen, G. and A. Iyer, "Backup Agreements in Fashion Buying - The Value of Upstream Flexibility," *Management Science*, 43, 11 (1997), 1469-1484.

Federgruen, A. and P. Zipkin, "Computational Issues in an Infinite-Horizon, Multiechelon Inventory Model," *Operations Research*, 32, 4 (1984), 818-836.

Fisher, M. and A. Raman, "Reducing the Cost of Demand Uncertainty through Accurate Response to Early Sales," *Operations Research*, 44, (1996), 87-99.

Fudenberg, D. and J. Tirole (1991). Game Theory. Cambridge: MIT Press.

Ha, A., "Supply Contract for a Short-Life-Cycle Product with Demand Uncertainty and Asymmetric Cost Information," Yale School of Management working paper, (1996).

Hausman, W. and N. Erkip, "Multi-echelon vs Single-echelon Inventory Control Policies," *Management Science*, 40, 5 (1994), 507-602.

Jeuland, A. and S. Shugan, "Managing Channel Profits," *Marketing Science*, 2, (1983), 239-272.

Kandel, E., "The Right to Return," *Journal of Law and Economics*, 39, (1996), 329-56.

Kouvelis, P. and M. Lariviere, "Decentralizing Cross-Functional Decisions: Coordination Through Internal Markets," Duke University working paper, (1997).

Lal, R. and R. Staelin, "An Approach for Developing an Optimal Discount Pricing Policy," *Management Science*, 30, 12 (1984), 1524-1539.

Lariviere, M. "Supply Chain Contracting and Coordination with Stochastic Demand," Fuqua School of Business working paper, (1998).

Lee, H., P. Padmanabhan and S. Whang, "Information Distortion in a Supply Chain: The Bullwhip Effect," *Management Science*, 43, 4 (1997), 546-558.

Lee, H. and M.J. Rosenblatt, "A Generalized Quantity Discount Pricing Model to Increase Supplier's Profits," *Management Science*, 32, 9 (1986), 1177-1185.

Lee, H. and S. Whang, "Decentralized Multi-Echelon Supply Chains: Incentives and Information," Stanford University working paper, (1996).

Mallik, S. and P. Harker, "Coordinating Supply Chains with Competition: Capacity Allocation in Semiconductor Manufacturing," University of Pennsylvania working paper, (1997).

Milgrom, P. and J. Roberts, "Rationalizability, Learning and Equilibrium in Games with Strategic Complementarities," *Econometrica*, 58, (1990), 1255-78.

Moorthy, K.S., "Managing Channel Profits: Comment," *Marketing Science*, 6, (1987), 375-379.

Muckstadt, J.A. and L.J. Thomas, "Are Multi-Echelon Inventory Methods Worth Implementing in Systems with Low-Demand Rates?," *Management Science*, 26, 5 (1980), 483-494.

Narayanan, V. and A. Raman, "Contracting for Inventory in a Distribution Channel with Stochastic Demand and Substitute Products," Harvard University working paper, (1997).

Padmanabhan, V. and I.P.L. Png, "Returns Policies: Make Money by Making Good," *Sloan Management Review*, Fall, (1995), 65-72.

Padmanabhan, V. and I.P.L. Png, "Manufacturer's Returns Policy and Retail Competition," *Marketing Science*, 16, 1 (1997), 81-94.

Pasternack, B., "Optimal Pricing and Returns Policies for Perishable Commodities," *Marketing Science*, 4, 2 (1985), 166-176.

Porteus, E., "Responsibility Tokens in Supply Chain Management," Stanford University working paper, (1997).

Porteus, E. and S. Whang, "On Manufacturing/Marketing Incentives," *Management Science*, 37, 9 (1991), 1166-1181.

Roundy, R., "98%-Effective Integer Ratio Lot-Sizing For One-Warehouse Multi-Retailer Systems," *Management Science*, 31, (1985), 1416-1430.

Saggi, K. and N. Vettas, "On the Strategic Role of Fees and Royalties in Vertical Contracting," Duke University working paper, (1998).

Spengler, J., "Vertical Integration and Antitrust Policy," *Journal of Political Economy*, (1950), 347-352.

Sterman, J., "Modeling Managerial Behavior: Misperceptions of Feedback in a Dynamic Decision Making Experiment," *Management Science*, 35, 3 (1989), 321-339.

Tirole, J. (1990). *The Theory of Industrial Organization*. Cambridge: MIT Press.

Tsay, A., "The Quantity Flexibility Contract and Supplier-Customer Incentives," Santa Clara University working paper, (1997).

Tsay, A. and W. Lovejoy, "Quantity Flexibility Contracts and Supply Chain Performance," Santa Clara Univeristy working paper, (1998).

Weng, Z. K., "Channel Coordination and Quantity Discounts," *Management Science*, 41, (1995), 1509-1522.

6 VEHICLE ROUTING AND THE SUPPLY CHAIN

Shoshana Anily

Faculty of Management
Tel-Aviv University
Tel-Aviv, Israel 69978

anily@post.tau.ac.il

Julien Bramel

406 Uris Hall
Columbia University
New York, NY 10027

jdb8@columbia.edu

6.1 INTRODUCTION

One of the central problems of supply chain management is the coordination of product and material flows between locations. A typical problem involves bringing product located at a central facility to geographically dispersed facilities at minimum cost. For example, a supply of product is located at a plant, warehouse, cross-docking facility or distribution center and must be distributed to customers or retailers. The task is often performed by a fleet of vehicles under direct control of the firm. Depending on the structure of the supply chain, a number of different distribution problems can be defined, e.g., where retailers/customers are *external*, i.e., not part of the same company, or where retailers are *internal* to the firm. In this chapter, we present models, analyses and approaches for solving a number of these distribution problems.

We first consider *single period* models where we assume that all delivery *sizes* have been determined exogenously. For instance, the deliveries may be actual customer orders, or retailer inventory replenishments whose sizes have been specified by some (possibly multi-period) inventory/distribution policy in use. In this case, the single period problem consists of minimizing the one-time cost of delivering the product to the retailers/customers. When the only route restriction is the vehicle capacity, this is a Capacitated Vehicle Routing Problem (CVRP). Vehicles of limited capacity are assumed to be initially stationed at the supply point. The problem is to design routes for the vehicles that deliver the required amounts to each customer, do not violate the capacity constraint and minimize the total distance traveled. The objective suggests that only the variable cost needs to be minimized (i.e., truck costs are sunk). Other objectives could be used to determine the number of trucks needed, the minimum time required to accomplish the delivery, etc.... We also investigate multi-depot versions of this delivery problem, as well as problems with both deliveries and backhauls.

We then consider *multi-period* models where deliveries are required a number of times over a planning horizon. This occurs when, e.g., the delivery points are retailers that face period specific demands over a given time-horizon. The costs now include transportation and inventory costs. A solution to the problem consists of a *distribution strategy* that specifies the sizes and timings of all deliveries over the given horizon. This problem has been called the One-Warehouse Multi-Retailer Problem (*with vehicle routing costs*) and also the Inventory-Routing Problem (IRP). We will consider a number of single-warehouse multi-retailer continuous-time models where demands are deterministic, constant (though retailer-specific) and must be met (i.e., no backlogging).

Distribution problems have been a fertile ground for research over the last three decades. Since most are $\mathcal{NP}$-Hard [41], researchers have long given up hope of developing optimization algorithms for these problems that will run in polynomial time. Therefore, *heuristic algorithms* or simply *heuristics* have received much attention. A heuristic's "reputation" can be established by a combination of empirical studies and whether solutions it creates are provably good, either from a worst-case or average-case perspective.

We present a number of heuristics and algorithms in this chapter. Particular note will be taken to describing worst-case and average-case results. We restrict ourselves to *deterministic* models. In addition, we do not consider complicating route constraints besides capacity, such as time windows or distance constraints.

A number of the problems, algorithms and analyses we consider here have recently appeared. Details can be found in the referenced papers and also in a number of general surveys on the analysis of distribution systems. This includes Christofides [26], Fisher [38], Toth and Vigo [71], Federgruen and Simchi-Levi [35] and Bramel and Simchi-Levi [19]. For a review of vehicle routing and scheduling we refer the reader to the survey of Bodin et al. [16], and for routing models with time windows to Desrosiers et al. [33].

In the next Section we present basic notation and assumptions that appear in most models we consider. In §6.3 we analyze single period deterministic delivery problems. In §6.4 we study multi-period deterministic inventory/distribution problems.

6.2 NOTATION AND PRELIMINARIES

We assume n retailers (or customers), indexed by the set $N = \{1, 2, \ldots, n\}$, are geographically dispersed in a given region. These retailers are supplied product by a single central warehouse (or plant, or distribution center). We will generically call this the *depot* and denote it by $\{0\}$. For any set $S \subseteq N$, let $S^0 \stackrel{\text{def}}{=} S \cup \{0\}$. (In some models we will assume there are a number of such supply points, see §6.3.5.) A fleet of capacitated vehicles is available and is assumed to be initially stationed at the depot. If the vehicles are identical, we use Q to denote their capacity (either volume or weight). We let d_{ij} denote the distance between any two points $i \in N^0$ and $j \in N^0$. We assume the distance matrix is symmetric and satisfies the triangle inequality: $\forall i, j, k \in N^0$, $d_{ij} \leq d_{ik} + d_{kj}$. We denote by (i, j) the undirected edge between $i \in N^0$ and $j \in N^0$. In the case of a single depot, we let $r_i \stackrel{\text{def}}{=} d_{i0}$ designate retailer i's *radial* distance from the depot. In addition, we define $r_{\max} \stackrel{\text{def}}{=} \max_{i \in N} \{r_i\}$.

Consider a generic distribution problem and an instance with n retailers. Let Z_n^* denote the cost of an optimal solution to the problem, and let Z_n^H be the cost of the solution constructed by a heuristic H on this same instance. In some cases, we consider the performance of heuristics in specific policy classes. For a policy class Ψ, let $Z_n^*(\Psi)$ denote the infimum of the cost of any policy in Ψ. A lower bound on Z_n^* is typically denoted $\underline{Z}_n$.

Much of our focus is on worst-case and average-case analytical results. A heuristic H is an $\alpha-approximation$ algorithm if for all instances (in particular, for all n) $Z_n^H \leq \alpha Z_n^*$. The number α is also called the heuristic's *worst-case bound*.

Average-case analysis characterizes the average performance of the heuristic under assumptions on the data. As noted in [29], calculating the expected performance of relatively simple heuristics on even small size instances presents a challenge. Therefore, at present, much of the analysis is confined to an *asymptotic* analysis of the performance. This, of course, tells us less about the performance on a "typical" instance, but in many cases it provides (i) insights into the ways in which the important cost components interact, (ii) the crucial ingredients necessary for an effective algorithm on large size problems, and (iii) expressions that can be used as approximations to the optimal cost for use in models that integrate several levels of hierarchical planning (see, e.g., [69, 70]). Formally, let φ be a probability measure on the set of all instances, and note that Z_n^* and Z_n^{H} are random variables. A heuristic H is an *asymptotic* $\alpha-$*approximation algorithm for* φ if almost surely (a.s.)

$$\varlimsup_{n\to\infty} \frac{Z_n^{\mathrm{H}}}{Z_n^*} \le \alpha. \tag{6.1}$$

In particular, if (6.1) holds with $\alpha = 1$, then the heuristic is said to be *asymptotically optimal for* φ. That is, under specific assumptions on the distribution of the data, H generates solutions whose relative error tends to zero as n, the number of retailers/customers, increases. We will see that in many cases the assumptions on φ are quite general. We also use the term *asymptotically accurate* (or *tight*) to describe a lower/upper bound which is asymptotically equal to the optimal solution value.

We now describe the general conditions we assume throughout the paper, specifically concerning average-case analysis. The analysis requires assumptions on retailer/customer locations, demand sizes or other problem parameters. The results we cite were made under a number of different conditions. For consistency in this presentation, we will use one set of conditions, which in all cases is equal to or stronger than the conditions described in the cited works. We feel the spirit of the analysis is not lost by making this stronger assumption.

Condition $\mathcal{A}$. Retailer/customer *locations* are independently drawn from a distribution μ with compact support $A \subset \mathbf{R}^2$, i.e., they are restricted to a bounded region of the plane. We also assume that the expected radial distance of the retailers/customers from the warehouse ($E_\mu(r)$ or simply $E(r)$) is strictly positive. If, in addition, other retailer/customer-dependent characteristics are modeled, such as demand sizes (alternatively, demand rates), inventory holding cost rates etc., then we assume that these characteristics are independently drawn from specified distributions, and moreover they are independent of the retailer/customer location or any of its other characteristics.

The compact support assumption, in particular, implies that the radial distances are uniformly bounded by some $\rho < +\infty$.

The Traveling Salesman Problem (TSP) plays an important role in some of the analyses. Let $TSP^*(S)$ denote the length of an optimal traveling salesman tour through a set $S \subseteq N^0$. If an $\alpha-$approximation algorithm for the TSP is used to construct the tour, then the length is denoted $TSP^\alpha(S)$. The TSP has been extensively analyzed (see [52] for an excellent survey). The polynomial-time heuristic with the smallest worst-case bound currently known is Christofides' heuristic [25] which is a $\frac{3}{2}$-approximation algorithm.

The Bin Packing Problem (BPP) also plays a central role in several of the algorithms and models we consider. The BPP is defined by a set of item sizes $\{w_1, w_2, \ldots, w_n\}$ and bins of capacity C (it is assumed that $w_i \leq C$, for $i = 1, 2, \ldots, n$). The problem is to assign the items to bins in such a way that the number of bins used is minimized. A set $S \subseteq \{1, 2, \ldots, n\}$ makes up a feasible bin if and only if $w(S) \stackrel{\text{def}}{=} \sum_{i \in S} w_i \leq C$. Excellent surveys of this fundamental problem appear in [29, 30]. An important result concerns the relationship between the number of items and the number of bins required (for large n). Let b_n^* be the number of bins used in the optimal solution to the problem defined by the item sizes $\{w_1, w_2, \ldots, w_n\}$, and assume these item sizes are drawn from a distribution ν on $(0, C]$ with mean $E(w)$. Clearly $b_n^* \geq \sum_{i=1}^{n} w_i/C$ which, by the Strong Law of Large Numbers, leads to the inequality: $\underline{\lim}_{n \to \infty} b_n^*/n \geq E(w)/C$, almost surely. Using the techniques of Rhee and Talagrand [63] (or Kingman [51]) it is possible to show that for each distribution ν there exists a constant $\gamma \in [E(w)/C, 1]$, called the *bin-packing constant associated with* ν, such that $\lim_{n \to \infty} b_n^*/n = \gamma$, almost surely. Distributions with $\gamma = E(w)/C$ are said to allow *perfect packing* since asymptotically the amount of wasted space becomes a smaller and smaller fraction of the total bin space used.

6.3 SINGLE PERIOD CAPACITATED DELIVERY PROBLEMS

6.3.1 Introduction

We start with a review of single period models, specifically the Capacitated Vehicle Routing Problem (CVRP) which has received much attention. We assume retailer i requires delivery of an integer number w_i of units (the *demand*) currently at the depot. The problem is to design routes, each starting and ending at the depot, so that each retailer receives its demand, the vehicle is never loaded with more than Q units and the total distance traveled is as small as possible.

When analyzing or constructing solutions for the CVRP it is important to distinguish between the two versions of the problem: with *split* or *unsplit* demands. In the split demand version (considered in §6.3.3), vehicles are allowed to deliver partial loads. That is, a retailer may receive multiple deliveries and only a portion of the retailer's demand is delivered each time. In this case, it is customary to replace retailer i (with demand w_i) with w_i demand points each with *unit* demand that are all placed at the same location (of zero distance apart). We therefore assume each retailer has a unit demand. This is equivalent to the case where all retailers have equal demand and hence is also called the *equal weight* CVRP. In many cases splitting customer demands among several vehicles is undesirable, such as when loading/unloading takes a substantial amount of time, or disrupts operations at the retailer. In the unsplit demand version (considered in §6.3.4) each retailer can receive at most one delivery.

To perform a worst-case, average-case (or asymptotic analysis), a lower bound is usually required. As described in §6.2, we define Z_n^* (Z_n^H) as the length of an optimal solution (a solution constructed by a heuristic H) on a specific instance. Lemma 6.3.1 (from [45]) provides a useful lower bound for either the split or unsplit demand versions:

Lemma 6.3.1 *For n retailers with demands $\{w_1, w_2, \ldots, w_n\}$,*

$$Z_n^* \geq \max \left\{ TSP^*(N^0), \frac{2}{Q} \sum_{i=1}^n w_i r_i \right\}.$$

Proof. First observe that the split demand version is a relaxation of the unsplit demand version. Thus, it is sufficient to prove the lemma for the split demand version. The first term is clearly a lower bound. For the second term, replace each retailer i with w_i unit-demand points, as explained above. Let S_j be the set of demand points served by the jth vehicle in an optimal solution to the problem. Then

$$TSP^*(S_j^0) = TSP^*(S_j \cup \{0\}) \geq 2 \max_{i \in S_j} \{r_i\} \geq \frac{2}{|S_j|} \sum_{i \in S_j} r_i \geq \frac{2}{Q} \sum_{i \in S_j} r_i,$$

since $|S_j| \leq Q$. Summing over all routes, we obtain

$$\sum_j TSP^*(S_j^0) \geq \frac{2}{Q} \sum_j \sum_{i \in S_j} r_i = \frac{2}{Q} \sum_{i=1}^n w_i r_i.$$

∎

In the next subsection we consider heuristics for the CVRP. In §6.3.3, we analyze the split demand version of the problem, and in §6.3.4, we consider the unsplit case. In §6.3.5, we analyze several versions with multiple sources.

6.3.2 Heuristics

A large number of heuristics have been proposed for the CVRP. We classify these heuristics (as in Christofides [26]) into four categories: (*i*) *constructive*, (*ii*) *route first-cluster second*, (*iii*) *cluster first-route second*, and (*iv*) *incomplete optimization* methods. We describe the main characteristics of these classes and give examples of each. Computational tests of some of these are reported in e.g. [38, 18, 43].

6.3.2.1 Constructive Methods. Constructive methods build routes on the basis of strictly *local* considerations and are not related to any specific mathematical programming formulation. As such they fail to provide the user with an *ex ante* or *ex post* bound on the optimality gap of the generated solution.

The well-known Savings Algorithm suggested by Clarke and Wright [28] is a prominent member of this class. Many heuristics make use of modifications to the Savings Algorithm, this includes the heuristics of Gaskel [42], Yellow [73] and Paessens [57]. Also included in this category are tabu search methods which have empirically proven to be very effective on the CVRP, see e.g. the implementations of Pureza and Franca [60], Semet and Taillard [68] and Gendreau et al. [43].

6.3.2.2 Route First-Cluster Second Methods. Heuristics of this class first determine an *ordering* of the retailers (independently of the demand sizes), and then *partition* this ordering into feasible demand *clusters*. Customers are then served within each cluster depending on the specific heuristic. An analysis is presented in §6.3.4.3. In *tour partitioning* heuristics, such as Optimal Partitioning of Beasley [12] and Iterated Tour Partitioning (ITP) of Haimovich and Rinnooy Kan [45], the initial sequence is chosen as the sequence in which retailers appear in an optimal (or near-optimal) traveling salesman tour. We consider such schemes in §6.3.3.1, §6.3.3.2 and §6.3.4.1. The Sweep Algorithm of Gillett and Miller [44] determines the initial ordering based on sorting the retailers' polar coordinates (where the depot is the origin).

6.3.2.3 Cluster First-Route Second Methods. In this class of heuristics, retailers are first clustered into feasible groups to be served by the same vehicle without regard to any preset ordering and then efficient routes are designed for each cluster. The main approaches for determining the clusters are via solution of a *mathematical programming problem* or via *region partitioning schemes*.

Mathematical programming based approaches tend to be more technically elaborate than the previous classes. These include the Two-Phase Method of Christofides et al. [27], the Matching Based Savings Algorithm of Desroshers and Verhoog [32], the Parallel Savings Algorithm of Altinkemer and Gavish [3], the Generalized Assignment Heuristic of Fisher and Jaikumar [37], the Location-Based Heuristic of Bramel and Simchi-Levi [18] and the TSSP+1 Method of Noon et al. [56]. The algorithms of [37] and [18] use in a first step the concept of *seed retailers*. The seed retailers' demands are in separate vehicles in the solution, and tours are constructed around them. In both cases, the performance of the algorithm depends highly on the determination of these seeds. The two algorithms are similar in spirit; their main difference is that [18] incorporate the seed selection problem directly into the optimization problem.

Region partitioning procedures were first proposed by Karp for the TSP (see [50]) as a way of more effectively taking advantage of the geometrical setting of the problem. Haimovich and Rinnooy Kan applied them to the CVRP with split demands (see [45]). In both of these contexts, the retailers are divided into smaller regions (clusters), traveling salesman tours are determined for each of the regions (clusters), and then these are transformed into a single tour (in the case of the TSP) or a set of routes (in the case of the CVRP). In addition, [45] developed easily computable bounds on the optimal solution value that are asymptotically accurate. We analyze these procedures in the case of the CVRP in §6.3.3.3. This general approach was also used successfully to design asymptotically optimal heuristics and bounds for a number of single and multi-period capacitated delivery problems discussed in §6.3.5.2 and §6.4.

6.3.2.4 Incomplete Optimization Methods. Incomplete optimization methods are optimization algorithms based on mixed integer programming formulations that due to the prohibitive computing time required in reaching an optimal solution, are terminated prematurely. These procedures usually have the advantage of generating bounds on the optimal solution value. Examples include the cutting plane approach of Cornuéjols and Harche [31], and the Minimum K-Tree approach of Fisher [39].

6.3.3 The Split Demand Case

In addition to the many heuristic approaches developed for it, the CVRP has also been extensively analyzed. We first present analyses of the split demands version. As explained in §6.3.1, without loss of generality, each retailer is assumed to request an equal amount (a *unit*) of product.

6.3.3.1   Worst-Case Analysis of Tour Partitioning Heuristics.   Worst-case analysis of heuristics for the CVRP with unit demands begins with the ITP described by Haimovich and Rinnooy Kan [45] and later analyzed by Altinkemer and Gavish [2].

The ITP(α) heuristic first constructs a traveling salesman tour through all the retailers and the depot using an α-approximation algorithm for the TSP. The unit-demand points are then labeled $1, 2, \ldots, n$ in the order they appear on the tour where the depot is connected to 1 and n. The heuristic constructs Q different solutions to the problem and picks the one with minimum length. For $i = 1, 2, \ldots, Q$, solution i is constructed as follows. Demand points $\{1, 2, \ldots, i\}$ are served by the first vehicle. Demand points $\{i+1, i+2, \ldots, i+Q\}$ are served by the second vehicle. In general, demand points $\{i+1+Q(k-2), i+2+Q(k-2), \ldots, i+Q(k-1)\}$ are served by the kth vehicle, for $k = 2, 3, \ldots, \lfloor n/Q \rfloor$. The remaining points, of which there are at most Q if any, are served by the last vehicle (vehicle $\lfloor n/Q \rfloor + 1$).

Superimposing all Q solutions reveals two things: each edge of the traveling salesman tour is used a total of $Q - 1$ times and, additionally, each edge $(0, i)$ is used twice, for each $i \in N$. Therefore the sum of the lengths of these Q solutions is:

$$2 \sum_{i=1}^{n} r_i + (Q-1)TSP^{\alpha}(N^0) \leq 2 \sum_{i=1}^{n} r_i + (Q-1)\alpha TSP^*(N^0).$$

Since the best solution is at most the length of the average solution:

$$Z_n^{\text{ITP}(\alpha)} \leq \frac{2}{Q} \sum_{i=1}^{n} r_i + \left(1 - \frac{1}{Q}\right)\alpha TSP^*(N^0). \tag{6.2}$$

Combining this upper bound with the lower bound of Lemma 6.3.1 (with unit demands), gives:

$$\frac{Z_n^{\text{ITP}(\alpha)}}{Z_n^*} \leq 1 + \left(1 - \frac{1}{Q}\right)\alpha.$$

Thus if the initial tour is an optimal traveling salesman tour ($\alpha = 1$), then the worst-case bound is $2 - \frac{1}{Q}$. For the particular case of $\alpha = 1$, the bound is shown to be asymptotically tight (as $Q \to \infty$) since [53] construct an example where the solution created by ITP(1) has length $2 - \frac{2}{Q+1}$ times the optimal length for any $Q \geq 1$. If one uses Christofides' heuristic [25] for the TSP, then the performance guarantee of ITP($\frac{3}{2}$) is $\frac{5}{2} - \frac{3}{2Q}$.

The performance of ITP(α) can be improved by observing that it is possible to optimally partition the initial tour so that total distance traveled is minimized. This problem can be formulated as a shortest path problem in an acyclic graph. This heuristic, called Optimal Partitioning (OP(α)), was first described by Beasley [12]. By construction its performance is no worse than ITP(α), and in fact its worst-case performance is identical to ITP(α).

6.3.3.2 Average-Case Analysis of Tour Partitioning Heuristics.

Haimovich and Rinnooy Kan [45] consider tour partitioning heuristics for the CVRP with unit demands where the route cost is assumed to be proportional to the Euclidean distance traveled. We note that the analyses of [45] were performed under conditions weaker than Condition $\mathcal{A}$ (in particular, they assume merely $E(r) < +\infty$).

Consider ITP(α) and its upper bound (6.2) as n increases. We observe that the first term $(2 \sum_{i=1}^{n} r_i/Q)$ grows like n, while the second term grows at the same rate as the length of an optimal traveling salesman tour.

An important result, first proved by Beardwood et al. [11], concerns the asymptotic growth of the length of an optimal traveling salesman tour through the n points of the set N ($TSP^*(N)$) in a region of $\mathbf{R}^2$. They show that $TSP^*(N)/\sqrt{n} \to \beta_{\text{TSP}} c_\mu$, where c_μ is a constant related only to the density of the points and β_{TSP} is the "TSP constant" (see [52]).

Therefore, the first term of (6.2) dominates the second for large enough n, and it coincides with the lower bound from Lemma 6.3.1 (with unit demands). Therefore:

Theorem 6.3.2 *Under Condition $\mathcal{A}$:*

$$\lim_{n \to \infty} \frac{1}{n} Z_n^* = \lim_{n \to \infty} \frac{1}{n} Z_n^{\text{ITP}(\alpha)} = \frac{2}{Q} E(r) \qquad \text{a.s.}$$

For the unequal (split) demand case, assume the retailer demands are drawn from a distribution ν on the integers $\{1, 2, \ldots, Q\}$ with mean $E(w)$. Then:

Corollary 6.3.3 *Under Condition $\mathcal{A}$, for any fixed $\alpha \geq 1$:*

$$\lim_{n \to \infty} \frac{1}{n} Z_n^* = \lim_{n \to \infty} \frac{1}{n} Z_n^{\text{ITP}(\alpha)} = \frac{2}{Q} E(w) E(r) \qquad \text{a.s.}$$

Therefore ITP(α), as well as OP(α), are *asymptotically optimal* for any fixed $\alpha \geq 1$.

6.3.3.3 Average-Case Analysis of Region Partitioning Schemes.
In region partitioning schemes, the retailers (or demand points) are divided into regions (clusters) each containing no more than a given number of points. Each

region is served according to an optimal (or near-optimal) traveling salesman tour in the region. When considering region partitioning schemes, we assume each region is served by a separate vehicle and therefore we use interchangeably *regions* and *vehicles*.

We consider the Circular Region Partitioning (CRP) scheme developed by [45]. Since subsequent models will make use of a partitioning scheme for problems with a more general cost structure, we present here a modification of CRP, called Modified Circular Region Partitioning (MCRP), due to Anily and Federgruen [4]. Like most (e.g., [45, 50, 54]), this scheme has complexity $O(n \log n)$.

The MCRP heuristic uses an important subroutine called MCRP(q), which depends on a parameter q. This subroutine is also used by other algorithms in subsequent sections. MCRP(q) partitions the demand points into subregions in such a way that each contains exactly q demand points (except possibly one subregion). The proposed region partition for the CVRP is constructed by MCRP through a single call to the subroutine MCRP(Q).

Subroutine MCRP(q): Modified Circular Partitioning Scheme with capacity q.
Step 1. Partition the circle with radius $r_{\max}$ into $\lfloor \sqrt{n} \rfloor$ sectors each containing $q\ell_0$ points, for $\ell_0 = \lfloor n/(q\lfloor \sqrt{n} \rfloor) \rfloor$, and potentially one additional sector containing $n - \lfloor \sqrt{n} \rfloor q\ell_0$ points. Let K denote the number of sectors generated, and index them $k = 1, \ldots, K$. (Note that each sector, except possibly sector K, contains an integer multiple of q points.)
Step 2. For $k = 1, 2, \ldots, K$, partition sector k by means of *circular cuts* such that each of the subregions, except possibly one, contains q points. Let $S_1, S_2, \ldots, S_L$ denote all subregions created, and note that each subregion $S_1, S_2, \ldots, S_{L-1}$ contains q points and S_L may contain less.
Step 3. For each subregion S_ℓ, for $\ell = 1, 2, \ldots, L$ determine an optimal traveling salesman tour through the subregion's points and the depot.

We note that in *Step 3*, optimal traveling salesman tours can be computed in constant time if q is constant. An alternative method, that does not affect the asymptotic optimality of the scheme proved below, is to use an α-approximation algorithm for the TSP.

Let $\mathcal{L}$ denote the sum of the lengths of the optimal traveling salesman tours in each subregion (not including the depot), i.e., $\mathcal{L} \overset{\text{def}}{=} \sum_{\ell=1}^{L} TSP^*(S_\ell)$. By using the triangle inequality, the total cost of the solution constructed by MCRP(Q) is:

$$Z_n^{\text{MCRP}(Q)} = \sum_{\ell=1}^{L} TSP^*(S_\ell^0) \leq \sum_{\ell=1}^{L} 2 \min_{i \in S_\ell} \{r_i\} + \mathcal{L}$$

$$\leq \sum_{\ell=1}^{L-1} \frac{2}{Q} \sum_{i \in S_\ell} r_i + 2r_{\max} + \mathcal{L}$$

$$\leq \frac{2}{Q} \sum_{i=1}^{n} r_i + 2r_{\max} + \mathcal{L}. \qquad (6.3)$$

We note that the first term on the right hand side of (6.3) coincides with a lower bound (see Lemma 6.3.1 with unit demands) which grows like n. We now consider $\mathcal{L}$. Using the following two *lemmata* (the first from Karp [50] and the second from [45]), Anily and Federgruen [4] prove that $\mathcal{L} = O(\sqrt{n})$.

Lemma 6.3.4 *Let* $\{X_1, \ldots, X_L\}$ *be a partition of the set* N *of retailers generated by a partitioning scheme. Let* Π *denote the total perimeter of the subregions. Then*

$$\sum_{\ell=1}^{L} TSP^*(X_\ell) \leq TSP^*(N) + \frac{3}{2}\Pi.$$

Lemma 6.3.5 *If* N *is contained in a connected planar region with area* A *and finite perimeter* $\Pi(A)$, *then*

$$TSP^*(N) \leq \sqrt{2nA} + \frac{3}{2}\Pi(A).$$

Let Π^{MCRP} be the total perimeter of the subregions generated by MCRP on N. Then:

$$
\begin{aligned}
\mathcal{L} \;&\leq\; TSP^*(N) + \frac{3}{2}\Pi^{\mathrm{MCRP}} && \text{(by Lemma 6.3.4)} \\
&\leq\; r_{\max}\sqrt{2\pi n} + 3\pi r_{\max} + \frac{3}{2}\Pi^{\mathrm{MCRP}},
\end{aligned}
$$

where this last inequality follows from Lemma 6.3.5 and since A can be chosen as the circle of radius $r_{\max}$ centered at the depot. Anily and Federgruen [4] prove that $\Pi^{\mathrm{MCRP}} = O(\sqrt{n})$, and therefore $\mathcal{L} = O(\sqrt{n})$. They then obtain the following theorem:

Theorem 6.3.6 *Under Condition* $\mathcal{A}$, *MCRP(Q) is asymptotically optimal for the CVRP with unit demands, i.e.,*

$$\lim_{n \to \infty} \frac{1}{n} Z_n^{\mathrm{MCRP}(Q)} = \frac{2}{Q} E(r).$$

In addition, MCRP(Q) is asymptotically optimal for the CVRP with unequal, but split, demands, i.e.:

$$\lim_{n \to \infty} \frac{1}{n} Z_n^{\mathrm{MCRP}(Q)} = \frac{2}{Q} E(w) E(r).$$

General Routing Costs

Theorem 6.3.6 above was proved by [4] under slightly different assumptions, but more importantly, for a much broader *class* of problems. They consider the problem where the cost of a route is dependent on the route's length ϑ as well as on m, the number of demand points visited, through a general function $f(\vartheta, m)$. They merely assume f to be nondecreasing and concave in ϑ. Details can be found in [4] and [35]. They also consider a non-homogeneous fleet, where the vehicles are numbered in ascending order of their capacities, i.e., $Q_1 \leq Q_2 \leq \cdots \leq Q_L$. The number of vehicles (L) may be fixed or a decision variable. Feasibility requires that if L is fixed, $n \leq \sum_{\ell=1}^{L} Q_\ell$.

We say a partition $\chi = \{X_1, \ldots, X_L\}$ of N is *feasible* if $|X_\ell| \leq Q_\ell$, for $\ell = 1, \ldots, L$. The cost of a feasible partition χ of N is therefore given by

$$U(\chi) = \sum_{\ell=1}^{L} f(TSP^*(X_\ell^0), |X_\ell|).$$

The optimal solution value is: $Z_n^* = \min\{U(\chi) : \chi \text{ is a feasible partition of } N\}$. By substituting the length of the routes in the cost function by a lower bound (see the proof of Lemma 6.3.1 with unit demands), the authors derive two lower bounds on the optimal solution. For that sake, let:

$$\underline{U}^{(1)}(\chi) = \sum_{\ell=1}^{L} f\left(\frac{2}{|X_\ell|} \sum_{j \in X_\ell} r_j, |X_\ell|\right) \quad \text{and} \quad \underline{U}^{(2)}(\chi) = \sum_{\ell=1}^{L} f(2 \max_{j \in X_\ell}\{r_j\}, |X_\ell|).$$

Clearly $\underline{U}^{(1)}(\chi) \leq \underline{U}^{(2)}(\chi) \leq U(\chi)$ since f is nondecreasing in ϑ. Define two partitioning problems for $j = 1, 2$:

$$(\underline{P}^{(j)}) : \underline{Z}_n^{(j)} = \min\{\underline{U}^{(j)}(\chi) : \chi \text{ is a feasible partition of } N\}. \tag{6.4}$$

One obviously concludes:

Lemma 6.3.7 $\underline{Z}_n^{(1)} \leq \underline{Z}_n^{(2)} \leq Z_n^*$.

Anily and Federgruen show that these lower bounds are asymptotically accurate. In addition, they can easily be computed (in at most $O(n \log n)$ operations, sometimes only $O(n)$). We now describe how these partitioning problems can be solved.

Solving $\underline{P}^{(1)}$ and $\underline{P}^{(2)}$

The partitioning problems $\underline{P}^{(1)}$ and $\underline{P}^{(2)}$ are known to be $\mathcal{NP}$-complete for general $f(\cdot, \cdot)$, see [23]. However, if we assume, without loss of generality, that the demand points are numbered $r_1 \leq r_2 \leq \ldots \leq r_n$, then the concavity of f in ϑ implies that an optimal *consecutive* partition exists for both $\underline{P}^{(1)}$ and $\underline{P}^{(2)}$. A consecutive partition is one in which the indices of each set of the partition are consecutive integers. $\underline{P}^{(1)}$ and $\underline{P}^{(2)}$ can thus be computed by solving a shortest path in an acyclic network; see Chakravarty et al. [23] or Anily and Federgruen [6] for details.

Thus, let χ_1 and χ_2 be optimal partitions for $\underline{P}^{(1)}$ and $\underline{P}^{(2)}$, respectively, and define, for $m = 1, 2, \ldots, Q$ and $j = 1, 2$, the sets

$$X_j^{(m)} = \{i : \chi_j \text{ assigns } i \in N \text{ to a set of cardinality } m\}.$$

Note that $X_j^{(m)}$ contains an integer multiple of m points. If f has antitone differences, an optimal *monotone* partition exists for both $\underline{P}^{(1)}$ and $\underline{P}^{(2)}$ (i.e., the partition $\{X_j^{(1)}, X_j^{(2)}, \ldots, X_j^{(Q)}\}$ is also consecutive), meaning that the sets $\{X_j^{(m)} : m = 1, \ldots, Q\}$ are contained in disjoint rings around the depot. (A function $\varphi : \mathbf{R}^2 \to \mathbf{R}$ has *antitone differences* if $\varphi(\vartheta + \Delta, m) - \varphi(\vartheta, m)$ is nonincreasing in m for $\Delta > 0$.) Anily and Federgruen [6] demonstrate that when the partitioning problem is *extremal*, a special algorithm, the Extremal Partitioning Algorithm (EPA), can be used to find an optimal partition in $O(n)$ operations. They show sufficient conditions with respect to f under which $\underline{P}^{(1)}$ and $\underline{P}^{(2)}$ are extremal in addition to possessing an optimal monotone partition. (One such condition is that f has the additional properties of antitone differences and is concave in m.) The EPA is as follows:

Extremal Partitioning Algorithm (EPA)

Step 1. Let $m = n$. And let $\ell = L$ if L is a given constant, otherwise let $\ell = 1$.

Step 2. If $1 < Q_\ell < m - \ell + 1$ then
  o add the set $\{m - Q_\ell + 1, \ldots, m\}$ to the partition.
  o Set $m \leftarrow m - Q_\ell$.
  o If $\ell > 1$ then $\ell \leftarrow \ell - 1$.
  o Repeat Step 2.

Else if $Q_\ell = 1$, add the sets $\{1\}, \ldots, \{m\}$ to the partition, stop.

Else if $Q_\ell \geq 2$ and $\ell > 1$, add the sets $\{1\}, \ldots, \{\ell - 1\}, \{\ell, \ldots, m\}$ to the partition.

Otherwise add the single set $\{1, \ldots, m\}$ to the partition, stop.

The EPA thus generates a partition *independent* of f, a robustness property suggested by the definition of extremality. The conditions established in [6] can also be used to establish extremality of a number of $\underline{P}^{(1)}$ problems discussed in §6.4.

The MCRP Heuristic

The MCRP heuristic can be applied to either the sets $X_1^{(m)}$, for $m = 1, 2, \ldots, Q$, or the sets $X_2^{(m)}$, for $m = 1, 2, \ldots, Q$. As a result we may obtain two different region partitions. For $m = 1, 2, \ldots, Q$, subroutine MCRP(m) is applied independently to each of the sets $X_j^{(m)}$, for either $j = 1$ or 2. MCRP(m) partitions the points in $X_j^{(m)}$ into regions consisting of exactly m demand points (since $|X_j^{(m)}|$ is a multiple of m). We let $Z_n^{\text{MCRP}(1)}$ ($Z_n^{\text{MCRP}(2)}$) denote the length of the solution obtained by the MCRP heuristic when applied to the sets $X_1^{(m)}$ ($X_2^{(m)}$), for $m = 1, 2, \ldots, Q$. The complexity of MCRP is again $O(n \log n)$ (see [6]).

The following theorem shows that MCRP heuristic is asymptotically optimal for this general class of problems.

Theorem 6.3.8 *Let $f(\vartheta, m)$ be nondecreasing and concave in ϑ, and define*

$$\underline{f}(\vartheta) \stackrel{\text{def}}{=} \min\{f(\vartheta, m) \mid 1 \le m \le Q\}.$$

Then, under Condition $\mathcal{A}$,

(a)

$$\lim_{n \to \infty} \frac{1}{n} \underline{Z}_n^{(2)} \ge \lim_{n \to \infty} \frac{1}{n} \underline{Z}_n^{(1)} \ge \frac{1}{Q} E(\underline{f}(2r)) \quad \text{a.s.}$$

(b) If, in addition, $E(\underline{f}(2r)) > 0$ then the MCRP heuristic is asymptotically optimal and the lower bounds $\underline{Z}_n^{(1)}$ and $\underline{Z}_n^{(2)}$ are asymptotically accurate, i.e.,

$$\overline{\lim_{n \to \infty}} \ Z_n^{\text{MCRP}(1)} / \underline{Z}_n^{(1)} = \overline{\lim_{n \to \infty}} \ Z_n^{\text{MCRP}(2)} / \underline{Z}_n^{(2)} = 1, \quad \text{a.s.}$$

6.3.4 The Unsplit Demand Case

We now turn our attention to the more restrictive version of the delivery problem: the unsplit demand case where each retailer can receive at most one delivery. As noted by Haimovich and Rinnooy Kan in [45], "this additional constraint introduces bin-packing features into the routing problem," which makes a direct extension of the above methods (e.g. ITP) non-trivial.

6.3.4.1 Worst-Case Analysis of Tour Partitioning Heuristics. Applying ITP to this version of the problem, by first replacing retailer i with demand w_i by w_i demand points each requiring delivery of one unit, fails to guarantee a feasible solution. A partition of the traveling salesman tour may split a number of customer demands into separate vehicles. Altinkemer and Gavish [1] show how ITP can be adapted to ensure that each partition does not split any customer demands. They partition the tour using a vehicle capacity of $Q/2$ (assuming without loss of generality that Q is even). This solution may also split demands. However, they present a simple rule to fix the partition and create a solution that is feasible for capacity Q and does not split demands. Their step does not increase the total distance traveled and therefore the analysis follows as in (6.2):

$$Z_n^{\text{ITP}(\alpha)} \le \frac{4}{Q} \sum_{i=1}^{n} w_i r_i + \left(1 - \frac{2}{Q}\right) \alpha TSP^*(N). \tag{6.5}$$

Combining this upper bound with the lower bound of Lemma 6.3.1 gives:

$$\frac{Z_n^{\text{ITP}(\alpha)}}{Z_n^*} \le 2 + \left(1 - \frac{2}{Q}\right) \alpha.$$

For the case of $\alpha = 1$, the bound is known to be tight (see [53]). For the case where Christofides' heuristic is used ($\alpha = \frac{3}{2}$), the worst-case bound is $\frac{7}{2} - \frac{3}{Q}$.

Again, the tour partitioning problem can be formulated as a shortest path in an acyclic network and thus the partitioning can be done optimally. This is exactly the Optimal Partitioning (OP) heuristic designed by Beasley [12]. The OP(α) heuristic has the same worst-case performance as ITP(α).

6.3.4.2 Average-Case Analysis of the Optimal Solution Value. An average-case analysis of the unsplit demand model begins with a determination of the asymptotic optimal solution value. Unfortunately, it cannot be derived from a simple extension of Corollary 6.3.3. However, the analysis of Bramel et al. [17], which we describe here, reveals its structure.

We assume customer demands $\{w_i\}$ (not necessarily integers) are drawn from a distribution ν on $(0, Q]$ with mean $E(w)$. Let $\gamma \in [E(w)/Q, 1]$ denote the *bin-packing constant associated with* ν. In [17], the following result is proven:

Theorem 6.3.9 *Under Condition* $\mathcal{A}$,

$$\lim_{n \to \infty} \frac{1}{n} Z_n^* = 2\gamma E(r), \qquad \text{a.s.}$$

Before proving this Theorem, we note that the structure of this asymptotic cost illustrates the two important cost components. The first is the customer's expected distance to the depot $E(r)$. The *shape* of the service region has no effect. The other determinant of the cost is the bin-packing constant γ. This is a number which is dependent on the types of items distributed. We also note that if the distribution of the demands allows for perfect packing, then $\gamma = E(w)/Q$, and the split and unsplit demand problems are asymptotically equivalent.

Proof. We now prove Theorem 6.3.9 by approximating Z_n^* from above by that of the following three-step cluster first-route second procedure. In the first step, using a simple grid we partition the region where the retailers are distributed into subregions. Then, for each of these subregions, we find the optimal packing of the retailer demands into bins of capacity Q. Finally, for each bin, we send one vehicle to serve the retailers in the bin using an efficient route.

For any fixed $\epsilon > 0$, let $G(\epsilon)$ be an infinite grid of squares of side $\epsilon/\sqrt{2}$ with edges parallel to the system coordinates. Recall A is the compact support of the distribution μ, and let $A_1, A_2, \ldots, A_{t(\epsilon)}$ be the subregions of $G(\epsilon)$ that intersect A and have $\mu(A_i) > 0$.

Let N_i be the set of retailers located in subregion A_i, and define $n_i = |N_i|$. For every $i = 1, 2, \ldots, t(\epsilon)$, let b_i^* be the minimum number of bins of capacity Q needed to pack the demands $\{w_j\}_{j \in N_i}$. Finally, for each subregion A_i and for each $j = 1, 2, \ldots, b_i^*$, let n_{ij} be the number of retailers in the jth bin of this optimal packing.

We now proceed to find an upper bound on the value of the procedure. For each bin, we send one vehicle to serve all the retailers in the bin. Let $\underline{r}_i$ be the distance from the depot to the closest point in subregion A_i. Note that the distance between any two retailers in A_i is no more than ϵ. Consequently, using any efficient routing method, the distance traveled by the vehicle serving the jth bin of subregion A_i is no more than $2\underline{r}_i + \epsilon(n_{ij} + 1)$.

Hence,

$$Z_n^* \leq \sum_{i=1}^{t(\epsilon)} \sum_{j=1}^{b_i^*} \left[2\underline{r}_i + \epsilon(n_{ij} + 1) \right] \leq 2 \sum_{i=1}^{t(\epsilon)} b_i^* \underline{r}_i + 2n\epsilon,$$

and thus

$$\varlimsup_{n\to\infty} \frac{1}{n} Z_n^* \leq \varlimsup_{n\to\infty} \frac{2}{n} \sum_{i=1}^{t(\epsilon)} b_i^* \underline{r}_i + 2\epsilon. \tag{6.6}$$

Consider the first term of this upper bound:

$$\varlimsup_{n\to\infty} \frac{2}{n} \sum_{i=1}^{t(\epsilon)} b_i^* \underline{r}_i = \varlimsup_{n\to\infty} 2 \sum_{i=1}^{t(\epsilon)} \underline{r}_i \left(\frac{b_i^*}{n_i} \cdot \frac{n_i}{n} \right) \leq 2 \sum_{i=1}^{t(\epsilon)} \underline{r}_i \varlimsup_{n\to\infty} \frac{b_i^*}{n_i} \varlimsup_{n\to\infty} \frac{n_i}{n} \qquad \text{a.s.}$$

For each $i = 1, 2, \ldots, t(\epsilon)$, the Strong Law of Large Numbers implies that $\lim_{n\to\infty} n_i/n = \mu(A_i) > 0$ almost surely, and therefore n_i grows to infinity almost surely, as n grows to infinity. Then:

$$\varlimsup_{n\to\infty} \frac{b_i^*}{n_i} = \varlimsup_{n_i\to\infty} \frac{b_i^*}{n_i} = \gamma, \qquad \text{a.s.}$$

Hence,

$$\varlimsup_{n\to\infty} \frac{2}{n} \sum_{i=1}^{t(\epsilon)} b_i^* \underline{r}_i \leq 2\gamma \sum_{i=1}^{t(\epsilon)} \underline{r}_i \mu(A_i) \leq 2\gamma E(r) \qquad \text{a.s.}$$

Using this in (6.6):

$$\varlimsup_{n\to\infty} \frac{1}{n} Z_n^* \leq 2\gamma E(r) + 2\epsilon \qquad \text{a.s.}$$

Since this inequality holds for arbitrary $\epsilon > 0$, we have

$$\varlimsup_{n\to\infty} \frac{1}{n} Z_n^* \leq 2\gamma E(r) \qquad \text{a.s.}$$

In [17], this upper bound is combined with a lower bound (derived using similar region partitioning ideas) to prove Theorem 6.3.9. ∎

6.3.4.3 Analysis of Route First-Cluster Second Heuristics. Having determined the asymptotic optimal cost, it is then possible to compare the performance of several classical heuristics to this benchmark. In particular, we consider the analysis of heuristics of the route first-cluster second class described in §6.3.2.2. This class includes, among others, the Sweep Algorithm [44], Iterated Tour Partitioning [45] and Optimal Partitioning [12]. We show that, as the optimal solution to the routing problem is asymptotically related to the optimal solution to the bin-packing problem (through Theorem 6.3.9), the performance of these heuristics is asymptotically related to the Next-Fit (NF) bin-packing heuristic. Given an instance of the BPP, Next Fit can be described as follows. For a given ordering of the items, place the first item in the first bin. If the next item fits in the currently opened bin, place it there otherwise close this bin, open a new bin and place the item in the new bin. Repeat this for each subsequent item.

To give the intuition behind how the route first-cluster second heuristics are related to Next Fit, note that a route first-cluster second heuristics first *sorts* the retailers into a fixed sequence. The sequencing is done independently of the demand sizes. For example, an efficient ordering of the demands (based on locations alone) might be: $9, 2, 9, 2, 9, 2, 9, 2, \ldots$. With a vehicle capacity of 10, any partition of this tour *must* assign one vehicle per customer. This is not an efficient strategy but it is exactly the solution provided by Next Fit on this sequence of item sizes (demands). By contrast, a more intelligent clustering algorithm would reduce this type of inefficiency.

We now more formally relate NF and route first-cluster second strategies. Let $\{w_i\}_{i=1,2,\ldots,n}$ be generated according to ν and define b_n^{NF} be the number of bins produced by NF on these set of item sizes. Rhee and Talagrand [63] show that, as in the case of an optimal packing, there exists a constant $\gamma^{\mathrm{NF}} > 0$ such that $\lim_{n\to\infty} b_n^{\mathrm{NF}}/n = \gamma^{\mathrm{NF}}$ almost surely. Note that $b_n^* \le b_n^{\mathrm{NF}} \le n$ implies that $\gamma^{\mathrm{NF}} \in [\gamma, 1]$, where γ is the bin-packing constant associated with ν. This is used in [14] to prove the following:

Theorem 6.3.10 *(i) Let H denote a route first-cluster second heuristic. Then, under Condition $\mathcal{A}$:*

$$\lim_{n\to\infty} \frac{1}{n} Z_n^{\mathrm{H}} \ge 2\gamma^{\mathrm{NF}} E(r) \qquad \text{a.s.}$$

(ii) Asymptotically, $OP(\alpha)$ is the best possible heuristic in this class, that is, for any fixed $\alpha \ge 1$, under Condition $\mathcal{A}$:

$$\lim_{n\to\infty} \frac{1}{n} Z_n^{\mathrm{OP}(\alpha)} = 2\gamma^{\mathrm{NF}} E(r) \qquad \text{a.s.}$$

This theorem along with Theorem 6.3.9 imply that heuristics in the route first-cluster second class *can never be asymptotically optimal*, except in some simple cases where $\gamma^{\mathrm{NF}} = \gamma$ (e.g., when demand sizes are equal.). In general, route first-cluster second heuristic (including ITP(α)) will be asymptotically ineffective for the CVRP with *unsplit* demands. In fact, for the case where the customer demand sizes are uniform on $(0, Q]$ (and not necessarily integer), $\gamma^{\mathrm{NF}} = \frac{2}{3}$ while $\gamma = \frac{1}{2}$ and thus any algorithm of the route first-cluster second variety is at best an asymptotic $\frac{4}{3}$-approximation algorithm.

6.3.4.4 Algorithmic Insights. An important byproduct of the analysis of §6.3.4.2 and §6.3.4.3 is that it suggests new ways to solve the problem. Bramel and Simchi-Levi [18] used the insights to develop an effective heuristic for the CVRP called the Location-Based Heuristic (LBH). Specifically, this heuristic was motivated by the following observations.

The grid region partitioning scheme used to find an upper bound on Z_n^* in the proof of Theorem 6.3.9 is accurate in an asymptotic sense. Unfortunately, the scheme is not polynomial (e.g. it requires solving the BPP). But the scheme suggests that, asymptotically, the tours in an optimal solution will be of a very simple structure consisting of two parts. The first is the round trip the vehicle makes from the depot to a subregion (where retailers are located); we call these the *stems*. The second is the additional distance (the *insertion cost*) accrued by visiting each of the retailers served in the subregion. The goal is therefore to construct a heuristic that assigns retailers to vehicles so as to minimize the sum of the length of all stems plus the total insertion costs of retailers into each cluster. If done carefully, the solution obtained should be asymptotically optimal.

The LBH of [18] uses a surrogate problem, called the Capacitated Concentrator Location Problem (see Pirkul [59]), to approximate this construction. In addition, under Condition $\mathcal{A}$, the LBH is *asymptotically optimal* (though non-polynomial). This heuristic also provides excellent solutions to the standard test problems of Christofides et al. [27] whose size range from 50 to 200 retailers. This case demonstrates that the insights obtained from the asymptotic analysis above can suggest new solution methods that are effective on problems of reasonable size.

6.3.5 Multi-Source Delivery Problems

So far we have restricted our attention to delivery problems where the product is assumed to be supplied from one location, such as a warehouse or plant. We now consider two models where product is available at a number of locations.

6.3.5.1 The Multi-Depot CVRP. Li and Simchi-Levi [53] consider the Multi-Depot CVRP (MCVRP) where a set of depots are assumed to have an unlimited number of vehicles and supply of product. The objective is to design routes serving all customers, each starting and ending at a depot, so that the total distance traveled is minimized. They distinguish between the *fixed destination* and *non-fixed destination* case. In the fixed destination problem, a vehicle leaving a depot must return to that same depot at the end of the route. In the non-fixed destination case, the vehicle can return to any depot.

In the non-fixed destination MCVRP and for unit demands (or general, but split, demands), the algorithm of [53] first condenses the depots into one super-depot. The distance from this super depot to each retailer is set to the distance the retailer is to *its* nearest depot. After recalculating the matrix of shortest distances, $\text{ITP}(\alpha)$ is run on this new single depot problem. In the solution, all edges connecting a retailer to the super-depot are replaced with a new edge connecting the retailer to *its* closest depot. The authors show that this solution is feasible and has length at most $1 + (1 - \frac{1}{Q})\alpha$ times the length of the optimal solution. A similar analysis and bound can be developed for $\text{OP}(\alpha)$. For the general unsplit demand version, these bounds become $2 + (1 - \frac{2}{Q})\alpha$. These bounds are proven to be asymptotically tight (as $Q \to \infty$) for $\alpha = 1$.

The fixed destination case is more involved. After deleting all edges connecting a retailer to the super-node, the resulting line segments are attached (from both ends) to the "best" depot (the one which minimizes the sum of the distances to each end). The solution is feasible and has length at most $1 + (2 - \frac{1}{Q})\alpha$

times the length of an optimal solution. For the general unsplit demand case, a similar approach leads to a solution at most $2 + (2 - \frac{2}{Q})\alpha$ times the optimal length. Again, for the specific case of $\alpha = 1$, the first bound is proven to be asymptotically tight (as $Q \to \infty$) and the second is close to asymptotically tight.

6.3.5.2 Problems with Deliveries and Backhauls.

A feature of many supply chains is the need to additionally transport items in the reverse direction, from customers (retailers) to the depot (or warehouse, or plant). Casco et al. [21] describe a problem of this type called the Vehicle Routing Problem with Delivery and Backhauls (VRPDB). Here, *demand* product must be brought from the depot to a subset of the customers and *supply* product currently at a subset of the customers needs to be brought to the depot. Two variants exist: *mixed* and *non-mixed* loads. The mixed version allows delivery and backhaul products to simultaneously be on the vehicle. In the non-mixed loads version, the delivery product cannot be mixed with the backhaul product (meaning in any route all deliveries must precede backhauls). As stated in [21], firms that can use backhauls effectively can significantly reduce their total transportation costs.

The VRPDB was analyzed by Anily [10]. She considers two sets of customers geographically dispersed in the plane. The first set consists of delivery customers that require goods from the depot. The second set consists of backhaul customers that need to have goods currently at their location brought to the depot. In order to avoid excessive notation, we assume a single commodity (the extension to multiple commodities can be found in [10]). The objective is to design a set of routes of minimum total length such that each customer is served (demand customers get product, supply customers are relieved of theirs), each route starts and ends at the depot and the vehicle capacity is not exceeded.

From here on, we assume demands can be split, or equivalently, all demands and backhauls are of unit size. We further assume there are $m \overset{\text{def}}{=} n/2$ unit-demand delivery customers and m unit-demand backhaul customers (additionally, m is assumed to be an integer multiple of Q). These assumptions are made without loss of generality since dummy delivery or backhaul customers (located at the depot) can always be added. The set D (B) is used to represent the set of delivery (backhaul) customers, and $N = D \cup B$.

Observe that in both mixed and non-mixed versions of the problem each vehicle can serve at most $2Q$ points on a single route: Q delivery and Q backhaul points. Thus, the lower bound for the CVRP proposed by Haimovich and Rinnooy Kan (see Lemma 6.3.1 with unit demands), can be applied to the VRPDB, giving $Z_n^* \geq \sum_{i=1}^{n} r_i/Q$. Unfortunately, it is simple to show that this bound is <u>not</u> asymptotically tight for the VRPDB under general conditions.

Anily first considers an improved lower bound. For that sake, define the $m \times m$ matrix $M = \{M_{ij}\}$ where $M_{ij} = d_{ij}$ if $i \in D$ and $j \in B$. Let $g(M)$ be the cost of a minimum cost bipartite matching between D and B. This can be computed in $O(m^3)$ operations; see [58], Chapter 11. The following lemma (proven in [10]) proposes a lower bound.

Lemma 6.3.11 *For either mixed or non-mixed versions,*

$$Z_n^* \geq \underline{Z}_n^{(1)} \stackrel{\text{def}}{=} \frac{1}{Q} \sum_{i=1}^{n} r_i + \frac{g(M)}{Q}.$$

Anily proposes the following heuristic for the VRPDB based on the MCRP(Q) subroutine introduced in §6.3.3.3. The subroutine is applied separately to D and to B. As a result we obtain two types of regions: delivery regions and backhaul regions, each consisting of Q points. By solving a bipartite matching problem on these two types of regions (see below) we determine which delivery and backhaul regions to serve together. To this end, the cost of matching two regions D_ℓ and B_k is defined as the minimum distance between a point in D_ℓ and a point in B_k. Define the $\frac{m}{Q} \times \frac{m}{Q}$ cost matrix M^R, i.e., $M_{\ell k}^R = \min\{d_{st} : s \in D_\ell, t \in B_k\}$. Then:

The MCRPDB Heuristic:
Step 1. Apply Steps 1 and 2 of MCRP(Q) to D and B separately, resulting in regions $D_1, \ldots, D_{m/Q}$ and $B_1, \ldots, B_{m/Q}$. In each of these regions find an optimal traveling salesman tour through the points in the region. In addition, define the matrix M^R as explained above.
Step 2. Compute an optimal bipartite matching of D to B, using the cost matrix M^R.
Step 3. Route construction. For each delivery region D_ℓ matched to a backhaul region B_k (say from $s \in D_\ell$ to $t \in B_k$), use the following route: the vehicle leaves the depot fully loaded and travels to the closest point in D_ℓ, serves all delivery points in D_ℓ by traveling along the optimal traveling salesman tour in the region. The vehicle then travels empty to $t \in B_k$ and continues to serve all backhaul points in B_k by traveling along the optimal traveling salesman tour in B_k, and finally returns fully loaded to the depot. As we remark in §6.3.3.3, instead of computing optimal traveling salesman tours within each of the regions, one can use any heuristic for the TSP with a bounded worst-case ratio.

The matrix M^R can also be used to construct an additional lower bound:

$$\underline{Z}_n^{(2)} \stackrel{\text{def}}{=} \frac{1}{Q} \sum_{i=1}^{n} r_i + g(M^R).$$

Lemma 6.3.12 $Z_n^* \geq \underline{Z}_n^{(1)} \geq \underline{Z}_n^{(2)}.$

We conclude with the following theorem (of [10]):

Theorem 6.3.13 *Under Condition $\mathcal{A}$, where m unit-delivery point locations are drawn from a distribution μ_D, and independently, m unit backhaul point* locations are drawn from a distribution μ_B with $E(r) = \frac{1}{2}[E_{\mu_D}(r) + E_{\mu_B}(r)]$, then:

(a)

$$\lim_{n\to\infty} \frac{1}{n} Z_n^* \geq \frac{1}{Q} E(r), \quad \text{a.s.}$$

(b) moreover, MCRPDB is asymptotically optimal and $\underline{Z}_n^{(1)}$ and $\underline{Z}_n^{(2)}$ are asymptotically accurate, i.e.,

$$\overline{\lim_{n\to\infty}} \, Z_n^{\text{MCRPDB}} / \underline{Z}_n^{(1)} = \overline{\lim_{n\to\infty}} \, Z_n^{\text{MCRPDB}} / \underline{Z}_n^{(2)} = 1 \quad \text{a.s.}$$

Note that this particular split demand version of the VRPDB allows the delivery and backhaul points to be from different locational distributions. If we make the assumption that the two types of points have the same locational distribution, then the cost of the matching (above) becomes asymptotically a negligeable fraction of the total cost. For this version, it is possible to determine the asymptotic optimal solution value of the VRPDB with *unsplit* demands. The result is as follows:

Theorem 6.3.14 *Under Condition $\mathcal{A}$, assume n locations are drawn from a distribution μ (with expected distance $E(r)$ from the depot). Assume each location is chosen independently as a delivery location with probability $p_D \in (0,1)$ and as a backhaul location with probability p_B (with $p_B = 1 - p_D$). For each delivery (backhaul) location, the demand (backhaul) quantity is drawn from a distribution ν_D (ν_B) on $(0, Q]$. Let γ_D (γ_B) designate the bin-packing constant associated with ν_D (ν_B). Then:*

$$\lim_{n\to\infty} \frac{1}{n} Z_n^* = 2E(r) \cdot \max\{p_D \gamma_D, p_B \gamma_B\}, \quad \text{a.s.}$$

Proof. Let n_D (n_B) be the number of delivery (backhaul) retailers. Let $Z_{n_D}^*$ ($Z_{n_B}^*$) designate the cost of serving the n_D delivery (n_B backhaul) retailers. We first show the lower bound:

$$\frac{1}{n} Z_n^* \geq \frac{1}{n} \max\{Z_{n_D}^*, Z_{n_B}^*\} = \max\left\{ \frac{n_D}{n} \cdot \frac{Z_{n_D}^*}{n_D}, \frac{n_B}{n} \cdot \frac{Z_{n_B}^*}{n_B} \right\}.$$

Taking the limit and applying the Strong Law of Large Numbers and Theorem 6.3.9, we get:

$$\lim_{n\to\infty} \frac{1}{n} Z_n^* \geq \max\{p_D \cdot 2\gamma_D E(r), p_B \cdot 2\gamma_B E(r)\} \quad \text{a.s.}$$

For the upper bound, we sketch here a proof using the grid partition (as in the proof of Theorem 6.3.9). In each subregion of the grid, two bin-packing problems are solved: one on the delivery loads and one on the backhaul loads. Each vehicle is assigned arbitrarily a *delivery* bin and a *backhaul* bin: it leaves the depot with an amount equal to the total delivery quantity in its delivery bin, serves these delivery retailers, then travels (empty) to one of the backhaul retailers in the backhaul bin, serves these in any order and returns to the depot. The analysis now follows. Asymptotically, there are $n_i p_D + o(n)$ $(n_i p_B + o(n))$ delivery (backhaul) retailers in region A_i, requiring $n_i p_D \gamma_D + o(n)$ $(n_i p_B \gamma_B + o(n))$ vehicles. Therefore the number of vehicles required to serve region A_i is the maximum between the two. The bound then follows as in the proof of Theorem 6.3.9. ∎

6.4 MULTI-PERIOD CAPACITATED DELIVERY PROBLEMS

6.4.1 *Introduction*

In many multi-period supply chain settings, important cost reductions and/or service improvements may be achieved by adopting an *integrated* approach towards various logistical planning functions. In particular the areas of *inventory control* and *transportation planning* need to be *closely coordinated*. For example, shipping in smaller quantities and with higher frequency generally leads to reductions in inventory investments but requires additional transportation costs. In this section, we consider models that attempt to balance these two important cost components by determining effective combined inventory and distribution strategies.

Specifically, this section deals with infinite horizon deterministic Inventory Routing Problems (IRP) where retailers face constant deterministic retailer-specific demand rates. All of the models considered here assume a single warehouse that serves as the supply point for the retailers. The objective is to determine long-term *integrated* replenishment strategies (i.e., inventory rules and routing patterns) enabling all retailers to meet their demands while minimizing long-run average system-wide transportation and inventory costs. A number of variants of the problem are considered. In particular, we consider systems where: (i) the retailers' holding costs are identical or retailer-specific, (ii) central inventories may be held at the warehouse (single versus two-echelon systems), (iii) there is a single product or multiple products, and (iv) the retailer demand rates can or cannot be split among different routes.

The notion of splitting an integer *demand rate* is different from that of splitting a demand (as in §6.3.3). Splitting a demand rate implies that a retailer of demand rate $m > 1$ (an outlet) is treated as m unit demand rate retailers (suboutlets). One result of this assumption is that if an outlet is split into a number of suboutlets they are treated as separate suboutlets each responsible for a specific fraction of the sales; it is therefore possible that a delivery is made to one suboutlet at an epoch at which another suboutlet continues to have stock. Another result is that splitting a demand rate is likely to result in non-stationary deliveries to the retailer.

The first studies in this field include Bell et al. [13] and Burns et al. [20]. Bell et al. [13] developed a computerized planning system based on a multi-period, combined inventory/vehicle scheduling model. Their system was implemented at several companies and was awarded the 1983 CPMS prize for its implementation at Air Products, Inc. A delivery plan is determined ensuring that the inventory levels stay above prespecified safety stock levels. The model allows for a variety of operational constraints with respect to acceptable time windows for the deliveries, vehicle capacity constraints, acceptable assignments of retailers to vehicles, etc.... A large scale mixed integer program is solved by a Lagrangean relaxation technique.

The first *integrated infinite horizon* model appears to be due to Burns et al. [20] and Blumenfeld et al. [15]. They assume that variable transportation cost is proportional to the Euclidean distance traveled. Confining themselves to the case where all retailers face *identical* demand rates, these authors obtain several solution heuristics and cost approximations, on the basis of a number of simplifying assumptions. In particular, they use information on the spatial density of the retailers rather than their exact locations. Neither of these studies consider vehicle routes as decision variables.

In this section, we describe region partitioning schemes for three different models; in §6.4.2 we introduce the common assumptions and notation used by these models and their analyses. In §6.4.3, §6.4.4 and §6.4.5 we consider distribution systems, in which all stock enters the system through the warehouse from where it is distributed to (some of) the retailers. Inventories are kept at the retailers but not at the warehouse. Such systems are called *coupled* systems, see [64]. In other words, the warehouse (depot) serves as a coordinator of the replenishment process and acts as a "breakbulk" or transshipment point from where all vehicles start and end their routes. In some applications the warehouse plays the role of an outside supplier. In §6.4.3 we consider the single commodity problem with split demands and where retailer holding costs are identical. In §6.4.4 we consider several models where holding costs are retailer-specific. In §6.4.5 we present an average-case analysis of a class of policies called Fixed Partition policies, for the version of the problem with identical retailer holding costs. We conclude with §6.4.6 where we consider two-echelon systems, i.e., systems where central inventories may be stored at the warehouse.

6.4.2 Notation and Assumptions for Region Partitioning Schemes

We review here the notation and assumptions shared by all three region partitioning schemes of this section.

All models consider a single warehouse with multiple retailers. Demand occurs at each retailer at a retailer-specific rate and may be split. Inventory carrying costs are incurred at a constant rate per unit of time per unit stored at the retailer. The transportation costs include a fixed cost per route driven (for leasing or renting) and variable costs proportional to the total (Euclidean) distance traveled. Delivery patterns are restricted by a truck capacity and possibly also by a maximum frequency with which a route may be driven. The frequency restriction may be due to the set-up time required for unloading at the retailers or other material handling constraints, see [22]. In addition, we allow for sales volume constraints that limit the total demand to be served by a route. In §6.4.3 and §6.4.4 we consider coupled systems, where in the first, we assume identical retailer holding cost rates, and in the second, we allow for general holding cost rates. In §6.4.6 we consider two-echelon systems with identical retailer holding cost rates.

Optimal policies for such models can be very complex. This complexity may make optimal policies difficult, if not impossible, to implement even if they could be computed efficiently. Instead the authors restrict themselves to a class of replenishment strategies denoted by Φ described below. A replenishment strategy in Φ specifies a collection of regions (clusters) covering all retailers. A retailer may belong to several regions (clusters) and in that case a specific fraction of its sales/operations is assigned to each of these regions. In particular, each time one of the retailers in a given region receives a delivery, this delivery is made by a vehicle which visits all other retailers in the region as well. Note that a large amount of flexibility is preserved within the class Φ by allowing regions to overlap. On the other hand, under a strategy in Φ, all regions are controlled independently of each other. However, the proposed region partitioning schemes generate regions in which only a *few* retailers are *split* among different (usually two) routes. Also note that under strategies in Φ deliveries are never coordinated among different regions and that retailers that are assigned to different regions are never served in a common route. As noted by Hall [46], in an optimal strategy any given retailer may be served with varying rather than constant combinations of other retailers. In the first two models (§6.4.3 and §6.4.4.3), the cost of the solution provided by the proposed region partitioning schemes asymptotically converge to a lower bound on the minimum cost among all strategies in Φ, but it is not yet known how close this is to the unrestricted minimum. In the last model (§6.4.6), the cost of the proposed region partitioning scheme is asymptotically at most 1.061 times the cost of the best possible policy in Φ. For a further discussion on this assumption see [7, 5].

We use here the same notation as in §6.3.3.3, and make the following assumptions:

- Retailer j's demand rate is an integer $1 \leq w_j \leq w_{\max} < +\infty$, and retailer j is replaced with w_j unit demand rate retailers.

- The unit demand rate retailers are indexed by the set $N = \{1, 2, \ldots, n\}$. The depot is denoted $\{0\}$.

- The vehicle capacity is Q (which may assume any positive value).

- r_i denotes retailer i's radial distance from the depot. The retailers are indexed $r_1 \leq r_2 \leq \ldots \leq r_n$.

- Retailers hold no initial stock.

- No backlogging is allowed.

- Inventory holding cost at each of the retailers is proportional to its time-average inventory level.

- The cost of a route is c, the fixed cost per route driven, plus a term proportional to the distance traveled. Without loss of generality, we set the cost per mile to one.

- The upper bound on the frequency with which a given route can be driven is f^*. (To ensure feasibility it is assumed that $f^* \geq 1/Q$).

- The upper bound on the sales volume per region is $M^{**} \geq 1$, i.e., no region can consist of more than M^{**} unit-demand retailers. We also assume $\min\{\lfloor f^*Q \rfloor, M^{**}\} < +\infty$.

- L denotes the number of regions (clusters) which may be fixed or variable.

- $\chi = \{X_1, X_2, \ldots, X_L\}$ denotes a partition of N, and X_ℓ ($\ell = 1, 2, \ldots, L$) denotes the set of retailers on route ℓ. Let also $m_\ell = |X_\ell|$ denote the number of retailers in region ℓ.

- Let T_ℓ denote the replenishment interval of region ℓ. Then T_ℓ satisfies the frequency and capacity constraints if and only if $1/f^* \leq T_\ell \leq Q/m_\ell$.

- $TSP^*(S)$ denotes the length of an optimal traveling salesman tour through $S \subseteq N^0$.

- $Z_n^*(\Phi)$ denotes the minimal long-run average cost among all strategies in the class Φ.

As a consequence of the frequency and sales volume constraints, we obtain that a region should consist of at most $M^* \overset{\text{def}}{=} \min\{\lfloor f^*Q \rfloor, M^{**}\} < +\infty$ retailers.

All region partitioning schemes proposed in this section are of complexity $O(n \log n)$. We note that in view of the assumption that $w_i \leq w_{\max}$, for each $i \in N$, n (the number of unit-demand retailers) in the complexity expression can be replaced with the number of original retailers.

6.4.3 Region Partitioning Schemes for One-Warehouse Multi-Retailer Coupled Systems

The first region partitioning schemes for the IRP were proposed by Anily and Federgruen [5] (see also [35]). They consider the one warehouse multi-retailer model with identical holding cost h at all (unit demand-rate) retailers. A policy of Φ consists of a partition of the retailers (N) into regions. For a *given* partition of N into regions, the remaining problem reduces to a separate (constrained) EOQ problem in each of the regions. It then follows that each strategy in Φ is dominated by one under which the retailers of each region receive deliveries at equidistant epochs which are therefore of *constant* size. (Note that an original retailer, now consisting of multiple unit-demand retailers, may be assigned to more than one region and therefore its delivery sizes may vary over time.)

Thus, for $\ell = 1, \ldots, L$, let Ω_ℓ be the total amount of product delivered to X_ℓ each time route ℓ is driven. As stated above, we consider here the basic model where backlogging is not allowed ([5] consider the backlogging option). The following Lemma was proven in [5].

Lemma 6.4.1 *(a)*

$$Z_n^*(\Phi) = \min\Big\{ \sum_{\ell=1}^{L} \min\Big[\frac{1}{2}h\Omega_\ell + \frac{m_\ell}{\Omega_\ell}(TSP^*(X_\ell^0) + c) : \chi = \{X_1, \ldots, X_L\}$$

$$\text{is a partition of } N, \ m_\ell \leq M^{**} \text{ and } \frac{m_\ell}{f^*} \leq \Omega_\ell \leq Q\Big]\Big\}. \quad (6.7)$$

(b) For a given route X_ℓ, the optimal delivery size is:

$$\Omega_\ell^* = \min\Big\{Q, \max\Big[\frac{m_\ell}{f^*}; \sqrt{2m_\ell(TSP^*(X_\ell^0) + c)/h}\Big]\Big\}. \quad (6.8)$$

(c) A partition $\chi = \{X_1, \ldots, X_L\}$ is feasible if $m_\ell \leq M^*$.

Substituting (6.8) into (6.6) we obtain:

$$Z_n^*(\Phi) = \min\Big\{ \sum_{\ell=1}^{L} f(TSP^*(X_\ell^0), m_\ell) : \chi = \{X_1, \ldots, X_L\}$$

$$\text{is a feasible partition of } N\Big\}$$

where

$$f(\vartheta, m) = \begin{cases} hm/(2f^*) + f^*(\vartheta + c) & \text{if } \vartheta + c < mh/(2f^{*2}), \\ \sqrt{2hm(\vartheta + c)} & \text{if } mh/(2f^{*2}) \leq \vartheta + c \leq Q^2 h/(2m), \\ hQ/2 + m(\vartheta + c)/Q & \text{otherwise.} \end{cases}$$

$$(6.9)$$

It follows that the problem of determining an optimal strategy within the class Φ reduces to the problem of partitioning the set N into L feasible routes with minimal total cost where the cost of a route with length ϑ and m unit demand rate retailers is given by $f(\vartheta, m)$. (A route is feasible if it covers no more than M^* unit demand retailers.) This class of routing problems with general route cost function is addressed in §6.3.3.3 (see also [4]). It is easily verified that f is nondecreasing in ϑ. Following the methodology in §6.3.3.3, we replace the route's length with a lower bound (i.e., defining $\underline{U}^{(1)}$ and $\underline{U}^{(2)}$). This results in two partitioning problems $\underline{P}^{(1)}$ and $\underline{P}^{(1)}$ as in (6.4) where feasibility of the partitions is stated as in Lemma 6.4.1, part *(c)*. Lemma 6.3.7 then implies that $\underline{Z}_n^{(1)} \leq \underline{Z}_n^{(2)} \leq Z_n^*(\Phi)$.

The results in [6] show that both $\underline{Z}_n^{(1)}$ and $\underline{Z}_n^{(2)}$ can be computed efficiently. Since f is nondecreasing, it follows from Theorem 7 (*ibid.*), that an optimal consecutive partition exists for $\underline{P}^{(2)}$ (recall $r_1 \leq r_2 \leq \cdots \leq r_n$). This allows us to solve (6.4) (for $j = 2$) by a shortest path algorithm on an acyclic graph, for details see [5]. Anily and Federgruen (in [5]) prove that $\underline{P}^{(1)}$ (in (6.4)) with $f(\vartheta, m)$ as defined in (6.9) is extremal which permits us to employ the simple Extremal Partitioning Algorithm (EPA) presented in §6.3.3.3 (this fails to be true of $\underline{P}^{(2)}$).

For $j = 1, 2$, let χ_j denote the optimal partition for $\underline{P}^{(j)}$, and define:

$$X_j^{(m)} = \{i : \chi_j \text{ assigns } i \text{ to a set of cardinality } m\}, \quad m = 1, \ldots, M^*.$$

As explained in §6.3.3.3, in view of the extremality of $\underline{P}^{(1)}$ the sets $X_1^{(m)}$, for $m = 1, \ldots, M^*$, define disjoint rings in the plane. This property does not necessarily hold for the sets $X_2^{(m)}$, for $m = 1, 2, \ldots, M^*$. The Combined Routing and Replenishment Strategies Algorithm (CRRSA) invokes the MCRP heuristic (as described in §6.3.3.3) on the rings $\{X_j^{(m)} : 1 \leq m \leq M^*\}$, for $j = 1, 2$, to determine the specific routes. We note, that as a result we may get two different heuristic solutions. Below we summarize the procedure for variable L when the MCRP heuristic is applied to the rings defined by $\underline{P}^{(1)}$. For that sake, we let $EOQ(\vartheta, m)$ be the optimal delivery size for a given region containing m unit-demand rate retailers when its retailers are replenished with a route of length ϑ, as described by (6.8).

CRRSA: The Combined Routing and Replenishment Strategies Algorithm

Step 1. Let $p_1 := \lfloor n/M^* \rfloor$; $p_2 := n - p_1 M^*$;

Step 2. Apply MCRP(M^*) to the set $\{p_2 + 1, \ldots, n\}$. Let $S_1, S_2, \ldots, S_{p_1}$ be an enumeration of the generated regions. Each of these has exactly M^* points. If $p_2 \neq 0$, let $S_0 = \{1, 2, \ldots, p_2\}$.

Step 3. For each region $\ell = 1, \ldots, p_1$, and if $p_2 \neq 0$ also for $\ell = 0$, let $\vartheta_\ell := TSP^*(S_\ell^0)$ and $\Omega_\ell := EOQ(\vartheta_\ell, |S_\ell|)$. The region S_ℓ is served once every $T_\ell := \Omega_\ell / |S_\ell|$ units of time and each time it is served the vehicle leaves the depot loaded with Ω_ℓ units.

Step 2 of CRRSA is of complexity $O(n \log n)$ while Step 3 is only $O(n)$. The asymptotic optimality of CRRSA (within the class Φ) follows now directly from Theorem 6.3.8.

6.4.4 One-Warehouse Multi-Retailer Systems with Retailer Specific Holding Costs

In this subsection, we consider generalizations of the One Warehouse Multi-Retailer System presented in §6.4.3.

6.4.4.1 Worst-Case Analysis of Direct Shipping Strategies.
Gallego and Simchi-Levi [40] characterize the effectiveness of the so-called *direct shipping* strategies. Such strategies have very simple vehicle routes: each route starts at the depot, delivers product to *one* retailer, and then returns to the depot.

For each retailer $i \in N$, we let w_i be the demand per unit of time (this may be non-integer), we let h_i be the holding cost per unit per unit of time and K_i the fixed ordering cost. Each time a vehicle is sent out to replenish inventory of a retailer $i \in N$, it incurs a cost proportional to the distance traveled, $2r_i$, plus the set up cost K_i. Note that this analysis does not allow for fixed cost per route but on the other hand allows retailer-specific set-up costs. For each $i \in N$, let $\hat{Q}_i \stackrel{\text{def}}{=} \sqrt{2K_iw_i/h_i}$.

Gallego and Simchi-Levi characterize the worst-case performance of a particular direct shipping policy called *fully loaded* direct shipping (FDS). The policy specifies that all vehicles dispatched are fully loaded. The author derive a simple lower bound $\underline{Z}_n$ on the optimal cost:

$$\underline{Z}_n = \sum_{i=1}^{n} \left(\sqrt{2K_ih_iw_i} + 2w_i/Q \right).$$

The first term in the parentheses is a lower bound on the average set-up and holding costs (as in the EOQ formula) and the second term is a lower bound on the average transportation cost due to retailer i.

The worst-case bound of [40] is parametrized by $\eta \stackrel{\text{def}}{=} \max\{\sqrt{2}, \max_{i=1}^{n}(Q/\hat{Q}_i)\}$. They show:

Theorem 6.4.2 *For any instance,* $Z_n^{\text{FDS}}/\underline{Z}_n^* \leq \frac{1}{2}(\eta + \eta^{-1})$.

The worst-case bound of FDS (relative to a lower bound on the optimal cost) is at most 1.061 whenever each retailer-specific economic lot size $(\hat{Q}_i)$ exceeds 71% of the vehicle capacity, that is, $Q/\sqrt{2}$. The worst-case ratio increases as the economic lot sizes decrease. For instance, if the minimum lot size is 50% (respectively, 33%) of the vehicle capacity, then the worst-case ratio is 1.25 (respectively, 1.68). Hall [47] explains why, in practice, direct shipping performs somewhat better than the bound indicated by Gallego and Simchi-Levi [40]: the authors underestimate, in their analysis, the unconstrained optimal replenishment quantities. They use the optimal EOQ for retailer i as given by $\hat{Q}_i = \sqrt{2K_iw_i/h_i}$, ignoring the transportation cost. In reality the policy would be implemented using the EOQ value $\sqrt{2(K_i + 2r_i)w_i/h_i}$, which may result in a lower value of η and hence better performance than suggested by this analysis.

6.4.4.2 Stationary Nested Joint Replenishment Policies. Viswanathan and Mathur [72] consider a one warehouse multi-retailer coupled system with a number of different products. Constant deterministic demands occur at n retailers for each of the m products. An *item* is defined as a combination of a product and a retailer, thus the system consists of nm items. The authors assume (*i*) a fixed cost per route driven; (*ii*) variable route costs that are proportional to the distance driven; (*iii*) retailer-specific fixed costs for placing an order (independent of the products ordered); (*iv*) product-specific fixed cost that have to be paid by each retailer placing an order for that product; and, (*v*) item-specific linear inventory holding cost for each item.

The heuristic generates a *stationary nested joint replenishment policy* (SNJRP) where all items are replenished in equi-distant intervals, and moreover, replenishment intervals of items that are ordered less frequently are integer multiples of replenishment intervals of items that are ordered at higher frequencies. In particular, they use *power-of-two* policies, where the replenishment intervals are power-of-two multiples of some base planning period. Power-of-two policies have been extensively used for multi-echelon inventory problems and for joint replenishment problems (see [55] for a survey).

The SNJRP heuristic does not split the service of any retailer to different routes, thus it can also be applied to the unsplit demand version of the problem. The SNJRP heuristic consists of two stages: in the first, the uncapacitated version is considered. All items then fall within a single cluster which is partitioned into *nested* sets $S_1 \subset S_2 \ldots \subset S_L = \{1, \ldots, nm\}$. The replenishment interval of S_ℓ is an integer (greater than one) power-of-two multiple of the replenishment interval of S_k, for $k < \ell$. For example, the items in S_1 are replenished once in a week, the items in S_2 are replenished once in two weeks, and the items in S_L are replenished once in 8 weeks. As a result, periodically, when S_L is replenished the vehicle visits <u>all</u> retailers in one single tour. Determining the best such policy requires solving a joint replenishment problem. This problem has been extensively studied, however, none of the existing solution methods deal with a general cost structure such as routing costs. The authors propose a greedy heuristic to solve this problem. In the second stage of the SNJRP heuristic, capacity restrictions are considered, and the cluster from the first stage is partitioned into a collection of disjoint clusters. A new cluster is created whenever the vehicle capacity is not sufficiently large to serve all items in the cluster. Each cluster is then partitioned into a sequence of nested sets, as explained above. In addition, a minimum total annual dollar demand restriction is imposed in order to avoid getting clusters that are too small when the vehicle's capacity is binding. I.e., when the capacity is binding for a given cluster, and the minimum sales volume is not achieved, items are added to the cluster and the respective replenishment interval is reduced. See [72] for details.

The authors compare, via a computational study, the effectiveness of the proposed SNJRP heuristic against the FDS policy (see §6.4.4.1 or [40]) and the two CRRSA heuristics of §6.4.3. Viswanathan and Mathur limit the test to single product problems since the other models are single commodity models. In view of the fact that asymptotic optimality of the CRRSA heuristics is guaranteed only when there exists a constant upper bound on the total demand of a region (a restriction that is not needed by the SNJRP heuristic) the authors implement the CRRSA heuristics by searching the best such bound. The traveling salesman tours in each of the regions created by the CRRSA heuristics were calculated by the *cheapest insertion* method (see [65]). A simplified version of the *cheapest insertion* method was used to calculate the traveling salesman tours for the SNJRP, as these have to be calculated repeatedly.

The computational study reveals that the SNJRP heuristic performs extremely well compared to FDS when the vehicle capacity is large. Also, the larger the fixed cost per route the better the performance of SNJRP compared to that of FDS. (Note that the FDS, when *implemented* with a fixed cost per route, no longer has the worst-case performance guarantee described in §6.4.4.1.) The SNJRP heuristic outperformed the CRRSA heuristics for most of the instances tested; however, for small capacities the performance of the CRRSA heuristics improved significantly as the problem size increased. For equal demand sizes the CRRSA heuristics out-performed SNJRP as the problem size increased. For the detailed computational study, see [72].

6.4.4.3 Average-Case Analysis of Region Partitioning Schemes.

Anily, see [9], generalizes the results of §6.4.3 to the case where holding costs are retailer-specific. Holding costs rates may vary substantially among the retailers in large distribution systems. Usually the holding cost is an increasing function of the item-value which tends to increase with the distance traveled. Other factors, such as proximity to a city-center also have an impact on the holding cost rates.

Anily considers the one-warehouse multiple-retailer system with unit demand rate retailers as in §6.4.3 with retailer-specific holding cost rates. All assumptions and notation of the model are stated in §6.4.2. The holding cost rate at retailer i is denoted h_i. These rates h_i are assumed to be uniformly bounded from below by a constant $h_{\min} > 0$. The cost of an optimal strategy in Φ is defined by:

$$
Z_n^*(\Phi) = \min \left\{ \sum_{\ell=1}^{L} \min \left[\frac{1}{2} \frac{\Omega_\ell \sum_{i \in X_\ell} h_i}{m_\ell} + \frac{m_\ell}{\Omega_\ell} (TSP^*(X_\ell^0) + c) : \right. \right.
$$

$$
\chi = \{X_1, \ldots, X_L\} \text{ is a partition of } N, \ m_\ell \leq M^{**} \text{ and}
$$

$$
\left. \left. \frac{m_\ell}{f^*} \leq \Omega_\ell \leq Q, \text{ for } \ell = 1, 2, \ldots, L \right] \right\}. \tag{6.10}
$$

Thus, similarly to Lemma 6.4.1, the optimal order quantity for a set of demand points X_ℓ is given by a constrained EOQ problem:

$$
\Omega_\ell^* = \min \left\{ Q, \max \left[\frac{m_\ell}{f^*} ; \sqrt{2m_\ell^2 (TSP^*(X_\ell^0) + c) / \sum_{i \in X_\ell} h_i} \right] \right\}. \tag{6.11}
$$

By substituting (6.11) into (6.9) we obtain an equivalent problem whose objective function is a minimization over partitions χ of a cost function that depends on certain problem parameters, subject to constraints: $m_\ell \leq M^*$, for $\ell = 1, \ldots, L$. The following notation will simplify the presentation:

$$
r_i^c = r_i + \frac{c}{2}; \quad \vartheta_\ell^c = TSP^*(X_\ell^0) + c; \quad R_\ell = \sum_{i \in X_\ell} r_i; \quad R_\ell^c = \sum_{i \in X_\ell} r_i^c; \quad \text{and } H_\ell = \sum_{i \in X_\ell} h_i.
$$

The next lemma provides an alternative expression for $Z_n^*(\Phi)$ and its proof is immediate from the above observations.

Lemma 6.4.3 *(a) The optimal average cost of a given region X_ℓ is $g(\vartheta_\ell^c, H_\ell, m_\ell)$ where*

$$g(\vartheta^c, H, m) = \begin{cases} H/(2f^*) + f^*\vartheta^c & \text{if } \vartheta^c/H \leq 1/(2f^{*^2}), \\ \sqrt{2H\vartheta^c} & \text{if } 1/(2f^{*^2}) \leq \vartheta^c/H \leq Q^2/(2m^2), \\ HQ/(2m) + m\vartheta^c/Q & \text{otherwise.} \end{cases}$$

(b) $\quad Z_n^*(\Phi) = \min\left\{ \sum_{\ell=1}^{L} g(\vartheta_\ell^c, H_\ell, m_\ell) \; : \chi = \{X_1, \ldots, X_L\} \text{ is a partition of } N \right.$

$$\left. \text{and } m_\ell \leq M^*, \text{ for } \ell = 1, \ldots, L \right\}.$$

The next theorem provides a lower bound (whose proof follows in view of $g(\vartheta^c, \cdot, \cdot)$ being strictly increasing in $\vartheta^c > 0$ and Lemma 6.3.1 with unit demands):

Theorem 6.4.4

$$Z_n^*(\Phi) \geq \underline{Z}_n \stackrel{\text{def}}{=} \min \quad \left\{ \sum_{\ell=1}^{L} g(2R_\ell^c/m_\ell, H_\ell, m_\ell) : \chi = \{X_1, \ldots, X_L\} \text{ is a} \right.$$

$$\left. \text{partition of } N \text{ and } m_\ell \leq M^*, \text{ for } \ell = 1, \ldots, L \right\}. \quad (6.12)$$

The lower bound, as given in (6.11), is a partitioning problem with a separable cost function. As mentioned in §6.3.3.3, general partitioning problems are NP-complete, see Chakravarty et al. [23]. Unfortunately, the partitioning problem defined by (6.11) is not known to be solvable in polynomial-time.

As an alternative, the author proposes a lower bound on $\underline{Z}_n$ which is obtained by relaxing the cardinality constraints. For that sake, we first define an equivalent set function G having as arguments R_ℓ^c, H_ℓ and m_ℓ. I.e, $G(R_\ell^c, H_\ell, m_\ell) \stackrel{\text{def}}{=} g(2R_\ell^c/m_\ell, H_\ell, m_\ell)$. Below, we write the function G explicitly.

$$G(R^c, H, m) = \begin{cases} H/(2f^*) + 2f^*R^c/m & \text{if } R^c/H < m/(4f^{*^2}), \\ 2\sqrt{HR^c/m} & \text{if } m/(4f^{*^2}) \leq R^c/H \leq Q^2/(4m), \\ HQ/(2m) + 2R^c/Q & \text{otherwise.} \end{cases}$$

As proven by Anily [9], the function G is nonincreasing in m. Let

$$G_{M^*}(R^c, H) \stackrel{\text{def}}{=} G(R^c, H, M^*),$$

substitute G_{M^*} in the definition of $\underline{Z}_n$ and relax the cardinality constraints to obtain:

$$\underline{Z}_n^{(M^*)} \stackrel{\text{def}}{=} \min\left\{ \sum_{\ell=1}^{L} G_{M^*}(R_\ell^c, H_\ell) : \chi = \{X_1, \ldots, X_L\} \text{ is a partition of } N \right\}.$$

Thus:

Theorem 6.4.5 $\underline{Z}_n^{(M^*)} \leq \underline{Z}_n \leq Z_n^*(\Phi)$.

Unlike $\underline{Z}_n$, $\underline{Z}_n^{(M^*)}$ and the corresponding partition can be easily obtained. In addition, as is shown below, $\underline{Z}_n^{(M^*)}$ is asymptotically accurate with the optimal average cost of all policies in Φ.

Lemma 6.4.6 *The partition* $\chi^* = \Big\{\{1\}, \{2\}, \ldots, \{n\}\Big\}$ *is optimal for* $\underline{Z}_n^{(M^*)}$.

In order to design the region partitioning scheme we classify the retailers into three (possibly empty) categories. Without loss of generality we assume the retailers are indexed in non-decreasing order of the r_i^c/h_i values, i.e., $r_1^c/h_1 \leq r_2^c/h_2 \leq \ldots \leq r_n^c/h_n$. Define the following three consecutive sets:

$$
\begin{aligned}
F &= \{i \in N : r_i^c/h_i < M^*/(4f^{*^2})\}, \\
S &= \{i \in N : M^*/(4f^{*^2}) \leq r_i^c/h_i \leq Q^2/(4M^*)\}, \\
C &= \{i \in N : Q^2/(4M^*) < r_i^c/h_i\}.
\end{aligned}
$$

The sets F, S and C are used as inputs to the heuristic proposed by [9]. Using MCRP(M^*), see §6.3.3.3, the heuristic partitions F and C separately into regions, each consisting of M^* retailers, except possibly one (per set) that contains less than M^* retailers. However, more care is needed with S as applying MCRP(M^*) to it does not result in an asymptotically optimal solution. For that sake, we let $\zeta_0 = \min_{i \in S}\{r_i^c/h_i\}$, $\zeta_m = \max_{i \in S}\{r_i^c/h_i\}$, and $\delta = \zeta_m/\zeta_0$. Anily [9] proposes a two-stage partitioning algorithm to S: first the points are aggregated into $O(\sqrt{|S|})$ clusters having similar r_i^c/h_i ratios, and second, MCRP(M^*) is applied on each of these clusters separately.

The MCRPGH Heuristic: The MCRP with General Holding Costs.

Step 0. Let $\Upsilon = \emptyset$ denote the index set of regions. Construct F, S and C as above.
Step 1. Apply MCRP(M^*) to F and C separately and add the regions obtained to Υ.
Step 2. Let $m := \lceil \sqrt{|S|} \rceil$; $\zeta_k := \delta^{k/m}\zeta_0$; for $k = 1, \ldots, m$;
Step 3. Define the clusters $S_k := \{i : i \in S$ and $\zeta_{k-1} < r_i^c/h_i \leq \zeta_k\}$, $k = 1, \ldots, m$; let $S_1 := S_1 \cup \{i : i \in S$ and $r_i^c/h_i = \zeta_0\}$;
Step 4. Apply MCRP(M^*) to each cluster S_k, $k = 1, \ldots, m$, separately. Add to Υ all regions obtained in this step that contain exactly M^* points. Let S_0 be the union of all regions created in this step that contain <u>less</u> than M^* points.
Step 5. Apply MCRP(M^*) to S_0 and add the regions obtained to Υ.
Step 6. For each region in Υ compute the optimal policy according to the EOQ formula.

We conclude this section with the following theorem (see the proof in [9]).

Theorem 6.4.7 *Under Condition* $\mathcal{A}$, *where retailers face identical demand rates:*

(a)

$$\lim_{n \to \infty} \frac{1}{n} \underline{Z}_n^{(M^*)} = E(G_{M^*}(r + c/2, h)) \quad \text{a.s.}$$

(b) If, in addition, $E(G_{M^*}(r + c/2, h)) > 0$ and $0 < h_{\min} \leq h_i \leq h_{\max}$, $\forall i$, for given constants $h_{\min}$ and $h_{\max}$ (and if $c = 0$ then $r_i \geq r_{\min} > 0$, $\forall i$, for given $r_{\min}$), then

$$\lim_{n \to \infty} Z_n^{MCRPGH} / \underline{Z}_n^{(M^*)} = 1 \quad \text{a.s.,}$$

i.e., the heuristic solution and the lower bound $\underline{Z}_n^{(M^*)}$ are asymptotically accurate.

In the general retailer demand-rate case (with split demand rates),

(c)

$$\lim_{n \to \infty} \frac{1}{n} \underline{Z}_n^{(M^*)} = E(w)E(G_{M^*}(r + c/2, h)) \quad \text{a.s.}$$

(d) If, in addition, r_i and h_i are as specified in (b), then (b) holds in this case as well.

6.4.5 An Average-Case Analysis of Fixed Partition Policies

We now consider an analysis of Fixed Partition (FP) policies, which were introduced in [18]. Under an FP policy one seeks a partition of the set of retailers N into a collection of regions (clusters) such that each retailer belongs to a *single* region, i.e., retailer demand rates are not split between different routes. Each region is then served using region-specific equi-distant intervals.

We present the analysis of Chan et al. [22]. Each retailer is associated with a demand D_i. We define a unit D (small enough) so that $D_i = w_i D$ and $w_i \geq 1$ is integer. As in §6.4.2, these authors assume frequency and capacity constraints on each of the routes, i.e., a retailer may be served at most f^* times per unit of time and the load on the vehicle should not exceed Q. Since sales volume constraints are not imposed here, both Q and f^* are assumed to be finite, i.e., the total sales volume served on a route cannot exceed f^*Q. In view of the frequency constraint, the minimum possible delivery quantity to retailer i is $w_i q$, where $q \stackrel{\text{def}}{=} D/f^*$. Thus, (since w_i is integer) in each visit a retailer can get a replenishment quantity no smaller than q. Since D can be chosen as small as desired, we can further assume that Q, the vehicle capacity, is an integer multiple of q, i.e., $\bar{b} \stackrel{\text{def}}{=} Q/q$ for some integer $\bar{b} \geq 1$. The cost structure is the same as in §6.4.2.

We assume the retailer multipliers $\{w_i\}$ are drawn from a distribution ν defined on $\{1, 2, \ldots, w_{\max}\}$ where $\bar{b} \geq w_{\max}$. Let γ denote the bin-packing constant associated with the distribution ν and bin capacity $\bar{b}$. It will be convenient to define $\beta \stackrel{\text{def}}{=} \bar{b}\gamma/E(w)$. If we consider an optimal packing of the multipliers $\{w_1, w_2, \ldots, w_n\}$ into bins of size $\bar{b}$ as $n \to \infty$, then β^{-1} is the average utilization of the bins. Recall that $\beta^{-1} \in (\frac{1}{2}, 1]$ since in an optimal packing all but possibly one bin is less than half full. Thus $\beta \in [1, 2)$.

We define the following: let Z_n^* (respectively, $Z_n^*(\text{ZIO})$, $Z_n^*(\text{FP})$) denote the infimum of the long-run average cost per unit time over all feasible policies (respectively, all zero-inventory ordering policies, all fixed partition policies).

Clearly, $Z_n^* \leq Z_n^*(\text{ZIO}) \leq Z_n^*(\text{FP})$ as a Fixed Partition policy is, in particular, a zero-inventory ordering policy. A problem of fundamental interest is finding relationships between these values. For example, zero-inventory ordering policies can be suboptimal (asymptotically), as demonstrated by the example of [22].

The following lower bound on the long-run average cost of any policy is from [22].

Lemma 6.4.8

$$Z_n^* \geq \underline{Z}_n \stackrel{\text{def}}{=} \sum_{i=1}^{n} w_i \left[\frac{D(2r_i + c)}{Q} + \frac{hq}{2} \right].$$

An FP policy is constructed which comes close to being optimal in a specific sense to be described below. In particular, the cost of this FP policy asymptotically (as $n \to \infty$) exceeds the lower bound $\underline{Z}_n$ by no more than $\sqrt{\beta}$. Since $\beta \in [1, 2]$, this implies that the cost of the FP policy is within a factor of $\sqrt{2}$ of the minimal policy's cost. Unfortunately, the construction is not polynomial in n (since it requires optimal solutions to bin-packing problems) and thus cannot serve directly as a heuristic.

We construct the FP policy using the following two-step procedure. In the first step, we partition the compact region A where the retailers are located, into subregions using a simple grid. The retailers in each such subregion are then partitioned into sets of retailers by solving the bin-packing problem defined by the multipliers of the retailers and bins of size $\bar{b}$. Each bin defines a region (or cluster) that is served by a separate vehicle. This packing ensures a feasible region in the case that the region is served at the maximum frequency allowed, i.e., f^* times per unit of time. Once the partition into bins is obtained, the actual replenishment intervals are calculated according to the region specifics and the vehicle's capacity, see below.

To construct the region partition, pick a fixed $\epsilon > 0$. Let $G(\epsilon)$ be an infinite grid of squares with edges parallel to the coordinate axes and side length $\frac{\epsilon}{\sqrt{2}}$. Let $\{A_1, A_2, \ldots, A_{t(\epsilon)}\}$ denote the subregions that intersect A and have $\mu(A_i) > 0$.

Let N_j, be the set of retailers in subregion A_j, for $j = 1, 2, \ldots, t(\epsilon)$. Let $\underline{r}_j$ be the distance from the warehouse to the closest point in A_j, for $j = 1, 2, \ldots, t(\epsilon)$.

Now group all the retailers in subregion A_j, $j = 1, 2 \ldots, t(\epsilon)$, into sets by solving the bin-packing problem defined by the multipliers $\{w_i : i \in N_j\}$ and bins of capacity $\bar{b}$. Consider a subregion A_j, and let $S \subseteq N$ denote a set of retailers that are packed together in a bin. The optimal reorder interval for S depends on $w(S) \stackrel{\text{def}}{=} \sum_{i \in S} w_i$ and the distance the subregion is from the depot. To determine this reorder interval, first calculate the value of t_S^o, the unconstrained minimizer of the approximate cost (the local cost of the route is ignored), i.e.,

$$t_S^\circ \stackrel{\text{def}}{=} \sqrt{\frac{2(2\underline{r}_j + c)}{w(S)Dh}}.$$

Since the vehicle is capacitated, the actual (feasible) reorder interval for S is:

$$t_S = \begin{cases} 1/f^*, & \text{if } 0 \le t_S^\circ < 1/f^* \\ t_S^\circ, & \text{if } 1/f^* \le t_S^\circ \le \frac{Q}{w(S)D} \\ \frac{Q}{w(S)D}, & \text{otherwise.} \end{cases}$$

That is, the reorder interval is chosen so that $q_S = w(S)Dt_S$ is the value of q achieving

$$\min_{w(S)\underline{q} \le q \le Q} \left\{ \frac{w(S)D(2\underline{r}_j + c)}{q} + \frac{hq}{2} \right\}.$$

Consequently, these reorder intervals satisfy the capacity as well as the frequency constraints.

For any set of retailers S, we use the following routing strategy. The vehicle travels from the warehouse to a retailer of S and then visits the other retailers in any order and then returns to the depot. The total distance traveled by this vehicle using this strategy is no more than $2\underline{r}_j + (|S| + 1)\epsilon$.

To analyze the effectiveness of this overall strategy, we need some additional notation. For each subregion A_j, let $b(N_j)$ be the number of bins used in an optimal solution to the bin-packing problem defined by the multipliers $\{w_i : i \in N_j\}$, $j = 1, 2, \ldots, t(\epsilon)$ and bins of capacity $\bar{b}$. Let $S_{j\ell}$, $\ell = 1, 2, \ldots, b(N_j)$ be the set of retailers in the ℓth bin of this optimal solution.

We are now able to derive an upper bound on the cost of the above defined FP policy. This bound depends on the number of bins (routes) $b(N_j)$ into which each of the regions is partitioned. For each subregion $j = 1, 2, \ldots, t(\epsilon)$, we express the number of routes generated in the subregion relative to the minimum possible number of routes, i.e., the number of routes required if the demand multipliers $\{w_i : i \in N_j\}$ (as item sizes) were allowed to be *split*; in other words, we express the number of routes employed by the FP policy in terms of

$$\beta_j \stackrel{\text{def}}{=} \frac{b(N_j) \cdot \bar{b}}{w(N_j)} \ge 1.$$

The main theorem of [22] is the following:

Theorem 6.4.9 *Under Condition $\mathcal{A}$,*

$$\varlimsup_{n \to \infty} \frac{Z_n^*(\text{ZIO})}{Z_n^*} \le \varlimsup_{n \to \infty} \frac{Z_n^*(\text{FP})}{Z_n^*} \le \sqrt{\beta} \le \sqrt{2} \qquad \text{a.s.}$$

Proof. Obviously $Z_n^*(\text{ZIO}) \le Z_n^*(\text{FP})$. Also the right-hand inequality follows since $\beta \le 2$. Therefore we need only show the middle inequality.

We bound $Z_n^*(\text{FP})$ from above by determining the cost of the particular FP policy constructed above. Using this policy, we derive:

$$
\begin{aligned}
Z_n^*(\text{FP}) \;\leq\; & \sum_{j=1}^{t(\epsilon)} \sum_{\ell=1}^{b(N_j)} \left[\frac{2\underline{r}_j + c + \epsilon(|S_{j\ell}| + 1)}{t_{S_{j\ell}}} + \frac{hw(S_{j\ell}) D t_{S_{j\ell}}}{2} \right] \\
\leq\; & \sum_{j=1}^{t(\epsilon)} \sum_{\ell=1}^{b(N_j)} \left[\frac{2\underline{r}_j + c}{t_{S_{j\ell}}} + \frac{hw(S_{j\ell}) D t_{S_{j\ell}}}{2} \right] + 2n\epsilon f^* \qquad (\text{since } t_{S_{j\ell}} \geq 1/f^*) \\
=\; & \sum_{j=1}^{t(\epsilon)} \sum_{\ell=1}^{b(N_j)} \min_{w(S_{j\ell})\underline{q} \leq q \leq Q} \left\{ \frac{w(S_{j\ell}) D(2\underline{r}_j + c)}{q} + \frac{hq}{2} \right\} + 2n\epsilon f^*.
\end{aligned}
$$

Now for any $r \geq 0$ and $b \in [1, \bar{b}]$, define the function

$$
F(b, r) = \min_{b\underline{q} \leq q \leq Q} \left\{ \frac{bD(2r + c)}{q} + \frac{hq}{2} \right\}.
$$

Then

$$
Z_n^*(\text{FP}) \leq \sum_{j=1}^{t(\epsilon)} \sum_{\ell=1}^{b(N_j)} F(w(S_{j\ell}), \underline{r}_j) + 2n\epsilon f^*. \tag{6.13}
$$

In [22], the authors prove that $F(b, \cdot)$ is concave in $b \in [1, \bar{b}]$, and therefore for every $j = 1, 2, \ldots, t(\epsilon)$, we have:

$$
\sum_{\ell=1}^{b(N_j)} F(w(S_{j\ell}), \underline{r}_j) \leq b(N_j) F\left(w(N_j)/b(N_j), \underline{r}_j \right) = b(N_j) F\left(\bar{b}/\beta_j, \underline{r}_j \right).
$$

Using this in (6.13),

$$
Z_n^*(\text{FP}) \leq \sum_{j=1}^{t(\epsilon)} b(N_j) F\left(\bar{b}/\beta_j, \underline{r}_j \right) + 2n\epsilon f^*.
$$

In [22] the authors also show that $F(b, r) \leq F(\bar{b}, r)\sqrt{b/\bar{b}}$, for any $b \in [1, \bar{b}]$, and that $F(\bar{b}, r) = \left[f^*(2r + c) + hQ/2 \right]$. Therefore:

$$
\begin{aligned}
Z_n^*(\text{FP}) \;\leq\; & \sum_{j=1}^{t(\epsilon)} b(N_j) F(\bar{b}, \underline{r}_j) \sqrt{\frac{1}{\beta_j}} + 2n\epsilon f^* \\
=\; & \sum_{j=1}^{t(\epsilon)} w(N_j) \sqrt{\beta_j} \frac{F(\bar{b}, \underline{r}_j)}{\bar{b}} + 2n\epsilon f^* \qquad (\text{by the def. of } \beta_j) \\
=\; & \sum_{j=1}^{t(\epsilon)} \sqrt{\beta_j} \sum_{i \in N_j} w_i \frac{F(\bar{b}, \underline{r}_j)}{\bar{b}} + 2n\epsilon f^*
\end{aligned}
$$

$$\leq \sum_{j=1}^{t(\epsilon)} \sqrt{\beta_j} \sum_{i \in N_j} w_i \frac{F(\bar{b}, r_i)}{\bar{b}} + 2n\epsilon f^* \qquad \text{(since } \underline{r}_j \leq r_i)$$

$$= \sum_{j=1}^{t(\epsilon)} \sqrt{\beta_j} \sum_{i \in N_j} \left[\frac{w_i f^*(2r_i + c)}{\bar{b}} + \frac{hQw_i}{2\bar{b}} \right] + 2n\epsilon f^*$$

$$= \sum_{j=1}^{t(\epsilon)} \sqrt{\beta_j} \sum_{i \in N_j} w_i \left[\frac{D(2r_i + c)}{Q} + \frac{h\underline{q}}{2} \right] + 2n\epsilon f^*,$$

where the last equation follows from the definitions of f^* and $\underline{q}$. Observe from Lemma 6.4.8 that $Z_n^* \geq n\left[\frac{Dc}{Q} + \frac{h\underline{q}}{2}\right]$, thus:

$$\frac{Z_n^*(\text{FP})}{Z_n^*} \leq \frac{\sum_{j=1}^{t(\epsilon)} \sqrt{\beta_j} \sum_{i \in N_j} w_i[D(2r_i + c)/Q + h\underline{q}/2]}{\sum_{j=1}^{t(\epsilon)} \sum_{i \in N_j} w_i[D(2r_i + c)/Q + h\underline{q}/2]} + \frac{2\epsilon f^*}{Dc/Q + h\underline{q}/2}.$$

For some $K \geq 0$, almost surely:

$$\varlimsup_{n \to \infty} \frac{Z_n^*(\text{FP})}{Z_n^*} \leq \varlimsup_{n \to \infty} \left\{ \max_{j=1}^{t(\epsilon)} \sqrt{\beta_j} \right\} + K\epsilon = \max_{j=1}^{t(\epsilon)} \left\{ \varlimsup_{n \to \infty} \sqrt{\beta_j} \right\} + K\epsilon.$$

Now, from our assumptions, for all $j = 1, 2, \ldots, t(\epsilon)$,

$$\lim_{n \to \infty} \frac{b(N_j)}{|N_j|} = \lim_{|N_j| \to \infty} \frac{b(N_j)}{|N_j|} = \gamma, \quad \text{a.s.}$$

while by the Strong Law of Large Numbers,

$$\lim_{n \to \infty} \frac{w(N_j)}{|N_j|} = E(w) > 0 \quad \text{a.s.}$$

We conclude that for all $j = 1, 2, \ldots, t(\epsilon)$,

$$\lim_{n \to \infty} \beta_j = \lim_{n \to \infty} \bar{b} \, \frac{b(N_j)}{|N_j|} \frac{|N_j|}{w(N_j)} = \frac{\bar{b}\gamma}{E(w)} = \beta,$$

and obtain the desired inequality. ∎

Following directly from Theorem 6.4.9, we have:

Corollary 6.4.10 *Under Condition* $\mathcal{A}$, *if the distribution of retailer multipliers* ν *allows for* perfect *packing, i.e.,* $\beta = 1$ *(or equivalently,* $\gamma = E(w)/\bar{b}$*):*

$$\lim_{n \to \infty} \frac{Z_n^*(\text{ZIO})}{Z_n^*} = \lim_{n \to \infty} \frac{Z_n^*(\text{FP})}{Z_n^*} = 1 \quad \text{a.s.,}$$

i.e., the FP policy is asymptotically optimal.

Since the FP policy described here cannot be constructed in polynomial time, Chan et al. [22] also present a heuristic, based on ideas similar to the LBH of [18]. The heuristic is shown to perform well on a set of test problems.

6.4.6 Two-Echelon Distribution Systems with Central Inventories

In all models discussed so far in this section we have assumed that the depot serves as a mere coordinator of the replenishment process or alternatively as a transshipment point at which no inventory is kept. Here we extend the analysis to the case where central inventories can be kept at the depot (warehouse), i.e., the warehouse places orders with an outside supplier. As a consequence, the above problems are compounded by that of determining a replenishment strategy for the warehouse, optimally coordinated with that of each retailer and synchronized with the transportation schedules.

6.4.6.1 Average Case Analysis of Region Partitioning Schemes.

Anily and Federgruen [8] extend the analysis of [5] presented in §6.4.3, to two-echelon systems (see also [35]). Following the assumptions and notation of §6.4.2 they assume, in addition, a fixed cost for placing an order from the outside supplier. Inventory holding costs are identical at the retailers but is different at the warehouse.

The classical two-echelon one-warehouse multi-retailer model (from Roundy [66]) assumes that each retailer is served on an *individual* basis, rather than deliveries being combined into efficient routes. Even with individual and unco-ordinated deliveries, optimal policies can be too complex to be implementable. The combinatorial nature of the routing costs compounds this complexity.

We present here an $O(n \log n)$ heuristic solution, based on region partitioning, that asymptotically comes within 6% of the minimal average system-wide cost under Φ, $Z_n^*(\Phi)$. This solution may combine deliveries to a number of retailers in each route.

Let h^+ be the retailers holding cost rate; h_0 the warehouse holding cost rate; and $h = h^+ - h_0$ the *echelon holding cost* rate. We assume that $h > 0$. We let T_0 denote the replenishment interval of the warehouse. We need the following parameter restrictions for the analysis: (i) Qf^* is integer (or $+\infty$), and (ii) if $M^{**} < Qf^* < \infty$ then Qf^* is some power-of-two multiple of M^{**}.

For a given partition $\chi = \{X_1, \ldots, X_L\}$ of N, the remaining problem reduces to identifying an optimal inventory replenishment strategy in a classical one-warehouse L-retailer system in which each set X_ℓ plays the role of a single "super retailer" with demand rate m_ℓ and a fixed procurement cost $TSP^*(X_\ell^0) + c$. No method is known for computing an optimal strategy, even in the uncapacitated version of this problem, but Roundy [66] has shown that for the latter a *close to optimal* simple strategy may be found of the following power-of-two structure: the warehouse (region ℓ) replenishes its inventory every T_0 (T_ℓ) time units when its inventory reaches zero (for $\ell = 1, \ldots, L$); also $(T_0, T_1, \ldots, T_L)$ are power-of-two multiples (greater, equal or less than 1) of a base planning period T^B. A power-of-two policy exists whose cost comes within 6% or 2% of the optimal cost depending upon whether the base planning period is fixed or variable.

For a given partition $\chi = \{X_1, \ldots, X_L\}$ of N, and a given power-of-two policy $\vec{T} = (T_0, T_1, \ldots, T_L)$, we denote the corresponding *average cost* by:

$$C_\chi(\vec{T}) = K_0/T_0 + \sum_{\ell=1}^{L} D_{T_0}(T_\ell, TSP^*(X_\ell^0), m_\ell). \qquad (6.14)$$

Here $D_{T_0}(T_\ell, TSP^*(X_\ell^0), m_\ell)$ represents the average cost per unit time of replenishing the ℓth region. This includes the transportation costs as well as the inventory carrying costs on *(i)* the retailers inventories in the region and *(ii)* the part of the warehouse inventory which is destined to be shipped to region ℓ. As in Roundy [66]:

$$D_{T_0}(T_\ell, TSP^*(X_\ell^0), m_\ell) = (TSP^*(X_\ell^0) + c)/T_\ell + m_\ell \left(\frac{h}{2} T_\ell + \frac{h_0}{2} \max(T_0, T_\ell) \right). \qquad (6.15)$$

The following lemma, proven in [8], provides a lower bound on the minimum long-run average costs over *all* feasible policies that employ the partition χ. The lower bound is obtained by relaxing the power-of-two constraints. This relaxation is not straightforward as (6.15) represents the average cost due to X_ℓ only for the case where T_0/T_ℓ or T_ℓ/T_0 is integer. For the case where neither upper nor lower bounds are imposed on the replenishment intervals, this lower bound follows directly from Roundy [66]. Anily and Federgruen [8] generalize the lemma to the case where such bounds are imposed.

Lemma 6.4.11 *For any given partition* $\chi = \{X_1, \ldots, X_L\}$ *of* N, $U(\chi) \overset{\text{def}}{=}$ $\inf\{C_\chi(\vec{T}) : T_0 > 0 \text{ and } 1/f^* \le T_\ell \le Q/m_\ell, \ \ell = 1, \ldots, L\}$, *is a lower bound for the minimum long-run average costs over all feasible policies that employ the partition* χ.

We now describe how $U(\chi)$ may be evaluated. For any partition $\chi = \{X_1, \ldots, X_L\}$ of N and $T_0 > 0$ we define

$$U_{T_0}(\chi) = K_0/T_0 + \sum_{\ell=1}^{L} f_{T_0}(TSP^*(X_\ell^0), m_\ell) \qquad (6.16)$$

where $f_{T_0}(\vartheta_\ell, m_\ell) = \inf_{1/f^* \le T_\ell \le Q/m_\ell} D_{T_0}(T_\ell, \vartheta_\ell, m_\ell)$. Note that D_{T_0} is convex in T_ℓ so $f_{T_0}(\vartheta, m)$ is easy to evaluate in closed form, see [8]. We note that $f_{T_0}(\vartheta, m)$ is convex in T_0, for given ϑ and m, thus $\inf_{T_0>0} U_{T_0}(\chi)$ is also easily obtained. We conclude from $U(\chi) = \inf_{T_0>0} U_{T_0}(\chi)$, (6.14), Lemma 6.4.11 and (6.16) that a lower bound on Z_n^* is given by:

$$\begin{aligned}
\underline{Z}_n &= \min\left\{ U(\chi) : \chi = \{X_1, \ldots, X_L\} \text{ is a partition of } N \text{ and } m_\ell \le M^* \right\} \\
&= \inf_{T_0>0} \left\{ K_0/T_0 + \min\left[\sum_{\ell=1}^{L} f_{T_0}(TSP^*(X_\ell^0), m_\ell) : \chi = \{X_1, \ldots, X_L\} \right. \right.
\end{aligned}$$

$$\left. \left. \text{is a partition of } N \text{ and } m_\ell \le M^* \right] \right\}. \qquad (6.17)$$

It follows that for any $T_0 > 0$, the minimization problem within the curled brackets in (6.16) reduces to the problem of partitioning the set N into L routes with minimal total cost, where the cost of a route with length ϑ and $m \leq M^*$ retailers is given by $f_{T_0}(\vartheta, m)$. This class of routing problems with general route cost function has been addressed in [4] and described in §6.3.3.3 and in the remainder we draw on these results. In view of the fact that $f_{T_0}(\vartheta, m)$ is nondecreasing in ϑ we can obtain a further lower bound by replacing $TSP^*(X_\ell^0)$ in (6.16) by twice the average radial distance of the points in the region, see Lemma 6.3.1 with unit demands. For that sake, we define for any $T_0 > 0$ the partitioning problem:

$$\underline{P}_{T_0}^{(1)} : \underline{Z}_n^{(1)}(T_0) = K_0/T_0 + \min \left\{ \sum_{\ell=1}^{L} f_{T_0}\left(\frac{2}{m_\ell} \sum_{i \in X_\ell} r_i, m_\ell\right) : \right.$$

$$\left. \chi = \{X_1, \ldots, X_L\} \text{ partitions } N \text{ and } m_\ell \leq M^* \right\},$$

and let $\underline{Z}_n^{(1)} \overset{\text{def}}{=} \inf_{T_0 > 0} \underline{Z}_n^{(1)}(T_0)$. Clearly, $Z_n^* \geq \underline{Z}_n \geq \underline{Z}_n^{(1)}$.

Note that the partitioning problem $\underline{P}_{T_0}^{(1)}$ depends on the parameter $T_0 > 0$. Thus, even if $\underline{P}_{T_0}^{(1)}$ falls in the small class of polynomially solvable partitioning problems, the optimal partition may be expected to vary with T_0, so that $\underline{Z}_n^{(1)}(T_0)$ as the finite minimum of convex functions may fail to be convex and the minimization over T_0 required to evaluate $\underline{Z}_n^{(1)}$ can be expected to be cumbersome. The next theorem is proved in [8].

Theorem 6.4.12 $\underline{P}_{T_0}^{(1)}$ *is extremal for all* $T_0 > 0$; *the same partition* χ^* *optimizes* $\underline{P}_{T_0}^{(1)}$ *for all* $T_0 > 0$.

As mentioned in §6.3.3.3, the partition χ^* which optimizes $\underline{P}_{T_0}^{(1)}$ for all $T_0 > 0$ is easy to construct by the EPA algorithm given in §6.3.3.3.

It follows from Theorem 6.4.12 that

$$\underline{Z}_n^{(1)} = \inf_{T_0 > 0} \underline{Z}_n^{(1)}(T_0) = \inf_{T_0 > 0} \left\{ K_0/T_0 + \sum_{\ell=1}^{L} f_{T_0}\left(\frac{2}{|X_\ell^*|} \sum_{i \in X_\ell^*} r_i, |X_\ell^*|\right) \right\}$$

and in view of the function $f_{T_0}(\cdot, \cdot)$:

$$\underline{Z}_n^{(1)}(T_0) = K_0/T_0 + \sum_{\ell=1}^{L} \left[\frac{\alpha_\ell(T_0)}{T_0} + \beta_\ell(T_0) + \gamma_\ell(T_0)T_0 \right]$$

where $\alpha_\ell(T_0)$, $\beta_\ell(T_0)$ and $\gamma_\ell(T_0)$ are piecewise constant functions for all $\ell = 1, \ldots, L$. It is easy to verify that the functions $\left[\frac{\alpha_\ell(T_0)}{T_0} + \beta_\ell(T_0) + \gamma_\ell(T_0)T_0\right]$, for $\ell = 1, \ldots, L$, are convex and continuously differentiable in $T_0 > 0$ except possibly at three (four) specific values of T_0 for L variable (fixed) as identified in [8]. Thus, $\underline{Z}_n^{(1)}(T_0)$ is strictly convex and continuously differentiable in T_0

everywhere, except possibly for the few points mentioned above. In addition, $\underline{Z}_n^{(1)}(T_0)$ is of the form $\alpha(T_0)/T_0 + \beta(T_0) + \gamma(T_0)T_0$ where the functions α, β and γ are piecewise constant functions changing values only when T_0 crosses one of at most $2n + 3$ ($2n + 4$ if L is fixed) specific values. Thus, $\underline{Z}_n^{(1)}$ is achieved at the unique point T_0^* where either $d\underline{Z}_n^{(1)}(T_0)/dT_0 = 0$ (i.e., $T_0^* = \left[(K_0 + \sum_{\ell=1}^{L} \alpha_\ell(T_0^*))/\sum_{\ell=1}^{L} \gamma_\ell(T_0^*)\right]^{1/2}$ or T_0^* is one of the few points where the function is not differentiable. In [8], an $O(n \log n)$ algorithm is presented for the minimization of $\underline{Z}_n^{(1)}(T_0)$. Queyranne [62] proposes an $O(n)$-time algorithm based on an alternative linear-time median finding procedure.

As in §6.3.3.3 we define the rings

$$X^{(m)} = \{i \in N : \chi^* \text{ assigns } i \text{ to a set of cardinality } m\}$$

for $m = 1, \ldots, M^*$. For simplicity we describe here the procedure for a variable fleet size. In this case, there are at most two non-empty rings $X^{(p_2)} = \{1, \ldots, p_2\}$ and $X^{(M^*)} = \{p_2+1, \ldots, n\}$ for $p_2 = (n) \bmod M^*$. The MCRP(M^*) algorithm, proposed in §6.3.3.3 is then applied on $X^{(M^*)}$. The MCRP generates a collection of $\lfloor n/M^* \rfloor$ regions each consisting of M^* retailers. To this list of regions we add, if $p_2 \neq 0$, the region $\{1, \ldots, p_2\}$ which will be served by a separate vehicle. Let $\chi^H = \{X_1^H, \ldots, X_L^H\}$ be the resulting collection of regions. Given this set of regions, as explained above, the system reduces to a one-warehouse multi-retailer system, with each region acting as a (super) retailer. An optimal power-of-two policy for this system is easy to determine following the procedure in Roundy [66] with appropriate modifications to capacitated systems. The first step in determining such a power-of-two policy is the determination of a (not necessarily power-of-two) vector $\vec{\tau}^H$ that achieves $\min_{\vec{\tau}>0} C_{\chi^H}(\vec{\tau})$. This can be done with the same $O(L \log L)$ or $O(L)$ procedure required to compute $\underline{Z}_n^{(1)} = \min_{\vec{T}>0} C_{\chi^*}(\vec{T})$.

Next a rounding procedure described in [8] is employed to round $\vec{\tau}^H$ to a feasible power-of-two policy $\vec{T}^H$ with respect to the base planning period $T^B = 1/f^*$. Let $Z_n^H \stackrel{\text{def}}{=} C_{\chi^H}(\vec{T}^H)$. The entire procedure required to obtain the lower bound and the heuristic solution is of complexity $O(n \log n)$. The following theorem, proved in [8], characterizes the asymptotic worst-case performance of the proposed heuristic.

Theorem 6.4.13 (a) $\underline{Z}_n^{(1)} \leq Z_n^* \leq Z_n^H$.
(b) Under Condition $\mathcal{A}$, where retailers face identical demand rates such that

$$E\left[\min_{m=1,\ldots,M^*} f_{T_0}(2r, m)\right] > 0, \qquad \text{for } T_0 > 0,$$

then:

$$\varlimsup_{n \to \infty} \frac{Z_n^H}{\underline{Z}_n^{(1)}} \leq 1.061, \quad \text{a.s.}$$

Remark: For uncapacitated systems, i.e., when $Q = f^* = \infty$ and thus only sales volume constraints are imposed on the routes (see the parameter restrictions), it is possible to guarantee that the heuristic solution is within 2% of optimality by optimizing over the base planning period T^B. This optimization can be performed in $O(n)$ time, see [66].

6.4.6.2 Uncapacitated Systems - Worst-Case Analysis of Nested Policies.

Herer and Roundy [49] investigate two-echelon systems where the cost of serving a subset $S \subseteq N$ of retailers on a single route consists of two components: the first component is a *nonnegative monotone submodular* function of S, and the other is a per-mile charge (which we assume to be one) times the distance traveled. Warehouse and retailer-specific holding cost rates are assumed. Backlogging is not allowed. No restrictions are imposed on the vehicles capacity (i.e., $Q = \infty$), the frequency routes are driven (i.e., $f^* = \infty$) or the sales volume per route ($M^{**} = \infty$). This means that the total demand served on a route is not bounded by a constant which was assumed in all regional partitioning schemes considered so far. The goal is to minimize long-run average cost.

The only known structural property of optimal solutions for this system is zero-inventory ordering (see [34, 49]). Power-of-two policies are proposed for this uncapacitated model, where all retailers scheduled for replenishment at a certain period are served on a single route. As a result, the proposed heuristic provides a nested sequence of N^0, similarly to the heuristic proposed by Vishwanathan and Mathur [72] for coupled systems, see §6.4.4.2.

A *submodular* set function G is a function which maps subsets of N^0 to $\mathbf{R}$ and satisfies the following inequality:

$$G(S_1 \cup S_2 \cup S_3) - G(S_1 \cup S_2) \leq G(S_1 \cup S_3) - G(S_1) \text{ for all disjoint } S_1, S_2, S_3 \subseteq N^0.$$

By *nonnegative monotone* we mean $G(S_1 \cup S_2) \geq G(S_1) \geq 0$ for all $S_1, S_2 \subseteq N^0$.

The joint replenishment problem with nonnegative monotone submodular costs was extensively studied, see [61, 36, 34, 55]. Federgruen et al. [34] show that the cost of an optimal power-of two policy is no more than 1.021 (or 1.061 when the base period is fixed) a lower-bound; the lower bound and the power-of-two policy can be computed in polynomial time given an oracle that can evaluate the ordering cost for any set S.

The joint ordering cost $K(S)$ for $S \subseteq N^0$ assumed in [49] is $K(S) = G(S) + TSP^*(S^0)$ where G is a nonnegative monotone submodular function. If $TSP^*(S)$ for $S \subseteq N^0$ were nonnegative monotone submodular and could be evaluated in polynomial-time, then the problem could be solved within the bounds described above. However, unless $\mathcal{P} = \mathcal{NP}$, $TSP^*(S)$ cannot be evaluated in polynomial time. Even if $\mathcal{P} = \mathcal{NP}$, the set function $TSP^*(S)$ is not submodular, see [5]. However, [49] show that the function TSP^* is *almost* submodular in the following sense. We say a set function F is η-*close* to being nonnegative monotone submodular, if there exists a nonnegative monotone submodular function $\overline{F}$ such that:

$$F(S) \leq \overline{F}(S) \leq \eta F(S), \text{ for all } S \subseteq N^0. \tag{6.18}$$

The smallest η that satisfies (6.18) determines how close $F(S)$ is to the submodular function.

For a given nonnegative monotone submodular function $\overline{TSP}$ that satisfies (6.18) with $F = TSP^*$, define two new nonnegative monotone submodular functions $\hat{K}^\ell(S) = G(S) + \overline{TSP}(S^0)/\eta$ and $K^u(S) = G(S) + \overline{TSP}(S^0)$ which are respectively, lower and upper bounds on the order cost function $K(S)$. We define ξ to be the smallest value such that $K^u(S) \leq \eta\xi\hat{K}^\ell(S)$ for all $S \subseteq N$. Note that $\eta \geq \eta\xi \geq 1$, and let $K^\ell(S) \overset{\text{def}}{=} K^u(S)/(\eta\xi)$ which is also a lower bound on the order cost. [49] show that a smaller value of $\eta\xi$ (meaning that the submodular component is increasingly dominating the order cost) results in tighter bounds for the proposed heuristic.

In the sequel we refer to the K^ℓ-*system* (K^u-*system*) as the system with an ordering cost for $S \subseteq N$ of $K^\ell(S)$ ($K^u(S)$). We denote the optimal cost of the K^ℓ-system (K^u-system) by Z^*_{n,K^ℓ} (Z^*_{n,K^u}). In addition, let T^B denote the base planning period. The optimal solution Z^*_n is related to the optimal solution of these two systems as follows (from [49]):

Theorem 6.4.14

$$Z^*_{n,K^\ell} \leq Z^*_n \leq Z^*_{n,K^u}.$$

Moreover,

$$Z^*_{n,K^u}/Z^*_{n,K^\ell} \leq 1.021\sqrt{\eta\xi} \qquad \textit{if } T^B \textit{ is variable, and}$$

$$Z^*_{n,K^u}/Z^*_{n,K^\ell} \leq 1.061\sqrt{\eta\xi} \qquad \textit{if } T^B \textit{ is fixed.}$$

In view of the fact that the K^ℓ- and K^u-systems are associated with an ordering cost that is nonnegative monotone submodular, a lower bound for Z^*_{n,K^ℓ} and an optimal power-of-two policy for the K^u-system can be obtained in a polynomial time. Herer and Roundy [49], propose to use this optimal power-of-two policy as the heuristic solution for the actual system. We let Z^H_n denote the average cost of this solution when applied on the actual system.

Corollary 6.4.15

$$Z^H_n/Z^*_n \leq 1.021\sqrt{\eta\xi} \qquad \textit{if } T^B \textit{ is variable, and}$$

$$Z^H_n/Z^*_n \leq 1.061\sqrt{\eta\xi} \qquad \textit{if } T^B \textit{ is fixed.}$$

Herer and Roundy [49] propose four polynomial-time methods for approximating $TSP^*(S^0)$ in the Euclidean plane, by a nonnegative monotone submodular function. The $\eta\xi$ values of these approximation methods were investigated theoretically and their average values were investigated computationally. In all four cases the $\sqrt{\eta\xi}$ value is an increasing function of n, where the best one is $O(\log n)$. For details see [48].

References

[1] Altinkemer, K. and B. Gavish. Heuristics for Unequal Weight Delivery Problems with a Fixed Error Guarantee. *Operations Research Letters* 6:149–158, 1987.

[2] Altinkemer, K. and B. Gavish. Heuristics for Delivery Problems with Constant Error Guarantees. *Transportation Science* 24:294–297, 1990.

[3] Altinkemer, K. and B. Gavish. Parallel Savings Based Heuristics for the Delivery Problem. *Operations Research* 39:456–469, 1991.

[4] Anily, S. and A. Federgruen. A Class of Euclidean Routing Problems with General Route Cost Functions. *Mathematics of Operations Research* 15:268–285, 1990.

[5] Anily, S. and A. Federgruen. One Warehouse Multiple Retailer Systems with Vehicle Routing Costs. *Management Science* 36:92–114, 1990.

[6] Anily, S. and A. Federgruen. Structured Partitioning Problems. *Operations Research* 39:130–149, 1991.

[7] Anily, S. and A. Federgruen. Rejoinder to "Comments on One-Warehouse Multiple Retailer Systems with Vehicle Routing Costs". *Management Science* 37:1497–1499, 1991.

[8] Anily, S. and A. Federgruen. Two-echelon distribution systems with vehicle routing and central inventories. In the special issue on Stochastic and Dynamic Models in Transportation, M. Dror (ed.), *Operations Research* 41:37–47, 1993.

[9] Anily, S. The general multi-retailer EOQ problem with vehicle routing costs. *European Journal of Operational Research* 79:451–473, 1994.

[10] Anily, S. The Vehicle Routing Problem with Delivery and Backhaul Options. *Naval Research Logistics* 43:415–434, 1996.

192

[11] Beardwood, J., J. L. Halton and J.M. Hammerseley. The Shortest Path Through Many Points. *Proc. Cambridge Phil. Soc.* 55:299–327, 1959.

[12] Beasley, J. Route First–Cluster Second Methods for Vehicle Routing. *Omega* 11:403–408, 1983.

[13] Bell, W., L. M. Dalberto, M. L. Fisher, A. J. Greenfield, R. Jaikumar, P. Kedia, R. G. Mack and P. J. Plutzman. Improving the Distribution of Industrial Gases with an On-line Computerized Routing and Scheduling Optimizer. *Interfaces* 13:4–23, 1983.

[14] Bienstock, D., J. Bramel and D. Simchi-Levi. A Probabilistic Analysis of Tour Partitioning Heuristics for the Capacitated Vehicle Routing Problem with Unsplit Demands. *Mathematics of Operations Research* 18:786–802, 1993.

[15] Blumenfeld, D. E., L. D. Burns and J. D. Diltz. Analyzing Trade-offs Between Transportation, Inventory and Production Costs on Freight Networks. *Transportation Research* 19B:361–380, 1985.

[16] Bodin, L., B. L. Golden, A. Assad and M. Ball. The State-of-the-Art in the Routing and Scheduling of Vehicles and Crews. *Computers and Operations Research* 10:63–212, 1983.

[17] Bramel, J., E. G. Coffman, Jr., P. Shor and D. Simchi-Levi. Probabilistic Analysis of Algorithms for the Capacitated Vehicle Routing Problem with Unsplit Demands. *Operations Research* 40:1095–1106, 1991.

[18] Bramel, J. and D. Simchi-Levi. A Location Based Heuristic for General Routing Problems. *Operations Research* 43:649–660, 1995.

[19] Bramel, J. and D. Simchi-Levi. The Logic of Logistics: Theory, Algorithms and Applications for Logistics Management. Springer-Verlag, New York, 1997.

[20] Burns, L. D., R. W. Hall, D. E. Blumenfeld and C. F. Daganzo. Distribution Strategies that Minimize Transportation and Inventory Costs. *Operations Research* 33:469–490, 1985.

[21] Casco, D. O., B. L. Golden and E. A. Wasil. Vehicle Routing with backhauls: Models, Algorithms, and Case Studies. *Studies in Management Science and Systems/Vehicle Routing Methods and Studies*, North-Holland, Amsterdam, pages 127–147, 1988.

[22] Chan, L. M. A., A. Federgruen and D. Simchi-Levi. Probabilistic Analyses and Practical Algorithms for Inventory-Routing Models. *Operations Research* 46:96–106, 1998.

[23] Chakravarty, A. K., J. B. Orlin and U. G. Rothblum. A Partitioning Problem with Additive Objective with an Application to Optimal Inventory

Grouping for Joint Replenishment. *Operations Research* 30:1018–1020, 1982.

[24] Chakravarty, A. K., J. B. Orlin and U. G. Rothblum. Consecutive Optimizers for a Partitioning Problem with Applications to Optimal Inventory Groupings for Joint Replenishment. *Operations Research* 33:820–834, 1985.

[25] Christofides, N. Worst-Case Analysis of a New Heuristic for the Traveling Salesman Problem. Report 388, Graduate School of Industrial Administration, Carnegie-Mellon University, Pittsburgh, PA, 1976.

[26] Christofides, N. Vehicle Routing. Lawler, E. L., J. K. Lenstra, A. H. G. Rinnooy Kan and D. B. Shmoys (eds.), *The Traveling Salesman Problem,* John Wiley & Sons Ltd., New York, pages 431–448, 1985.

[27] Christofides, N., A. Mingozzi and P. Toth. The Vehicle Routing Problem, in *Combinatorial Optimization,* Christofides, N., A. Mingozzi, P. Toth and C. Sandi (eds.), John Wiley & Sons Ltd., New York, pages 318–338, 1978.

[28] Clarke, G. and J. W. Wright. Scheduling of Vehicles from a Central Depot to a Number of Delivery Points. Operations Research 12:568–581, 1964.

[29] Coffman, E. G., Jr., and G. S. Lueker. *Probabilistic Analysis of Packing and Partitioning Algorithms,* John Wiley & Sons Ltd., New York, 1991.

[30] Coffman, E. G., Jr., M. R. Garey and D. S. Johnson. Approximation Algorithms for Bin-Packing–An Updated Survey. G. Ausiello, M. Lucertini and P. Serafini (eds.), *Algorithm Design for Computer System Design,* Springer-Verlag, pages 49–106, 1984.

[31] Cornuéjols, G. and F. Harche. Polyhedral Study of the Capacitated Vehicle Routing Problem. *Mathematical Programming* 60:21–52, 1993.

[32] Desrochers, M. and T. W. Verhoog. A Matching Based Savings Algorithm for the Vehicle Routing Problem. Cahier du GERAD G-89-04, École des Hautes Études Commerciales de Montréal, Québec, 1989.

[33] Desrosiers, J., Y. Dumas, M. M. Solomon and F. Soumis. Time Window Constrained Routing and Scheduling. Handbooks in Operations Research and Management Science, "Networks Routing", M.O. Ball, T.L. Magnanti, C.L. Monma and G.L. Nemhauser, eds., pages 35–139, 1992.

[34] Federgruen, A., M. Queyranne and Y.S. Zheng. Simple Power-of-Two Policies Are Close to Optimal In a General Class of Production/Distribution Networks with General Joint Setup Costs. *Mathematics of Operations Research* 17:951–963, 1992.

[35] Federgruen, A. and D. Simchi-Levi. Analysis of Vehicle Routing and Inventory-Routing Problems. Handbooks in Operations Research and

Management Science, "Networks Routing", M.O. Ball, T.L. Magnanti, C.L. Monma and G.L. Nemhauser, eds., pages 297–373, 1992.

[36] Federgruen, A. and Y.S. Zheng. The Joint Replenishment Problem with General Joint Cost Structures. *Operations Research* 40:384–403, 1992.

[37] Fisher, M. L. and R. Jaikumar. A Generalized Assignment Heuristic for Vehicle Routing. *Networks* 11:109–124, 1981.

[38] Fisher M. L. Vehicle Routing. *Handbooks in Operations Research and Management Science*, the volume on Network Routing, M. Ball, T. Magnanti, C. Monma and G. Nemhauser, eds. pages 1–33, 1992.

[39] Fisher, M. L. Optimal Solution of Vehicle Routing Problems Using Minimum K-Trees. *Operations Research* 42:626–642, 1994.

[40] Gallego, G. and D. Simchi-Levi. On the Effectiveness of Direct Shipping Strategy for the One Warehouse Multi-Retailer R-Systems. *Management Science* 36:240–243, 1990.

[41] Garey, M. R. and D. S. Johnson. *Computers and Intractability*. W. H. Freeman and Company, New York, 1979.

[42] Gaskel, T. J. Bases for Vehicle Fleet Scheduling. *Oper. Res. Quart.* 18:281–295, 1967.

[43] Gendreau, M., A. Hertz and G. Laporte. A Tabu Search Heuristic for the Vehicle Routing Problem. Working Paper, Centre de recherche sur les transports, Université de Montréal, Québec, 1991.

[44] Gillett, B. E. and L. R. Miller. A Heuristic Algorithm for the Vehicle Dispatch Problem. *Operations Research* 22:340–349, 1974.

[45] Haimovich, M. and A. H. G. Rinnooy Kan. Bounds and Heuristics for Capacitated Routing Problems. *Mathematics of Operations Research* 10:527–542, 1985.

[46] Hall, R. W. Comments on One-Warehouse Multiple Retailer Systems with Vehicle Routing Costs. *Management Science* 37:1496-1497, 1991.

[47] Hall, R. W. A Note on Bounds for Direct Shipping Cost. *Management Science* 38:1212-1214, 1992.

[48] Herer, Y.T. Submodularity and the Traveling Salesman Problem. Working Paper, Technion - Israel Institute of Technology. To appear in *European Journal of Operational Research*, 1996.

[49] Herer, Y.T. and R. Roundy. Heuristics for a One-Warehouse Multiretailer Distribution Problem with Performance Bounds. *Operations Research* 45:102–115, 1997.

[50] Karp, R. M. Probabilistic Analysis of Partitioning Algorithms for the Traveling Salesman Problem. *Mathematics of Operations Research* 2:209–224, 1977.

[51] Kingman, J. F. C. Subadditive Processes. *Lecture Notes in Math. 539*, Springer-Verlag, Berlin, pages 168–222, 1976.

[52] Lawler, E. L., J. K. Lenstra, A. H. G. Rinnooy Kan and D. B. Shmoys. *The Traveling Salesman Problem: A Guided Tour of Combinatorial Optimization*, John Wiley & Sons Ltd., New York, 1985.

[53] Li, C.-L. and D. Simchi-Levi. Analysis of Heuristics for the Multi-depot Capacitated Vehicle Routing Problems. *ORSA Journal on Computing* 2:64–73, 1990.

[54] Marchetti Spaccamela, A., A.H.G. Rinnooy Kan and L. Stougie. Hierarchial Vehicle Routing Problems. *Networks* 14:571–586, 1985.

[55] Muckstadt, J. and R. O. Roundy. Analysis of Multistage Production Systems. *Handbooks in Operations Research and Management Science*, the volume on Logistics of Production and Inventory, S. C. Graves, A. H. G. Rinnooy Kan and P. H. Zipkin, eds., pages 59–131, 1993.

[56] Noon, C. E., J. Mitthenthal and R. Pillai. A TSSP+1 Decomposition Approach for the Capacity Constrained Vehicle Routing Problem. Working Paper, Management Science Group, The University of Tennessee, Knoxville, TN, 1991.

[57] Paessens, H. The Savings Algorithm for the Vehicle Routing Problem. *European Journal of Operations Research* 34:336–344, 1988.

[58] Papadimitriou, C.H. and K. Steiglitz. *Combinatorial Optimization Algorithms and Complexity*, Prentice Hall, Inc., Englewood Cliffs, NJ, pages 247–270, 1982.

[59] Pirkul, H. Efficient Algorithms for the Capacitated Concentrator Location Problem. *Computers and Operations Research* 14(3):197–208, 1987.

[60] Pureza, V. M. and P. M. Franca. Vehicle Routing Problems via Tabu Search Metaheuristics. Publication CRT-747. Centre de Recherche sur les Transports, Montréal, 1991.

[61] Queyranne, M. A Polynomial-Time, Submodular Extension to Roundy's 98% -Effective Heuristic for Production/Inventory Systems. Working Paper No. 1136, Faculty of Commerce and Business Administration, University of British Columbia, Vancouver, 1985.

[62] Queyranne, M. Finding 94% Effective Policies in Linear Time for Some Production/Inventory Systems. Working Paper, Faculty of Commerce, University of British Columbia, Vancouver, Canada, 1987.

[63] Rhee, W. T. and M. Talagrand. Martingale Inequalities and $\mathcal{NP}$-Complete Problems. *Mathematics of Operations Research* 12:177–181, 1987.

[64] Rosenfield, D. and M. Pendrock. The Effects of Warehouse Configuration Design on Inventory Levels and Holding Costs. *Sloan School of Management Review* 21:21-33, 1980.

[65] Rosenkrantz, D. J., R. E. Sterns and P. M. Lewis II. An Analysis of Several Heuristics for the Traveling Salesman Problem. *SIAM J. on Computing* 6:563–581, 1977.

[66] Roundy, R. O. 98%-effective Integer-Ratio Lot-Sizing Rule for One-Warehouse Multi-Retailer Systems. *Management Science* 31:1416–1430, 1985.

[67] Roundy, R.O. A 98%-effective Lot Sizing Rule for a Multi-Product, Multi-Stage Production/Inventory System. *Mathematics of Operations Research* 11:699–727, 1986.

[68] Semet, F. and E. Taillard. Solving Real-Life Vehicle Routing Problems Efficiently Using Taboo Search. Working Paper, Départment de Mathématiques, École Polytechnique Fédérale de Lausanne, Switzerland, 1991.

[69] Simchi-Levi, D. Hierarchical Planning for Probabilistic Distribution Systems. *Management Science* 38:198–211, 1992.

[70] Spaccamela, A. M., A. H. G. Rinnooy Kan and L. Stougie. Hierarchical Vehicle Routing Problems. *Networks* 14(4):571–586, 1984.

[71] Toth, P. and D. Vigo. The Vehicle Routing Problem. SIAM Monographs in Discrete Mathematics and Applications, New York. In Preparation, 1998.

[72] Viswanathan S. and K. Mathur. Integrating Routing and Inventory Decisions in One Warehouse Multi-Retailer Multi-Product Distribution Systems. *Management Science* 43:294–312, 1997.

[73] Yellow, P. A Computational Modification to the Savings Method of Vehicle Scheduling. *Oper. Res. Quart.* 21:281–283, 1970.

7 SUPPLY CONTRACTS WITH QUANTITY COMMITMENTS AND STOCHASTIC DEMAND

Ravi Anupindi

J.L. Kellogg Graduate School of Management
Northwestern University, Evanston, IL 60208

Yehuda Bassok

School of Business Administration
University of Washington, Seattle, WA 98195

7.1 INTRODUCTION

Supply chain management deals with the management of material, information, and financial flows in a network consisting of vendors, manufacturers, distributors and customers. Managing flows in this network is a major challenge due to the complexity (in space and time) of the network, the proliferation of products (often with short life cycles) that flow through this network, and the presence of multiple decision makers who each own and operate a piece of this network and optimize a private objective function. Supply chain management clearly involves a variety of issues including product/process design, production, third party logistics and outsourcing, supplier contracting, incentives and performance measures, multi–location inventory coordination, etc.

For each (set of) node(s) in the network – e.g., a manufacturer, an assembler, etc. – one could identify supplier(s) and buyer(s). Exchange of flows is a routine transaction that occurs between any pair of supplier and buyer nodes in the network. For example, material usually flows from a supplier to a buyer while information and financial flows are bi-directional. The entities that comprise the network are businesses except for the last set of links who may be individual customers. Therefore for most part, the exchange of flows occur between businesses. Almost all business–to–business transactions are governed by contracts.[1] An important criteria that we will use to measure the performance of the supply chain is *Coordination*. We say that a network (or channel) is coordinated when a *single* decision maker optimizes the network with the union of information that the various decision makers have. Lack of coordination occurs due to the existence of multiple decision makers in the network who may have different *information* and *incentives*. Even under information symmetry (all parties are equally informed) the performance of the network may be suboptimal since each decision maker optimizes a private objective function and local optima need not be globally optimal. A classical example is that of double–marginalization [26]. Often, however, decision makers have private information which they may not share with others. This too leads to sub-optimal performance. While market mechanisms may be used to achieve coordination, this is not always feasible. Hence the supply chain resorts to contracts. Contracts could be written that may ensure coordination through appropriate provisions for information and incentives such that channel performance can be optimized. Even when coordination is not achieved, contracts may provide Pareto optimal solutions. Therefore, the premise of this chapter is that the management of supplier–buyer interactions within a supply chain is governed by formal contracts. We refer to them as **Supply Contracts** since our primary focus here is in the material and information flows.

---

[1]The economics literature discusses in detail contracting issues between businesses, between individuals, and between businesses and individuals.

7.2 SUPPLY CONTRACTS: A GENERAL FRAMEWORK

In general the parameters over which two parties may contract upon will be context specific and also (under asymmetric information) a function of the information structure available to each. Broadly a supply contract should capture the three types of flows, namely, *material, information,* and *financial.* In practice, the set of parameters over which supply contracts are observed can be classified into the following categories:

- **Horizon Length**: This specifies the duration for which the contract is valid.

- **Pricing**: This is interpreted broadly to incorporate all financial flows. Clearly, one component of this is the purchase price. This could take several forms. For example, it could be linear (proportional) or non–linear (e.g., two–part tariff). In addition, other types of payment structures can be set up. For example, credit for return of goods by a buyer, holding cost subsidies from a supplier to a buyer, payments for inability to supply (e.g., due to stock-out), etc. Thus pricing may also depend on other parameters of the contract.

- **Periodicity of Ordering**: This specifies how often a buyer can place orders. It could be *fixed* (for example, a buyer may be asked to place orders every Monday), or *random* (any day of the week). It is not essential that non–zero orders are always placed. If the contract calls for fixed ordering (every Monday), then a buyer *may* place orders only on a Monday.

- **Quantity Commitment**: Quantity Commitments by a buyer could be on orders, its demand, or capacity of the supplier. **Order Commitments** take two generic forms.

  - **Total Minimum Commitment**: For single products, this implies that a buyer commits to cumulative purchases of at least a certain quantity; this is referred to as *Total Minimum Quantity Commitment*. For multiple products, this usually takes the form of commitments to purchase at least a certain minimum dollar value of goods, referred to as *Total Dollar Volume Commitment*.

  - **Periodical Commitment**: A buyer makes a commitment to purchase a certain quantity every period.

Under **demand commitment** a buyer commits to source a fraction of all his demand from a specific supplier. Observe that under such commitment, the uncertainty of the demand process is shared with the supplier. Finally, under **capacity commitment** a buyer usually reserves a fraction of the supplier's capacity. For example, in agricultural contracts, a buyer commits to buy all production from a certain acreage; in the semiconductor industry, often buyers purchase a certain fraction of the capacity of a supplier's foundry wafer fabrication facility.

- **Flexibility**: Whenever a buyer is required to make some commitments on the quantities to be purchased, often a supplier provides some flexibility to adjust these quantities. The contract may specify the *magnitude* and *frequency* of adjustments. For example, a supplier may specify that additional (unlimited) quantities may be ordered but no more than two times during the contract horizon. In contracts with restrictions on the magnitude of commitments, the extent of adjustments allowed may or may not be a function of the commitment made. Furthermore, the additional flexibility may come at extra cost to the buyer.

- **Delivery Commitment**: A supplier usually makes a commitment for the material delivery process. A commitment on the **lead time** would specify delay in delivery of the material. Service level agreements on lead time for the entire order or on fraction of the order are common. For example, "delivery of entire order 90% of the time within 2 weeks" or "delivery of 90% of the order within 2 weeks". Of course, this is usually coupled with a mutually agreed upon **shipment policy**. A shipment policy will specify if a buyer accepts multiple shipments for the same order.

- **Quality**: Quality restrictions could come in terms of defects rates, specifications, etc.

- **Information Sharing**: This characterizes the information flow between a buyer and a supplier. Specifically, it outlines what type of information will be shared between buyer and a supplier. For example, a buyer (retailer) may pass on the sales data to its supplier.

While pricing incorporates all *financial* flows, aspects of *material* flows are captured through periodicity of ordering, commitments (quantity and delivery), flexibility, and quality, and *information* flows are captured by the type of information shared. With this general framework for supply contracts in mind, from a research perspective there are two types of questions that need to be addressed:

- Given that several types of contracts are observed in practice. what is a decision maker's (buyer or supplier) optimal policy under these different contracts? What are the implications of this policy for the other parties? This stream of work may be referred to as *Analysis of Contracts*. It is primarily a single decision maker analysis of given contracts and their implications. Analysis of contracts serves two purposes:

  - It is a first step towards the design of contracts. Since it involves identifying the effect of various contract parameters analysis of contracts are useful in restricting the contract parameter space to a few desirable attributes, and thus indirectly assist in the design of contracts.

- They form building blocks for development of complex decision support systems.

- What types of supply contracts are written and why? We have already argued earlier the two basic problems for lack of coordination are information and incentives. Even when information known to the two players is symmetric, incentive problems may prevent coordination. Asymmetric information, in addition, may complicate the matter further. Usually, the asymmetry is about demand, cost structure, policies followed, etc. Several researchers have begun working in the area of design of contracts. This stream of work may be referred to as *Design of Contracts*.

Before we present specific examples of analyses and design of contracts, we will first argue that all of the extant inventory literature (see [17] for a review) can also be viewed as analyses of specific types of supply contracts. This comparison allows us to motivate other types of supply contracts that involve quantity commitments.

7.3 CLASSICAL INVENTORY THEORY AS ANALYSIS OF SUPPLY CONTRACTS

Consider the classical finite horizon periodic review newsvendor problem. The traditional view is that the model assumes either perfectly competitive marketplace or that prices are rigid [6] and hence the assumption of fixed proportional price of the input is appropriate without imposing any constraints on the order process. The imperfectly competitive case has received scant attention. Alternately, the newsvendor problem could also be viewed as analysis of a contract (say, *newsvendor supply contract*) with a given horizon length, proportional purchase price (often considered fixed), fixed periodicity of ordering (given by the period length), no commitments, unlimited flexibility (there are usually no limits on the order quantity), and a delivery commitment with a fixed lead time of supply and a single shipment of an order. Similarly, from a contractual viewpoint, two views of the (s, S) policy are possible. First, it can be considered as the outcome of an analysis of a newsvendor supply contract for a buyer who has some transaction costs. Alternately, (s, S) policy could also be considered as an outcome of an analysis of a contract similar to the newsvendor contract except that the pricing policy is a two–part tariff (fixed fee + linear) and in which a buyer has no internal transaction costs. A (Q, R) policy could be viewed similarly but for a system where the periodicity of ordering is random.

If we take a contractual view of these pieces of work, then from a channel perspective it is important to understand the implications of the contracts on the other parties in the supply chain. Consider the newsvendor supply contract. Under the assumption of stationary demands faced by a buyer, it is well known that the buyer's policy structure is such that the order process closely tracks the demand process[2] (is identical when the buyer's terminal salvage value equals the purchase cost) thus passing on the uncertainty in the demand process to the supplier.[3] If the buyer faces non-stationary demand, then the order process may exhibit a higher variability that the demand process [21], prominently known as the *bull-whip effect.*

7.3.1 Commitment Type Contracts

Based on the discussion on the nature of the order process under a *newsvendor* supply contract, it could be argued that such a contract may not be acceptable to a supplier, especially, if say, her costs are affected by the uncertainty in the order process (its demand process). Thus if the supplier's cost due to this uncertainty is high enough, she may not offer a newsvendor supply contract. Instead, she may require a buyer to make periodical quantity commitments. Furthermore, if her costs are affected by periodic variability in the order process, then she may even restrict the commitments to be stationary. If, in addition, she offers no flexibility to the buyer to adjust these commitments, then the implications of demand uncertainty is entirely borne by the buyer. Clearly, the newsvendor supply contract and a contract with stationary commitments and no flexibility are two extremes. At an intermediate level, a supplier may require a buyer to make commitments but allow him to adjust these quantities in some limited fashion. Thus supply contracts with periodical commitments and flexibility allow a supplier and a buyer to *share the burden of demand uncertainty*[4] and reduce the uncertainty in the order process.

Other forms of quantity commitments are also observed in practice. One reason for the existence of total (quantity or dollar volume) minimum commitment contracts is to *ensure markets.* In a competitive environment, when buyers have alternative sources of supply, suppliers prefer to lock–in buyers by providing them with an incentive to commit to purchase goods for a long period of time. Alternately, when capacity investment is expensive and demands highly uncertain, a supplier may wish to lock–in a buyer for a specific duration with a total minimum commitment contract.

The above two arguments posit when a supplier will offer a commitment contract. Alternately, if there is any uncertainty in the supply process, then a buyer may wish to offer such a contract to a supplier to *ensure supply.* For example, if a supplier needs to make a capacity investment, to ensure supply a

---

[2]Under an infinite horizon policy, order process is identical to the demand process. We assume that all contract durations are of finite horizon. The buyer, however, may use an infinite horizon policy as an approximation.

[3]In the classical newsvendor setting, a buyer always orders before observing demand in a period and hence carries some safety stock. Thus he bears some burden for the uncertainty in demand. If we allow the buyer to order after observing demand, then all of the uncertainty in demand process is borne by the supplier.

[4]We deliberately avoid the phrase "risk–sharing" since usually risk is associated with uncertainty in cash flows and for the purpose of our discussion we assume that all parties are risk neutral.

buyer may offer to commit to purchase certain minimum quantity. He may retain the flexibility of not purchasing the committed quantity exactly by paying, for example, a penalty for unused capacity [12].

Among the various elements in the general supply contract, we will particularly focus attention on the issue of *quantity commitments* and *flexibility*. Several types of contracts can be structured using quantity commitments and flexibility. For example,

- **Total Minimum Quantity Commitment** contract in which the buyer, at the beginning of the horizon, commits to purchase, during the entire horizon, at least a minimum quantity. Usually, a supplier gives a discount based on the level of commitment. The unit cost decreases with the magnitude of this commitment. There is no restriction on the maximum amount that can be purchased during the horizon. Minimum commitment contracts are a common practice in the electronic industry. Usually the buyer makes a commitment to purchase a minimum quantity during the coming year, and the supplier agrees to sell the product (integrated circuits) for a discounted price. A **Total Minimum Quantity Commitment with Flexibility** contract is similar to the Total Minimum Quantity Commitment except that the supplier imposes a maximum limit on the total quantity that can be purchased at the discounted price. This maximum is specified as a percentage above the total minimum commitment.

- **Total Minimum Dollar Volume Commitment** contract is a multi-product equivalent of the Total Minimum Quantity Commitment contract where a buyer, instead of making commitments on individual component purchases, commits to a minimum dollar volume of business for the horizon. Again, a supplier could offer discounts based on the dollar volume of commitment.

- **Periodical Commitment with Flexibility** contract. First, unlike the total minimum commitment contracts, such contracts impose restrictions on the periodical purchases made. These contracts take various forms depending on the nature of periodical commitments and the flexibility offered. Broadly, the commitments could be static (stationary or non-stationary) or dynamic and thus allowed to be updated periodically in a rolling horizon manner. The magnitude of the flexibility offered could either be unlimited (that is, any amount of additional units could be ordered) or limited. Usually when unlimited flexibility is allowed, it comes at an extra cost; that is, a supplier specifies that additional units will be supplied at a higher price. Limited flexibility could be specified to be either dependent or independent of the commitments made. The following are two examples of such contracts.

  - **Rolling Horizon Flexibility (RHF)**. At the beginning of the horizon the buyer commits to purchase a certain quantity every period. The buyer has limited flexibility to purchase quantities that are somewhat different than the original commitments, and is also allowed to update the previously made commitments. Clearly, there is some limitation on the flexibility to update the previously made commitments. In general, the per unit wholesale price may decrease as the magnitude of flexibility increases.

- **Periodical Commitments with Options.** In this setting a buyer commits to purchase some quantities in future periods. In addition, he purchases options (at unit option price) from the supplier that allows him to buy additional units of the good, if necessary, by paying an exercise price. Thus using options a buyer buys the right to adjust order quantities upwards.

There are several similarities between the RHF contract and the supply contracts with options. In both types of contracts the buyer makes periodical commitment ahead of time. These commitments provide the supplier with valuable and relatively accurate information of the buyer's future needs. Both contracts provide the buyer with a mechanism to adjust to changes in demand or inventory information. While the RHF contract allows for (upward and downward) adjustments that are proportional to the original commitment, the supply contract with options allows for "unlimited" (ex ante) upward adjustments. Also, in both contracts there is a notion that flexibility is a commodity that may be purchased at a cost. In the RHF contract it is implicitly assumed that the buyer can obtain higher flexibility by paying a higher purchasing cost per unit. The cost of flexibility is explicitly modeled in supply contracts with options where a buyer buys options to have the right to adjust quantities later. In addition, the supplier charges an exercise price when the buyer actually purchases an additional quantity.

In the remainder of this chapter we will first analyze in detail the *Total Minimum Quantity Commitment Contract (with Flexibility)* (sections 7.4 and 7.5), the *Total Minimum Dollar Volume Contract* (section 7.6), and the *Rolling Horizon Flexibility* (section 7.7) contracts. Subsequently, in the spirit of *design of contracts*, we consider a context most realistic of a fashion industry to demonstrate how a *Periodical Commitments with Options* (section 7.8) contract with appropriate pricing strategy allows the channel to be coordinated. We will assume that the decision makers have full information and focus on the incentive issues. A detailed literature review of supply contracts in supply chain management appears elsewhere in this book [1].

7.4 TOTAL MINIMUM QUANTITY COMMITMENT CONTRACTS

In a *total minimum quantity commitment* contract a buyer guarantees that his cumulative orders across all periods in the planning horizon will exceed a specified minimum quantity. The buyer has the flexibility to order any amount in specific periods as long as the cumulative quantity restriction is met at the end of the horizon. In return, a supplier, may offer price discounts. In practice, a supplier provides a menu of (per unit price, total minimum commitment) pairs from which the buyer chooses a commitment at the corresponding price. It is reasonable to expect that as the total minimum commitment increases, the per unit price would decline; in that sense, this menu is a price discount scheme for commitments. In return for the buyer's commitment, the supplier provides a price discount. The description below follows the analysis of a total minimum commitment contract by Bassok and Anupindi [9].

7.4.1 Model and Analysis

Consider a classical periodic review finite horizon stochastic demand inventory problem. Let the horizon length be N periods. The sequence of events and actions taken by the buyer are as follows. At the beginning of the horizon the buyer makes a commitment to buy a minimum quantity, say K_1 for the entire horizon. At the beginning of each period t, the starting inventory I_t and the remaining commitment quantity K_t are observed. Thus K_1 can be viewed as the minimum remaining commitment in periods $1, \ldots, N$. An order of size Q_t is placed which is delivered instantaneously.[5] Demand D_t is observed and is satisfied as much as possible from the available stock of $I_t + Q_t$. Excess demand is backlogged and excess inventory is carried to the next period. The marginal discounted purchasing, holding, and backlog penalty costs are c, h, and π respectively. We assume that the demands are independent and identically distributed across periods with a probability density function of $f(\cdot)$ and a cumulative density function of $F(\cdot)$.

The optimization problem is formulated as a stochastic dynamic program with two state variables - the on-hand inventory (I_t) and the remaining commitment (K_t). The periodical purchase quantities, Q_t, $t = 1, \ldots, N$ are the only decision variables. Let $C_t(I_t, K_t, Q_t)$ be the total expected cost from period t through N and $J_t(I_t, K_t)$ be the period t value function. That is,

$$J_t(I_t, K_t) = \min_{Q_t} C_t(I_t, K_t, Q_t).$$

Define $J_{N+1}(I_{N+1}, 0) \equiv 0$. Then the dynamic programming recursion is described as follows:

$$C_t(I_t, K_t, Q_t) = cQ_t + L(I_t + Q_t) + E_{D_t}\{J_{t+1}(I_{t+1}, K_{t+1})\}$$

---

[5]This assumption can be relaxed. It is well known that a model with a known positive lead time can be reduced, by appropriately adjusting the demand, to a model with instantaneous deliveries.

where,

$$I_{t+1} = I_t + Q_t - D_t, \quad t = 1, \ldots, N$$
$$K_{t+1} = (K_t - Q_t)^+, \quad t = 1, \ldots, N$$

and,

$$L(I + Q) = h\mathbf{E}[\max\{I + Q - D, 0\}] + \pi\mathbf{E}[\max\{D - (I + Q), 0\}]$$

gives the single period expected holding and penalty costs. The optimization problem is defined as follows:

$$\min_{Q_1} \quad C_1(I_1, K_1, Q_1)$$

$$\text{s.t.} \quad \sum_{i=1}^{N} Q_i \geq K_1$$

$$Q_i \geq 0, \quad i = 1, \ldots, N.$$

It can then be shown that,

Proposition 1   1. $C_t(I_t, Q_t, K_t)$ is convex in I_t, Q_t, and K_t.

2. *The optimal order policy for period t is:*

$$Q_t^* = \begin{cases} 0 & \text{if } I_t \geq S_\infty \\ S_\infty - I_t & \text{if } S_\infty - K_t \leq I_t < S_\infty \\ K_t & \text{if } S_t - K_t \leq I_t < S_\infty - K_t \\ S_t - I_t & \text{if } I_t \geq S_t - K_t \end{cases}$$

where S_t is the base stock level of a $(N - t + 1)$–period standard newsvendor problem without commitment and $F(S_\infty) = \frac{\pi}{\pi + h}$.

3. $J_t(I_t, K_t)$ *is convex in I_t and K_t.*

Observe that to implement the order policy we only need S_∞ and S_t which are easily computed. Using the order policy, until the minimum commitment is exhausted we order up to S_∞ (that is, order as if the purchase cost were sunk). Subsequently, we order like in a classical newsvendor problem. Graphically, the structure of the above solution is presented in Figure 7.1.

The importance of this solution is two fold. First it allows the buyer to evaluate his optimal costs as a function of the commitment and the purchase price, and to thus choose the best (purchase price, total minimum commitment) pair from the menu offered. Second, and related, is the fact that it is very simple and easy to calculate the base stocks of the optimal policy. Observe that in order to implement the optimal purchasing policy it is only necessary to calculate the base stock levels $S_1, \ldots, S_N$ and S_∞.

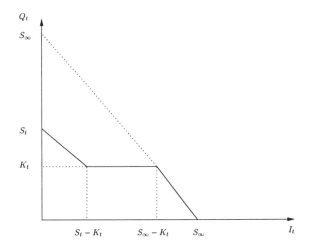

Figure 7.1 Optimal order quantity as a function of initial inventory.

7.4.2 Example

We now describe a numerical example to illustrate the benefits to the buyer of signing a contract for total minimum commitments. We evaluate such contracts against a base case in which no contract is signed; that is, in the base case, the buyer purchases quantities at a regular price with no restriction on total quantities purchased. Any discounted contracted price is specified as a percentage discount over the regular price. We then define percentage savings as the savings in total costs of the contract over that of the base case. The mean demand is 100.0 units with a coefficient of variation (CV) of 0.25. The unit regular price is $1.00. The unit penalty cost was set to $25.0 and the unit holding cost was 25% of purchase price. The horizon length is 10 periods. Demand distributions are assumed to be normal, truncated at three standard deviations.

Figure 7.2 demonstrates the effect of price discounts of 5%, 10%, 15% on the percentage savings. Clearly the higher the percentage discount the higher the percentage savings. The graphs of Figure 7.2 can be used to determine the acceptance or rejection of a specific contract. For example, if a 5% discount is offered, the buyer should never commit to purchase more than (approximately) 900 units. Any commitment below 920 units results in expected savings, and commitments above 920 units results in expected loss. Therefore a commitment for mean demand over the horizon (in this case 1000 units) may not be advantageous. It depends on the discount offered and the length of the horizon. Again, as expected, the buyer may commit to higher quantities if larger discounts are offered. Finally, observe that the total savings cannot exceed the percentage discount offered.

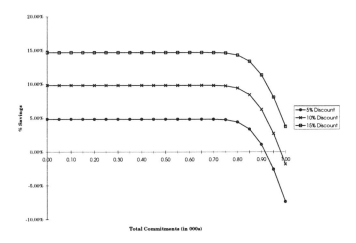

Figure 7.2   Effect of price discount.

Since the policy structure is simple, it is easy to compare the various (discount, total minimum commitment) contract tuples offered and select the best. Finally, since the policy structure is similar to the newsvendor supply contract, we can see that the order process will closely track the demand process when the latter is stationary. We can conjecture that under non–stationary demands, the bull-whip effect will also be observed. Using such a contract, however, a supplier is able to *ensure markets* for herself.

7.5   TOTAL MINIMUM QUANTITY COMMITMENT CONTRACTS WITH FLEXIBILITY

In the previous section we assumed that the buyer is able to purchase any additional quantity above the minimum committed at the same discounted price. Often a supplier, however, may impose restrictions on the total purchases at the discounted price. This happens, for example, when supplier's production capacity is restricted. In such instances, she may offer to supply quantities larger than a certain amount at a higher price (or the buyer may source these from another supplier at a higher price). Formally such contracts are structured as follows. A buyer is required to commit to a total minimum quantity (say, K_1^L) of purchases to avail of a discounted price c. The supplier extends this discounted price to purchases up to a fixed fraction (say, a) above the minimum; that is, purchases up to $K_1^U = (1+a)K_1^L$ are available at discounted price. Any further purchases above K_1^U are available at regular price c_r. The fraction a is the flexibility that a supplier offers to the buyer to adjust his total purchases over the horizon of the contract. A menu of (regular price, discounted price, total minimum commitment, flexibility) can then be offered to a buyer.

Anupindi and Bassok [4] analyze such contracts. Observe that under such a contract, there are two types of purchases: one at discounted price denoted by Q_t and the other at regular price denoted by M_t. It can be shown that the expected cost and the corresponding value function are convex in their arguments [5]. Then,

Proposition 2 *The optimal policy structure is defined by three critical levels S_∞, S_t, and S_t^m.*

$$
(Q_t, M_t) = \begin{cases}
(0,0) & \text{if } I_t \geq S_\infty \\
(S_\infty - I_t, 0) & \text{if } S_\infty - K_t^L < I_t < S_\infty \\
(K_t^L, 0) & \text{if } S_t - K_t^L < I_t \leq S_\infty - K_t^L \\
(S_t - I_t, 0) & \text{if } S_t - K_t^U < I_t \leq S_t - K_t^L \\
(K_t^U, 0) & \text{if } S_t^m - K_t^U < I_t \leq S_t - K_t^U \\
(K_t^U, S_t^m - K_t^U - I_t) & \text{otherwise}
\end{cases}
\tag{7.3}
$$

where S_∞ satisfies

$$
F(S_\infty) = \frac{\pi}{\pi + h}
$$

and, S_t is the base stock level in period t for a standard $(N - t + 1)$ period newsboy problem with per unit order cost at c where the periods are numbered $t, \ldots, N$; and, S_t^m is the base stock level in period t for a standard $(N - t + 1)$ period newsboy problem with per unit order cost at c_r where the periods are numbered $t, \ldots, N$.

The policy structure is diagrammed in Figure 7.3.

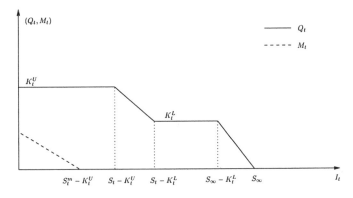

Figure 7.3 Optimal Order policy for a single product with total minimum commitment and flexibility.

The parameters that characterize the policy described above are simple to compute since they are solutions of unconstrained single product newsvendor problem with appropriate purchase costs. The implications on the buyer and supplier are similar to those observed under the Total minimum quantity contract without flexibility.

7.6 MINIMUM DOLLAR VOLUME COMMITMENT CONTRACTS

The total minimum quantity contracts discussed earlier are usually written for single products. For multiple products, it is common practice to have contracts on the total dollar volume of purchases. Under such a contract requires a buyer, to be eligible for a discounted price, needs to commit to a minimum cumulative dollar value of purchases during a specified time horizon. Furthermore, the supplier usually extends this discounted price for cumulative dollar volume purchases up to a certain fraction above this minimum commitment; this is a form of flexibility offered by the supplier. Any purchases in excess of this upper bound are charged a regular (market) price; the purchases at the regular price may be from a different supplier.

Business volume discounts offer several advantages to both the buyer and the supplier (see [24] for a detailed discussion). By pooling his purchases, a buyer is in a better position to negotiate higher discount rates. A supplier, by offering such discounts, increases her total volume of business. In addition, she can benefit from increased market presence for higher priced products. For example, a higher priced product from a supplier may not ordinarily be purchased if there is a competitor offering the same product at a lower price. However, by combining this higher priced product with other products in a business volume discount structure, the supplier increases the likelihood of selling this product. If the buyer, by including this higher price product exceeds the business volume cut-off point, he achieves a greater overall savings. Finally, from a production economics perspective, a supplier may wish to offer such contracts for multiple products whenever she has significant product / process commonality in her production; for example, if she uses a postponement or late customization strategy in producing the components. Supply contracts with business volume discounts for multiple products are commonly used in the electronics industries [18].

The description below is extracted from a model and detailed analysis of a multi-product contract with dollar volume commitments presented in Anupindi and Bassok [5].

7.6.1 Model

The multi-period contract is for a family of products/components[6] and takes the following form: The buyer agrees to buy a total (across all components)

---

[6]We will use products and components interchangeably.

minimum dollar volume, K_1^L, of components from the supplier over the length of the horizon. For consistency, we use similar notation as for the single product total minimum quantity contracts. The commitments for the multi-product case are specified for dollar volumes whereas in the single product case they are specified on quantities. The supplier guarantees a unit discounted price of c^p per component p.[7] The supplier provides the flexibility to adjust the total minimum dollar volume upwards but within limits, of say $a\%$, at the same price of c^p per unit; that is, the total dollar value supplied at price c^p from the supplier cannot exceed $K_1^U = (1+a)K_1^L$. Request for quantities of component p beyond this upper limit will be supplied at the regular price of $c_r^p > c^p$.

We make assumptions similar to those for the single product case. Demand for each component are independent and identically distributed across time; demands for components are independent of each other. Lead times of supply (of quantities purchased at the discounted or regular prices) are assumed to be zero.[8] Orders in a period are placed before observing demand for that period. Excess demands are backlogged. For notational simplicity, the salvage value of excess inventory (in the last period) is assumed to be zero. This assumption can be easily relaxed. The total dollar volume of purchase by buyer from the supplier at the discounted prices must exceed K_1^L but should be no more than $K_1^U = (1+a)K_1^L$.

We will use the following notation to model the problem. The horizon length is N periods with each period indexed by t. Let K_t^L and K_t^U be, respectively, the minimum and maximum dollar volume to be spent in purchasing components at discounted price from period t onwards till the end of the horizon. Let P be the total number of components in the family, indexed by p. We will modify the notation used for single product analysis to the multi–product case by using the superscript p for product index on the relevant parameters and quantities. Thus I_t^p denotes the initial inventory of component p at the beginning of period t, and Q_t^p and M_t^p, respectively, the order quantities of component p purchased in period t at the discounted and regular prices. As before all costs are assumed to be proportional. The per unit holding and penalty cost of component p per period is given by h^p and π^p respectively. Holding costs are assumed to be a function of the discounted prices only. That is, they do not change when components are purchased at the regular prices. We believe that errors due this approximation are usually insignificant. Finally, let D^p be the per period demand for component p with a probability distribution function of $f^p(\cdot)$ and a cumulative distribution function of $F^p(\cdot)$.

We use boldface letters to denote vectors. For example, $\mathbf{I}_t = (I_t^1, \ldots, I_t^P)$. All vectors are of size P unless specified otherwise. Let $L(\mathbf{I}_t, \mathbf{Q}_t, \mathbf{M}_t)$ be the single period expected holding and penalty costs, $C_t(\mathbf{I}_t, \mathbf{Q}_t, \mathbf{M}_t, K_t^L, K_t^U)$ the total expected costs for a $N - t + 1$ period problem starting in period t, and

---

[7]The contract may specify a percentage discount and a regular price. The discounted price is then easily inferred.

[8]It is easy to modify the model to incorporate fixed lead times.

$J_t(\mathbf{I}_t, K_t^L, K_t^U)$ the optimal expected costs for a $N - t + 1$ period problem starting in period t with initial inventory $\mathbf{I}_t$ and the minimum and maximum dollar volume to be purchased through the end of the horizon of K_t^L and K_t^U respectively. Then,

$$J_t(\mathbf{I}_t, K_t^L, K_t^U) = \min_{\mathbf{Q}_t, \mathbf{M}_t} C_t(\mathbf{I}_t, \mathbf{Q}_t, \mathbf{M}_t, K_t^L, K_t^U)$$

and $J_{N+1}(\mathbf{I}_{N+1}, 0, 0) \equiv 0$. Furthermore,

$$C_t(\mathbf{I}_t, \mathbf{Q}_t, \mathbf{M}_t, K_t^L, K_t^U) = \mathbf{c} \cdot \mathbf{Q}_t + \mathbf{c}_r \cdot \mathbf{M}_t + L(\mathbf{I}_t, \mathbf{Q}_t, \mathbf{M}_t)$$
$$+ \mathbf{E}_{D_t}[J_{t+1}(\mathbf{I}_{t+1}, K_{t+1}^L, K_{t+1}^U)]$$

To evaluate the contract, a buyer needs to solve

$$\min_{\mathbf{Q}_1, \mathbf{M}_1} C_1(\mathbf{I}_1, \mathbf{Q}_1, \mathbf{M}_1, K_1^L, K_1^U) = \mathbf{c} \cdot \mathbf{Q}_1 + \mathbf{c}_r \cdot \mathbf{M}_1 + L(\mathbf{I}_1, \mathbf{Q}_1, \mathbf{M}_1)$$
$$+ \mathbf{E}_{D_1}[J_2(\mathbf{I}_2, K_2^L, K_2^U)]$$

s.t.,

$$K_1^L \leq \sum_{p=1}^{P} \sum_{t=1}^{N} c^p Q_t^p \leq K_1^U$$

$$I_{t+1}^p = I_t^p + Q_t^p + M_t^p - D^p \qquad \text{for } t = 1, \ldots, N$$

$$K_{t+1}^L = \left(K_t^L - \sum_{p=1}^{P} c^p Q_t^p \right)^+ \qquad \text{for } t = 1, \ldots, N-1$$

$$K_{t+1}^U = K_t^U - \sum_{p=1}^{P} c^p Q_t^p \qquad \text{for } t = 1, \ldots, N-1$$

$$M_t^p \geq 0 \qquad \text{for } t = 1, \ldots, N$$
$$Q_t^p \geq 0 \qquad \text{for } t = 1, \ldots, N$$

Assuming proportional holding and penalty costs, we write

$$L(\mathbf{I}, \mathbf{Q}, \mathbf{M}) = \sum_{p=1}^{P} \left\{ h^p \mathbf{E}_{D_p}[\max\{I^p + Q^p + M^p - D^p, 0\}] \right.$$
$$\left. + \pi^p \mathbf{E}_{D_p}[\max\{D^p - (I^p + Q^p + M^p), 0\}] \right\}.$$

Finally, observe that under a business volume contract a supplier offers discounts for the total dollar volume of purchases. This equivalently implies that the price discount across all products in the contract is identical.[9] That is, for all p, there exists an $m > 0$ such that $c_r^p = m \cdot c^p$. We shall call this the **uniform discount policy property**.

---

[9] A supplier could construct a more sophisticated contract in which, in addition, to discounts for total dollar volume, she specifies a discount policy which is product specific. We do not consider such cases here.

7.6.2   Analysis

The optimization problem for a buyer is a complex multi–period, multi–product constrained dynamic program. The optimal solution will also likely be complex. Unfortunately, a generalization of the policy structure for the single product case with flexibility discussed in the previous section is not optimal. For the allocation policy, however, it can be shown that the buyer will first purchase goods at the discounted price, and only when K_1^U is exhausted will he purchase, if necessary, at the regular price. For the order policy, we suggest an approximation.

Order Policy Assumption.

- Order each product p up to S_∞^p, where $F^p(S_\infty^p) = \frac{\pi^p}{\pi^p + h^p}$, until the last period. In doing so, first order components at the discounted prices and only then, if necessary, order them at the regular prices.

- In the last period, solve the last (single) period problem optimally.

Thus under the order policy assumption, the base stock levels for a product p are identical for periods $1, \cdots, N - 1$ and independent of its purchase price (either discounted contract or regular). This is clearly not optimal and gives an upper bound on the total expected costs. Furthermore, given the **Order Policy Assumption** it can be shown that the allocation problem (of allocating dollar volume across various products) need to be solved only for the last period. Thus the only optimization problem that needs to be solved is for the last period. It is easily seen that the last period problem is a multi-product newsvendor problem with lower and upper bounds on the budget. Standard resource allocation algorithms can be used to solve this problem to optimality.

A Lower Bound.   A lower bound is derived from the following modification to the original model. Suppose goods can be salvaged in any period at their purchase price. This includes two cases: (a) either goods are first salvaged to the next period at its purchase price, if the next period is willing to purchase them, and (b) if necessary, dispose off the remaining (after salvaging to next period) at their purchase price. Now consider the **Order Policy Assumption** and, in addition, allow for units to be salvaged in any period as discussed; call this the LB policy. Then it can be shown that the LB policy gives a lower bound on the optimal expected costs.

   To compute the gap between the upper and lower bound, we compute the optimal dollar volume commitment for a given discount. Extensive computational studies indicate that the gap between the upper and lower bound is usually very small and does not exceed 3.03% for coefficient of variation (CV) of demand equal to 1.0; the gap is smaller for lower CVs. Furthermore, computational studies indicate that there is little loss in committing to the mean dollar volume and the value of flexibility (given by a) is small even when the number of components in the supply contract is small. An intuitive argument for this observation is that risk pooling of demands over multiple periods and multiple products by itself offers enough flexibility. In reality, a supplier would offer a menu of (discount, regular price, dollar volume commitment, flexibility) contracts and hence the approximation procedure suggested allows a buyer to efficiently evaluate such contracts.

The approximation method suggested is demonstrated to be quite effective and easy to compute. Since the approximation is based on intuition developed for a single product quantity commitment contract, the implications on the buyer and supplier are similar to those observed under those contracts.

7.7 PERIODICAL COMMITMENT CONTRACTS WITH FLEXIBILITY

Thus far we have presented contracts for single and multiple products that require a buyer to make commitments on the total quantity (dollar volume) over the horizon. No specific restrictions were imposed on the exact quantities purchased every period. A buyer was free to order any quantity in the various periods. For single product problems, observe that the policy structure implied that the buyer orders exactly the demand in the previous period until the total minimum commitment is exhausted. Subsequently, the orders may differ from demand due to the finite horizon nature of the remaining problem. Thus, like in the newsvendor supply contract discussed earlier, the uncertainty in the demand process is passed on to the supplier. The buyer, however, does bear some of the uncertainty in the demand since he has to place an order before observing the demand. The demand uncertainty "pass-through" to the supplier may not be optimal for the entire supply chain, especially, if the supplier's production costs are affected by it. This is the primary motivation for contracts with periodical commitments and flexibility.

In section 7.1, we outlined two examples of such contracts. We will, shortly discuss in detail the RHF contract. Before we do so, we briefly discuss a few periodical commitment contracts in which the commitments are restricted to be stationary.

- **Stationary Commitments.** In supply contracts with stationary commitments, a buyer is required to purchase a fixed minimum amount in each period. Discounts are given based on the level of minimum commitment. Additional units can be purchased but at an extra cost. Such contracts are analyzed by Anupindi and Akella [3] and Moinzadeh and Nahmias [23]. In Anupindi and Akella, the unit price of additional units is more than the discounted price. In addition, the additional units may not be delivered immediately, whereas, the supplier commits (a form of *Delivery Commitment*) to deliver the previously committed quantities. Moinzadeh and Nahmias do not allow for lead time uncertainty in the adjusted quantities but model a fixed cost of adjustments.

- **Order Bands.** Under such a contract a buyer is required to restrict all order quantities to be within an exogenously specified lower and upper limit that are stationary over time. The unit price could depend on the difference between the upper and lower limits (band–width), increasing with the band–width. Clearly, a larger band–width offers greater flexibility to the buyer. The concept of order-bands were initially studied by Kumar [19] and Anupindi [2]. Both these studies showed the existence of order band contracts in a game–theoretic setting. Analysis of a contract with order bands and two suppliers under a Markovian demand setting is carried out by Scheller–Wolf and Tayur [25].

A general model of an RHF contracts with detailed analysis is presented in Bassok and Anupindi [8]. The discussion here follows from their model and analysis. Tsay and Lovejoy [22] analyze a RHF contract (they call them Quantity Flexibility contracts) for a multi-echelon system. We first illustrate the RHF contract using an example. Suppose that a supplier allows 5%, 10%, and 20% flexibility in adjusting the current period's order, the commitment one-period from now, and the commitment two periods from now, respectively. We illustrate the dynamics of a RHF model in Figure 7.4. The current period is represented by a diamond and future periods by a rectangle. Suppose that the commitment for period 3 in period 0 was 150. In period 1, this commitment can be adjusted by 20%. Thus the period 3 commitment made in period 1 could be between [120,180]. Say, the commitment chosen is 130. In period 2, the commitment made for period 3 can be adjusted by 10%; that is, the new commitment could be between [117,143]. Say the buyer chooses 110. In period 3, the buyer is allowed to actually purchase within 5% of previous commitment of 110, i.e., between 105 and 116.

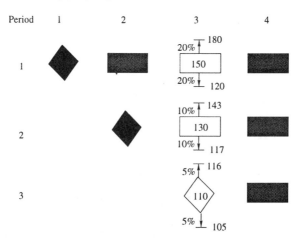

Figure 7.4 The Dynamics of a RHF Contract

Usually these types of contracts provide the manufacturer with a high level of long–term and low level of short–term flexibility. The short–term flexibility is low since the supplier may have limited means to adjust quantities within the production lead-time. Such contracts are popular in the electronics industry [10] where the production lead time is long. The supplier has very limited flexibility during the production lead time, and as a result usually offers no more than 1% or 2% flexibility within the production lead-time. The flexibilities beyond the production lead-time may be larger.

7.7.1 Model and Measurements

A finite horizon RHF contract is described as follows. At the beginning of the horizon, the buyer gives an initial commitment $Q_{1,i}$ for all the periods $i = 1, \ldots, N$ in the horizon. The supplier offers flexibilities to adjust the commitments for future periods and actual orders, upwards or downwards, in any period. Let $\overline{\alpha}_{i,j}$, $i \leq j$, be the fraction by which the buyer may adjust period j commitment (or actual order if $j = i$) *upward* at the beginning of period i. Similarly, let $\underline{\alpha}_{i,j}$, $i \leq j$, be the fraction by which the buyer may adjust period j commitment (or actual order if $j = i$) *downward* at the beginning of period i. Clearly, $\overline{\alpha}_{i,j} = 0$ and $\underline{\alpha}_{i,j} = 0$ for $i > j$. Let $Q_{t,t+i}$ denote the commitment for period $t+i$ made in period t for $i > 0$. $Q_{t,t}$ denotes the actual order quantity in period t. Then, $Q_{t,t+i}$ for $i \geq 0$ are constrained as follows:

$$(1 - \underline{\alpha}_{t,t+i})Q_{t-1,t+i} \leq Q_{t,t+i} \leq (1 + \overline{\alpha}_{t,t+i})Q_{t-1,t+i}. \tag{7.8}$$
$$(1 - \underline{\alpha}_{t,t})Q_{t-1,t} \leq Q_{t,t} \leq (1 + \overline{\alpha}_{t,t})Q_{t-1,t}. \tag{7.9}$$

The upward and downward flexibilities may specified as two matrices $[\overline{\alpha}]$ and $[\underline{\alpha}]$ with entries $\overline{\alpha}_{i,j}$ and $\underline{\alpha}_{i,j}$ respectively. This is the most general characterization of an RHF contract. Usually, most RHF contracts have the property that the flexibility offered for a future period depends only in the number periods into the future; formally $\overline{\alpha}_{t_1,t_1+j} = \overline{\alpha}_{t_2,t_2+j}$, and $\underline{\alpha}_{t_1,t_1+j} = \underline{\alpha}_{t_2,t_2+j}$ for all t_1, t_2, and j.

In each period t the buyer faces a stochastic demand represented by D_t with a probability density function of $f_t(\cdot)$ and a cumulative density function of $F_t(\cdot)$. Observe that we allow for demands to be non-stationary. We only assume that they be independent across periods. At the beginning of each period t, he observes his inventory position I_t and orders quantity $Q_{t,t}$ for the current period (constrained as in (7.8)). In addition, he updates the commitment for the future periods (constrained as in (7.9)). The quantity ordered for the current period is delivered immediately. Subsequently, the buyer sees the demand and satisfies it from stock on–hand as much as possible. Any excess demand is back ordered. As usual the marginal holding, penalty and purchase costs per period are h, π, and c respectively. In the last period, any excess inventory may be salvaged at a unit price of s.

In every period t the buyer needs to determine the actual order quantity to purchase and also update the commitments for future periods so as to minimize total expected costs. This can be formulated as a stochastic dynamic program. The single period expected holding and penalty costs when the total quantity on hand at the beginning of period t after receiving the orders is $I_t + Q_{t,t}$ is written as:

$$L_t(I_t + Q_{t,t}) = h\mathbf{E}[\max\{I_t + Q_{t,t} - D_t, 0\}] + \pi\mathbf{E}[\max\{D_t - I_t - Q_{t,t}, 0\}].$$

Define the expected costs from period i through N given a starting inventory of I_t and commitments $Q_{t-1,t}, \ldots, Q_{t-1,N}$ made in period $t-1$ as:

$$C_t(Q_{t,t}, \ldots, Q_{t,N} | I_t, Q_{t-1,t}, \ldots, Q_{t-1,N}) =$$
$$cQ_{t,t} + L_t(I_t + Q_{t,t}) + \mathbf{E}_{D_t}\{J_{t+1}(I_{t+1}, Q_{t,t}, \ldots, Q_{t,N})\}$$

where, $I_{t+1} = I_t + Q_{t,t} - D_t$ for $t = 1, \ldots, N$ is the inventory balance equation. Define the terminal cost $J_{N+1}(I_{N+1}) = -sI_{N+1}$. Then the buyer's problem in period $t = 2, \ldots, N$ is:

$$\min_{Q_{t,t}, \ldots, Q_{t,N}} C_t(Q_{t,t}, \ldots, Q_{t,N} | I_t, Q_{t-1,t}, \ldots, Q_{t-1,N}) \qquad (7.12)$$

subject to

$$(1 - \underline{\alpha}_{t,i})Q_{t-1,i} \leq Q_{t,i} \leq (1 + \overline{\alpha}_{t,i})Q_{t-1,i} \quad i = t, \ldots, N. \qquad (7.13)$$

For $t = 1$, the buyer solves (7.12) without the constraint (7.13).

Order Process Measurements. Recall that under the newsvendor supply contract with stationary demand the order process is almost identical to the demand process. Hence any variability in the demand is passed on directly to the supplier through the orders. If the demand process is non–stationary, the uncertainty in the order process may even be larger [21]. In addition, under such contracts the supplier never has any information regarding future orders which may assist her in better production planning. RHF contracts attempt to bridge these two gaps. We propose two metrics to measure the effectiveness of RHF contracts for this purpose.

- *Order Process Variability:* RHF contracts restrict the variability in the order process to be less than the variability in the demand process. Recall that $Q_{t,t}$ is the order quantity in period t under the RHF contract. It should be clear that in a finite horizon problem, the average quantity ordered will vary depending upon the various parameters (purchase cost, flexibilities, etc.) of the contract. Therefore, we focus on the coefficient of variation (CV) of period t order, $ORCV_t$, defined as:

$$ORCV_t = \frac{\sqrt{\mathbf{E}[Q_{t,t}^2] - \{\mathbf{E}[Q_{t,t}]\}^2}}{\mathbf{E}[Q_{t,t}]}.$$

Observe that in a finite horizon model, the CV of orders in the early periods will be small. This is basically due to the fact the effective flexibility in the earlier periods is small. In the computational studies with stationary demands we observe that the CV of orders converge as the horizon length increases. We therefore compare this converged value of the CV of orders to the CV of demand and show that the CV of orders in a RHF contract is significantly lower than the CV of demand.

■ *Advance Information*: Under an RHF contract information regarding future orders is conveyed to the supplier through the commitments made for the future by the buyer. By making commitments for the future using RHF contracts a buyer also provides advance information regarding (future) orders. A supplier would be interested in knowing how good this advance information is. We measure the mean absolute deviation (MAD) between the commitment made in period t for period $t+i$ and the actual order placed in period $t+i$.[10] That is,

$$MAD_{t,t+i} = \mathbf{E}|Q_{t,t+i} - Q_{t+i,t+i}|$$

7.7.2 Analysis

It should be clear that an optimal policy for problem RHF will be extremely complex and perhaps unattractive for implementation. This motivates us to develop some approximations. One solution mechanism is the Open Loop Feedback Control (OLFC). The basic idea is to determine the optimal periodical commitments at the beginning of period t, *assuming that these commitments will not be changed at a future period*. While OLFC policies are not optimal, they provide results that are close to optimal [11].

Let $Q_t(i)$ denote the sum $Q_{t,t} + \cdots + Q_{t,i}$ and $D_t(i)$ denote the sum $D_t + \cdots + D_{t+i}$. Let $G(Q_t(t), \ldots, Q_t(N))$ be the total expected costs for periods t through N. We write,

$$G(Q_t(t), \ldots, Q_t(N)) = cQ_t(N) + \sum_{i=t}^{N} M_{t,i}(Q_t(i)),$$

where

$$M_{t,i}(Q_t(i)) = h\mathbf{E}[\max\{Q_t(i) - D_t(i), 0\}] + \pi\mathbf{E}[\max\{D_t(i) - Q_t(i), 0\}]$$

is the expected holding and penalty costs in period $i < N$ for a problem starting in period $t \leq i$. For $i = N$, we write,

$$M_{t,N}(Q_t(N)) = (h-s)\mathbf{E}[\max\{Q_t(N) - D_t(N), 0\}] + \pi\mathbf{E}[\max\{D_t(N) - Q_t(N), 0\}]$$

The OLFC solution is obtained as follows. For period one solve

$$\min_{Q_1(1), \ldots, Q_1(N)} G(Q_1(1), \ldots, Q_1(N))$$

$$\text{s.t.} \quad Q_1(i) \geq Q_1(i-1) \quad \forall i = 1, \ldots, N$$

---

[10] Observe that this is different from the usual MAD measure for a random variable where the average absolute deviation from the mean of the random variable is taken. The expression presented is more appropriate for the current situation.

For period $t = 2, ...N$, solve

$$\min_{Q_t(t),...,Q_t(N)} G(x_t + Q_t(t), ..., x_t + Q_t(N))$$

s.t.

$$Q_t(t+i) \geq Q_t(t+i-1), \qquad \forall i = 0, ..., N-t.$$

$$(1 - \underline{\alpha}_{t,t+i})Q_{t-1,t+i} \leq Q_{t,t+i} \leq (1 + \overline{\alpha}_{t,t+i})Q_{t-1,t+i}, \qquad \forall i = 0, ..., N-t.$$

A Lower Bound. Consider the following specialization of the RHF contract. Set $\underline{\alpha}_{t,i} = 0$ and $\overline{\alpha}_{t,i} = 0$ for $i > t$. That is, a buyer makes a commitment in the first period for quantities to be purchased in each period. While the actual order quantity can be adjusted, the future commitments may not be. Since the buyer is only allowed to adjust the order quantity in the current period, we call this a Zero Lead time Flexibility (ZLF) contract. Thus a ZLF contract is a special case of the RHF contract with $\underline{\alpha}_{t,i} = 0$ and $\overline{\alpha}_{t,i} = 0$ for $t > i$. Furthermore, define matrices $[\underline{\gamma}]$ and $[\overline{\gamma}]$ such that,

$$\underline{\gamma}_{i,j} = \begin{cases} 0 & \text{if } i \neq j \\ 1 - \prod_{k=1}^{j}(1 - \underline{\alpha}_{k,j}) & \text{if } i = j \end{cases}$$

and,

$$\overline{\gamma}_{i,j} = \begin{cases} 0 & \text{if } i \neq j \\ \prod_{k=1}^{j}(1 + \overline{\alpha}_{k,j}) - 1 & \text{if } i = j \end{cases}$$

Consider a ZLF contract with flexibilities $[\underline{\gamma}]$ and $[\overline{\gamma}]$. It can then be shown that a solution to this ZLF contract gives a lower bound to the original RHF contract.

7.7.3 Performance Evaluation

In Bassok and Anupindi [8], we perform extensive computational testing to evaluate the performance of the RHF contracts. From a managerial perspective, we are primarily interested in answers to the following types of questions: (i) How much flexibility of any given type is sufficient? (ii) What is the value of an additional (say, 5%) flexibility? (iii) What is the impact of various types of contracts on coefficient of variation of the order process? (iv) What is the nature of advance information provided by an RHF contract?

The main conclusions from our computational studies with stationary demand can be summarized as follows:

- We observe that the OLFC heuristic performs well when the salvage value of the product can be assumed to be equal to its purchase price. This is a reasonable assumption for most products that are not at the end of their product life cycles. The percentage gap between the OLFC and the suggested lower bound increases with the coefficient of variation (CV) of demand and decreases with flexibility. For CV = 0.5 and flexibility = 5%, the gap was less than 1.6%.

- The amount of flexibility required to get close (within 1.0%) to the newsvendor solution (ideal from a buyer's perspective) is usually low, ranging from 10% for CV of 0.25 to 25% for CV of 0.5.

- There are decreasing returns to the value of additional (say, 5%) flexibility to the buyer.

- Increased flexibility allows a buyer to offer better service as measured by fill-rates. Thus negotiating flexibility (traditionally a purchasing function) impacts customer service (marketing function).

- The variability in the order process is significantly reduced under the RHF contract. For example, when the demand process CV = 0.5 the order process CV is 0.12 at 5% flexibility increasing to 0.30 at 30% flexibility.

- Finally, we observe that the process of generating and updating the commitments for the future merely creates information regarding the future orders. We call this "information processing" since a buyer processes information regarding current demands and available inventory to create new information regarding future orders. Extensive computations suggest that the information that is withing 1–3 periods of the actual order is more "reliable" in the sense that it has less variability (between the information and the order) that that the variability in the order (between the order and mean demand).

To summarize, we have studied two classes of commitment contracts – total volume commitments and periodic commitments. While the former class of contracts effectively ensure a market for the supplier, the later is demonstrated to be effective in reducing the variability in the order process and hence counteract the bull–whip effect.

7.8 COORDINATION AND FLEXIBILITY USING SUPPLY CONTRACTS WITH OPTIONS

In the previous sections we described supply contracts in which the buyer makes periodical commitments and the supplier provides some flexibility for future adjustments of the commitments. The focus of these sections was on analysis of contracts from a buyer's perspective and their implications on the supplier. In contrast in this section, we explore if a certain class of supply contracts with periodical commitments can be implemented in a decentralized system. That is, can the decentralized system achieve channel coordination? Lariviere [20] summarizes a large body of developing literature in the area of supply chain contracting with stochastic demand that seeks to address this issue in a single period setting. In contrast, the model we develop here is for two–period with correlated demands.

The supply contract in this section is similar to those in the previous in the sense that they involve quantity commitments by a buyer. As a mechanism to provide flexibility, however, the buyer may purchase ahead of time some options. Of course, he pays an option price for each unit of option purchased. Options enable the buyer to procure, in a future period, additional quantity of goods at the exercise price. To facilitate discussion of coordination issues in the supply chain, we consider three channel structures:

Centralized System (CS): In this system, the buyer and the supplier cooperate to optimize the total system profits The optimal solution to the CS is called the *first best* solution.

Decentralized System (DS): In this system, the buyer and supplier play a Stackelberg game where the supplier is the leader and the buyer the follower.

Decentralized System, No Options (DSNO): This is similar to the DS but without options. Since there are no options, the buyer has no flexibility to order additional quantities, if needed.

The CS is used as a benchmark case to investigate if there exists an appropriate decentralization mechanism (through prices and/or quantities) such that the DS will achieve the first best solution. That is, we investigate if *channel coordination* is feasible. In addition, we use the DSNO as a benchmark case to (i) determine if both the supplier and the buyer increase their profits from the use of options, and (ii) numerically quantify the benefits of using options.

7.8.1 A General Model

In what follows we describe a general structure of supply contract with options. This framework was suggested and analyzed by Barnes-Schuster et. al. [7].

Consider a supply chain with a single buyer and a single supplier. The buyer sells a single product to the external market. We assume that the product life cycle (season) is short and the buyer has limited ability, arising due to the supplier's production constraints (discussed later), to react to changing market conditions. Therefore, there may be limited opportunity to replenish the stock of goods *during* the season. We model the season as consisting of two periods (possibly of unequal lengths). A two period model allows us to have one replenishment during the season in addition to the one at the beginning of the season.[11] The demand D_t in period $t \in \{1, 2\}$, is assumed to be normally distributed with mean μ_t, standard deviation σ_t with conditional density and distribution functions of $f_{D_t}(\cdot)$ and $F_{D_t}(\cdot)$ respectively. Demands in the two periods are correlated with a correlation coefficient of ρ. We assume that the demand distributions are common knowledge. Since demands are correlated between periods, the buyer will gain some information regarding the second period demand after observing the first period demand. Options allow a buyer to exploit value of this information effectively.

---

[11] In the textile garment industry the ability to place more than one order in a season is increasingly becoming common; see Fisher and Raman [16] and Eppen and Iyer [14].

We assume that the supplier sources the components / material needed to produce the product from its upstream supplier that requires a long lead-time. Hence the supplier has no opportunity to reorder raw material during the season. Since the supplier needs to procure raw material all at once at the beginning of the season, she takes on a risk of buying perhaps too much or too little material. Requiring the buyer to commit to order quantities which cannot be changed is a simple solution which eliminates this risk. However, this solution may not be efficient from a channel perspective, since there might be opportunities (e.g., after updating demand at the end of the first period) to sell more products. Since input raw material is the primary constraint in the channel, the supplier mitigates the risk of raw material purchase by offering options to the buyer at a price. Specifically, we assume that each unit of option allows a buyer to procure an additional unit of the product, if necessary, in the second period at a price. The supplier, however, is now liable to produce products for the second period, up to the number of options purchased by the buyer. Thus a supplier makes a *delivery commitment*.

Before the beginning of the horizon the buyer makes three decisions. He places *firm* orders (commitments) for goods to be delivered at the beginning of periods one and two; these are denoted by Q_1 and Q_2 respectively. In addition, he purchases M *options* from the supplier which give him an opportunity to order additional units of the product (up to the number of options purchased) before the start of period two but after observing demand in period one. The total procurement costs consist of the following components: (i) a unit *wholesale price* (c) for goods procured against firm orders, (ii) an *option price* (c_o) for every unit of option purchased, and (iii) an *exercise price* (c_e) for options exercised to obtain additional products in the second period. Excess demand in the first period is back-ordered. Any excess demand in the second period is considered lost sales. The buyer incurs the standard holding (h_t) and penalty costs (p_t) for the two periods and earns revenues for the products sold. We assume that the product is sold to the consumer at a fixed (exogenously specified) market price r. Any leftover finished goods at the end of the second period may be salvaged at a per unit price of $\tilde{v}_f^b$. We define the *effective salvage value* of finished goods for the buyer to be $v_f^b = \tilde{v}_f^b - h_2$.

The supplier produces goods for the buyer. The supplier must purchase sufficient raw material to produce the maximal quantity possibly requested (that is, $Q_1 + Q_2 + M$, the sum of the committed orders and the options purchased). The per unit procurement cost of raw material is c_r. The supplier has two modes of production. Under the cheaper mode of production (at a per unit labor cost of c_L) she will produce at least enough goods to satisfy the firm orders. Let X_L be the total units produced using the cheaper production mode; clearly $Q_1 + Q_2 \leq X_L \leq Q_1 + Q_2 + M$. She uses the expensive production mode (at a per unit labor cost of $(1 + \gamma)c_L$), if necessary, to satisfy additional goods requested by the buyer against exercised options. We assume that additional production in this period is expensive due to increased labor costs (for example, due to overtime costs). We assume that capacity in both the production modes is unrestricted. Any raw material leftover after satisfying the goods needed against the options exercised by the buyer is immediately salvaged for v_r^s. Hence, the supplier does not hold any raw material inventory in the second period. Any leftover finished goods are also immediately (at the beginning of second period, after satisfying exercised options) salvaged for v_f^s per unit. Since there are no constraints on amount of capacity or raw material availability, the supplier will be able to produce exactly what the buyer requests, and hence experiences no shortage cost. The supplier needs to determine the optimal wholesale, option, and exercise prices, and in addition determine the optimal production quantities for both periods. We make the following assumptions:

- $v_f^s < c_L + c_r$ and $v_r^s < c_r$: This is required to ensure that the profits in all the channel structures considered are finite.

- $v_f^b < c_r + c_L$: This is required to ensure that the total profits in the CS are finite.

The above framework is quite general and can be used to model many different situations. For example, Eppen and Iyer (1997) present a model in which at the beginning of the horizon the buyer makes a firm commitment to purchase during the horizon Q units. At the first period she purchases $Q_1 = Q(1 - \beta)$ units, at the price c per unit. At the second period the buyer may purchase up to $Q\beta$ units at the price c per unit. If the buyer decides to purchase m units, where $m < Q\beta$ then she pays a penalty, b per unit, (for not purchasing $Q\beta - m$ units). The reader may notice that this model is actually a special case of the model presented above. The correspondence between the general model parameters and those in Eppen and Iyer are shown in Table 7.1.

Table 7.1 Correspondence between Eppen & Iyer and the General Model Parameters.

General Model	Q_1	Q_2	M	c	c_o	c_e
Eppen & Iyer	$Q(1 - \beta)$	0	$Q\beta$	c	b	$c - b$

7.8.2 Analysis

A primary goal of our analysis is to explore if channel coordination is achieved. To do this we need to first develop the policy structures and optimal profits for the CS and the DS.

Observe that the conditional mean and standard deviation of a normally distributed and correlated random variable are as follows:

$$\mu_{D_2|d_1} = \mu_2 + \rho\frac{\sigma_2}{\sigma_1}(d_1 - \mu_1) \qquad \text{and} \qquad \sigma_{D_2|d_1} = \sigma_2\sqrt{1 - \rho^2}.$$

Decentralized System. It can be shown that the buyer's profit function is a separable function of (X_1, X_2, X_3) as follows:

$$\Pi^b(X_1, X_2, X_3) = J_1(X_1) + J_2(\dot{X_2}) + J_3(X_3),$$

where $X_1 = Q_1$, $X_2 = Q_1 + Q_2$ and $X_3 = Q_1 + Q_2 + M$ and $0 \leq X_1 \leq X_2 \leq X_3$. $J_i(X_i)$ for $i = 1, 2, 3$ is derived in Barnes–Schuster et. al. [7]. Furthermore,

Proposition 3   *1. The optimal number of options m^* to be exercised in period two is:*

$$m^* = \begin{cases} G(\delta^d, X_2) - G(\delta^d, X_3) & \text{if } c_e > v_f^b \\ M & \text{if } c_e \leq v_f^b \end{cases}$$

where

$$G(x, y) = \max\{x + \eta d_1 - y, 0\},$$

and

$$\delta^d = \mu_2 - \rho\frac{\sigma_2}{\sigma_1}\mu_1 + k_m\sigma_2\sqrt{1 - \rho^2} \quad \text{with} \quad \Phi(k_m) = \frac{p_2 + r - c_e}{p_2 + r - v_f^b},$$

$$\eta = 1 + \rho\frac{\sigma_2}{\sigma_1}$$

2. $J_i(X_i)$ is concave in X_i for $i \in \{1, 2, 3\}$. Hence the buyer's profit function in the DS, $\Pi^b(X_1, X_2, X_3)$, is jointly concave in X_i, $i = 1, 2, 3$.

The supplier's profit function is then derived as a function of the buyer's decisions X_1, X_2, X_3, and m^* to give:

$$\Pi^s(X_L, c, c_o, c_e) = X_2(c - c_o - v_f^s) - X_L(v_r^s - v_f^s + c_L) + X_3(c_o + v_r^s - c_r)$$
$$+ (c_e - v_f^s)\int_{\frac{X_2 - \delta^d}{\eta}}^{\infty}(\delta^d - X_2 + \eta d_1)dF_{D_1}(d_1)$$
$$+ (v_f^s - v_r^s - c_L(1 + \gamma))\int_{\frac{X_L - \delta^d}{\eta}}^{\infty}(\delta^d - X_L + \eta d_1)dF_{D_1}(d_1)$$
$$+ (v_r^s + c_L(1 + \gamma) - c_e)\int_{\frac{X_3 - \delta^d}{\eta}}^{\infty}(\delta^d - X_3 + \eta d_1)dF_{D_1}(d_1).$$

The supplier then solves the following problem:

$$\max_{X_L, c, c_o, c_e} \Pi^s(X_L, c, c_o, c_e) \quad \text{s.t.} \quad X_2 \leq X_L \leq X_3.$$

Centralized System. Similar results can be developed for the CS. Note that c, c_o and c_e play no role in the expression for the joint profits since they merely affect transfer payments between the two parties. As a result, the only decisions which affect system profit are decisions concerning order and production quantities. Furthermore, the second period committed order (Q_2) used in the DS plays no role in the CS. The amount shipped to the buyer's location (to satisfy demand) in the second period is a function of the total production (using both modes of production) and the number of options exercised. Let X_1^c, X_L^c, and X_3^c represent the respective quantities in the CS corresponding to X_1, X_L, and X_3 in the DS.

It can then be shown [7] that the joint profit function in the CS is also separable in X_i^c for $i = 1, L, 3$

$$\Pi^c(X_1^c, X_L^c, X_3^c) = J_1^c(X_1^c) + J_L^c(X_L^c) + J_3^c(X_3^c),$$

where $0 \leq X_1^c \leq X_L^c \leq X_3^c$ and X_1^c is the quantity available to satisfy first period demand, X_L^c is the total quantity produced using the cheaper mode, and X_3^c is the total amount of raw material purchased. Barnes–Schuster et. al. then derive the policy structure for the number of options to be exercised and show that the profit function is concave in X_1^c, X_L^c, and X_3^c.

7.8.3 Channel Coordination

It is can be shown that the optimal profits in the CS is, in general, larger than in the DS and hence the channel is not coordinated. One general way to achieve coordination is if the buyer "internalizes" the costs of the supplier. This occurs, for example, when the supplier sells the "firm" to the buyer implemented by charging marginal costs for the various transactions. There are three price instruments available to the supplier, namely, the wholesale, option, and exercise prices. All are assumed to be linear. In the CS, the marginal cost of providing goods against options is not constant due to the fact that $X_L^c - X_1^c$ units of goods are produced using the cheaper production mode and the rest using the expensive production mode. Since c_e is assumed to be linear, in general, the number of options exercised in the CS and the DS will differ. This appears to the be the main difficulty in achieving coordination. However, sufficient conditions can be developed under which channel coordination is achieved with linear prices. For example, it can be shown that if $v_f^s \leq v_f^b$ and $v_r^s + c_L > v_f^b$ then channel coordination is achieved with $c = c_r + c_L$, $c_o = c_r - v_r^s$, and $c_e = v_r^s + c_L(1 + \gamma)$. Barnes–Schuster et. al. develop several other sufficient conditions under which channel coordination is achieved. These conditions depend on the salvage values of the finished goods for the buyer and the supplier, the procurement cost and the salvage value of raw material for the supplier, the labor cost of production, and the correlation coefficient of demand.

Even though channel coordination is achieved the profits of the buyer and supplier may be affected differently. For example, under the sufficient conditions described and the corresponding prices, the supplier makes zero profits. This is not surprising since the supplier resorted to marginal cost pricing. Thus the prices that lead to the coordinated solution, however, is not *individually rational* and hence may not be implementable. Individual rationality can be restored using non–linear prices; e.g., a two–part tariff with a fixed fee of A and the linear prices obtained earlier. Alternately, quantity discount schemes

could be used. Barnes–Schuster et. al. propose two types of discount schemes
– *simple* (or unbundled) and *bundled* – each of which achieves channel coordination.

Let Π^{c*} be the optimal profits in the CS. Let $\hat{X}_j^*$ be the unconstrained optimal quantities in the DS, that is, without the restriction that $X_1 \leq X_2 \leq X_3$. Consider the prices that lead to channel coordination. For these prices let $\tilde{X}_j^*$, $j \in \{1,2,3\}$ be the optimal quantities with the restriction that $X_2 = X_3$. Thus $\tilde{X}_j^*$ is the optimal solution when no options are purchased. The all–unit discount schemes are then described as follows. Consider the set of wholesale, option, and exercise prices that give channel coordination (but is not individually rational).

Simple Discount Scheme:

$$c^{new} = c + A_w/X_2 \qquad \text{if } X_2 > 0$$
$$c_o^{new} = c_o + A_o/M \qquad \text{if } M > 0$$

for constants A_w and A_o such that $A_w \leq \Pi^{c*} - \Pi^b(0,0,\hat{X}_3^*)$, $A_o \leq \Pi^{c*} - \Pi^b(\tilde{X}_1^*, \tilde{X}_2^*, \tilde{X}_2^*)$, and $A_w + A_o \leq \Pi^{c*}$.

Bundled Discount Scheme:

$$c^{new} = c + A_w/X_2; \qquad c_o^{new} = c_o + A_o/M \qquad \text{if } X_2 > 0 \text{ and } M > 0$$
$$c^{new} = c + \Pi^{c*}/X_2; \qquad\qquad\qquad\qquad \text{if } X_2 > 0 \text{ and } M = 0$$

for constants A_w and A_o such that $A_w \leq \Pi^{c*} - \Pi^b(0,0,\hat{X}_3^*)$, and $A_o = \Pi^{c*} - A_w$.

Suppose, in addition, that the holding and penalty cost of the buyer do not depend on the wholesale price c^{new} and/or option price c_o^{new}. Then, Barnes–Schuster et. al. show that the supplier can achieve channel coordination with the new wholesale and option prices, c^{new} and c_o^{new}, and with c_e unchanged. For these sets of prices, the supplier's expected profits is $A_w + A_o$. A key difference between the two discount schemes is that while a supplier can extract all of the channel profits under the bundled discount scheme, she may not be able to do so under the simple discount scheme.

Use of non–linear prices, while prevelant, is not universal [15]. When non–linear prices are not used, ensuring individual rationality implies that channel coordination is not achievable. Thus in the DS, the supplier optimizes her profits within the class of linear wholesale, option, and exercise prices. Several structural properties of these systems are derived in Barnes–Schuster et. al. In addition, numerical studies indicate that both the buyer and the supplier gain from the use of options, though the supplier usually gains more. Furthermore, as expected, the value of options increases with correlation in demand.

7.9 SUMMARY AND CONCLUSION

In this chapter, we first presented a general framework for supply contracts. The key objective of supply contracts is to strive for coordination between a supplier and a buyer. The main problems that need to be addressed are those of incentives and information asymmetries. We argued that there are two main streams of research – *analysis* and *design* of contracts. The chapter addresses these two issues. The first part of the chapter discusses analysis of contracts. We suggest that the extant inventory literature could be viewed as analyses of specific types of supply contracts. A contractual view of classical inventory theory raises some issues not addressed in the past and thus questions whether the contract structure assumed (in this literature) will be observed in practice. We suggest a class of "quantity commitment contracts" and present analyses of several such contracts. Under such contracts a buyer agrees to either purchase a certain quantity during the entire horizon or to purchase a certain quantity every period. It is important to note that the buyer's commitment is always followed by a delivery commitment by the supplier.

Quantity commitment contracts usually provide some *flexibility* to the buyer, perhaps at a cost. Flexibility may take different forms. These include flexibility to adjust commitments upwards only and both upwards and downwards. Furthermore, these adjustments may or may not be dependent on the specific commitments made and their magnitude may be limited. Usually there is a cost associated with the adjustments and a supplier could price based on the flexibility offered.

In each of the contracts discussed, quantity commitments have several important roles. One of the main purposes of quantity commitments is to provide the supplier with reliable information with respect to the buyer's overall demand and specific future orders. This information is used by the supplier to prepare enough capacity and raw materials to meet the buyer's requirements. Another important role of (periodical) quantity commitments is to reduce the uncertainty passed onto the supplier, and to share the risks, due to uncertainty, between the two parties. Finally, commitments ensure markets for the supplier.

The second part of the chapter focuses on the design of supply contracts. Clearly, the design of an "optimal" contract will be context specific. The two problems of incentives and information need to be considered. Information asymmetries could be due to private information regarding demands, costs, policy structures, inventories, etc. In this chapter, we concentrate on incentive contracts and assume symmetric information between the players. We describe context most suitable for a fashion (short product life-cycle) industry and suggest commitments together with options as a mechanism to achieve coordination of the channel.

The area of supply contracts is new and provides several opportunities for further research. Once again the research could be either analysis or design. Future research in analysis of contracts could be to incorporate other parameters of the supply contract to study a decision maker's behavior and implications on the other player. In addition, most analyses thus far has considered the buyer. Analysis of contracts from a supplier's perspective will be another useful area of research. Finally, analysis of contracts for complex settings – multiple players, multiple products – need to be addressed. Analysis of richer environments with several buyers and suppliers that compete for markets and resources will be a helpful tool to most practitioners and is crucial to the understanding of the relationship between market structures and the form of the different types of supply contracts. Future research in design of supply contracts should include information asymmetries. Several researchers are working to incorporate asymmetric information regarding demand, costs, and policy structure are in progress [13]. Most of these consider a single supplier and a single buyer, assume players to be risk neutral, and restrict to single period models. Relaxing these constraints to incorporate more realistic settings will be very useful in understanding how to structure the material, information, and financial flows in a supply chain.

Acknowledgements

We would like to thank Krishnan Anand for his helpful comments on an early draft of this chapter.

230

References

[1] Agrawal, N., Nahmias, S., and Tsay, A. (1998). A review of literature on contracts in supply chain management. In Tayur, S., Magazine, M., and Ganeshan, R., editors, *Quantitative Models for Supply Chain Management* (Chapter 10). Kluwer Academic Publishers..

[2] Anupindi, R. (1993). *Supply Management Under Uncertainty.* PhD thesis, Graduate School of Industrial Administration, Carnegie Mellon University.

[3] Anupindi, R. and Akella, R. (1997). An inventory model with commitments. October 1993. Under Revision.

[4] Anupindi, R. and Bassok, Y. (1995). Analysis of supply contracts with total minimum commitment and flexibility. In *Proceedings of the 2nd International Symposium on Logistics.* University of Nottingham, UK.

[5] Anupindi, R. and Bassok, Y. (1997). Approximations for multiproduct contracts with stochastic demands and business volume discounts: Single supplier case. Accepted for publication in *IIE Transactions.*

[6] Arrow, K., Karlin, S., and Scarf, H. (1958). *Studies in the Mathematical Theory of Inventory and Production.* Stanford University Press.

[7] Barnes-Schuster, D., Bassok, Y., and Anupindi, R. (1998). Supply contracts with options: Flexibility, information, and coordination. Working Paper; J.L. Kellogg Graduate School of Management, Northwestern University, Evanston IL 60208.

[8] Bassok, Y. and Anupindi, R. (1997). Analysis of supply contracts with commitments and flexibility. March 1995. Revised August 1997. Northwestern University, Evanston, IL 60208.

[9] Bassok, Y. and Anupindi, R. (1997b). Analysis of supply contracts with total minimum commitment. *IIE Transactions*, 29(5):373–381.

[10] Bassok, Y., Srinivasan, R., Bixby, A., and Wiesel, H. (1995). Design of component supply contracts with forecast revision. *IBM Journal of Research and Development*, 41(6).

[11] Bertsekas, D. P. (1987). *Dynamic Programming: Deterministic and Stochastic Models*. Prentice Hall, Inc.

[12] Cachon, G. and Lariviere, M. (1997). Contracting to assure supply or what did the supplier know and when did he know it? Working paper; Fuqua School of Business, Duke University, Durham, NC.

[13] Corbett, C. and Tang, C.(1998). Designing supply contracts: Contract type and information asymmetry. In Tayur, S., Magazine, M., and Ganeshan, R., editors, *Quantitative Models for Supply Chain Management* (Chapter 9). Kluwer Academic Publishers.

[14] Eppen, G. and Iyer, A. (1997). Backup agreements in fashion buying – the value of upstream flexibility. *Management Science*, 43(11):1469–1484.

[15] Felstead, A. (1993). *The Corporate Paradox: Power and Control in the Business Franchise*. Routledge, London and New York.

[16] Fisher, M. and Raman, A. (1996). Reducing the cost of demand uncertainty through accurate response to early sales. *Operations Research*, 44(1):87–89.

[17] Graves, S. et al., editors (1992). *Handbooks in Operations Research and Management Science, Volume 4 : Logistics of Production and Inventory*. Elsevier Science Publishers.

[18] Katz, P., Sadrian, A., and Tendick, P. (1994). Telephone companies analyze price quotations with bellcore PDSS software. *Interfaces*, 24:50–63.

[19] Kumar, A. (1992). *Supply Contracts and Manufacturing Decisions*. PhD thesis, Graduate School of Industrial Administration, Carnegie Mellon University, Pittsburgh, PA 15213.

[20] Lariviere, M. (1998). Supply chain contracting and coordination with stochastic demand. In Tayur, S., Magazine, M., and Ganeshan, R., editors, *Quantitative Models for Supply Chain Management* (Chapter 8). Kluwer Academic Publishers.

[21] Lee, H. L., Padmanabhan, V., and Whang, S. (1997). Information distortion in a supply chain: The bullwhip effect. *Managament Science*, 43:546–558.

[22] Tsay, A., and Lovejoy, W. (1998). Quantity Flexibility Contracts and Supply Chain Performance. Working Paper, Santa Clara University, Santa Clara, CA.

232

[23] Moinzadeh, K. and Nahmias, S. (1996). Adjustment strategies for a fixed delivery contract. Working Paper, #8-96, School of Business Administration, University of Washington, Seattle.

[24] Sadrian, A. and Yoon, Y. (1992). Business volume discounts: A new perspective on discount pricing strategy. *International Journal of Purchasing and Materials Management*, 28(2):43–46.

[25] Scheller-Wolf, A. and Tayur, S. (1998). A Markovian Dual–Source Production Inventory Model with Order Bands. Working Paper, #1998–E200, Graduate School of Industrial Administration, Carnegie Mellon University, Pittsburgh, PA 15213.

[26] Tirole, J. (1990). *The Theory of Industrial Organization*. The MIT Press, Cambridge, Massachusetts.

8 SUPPLY CHAIN CONTRACTING AND COORDINATION WITH STOCHASTIC DEMAND

Martin A. Lariviere

Fuqua School of Business
Duke University
Durham, NC 27708

Recent years have seen a growing interest among both academics and practitioners in the field of supply chain management. With that has come a growing body of work on supply chain contracts. Few firms are so large and few products so simple that one organization can manage the entire provision of the good. Rather, most supply chains require the coordination of independently managed entities who seek to maximize their own profits. Issues of who controls what decisions and how parties will be compensated become critical. An understanding of contractual forms and their economic implications is therefore an important part of evaluating supply chain performance.

The literature on supply chain contracts can be roughly split into two classes. The first takes a particular contract and determines what optimal actions are assuming that the contract terms are fixed. Some attention may be given to the impact of contract parameters on agents' profits or costs, but there is generally no attempt to establish systematically whether the contract allows the decentralized system to perform as well as a centralized one (i.e., whether the contract can *coordinate* the system). Eppen and Iyer (1997) and Brown and Lee (1997) are two recent examples that fall within this class.

The second class takes agents' optimal policies under a contract as given and considers whether the terms of trade can be adjusted to at least improve, if not coordinate, the supply chain. Generally, one supposes that a player may propose the specific terms within a contractual form and asks what contract she would offer. This line of work is closely related to the work on vertical restraints in the economics literature (Mathewson and Winter, 1984) and channel coordination in marketing (Jeuland and Shugan, 1983; Moorthy, 1987). In this chapter, we essentially ignore the first class of papers and focus on a subset of the second. Our basic model is a one period setting in which a manufacturer sells to a retailer facing a newsvendor problem.

Most stochastic inventory models rest in part on intuition gained from the single period newsvendor problem. It is our contention that the same holds true for supply chain contracting under stochastic demand. Under the contracts we consider here, the fundamental inventory problem remains sufficiently simple that one can characterize its solution with some precision. Because the solution to the inventory problem is well understood, we can develop a detailed analysis of the economic incentives the contracts provide. In particular, we will show that contracting on excess inventory in the form of returns policies can greatly improve channel performance. Returns policies allow for the parties to place "bets" on how realized demand compares to the chosen stocking level. Improved channel performance follows from the manipulation of these wagers.

In what follows, we first present the basic assumptions of the model and evaluate the performance of the integrated channel. Section 8.2 presents the simplest terms of trade, a price-only contract. Subsequent sections examine richer contractual forms that allow the channel to be coordinated. Section 8.3 considers buy back contracts while section 8.4 analyzes quantity flexibility contracts. Section 8.5 briefly discusses other coordination schemes found in the literature. Section 8.6 offers concluding remarks and directions for future research.

8.1 MODEL ASSUMPTION AND INTEGRATED CHANNEL PERFORMANCE

We consider a one-period setting in which a manufacturer sells to a retailer facing demand from consumers. If the channel fails to provide sufficient stock, unmet demand is lost. The retail price is fixed at r per unit regardless of the terms the manufacturer offers. The manufacturer can produce the good at a constant marginal cost c. Any unsold inventory can be salvaged at a value v per unit. To avoid trivial problems, assume $r > c > v$. Obviously, the setting can be characterized as a newsvendor problem.

We assume that demand ξ has a continuous distribution $F(\xi)$ on the non-negative reals with density $f(\xi)$. In addition it will be convenient to assume that $F(\xi)$ is invertible and that $f(\xi)$ has a continuous derivative $f'(\xi)$. Also let $\bar{F}(\xi) = 1 - F(\xi)$. The demand distribution, as well as all cost and revenue information, are common knowledge.

It is straightforward to verify that the profits of an integrated firm (i.e., one that controls both manufacturing and sales to the public) for stocking level y are:

$$\Pi_I(y) = (r - c) y - (r - v) \int_0^y F(\xi) \, d\xi. \tag{8.1}$$

The problem is concave in the stocking level, and the optimal solution is given by:

$$y_I = F^{-1} \left(\frac{r - c}{r - v} \right)$$

where F^{-1} is the inverse of the cumulative distribution. Denote the maximum system profits by $\Pi_I^* = \Pi_I(y_I)$. Note that system profits are completely determined by the stocking level. Looking ahead to when we consider contracting between independent parties below, we can prove that terms coordinate the channel if we can show that they induce the choice of the centralized system's optimal stocking level, y_I.

As we consider contracting in this environment, we assume the manufacturer acts as a Stackelberg leader. She offers the terms of trade as a take-it-or-leave-it proposition to the retailer, which he can only accept or reject. We assume he accepts the terms if they allow him to earn a non-negative return (perhaps net of some opportunity cost). Clearly, this is a gross simplification; it would be more natural to assume that the parties negotiate over the terms. Unfortunately, properly modeling bargaining is sufficiently complex that no consensus exists regarding the appropriate equilibrium concept (Salanié, 1997). We skirt the issue by assigning all the bargaining power to one player. Where appropriate, we will comment on how the results would differ if the retailer were able to offer the terms of trade.

8.2 PRICE-ONLY CONTRACTS

Here we consider price-only contracts; the manufacturer offers the good at a per unit wholesale price w, and the retailer retains possession of any excess stock. We first examine the retailer's problem under a price-only contract before turning to the manufacturer's problem. We present sufficient conditions for the manufacturer's profits to be unimodal and characterize the optimal wholesale price (where "optimal" should be interpreted as maximizing the manufacturer's profits). We show that whenever the manufacturer proposes the terms of trade, a price-only contract fails to coordinate the channel. We close the section with a brief discussion of standard remedies from economics for improving channel performance.

8.2.1 The retailer's problem

The retailer faces a problem analogous to that of the integrated channel given in (8.1). The principal difference is that the retailer must buy stock at price w instead of producing it at cost c. If we assume that the retailer has the same salvage opportunities as the integrated channel, we then have:

$$\Pi_R(y) = (r - w)y - (r - v)\int_0^y F(\xi)\,d\xi.$$

Clearly, $\Pi_R(0) = 0$ and $\Pi'_R(0) > 0$ if $w < r$. Thus if we normalize the retailer's opportunity cost to zero, he will accept any contract such that the wholesale price is less than the retail price and order a positive quantity from the manufacturer.[1] The retailer's problem is concave and the optimal solution is given by:

$$y(w) = F^{-1}\left(\frac{r - w}{r - v}\right).$$

8.2.2 The manufacturer's problem

Acting as a Stackelberg leader, the manufacturer correctly anticipates how the retailer will order for any wholesale price. She therefore anticipates facing a demand curve $y(w)$, yielding a profit function of:

$$
\begin{aligned}
\Pi_M(w) &= (w - c)y(w) && (8.2)\\
&= (w - c)F^{-1}\left(\frac{r - w}{r - v}\right).
\end{aligned}
$$

There are a few points worthy of mention from (8.2). First, under a price-only contract, the manufacturer's profits are deterministic. She knows exactly what the retailer will order at every wholesale price and bears no responsibility for

---

[1]This will not necessarily hold if there is a positive stock out penalty. See Lariviere and Porteus, 1997.

the goods once the retailer takes possession. All uncertainty regarding channel profits is foisted onto the retailer. The richer contracts we consider in subsequent sections differ from price-only contracts by allowing the manufacturer to assume some of the risk arising from random demand.

A second observation is that (8.2) is simply not the most convenient expression with which to work. We would like to develop conditions for the manufacturer's profit function to be well behaved. When, for example, is it unimodal? Answering such questions could be exceedingly difficult since we lack general statements about the inverse of the cumulative density function. We would be forced to evaluate (8.2) distribution by distribution.

To allow for broader statements, we follow Lariviere and Porteus (1998) to develop an alternative expression for manufacturer profits. First, instead of working with the demand curve $y(w)$, we work with the inverse demand curve $w(y)$, where

$$w(y) = (r - v)\bar{F}(y) + v.$$

The change may be interpreted as follows. Previously, we had assumed the manufacturer chose her wholesale price while anticipating selling the most the retailer would freely take at that price. Now, we are assuming that manufacturer chooses how much she wants to sell anticipating receiving the most per unit that the retailer would freely pay to take all of the offered stock. In a competitive setting, choosing quantities instead of price can lead to markedly different outcomes. Here, however, the approaches are equivalent since the manufacturer holds a monopoly position.

The second change we make is to alter the costing conventions. Let $\hat{w} = w - v$ and $\hat{r} = r - v$. The inverse demand curve can now be written as:

$$\hat{w}(y) = \hat{r}\bar{F}(y). \tag{8.3}$$

Under the revised accounting structure, the retailer immediately credits himself for salvaging the goods upon receipt, thus lowering the effective wholesale price. To keep himself from booking deceptively high profits, he must also lower the retail price.[2] While in what follows we will refer to $\hat{w}$ and $\hat{r}$ as the wholesale and retail prices, respectively, one should keep in mind that they are, in fact, mark ups over the salvage value.

An immediate observation following from (8.3) is that the "market-clearing" wholesale price $\hat{w}(y)$ is proportional to the retail price $\hat{r}$. Another consequence is a simpler expression for manufacturer profits:

$$\begin{aligned} \Pi_M(y) &= y(\hat{w}(y) - \hat{c}) \\ &= y(\hat{r}\bar{F}(y) - \hat{c}) \end{aligned} \tag{8.4}$$

where $\hat{c} = c - v$. Expressing the marginal cost of production as a mark up over the salvage value assures that (8.2) and (8.4) yield the same results for a given

---

[2]The interpretation requires an obvious modification if v is negative and the retailer pays to dispose of stock.

$(y, \hat{w}(y))$ pair (or, equivalently, a $(\hat{w}, y(\hat{w}))$ pair). We also have an alternative statement of the manufacturer's problem: Choose y to maximize $\Pi_M(y)$.

8.2.3 Characterizing the optimal solution

In examining the manufacturer's problem, a few points are obvious. First, she will never sell more than y_I since that would necessitate a wholesale price below the marginal cost of production. Second, if $F(\xi) > 0$ for all $\xi > 0$, she will choose a sales quantity strictly between 0 and y_I which will result in a wholesale price between $\hat{c}$ and $\hat{r}$. The optimal sales quantity will be an interior solution, and first order conditions must hold.[3]

But are first order conditions sufficient? Is it possible for the manufacturer's problem to have multiple local maxima? Alternatively, is it possible that there is no solution to the manufacturer's problem? Both are possible; the manufacturer's first order conditions may have multiple or no solution. To see the former, consider the following density defined for $\xi \in (0,1)$:

$$f(\xi) = Cos(32\pi\xi) + 1. \tag{8.5}$$

The corresponding profit function exhibits multiple local maxima and minima. While first order conditions must hold at the optimal solution, they may hold at multiple points, some of which may be local minima.

For an example of a distribution for which first order conditions never hold, consider the Pareto distribution with $f(\xi) = k\theta^k/\xi^{k+1}$ for $\xi \geq \theta$. The corresponding inverse demand curve,

$$\hat{w}(y) = \hat{r}\bar{F}(y) = \hat{r}\theta^k y^{-k},$$

is isoelastic with elasticity $1/k$. For the manufacturer's profits to be concave we require that the elasticity be greater than one or, equivalently, that $k < 1$. However, for $k < 1$, the mean of the Pareto, and hence the retailer's problem, is undefined. One can show that for any problem with finite expected demand the manufacturer's problem is convex; she will charge $\hat{w} = \hat{r}$ and sell the minimum quantity θ.

The density given in (8.5) is admittedly perverse and an unrealistic model of demand. The Pareto example is a little more troubling; while not commonly used as a demand distribution, it is not a completely absurd choice for a modeler. It forces one to question what other distributions fail to yield sensible solutions to the manufacturer's problem.

---

[3]There are two exceptions to note. If $F(\xi) = 0$ for all $\xi \in [0, \bar{\xi}]$, the manufacturer may sell $\bar{\xi}$ units at a wholesale price of r. We cannot rule out the possibility that she prefers the corner solution of extracting all profits on the certain sales of $\bar{\xi}$. In addition, if the retailer has a positive opportunity cost that exceeds his profits under the manufacturer's optimal retail price, the manufacturer will deviate from the price found from the first order conditions. It is straightforward to show that the manufacturer prefers cutting her wholesale price to charging the unconstrained price and paying a lump sum to gain participation (Lariviere and Padmanabhan, 1997).

In order to assure that the manufacturer's first order conditions have a unique solution, we must place limits on the demand distribution that rule out the likes of (8.5) and the Pareto. To that end, define $h(\xi)$ as the failure, or hazard, rate function of the demand distribution.

$$h(\xi) = \frac{f(\xi)}{\bar{F}(\xi)}.$$

In our setting, $h(\xi)\, d\xi$ may be interpreted as the probability that demand will lie in the interval $[\xi, \xi + d\xi]$ given that demand is at least ξ. A distribution is said to have an increasing failure rate (IFR) if $h'(\xi) > 0$ for all ξ.

Define $g(\xi) = \xi h(\xi)$ as the *generalized failure rate*. We say that a distribution has an increasing generalized failure rate (IGFR) if $g'(\xi) > 0$ for all ξ. Clearly, a distribution that is IFR is also IGFR, but the distinction is not vacuous. Consider the Weibull distribution,

$$f(\xi) = \theta k \xi^{k-1} e^{-\theta \xi^k},$$

for $k > 0$ and $\theta > 0$, or the gamma distribution,

$$f(\xi) = \theta^k \xi^{k-1} e^{-\theta \xi} / \Gamma(k),$$

for $k > 0$ and $\theta > 0$. Both are IFR for a restricted set of parameters (i.e., $k > 1$) but are IGFR for all parameters. IGFR distributions offer additional flexibility over IFR ones that are particularly relevant in our setting. For example, IFR distributions must have a coefficient of variation of less than one (Barlow and Proschan, 1965) while IGFR distributions are not similarly restricted.

The following theorem shows that an IGFR demand distribution is also sufficient for manufacturer profits to be well-behaved.[4]

Theorem 1 *Let $\nu(\xi)$ denote the own-price elasticity of the retailer's orders to the manufacturer, and let y^1 be the smallest value of ξ such that $g(\xi) = 1$. If no such ξ exists, let $y^1 = \infty$.*

1. *The elasticity of retailer's orders is given by*

$$\nu(\xi) = 1/g(\xi). \tag{8.6}$$

2. *The manufacturer's first order conditions may be written as*

$$\hat{w}(y)(1 - 1/\nu(y)) = c. \tag{8.7}$$

---

[4]This and most of the results in the current section are from Lariviere and Porteus, 1998.

3. *If the demand distribution is IGFR, then the manufacturer's profits are unimodal on $[0, \infty)$, concave on $[0, y^1]$, and strictly decreasing on $[y^1, \infty)$. Any solution to (8.7) is a unique global maximum and must lie in the interval $[0, y^1]$.*

Proof: For the first part of the theorem, note that $\nu(\xi)$ is defined as $\nu(\xi) = -\frac{1}{\hat{w}'(\xi)} \frac{\hat{w}(\xi)}{\xi}$. The result is then immediate from the definition of $\hat{w}(\xi)$. The second part follows from a standard microeconomic result (Kreps, 1990). Differentiating revenue, $y\hat{w}(y)$, yields marginal revenue, $MR(y)$:

$$MR(y) = \hat{r}\bar{F}(y) - y\hat{r}f(y) = \hat{r}\bar{F}(y)(1 - g(y)).$$

The condition (8.7) thus reduces to setting marginal revenue equal to marginal cost. For the final part of the theorem, note that IGFR implies that $\bar{F}(y)$ and $1 - g(y)$ are decreasing on $[0, \infty)$. Both are positive on $[0, y^1]$ so marginal revenue is positive and decreasing there. Hence revenue and profits are concave on $[0, y^1]$. As $MR(y^1) = 0$, any solution to (8.7) must lie in $[0, y^1]$ and from concavity must be unique. Profits are falling on $[y^1, \infty)$ because revenue is strictly decreasing while costs are strictly increasing over this region. $\square$

Note that $\nu(\xi)$ is the elasticity of the orders the retailer places with the manufacturer, not the elasticity of end consumer demand. We have not modeled consumer behavior and, indeed, have assumed a fixed retail price. On the other hand, by modeling the retailer's ordering policy, we have defined an induced demand curve (and the equivalent induced inverse demand curve) that the manufacturer faces. $\nu(\xi)$ measures the percent change in the retailer's orders for a percent change in the wholesale price. Thus, if $\nu(\xi) > 1$, a wholesale price cut is offset by such an increase in volume that total revenue increases. Conversely, if $\nu(\xi) < 1$, a price increase will boost total revenue. An immediate consequence of the theorem is the standard result that the optimal sales quantity will lie on the elastic portion of the demand curve (i.e., $\nu(\xi) > 1$).

The relationship between the elasticity of orders and the generalized failure rate captured by (8.6) offers some intuition for why IGFR is the correct sufficient condition for a well-behaved profit function. IGFR implies that the elasticity of orders falls monotonically with the sales quantity. If the elasticity of orders falls monotonically, then marginal revenue falls monotonically and can equal the marginal cost of production at only one point.

The two poorly-behaved examples we considered above fail to be IGFR. The generalized failure rate for the density in (8.5) is not monotonic since the density itself drops to zero periodically. The Pareto, on the other hand, has a constant generalized failure rate. For meaningful parameter values, the generalized failure rate is greater than one and marginal revenue is everywhere negative.

While the theorem limits the distributions that one can be assured are well-behaved, the IGFR class is broad enough to capture most of the distribution a modeler would choose to employ. In addition to the distributions discussed above, the uniform and the normal are both IFR and hence IGFR. In addition to assuring a well behaved profit function, IGFR gives us some additional attractive properties.

Theorem 2 *If the demand distribution is IGFR, then the optimal sales quantity y^* is increasing $\hat{r}$ and decreasing in $\hat{c}$. Additionally, $\hat{w}(y^*) \geq \hat{w}(y^1)$.*

Proof: The first part follows immediately from noting that IGFR is sufficient for marginal revenue to be monotonically decreasing on $[0, y^1]$. The second part follows from the fact that y^* must lie in $[0, y^1]$. $\square$

Thus an IGFR distribution assures that the optimal sales quantity changes as one would expect with respect to cost and revenue parameters. The lower bound on the manufacturer's profit maximizing wholesale price $\hat{w}(y^*)$ given by her revenue maximizing price $\hat{w}(y^1)$ is useful since y^1 may be found simply from the generalized failure rate. For example, the generalized failure rate for the Weibull is $g(\xi) = \theta k \xi^k$. Solving for the sales quantity such that $\nu(y) = 1$ yields $y^1 = \left(\frac{1}{\theta k}\right)^{1/k}$. The lower bound on the wholesale price is then

$$\hat{w}(y^1) = \hat{r} e^{-1/k}. \tag{8.8}$$

We are consequently able to conclude that the optimal wholesale price for the Weibull goes to the retail price as the parameter k goes to infinity.

8.2.4 Special cases

The preceding results give a general characterization of the optimal solution. Further assumptions regarding the demand distribution allow some additional analysis. First we require some definitions. We say a distribution is from a "scaled" family if the distribution depends on a parameter θ and there exists an increasing, positive function $\tau(\theta)$ such that

$$F(\xi|\theta) = F\left(\frac{\xi}{\tau(\theta)} \Big| 1\right).$$

We immediately have that for scaled families $f(\xi|\theta) = \frac{1}{\tau(\theta)} f\left(\frac{\xi}{\tau(\theta)} \Big| 1\right)$, $h(\xi|\theta) = \frac{1}{\tau(\theta)} h\left(\frac{\xi}{\tau(\theta)} \Big| 1\right)$, and $g(\xi|\theta) = g\left(\frac{\xi}{\tau(\theta)} \Big| 1\right)$. The most obvious examples of scaled families are uniform $[0, \theta]$ random variables or exponential random variables with mean θ. Less obvious examples are the Weibull and gamma distributions.

We say that a family of distributions is "shifted" if, for some parameter $\theta \geq 0$,

$$F(\xi|\theta) = F(\xi - \theta|0).$$

This implies that $f(\xi|\theta) = f(\xi - \theta|0)$, $h(\xi|\theta) = h(\xi - \theta|0)$, and $g(\xi|\theta) \geq g(\xi - \theta|0)$. The normal distribution with mean θ and fixed standard deviation σ is an example of a shifted family.

Theorem 3 *Suppose that the demand distribution $F(\xi|\theta)$ is IGFR for all values of θ.*

1. *If $F(\xi|\theta)$ is from a scaled family, the optimal order quantity is proportional to $\tau(\theta)$ and the resulting wholesale price is independent of θ. That is:*

$$y^*(\theta) = \tau(\theta) y^*(1) \quad and \quad \hat{w}^*(\theta) = \hat{w}^*(1)$$

   where $y^*(\theta)$ and $\hat{w}^*(\theta)$ are the optimal values of y and $\hat{w}$ given θ.

2. *If $F(\xi|\theta)$ is a shifted family, then for $\Delta > 0$:*

$$y^*(\theta + \Delta) < y^*(\theta) + \Delta \quad and \quad \hat{w}^*(\theta) < \hat{w}^*(\theta + \Delta).$$

   If in addition $F(\xi|\theta)$ is IFR, then $y^*(\theta) < y^*(\theta + \Delta)$.

Proof: For the first part, the manufacturer's first order conditions for an arbitrary value of θ can then be written as:

$$\hat{r}\bar{F}(y^*(\theta)|\theta)(1 - g(y^*(\theta)|\theta)) =$$
$$\hat{r}\bar{F}(y^*(\theta)/\tau(\theta)|1)(1 - g(y^*(\theta)/\tau(\theta)|1)) = c. \qquad (8.9)$$

However, for $\theta = 1$, first order conditions are:

$$\hat{r}\bar{F}(y^*(1)|1)(1 - g(y^*(1)|1)) = c.$$

Thus (8.9) must be satisfied by $y^*(\theta) = \tau(\theta) y^*(1)$. The corresponding wholesale price is found from:

$$\hat{w}^*(\theta) = \hat{r}\bar{F}(y^*(\theta)|\theta) = \hat{r}\bar{F}\left(\frac{\tau(\theta) y^*(1)}{\tau(\theta)}\Big|1\right) = \hat{w}^*(1).$$

For the second part, we first show that $y^*(\theta + \Delta) < y^*(\theta) + \Delta$. Since F is a shifted family:

$$\bar{F}(y^*(\theta) + \Delta|\theta + \Delta) = \bar{F}(y^*(\theta) - \theta|0) = \bar{F}(y^*(\theta)|\theta)$$
$$h(y^*(\theta) + \Delta|\theta + \Delta) = h(y^*(\theta) - \theta|0) = h(y^*(\theta)|\theta)$$

which gives that $g(y^*(\theta) + \Delta|\theta + \Delta) > g(y^*(\theta)|\theta)$. We then have:

$$\hat{r}\bar{F}(y^*(\theta) + \Delta|\theta + \Delta)(1 - g(y^*(\theta) + \Delta|\theta + \Delta))$$
$$< \hat{r}\bar{F}(y^*(\theta)|\theta)(1 - g(y^*(\theta)|\theta)) = \hat{c}.$$

The final equality follows from the definition of $y^*(\theta)$. That $y^*(\theta + \Delta) < y^*(\theta) + \Delta$ follows from the IGFR property which implies that marginal revenue is a decreasing function. That $\hat{w}^*(\theta) < \hat{w}^*(\theta + \Delta)$ immediately follows from $y^*(\theta + \Delta) < y^*(\theta) + \Delta$.

For $y^*(\theta) < y^*(\theta + \Delta)$, note that

$$\bar{F}(y^*(\theta)|\theta + \Delta) = \bar{F}(y^*(\theta) - (\theta + \Delta)|0) > \bar{F}(y^*(\theta) - \theta|0) = \bar{F}(y^*(\theta)|\theta)$$
$$h(y^*(\theta)|\theta + \Delta) = h(y^*(\theta) - (\theta + \Delta)|0) < h(y^*(\theta) - \theta|0) = h(y^*(\theta)|\theta),$$

The second inequality follows from the IFR assumption and implies that

$$g\left(y^{*}\left(\theta\right)|\theta+\Delta\right) < g\left(y^{*}\left(\theta\right)|\theta\right).$$

The remainder of the proof is essentially the reverse of the previous case. □

The first part of the theorem states that for many distributions the optimal wholesale price never depends on some distribution parameter. For example, the same wholesale price is optimal for all markets with exponential demand. More generally, one can show that a scaled family has a constant coefficient of variation, and it is the coefficient of variation that determines the optimal price when demand is gamma or Weibull.

An alternative interpretation of the result is that the manufacturer induces the retailer to serve the same fraction of demand for all distributions in the family. Recall that the elasticity of retailer $\nu\left(\xi\right)$ gives the percentage change in retailer's orders for a percentage change in the wholesale price. With a scaled family we have $\nu\left(\xi|\theta\right) = \nu\left(\frac{\xi}{\tau(\theta)}|1\right)$. The responsiveness of retailer orders to a price change at a given fractile of the demand distribution is the same for all members of the family. Thus the optimum is always found at the same fractile.

The second part of the theorem relates the optimal sales quantity and wholesale price to an additive parameter. As with a scaled distribution, a higher θ for a shifted family corresponds to a larger market, and the manufacturer responds to an increased market size by increasing her sales quantity (although the increase is less than the increase in the market size). Unlike a scaled family, a shifted distribution leads to the manufacturer charging a higher price. These observations immediately give the following.

Theorem 4 *If the demand distribution $F\left(\xi|\theta\right)$ is from either a shifted or a scaled family, then the manufacturer's profits are increasing in θ.*

Note that for a shifted family, varying θ changes the mean but has no impact on the variance of the distribution. Thus while scaled families have constant coefficients of variations, shifted families have coefficients of variations that decrease as θ increases. Further, the optimal price is higher for lower coefficients of variation. This logic yields the following:

Theorem 5 *Suppose demand is normally distributed. The optimal wholesale price is determined by the coefficient of variation and the manufacturer charges more the smaller the coefficient of variation is.*

Proof: We first show that the normal can be expressed as a scaled family. Let θ denote the mean, γ denote the coefficient of variation (giving a standard deviation of $\theta\gamma$), and $f\left(\xi|\theta,\theta\gamma\right)$ denote the density. We have:

$$
\begin{aligned}
f\left(\xi|\theta,\theta\gamma\right) &= \frac{1}{\theta\gamma\sqrt{2\pi}} \exp\left[-\left(\xi-\theta\right)^{2}/2\gamma^{2}\theta^{2}\right] \\
&= \frac{1}{\theta\gamma\sqrt{2\pi}} \exp\left[-\left(\xi/\theta-1\right)^{2}/2\gamma^{2}\right] \\
&= \frac{1}{\theta}f\left(\xi/\theta|1,\gamma\right).
\end{aligned}
$$

The normal is therefore a scaled family, and the optimal wholesale price depends only on the coefficient of variation. To see that the wholesale price must decrease in the coefficient of variation, consider two markets, one with mean one and coefficient of variation γ_1 and the other with mean γ_1/γ_2 and coefficient of variation γ_2. Assume $\gamma_1 > \gamma_2$. The demand distributions are now from a shifted family since they have the same standard deviation but different means. As $\gamma_1/\gamma_2 > 1$, market two must have the higher price, but it is also the market with the smaller coefficient of variation. $\square$

We derived a similar result for the Weibull using the lower bound on the optimal wholesale price given in (8.8) above. The coefficient of variation for the Weibull is decreasing in the parameter k while the lower bound increases in k. To gain some intuition to support these results, suppose the coefficient of variation were zero and demand were known with certainty. The retailer's order is then completely inelastic for any wholesale price less than or equal to the retail price. The manufacturer responds by pushing the wholesale price up to $\hat{r}$, capturing all channel profits for herself. This observation suggests if the optimal wholesale price is monotonic in the coefficient of variation (as it is for the normal), it must fall as variability increases.

8.2.5 Supply chain performance

While we have determined the best the manufacturer can do for herself, we must acknowledge that it will not be the best outcome for the supply chain as a whole. The manufacturer will always choose to sell a quantity less than y_I (or, equivalently, will always charge a wholesale price above the marginal cost of production). Total profits in the decentralized system will thus be less than the profits of a centralized system.

This phenomenon of double marginalization has long been known in the study of industrial organization (Spengler, 1950). The standard presentation has a manufacturer selling to a retailer facing a known, downward-sloping demand curve. By selling at a wholesale price above her marginal cost of production, the manufacturer induces the retailer to set a retail price above what an integrated firm would charge. Sales, and system profits, are thus below what an integrated channel could achieve. In our setting, the retail price is fixed. Double marginalization consequently appears as inadequate stocking levels. Insufficient inventories, not overly high retail prices, are the source of the problem. In an applied setting, this issue is explored by Pasternack (1980).

Economists have suggested a number of possible remedies to improve supply chain performance. These generally fall under the heading of "vertical restraints." The most common suggestions are franchising, quantity forcing, resale price maintenance (RPM) and exclusive territories. Tirole (1988) provides an introduction to the topic while Mathewson and Winter (1984) evaluate their relative merits in a number of environments.

RPM (i.e., allowing the manufacturer to dictate the retail price) and exclusive territories provide little help in our setting. We have explicitly assumed that the retail price is fixed regardless of the terms offered and have implicitly assumed that the retailer already enjoys a local monopoly. The impact of RPM when the market size is uncertain and the retailer has control over the retail price is considered in Deneckere, Marvel, and Peck (1996 and 1997) and Butz (1997). A comparison of RPM and exclusive territories in such a setting is offered by Rey and Tirole (1986).

On the other hand, franchising and quantity forcing are capable of coordinating our system. That is, allowing the decentralized system to earn the same profits as a centralized one. Further, both allow for an arbitrary split of the system's profits. Under franchising, the manufacturer charges an upfront fee A to carry the product (regardless of the stocking level) and then sells the product at a wholesale price of w.[5] Since the size of the lump sum payment is independent of the order quantity, the retailer will willingly pay any quantity such that

$$A \leq \Pi_R (w) - \kappa,$$

where $\Pi_R (w) = \Pi_R (y(w))$ and $\kappa \geq 0$ is the retailer's opportunity cost for carrying the product. It is straightforward to show that $\Pi_R (w)$ is decreasing in w. The largest fee the manufacturer can extract without subsidizing the retailer (i.e., selling the product at a loss) is

$$A = \Pi_R (c) - \kappa = \Pi_I^* - \kappa.$$

She can thus capture all channel profits except for the minimum amount the retailer requires to participate in the system. Note that system profits are independent of A. It is transferring at marginal cost that eliminates double marginalization; the franchise fee serves only to redistribute profits and consequently enables an arbitrary split of profits.

Franchising works by eliminating distorted incentives that affect the retailer's choice. Quantity forcing works by simply eliminating the retailer's choice. The manufacturer offers the product a wholesale price of w and insists that the retailer take some quantity Q. Clearly, if $Q = y_I$, inventory in the system is equal to the integrated channel's quantity, so system profits must equal Π_I^*. It is now the wholesale price that serves no purpose in coordination and is freed to redistribute profits.[6] The retailer's profits are

$$\Pi_I^* - (w - c) y_I.$$

Hence he accepts any contract such that

$$w \leq \frac{\Pi_I^* - \kappa}{y_I} + c.$$

Quantity forcing is therefore also able to support an arbitrary split of profits and, in particular, can be structured to drive the retailer to indifference.

---

[5]Note that here and for the rest of the chapter we return to our original costing convention.
[6]See Cachon and Lariviere (1997) for ways in which this pricing flexibility can be exploited.

8.3   RETURNS POLICIES: BUY BACK CONTRACTS

The nostrums economists favor for improving supply chain performance have generally been developed for settings in which demand is deterministic. Consequently, no sale is ever lost and no unit ever goes unsold. That is not so in our setting, and that change allows for the consideration of some additional contracts. In this and the next section we focus on returns policies. See Padmanabhan and Png (1995) for a managerial discussion of such programs.

Given a stocking level, some demand realizations will result in excess stock. We show that by judiciously assigning responsibility for that surplus the system can be coordinated. Accountability can be adjusted through a price mechanism (buy backs) or a quantity scheme (quantity flexibility contracts). Further, either approach can support an arbitrary split of profits without any ancillary payments. Surprisingly, a player's profits are increasing in the burden she bears for excess stock. Either method is consequently capable of performing as well as franchising or quantity forcing without being as heavy-handed as those schemes.

Here we consider buy back contracts, postponing quantity flexibility until the next section. We first present the basics of the contract, the retailer's optimal policy under the contract and some properties of the resulting manufacturer's profits. We then determine the coordinating contract and demonstrate the flexibility of buy backs by considering the various settings in which they have been applied.

8.3.1   Contract basics

We now assume that in addition to posting a wholesale price w, the manufacturer stands ready to buy back any unsold stock from the retailer at a per unit rate $b < w$. If realized demand ξ is less than the order quantity y, the retailer receives $b(y - \xi)$. Restricting b to be less than w ensures that the manufacturer does not create an arbitrage opportunity for the retailer, allowing him to buy stock in order to return it for a profit. Additionally, for the deal to be attractive to the retailer, b must be greater than what the retailer can achieve by salvaging the stock himself, which raises an important issue: To evaluate the effectiveness of such contracts, we must make some assumption about the salvage opportunities open to the players. It will obviously be impossible to replicate the outcome of the integrated channel if our buy back scheme results in some stock being salvaged at unfavorable terms.

There are a number of assumptions that will prevent this outcome. First we can assume that the manufacturer can salvage the product at the integrated channel rate v while the retailer can only salvage the product for an amount less than or equal to v. Offering a buy back of b to the retailer costs the manufacturer $b - v$ per unit.[7] If, on the other hand, only the retailer can salvage at the integrated channel level, we can assume that the manufacturer can verify the retailer's excess inventory. The manufacturer can then pay the retailer $b - v$ per unit, leaving the retailer with an effective salvage rate of b. We shall see that the power of returns policies is that they allow payments that are conditional on both the chosen stocking level and realized demand. This second assumption recognizes that these payments can be divorced from the actual movement of goods as long as the parties can verify that the payments are warranted.

Both alternatives must be recognized as simplifications of reality. Moreover, they both rest on a larger assumption: Imposing a returns policy on the decentralized channel introduces no additional cost beyond that incurred by the centralized system. Thus in the first case, returning stock to the manufacturer requires no transportation or handling costs that the centralized system could avoid. In the second, verifying the retailer's excess inventory must be costless.

8.3.2 The retailer's problem

With these assumptions, we can write the retailer's objective as:

$$\Pi_R\left(y\right) = \left(r - w\right)y - \left(r - b\right)\int_0^y F\left(\xi\right)d\xi.$$

The retailer still faces a newsvendor problem. The optimal solution is now

$$y\left(w, b\right) = F^{-1}\left(\frac{r - w}{r - b}\right).$$

It is straightforward to verify that the retailer's optimal order quantity and profits for a fixed wholesale price are increasing in b. Holding the manufacturer's wholesale price constant, he prefers a generous returns policy.

8.3.3 The manufacturer's problem

Under price only contracts, the manufacturer's profits were deterministic. For any wholesale price, she knew exactly what she would sell, and the retailer's presence sheltered her from market uncertainty. With a buy back contract, she is now exposed to the possibility of a poor demand outcome. Her profits are given by:

$$\Pi_M\left(w, b\right) = \left(w - c\right)y\left(w, b\right) - \left(b - v\right)\int_0^{y(w,b)} F\left(\xi\right)d\xi. \tag{8.10}$$

---

[7]Note that b does not have to be positive. For example, the retailer may pay the manufacutrer to take excess stock off his hands as long as b is less costly than his next-best disposal opportunity.

For price-only contracts, we could develop sufficient conditions on the demand distribution for the manufacturer's problem to be well-behaved. As the following shows, buy backs are more complicated, and her objective is never well-behaved.

Theorem 6 *The Hessian of* $\Pi_M(w, b)$ *is nowhere negative definite. Thus second order conditions for a local maximum are never satisfied.*

Proof: For second order conditions to be satisfied, the Hessian must be negative definite, which requires that its determinant be positive. Its determinant, however, is

$$-\frac{(r(w-c) + v(r-w) - b(r-c))^2}{(r-b)^6 f(y(w,b))^2},$$

which is always negative. □

For price only contracts we could place limits on the demand distribution that assured that first order conditions were sufficient to determine the optimal contract. While intuitively appealing, that approach will not work here. We require an alternative approach to determine the equilibrium contract.

8.3.4 Coordinating contracts

While we cannot identify the optimal contract through standard optimization approaches, some form of returns policy is beneficial to the system. Pasternack (1985) shows that having the manufacturer accept no returns leads to a suboptimal outcome for the system. Similarly a policy of full returns (i.e., fully refunding the wholesale price on all unsold items) is also suboptimal. An intermediary policy, however, results in channel coordination, as the following theorem (from Pasternack, 1985) shows.

Theorem 7 *Suppose that the manufacturer offers a contract* $(w(\varepsilon), b(\varepsilon))$ *for* $\varepsilon \in (0, r - c]$ *where*

$$w(\varepsilon) = r - \varepsilon \quad and \quad b(\varepsilon) = r - \frac{\varepsilon(r-v)}{r-c}.$$

1. *The retailer orders the integrated channel quantity (i.e.,* $y(w(\varepsilon), b(\varepsilon)) = y_I$) *and system profits are equal to the integrated channel profits.*

2. *Retailer profits are increasing in* ε. *Specifically,* $\Pi_R(w(\varepsilon), b(\varepsilon)) = \frac{\varepsilon}{r-c}\Pi_I^*$.

3. *Manufacturer profits are decreasing in* ε. *Specifically,* $\Pi_M(w(\varepsilon), b(\varepsilon)) = \left(1 - \frac{\varepsilon}{r-c}\right)\Pi_I^*$.

Proof: Note that for all allowed ε, $r > w(\varepsilon) > b(\varepsilon) \geq v$. So the contract is feasible even if the retailer may salvage excess items at rate v. Next observe that for all ε:

$$\frac{r - w(\varepsilon)}{r - b(\varepsilon)} = \frac{r - c}{r - v}.$$

The retailer faces the same critical fractile as the integrated channel and thus orders the same amount. As channel profits only depend on the total stock in the system, the decentralized system's profits must equal integrated system profits. To determine retailer profits, we have:

$$\Pi_R\left(w(\varepsilon), b(\varepsilon)\right) = (r - w(\varepsilon))y^* - (r - b(\varepsilon))\int_0^{y^*} F(x)dx$$

$$= \frac{\varepsilon}{r - c}\left((r - c)\,y^* - (r - v)\int_0^{y^*} F(x)dx\right)$$

$$= \frac{\varepsilon}{r - c}\Pi_I^*.$$

System profits are fixed, so the manufacturer earns $\left(1 - \frac{\varepsilon}{r-c}\right)\Pi_I^*$. □

The coordinating contract is therefore not unique. Rather a continuum of contracts exists. Possible contracts differ in how they divide the channel profits. A few observations add insight to this outcome. First, if the manufacturer wishes to sell at a price above the marginal cost of production, the retailer faces a higher acquisition cost than the integrated channel does. Double marginalization suggests that the retailer will order less than the integrated channel unless the manufacturer can simultaneously raise the retailer's marginal revenue. Granting a higher pay out in low demand states accomplishes this. Not surprisingly, the higher the wholesale price, the more the manufacturer must compensate the retailer in the event of low demand.

Second, what the retailer orders depends on a single value, the critical fractile. Under price only contracts, the manufacturer has a single contract parameter that affects the fractile, and the only way to coordinate the channel is to transfer at marginal cost. With a buy back contract, on the other hand, she may manipulate two parameters to have one value come out correctly. The system is under-determined, and a continuum of solutions exists.

This observation together with the third part of the theorem suggests why Theorem 6 holds. The manufacturer would prefer to push ε as low as possible and capture all channel profits. However, she cannot set ε to zero since this would result in a full returns policy which Pasternack (1985) has shown to be suboptimal. She consequently does not have an optimal contract; for any feasible, coordinating contract she will prefer to cut ε even further to grab a larger share of the profits.

The coordinating contracts have a number of additional properties. First, it is good to be responsible. Either party would prefer to take greater responsibility for excess stock as this leads to higher profits. This holds despite the fact that for any fixed wholesale price, the retailer would prefer the manufacturer offer an easy returns policy. But there is, of course, a difference between being easy and being cheap. Under coordinating contracts, generous return rates are paired with high wholesale prices. The price hikes are more than enough to offset any gains from the higher buy back rate. The manufacturer is in essence bundling insurance with the physical good. The more generous the insurance is, the higher the bundle is priced. The retailer, however, is risk neutral and would not pay for the indemnity if it were offered separately. It is the bargaining power we have assigned to the manufacturer that allows her to force the retailer to purchase the bundle.

Another interesting feature is that the coordinating contracts $(w(\varepsilon), b(\varepsilon))$ are independent of the demand distribution. If the manufacturer were to contract with a second retailer who faced a different demand distribution but had an identical cost structure, she could offer the same contract and capture the same fraction of total channel profits. While this independence is generally useful, one must be careful in interpreting the result. For example, one might be tempted to infer that the manufacturer does not need to know the demand distribution to design a coordinating contract; as long as the retailer knows the demand distribution, the correct actions will be taken. Mathematically, this is true, but it makes little economic sense. If the manufacturer does not know the demand distribution, she cannot know Π_I^*. We must then suppose that a rational agent is willing to enter into a transaction even though she cannot estimate her expected profits from the deal. Further, if the retailer were willing to take any contract the manufacturer might propose, we must infer that he has no opportunity cost. If more realistically the retailer has a positive opportunity cost κ, the manufacturer would want to offer a contract that assure acceptance without leaving any excess rents for the retailer. That is, she would choose ε such that:

$$\varepsilon = \frac{\kappa (r - c)}{\Pi_I^*}.$$

Clearly to define an acceptable contract, she must have enough information to calculate Π_I^*. Note that the presence of a positive retailer opportunity cost resolves the indeterminacy inherent in Theorem 7. Where before we said that for any positive ε the manufacturer could find a better contract, we now have a specific value of ε defining the best she can do for herself.

A final property to note is that different contracts result in the parties having wildly different variability in their profits. If ε equals $r - c$, the manufacturer transfers at cost and has profits that are identically zero. As ε falls to zero, both her profits and the variability of her earnings increase. The outcomes are reversed for the retailer. Again, the result should be evaluated with care. Both players are assumed to be risk neutral; they are concerned only with their mean earnings and not the variance of those earnings. A buy back contract is not a risk-sharing device in the economic sense. Risk sharing is usually interpreted as a means of maximizing some measure of the total utility of a group of risk averse agents. Common results are, for example, that if one agent is risk neutral he should bear all the risk (see Kreps, 1990, or Salanié, 1997). Here, both players are risk neutral and we seek to maximize the sum of their profits. From a social welfare perspective, all of the coordinating contracts lie on the Pareto frontier and are equally effective. Shifting the variability in profits is just a second-order consequence of using the terms of trade to both assure the stocking of the system-optimal quantity and allocate profits among the players.

8.3.5 Applications of buy backs

The coordinating contracts of Pasternack (1985) are surprisingly general and robust. The result can be interpreted as saying that if a setting can be manipulated to look like a newsvendor problem, it can be successfully decentralized through a system of linear prices. The usefulness of this approach can be seen by the number of researchers who have applied or extended the result. In general, a returns policy can have a significant impact on system performance even if it fails to replicate the outcome of the integrated firm.

Kandel (1996) covers much of the same ground as Pasternack (1985) but from an economist's perspective. In particular, he emphasizes the incentive for a manufacturer to implement a consignment policy. He also notes that if the demand distribution depends on the retail price, coordination cannot be achieved through buy backs unless the manufacturer can impose resale price maintenance. Although he does not formally model them, he does consider risk aversion and the provision of promotional effort and customer service.

Donohue (1996) considers two settings. In the first, the manufacturer has two production technologies and the retailer learns additional information prior to the demand realization. The first production technology has a low marginal cost but a long lead time while the second has a high marginal cost but short lead time. The retailer's additional information allows for a better estimate of realized demand but is learned at point at which only the high cost technology is feasible. She shows that the system can be coordinated by allowing the retailer to place a supplemental order after observing the information. The wholesale price for items purchased in the supplemental order is higher than for those purchased in the initial order, but any leftover stock may be returned at a single return rate. Again, a continuum of coordinating contracts exists and is independent of the demand distribution.

In her second model, there is a single production opportunity which may be initiated after the retailer has gained his information but production requires a specific component with a long lead time; there is no substitute for the component available after the retailer learns his refined information. Thus the total production quantity is limited by the amount of the component purchased. The coordinating contract now includes an option. The retailer pays the manufacturer o per unit to reserve the production of a finished good but must pay w per unit when exercising the option. The manufacturer again offers a buy back rate b. Once more we have a continuum of contracts that is independent of the demand distribution that will coordinate the system.

Note that in both models, there are two decisions that must be coordinated. In the first, it is the production quantities for the two technologies. In the second, it is the amount of the component to purchase and the final production quantity. The contracts consequently have three parameters. The extra degree of flexibility allows for the coordination of both actions.

Emmons and Gilbert (1998) also have a system in which two decisions must be made. Their model is similar to what we have considered here except that the retailer sets both the stocking level and the retail price. The demand distribution is a function of the retail price. Its density $f(\xi|r)$ satisfies

$$f(\xi|r) = \frac{1}{D(r)}\hat{f}\left(\frac{\xi}{D(r)}\right),$$

where $D(r)$ is a decreasing function of the retail price and $\hat{f}(\xi)$ is a probability density with mean one. While buy back contracts cannot coordinate the channel, there does exist a range of wholesale prices such that for any price in that range, the manufacturer and the retailer are both better off when the manufacturer offers a positive buy back rate.

In Ha (1998), a supplier proposes a contract to a manufacturer for the supply of a component. The manufacturer incurs a cost to turn the component into a finished good and must choose the number of units to produce as well as the retail price. The retail price affects demand in an additive fashion. For a given retail price r, demand is given by $D(r)+\xi$, where $D(r)$ is a decreasing function and ξ is a mean zero random variable. If the supplier knows the manufacturer's cost to complete production, the system can be coordinated through franchising, quantity forcing, or resale price maintenance combined with a buy back policy. Obviously, the latter reduces to our setting. If there is asymmetric information regarding the manufacturer's cost, the supplier can design a menu of incentive compatible contracts. For each possible cost realization, the manufacturer chooses a unique contract. The menu relies on resale price maintenance and a non-linear price schedule. The supplier is worse offer under asymmetric information since she must leave higher returns for manufacturers with lower costs to induce self revelation.

Cachon and Lariviere (1997) also consider alternative information structures. Like Ha (1998), they have a manufacturer contracting with an upstream supplier except now it is the manufacturer who proposes the contract. They show that the effectiveness of different contractual forms depends on the system's enforcement mechanism and information structure. If the supplier is obliged to cover the manufacturer's order and there is symmetric information, then offering to pay a termination fee for each item in the original order not taken allows the manufacturer to coordinate the system. It is the quantity forcing aspect of the enforcement mechanism that produces the result, not the marginal incentives of the price and termination fee. If enforcement is not possible, the manufacturer never offers a termination fee under full information. However, if the manufacturer is privately informed about the demand distribution, a manufacturer expecting a large market offers a termination fee to signal her information to the supplier. The fee plays only an information role. A high wholesale price is still required to induce the supplier to provide adequate capacity.

Narayanan and Raman (1997) note that most retailers carry multiple products in each category. Thus while a manufacturer loses sales following a stock out, the retailer might capture sales of a competing good. They consider a manufacturer selling to a retailer who also carries another product (say, a private label) that is never out of stock. If the manufacturer's good stocks out, a fraction of its customer p switch to the other good. In addition, the manufacturer can exert effort to increase demand for the product. They consider three channel management structures: retailer managed inventory (similar to our price-only contacts); vendor managed inventory (in which the manufacturer sets the stocking level of her product); and contracting on unsold inventory (similar to our buy back contracts). They show that double marginalization will result in too little stock and too little effort relative to the integrated channel when retail managed inventory is used. Vendor managed inventory results in the integrated channel effort level but too much stock since the vendor does not internalize the sales of the other product. Which management structure leads to better channel performance depends on the parameters of the problem. When the manufacturer implements a buy back, the channel outperforms simple retailer managed inventory although it does not perfectly coordinate the channel. How buy backs stack up to vendor managed inventory depends on the parameters of the problem.

While Narayanan and Raman (1997) have one retailer carrying two competing products, Padmanabhan and Png (1997) have two competing retailers carrying one product that must be purchased from the same manufacturer. The retailers face a linear demand curve with an uncertain intercept. The sequence of play has the manufacturer posting the terms of trade and the retailers ordering. Only after ordering do the retailers observe the realized intercept and determine which price to charge. Absent a returns policy, the retailers will have so much stock that they will engage in intense competition when the demand realization is low. Fearing a disastrous outcome in a small market, they will order relatively little stock and not be able to take full advantage of large market realizations. By offering a returns policy, the manufacturer mitigates "excessive" competition in small markets and encourages higher stocking levels that allow the channel to take advantage of large market outcomes.[8] The authors limit the analysis to full returns policies, so in some instances the manufacturer may prefer a price-only contract. Intuitively, allowing for partial returns should make returns more attractive.

Finally, a pair of papers suggest decentralization schemes related to buy backs for multiechelon inventory systems similar to Clark and Scarf (1960). Chen (1997) supposes that the supply chain is owned by a principal who hires a manager to run each echelon. The echelons are run as cost centers evaluated on long run average costs. If the costs charged to each echelon are based on accounting inventory (which measures what net inventory would have been if the upstream supply had been perfectly reliable), each manager's problem reduces to setting a base stock level to a critical fractile. By setting the holding and back order cost for each echelon appropriately (the solution is not unique), the principal can induce each manager to set the system optimal base stock level for his echelon.

---

[8]Deneckere, Marvel, and Peck (1996 and 1997) present related arguments for resale price maintenance.

Cachon and Zipkin (1997) consider a two echelon supply chain in which the upstream manufacturer and the downstream retailer set their base stock levels in a competitive fashion. They consider the difference in outcome if the players choose local or echelon base stock levels and show that the system optimal solution is rarely a Nash equilibrium. They also show that a simple linear payment scheme can result in the competitive system matching the performance of the centralized. Under this system, the manufacturer subsidizes the retailer's holding of inventory and pays a penalty when she is back ordered. The retailer, on the other, pays a penalty to the manufacturer when he is back ordered. Again, the coordinating contract is not unique. As in Donohue (1996) two decisions must be coordinated, and the contract must involve three parameters.

Together, these last two papers show the power of buy back contracts. In comparison to other decentralized schemes that have been suggested for multiechelon systems (e.g., Lee and Whang, 1996), these are simple and intuitive. The Chen approach is somewhat more robust. The resulting equilibrium under his scheme is supported by iterated dominance; for the manager of the n-th echelon, picking the system optimal is a dominant strategy assuming that all managers below him follow the optimal policy. In contrast, Cachon and Zipkin cannot rule out that the existence of multiple Nash equilibria under a coordinating contract. The robustness of Chen's scheme, however, comes at a significant cost: All managers must contract individually with a principal, which essentially limits the approach to intrafirm transactions. In Cachon and Zipkin, the parties contract directly with each other, which potentially allows for greater applicability.

8.4 RETURNS POLICIES: QUANTITY FLEXIBILITY CONTRACTS

As the previous section has shown, a returns policy in the form of a buy back rate can coordinate a system in which a manufacturer sells to a retailer facing a newsvendor problem. We now examine an alternative way of implementing returns, quantity flexibility (QF) contracts. QF contracts are frequently used for components in the electronics and computer industry and have occasionally been used in the automotive industry. They also bear some resemblance to "take-or-pay" contracts used in natural resource markets. Tsay and Lovejoy (1998) provides a discussion of these points and a detailed analysis of the contract in a multiperiod setting.

Here we limit the analysis to a single period, so our newsvendor model applies. A QF contract is specified by three parameters: a wholesale price w, a downward adjustment parameter $d \in [0,1)$, and an upward adjustment parameter $u \geq 0$. The sequence of play has the manufacturer offering the terms of trade and the retailer placing an initial order y. The manufacturer commits to providing $y(1+u)$ units to the system. If realized demand ξ is between $y(1-d)$ and $y(1+u)$, the retailer buys ξ units at price w from the manufacturer to sell to the market. The manufacturer thus provides an "upside" coverage to the retailer of u-percent above his initial order. The QF contract also requires a "downside" commitment from the retailer in the form of a minimum purchase requirement; he may cancel d-percent of his order but must take the remainder.[9] If realized demand is below $y(1-d)$, the retailer must take $y(1-d)$ units at a wholesale price of w, filling demand and salvaging the rest.

It will be convenient to define $\chi = \frac{1-d}{1+u}$. Since total channel stock will be $y(1+u)$, χ represents the fraction of channel stock for which the supplier is responsible and is thus a measure of the flexibility offered the retailer with lower values of χ reflecting greater flexibility.

To see that relationship between QF contracts and buy backs, suppose that under a QF contract the retailer had to pay for all $y(1+u)$ units up front but could still cancel his commitment down to $y(1-d)$ units. For each unit canceled, the manufacturer would refund the wholesale price. Interpreted this way, a QF contract allows the retailer to cancel or return a fraction of his order for a full refund of the wholesale price. In contrast, a buy back contract allows the retailer to cancel his full order for a fractional refund of the wholesale price.

In the remainder of this section, we first present the retailer's problem and then consider coordinating contracts. We close with a comparison of QF contracts with buy backs.

8.4.1 The retailer's problem

The retailer's objective under a QF contract is given by:

$$
\begin{aligned}
\Pi_R(y) \;=\; & (r-w)y(1+u) - (r-w)\int_{y(1-d)}^{y(1+u)} F(x)dx \qquad (8.11) \\
& -(r-v)\int_0^{y(1-d)} F(x)dx.
\end{aligned}
$$

The first term represents the retailer's profits if he were to sell everything. The subsequent terms represent adjustments for lower demand realizations. If demand falls between $y(1-d)$ and $y(1+u)$, he loses revenue of r but saves w. However, when demand is below $y(1-d)$, he only saves v since demand is below his minimum purchase commitment and he is obliged to salvage some units. Note that for a fixed y and w, the retailer's profits are increasing in d and u. That is, if at a given order quantity and fixed wholesale price the manufacturer offers greater upside coverage or demands less downside commitment, the retailer is better off. The following theorem (largely from Tsay, 1997), establishes some properties of the $\Pi_R(y)$ and the optimal retailer order y^*.

Theorem 8 *Suppose that the manufacturer offers a QF contract with parameters (w, d, u).*

---

[9]Many of the result of this section go through if the total stock in the system is given by $y + \hat{u}$ and the retailer's minimum purchase is $y - \hat{d}$. Such terms would allow the retailer a fixed range $(\hat{d} + \hat{u})$ of flexibility instead of a fixed percentage of flexibility. The QF formulation is preferable since a percentage commitment will always be positive while an additive commitment will not if $y < \hat{d}$.

1. The retailer's profits $\Pi_R(y)$ are concave in y.

2. The retailer's optimal order y^* unique and is implicitly defined by

$$\Pi'_R(y^*) = (r - w)(1 + u)\bar{F}(y^*(1 + u)) \qquad (8.12)$$
$$-(w - v)(1 - d)F(y^*(1 - d)) = 0.$$

3. The retailer's optimal order y^* is decreasing in the wholesale price and increasing in the downward adjustment parameter d. It is increasing in the upward adjustment parameter u if

$$g(y^*(1 + u)) < 1, \qquad (8.13)$$

where $g(\xi)$ is the generalized hazard rate of the demand distribution.

4. The total amount of stock in the system $y^*(1 + u)$ is increasing in u.

Proof: Differentiating $\Pi_R(y)$ twice yields

$$\Pi''_R(y) = -(r - w)(1 + u)^2 f(y(1 + u)) - (w - v)(1 - d)^2 f(y(1 - d)),$$

which is always negative. As profits are concave, first order conditions are sufficient for a unique solution. The third part of the theorem follows from applying the implicit function theorem to (8.12). In particular we have:

$$\frac{\partial y^*}{\partial u} = -(r - w)\left(\bar{F}(y^*(1 + u)) - (1 + u)y^* f(y^*(1 + u))\right)/\Pi''_R(y^*)$$
$$= -(r - w)\bar{F}(y^*(1 + u))(1 - g(y^*(1 + u)))/\Pi''_R(y^*).$$

As $\Pi''_R(y^*)$ is negative, we require that $1 - g(y^*(1 + u))$ be positive for y^* to be increasing in u. The final part of the theorem follows from differentiating $y^*(1 + u)$ and using the expression for $\frac{\partial y^*}{\partial u}$. □

Note that if $d = u = 0$, than the retailer's problem reduces to a newsvendor problem. Otherwise it will not be generally possible to determine an explicit solution to (8.12) even if $F(\xi)$ is easily invertible. In the standard newsvendor problem, one balances the chance of being over and under realized demand at one point. Under a QF contract, that balance must be struck at two distinct points. One point corresponds to the manufacturer's upside commitment and the other to the retailer's downside minimum purchase. An explicit solution is hence not readily possible unless either $F(\xi)$ or $\bar{F}(\xi)$ is a homogeneous function so that either $F(\lambda\xi) = \lambda^\rho F(\xi)$ or $\bar{F}(\lambda\xi) = \lambda^\rho \bar{F}(\xi)$ for $\lambda > 0$ and some ρ. For example, if $F(\xi)$ is homogeneous, the solution to (8.12) has a modified critical fractile structure:

$$F(y^*) = \frac{(r - w)(1 + u)}{(r - w)(1 + u)^{\rho+1} + (w - v)(1 - d)^{\rho+1}}.$$

The alternative case has similar results. It is straightforward to show that $F(\xi)$ is homogeneous for the uniform and that $\bar{F}(\xi)$ is homogeneous for the Pareto distribution.

The comparative statics are largely as one would expect. The retailer orders less when he has a slimmer margin. Increasing d lowers both the absolute level of the minimum purchase requirement and the marginal rate at which the minimum purchase obligation is increasing. Hence he orders more. The case for the upward adjustment is not as clear. On the one hand as u increases, the retailer has greater upside coverage for any order level. He could reduce his current order, lower his minimum purchase requirement, and still enjoy a larger upside coverage. On the other hand, the rate at which an increase in his order extends his upside coverage has increased – arguing for a boost in his order.

The outcome depends on how likely an incremental increase of upside coverage is going to be used. That, in turn, depends on the generalized hazard rate. Note that (8.13) corresponds to the elasticity of orders under a price-only contract being greater than one, which will be so under the optimal price-only contract. Consequently, if a manufacturer moves from the optimal price-only contract to a QF contract with upside flexibility without changing the wholesale price, the retailer's order will increase. If $F(\xi)$ is IGFR, however, the retailer's order will at some point be decreasing in u. The final part of the theorem establishes that even if the retailer reduces his order in response to an increase in the upward adjustment parameter, he does not reduce it so much that the total amount of stock in the system falls.

8.4.2 Coordinating contracts

We now seek coordinating contracts similar to those we had for buy backs. To begin, note that if we want the system to stock y_I units, we must induce the retailer to order $\frac{y_I}{1+u}$. Following Tsay (1997), we substitute $\frac{y_I}{1+u}$ for y in (8.12) and perform a few manipulations to yield:

$$
\begin{aligned}
w^*(d, u) &= v + \frac{(r-v)(1+u)\frac{c-v}{r-v}}{(1+u)\frac{c-v}{r-v} + (1-d)F\left(y_I \frac{1-d}{1+u}\right)} \\
&= v + \frac{c-v}{\frac{c-v}{r-v} + \chi F(\chi y_I)}. \qquad (8.14)
\end{aligned}
$$

Note that for the allowed ranges of d and u, $0 < \chi F(\chi y_I) \leq \frac{r-c}{r-v}$, which in turn implies that $c \leq w^*(d, u) < r$. The contract is thus economically sensible. The optimal wholesale price depends on the adjustment parameters only through the ratio χ, the fraction of the total system inventory for which the retailer is ultimately responsible. Although a QF contract generally has three parameters, a coordinating QF contract in reality only has two: the wholesale price and the flexibility offered the retailer, χ. For simplicity we will write the coordinating wholesale price simply as a function of χ, i.e., $w^*(\chi)$.

Given our restrictions on d and u, it must be that $0 < \chi \leq 1$. We again have that a continuum of coordinating contracts exists, and they differ in the flexibility they offer the retailer. As before, if the retailer is given no flexibility, the product must be transferred at marginal cost ($w^*(1) = c$). At the other extreme, the retailer cannot be offered complete flexibility. He must remain responsible for some fraction (even if trivially small) of the system's inventory. It is simple to show that the coordinating wholesale price is decreasing in the flexibility parameter χ and that the limiting price as χ goes to zero is the retail price r.

The remaining issue is how system profits are divided. We again have that both parties would prefer to bear greater responsibility for excess stock.

Theorem 9 *If the manufacturer offers a coordinating contract $(w^*(\chi), \chi)$, the retailer's profits $\Pi_R(w^*(\chi), \chi)$ are increasing in χ.*

Proof: Under a coordinating contract, the derivative of the retailer's profits with respect to χ can be written as:

$$\frac{\partial \Pi_R(w^*(\chi), \chi)}{\partial \chi} = -w^{*\prime}(\chi)\left(y_I - \int_{\chi y_I}^{y_I} F(\xi)\,d\xi\right) - y_I F(\chi y_I)(w^*(\chi) - v)$$

$$= \frac{(w^*(\chi) - v)}{\frac{c-v}{r-v} + \chi F(\chi y_I)}\left((F(\chi y_I) + \chi y_I f(\chi y_I))\right)$$
$$\left(y_I - \int_{\chi y_I}^{y_I} F(\xi)\,d\xi\right) - y_I F(\chi y_I)\left(\frac{c-v}{r-v} + \chi F(\chi y_I)\right)\right),$$

where the second equality follows from noting that:

$$w^{*\prime}(\chi) = -\frac{(w^*(\chi) - v)((F(\chi y_I) + \chi y_I f(\chi y_I)))}{\left(\frac{c-v}{r-v} + \chi F(\chi y_I)\right)}.$$

To prove the result, it is sufficient to show that

$$F(\chi y_I)\left(y_I - \int_{\chi y_I}^{y_I} F(\xi)\,d\xi\right) > y_I F(\chi y_I)\left(\frac{c-v}{r-v} + \chi F(\chi y_I)\right).$$

This follows from noting that

$$y_I - \int_{\chi y_I}^{y_I} F(\xi)\,d\xi = y_I\left(\frac{c-v}{r-v} + \chi F(\chi y_I)\right) + y_I \int_{\chi y_I}^{y_I} f(\xi)\,d\xi,$$

where we use the fact that $F(y_I) = \frac{r-c}{r-v}$. $\square$

With the theorem, we have that QF contracts are almost as flexible as buy backs. Under either, a range of coordinating contracts exists, and the more flexibility the manufacturer offers, the higher the wholesale price and manufacturer profits are. The only reason to qualify the comparison is that the under QF contracts, the coordinating wholesale price depends on the demand distribution. The manufacturer can no longer rely on one contract to coordinate all markets. This is hardly surprising since the retailer's optimal order balances overage and underage costs at two distinct points of the demand distribution. It is also as we argued above not that dramatic a shortcoming since it often makes no economic sense to assume that the manufacturer is uninformed about the demand distribution. Nevertheless, one can show that for some demand distributions the coordinating contract is independent of some parameter.

Theorem 10 *If $F(\xi|\theta)$ is from a scaled family, the coordinating wholesale price $w^*(\chi|\theta)$ is independent of θ for all values of χ. That is:*

$$w^*(\chi|\theta) = w^*(\chi|1).$$

Proof: For a scaled family, we have that $y_I(\theta) = \tau(\theta) y_I(1)$ and thus that $F(\chi y_I(\theta)|\theta) = F(\chi y_I(1)|1)$. Therefore, $w^*(\chi|\theta) = w^*(\chi|1)$. $\square$

The theorem immediately suggests that the same coordinating contract can be used in multiple markets if the demand distribution for each market is from the same scaled family. In particular, if demand in each market is normal with the same coefficient of variation, then one coordinating QF contract suffices.

8.4.3 Quantity flexibility vs. buy backs

The remaining issue to consider is whether a manufacturer offering a returns policy has any reason to favor a QF contract over a buy back. Relatively little research has been done on QF contracts, but there are reasons to believe that they will generally prove as useful as buy backs. For example, Tsay (1997) has extended their application to a setting similar to Donohue (1996). Further, it is simple to show that QF contracts, like buy backs, will fail to coordinate the channel when the demand distribution depends on the retail price as in Emmons and Gilbert (1998).

Hence, the sole reason we have established for favoring buy backs is that coordinating buy back contracts do not depend on the demand distribution. Even this property is arguably over-valued, and the QF dependence on the distribution can be relaxed for some distributional families. Indeed, one may argue that the distributional dependence benefits QF contracts since the manufacturer cannot induce self selection through coordinating buy back contracts.

To see this point, consider a manufacturer selling to two retailers who face different demand distributions and have positive opportunity costs. The manufacturer knows that each market is characterized by one of two demand distributions but cannot verify a particular market's prevailing distribution. Ideally, the manufacturer would like to offer a menu of contracts that induces retailers with different demand distributions to choose different contracts for themselves. Further, she would like to avoid sacrificing any channel profits and grab as large a share of profits as possible. Coordinating buy backs simply cannot do this. Because the coordinating contracts are independent of the demand distribution, all retailers will prefer the one that offers the least flexibility regardless of their demand distribution. Self selection necessarily requires deviating from the coordinating contract and sacrificing channel profits. In any given setting, self selection through coordinating QF contracts might also require sacrificing coordination, but there is at least a possibility of achieving both goals.

8.5 ALTERNATIVE CONTRACTS

Returns policies are an effective means of improving system performance, but other possibilities exists. Here we consider two. The first approach can be classified as penalty schemes. In lieu of offering a safety net to the retailer, the manufacturer brandishes a stick. The second contract is the standard-setting scheme of Atkinson (1979). Unlike returns policies, both penalties and standards are best suited for intrafirm coordination.

8.5.1 Penalty methods

Suppose that the manufacturer offers the good to the retailer at a wholesale price w with no return policy but with a demand for a payment of p per unit for any missed sales. The retailer's objective becomes:

$$\Pi_R(y) = (r - w)y - (r - v)\int_0^y F(\xi)\,d\xi - p\int_y^\infty \bar{F}(\xi)\,d\xi,$$

and his optimal order is:

$$y(w, p) = F^{-1}\left(\frac{r + p - w}{r + p - v}\right).$$

It is straightforward to show that if the manufacturer imposes a penalty of

$$p(w) = \frac{(w - c)(r - v)}{c - v}$$

with a wholesale price of w then the retailer orders the integrated channel quantity (i.e., $y(w, p(w)) = y_I$).

It is straightforward to see that $p(c) = 0$ and that $p(w)$ is increasing in w. We also have that the retailer's profits are decreasing in w since this decreases his expected revenue and increases his expected penalty payments. Consequently, the manufacturer can capture all channel profits at a wholesale price strictly below the retail price. If she were to push the wholesale price to the retail price (as she can under returns policies), the retailer would rationally refuse the contract since it would saddle him with an uncompensated loss.

Where a returns policy works by manipulating the consequence of having excess stock, a penalty method alters the consequences of being short. As such, its implementation may be difficult. In particular, it must be possible to observe lost sales. This can be relaxed slightly. Suppose that instead of imposing a linear payment rate, the manufacturer charges a lump sum in the event of a stock out and offers the retailer a menu of contracts in which the lump sum depends on the retailer's initial order. In particular suppose that for a given wholesale price w, the penalty payment is

$$P(y|w) = p(w)\frac{\int_y^\infty \bar{F}(\xi)\,d\xi}{\bar{F}(y)}.$$

For a stocking level of y, the retailer incurs the penalty with probability $\bar{F}(y)$, and his objective becomes equivalent to that under the linear payment scheme.

The quantity $\int_y^\infty \bar{F}(\xi)\, d\xi / \bar{F}(y)$ is known as the mean residual life in the reliability literature (Barlow and Proschan, 1965) and may be interpreted here as the expected number of items short given that demand is greater than the stocking level. Instead of basing the charges on the realized shortfall, the manufacturer exploits the retailer's risk neutrality and bases the charges on the expected shortfall given the stocking level. The information requirements have been significantly reduced. The manufacturer does not have to observe how many sales were lost, only that a stock out occurred.

Implementation may still be difficult. Under a returns policy, the retailer has an incentive to cooperate and allow the manufacturer to audit his sales since demand information is tied to his receiving compensation in a down market. Here, demand information is tied to being penalized and the retailer has every reason to make the auditing process difficult. Consequently, a penalty scheme may be more suitable for use within a firm instead of between firms since verifying information would hopefully be simpler in an intrafirm setting.

Intrafirm coordination through an alternative penalty scheme is considered by Celikbas, Shanthikumar, and Swaminathan (1997). Here, marketing provides a forecast of demand to manufacturing who must then produce the good prior to the realization of demand. The two functions are run by separate managers. They propose coordinating the system by evaluating manufacturing as a cost center and having the manufacturing manager pay a per unit penalty to the principal if the production quantity is less than both realized demand and marketing's forecast. The penalty induces manufacturing to internalize the cost of a lost sale. The marketing manager is sold the rights to the market but must pay a per unit penalty to the principal for each unit that his forecast exceeds the demand. As excess inventory remains on the books of manufacturing, the penalty forces the marketing manager to internalize the cost of excess production.

The scheme is somewhat more complicated than establishing an internal pricing or returns policy between the functions since it requires the parties to contract with the principal instead of each other. Also note that the data point marketing provides is not what one would usually term a "forecast." For the scheme to work, the complete demand distribution must be common knowledge so the data transferred from marketing does not lead anyone to alter their beliefs regarding the market. The forecast plays no informational role but is a useful contracting device for manipulating incentives.

8.5.2 Standard setting

Atkinson (1979) also considers intrafirm coordination but has a more classical principal-agent format. A risk neutral principal hires a risk averse agent to set the stocking level in a newsvendor problem. He shows that if the agent and the principal have the same evaluation of the demand distribution, the agent will set the stocking level below what the principal would choose.

He further argues that the manager through greater involvement in the daily operations of the business may have information about market demand unavailable to the principal. That is, between the time the parties enter into their contract and the time at which the stocking decision must be made, the manager is able to revise his estimate of the demand distribution but the principal is not. This leads to a standards-based contract. The principal offers a fixed wage ω and a gains sharing parameter ϕ (assumed to be between zero and one) while posting a standard y_0. Letting $\pi(y)$ represent realized system profits when the initial stocking level is y, the retailer's total compensation can be written as

$$\omega + \phi\left(\pi\left(y\right) - \pi\left(y_0\right)\right)$$

The agent receives ω for certain and shares in a fraction of the gains or loss that result from deviating from the principal's standard.

Suppose the principal sets the standard optimally given her initial information (i.e., so that y_0 is the critical fractile of the initial distribution). Atkinson shows that the agent only deviates from the standard if his revised information leads him to believe the critical fractile has shifted. Further, the stock level moves in the correct direction; the agent's choice always lies between the standard and what the risk neutral principal would have chosen given the agent's revised beliefs. If the players start with a common prior and update beliefs in a Bayesian fashion, the principal is better off implementing the standard based scheme than paying a fixed wage and insisting on implementing y_0.

The standard scheme is intended to be used within a firm and is not necessarily optimal. Dealing with a risk averse agent is going to impose some loss on the system, and there is no guarantee that this scheme minimizes that loss. Nonetheless, the paper is of interest. Contracting with a risk averse manager is an important topic to consider, and the proposed contract is simple and intuitive. It manages to balance the need for risk sharing (by providing a fixed wage) with the need to induce the manager to act on his revised information.

8.6 SUMMARY AND DIRECTIONS FOR FUTURE RESEARCH

We have reviewed a series of results related to perhaps the simplest supply chain contracting model with stochastic demand: a manufacturer selling to a newsvendor. We have presented conditions for the manufacturer's problem under price-only contracts to be well-behaved and shown that the optimal wholesale price is closely related to a generalization of the failure rate of the demand distribution. We have also shown that returns policies of various forms are a powerful tool for improving supply chain performance. Returns policies allow for payments that are conditional on how realized demand compares to the chosen stocking level. As such, they alter the marginal incentive to hold inventory and can reduce – or eliminate – the impact of double marginalization.

Given the apparent power of returns policies, it is not surprising that they are common in industries such as publishing. Indeed, one may wonder why they are not even more common. Relatively little work has examined this issue, but the fact that returns policies are not ubiquitous suggests that in some environments one or more of our assumptions fail to hold. Two obvious considerations are enforceability and cost. To begin with the latter, we assumed that offering a returns policy imposed no additional costs on the system relative to the integrated channel. That is obviously a simplification; merely accounting for the amount returned and providing compensation adds some costs. If system performance under a price-only contract is sufficiently good, it may not be worth incurring the additional cost of running a returns policy. Lariviere and Porteus (1998) offer some evidence that performance under price-only contracts improves as the coefficient of variation falls. It may be that demand in some markets is not sufficiently variable to warrant a returns policy.

The relevance of contract enforceability is best illustrated by quantity flexibility contracts. A manufacturer who offers a QF contract leaves herself facing a newsvendor with demand distributed between $y(1-d)$ and $y(1+u)$. Building sufficient stock to cover all demand is not necessarily optimal if the manufacturer can freely choose her production quantity. If the retailer anticipates that the manufacturer is not committed to providing the full upside coverage, his optimal order will change. Thus our analysis of QF contracts implicitly assumes an unmodeled enforcement mechanism that compels the manufacturer to fulfill her contractual obligation. Cachon and Lariviere (1997) show that absent such a mechanism returns policies can collapse under symmetric information.

An alternative explanation for not employing returns is offered by Marvel and Peck (1995). They note that there are two forms of market uncertainty: how many customers will show up and how a typical customer will value the product. The problem we have considered here is an extreme representation with the number of arrivals unknown but every arrival's valuation known to be r. As we have seen, returns are valuable in this setting. At the other extreme with the number of arrivals certain but the price they are willing to pay unknown, returns can hurt the manufacturer; the retailer sets a higher price in the presence of a returns policy and sales fall. In most markets, the manufacturer must balance both concerns.

While exploring the limits on the applicability of returns policy is one possible research direction, another fruitful avenue is to extend the lessons learned here to other settings. We argued in the introduction that the basic model we have studied provides an intuitive foundation for designing contracts to improve the performance of more complicated supply chains. This contention is supported by work such as Donohue (1996) and especially Chen (1997) and Cachon and Zipkin (1997), which develop schemes for multiechelon inventory systems.

An important extension to consider is the coordination of multiple actions. As we noted above, a returns policy cannot coordinate the system when the retailer controls the retail price. Finding a modification to returns policies to coordinate such a setting would be very useful. Similar statements could be made for actions such as manufacturer or retailer promotional effort. One might also want to consider the role of forecasting. Atkinson (1979), Donohue (1996), and Tsay (1997) all partially address this issue by considering systems in which some informational refinement occurs over time, but none of these authors explicitly models how that revision occurs. In particular, they do not consider that gathering information might be costly for the retailer and thus that the contract terms might affect the quality of information available to the supply chain. All of these would be worth pursuing, and likely involve some variant of the schemes considered here.

Acknowledgments

I am grateful for the helpful comments of Gerard Cachon, Evan Porteus, Andy Tsay, and Paul Zipkin.

266

References

Atkinson, A. A., "Incentives, Uncertainty, and Risk in the Newsboy Problem," *Decision Sciences*, 10 (1979), 341-353.

Barlow, R. E. and F. Proschan, *Mathematical Theory of Reliability* Cambridge: Princeton University Press, 1965.

Butz, D., "Vertical Price Controls with Uncertain Demand," *Journal of Law and Economics*, 40 (1997), 433-459.

Brown, A. O. and H. L. Lee, "Optimal Pay to Delay Capacity Reservation with Applications to the Semiconductor Industry," Stanford University working paper, 1997.

Cachon, G. P. and M. A. Lariviere, "Contracting to Assure Supply, or What did the Supplier Know and When did He Know It?," Duke University working paper, 1997.

Cachon, G. P. and P. H. Zipkin, "Competitive and Cooperative Inventory Policies in a Two-Stage Supply Chain," Duke University working paper, 1997.

Celikbas, M., J. G. Shanthikumar, and J. M. Swaminathan, "Coordinating Production Quantities and Marketing Efforts through Penalty Schemes," University of California at Berkeley working paper, 1997.

Chen, F., "Decentralized Supply Chains Subject to Information Delays," Columbia University working paper, 1997.

Clark, A.J. and H.E. Scarf, "Optimal Policies for a Multi-Echelon Inventory Problem," *Management Science*, 6 (1960), 475-490.

Deneckere, R., H. Marvel, and J. Peck, "Demand Uncertainty, Inventories, and Resale Price Maintenance," *Quarterly Journal of Economics*, 111 (1996), 885-913.

Deneckere, R., H. Marvel, and J. Peck, "Demand Uncertainty and Price Maintenance: Markdowns as Destructive Competition," *American Economic Review*, 87, 4 (1997), 619-641.

Donohue, K., "Supply Contracts for Fashion Goods: Optimizing Channel Profits," Wharton School working paper, 1996.

Emmons, H., and S. Gilbert, "Returns Policies in Pricing and Inventory Decisions for Catalogue Goods," *Management Science*, 44, 2 (1998), 276-283.

Eppen, G. and A. Iyer, "Backup Agreements in Fashion Buying - The Value of Upstream Flexibility," *Management Science*, 43, 11 (1997), 1469-1484.

Ha, A., "Supply Contract for a Short-Life-Cycle Product with Demand Uncertainty and Asymmetric Cost Information," Yale School of Management working paper, 1998.

Jeuland, A. and S. Shugan, "Managing Channel Profits," *Marketing Science*, 2 (1983), 239-272.

Kandel, E., "The Right to Return," *Journal of Law and Economics*, 39 (1996), 329-56.

Lariviere, M. A. and V. Padmanabhan, "Slotting Allowances and New Product Introductions," *Marketing Science*, 16, 2 (1997), 112-128.

Lariviere, M. A. and E. L. Porteus, "Stalking Information: Bayesian Inventory Management with Unobserved Lost Sales," Stanford University working Paper, 1997.

Lariviere, M. A., and E. L. Porteus, "Selling to the Newsvendor," Duke University working paper, 1998.

Lee, H. and S. Whang, "Decentralized Multi-Echelon Supply Chains: Incentives and Information," Stanford University working paper, 1996.

Kreps, D. M., *A Course in Microeconomic Theory* Cambridge: Princeton University Press, 1990.

Marvel, H. P. and J. Peck, "Demand Uncertainty and Returns Policies," *International Economics Review*, 36, 3 (1995), 691-714.

Mathewson, G. F. and R. A. Winter, "An Economic Theory of Vertical Restraints," *The Rand Journal of Economics*, 15 (1984), 27-38.

Moorthy, K. S., "Managing Channel Profits: Comment," *Marketing Science*, 6 (1987), 375-379.

Narayanan, V. G. and A. Raman, "Assignment of Stocking Decision Rights under Incomplete Contracting," Harvard University working paper, 1997.

Padmanabhan, V. and I.P.L. Png, "Returns Policies: Make Money by Making Good," *Sloan Management Review*, Fall (1995), 65-72.

Padmanabhan, V. and I.P.L. Png, "Manufacturer's Returns Policy and Retail Competition," *Marketing Science*, 16, 1 (1997), 81-94.

Pasternack, B. A., "Filling Out the Doughnuts: The Single Period Inventory Model in Corporate Pricing Policy," *Interfaces*, 10, 5 (1980), 96-100.

Pasternack, B. A., "Optimal Pricing and Returns Policies for Perishable Commodities," *Marketing Science*, 4, 2 (1985), 166-176.

Rey, P. and J. Tirole, "The Logic of Vertical Restraints," *The American Economics Review*, 5, 76 (1986), 921-939.

Salanié, B., *The Economics of Contracts: A Primer* Cambridge: MIT Press, 1997.

Spengler, J. J., "Vertical Integration and Antitrust Policy," *Journal of Political Economy*, 58 (1950), 347-352.

Tirole, J., *The Theory of Industrial Organization* Cambridge: MIT Press, 1988.

Tsay, A. A., "The Quantity Flexibility Contract and Supplier-Customer Incentives," Santa Clara University working paper, 1997.

268

Tsay, A. A. and W. S. Lovejoy, "Quantity Flexibility Contracts and Supply Chain Performance," Santa Clara University working paper, 1998.

9 DESIGNING SUPPLY CONTRACTS: CONTRACT TYPE AND INFORMATION ASYMMETRY

Charles J. Corbett* and Christopher S. Tang[†]

The Anderson School at UCLA
110 Westwood Plaza, Box 951481
Los Angeles, CA 90095-1481, U.S.A.

*This research is partially supported by the Center for Operations and Technology Management of The Anderson School at UCLA.
†This research is partially supported by UCLA Committee on Research Grant 92 and UCLA James Peters Research Fellowship.

9.1 INTRODUCTION

In designing supply contracts, a supplier has to consider the type of contract he can offer and the information he has about the buyer's cost structure. In this paper we provide a framework for fleshing out these two effects in the context of a simple single-supplier single-buyer supply chain facing price-sensitive deterministic demand. There are two well-known reasons for the resulting suboptimality (from both the supplier's and the joint perspectives):

- *Double marginalization:* because the buyer and the supplier only receive a portion of the total contribution margin, their decisions do not reflect the supply-chain wide incentive structure. As a result of receiving less than the full margin at any given quantity, they will produce less than a vertically integrated monopolist.

- *Asymmetric information:* the supplier rarely has complete information about the buyer's cost structure. However, the quantity the buyer will purchase (and therefore the supplier's profits) depend on that cost structure. Somehow, the supplier will have to take this information asymmetry into account.

In this paper we provide a simple but effective framework that allows us to study the following questions:

- If the supplier is faced with decreased buyer demand (due to a buyer cost increase), should the supplier sacrifice unit margin so as to maintain volume or should he do the opposite?

- What is the value to the supplier of obtaining additional information about the buyer's cost structure?

- What is the value to the supplier of being able to offer progressively more sophisticated supply contracts, eg. contracts with side payments or nonlinear contracts as opposed to merely specifying a constant unit wholesale price?

- Combining the previous two questions, when should a supplier focus on obtaining additional information and when should he focus on offering more sophisticated contracts?

- Under what circumstances can the double marginalization problem be overcome?

To study these questions, we examine the interaction between two key issues in designing supply contracts. The first deals with the type of contract that the supplier can offer while the second deals with the supplier's knowledge about the buyer's cost structure. We consider three types of contracts: one-part linear contracts, two-part linear contracts, and two-part nonlinear contracts. Under the one-part linear contract, the supplier charges a constant unit wholesale price. Under the two-part linear contract, the supplier charges a constant unit wholesale price but offers a fixed lump sum side payment to the buyer. Under a two-part nonlinear contract, the supplier offers a "menu" of contracts, where each item on that menu consists of a pair of unit wholesale price and lump sum-side payment, leaving it to the buyer to select the pair of his choice. In all cases, the buyer chooses the order quantity based on the wholesale price and side payment specified in the contract selected. Clearly, the design of an optimal supply contract requires a good understanding of how the buyer will choose an order quantity under different types of contract. However, the buyer's order quantity will depend on the buyer's internal cost structure, which may not be known to the supplier. This leads us to consider the two situations in which the supplier has complete or incomplete information about the buyer's cost structure. Therefore, there are six possible scenarios (see Table 9.1 at the end of this chapter) to be examined.

In this paper, we determine the optimal supply contract, the buyer's optimal order quantity, and the corresponding profits for supplier and buyer under each of the six scenarios depicted in Table 9.1. By comparing the profits for different scenarios, we aim to examine the value of offering successively more complex contracts (from one-part linear to two-part linear to two-part nonlinear) and the value to the supplier of getting better information about the buyer's cost structure. These comparisons enable us to gain a better understanding about the impact of different types of contracts and information asymmetry on supplier's and buyer's profits.

We assume that demand is deterministic and decreases linearly in price. This case captures the situation in which the product market is relatively mature and demand for any given price level is known. In this case, the supplier specifies the supply contract while the buyer determines the order quantity and the price at which he resells the product. For a more general demand function, the reader is referred to the fundamental work of Ha (1997) for details. Ha (1997) analyzes two-part nonlinear contracts under complete and incomplete information about the buyer's cost structure, in the case of stochastic price-sensitive demand. His results are therefore more general; however, his results do not lend themselves to simple interpretations. By focusing on this simpler case, we are able to obtain more qualitative insights into the value of information and the value of contracting flexibility.

Let us briefly summarize our findings. First, when the buyer's cost increases, we find it is optimal for the supplier to sacrifice part of his margin to maintain volume when he can observe the buyer's costs. However, when the supplier cannot observe the buyer's costs, the opposite may be true: the supplier will sometimes end up *increasing* his margin and thus sacrifice volume. Second, under the one-part linear contract, we find that the value to the supplier of information about the buyer's cost structure increases with the price-sensitivity of demand and with the uncertainty about the buyer's cost structure (measured by the variance of the supplier's prior distribution). Under the two-part linear contract, the value of information to the supplier also increases with the difference between the expected and worst-case value for the buyer's retail cost. Third, we find that the value of offering two-part contracts instead of one-part decreases with price-sensitivity; ie., when demand is more price sensitive, the supplier loses less when forced to contract on wholesale price alone.

The organization of this paper is as follows. Below, we first discuss the relevant literature in supply-chain management and in economics. In Section 9.3, we present the model and some of the underlying assumptions. In Section 9.4, we analyze the optimal contracts for each of the six scenarios for the case in which the demand is decreasing linearly in the selling price. In Section 9.5, we compare the supplier's and the buyer's profits and profit margins for different scenarios and comment on the value of more sophisticated contracts and the value to the supplier of obtaining information about the buyer's cost structure. Section 9.6 contains numerical examples, and Section 9.7 concludes the paper and provides some suggestions for future research.

9.2 LITERATURE

Although the learnings from this paper are targeted at researchers in the operations management community, the paper draws on both the supply-chain literature in operations and on the economics literature. We briefly review some of the key papers leading to this work below. (Any omission is our oversight.)

Most of the supply-chain literature on contracting has focused either on deriving optimal ordering policies in the context of a given contract, or on deriving optimal contract parameters given the functional form of that contract. Some exceptions have started studying questions related to coordination within supply chains, the value of information and various alternative contracting schemes. Lee, So and Tang (1998) quantify the value of sharing demand information to retailers and manufacturers in a two-level supply chain. Their work examines the case in which the demand follows an AR(1) process and is not directly observed by the manufacturer. Other recent work that examines the benefits of information sharing when demand is i.i.d. include Bourland, Powell and Pyke (1996), Cachon and Fisher (1997), and Gavirneri, Kapuscinski and Tayur (1996). Lee and Whang (1996) derive an incentive scheme for a multi-echelon supply chain that can be implemented by a central planner where each echelon uses local information only and which leaves each party with at least the same expected profit as the classic Clark and Scarf (1960) decomposition scheme. Other recent work that examines various incentive schemes includes Cachon and Zipkin (1997) and Chen, Federgruen and Zheng (1997). In contrast, Corbett (1996,1998) shows how asymmetric information (about upstream setup costs or downstream backorder penalty) in the absence of a central planner generally does lead to suboptimal outcomes. Weng (1995) quantifies the value of channel coordination (eg. using quantity discounts) in a two-level supply chain where both retailer and manufacturer face setup costs, and finds that quantity discounts (ie. one-part nonlinear contracts) alone are not sufficient to achieve coordination. Corbett and de Groote (1997) compare various coordination schemes for a two-level deterministic-demand supply chain with setup costs at both levels, and show how the equivalence of these schemes under full information breaks down when one party holds private information; that paper also derives preference orderings of these schemes for the supplier, buyer, and vertically-integrated firm. The current paper adds to this literature by explicitly quantifying the value of information and the value of more complex contracts.

The economics literature has a rich history of studying vertical contracting relationships, a brief introduction to which can be found in eg. Tirole (1988) and the references therein. The basic problem is that of two successive monopolists, where the downstream one faces a price-sensitive demand curve (often taken to be linear). Demand is generally deterministic or, if it is stochastic, the uncertainty is resolved before the buyer places his order, so that safety stock is not an issue. Left to their own devices, both parties add a markup to their costs, leading to the classic "double marginalization" phenomenon of higher prices and lower output, and lower hence profits, than an integrated monopolist would offer. Most work in this area focuses on comparing total surplus under various schemes and on finding what type of contract a manufacturer should offer to mitigate the double marginalization issue. A comprehensive and critical discussion of this issue, including an overview of the three schools of thought on bilateral monopoly in economics up to then, is given by Machlup and Taber (1960). The work by Gal-Or (1991a) is probably the most closely related to the current paper. Her paper focuses on a single-manufacturer single-retailer situation, where the retailer has private information about demand and about retail costs. She finds that, in general, neither franchise fees (wholesale price plus fixed side payment) nor retail price maintenance (supplier forces a particular retail price as part of the contract) can achieve the vertically-integrated solution under asymmetric information. She does not go beyond showing suboptimality to actually quantifying that suboptimality as we do here. Gal-Or (1991b) studies the situation with two suppliers but complete information, and finds that equilibrium is sometimes achieved with linear pricing and sometimes with a franchise fee contract. Bresnahan and Reiss (1985) look at the ratio of the profit margins of manufacturer and retailer under simple wholesale pricing with full information, and show how this ratio (which can also be seen as a measure of relative power) depends on the convexity of the demand function.

The current paper attempts to combine these two strands of theory, by building on the basic bilateral monopoly framework offered in economics and asking the normative and more micro-level questions more typical of the supply-chain literature. For instance, rather than focusing on achieving first-best outcomes and profits or on showing suboptimality of certain contract types, we explicitly measure the cost of that suboptimality. The contribution of this paper lies not in the analysis of the individual cases (which is either well-known or trivial), but in the comparisons between the cases, the quantification of the differences between them, and the insights that result.

9.3 THE MODEL

Consider a supply chain that consists of a single supplier and a single buyer. The supplier is a manufacturer who provides an important product to the buyer, who in turn resells it to final consumers. Alternatively, the supplier can provide a critical component to the buyer who in turn integrates this component with other components to form the finished products. For mathematical convenience, we assume that each unit of finished product requires one unit of the component. Demand per period for the finished product is denoted by q, and the selling price is denoted by p. We assume demand decreases linearly in price; i.e., $q = a - bp$, where $a \geq 0$ and $b \geq 0$ are known parameters. In addition, since q is a linear function of price p, the buyer's order quantity q and profit Π_b can be uniquely determined by the selling price p. Thus, it is sufficient for the buyer to select either the quantity q or the selling price p that maximizes his profit.

The supplier's marginal costs (including production costs) are given by s, the buyer's internal marginal costs (i.e. excluding the part cost) by c. When there is complete information, the supplier knows the actual value of the buyer's internal marginal cost c. However, when there is information asymmetry, the supplier does not know the actual value of the buyer's internal marginal cost c but holds a prior distribution $F(c)$ (with continuous density function $f(c)$), where the distribution is defined on a finite interval $[\underline{c}, \overline{c}]$. To ensure nonnegative demand, we assume that $a - b(s + \overline{c}) \geq 0$, and to establish tractable results we also assume that $\frac{F(c)}{f(c)}$ is increasing in c. When this assumption does not hold, tractable results are essentially impossible to obtain. The reader is referred to the work of Ha (1997) that examines various general cases. (To justify that this assumption is not too restrictive, let us examine the conditions under which it does hold. Consider a random variable $k = \overline{c} - c$, where k has a probability distribution $G(k)$ and density function $g(k)$. In this case, it can be easily shown that $\frac{F(c)}{f(c)}$ is increasing in c if and only if $\frac{g(k)}{1 - G(k)}$ is increasing in k. Notice that $\frac{g(k)}{1 - G(k)}$ is increasing in k so long as G is an IFR (increasing failure rate) distribution, which includes the exponential, normal and truncated normal.)

We consider three types of contract: a basic one-part linear contract, in which wholesale price w does not depend on the order quantity q, a two-part linear contract (w, L) consisting of wholesale price w and a per-period lump-sum side payment L from the supplier to the buyer, where both w and L are independent of order quantity q. When $L > 0$, the side payment can be interpreted as a slotting fee, which is common in dealing with large retailers. When $L < 0$ this lump sum can be seen as a franchise fee, more common when the supplier's product has a strong brand. The most complex type of contract we consider is that of two-part nonlinear contracts $\{(w(q), L(q)\}$, where actual wholesale price and side payment both depend on the order quantity selected by the buyer. Clearly, the two-part nonlinear contracts generalize the one-part linear contract and the two-part linear contract. As such, it is sufficient to formulate the supplier's problem for the two-part nonlinear contract first, and analyze different types of contracts subsequently as special cases. To develop an optimal supply contract that maximizes his expected profit for the two-part nonlinear contract, the supplier solves the problem below; if the supplier is restricted to offering simpler contracts, the problem becomes correspondingly more constrained.

$$S \quad \max_{\{w(q), L(q)\}} \quad \Pi_s(w, L) \quad := \quad E_c[(w(q(c)) - s)q(c) - L(q(c))] \qquad (9.1)$$

$$\text{s.t.} \quad \Pi_b(c, q) \quad := \quad (p(q) - w(q) - c)q + L(q) \quad \geq \Pi_b^- \quad \forall \ c \ (9.2)$$

$$D_q \Pi_b(c, q) \quad = \quad 0 \qquad\qquad\qquad \forall \ c \ (9.3)$$

The supplier's expected net profits Π_s in (9.1) depend on the quantity $q(c)$ ordered by the buyer. This in turn depends on the buyer's internal cost structure c, which may be unknown to the supplier, hence the expectation $E_c[\cdot]$ using the prior distribution $F(c)$. Depending on the type of contract, the supplier may also offer a side payment $L(q)$. With $p(q)$ as the selling price, inequality (9.2) represents the buyer's individual rationality constraint: the buyer's net profits Π_b (which include the side payment $L(q)$), the term to the left of the inequality in (9.2), must exceed his reservation profit level Π_b^-. (For a retailer, Π_b^- could be the profit that the buyer could obtain from allocating the shelf space to another supplier's product.) Finally, condition (9.3) is the buyer's incentive-compatibility constraint: presented with a menu $\{w(q), L(q)\}$, a buyer with cost c will choose $q(c)$ so as to maximize his net profit, denoted by $\Pi_b(c, q)$. Throughout this paper, we reserve the notation $D_x H$ to denote the derivative of any given function H with respect to x.

We shall consider the following six cases:

1. Case F1: the supplier offers a one-part linear contract on wholesale price w only, but has full information about c.

2. Case F2: the supplier offers a two-part contract (w, L), and has full information about c.

3. Case F3: the supplier offers a two-part nonlinear menu of contracts $\{w(q), L(q)\}$, and has full information about c.

4. Case A1: the supplier offers a one-part linear contract on w only, and does not know c.

5. Case A2: the supplier offers a two-part contract (w, L), but does not know c.

6. Case A3: the supplier offers a two-part menu of contracts $\{w(q), L(q)\}$, but does not know c.

Table 9.1 depicts the cases to be analyzed in Section 9.4. The sequence of events is always:

1. The supplier offers one type of contract (one-part linear contract, two-part linear contract, or two-part nonlinear contract).

2. The buyer (who has internal cost c) chooses a specific contract by selecting the order quantity q (when a one-part linear or two-part linear contract is offered) or by selecting $(w(q(c)), L(q(c)))$, and the corresponding order quantity q (when a two-part nonlinear contract is offered).

3. All sales and financial transactions take place simultaneously.

When the buyer's marginal cost is unknown to the supplier, we consider the case in which the supplier treats the buyer's marginal cost as a random variable C and imposes a prior probability distribution $F(c)$ over the buyer's marginal cost C. During the analysis of case A3 (asymmetric information, two-part nonlinear contract), we transform our model into an equivalent but mathematically more convenient form. This transformation utilizes the revelation principle from economics (see Fudenberg and Tirole 1991 for more detail or Corbett 1998 for a simple proof in a similar context). The revelation principle can be explained as follows. Intuitively, for any given order quantity q selected by the buyer, the supplier can deduce the buyer's corresponding cost c. Hence, the buyer's selecting q is essentially equivalent to his announcing a cost parameter c. This implies that the supplier can reformulate the contracts in terms of c, i.e. optimizing over $\{w(c), L(c)\}$ rather than $\{w(q), L(q)\}$. Although the buyer could announce a cost parameter $\hat{c}$ where $\hat{c} \neq c$, the revelation principle assures us that there is an optimal contract under which the buyer will reveal truthfully, i.e. $\hat{c} = c$. Throughout this paper, our analyses will be based on $\{w(c), L(c)\}$ and $\{w(q), L(q)\}$ interchangeably. Note, however, that the contribution of this paper lie in the framework and the insights derived from comparing the six scenarios, not in the application of the revelation principle which has been done previously.

For any function $H(\cdot)$, we reserve $\dot{H}$ to denote the derivative of the function H with respect to the variable $\hat{c}$ and evaluated at $c = \hat{c}$, i.e. $\dot{H} = D_{\hat{c}}H$. The notation is summarized in Table 9.2.

9.4 THE SUPPLIER'S OPTIMAL SUPPLY CONTRACTS

The basic steps for determining the optimal supply contract for each of the six cases are as follows. We first solve the buyer's optimization problem and determine the buyer's optimal order quantity when either the one-part contract or the two-part linear or nonlinear contract are being offered by the supplier. Then we solve the supplier's optimzation problem that accounts for the buyer's optimal order quantity and determine the optimal supply contract. After we complete the analysis for those six possible scenarios, we examine the impact of the contract type and of information asymmetry on the supplier's and the buyer's profits. Though the first parts below (the buyer's optimization under a given contract, and the supplier's problem under full information) are already well-established, we briefly review them below for the sake of completeness.

9.4.1 The buyer's optimization problem

9.4.1.1 The buyer's optimization problem under a one-part or two-part linear contract.
Under the one-part linear contract, the supplier selects the wholesale price w. For any given w, the buyer solves:

$$\mathcal{B}_{1,2} \qquad \max_{q} \qquad \Pi_b := (p(q) - w - c)q$$

Since $q = a - bp$, it is well-known that the buyer will select

$$p^* = \frac{a + b(w + c)}{2b} \tag{9.4}$$

$$q^* = \frac{1}{2}[a - b(w + c)] \tag{9.5}$$

Under the two-part linear contract, the supplier also selects a lump-sum side payment L, but as L is independent of q it will clearly not affect the buyer's order quantity selection. Hence, when the supplier offers a two-part linear contract, it is optimal for the buyer to order q^* as given in (9.5).

9.4.1.2 The buyer's optimization problem under a two-part nonlinear contract.
Under the two-part nonlinear contract, the supplier offers a menu of contracts $\{w(q), L(q)\}$, and the buyer's decision is to choose the order quantity q. Or, equivalently (as explained above), the supplier offers a menu of contracts $\{w(\hat{c}), L(\hat{c})\}$, and the buyer announces a cost parameter $\hat{c}$. For a given menu of contracts $\{w(\hat{c}), L(\hat{c})\}$, the buyer would choose to announce whatever level of cost $\hat{c}$ will maximize his profit. Associated with the cost $\hat{c}$ announced by the buyer to the supplier, the buyer would pay wholesale price $w(\hat{c})$, and would order $q^*(w(\hat{c}))$ and set the selling price $p^*(w(\hat{c}))$, where p^* and q^* are as given in (9.4) and (9.5), respectively. Thus, the buyer solves:

$$\mathcal{B}_3 \qquad \max_{\hat{c}} \qquad \Pi_b(c, \hat{c}) := L(\hat{c}) + [p^*(w(\hat{c})) - w(\hat{c}) - c]q^*(w(\hat{c})) =$$

$$= L(\hat{c}) + b[\tfrac{a}{2b} - \tfrac{1}{2}(w(\hat{c}) + c)]^2 \tag{9.6}$$

We know, from the revelation principle, that there is an optimal menu of contracts under which it is optimal for the buyer to reveal his true costs so that $\hat{c}^* = c$. This implies that at the optimum, the first-order condition is solved at $\hat{c} = c$. The second-order condition is verified below. By considering the first-order condition, the buyer's incentive-compatibility constraint is given as:

$$\dot{L}(c) = \frac{1}{2}[a - b(w + c)]\dot{w}(c) \tag{9.7}$$

In addition, the supplier will need to set the side payment L so as to meet the buyer's individual-rationality constraint (9.2); i.e., $\Pi_b(c) \geq \Pi_b^-$ for all $c \in [\underline{c}, \overline{c}]$.

To verify sufficiency of the first-order condition for the buyer's optimization problem, first note that for given w, the buyer's price selection or order quantity problem is concave in p or q respectively. Now, to verify sufficiency, we examine the second-order condition. Specifically, for any given menu $\{L(\cdot), w(\cdot)\}$, the second-order condition is given as:

$$D_{\hat{c}}^2 \Pi_b = -b[\frac{a}{2b} - \frac{1}{2}(w + c)]\ddot{w} + \frac{1}{2}b\dot{w} + \ddot{L} \tag{9.8}$$

$$= \frac{1}{2}a\ddot{w} + \frac{1}{2}b(w + c)\ddot{w} + \frac{1}{2}b\dot{w} + \ddot{L} \tag{9.9}$$

Substituting

$$D_{\hat{c}}^2 L = \frac{1}{2}a\dot{w} - \frac{1}{2}b(w + \hat{c}))\dot{w} \tag{9.10}$$

gives

$$D_{\hat{c}}^2 \Pi_b = -\frac{1}{2}b\dot{w} \leq 0 \tag{9.11}$$

In the analysis below, we will see that the optimal wholesale price in case A3 w_{A3}^* is increasing in c, so that Π_b is concave in c, which verifies sufficiency.

9.4.2 Optimal supply contracts under complete information

By using the information about the buyer's optimal order quantity (given in (9.5)) when a one-part or two-part linear contract is offered and the information about the buyer's cost announced to the supplier, we can derive the optimal supply contract for each of the six cases below. The characterization of the optimal supply contract, the buyer's optimal order quantity, and the optimal supplier's and buyer's profit margins for each of these six cases are summarized in Table 9.3.

9.4.2.1 Case F1: one-part linear contract with complete information. When the supplier has complete information about the buyer's marginal cost c, the supplier knows the buyer's optimal order quantity q^* as given in (9.5) for any wholesale price w. Thus, the supplier needs to determine the wholesale price w so that his profits $\Pi_{s,F1}$ are maximized. In this case, the supplier solves:

$$S_{F1} \quad \max_w \quad \Pi_{s,F1}(w) := (w - s)q^* = (w - s)\frac{1}{2}[a - b(w + c)] \tag{9.12}$$

It is easy to check from the first and second order conditions that the problem is concave and that the optimal wholesale price w^* can be written as:

$$w_{F1}^* = \frac{a}{2b} + \frac{1}{2}(s - c) \tag{9.13}$$

Substituting the optimal wholesale price into the objective function of problem S_{F1}, one can show that the supplier's profits satisfy

$$\Pi_{s,F1}(w_{F1}^*) = \frac{b}{2}(\frac{a}{2b} - \frac{(s + c)}{2})^2 \tag{9.14}$$

and that the supplier's profit margin is equal to $w^* - s = \frac{1}{2}[\frac{a}{b} - (s + c)]$.

Similarly, we can substitute w^* into the expressions for p^* and q^* given in (9.4) and (9.5) to determine the buyer's profit. In this case, it is easy to show that the buyer's profit for case F1, denoted by $\Pi^*_{b,F1}$, is $\Pi^*_{b,F1} = (p^* - w^* - c)q* = \frac{b}{4}(\frac{a}{2b} - \frac{(s+c)}{2})^2$. Also, notice that the buyer's profit margin is equal to $p - w^* - c = \frac{1}{4}[\frac{a}{b} - (s + c)]$.

By comparing the supplier's and the buyer's profit margins and profits, it is clear that the supplier's profit and profit margin are double those of the buyer. This well-known result corresponds to the special linear demand case in Bresnahan and Reiss (1985); in general, this ratio is equal to $\frac{1}{2+\eta}$, where $\eta = q\frac{D^2_q p}{D_q p}$, a local measure of the curvature of the demand curve. Recall that, throughout this paper, the supplier is the party with the initiative to propose contract terms, so it is not surprising to find the supplier capturing a larger proportion of total profits.

9.4.2.2 Case F2: two-part linear contract with complete information.
Two-part contracts are often referred to as franchise fee (FF) contracts in the literature. Using the same argument as presented for case F1, it is easy to see that the supplier can determine the contract of the form $\{w, L\}$ that optimizes his profits by solving the following problem:

$$\mathcal{S}_{F2} \quad \max_{\{w,L\}} \Pi_{s,F2}(w, L) \quad := \quad (w - s)q^* - L = (w - s)\frac{1}{2}(a - b(w + c)) - L$$

$$\text{s.t.} \quad \Pi_{b,F2}(c, c) \geq \Pi^-_b \tag{9.15}$$

The problem can be solved by using two observations. First, because the supplier has complete information, he can set inequality (9.15) to be binding. Second, the supplier's profits $\Pi_{s,F2} = \Pi_{j,F2} - \Pi_{b,F2} = \Pi_{j,F2} - \Pi^-_b$ where Π_j denotes joint profits. Hence, problem $\mathcal{S}_{F2}$ is equivalent to maximizing joint profits $\Pi_{j,F2}$. It is easy to verify that

$$L^*_{F2} = \Pi^-_b - \frac{1}{4b}(a - b(w + c))^2 \tag{9.16}$$

and

$$w^*_{F2} = s \tag{9.17}$$

Hence, when the supplier has complete information about the buyer's marginal cost information, it is optimal for the supplier to set the wholesale price equal to the supplier's marginal cost and use the lump sum side payment to extract all profits from the buyer in excess of his reservation profit level Π^-_b. Note that this means that L^*_{F2} will be negative whenever the buyer's reservation profit level is not too high. This corresponds to a franchise fee paid by the buyer to the supplier. In case F2, the supplier's profits satisfy

$$\Pi_{s,F2}(w^*_{F2}, L^*_{F2}) = -L^*_{F2} = \Pi^-_b - \frac{1}{4b}(a - b(w + c))^2 \tag{9.18}$$

and the buyer's profit $\Pi^*_{b,F2} = \Pi^-_b$. In this case, the sum of the supplier's and the buyer's profits is equal to $\Pi_{j,F2} = \frac{1}{4b}(a - b(s + c))^2$.

9.4.2.3 Case F3: two-part nonlinear contract with complete information.

Although the supplier now has the added flexibility of being able to offer nonlinear contracts, it is clear that in case F2 he is already extracting all profits beyond the minimum level Π^-_b from the buyer, so this additional flexibility has no value to the supplier in the complete information case. All results for case F2 carry over directly to case F3. Below, though, we will find that in the case of asymmetric information this equivalence no longer holds.

Later, in Section 9.5, we shall compare these three contracts, to examine the value of two-part linear and nonlinear contracts vs. one-part contracts under full information. First, though, let us analyze the asymmetric information cases.

9.4.3 Optimal supply contracts under asymmetric information

Most models that explicitly include asymmetric information assume the supplier can offer two-part nonlinear contracts (or even more sophisticated contracts, including resell price maintenance). We do not make this assumption, and in doing so we can precisely quantify the value of information and the value of contracting sophistication to the supplier.

9.4.3.1 Case A1: one-part linear contract with asymmetric information.

In this case, the supplier holds a prior probability distribution $F(c)$ over the buyer's marginal cost c, and the supplier needs to specify a wholesale price so as to maximize his expected profit. We can utilize the buyer's optimal quantity q^* in (9.5) to formulate the supplier's optimization problem as follows:

$$
\mathcal{S}_{A1} \quad \max_{w} \; E_c[\Pi_{s,A1}(w)] \; := \; E_c[(w - s)q^*] =
$$

$$
= \int_{\underline{c}}^{\overline{c}} \frac{1}{2}(w - s)(a - b(w + c))dF(c) \tag{9.19}
$$

The first-order condition $D_w E_c[\Pi_{s,A1}(w)] = 0$ yields:

$$
D_w E_c[\Pi_{s,A1}(w)] \;=\; \frac{1}{2}a - \frac{1}{2}b(2w - s) - \frac{1}{2}bE[c] = 0
$$

which is solved at

$$
w^*_{A1} \;=\; \frac{a}{2b} + \frac{1}{2}(s - E[c]) \geq 0 \tag{9.20}
$$

The inequality follows from our assumption that $a - b(s + \overline{c}) \geq 0$. Substituting the optimal wholesale price in to the objective function of problem $\mathcal{S}_{A1}$, one can show that the supplier's expected profits satisfy

$$
E_c[\Pi_{s,A1}(w^*_{A1})] \;=\; \frac{1}{8b}(a - b(s + E[c]))^2 \tag{9.21}
$$

and that the supplier's profit margin is equal to $w^*_{A1} - s = \frac{a}{2b} - \frac{(s+E[c])}{2} \geq 0$.

Similarly, we can substitute w^*_{A1} into the expressions for p^* and q^* given in (9.4) and (9.5) to determine the buyer's profit. In this case, it is easy to show that the buyer's profit $\Pi^*_{b,A1} = (p^* - w^* - c)q* = \frac{b}{4}(\frac{a}{2b} - \frac{(s+c)}{2} + \frac{E[c]-c}{2})^2$, and that the buyer's profit margin is equal to $p - w^* - c = \frac{a}{4b} - \frac{(s+c)}{4} + \frac{E[c]-c}{4}$. Observe that the supplier's and the buyer's profits and profit margins depend on the accuracy of the estimated buyer's marginal cost $E[c]$. Therefore the supplier has the incentive to induce the buyer to reveal his true cost c so as to gain a higher profit; however, within the limited flexibility allowed by the one-part linear contract, the supplier cannot achieve this. We shall show how the supplier can induce the buyer to reveal his true cost in case A3, i.e., the two-part nonlinear contract case.

9.4.3.2   Case A2: two-part linear contract with asymmetric information.   By following the same approach as in case A1, we can formulate the supplier's optimization problem for case A2 as:

$$S_{A2} \quad \max_{w,L} E_c[\Pi_{s,A2}(w, L)] := E_c[(w - s)q^* - L] =$$

$$= \int_{\underline{c}}^{\overline{c}} [\frac{1}{2}(w - s)(a - b(w + c)) - L]dF(c) \quad (9.22)$$

For any given w, the supplier will always choose the lowest L that still satisfies the buyer's individual rationality constraint $\Pi_b(c) \geq \Pi_b^-$ for all c. Since $\Pi_b(c)$ is decreasing in c (because, by assumption, $a - b(w+c) = q \geq 0$), this constraint holds for all c if it holds at $c = \overline{c}$. thus, the buyer's profits can be written as

$$\Pi_{b,A2}(c, q) = \frac{1}{4b}(a - b(w + c))^2 + L \quad (9.23)$$

Setting $\Pi_b(\overline{c}, q) = \Pi_b^-$, we can determine the optimal side payment L^*_{A2}, where

$$L^*_{A2} = \Pi_b^- - \frac{1}{4b}(a - b(w + \overline{c}))^2 \quad (9.24)$$

Substituting this expression for L into $\Pi_{s,A2}$ as given in (9.22) and noting that the resulting expression is concave in w, we can solve the first-order condition for w to find that:

$$w^*_{A2} = s + \overline{c} - E[c] \quad (9.25)$$

Substitute (9.25) into (9.24) and (9.22), to get:

$$L^*_{A2} = \Pi_b^- - \frac{1}{4b}(a - b(s + 2\overline{c} - E[c]))^2 \quad (9.26)$$

$$E_c[\Pi_{s,A2}(w^*_{A2}, L^*_{A2})] = -\Pi_b^- + \frac{1}{2}(\overline{c} - E[c])(a - b(s + \overline{c})) +$$

$$+ \frac{1}{4b}(a - b(s + 2\overline{c} - E[c]))^2 \quad (9.27)$$

Clearly, under complete information, $\bar{c} = E[c] = \underline{c}$ and case A2 reduces to case F2. The information asymmetry means the supplier must now offer a larger side payment than before, i.e. $L^*_{A2} \geq L^*_{F2}$, to meet the "worst-case" buyer's minimum profit requirements. To compensate, the supplier adds a markup $\bar{c} - E[c]$, based on how far removed the "worst-case" buyer is from the mean, to his marginal cost s. We will discuss the qualitative differences in more depth below, after analyzing the final case A3.

9.4.3.3 Case A3: two-part nonlinear contract with asymmetric information.

In this case the supplier has the flexibility to offer a two-part non-linear menu of contracts $\{w(\hat{c}), L(\hat{c})\}$. By selecting any specific pair $(w(\hat{c}), L(\hat{c}))$ the buyer is essentially revealing a marginal cost $\hat{c}$ which, by the revelation principle explained earlier, will be his true marginal cost c. In this case, we can utilize the buyer's optimal quantity q^* in (9.5) and requirement (9.7) that the optimal lump sum payment satisfies the first order condition for problem $\mathcal{B}_3$ to formulate the supplier's optimization problem as follows:

$$\mathcal{S}_{A3} \quad \max_{\{w(\cdot), L(\cdot)\}} \quad E_c[\Pi_{s,A3}] \quad := \quad E_c[(w-s)q^* - L] =$$

$$= \quad E[(w-s)\frac{1}{2}(a - b(w+c)) - L] \qquad (9.28)$$

$$\text{s.t.} \qquad \dot{L} \quad = \quad \frac{1}{2}[a - b(w+c)]\dot{w} \qquad \forall\, c\,(9.29)$$

$$\Pi_b(c,c) \quad \geq \quad \Pi_b^- \qquad \forall\, c\,(9.30)$$

The solution procedure is given in the Appendix; the optimal wholesale price for case A3, w^*_{A3}, has the following form that is based on Laffont and Tirole (1993) and other related work:

$$w^*_{A3}(c) \quad = \quad s + \frac{F(c)}{f(c)} \qquad \forall\, c\ (9.31)$$

Recall that the function $\frac{F}{f}$ is assumed to be increasing in c, so one immediately sees from (9.31) that the optimal wholesale price is increasing in c. This is in contrast to the earlier cases, where wholesale price was decreasing in c (case F1), constant in c (case F2), or decreasing in $E[c]$ (cases A1 and A2). We further discuss this contrast in Section 9.5.1. As we had also assumed that $q^* = \frac{1}{2}a - \frac{1}{2}b(w+\hat{c})) \geq 0$, we can verify from (9.7) that the lump sum payment is also increasing in c. In this case, the buyer faces a tradeoff: accepting a higher lump sum payment and a higher unit wholesale price versus accepting a lower lump sum payment and a lower unit wholesale price.

9.4.3.4 Special case A3: uniform prior distribution.

To generate some managerial insights for case A3, let us consider the case when the prior distribution is uniformly distributed on $[\underline{c}, \bar{c}]$. In this case, it follows from (9.31) and (9.7) that:

$$w^*_{A3}(c) \quad = \quad s + c - \underline{c} \qquad (9.32)$$

$$L_{A3}^*(c) \;=\; \frac{1}{2}c[a - b(s + c - \underline{c})] + k \tag{9.33}$$

where k depends on Π_b^- and other parameters but not on c.

In this case, one can write w and L explicitly as function of q by inverting c out of $q^* = \frac{1}{2}[a - b(s + 2c - \underline{c})]$ from (9.5). This gives

$$w_{A3}^*(q) \;=\; \frac{1}{2}(s + \frac{a - 2q}{b} - \underline{c}) = \frac{1}{2}(s - \underline{c} + \frac{a}{b}) - \frac{1}{b}q \tag{9.34}$$

$$L_{A3}^*(q) \;=\; \Pi_b^- + \frac{1}{8b}[(a - b(s - \underline{c}))^2 - 4q^2] + k \tag{9.35}$$

One can easily verify that when $c = \underline{c}$, $q^* = \frac{1}{2}[a - b(s + \underline{c})]$ so that (9.34) reduces to $w_{A3}^*(q^*) = s$ (as indeed it must from (9.31)). The structure of w_{A3}^* in (9.34) is interesting: the unit wholesale price can be interpreted as the average of a constant part and a part decreasing in q, illustrating how unit wholesale price decreases with quantity.

9.5 COMPARISON OF THE SIX CASES

Using the analysis above and the results summarized in Table 9.3, we now compare the six cases (three contract types, full or asymmetric information) along various dimensions:

- Wholesale prices w_j and (if applicable) lump-sum side payments L_j for case j.

- Supplier's and buyer's profit margin for case j, denoted by $m_{s,j}$ and $m_{b,j}$ respectively, where $m_{s,j} := w_j - s$ and $m_{b,j} := p_j^* - w_j - c$.

- "Effective" wholesale prices $w_j^e := w_j - L_j/q_j^*$ for case j, the average unit wholesale price taking the lump-sum rebate L_j into account. Although ex post, w_j^e is the averge wholesale price paid by the buyer, we prefer not to call it that because ex ante the buyer's ordering behavior is based on w_j, not on w_j^e.

- Supplier's and buyer's "effective" unit profit margin for case j, denoted by $m_{s,j}^e := w^e - s$ and $m_{b,j}^e := p - w^e - c$ respectively.

- Profits of supplier, buyer, and joint profits, for case j.

9.5.1 The impact of the buyer's cost on the supplier's profit margin

One of the questions we are now able to answer is the following. As the buyer's cost c increases, the buyer's unit profit margin m_b decreases, leading the buyer to order less, which in turn decreases the supplier's profit. How should the supplier respond to this? The answer depends on the information structure and contract type allowed. In case F1, the supplier's margin $m_{s,F1} = \frac{1}{2}\frac{a}{b} - \frac{1}{2}(c + s)$ decreases with c. This means that a supplier faced with a buyer cost increase should accept a smaller profit margin in order to maintain volume. Similarly, in case A1, the supplier's margin $m_{s,F1} = \frac{1}{2}\frac{a}{b} - \frac{1}{2}(E[c]+s)$ decreases with $E[c]$, leading to the analogous insight that a supplier should respond to an increase in buyer's expected cost by decreasing his own margin. In cases F2 and F3, the supplier always sets wholesale price equal to marginal cost s, leading to $m_{s,F2} = m_{s,F3} = 0$, regardless of buyer cost c. In case A2, however, the supplier's margin $m_{s,A2} = \bar{c} - E[c]$; this is because the supplier needs to offer a side payment L that will satisfy even a buyer with the highest possible cost $\bar{c}$, so the supplier wishes to recoup part of that relatively high side payment by charging a higher unit profit margin than in the complete information case. The unit margin increases with $\bar{c}$, and decreases in $E[c]$ as in case A1. In all these cases, the supplier sacrifices margin for volume.

Interestingly, though, in case A3, the supplier should do the reverse: his margin is $m_{s,A3} = \frac{F(c)}{f(c)}$, which is increasing in c. This seems to suggest that the supplier can use the additional flexibility offered by nonlinear contracts to maintain a high unit profit margin in a way that he cannot do when restricted to offering linear contracts (with or without side payment).

To examine this phenomenon more precisely, one should take the side payment into account and evaluate the "effective" unit wholesale price $w^e = w - \frac{L}{q}$ and profit margins. Assume the buyer incurs a cost increase but the supplier cannot observe it, so his prior $F(c)$ remains unchanged. What will this cost increase do to the supplier? Although we do not yet have analytical results for this case, the numerical example in Section 9.6 and Figure 9.2 provides an instance where under full information, the supplier will lower his effective margin, but will increase it under asymmetric information (except in case A1 where average margins depends only on $E[c]$).

9.5.2 The value of information to the supplier

How much can the supplier gain from obtaining better information about the buyer's cost structure, without changing the type of contract he can offer? We answer this question for one-part and two-part linear contracts. First, introduce the difference operator $\Delta_{ij}\Pi_s := \Pi_{s,i} - \Pi_{s,j}$ where i and j denote the cases being compared.

Starting with the one-part linear contracts, one can easily verify from the expressions for supplier's (expected) profits in case F1 (9.14) and A1 (9.21) respectively, that the expected difference is equal to:

$$\Delta_{F1,A1}\Pi_s \;=\; E_c[\Pi_{s,F1}(w^*_{F1}) - \Pi_{s,A1}(w^*_{A1})] = \frac{b}{8}\text{Var}(c) \qquad (9.36)$$

This implies that the supplier would increase his expected profit by $\frac{b}{8}\text{Var}(c)$ if he had complete information about the buyer's marginal cost. In addition, the more price-sensitive the demand (i.e. the greater b), the more valuable is the information to the supplier; this confirms what one would expect intuitively.

Now turning to the two-part linear contract cases, similar analysis based on the equations for case F2 (9.18) and A2 (9.27) respectively, one finds

$$\Delta_{F2,A2}\Pi_s = E_c[\Pi_{s,F2}(w_{F2}^*, L_{F2}^*) - \Pi_{s,A2}(w_{A2}^*, L_{F2}^*)] =$$
$$= \frac{b}{4}\text{Var}(c) + \frac{1}{2}(\bar{c} - E[c])(a - b(s + \bar{c})) \qquad (9.37)$$

So now the value of information still depends on price-sensitivity b and on $\text{Var}(c)$, but also on the worst-case deviation from the supplier's expected value $E[c]$. This is because in cases F2 and A2 the supplier offers a side payment which the supplier can use to make the buyer's individual rationality constraint binding at $c = \bar{c}$; he can not do so in cases F1 and A1. We also see that $\Delta_{F2,A2}\Pi_s \geq 2\Delta_{F1,A1}\Pi_s$ because $\bar{c} \geq E[c]$ and $a - b(s + \bar{c}) \geq 0$ by assumption (to ensure nonnegative quantities). This leads to the important and intuitive finding that the value of information to the supplier is (significantly) greater when the supplier has the flexibility to offer two-part contracts.

9.5.3 The value to the supplier of offering side payments

Here we ask ourselves, how much can the supplier gain from offering more sophisticated contracts, without changing the information structure? In the full information case, we use the expressions for case F2 (9.18) and F1 (9.14) respectively, to find

$$\Delta_{F2,F1}\Pi_s = \Pi_{s,F2}(w_{F2}^*, L_{F2}^*) - \Pi_{s,F1}(w_{F1}^*) =$$
$$= -\Pi_b^- + \frac{1}{8}\frac{(a - b(s + c))^2}{b} \qquad (9.38)$$

This expression is exactly as one would expect: the second term is the difference between the supplier's profits in case F1 and an integrated firm's profits, the first term reflects the fact that the supplier cannot extract an arbitrary level of profit from the buyer. We see that the value of offering two-part contracts instead of one-part contracts decreases with the buyer's reservation profit level Π_b^- and with price-sensitivity b. The latter is perhaps surprising: as demand becomes more price-sensitive, the absolute penalty from using only wholesale price without side payments decreases. Measuring the penalty in relative terms by looking at $\frac{\Pi_{s,F2}(w_{F2}^*, L_{F2}^*)}{\Pi_{s,F1}(w_{F1}^*)}$ gives the same result: the relative penalty decreases in b (for $\Pi_b^- \geq 0$).

Moving to the asymmetric information case, we use the expressions for case A2 (9.27) and A1 (9.21) respectively, to find

$$\Delta_{A2,A1}\Pi_s = E_c[\Pi_{s,A2}(w_{A2}^*, L_{A2}^*) - \Pi_{s,A1}(w_{A1}^*)] =$$
$$= -\Pi_b^- + \frac{1}{8}\frac{(a - b(s + 2\bar{c} - E[c]))^2}{b} \qquad (9.39)$$

The comparative statics are as above: the value of contracting flexibility decreases with the buyer's reservation profit level Π_b^- and with price-sensitivity b. Moreover, because $\Delta_{A2,A1}\Pi_s \leq \Delta_{F2,F1}\Pi_s$, the value of contracting flexibility is greater under full information.

9.5.4 The value of information versus the value of contracting flexibility

Many suppliers in practice find themselves in case A1, with the simplest possible type of contract and incomplete information about the buyer's cost structure. Should such a supplier focus on offering more sophisticated contracts or on obtaining better information about the buyer's costs? Two observations are of interest here:

- The value of information increases with b, while the value of contracting flexibility decreases with b. This suggests that in more price-sensitive environments, the supplier should focus more on obtaining information about the buyer's costs.

- Whichever step the supplier takes first (A1 to F1 or A1 to A2), the value of that step would have been greater had he taken the other step first. In other words, a supplier investing in efforts to reduce his uncertainty about the buyer's costs should realize that, without increases contracting flexibility, he will not realize the full value of those efforts.

9.6 NUMERICAL EXAMPLES

To illustrate the behaviour of the various types of contract we provide a simple numerical example. Assume the supplier's prior distribution over c is uniformly distributed on $[\underline{c}, \overline{c}] = [10, 20]$, that the demand function is $q = a - bp = 200 - 2p$, that the manufacturing cost is $s = 50$, and that the buyer's reservation profits $\Pi_b^- = 0$. The figures below show how "effective" unit wholesale prices, "effective" margins for supplier and buyer, and both parties' profits, depend on c. Clearly, when the supplier cannot observe c, unit wholesale price w cannot depend directly on c (other than through a revelation mechanism), but w^e and the other variables displayed will depend on c through their dependence on q which in turn depends on c.

Figure 9.1 shows how the effective unit wholesale price, which we defined as $w^e = w - L/q$, depends on c. The figure is in accordance with our earlier suggestion that that in all full-information cases, the effective wholesale price decreases with c, which means the supplier is sacrificing margin to maintain volume; in the asymmetric information cases A2 and A3, the opposite occurs. (In case A1, w_{A1}^* cannot depend on c and there is no side payment which could introduce dependence on c.) Figure 9.2 shows the effective unit margins obtained by the supplier, and is similar to Figure 9.1. In Figure 9.3 we see that the buyer's unit margin always decreases in c, which is as one would expect. Figures 9.4 and 9.5 show which cases the supplier and buyer respectively will prefer.

9.7 CONCLUSIONS

In this paper we have used the simple bilateral monopoly framework with price-sensitive demand to study the interactions between information structure and contracting sophistication. We observed that, under full information, a supplier

288

Figure 9.1 "Effective" (average) unit wholesale prices

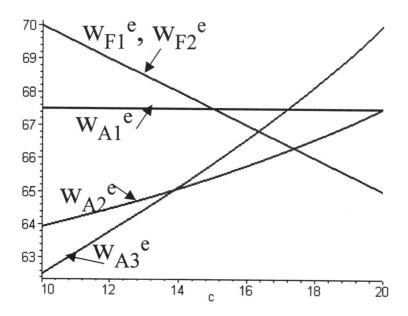

will decrease his wholesale price in reaction to a buyer cost increase, maintaining volume while sacrificing margin. Under asymmetric information, however, the supplier may do the opposite: increase average wholesale price, thus maintaining margin but sacrificing volume. We found that the value to the supplier of obtaining better information about the buyer's cost structure increases with the variance of the supplier's prior distribution about that cost parameter and with price-sensitivity of demand. We also found that the value of better information is greater when the supplier can offer two-part contracts rather than only one-part contracts. We saw that the value of being able to offer two-part contracts rather than one-part contracts is decreasing in price-sensitivity b.

Clearly, even in such a simple contracting framework as bilateral monopoly with asymmetric information, non-intuitive behaviour can occur. Clearly, many questions remain to be addressed, both within the framework presented here and by expanding the framework. In many contracting situations, the supplier starts in case A1: offering a simple linear wholesale price with no side payment, without knowing the buyer's cost structure. When should the supplier focus

Figure 9.2 Supplier's "effective" (average) unit margins

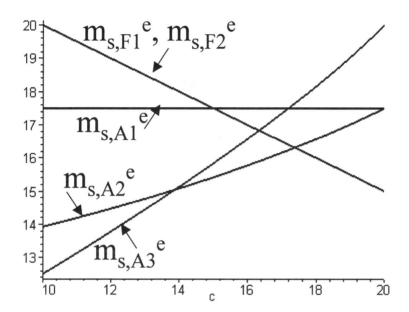

on obtaining better information about the buyer's cost structure, and when should he offer more sophisticated contracts? How would our results change if we introduce stochastic price-sensitive demand, as in Ha (1997) and Gal-Or (1991a)? What changes if the supplier cannot observe the price-sensitivity parameter b? We plan to investigate these questions in our future research.

9.8 REFERENCES

Bourland, K., S. Powell and D. Pyke, "Exploring timely demand information to reduce inventories", *European Journal of Operational Research*, 92 (1996), 239-253.

Bresnahan, T.F. and P.C. Reiss, "Dealer and manufacturer margins", *Rand Journal of Economics*, 16:2 (1985), Summer, 253-268.

290

Figure 9.3 Buyer's "effective" (average) unit margins

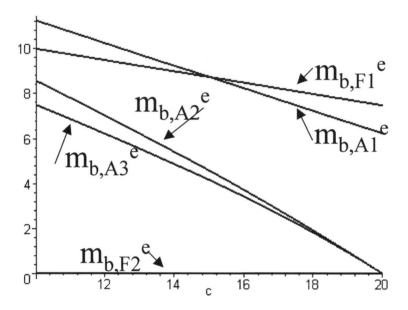

Cachon, G. and M. Fisher, "Supply chain inventory management and the value of shared information", working paper, Wharton School, University of Pennsylvania (1997).

Cachon, G. and P. Zipkin, "Competitive and cooperative inventory policy in a two stage supply chain", working paper, Fuqua School of Business, Duke University (1997).

Chen, F., A. Federgruen and Y. Zheng, "Coordination mechanisms for decentralized distribution systems", working paper, Wharton School, University of Pennsylvania (1997).

Clark, A.J. and H. Scarf, "Optimal Policies for a Multi-Echelon Inventory Problem", *Management Science*, Vol. 6, No. 4 (1960), pp. 475-490.

Corbett, C.J., "Supply chain logistics in an imperfect partnership (Inventory management and incentive and information asymmetries)", unpublished PhD dissertation, INSEAD (1996).

Figure 9.4   Supplier's profits

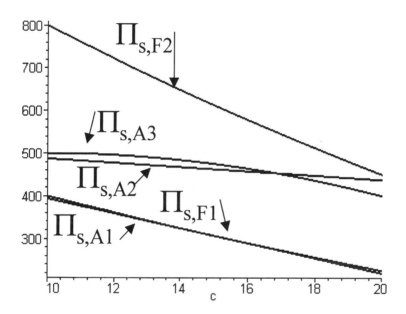

Corbett, C.J., "Stochastic inventory systems under asymmetric information", working paper, The Anderson School at UCLA (1998).

Corbett, C.J. and X. de Groote, "Integrated supply-chain lot sizing under asymmetric information", working paper, The Anderson School at UCLA; presented at First Xavier de Groote Memorial conference at INSEAD, May 16, 1997.

Fudenberg, D. and J. Tirole, *Game Theory*, The MIT Press, Cambridge, Mass., 1991.

Gal-Or, E., "Vertical restraints with incomplete information", *The Journal of Industrial Economics*, Vol. 39, September (1991a), 503-516.

Gal-Or, E., "Duopolistic vertical restraints", *European Economic Review*, Vol. 35 (1991b), 1237-1253.

292

Figure 9.5 Buyer's profits

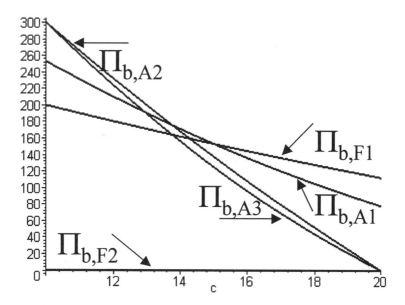

Gavirneri, S., R. Kapuscinski and S. Tayur, "Value of information in capacitated supply chains", working paper, Carnegie Mellon University (1996).

Ha, A.Y., "Supply contract for a short-life-cycle product with demand uncertainty and asymmetric cost information", working paper, Yale School of Management (1997).

Kamien, M.I. and N.L. Schwartz, *Dynamic Optimization (The Calculus of Variations and Optimal Control in Economics and Management)*, North-Holland, New York, 1981.

Laffont, J.-J. and J. Tirole, *A Theory of Incentives in Procurement and Regulation*, The MIT Press, Cambridge, Mass., 1993.

Lee, H.L., Kut C. So and Christopher S. Tang, "The Value of Information Sharing in a Two-Level Supply Chain", working paper, The Anderson School at UCLA (1998).

Lee, H. and S. Whang, "Decentralized Multi-Echelon Inventory Control Systems: Incentives and Information", working paper, Stanford University, (1996).

Machlup, F. and M. Taber, "Bilateral monopoly, successive monopoly, and vertical integration", *Economica*, May (1960), 101-119.

Tirole, J., *The Theory of Industrial Organization*, The MIT Press, Cambridge, Mass., 1988.

Proofs for case A3

Writing $u := \dot{w}$, formulate the Hamiltonian:

$$H = -fL + \frac{1}{2}f(w-s)(a-b(w+c)) + \lambda\frac{1}{2}b[\frac{a}{b}(w+c)]u + \mu u$$

Necessary optimality conditions:

$$\dot{\lambda} = -D_L H$$
$$\dot{\mu} = -D_w H$$
$$D_u H = 0$$

The first gives $\dot{\lambda} = -D_L H = f$, so that (combined with the transversality condition) $\lambda = F$. The second gives

$$\dot{\mu} = -D_w H = -\frac{1}{2}f(a-b(w+c)) + \frac{1}{2}f(w-s)b + \frac{1}{2}\lambda ub =$$
$$= -\frac{1}{2}f[a-b(w+c)-b(w-s)] + \frac{1}{2}Fub$$

The third gives

$$\frac{1}{2}\lambda b[\frac{a}{b} - (w+c)] + \mu = 0$$
$$\mu = -\frac{1}{2}Fb[\frac{a}{b} - (w+c)]$$
$$\dot{\mu} = -\frac{1}{2}fb[\frac{a}{b} - (w+c)] + \frac{1}{2}Fb(u+1)$$

Equating the two expressions for $\dot{\mu}$ gives:

$$\tfrac{1}{2}Fb = \frac{1}{2}fb(w-c)$$

from which the desired expression for w_{A3}^* follows immediately. We still need to verify several assumptions that we had temporarily set aside:

1. the buyer's individual rationality constraint;

2. sufficiency for supplier's optimization problem $\mathcal{S}_{A3}$.

Individual rationality: set the individual rationality constraint to equality for $c = \bar{c}$. A sufficient condition for it to hold for all c is that $D_c \Pi_b \leq 0$.

$$
\begin{aligned}
D_c \Pi_b &= -b\left[\frac{a}{2b} - \frac{1}{2}(w+c)\right](\dot{w}+1) + \dot{L} \\
&= -\frac{1}{2}[a - b(w+c)]
\end{aligned}
$$

As we had assumed $a - b(w + c) \geq 0$ (i.e. demand q is always nonnegative), we need only check the buyer's individual rationality constraint for $c = \bar{c}$ to ensure it is satisfied for all possible c.

Sufficiency, supplier: unfortunately with this type of analysis, this condition is generally impossible to verify, and we have indeed been unable to prove sufficiency in this case. However, the results are fully in line with what one would expect based on e.g. Laffont and Tirole (1993).

Table 9.1   The six combinations of contract type and information structure

| *Type of contract offered by supplier:* | *Does the supplier know the buyer's cost structure?* | |
	yes (Full information)	*no (Asymmetric information)*
1: one part, linear contract: w	Case F1: ■ supplier offers a one-part contract w, where w is the unit wholesale price, *independent* of quantity purchased q ■ buyer selects order quantity q ■ buyer pays supplier wq	Case A1: ■ supplier offers a one-part contract w, where w is the unit wholesale price, *independent* of quantity purchased q ■ buyer selects order quantity q ■ buyer pays supplier wq
2: two-part, linear contract: (w, L)	Case F2: ■ supplier offers a two-part contract (w, L), where w is the unit wholesale price and L the lump-sum side payment, both *independent* of q ■ buyer selects order quantity q ■ buyer pays supplier $wq - L$	Case A2: ■ supplier offers a two-part contract (w, L), where w is the unit wholesale price and L the lump-sum side payment, both *independent* of q ■ buyer selects order quantity q ■ buyer pays supplier $wq - L$
3: two-part, nonlinear contract: $\{w(q), L(q)\}$	Case F3: ■ supplier offers a two-part menu of contracts $\{w(q), L(q)\}$, where $w(q)$ is the unit wholesale price and $L(q)$ the lump-sum side payment, both *dependent* on q ■ buyer selects order quantity q ■ buyer pays supplier $w(q)q - L(q)$	Case A3: ■ supplier offers a two-part menu of contracts $\{w(q), L(q)\}$, where $w(q)$ is the unit wholesale price and $L(q)$ the lump-sum side payment, both *dependent* on q ■ buyer selects order quantity q ■ buyer pays supplier $w(q)q - L(q)$

Table 9.2 Notation

p	final selling price, selected by the buyer
w	wholesale price
L	lump sum payment per period from supplier to buyer
c	buyer's unit manufacturing costs
s	supplier's unit manufacturing costs
m_s	supplier's unit profit margin, $m_s := w - s$
m_b	buyer's unit profit margin, $m_b := p - w - c$
w^e	"effective" (or average) unit wholesale price, $w^e := w - L/q$
m_s^e	supplier's "effective" unit profit margin, $m_s^e := w^e - s$
m_b^e	buyer's "effective" unit profit margin, $m_b^e := p - w^e - c$
Π_b	buyer's profits
Π_b^-	buyer's reservation profit level
Π_s	supplier's profits
a, b	parameters of demand function $q = a - bp$
q	buyer's order quantity
$\hat{c}$	buyer's (announced) marginal unit manufacturing cost
$F(c), f(c)$	supplier's prior distribution over the buyer's marginal costs C, where F and f are defined on the finite interval $[\underline{c}, \bar{c}]$
$D_x H$	derivative of function H with respect to x
$\dot{H}$	derivative of a function H with respect to $\hat{c}$ evaluated at $c = \hat{c}$, ie. $\dot{H} = D_{\hat{c}} H$
$\Delta_{ij}\Pi_k$	increase in profits of player k going from case j to case i, ie. $\Delta_{ij}\Pi_k := \Pi_{k,i} - \Pi_{k,j}$, where $i, j \in \{F1, F2, F3, A1, A2, A3\}$

Table 9.3 Summary of optimal supply contracts

Type of contract offered by supplier:	*Does the supplier know the buyer's cost structure?*	
	yes (full information, F)	no (asymmetric information, A)

1: one part, linear contract: w

yes (full information, F):

$$w^*_{F1} = \frac{a}{2b} + \frac{1}{2}(s - c)$$

$$q^*_{F1} = \frac{1}{4}a - \frac{1}{4}(s + c)$$

$$p^*_{F1} = \frac{3}{4}a + \frac{1}{4}(s + c)$$

$$m^*_{s,F1} = \frac{a}{2b} - \frac{1}{2}(s + c)$$

$$m^*_{b,F1} = \frac{a}{4b} - \frac{1}{4}(s + c)$$

$$\Pi_{s,F1}(w^*_{F1}) = \frac{1}{8b}[a - b(s + c)]^2$$

$$\Pi^*_{b,F1} = \frac{1}{16b}[a - b(s + c)]^2$$

no (asymmetric information, A):

$$w^*_{A1} = \frac{a}{2b} + \frac{1}{2}(s - E[c])$$

$$q^*_{A1} = \frac{1}{4}a - \frac{1}{4}(s + c) - \frac{1}{4}b(c - E[c])$$

$$p^*_{A1} = \frac{3}{4}a + \frac{1}{4}(s + c) + \frac{1}{4}(c - E[c])$$

$$m^*_{s,A1} = \frac{a}{2b} - \frac{1}{2}(s + E[c])$$

$$m^*_{b,A1} = \frac{a}{4b} - \frac{1}{4}(s + c) - \frac{1}{4}(c - E[c])$$

$$E_c[\Pi_{s,A1}(w^*_{F1})] = \frac{1}{8b}[a - b(s + E[c])]^2$$

$$\Pi^*_{b,A1} = \frac{1}{16b}[a - b(s + 2c - E[c])]^2$$

2: two-part, linear contract: (w, L)

yes (full information, F):

$$w^*_{F2} = s$$

$$L^*_{F2} = \Pi^-_b - \frac{1}{4b}[a - b(s + c)]^2$$

$$q^*_{F2} = \frac{1}{2}[a - b(s + c)]$$

$$p^*_{F2} = \frac{1}{2b}[a + b(s + c)]$$

$$m^*_{s,F2} = 0$$

$$m^*_{b,F2} = \frac{a}{2b} - \frac{1}{2}(s + c)$$

$$\Pi_{s,F2}(w^*_{F2}, L^*_{F2}) = -\Pi^-_b + \frac{1}{4b}[a - b(s + c)]^2$$

$$\Pi^*_{b,F2} = -\Pi^-_b$$

no (asymmetric information, A):

$$w^*_{A2} = s + \bar{c} - E[c]$$

$$L^*_{A2} = \Pi^-_b - \frac{1}{4b}[a - b(s + \bar{c}) - b(\bar{c} - E[c])]^2$$

$$q^*_{A2} = \frac{1}{2}[a - b(s + \bar{c} + c - E[c])]$$

$$p^*_{A2} = \frac{1}{2b}[a + b(s + \bar{c} + c - E[c])]$$

$$m^*_{s,A2} = \bar{c} - E[c]$$

$$m^*_{b,A2} = \frac{a}{2b} - \frac{1}{2}(s + \bar{c} + c - E[c])$$

$$E_c[\Pi_{s,A2}] = -\Pi^-_b + \frac{1}{4b}[a - b(s + 2\bar{c} - E[c])]^2 + \frac{1}{2}(\bar{c} - E[c])[a - b(s + \bar{c})]$$

$$\Pi^*_{b,A2} = \Pi^-_b + \frac{1}{4b}[a - b(s + \bar{c} + c - E[c])]^2 - \frac{1}{4b}[a - b(s + 2\bar{c} - E[c])]^2$$

3: two-part, nonlinear contract: $\{w(q), L(q)\}$

Case F3: results are the same as case F2

Note: most results assume uniform prior

$$w^*_{A3} = s + \frac{F(c)}{f(c)} = s + c - \underline{c}$$

$$L^*_{A3} = \frac{1}{2}c[a - b(s + c - \underline{c})] + k$$

$$q^*_{A3} = \frac{1}{2}[a - b(s + c) - b(c - \underline{c})]$$

$$p^*_{A3} = \frac{1}{2b}[a + b(s + c) + b(c - \underline{c})]$$

$$m^*_{s,A3} = \frac{F(c)}{f(c)} = c - \underline{c}$$

$$m^*_{b,A3} = \frac{a}{2b} - \frac{1}{2}(s + c) - \frac{1}{2}(c - \underline{c})$$

10 MODELING SUPPLY CHAIN CONTRACTS: A REVIEW

Andy A. Tsay

Steven Nahmias

Narendra Agrawal

Department of Operations & Management Information Systems
Leavey School of Business
Santa Clara University
Santa Clara, CA 95053

10.1 INTRODUCTION

In this review, we summarize model-based research on contracts in the supply chain setting and provide a taxonomy for work in this area. During our discussions it became clear that the field has developed in many directions at once. Furthermore, as we surveyed the literature, it was not obvious what constitutes a contract in this context. While the nomenclature "supply chain management" is relatively new, many of the problems that are addressed are not. In particular, mathematical models for optimizing inventory control have a long history as a significant part of the mainstream of operations research and operations management. Inventory modeling, per se, dates to the early part of the century and the ideas of a Westinghouse engineer named Ford Harris (1915). A natural issue to address first is what is meant by supply chain management (SCM) research and how it relates to the vast body of work constituting classical inventory theory.

Modern usage of the term seems to be consistent with the following definition: *a supply chain is two or more parties linked by a flow of goods, information, and funds*. One way to interpret this is that SCM research is essentially the same as multi-echelon inventory theory, introduced by Clark and Scarf (1960) to model logistics problems encountered by the military. Using a dynamic programming formulation, the authors derived the optimal ordering and transshipment policy for a single-product serial system facing independent, identically distributed demand. The optimal policy is characterized by a critical "order-up-to" echelon inventory target for each installation, where echelon inventory counts not only the local stock but also the inventory downstream in the system. On the whole, much of the work on multi-echelon inventory theory relies on relaxing assumptions in the basic Clark-Scarf formulation. Experience thus far suggests that optimal control policies can be calculated only in limited settings. We direct the reader to Federgruen (1993) for a recent review of multi-echelon inventory models. Clark (1972) is also recommended for its excellent coverage of the earlier work.

While the basic structure of the systems studied can be similar, we believe that SCM research encompasses a much broader scope of issues than does multi-echelon inventory theory. That is, while multi-echelon inventory theory is primarily about controlling the timing and quantity of material flows, SCM studies this and more. For instance, SCM treats environments in which there are multiple decision makers, which may be different firms or different divisions within a single firm. Behavior that is locally rational can be inefficient from a global perspective (cf. Whang 1995), so attention turns to methods for improving system efficiencies. Some mechanisms for making these improvements are the contractual arrangements surveyed here. These include the reallocation of decision rights, rules for sharing the costs of inventory and stockout, and policies governing pricing to the end-customer or between supply chain partners. Representation of the information structure and rules for information sharing are also important, since the assumption of common information usually made in multi-echelon inventory theory may be inaccurate in real supply chains.

SCM also considers the topology of the system; that is, the number of suppliers, distributors or retailers. While traditional multi-echelon inventory theory tends to deal with issues of transportation delays via lead times, modern SCM studies the implications of alternative modes of logistics. One example is the current trend towards the outsourcing of the logistics function to external parties whose comparative advantage derives from scale and focus. Even the product itself may be redesigned to affect supply chain performance improvement. For example, several studies consider the extent to which delaying product differentiation ("postponement") affects volume and inventory pooling efficiencies within the supply chain (Lee et al 1993, Lee 1996). Of course, aspects of these problem areas have been considered in other fields, which suggests a distinguishing characteristic of SCM research: like supply chains themselves, the research questions and methodologies tend to cross traditional functional lines.

As suggested above, the management of supply chains consisting of multiple agents with possibly conflicting objectives requires consideration of the relationships among the parties. Recent research explores arrangements not considered in traditional inventory theory. We will refer to these structures as contracts. Because much of the research in this area is quite recent, many of the papers discussed have not yet appeared in the open literature, and are available as working papers only. Our including a paper in this review does not mean we believe it to be completely correct, nor that it will eventually pass peer review and appear in print. We believe this is the first review to try to provide a comprehensive description and classification of the model-based analyses of contracts in supply chain management.

10.2 SCOPE OF THE REVIEW

Before beginning our review, we note that contracts are an important area of study in disciplines other than SCM. Contracts are, of course, a major consideration in law, and the literature in this area is enormous. There is also a substantial literature on contracts in the economics literature (Tirole 1988 and Katz 1989 provide excellent reviews). These bodies of work contribute deep understanding of basic issues of motivation for a broad variety of contractual structures, with particular attention to their legality, enforceability, and ramifications for public policy and social welfare. Furthermore, they are the source of many concepts and techniques which SCM researchers make use of and build upon. What distinguishes SCM contract analysis may be its focus on operational details, requiring more explicit modeling of materials flows and complicating factors such as uncertainty in the supply or demand of products, forecasting and the possibility of revising those forecasts, constrained production capacity, and penalties for overtime and expediting. Due to space limitations, we review only those papers from other fields which are considered foundational to a stream of SCM research.

Even after restricting our search in this way, we were faced with the challenge of defining what constitutes a "contracts" paper in the SCM context.

This challenge arises because, broadly speaking, all literature on inventory theory could qualify. Certainly, the purchase of materials implies an agreement between two parties even if the behavior and preferences of one party are suppressed. We adopt the more narrow definition that the analysis must explicitly offer guidance in negotiating the terms of the relationship between buyer and seller. This would mean that classical inventory theory, in which the supply parameters (e.g. price, lead time, bounds on order size) are typically treated as exogenous environmental characteristics, would not satisfy this test. Thus, we review in depth only those papers which treat the terms of the contractual relationship as decision variables, or at the very least investigate the behavioral and performance consequences of changing these terms. The specific terms on which we focus are described later in §10.3.3.

10.3 FUNDAMENTALS OF CONTRACT ANALYSIS

In this section we describe the structure of a basic supply chain to provide a context for the subsequent discussion. Specifically, we use this to identify the decisions to be made, the relevant environmental components, and the behavioral dynamics which underlie the contracts discussed.

10.3.1 Supply Chain Structure

Consider a supply chain in which an upstream party (which we refer to as a manufacturer) provides a single product to a downstream party (which we refer to as a retailer), who in turn serves market demand, as shown in Figure 10.1. This scenario could describe the link between any two consecutive nodes in a supply chain, and indeed on occasion the tandem may be referred to as supplier and manufacturer, manufacturer and distributor, or, most generally, supplier and buyer.

Researchers commonly make the following simplifying assumptions to render the analysis more tractable. The manufacturer produces (or acquires) the product at a constant unit cost of c and charges the retailer a *wholesale/transfer payment* of $W(Q)$ for Q units. $W(Q)$ may either be exogenous, or a decision variable under the control of one of the parties. The retailer, in turn, sells the product at a price of p per unit. Market demand, denoted as $D(p)$, in reality is both price-sensitive and uncertain. While some models include both these features, it is more common to either take the retail price as fixed and represent market demand as a random variable (as in the operations research literature), or assume a deterministic, downward-sloping demand curve (as in the economics and marketing literatures). In the latter case the retailer's decision is primarily p, whereas in the former it is Q. A simpler underlying structure allows traditional inventory models to treat more complex problem settings including multiple periods, continuous review, and finite and infinite horizons. However, most contract papers assume only a one-period problem (i.e., a newsvendor setting), since the resulting models are often too complex to be amenable to multi-period analysis.

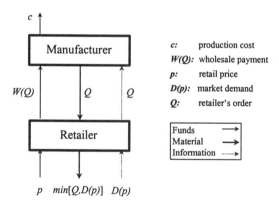

Figure 10.1 The Supply Chain

10.3.2 *Purposes of Contracts*

To understand what motivates the parties to pursue certain contract structures, consider the simple supply chain pictured in Figure 10.1. In the best of all possible worlds, total expected supply chain profits would be maximized if all decisions are made by a single decision maker with access to all available information. This is referred to as the *first-best case*, and is often associated with *central control*. Denote the resulting expected profit as Π^c in this case. However, in general neither the manufacturer nor the retailer is in a position to control the entire supply chain, and each has his own incentives and state of information. Refer to this as a *decentralized control* structure, with resulting total expected profit Π^d. It is *inefficient* if $\Pi^d < \Pi^c$.

Some contracts focus on how Π^d is to be split between the two parties. We call this the *risk-sharing* objective in that it provides a means for the buyer and supplier to share the risks arising from various sources of uncertainty, e.g. market demand, selling price, process yield, product quality, delivery time, and exchange rates. The risk sharing motive is common in the contracts reviewed here. As an example, suppose that the retailer is required to transmit sales forecasts upstream to manufacturers. These forecasts are intended to assist the manufacturer to make capacity and materials purchasing decisions. However, in most cases no commitment is attached to these forecasts. As a result, the manufacturer assumes a large portion of the risk of demand uncertainty. Not only might the retailer cancel orders if demand is lower than anticipated, there is an incentive for the retailer to deliberately inflate forecasts as a form of insurance. Minimum purchase agreements or penalties for returns are often included in contracts to protect the manufacturer against this behavior. Of course, a necessary condition for the adoption of any contractual agreement is that both parties ultimately benefit.

Contracts also provide a means for bringing Π^d closer to Π^c, which we call the *system-wide performance improvement* objective. This is also referred to as the *channel coordination* objective (a phrase coined in the marketing literature). Channel coordination may be achieved by first identifying the intra-chain dynamics which cause the inefficiency, then modifying the structure of the relationship to more closely align individual incentives with global optimization. For example, the existence of two separate entities within the channel can lead to *double marginalization*, a well-known cause of supply chain inefficiency. This notion apparently originated in the economics literature (Spengler 1950), where the fundamental decision is usually p, and the tradeoff to be made is between the unit profit margin (which favors higher p) and the size of the market (which favors lower p). If the retailer pays a unit price $t > c$ to the manufacturer, hence creating the two distinct profit margins referred to by the name of this phenomenon, the retailer's choice of p will be consistent with a profit margin of $(p - t)$ rather than the $(p - c)$ that would be appropriate from a central planner's perspective. Double marginalization can also occur when p is fixed, demand is stochastic and Q is the decision variable (more commonly assumed in the inventory management literature). Here, the optimal Q is determined by considering the overstocking and understocking costs. While the consequence of overstocking may be the same whether control is centralized or decentralized, the retailer perceives an overstock cost of $(p - t)$ rather than $(p - c)$. The problem does not go away if the manufacturer controls the order quantity decision, since the manufacturer's cost of overstocking is $(t - c)$. Many of the contractual structures recently studied (cf. Pasternack 1985, Donohue 1996, Tsay 1996, Ha 1997a) attempt to remedy some variant of this basic problem.

Contracts also facilitate *long-term partnerships* by delineating mutual concessions that favor the persistence of the business relationship, as well as specifying penalties for non-cooperative behavior. The lengthening of the time horizon may encourage parties to engage in activities that are unfavorable in the short term but have substantial payoffs over time. For example, chip vendor Intel might be willing to consign a large portion of its production of a new generation of microprocessors to a single computer maker, such as Dell, even though the chips would fetch more on the open market. Intel's motivation would be to build a long-term relationship in the hope that Dell would be a volume purchaser for many years to come. In addition to providing a reliable supply for a buyer and demand volume for a supplier, stable partnerships can reduce transactions costs and allow for greater cooperation (e.g. information sharing and collaborative product and process improvement). One effort to study some of these issues is Cohen and Agrawal (1998), which evaluates the impact of various contractual arrangements on total costs incurred by a risk-averse buyer firm. In particular, they analyze the tradeoffs between the flexibility offered by short-term contracts and the fixed investments, improvement opportunities, and price certainty associated with long-term contracts. However, few of the models we have encountered consider the time horizon of the contract as a decision variable.

Another important rationale for a contract that is not typically modeled is that it *makes the terms of a relationship explicit.* In fact, in the course of making legally concrete the expectations of each party, a contract can suggest and unambiguously define quantifiable performance metrics which are prerequisite to any systematic process improvement effort. Lead times, on-time delivery rates, and conformance rates are among the metrics commonly specified in supply contracts.

10.3.3 A Classification Scheme

Having restricted our attention only to papers on "contracts" in the SCM context, we still needed to determine a suitable scheme for classifying the resulting set of papers. This was complicated by the fact that no commonly accepted taxonomy appears to exist, and papers that purport to study contracts in supply chains consider a wide range of problems and issues. After experimenting with several schemes, we decided to classify the literature by contract clauses. These include:

> (a) *specification of decision rights,*
> (b) *pricing,*
> (c) *minimum purchase commitments,*
> (d) *quantity flexibility,*
> (e) *buyback or returns policies,*
> (f) *allocation rules,*
> (g) *lead time, and*
> (h) *quality.*

Several papers were candidates for more than one of these categories. For example, a minimum purchase commitment may also require special pricing terms to attract the buyer. Also, virtually all contracts must explicitly specify decision rights in order to be executable. However, classification in the "specification of decision rights" category suggests that a paper's main thrust is to investigate the relative desirability of alternative configurations of decision authority. Naturally, there is an element of subjectivity in this categorization, and there may be some clauses discussed in the literature or used in practice that simply do not fit well into this classification scheme. However, this classification covers the majority of the research that we have seen. Below we briefly describe the above contract clauses as they relate to the supply chain in Figure 10.1.

(a) Specification of Decision Rights

Here the goal is to achieve specific objectives by reassigning control of the decision variables. For example, although the retailer typically chooses Q and p given the $W(Q)$ specified by the manufacturer, there is an arrangement called *Resale Price Maintenance* (RPM) in which the manufacturer is allowed to dictate conditions on the p that the retailer may charge. Likewise, *Quantity Fixing* is in effect when the manufacturer exerts control over Q. The issue of local versus global control may also be interpreted as an issue of decision rights.

(b) Pricing

$W(Q)$ is the component of the contract between the manufacturer and retailer that defines the financial terms of the supply relationship. Typically, $W(Q) = F + tQ$ for constants F and t. $F = 0$ results in linear pricing, perhaps the most commonly assumed pricing structure. A positive F, often referred to as a franchise fee, results in "two-part tariff" pricing. Some researchers also treat more complex pricing schemes, such as quantity discounting.

(c) Minimum Purchase Commitments

Such an agreement requires the retailer to purchase a minimum quantity, either within each single transaction, or cumulatively over a specified time horizon. The manufacturer may reduce $W(Q)$ to provide an incentive to the retailer to agree to this arrangement.

(d) Quantity Flexibility

In a quantity flexibility clause, the quantity the retailer ultimately purchases may deviate from a previous planning estimate, subject to certain constraints and/or financial consequences. To properly represent such a setting requires a stochastic demand model in which some event (such as a forecast update) occurs within the time frame of the model to motivate the exercise of flexibility.

(e) Buyback or Returns Policies

A buyback clause is one that specifies that the retailer may return some or all unsold product to the manufacturer, possibly for only partial credit. Naturally, mismatches between the retailer's purchase and the market demand are only an issue when demand is assumed random.

(f) Allocation Rules

Allocation rules specify how the manufacturer's available stock or production capacity is to be distributed among multiple retailers in a shortage scenario.

(g) Lead Times

The lead time for delivery of product from the manufacturer to the retailer is treated in traditional inventory models as either a fixed constant (one special case being zero), or a realization of a random variable. This clause highlights the possible benefits of adjusting that lead time via a contractual agreement.

(h) Quality

Any supply relationship is premised on the quality of the delivered product. The specific notion of quality may be formalized within the contract.

10.4 REVIEW OF THE LITERATURE

In this section we review the current literature on contracts in supply chains. Each paper will be classified into one of the above categories and briefly summarized.

10.4.1 *Specification of Decision Rights*

While most of the vast multi-echelon inventory literature presumes central control (e.g. Clark and Scarf 1960, Bessler and Veinott 1966, Eppen and Schrage 1981, Federgruen and Zipkin 1984, Rosling 1989), some recent works have proceeded under the realization that localization of decision-making authority may be necessary for large enterprises operating in complex environments. This creates a different set of challenges. Conflicts in the interests of the different parties may lead to inefficiency. Double marginalization, as described earlier, can be interpreted in this way. Or, in a setting of information asymmetry, a key decision may be under the purview of a party with inferior information. Yet shifting control to the better-informed party may engender opportunistic behavior since self-interested actions might be undetectable. Hence, careful consideration of information and incentives is central to all attempts to improve supply chain performance by reconfiguration of decision rights. We note that the economics literature has contributed a great deal to the understanding of such possibilities, including the phenomena of Resale Price Maintenance (RPM) and Quantity Fixing. Rather than summarize this vast literature, we refer the reader to Katz (1989) for a review. In this section we first describe four papers that propose ways to facilitate the shift from centralized to decentralized control of a supply chain, followed by those which focus on solutions which transfer decision rights among the various independent agents.

Lee and Whang (1997) consider the problem of coordinating a decentralized version of the Clark-Scarf serial system. While each site incurs a holding cost, only the site serving the end customer faces a shortage cost. As a result, the upstream sites will carry less buffer inventory than may be best for the system as a whole. Realizing this, the site furthest downstream tends to carry extra inventory, which is inefficient since finished goods usually are most costly to hold. The authors propose rules for performance measurement and accountability, which have the following desirable properties: (i) *cost conservation* (i.e., all costs can be traced to individual sites without the need for subsidies or taxes from the central planner), (ii) *incentive compatibility* (i.e., what is optimal for each individual manager is also optimal for the system as a whole), and (iii) *information decentralizability* (i.e., the rules can be implemented using local information only). Their scheme involves a consignment policy for redistributing inventory carrying costs among the sites, an additional backlog penalty paid to an upstream site by its direct customer, and a shortage reimbursement paid to a downstream site by its direct supplier. No markups are added to the transfer prices. However, implementing this scheme requires common knowledge about the demand distribution. Kumar et al (1996) consider a similar problem but use a different set of control mechanisms for coordination: (i) an internal transfer price and a shortage penalty imposed on the upstream site, and (ii) a service level constraint imposed on the upstream site. For the same model as Lee and Whang (1997), except with delays between the transmission and receipt of orders between sites, Chen (1997) restores system-optimal performance using a measurement scheme based on "accounting inventory". A division's accounting inventory is its on-hand stock minus backlogs of orders placed by the downstream site, under the assumption that the upstream division is perfectly reliable. Porteus and Whang (1991) describe the conflicts of interest between the manufacturing and marketing divisions of a firm (which can be interpreted loosely as a "supply chain") over the production capacity decision. These conflicts can be reconciled by establishing an internal futures market for capacity. While the discussion in each of these papers concerns the coordination of individual agents within a single organization, we include them here because with merely a slight change in interpretation the mechanisms are equivalent to contracts. We now turn our attention to models in which control of certain decisions is transferred from one independent party to another.

The focus of Blair and Lewis (1994) is on designing efficient contracts for a one-manufacturer, one-retailer system, where the manufacturer cannot observe the demand or the retailer's service effort. This introduces both information asymmetry and moral hazard. Market demand (Q) is assumed to be a function of the retail price (p), the retailer's service effort (X) and some exogenous parameter (θ), which is known only to the retailer and reported to the manufacturer at the time of contracting. The retailer's cost for service is assumed to be X. Demand is deterministic, and inventory considerations are ignored. The retailer learns about θ and reports it to the manager (not necessarily truthfully), who in turn offers a menu of contracts from which the retailer selects one. The retailer then chooses his service effort. Assuming full information, first-best performance can be attained when the manufacturer charges a franchise fee and the retailer sets the price, service effort and quantity. However, under information asymmetry and moral hazard, the efficient contracts require some form of resale price maintenance and quantity fixing. The structure is shown to depend upon how the impact of market demand on the retailer's optimal promotion varies with price and quantity, i.e., upon the signs of $\partial^2 X / \partial p \partial \theta$ and $\partial^2 X / \partial Q \partial \theta$. When both these derivatives are zero, there exists a contract that will induce the retailer to select the first-best price, quantity and effort level. Otherwise, the manufacturer must impose a ceiling or floor on the retailer's price, and use quantity fixing/rationing.

Ha (1997a) studies the effect of decision rights in the presence of information asymmetry regarding the downstream member's production cost in an environment of demand uncertainty. He considers a two-member channel similar to that of Blair and Lewis (1994), but allows customer demand to be random and price-dependent, modeled as $D = \mu(p) + Y$, where $\mu(p)$ is deterministic and Y is random and independent of p. First, the author argues that under complete information, coordination can be achieved by various contractual arrangements: quantity fixing, franchising, or a returns policy with price fixing. Then, he considers the case in which the downstream member's variable production cost is private information (the upstream member has a probability distribution for this cost). Here, coordination can be achieved with a nonlinear pricing scheme with price fixing, in which the upstream member offers a menu of contracts that specifies the retail price, order quantity and a fixed fee.

Narayanan and Raman (1997) focus on the discrepancy between the impacts perceived by the manufacturer and the retailer when the retailer stocks out of a product. Typical retailers carry products that substitute for that of a given manufacturer, so tend to stock less of that product than may be optimal for the manufacturer. In a newsvendor framework with complete information, the authors assume a market demand consisting of a deterministic component (a linear function of the manufacturer's promotional effort) and a component that is uniformly distributed between 0 and 1. The cost of the effort is modeled as the square of the effort level. The retail price is fixed. They compare the optimal stock levels and channel profit for three scenarios: (i) the first-best case (a vertically integrated channel that manufactures and retails the product and its substitutes), (ii) "retailer managed inventory" (RMI), in which the retailer controls the inventory decisions, and (iii) "vendor managed inventory" (VMI), in which the manufacturer has the control. This abstraction of VMI is analogous to what is known in the economics literature as Quantity Fixing. The authors show that VMI may or may not do better than RMI, but VMI's relative performance improves as the manufacturer's influence over demand increases.

Finally, Agrawal and Tsay (1998) attempt to generalize the discussion on decision rights by focusing on the effect of intra-organizational goal incongruence on contract efficiency. They consider a single-period model of a supply chain consisting of an independent manufacturer and retailer. The retailer serves a price-sensitive stochastic demand of the form $D = N \cdot g(p)$, where N is random and $g(p)$ is a deterministic downward-sloping function of the retail price, p. Each party seeks to maximize its individual expected profits. However, the operational-level decisions in the retail organization (pricing and stocking level) are delegated by the owner to a manager whose compensation scheme induces behavior inconsistent with this objective. In particular, this manager is interested in maximizing the probability of meeting or exceeding a target profit level specified by the owner. The authors study how this impacts the behaviors and outcomes on both sides of the supply contract. Further, in order to determine the consequences for the welfare of end customers, they define the notion of *expected consumer surplus*. Their analysis evaluates the preferences of the various parties for wholesale and retail prices, inventory levels, and product and customer types. Certain implications for supply chain strategy are suggested: increasing the profitability of a supply chain need not be at the expense of the end customer, and coordination of goals within an organization does not necessarily improve the supply chain efficiency.

Industry case studies in which significant supply chain benefits have accrued in conjunction with the reallocation of decision rights include the Vendor Managed Inventory agreements between Proctor & Gamble and WalMart and between Levi Strauss and some of its retail partners, the Efficient Consumer Response initiative in the grocery industry, and the JIT II movement pursued by some manufacturing companies. On the other hand, obstacles to such programs are also well documented, as in the case of Barilla SpA, the world's largest pasta manufacturer. Its Just-in-Time Distribution program was resisted by its distributors as well as its own personnel (Hammond 1994). The transfer of inventory stocking decision rights to Barilla was perceived by the distributors as a strategic threat. Internally, the sales organization also opposed this program out of fear that it might adversely affect their incentive systems. Modeling of such implementation issues is an area for future research.

10.4.2 Pricing

In most of the traditional inventory models, $W(Q)$ is specified and is not subject to negotiation between the buyer and the seller. More recently, some researchers have considered this cost schedule as a means to modifying the behavior of one or both parties. In particular, they have considered the use of quantity discounting as a coordination mechanism.

One stream of such research includes Monahan (1984), Rosenblatt and Lee (1985), Lee and Rosenblatt (1986), and Banerjee (1986). Monahan (1984) considers the economic implication of offering quantity discounts to the (single) buyer from the supplier's point of view. Assuming that the supplier follows a lot-for-lot policy, he shows that a sufficient discount can induce the buyer to order a quantity that increases the supplier's net profit. This modified quantity is related to the buyer's original economic order quantity by a factor that depends only upon the ratio of the fixed ordering costs of the two parties. Lee and Rosenblatt (1986) extend Monahan's model to include the supplier's lot sizing decision, by considering the inventory carrying and fixed costs incurred by the supplier. Instead of the all-units discount schedule featured in both these papers, Rosenblatt and Lee (1985) study a linear discount schedule in the same setting as Lee and Rosenblatt (1986). Interestingly, here the benefits do not all accrue to the supplier (as in Monahan 1984 and Lee and Rosenblatt 1986), but instead both parties can benefit. Bannerjee (1986) takes the perspective of a central decision maker who can jointly optimize the total of both parties' costs. He computes the joint economic lot-size, quantifies the resulting benefit to the buyer and the supplier, and determines the optimal quantity discount schedule. The preceding papers all assume deterministic and price-independent demand,

and allow no shortages. The papers reviewed next relax some of these earlier assumptions.

A seminal paper considering the role of pricing in channel coordination is Jeuland and Shugan (1983), which appears in the marketing literature. They primarily consider a two-member channel facing deterministic market demand, assumed to be a function of the retailer's selling price (p) and service (s), where the latter costs s. The production cost is $C(Q)$, a function of the manufacturer's product quality (Q). By comparing the optimality conditions with and without coordination, they show that absent coordination, the manufacturer has the motive to raise his price markup above first-best level. At the same time, he will set levels of quality and all other promotional variables that are below the first-best level. The same is true of the retailer. The authors then go on to show how a quantity discount scheme can coordinate the channel, and also share the efficiency gains. However, they acknowledge practical and legal barriers to implementation: in particular, determining the parameters of their scheme will require all costs to be common knowledge of both parties. Moorthy (1987) offers other schemes which can coordinate the channel of Jeuland and Shugan (1983) and are easier to implement. In particular, he argues that a two-part tariff is superior to quantity discounting because it is simpler, it separates the coordination problem from the profit-sharing problem (coordination is achieved by setting the wholesale price equal to the manufacturer's variable cost, and the lump-sum transfer allows the profits to be split arbitrarily), and leads to fewer legal problems (charging all retailers the same wholesale cost is more consistent with the Robinson-Patman Act). Many subsequent papers in the marketing literature have considered variants of this model.

Weng (1995) extends Jeuland and Shugan (1983) by specifically incorporating the mechanisms that determine the relationship between operating decisions (such as order quantities and selling prices) and profits (revenues less inventory and setup costs, calculated using the typical EOQ framework) for a buyer and a supplier, and deriving pricing policies between them that can coordinate the channel's activities. The buyer determines the size of the order placed with the supplier and the selling price charged to customers. Customer demand is a known, deterministic function of that selling price. The supplier, whose profits depend upon the buyer's decisions, controls the transfer price and his own production lot size. All data are assumed to be common knowledge. The focus of the analysis is on determining how to implement a mechanism to divide the additional profits generated through coordination. Under the assumption that the buyer will receive a fixed fraction of the incremental profit, the author shows that a quantity discount for the buyer along with a franchise fee paid to the supplier is sufficient to induce the buyer to make decisions that lead to joint profit maximization. Further, he shows that the form of the quantity discount scheme (all-units vs. incremental quantity discount) is not critical to achieving channel coordination. The dependence of customer demand on price and of operating costs on order quantity are the critical factors.

Weng (1997a) also treats a newsvendor model in which the goal is to coordinate the decisions of the manufacturer and the retailer (which he calls the distributor) via pricing. The one-period demand is assumed to follow a phase-type distribution and includes a functional form that is an explicit function of the distributor's selling price. He compares the system performance when the manufacturer and retailer independently choose prices to that when production and ordering is jointly coordinated. He shows how coordination increases system profits and how the magnitude of the increase depends on the various system parameters. Weng (1997b) treats a problem similar to that considered in Weng (1997a), except with somewhat different assumptions regarding the costs and objectives. He assumes a quadratic manufacturer's cost function, and a service level constraint at the retailer level rather than a stockout cost. He derives the optimal system-wide policy and the transfer price between the manufacturer and retailer that achieves that policy.

Weng (1997c) includes demand uncertainty in the context examined in Weng (1995). However, a number of simplifying assumptions have been made to facilitate the analysis. A single-period model (newsvendor type) is assumed, and both the transfer price and retail price are exogenous. The manufacturer produces to order (with a fixed setup cost), and produces a second run at the end of the season to cover unmet demand (at a higher cost). Thus, the only decisions are the buyer's two order quantities. As in Weng (1995), the objective of the paper is to determine a pricing scheme to allocate the additional profits that result from coordination. Assuming an all-units discounting policy, the author shows that coordination increases the number of units produced by the manufacturer and held by the retailer. However, the increase in the system profit may not always be substantial.

Corbett and de Groote (1997) consider a two-member channel in a deterministic EOQ-type of environment, where all information is common except the buyer's holding cost. The supplier assumes a probability distribution on this holding cost. The authors compare the supplier's optimal contracts under full information to those under asymmetric information, both with and without coordination. These include contracts with a fixed cost transfer, one-step quantity discount, and with full cost transfer. As expected, all these contracts are equivalent under full information. However, under asymmetric information, only full cost transfer to the buyer is system-optimal. It is also the buyer's preferred scheme, while the supplier is indifferent between it and a one-step quantity discount.

10.4.3 Minimum Purchase Commitments

Traditional inventory theory generally assumes that the buyer can order any quantity from the supplier at any time. Certainly the buyer prefers to avoid any constraints on his ability to meet his own customers' demand in an economical way. However, this may be undesirable from the supplier's point of view for a variety of reasons. For instance, suppose a buyer facing uncertain demand places orders according to an (s, S) policy. The variance of the resulting orders will exceed the demand variance since the buyer makes no order in many periods, waiting until cumulative demand over a period of time is sufficiently large. Amplification of demand variance is referred to as the "bullwhip effect" (see Lee et al 1997), and is undesirable upstream since, generally speaking, the supplier's costs increase with the order variance. This is due, for instance, to the increased need for inventory buffers and/or the more tentative scheduling of machine and labor capacity that may result. Even when the buyer's demand is certain, the supplier's production costs may be lower when the order is larger than what the buyer may consider to be optimal.

One response to this conflict is an agreement in which the buyer agrees in advance to accept delivery of at least a certain quantity of stock, either in each individual order or cumulatively over some period of time. Depending on the relative strategic power of the parties, the seller may offer the buyer some forms of inducement, the most natural of which is a lower unit cost on items purchased under the contract.

Incorporation of a lower bound on the buyer's purchase quantity is fairly straightforward in many settings, for instance when the problem can be specified in the EOQ framework (cf. Nahmias 1997) or as a single-period newsvendor model (cf. Porteus 1990). We refrain from reviewing such models. Additionally, while some quantity flexibility contracts contain components which resemble minimum purchase commitments, we present them separately since the emphasis in modeling those structures is on the revision of orders. Here we discuss some recent papers that consider the choice to install a minimum purchase agreement in multi-period settings with demand uncertainty.

Anupindi and Akella (1993) cast the problem as a stochastic finite horizon model in which the buyer agrees to accept delivery of a fixed quantity of goods in each period. Discounts are given based on the level of this fixed commitment. The buyer can purchase extra units at a price premium. While the supplier guarantees the availability of the previously committed quantities, the additional units might not be delivered immediately. Thus, delivery of the additional units is expensive and uncertain. The authors prove that a modified S-type policy is optimal for this problem.

Moinzadeh and Nahmias (1997) treat the same general problem (the minimum commitment per period, Q, is given), but with both fixed and proportional penalties for adjustments, and over an infinite rather than finite horizon. Delivery of the additional units is assured. The authors contend (but do not formally prove) that a type of (s, S) policy is optimal: if the inventory on-hand just prior to a delivery is less than s, adjust so that the total inventory after delivery is S. Assuming normal demand, they develop a diffusion approximation of the system, which allows efficient approximation of both s and S. They show that this type of contract results in order variance lower than under conventional S or (s, S) policies. Hence, the fixed delivery contract serves as a risk sharing mechanism.

While the preceding two papers assume a constraint on every period's purchase, the agreement in Bassok and Anupindi (1997a) applies to the cumulative purchase (at least K_N units) over a given planning horizon of N periods. Demand is assumed to be a sequence of independent, identically distributed random variables. The authors prove inductively that the buyer's optimal policy in each period is a modified S policy. Letting (I_n, K_n) be the on-hand inventory and the minimum remaining purchase commitment when n periods are left, the optimal policy is defined by constants S_n and S^M $(S_n < S^M)$ in the following way:

if $I_n < (S_n - K_n)$, then order up to S_n;
if $(S_n - K_n) \leq I_n \leq (S^M - K_n)$, then order exactly K_n;
if $(S_M - K_n) \leq I_n \leq S_M$, then order up to S_M; and
if $S_M \leq I_n$, then do not order.

The impact of unit cost discounts on the buyer's total cost is described. The computations show that the advantages of the unit discount drop off quickly on approaching a particular quantity, which depends on the parameters of the problem. The buyer should never commit to more than this quantity.

Cachon and Lariviere (1997a) consider the problem faced by a manufacturer who faces uncertain demand for a component that is purchased from an external supplier. The manufacturer must offer a contract to the supplier for production capacity that must be built before the uncertain demand is observed. Obviously, the manufacturer would like the supplier to build ample capacity, but wishes to avoid the cost of excess. Under the simplifying assumption that the demand is a Bernoulli random variable, the authors show that termination fee contracts are equivalent to minimum purchase contracts in this setting. Under the former, in addition to a unit cost, the manufacturer pays a cancellation fee per unit for all units not purchased if he takes delivery of fewer than the agreed-upon number of units. Under the latter, the manufacturer guarantees a minimum purchase amount and pays a penalty per unit if he takes delivery of fewer than the guaranteed number of units. Further, the appropriate contract terms depend upon both the information available to the parties as well as the enforceability of the contract. When the demand distribution is known to all, the manufacturer offers such contracts only if he is able to force compliance and verify the supplier's capacity choice. Since such contracts require the manufacturer to share some risk, they simply add to his costs in this case. The manufacturer should, therefore, simply offer a price-only contract. However, when the manufacturer is privately informed about the demand distribution, he may offer a cancellation fee or a minimum purchase contract to convey information credibly, even when he must rely on the supplier's voluntary compliance.

10.4.4 Quantity Flexibility

A Quantity Flexibility (QF) clause defines terms under which the quantity a buyer ultimately obtains may deviate from a previous planning estimate. The conditions can include limits on the range of allowable changes, pricing rules, or both.

The motivation for each party to agree to such a clause depends on the nature of the alternative. The benefit to the buyer is clear if the clause ensures a degree of flexibility where previously there was none. This implies a willingness to pay a higher cost on all units and/or for each unit of deviation from the initial estimate. Indeed, some premium would be necessary to compensate the supplier for the increased exposure to demand risk. On the other hand, some buyers enjoy the strategic power to affect flexibility without the authority of a formal agreement. As noted earlier, when a buyer's estimate does not entail enforceable commitment, buyers commonly overstate their intended purchase, only to refuse undesired product later on. In this context, the clause is a way to encourage the buyer to forecast and plan more deliberately and honestly. In exchange, the supplier might need to provide a price break to give the buyer an incentive to participate. Either way, a QF clause has risk-sharing intent, and the hope is that the agreement can make both parties better off.

Typical research questions to ask of QF settings include: (1) how should the buyer behave (i.e. forecast and purchase) given the available flexibility, (2) how should the supplier behave given the flexibility promised the buyer, (3) what would be the cost or benefit to each party of changes to parameters of the agreement, e.g. a flexibility bound or pricing term.

Attempts to address these questions rigorously for the various types of QF clauses share a number of modeling challenges. Because the exercise of flexibility implies reconsideration of a prior decision, even the simplest model requires at least two decisions on the part of the buyer for each purchase: there is an initial inventory decision, and then a revision conditional on whatever new information about demand should become available. The first decision must properly anticipate the second for all contingencies, since the two are linked by the terms of the QF clause. Dimensionality of the decision space is an obstacle to analytical solution, as is the machinery required to represent updating of a stochastic demand. For these reasons, the available models tend to be very stylized, e.g. a selling season represented by a single random variable with a distribution amenable to Bayesian updating. Furthermore, most models focus only on the buyer's perspective, since the supplier's problem is even more complex. As is well known from multi-echelon inventory theory, fairly simple multi-level systems using relatively straightforward operating policies often yield demand processes at upper levels which are analytically quite difficult to characterize (cf. Schwarz 1981). Here, the informational dynamics of market demand are already an issue, and the buyer's ability to revise purchase quantities only adds complexity. Below we describe efforts to model contracts containing variants of the QF structure.

"Backup agreements," which have been observed in the fashion apparel industry, are the focus of Eppen and Iyer (1997). Such agreements are parametrized by (ρ, c). Prior to the selling season, the buyer (e.g. a mail-order retailer) commits to y units for the season, and takes immediate ownership of $(1 - \rho)y$ at unit price c. After observing the first two weeks of sales data (approximately 10% of total sales), which is used to perform a Bayesian update on the prior distribution of the total season's demand (modeled as a negative binomial random variable), the buyer can order up to the remaining ρy for the original price and receive quick delivery. There is a penalty cost of b for any of the backup units not purchased. The analysis suggests that the buyer's optimal strategy has "order-up-to" structure, meaning that each of the two decisions has some target threshold. The second is obtained as a critical fractile of the distribution of remaining demand conditional on the early sales, while the first does not have a simple solution. Some comparative statics properties of these thresholds are provided. While the bulk of the analysis focuses on the buyer's side, this paper reports that for certain parameter combinations, the backup agreement contract can improve profits for both parties relative to a setting with no backup agreement.

In Anupindi and Bassok (1995), the supply contract commits the buyer to purchase at least a given total quantity of a single product over a finite time horizon. A certain additional volume is also available at the same price, beyond which a higher price is charged. This paper derives the purchase policy that minimizes the buyer's total costs (purchase, holding, and backorder) under stationary and random demand, which is shown to be of modified order-up-to structure. The critical values are related to those obtained by solving a standard finite horizon model with no commitments and a single price.

Bassok and Anupindi (1995) consider forecasting and purchasing behavior in an arrangement in which the buyer initially forecasts its period-by-period purchases over a T-period horizon, then may revise each period's purchase one time within specified percentage bounds. Demands are non-stationary but independent. Because information about that demand never changes, the adjustment in period n is a response to the demand realizations and purchases made in periods 1 through $(n-1)$. With linear holding and shortage costs, and complete backordering from month to month, their objective is to solve for: (i) the purchase estimates prior to period 1, and (ii) each period's actual purchase. The authors discovered this problem to be very complex, and ultimately proposed a heuristic policy, constructed numerically. No special structure to this policy was reported.

Milner and Rosenblatt (1997) analyze a setting in which the buyer places orders for two periods, and may then adjust the second order after observing demand in the first period. This differs from the contract in Bassok and Anupindi (1995) in that there is per-unit penalty for any adjustments. They describe the optimal behavior of the buyer, both in the initial orders and the subsequent adjustment. As in Bassok and Anupindi (1995), beliefs about the second period's demand do not change between the time of the initial order and the adjustment, so what drives the use of flexibility in period 2 is purely the discrepancy between the purchase and the demand outcome in period 1. The optimal adjustment is characterized by a range $[L, U]$, whose endpoints are simple functions of the cost parameters and the demand distributions. If the pre-adjustment inventory position on entering the second period falls in this interval, no adjustment should be made; otherwise, adjust to get to the closest boundary of the interval. Closed forms are not available for the optimal initial orders, but some structural properties and comparative statics results are presented, all of which are consistent with intuition. Finally, parameter combinations are derived which characterize the buyer's preference for either the flexible contract, a non-flexible contract, or no contract at all (assumed to allow adjustment without financial consequence). The supplier's preferences are not considered.

Barnes-Schuster et al (1998) examine the use of options for supplier capacity as a means of affecting flexibility for the buyer. Here, a selling season is divided into two periods of correlated demands. Excess demand is backordered after period 1, and lost after period 2. Prior to the first period, the buyer places firm orders for both periods, and also purchases options which reserve additional supply in period 2. After observing the period 1 demand, the buyer has the prerogative to exercise some number of the options (at an additional fee), or let them expire, thereby losing the reservation fee but avoiding any additional cost that would have been incurred had actual production been commissioned initially. The supplier is obligated to position raw materials to the maximum buyer request (firm orders + options), but may convert these to finished product at two different points in time at different costs. A cheaper mode of production is available only immediately after the buyer's initial orders, while more costly production, if necessary, occurs after the buyer has exercised any options. The supplier dictates the price terms of the contract (wholesale price, option price, exercise price). Examining the efficiency of the contract form, this paper concludes that linear prices cannot coordinate the channel in a way that offers the supplier positive profits, and then proposes various quantity discount schemes that achieve this goal. The cost ramifications for both parties under linear pricing are characterized numerically. Finally, the decision of when to update demand is analyzed numerically by dividing the time horizon into several subintervals.

In contrast to the more common assumption in which the procurement price of a required component is either a decision variable or an exogenous parameter, in Li and Kouvelis (1997) what motivates a buyer's desire for flexibility is uncertainty in that price (modeled as a geometric Brownian motion with drift). The main thrust of this paper is to evaluate the value of flexibility in quantity and time, and determine when each form might be the more desirable contract structure conditional on the cost parameters, the design of the risk-sharing price arrangement, and the dynamics of the anticipated price fluctuations. In the model setting, the buyer must procure exactly D units by time T under pricing terms specified in the supply contract at time 0, and is interested in selecting a favorable flexibility arrangement to minimize his own expected purchasing and inventory costs. Possible arrangements that are considered in various combinations are: (1) a "time-inflexible contract", in which the buyer must state up front the purchase times, (2) a "time-flexible contract", in which the buyer may observe price movements and decide dynamically when to buy, and (3) a "quantity flexible contract", in which the buyer chooses a Q at time 0 that entails commitment to purchase an amount in the window $[(1 - \alpha) Q, Q]$ for a given α. The analysis determines the timing and quantity of purchases that minimize the buyer's expected net present value of purchase and holding costs. Given a single supplier that offers no quantity flexibility and sells at exactly the market price, the buyer's optimal strategy is shown to be invariant to the existence of any time flexibility. The buyer will purchase the full amount either at time 0 or time T, the choice of which depends on the holding cost, the discount rate, the drift parameter of the price process, and the value of T. However, if the contract price is distinguished from the market price, for example through an agreement that transfers some of the fluctuation risk to the supplier, then the optimal strategy must be obtained computationally. This is done via a discretization of the geometric Brownian motion into a random-walk type of process. Quantity flexibility is analyzed in the two-supplier case, where the suppliers are assumed to exist in different markets with separate but correlated prices.

Tsay (1996) models the incentives of both the supplier and the buyer in a setting in which the buyer first estimates a purchase quantity for a given selling season, the supplier then commits to production, and finally the buyer makes his actual purchase (which may differ from the estimate) in light of updated information about a stochastic market demand. System inefficiency can result in such an environment because, as noted earlier, the buyer might be expected to inflate the initial estimate. The root cause of this behavior can be linked to the phenomenon of double marginalization. This paper considers a QF contract which couples the buyer's commitment to purchase no less than a certain percentage below the initial estimate with the supplier's guarantee to deliver up to a certain percentage above. There is no penalty for making adjustments within the defined range. In conjunction with an appropriately chosen unit price, this contract structure is shown to be able to allocate the costs of market demand uncertainty so as to coordinate the individually motivated supplier and buyer to the system-wide optimal outcome. The structural properties of efficient QF contracts are characterized, as is the mechanism for allocating efficiency gains.

Bassok and Anupindi (1997b) analyze the buyer's side of the percentage flexibility contract studied in Tsay (1996), generalizing to an ongoing supply relationship in which planning for multiple future periods is performed in a rolling-horizon fashion. At each period, the buyer makes a purchase, and also provides estimates of purchases to be made in subsequent periods. The contract defines percentage limits on how these estimates may be revised from one planning iteration to the next. Demand is assumed to be independent and stationary, with known distributions. Shortages are completely backordered, and holding and backordering costs are linear. Because the decision space becomes substantially more complex, a heuristic forecast and purchase algorithm is proposed and defended by simulation analysis. Simulation experiments are also used to characterize the value to the buyer of a certain amount of flexibility and the variability of buyer purchases as a function of flexibility.

Tsay and Lovejoy (1998) consider the rolling-horizon QF contract in a multi-echelon setting, allowing for non-stationary demand with information updating. At period t, the buyer states to the supplier the vector

$$\{r(t)\} = [r_0(t), r_1(t), r_2(t), ...],$$

where

$r_0(t)$ = actual purchase in period t
$r_j(t)$ = estimate of purchase to be made in period $(t + j)$, for each $j \geq 1$.

The QF contract between the buyer and the supplier is parametrized by (α, ω), where $\alpha = [\alpha_1, \alpha_2, ...]$ and $\omega = [\omega_1, \omega_2, ...]$. This describes how much flexibility the buyer enjoys in revising $\{r(t)\}$ going forward in time. Specifically, for each t and $j \geq 1$:

$$[1 - \omega_j] r_j(t) \leq r_{j-1}(t+1) \leq [1 + \alpha_j] r_j(t).$$

$\{r(t)\}$ is the only information available to the supplier concerning the buyer's purchases, and any revisions are allowed provided that they observe the stated bounds. The analysis provides heuristics based on Open-Loop-Feedback-Control logic indicating how the buyer should construct $\{r(t)\}$ in light of the statistics of market demand and the flexibility parameters, as well as how the supplier should behave (i.e. submit orders and forecasts to its own upstream supplier, with whom there may be a separate QF contract) in order to fulfill its contractual commitment to support the buyer's order sequence. Attention is then given to how the flexibility characteristics of the system impact the inventory and service patterns, as well as how order variability propagates along a multi-level supply chain. Simulation experiments suggest that QF contracts can dampen the transmission of order variability throughout the chain, thus potentially retarding the well-known "bullwhip effect". This paper also addresses the issue of contract specification. A nonlinear programming formulation is provided for answering the channel coordination question: if the supplier and buyer are both independently managed units of the same firm, how might a central planner specify the internal contract to achieve good joint performance? Also, an investigation of the value of flexibility to a buyer indicates that flexibility increases in value as the market environment becomes more volatile, and that this value observes a principle of diminishing returns.

Anupindi and Bassok (1997) is the only paper among those reviewed here which analyzes flexibility contracts in a multi-product context. The contract requires the buyer to commit to a minimum cumulative dollar value of purchases during a specified time horizon to be eligible for a percentage discount off regular prices. The discount is available only for purchase volumes up to a certain fraction above the minimum commitment, with the regular price charged for any additional purchases. Again, because of problem complexity, a heuristic algorithm is used by the buyer. Numerical studies indicate that a policy which commits to a total dollar value that is the sum of the mean dollar volumes for the individual products performs relatively well. Moreover, the flexibility to increase purchases at the lower price is not particularly critical. This is because risk pooling across demands in different periods for various products already provides sufficient flexibility.

10.4.5 Buyback or Returns Policies

A buyback or return clause establishes who bears responsibility for unsold inventory, and to what extent. One can make an analogy between buyback clauses and QF clauses, in that both structures lay out ground rules for the buyer to end up with an amount that is potentially less than his prior estimate. A subtle difference is that buybacks generally take place after demand has been observed, whereas QF-style order reductions may be executed while demand uncertainty remains. Nevertheless, many of the modeling issues discussed in the previous section also apply here.

Analytical treatment of buybacks first appeared in the marketing literature, in Pasternack (1985). In this paper a manufacturer produces a commodity for sale to a retailer. The item has a relatively short shelf or demand life, and the retailer places only one order with the manufacturer. The manufacturer sets the wholesale price and the market selling price is fixed, so the only decision for the retailer is the order quantity. Using a single-period, newsvendor-style model, the author characterizes potential inefficiencies within this channel that are essentially due to double marginalization, and finds that neither a policy of allowing for unlimited returns at full credit nor one which allows for no returns is efficient. He determines that coordination of the channel can be achieved by a buyback clause which allows full return at a partial refund, and that the efficient prices (wholesale and buyback) can be set in a way that guarantees Pareto improvement. A key result is that the channel-coordinating prices are *independent of the market demand distribution*. This is significant in that the manufacturer need not know the market demand distribution in order to implement an efficient contract, although this remains necessary in order to properly value and allocate efficiency gains in a way that will insure the retailer's participation.

In the economics literature, Kandel (1996) extends Pasternack (1985) by modeling price-sensitivity in end customer demand, with a general downward-sloping expression for the expected demand. Hence, the retailer must also set the retail price. The author confirms that Pasternack's results proving the inefficiency of certain decision structures generalize to this setting, and proposes two arrangements under which coordination can be achieved. The first assumes that the manufacturer can additionally impose a resale price maintenance contract on the retailer, in which case a consignment agreement (full returns for full credit) is efficient. This allows the manufacturer to impose the channel-optimal retail price, and leads the retailer to choose the channel-optimal quantity. Here, the retailer makes zero profit and the manufacturer retains all profits of the coordinated channel. Alternatively, the manufacturer can charge a wholesale price of exactly the marginal production cost, rendering the retailer's decision problem identical to that of the integrated channel.

The profits of a coordinated channel accrue entirely to the retailer in this case. The author concludes with conjectures regarding the impacts of risk-aversion of either party, consideration of additional variables such as the manufacturer's choice of product quality and the retailer's promotions effort, information asymmetry, the elasticity of market demand, and the stability of market demand over time. Emmons and Gilbert (1998) generalize Pasternack (1985) in much the same way as does Kandel (1996), however assuming a specific multiplicative form of demand model. One result that is enabled by the functional assumption is that for a given transfer price, the offer to buy back excess stock tends to increase the total profits of the channel.

Donohue (1996) studies buyback contracts within two different two-stage production environments. In the first, the supplier offers the product for delivery at two different lead times. The buyer commits in advance to a quantity of the long-lead time item at a given wholesale price. After revising his demand forecast to reflect information gathered prior to the season, the buyer can place an additional order for short-lead time delivery at a different wholesale price. At the end of the season, the manufacturer takes back any unsold items at a third price. The author first finds that the buyer's optimal ordering policy has order-up-to structure, then determines the three price parameters that will result in the same system profit as the optimal centralized solution. Similar analysis is performed for the second setting, an assemble-to-order context in which some critical component must be prepositioned. Whereas coordination in the first environment entails a sort of minimum purchase commitment in the initial purchase, in the second environment there is an option-like arrangement which communicates a maximum purchase commitment. As in Pasternack (1985), the prices that coordinate the channel turn out to be independent of the distribution of market demand. There is also some discussion of how the efficiency gains might be split between the buyer and manufacturer, and the implications of the contract structure for the variability in each party's profits.

10.4.6 Allocation Rules

Allocation issues arise when multiple retailers compete for a product that is rendered scarce by some limit on the manufacturer's production capacity or stock availability. Such concerns are not considered in most of the papers in this review since they model only supply chains with a single retailer and a manufacturer with unlimited capacity. However, as suggested in Lee et al (1997), the possibility of rationing by the manufacturer can induce competition between retailers and, therefore, lead to strategic behavior. In particular, retailers will tend to inflate their orders, which distorts the flow of information. While insightful, Lee et al (1997) does not model the effect of alternative allocation policies. We are aware of only two papers on supply contracts that consider the design of allocation policies, as described below.

Cachon and Lariviere (1996) model a single-period, single-supplier, multi-retailer supply chain where the supplier's production capacity is limited and each retailer's stocking level is private information. The wholesale price is fixed. In case of shortage, the supplier's capacity is allocated using some fairly general allocation scheme, which is required only to be (i) "efficient", meaning that the available capacity is never wasted, (ii) "insured", i.e. if a retailer desires a positive quantity of stock, he will receive at least some stock, and (iii) "individually responsive", i.e. if a retailer wants more stock, he gets more as long as capacity is available. The main result is that Pareto optimal allocation mechanisms are easily manipulated by the retailers. At the same time, truth inducing mechanisms lead to system inefficiency. The authors also show that some mechanisms may lead the supplier to choose a higher capacity.

Cachon and Lariviere (1997b) consider a one-supplier, two-retailer supply chain in a two-period context. The supplier's production capacity and the wholesale price to the retailer are fixed, but retailers can set their own price. Demand in the second period can be high or low, and is a simple linear function of price. Common information is assumed. In case of shortage, capacity is allocated using a publicly known allocation scheme, of which they consider two versions: "even allocation" and "turn-and-earn". "Even allocation" means that the available capacity is divided evenly among all retailers placing orders. In "turn-and-earn", allocation is a function of past sales (for example, a retailer with higher past sales may receive a more favorable allocation). The main result of their analysis is that under turn-and-earn, the supplier's profits will increase but the retailers' profits may not. Since the retailers are identical, they end up just selling greater volume at a lower price to protect their allocation.

While both papers illustrate the effect of allocation on supply chain behavior and performance, they offer little by way of specifying what allocation policies might be optimal. Indeed, this is a difficult problem even when system control is centralized, as has become evident from extensive studies in a variety of contexts. Some examples include single-warehouse/multi-retailer settings (e.g. Eppen 1979, Eppen and Schrage 1981, Jonsson and Silver 1986, Jackson 1988), manufacturing systems with component commonality (e.g. Baker 1985, Gerchak and Henig 1989, Agrawal and Cohen 1997), inventory systems with different customer classes (e.g. Nahmias and Demmy 1981, Cohen et al 1988, Ha 1997b), and queuing systems with multiple servers or customer classes (e.g. Gilbert and Weng 1997). To understand the source of the difficulty, consider an assembly system with component commonality. The optimal allocation policy must depend on the inventory levels of all items in the system, which is a function of the joint distribution of demand for all end items. This makes the problem analytically and computationally complex. Also, in practice, when capacity shortages occur, suppliers may vary pricing to mitigate the effects of the problem and buyers may turn to alternate sources of supply. The models described in this section do not allow such possibilities.

10.4.7 Lead Times

In classical inventory models, delivery lead time is either fixed (at zero in some cases), or a realization of a random variable. Additionally, some researchers have allowed the buyer a choice from multiple (but fixed) lead times that result from multi-sourcing or the option to expedite material from a single supplier, two practices frequently observed in industry. For example, Hausman et al (1994) build on the efforts of Daniel (1963) and Fukuda (1964) and others in analyzing a buyer's optimal purchasing behavior given an exogenous menu of various lead time and unit cost combinations. Lawson and Porteus (1996) perform similar analysis on a multi-echelon system under central control. Agrawal et al (1998) model the contracting of capacity with suppliers which differ in commitment dates as well as other structural features, from the perspective of a retailer of multiple products. We do not present these works in detail since they take the characteristics of the supply base as given, and therefore fall beyond the scope of what we define to be "contracts" papers. Below we discuss papers that consider issues that arise when the terms of timing in delivery are control variables.

Barnes-Schuster et al (1997) consider a system composed of a supplier and one or more buyers in which the supplier faces a known production lead time (l_p), while the lead time for delivery to the buyer (l_d) is a decision variable. The buyer faces a stationary periodic review problem for which a static base-stock inventory policy is known to be optimal, and the appropriate base-stock level is an increasing function of l_d. The supplier faces an analogous problem, except that increasing l_d reduces its required safety stock. This paper shows that in the single-buyer case the l_d that is optimal from the system's point of view is either zero, which has the supplier hold all the system safety stock, or equal to $(l_p + 1)$, in which case the supplier produces to order and hence holds no safety stock. Conditions on the cost and demand parameters are provided that determine which lead time to use. In the case of multiple buyers with identical cost parameters, the supplier should hold the safety stock for buyers with "sufficiently low" standard deviations of demand, while the remaining buyers hold their own.

Iyer and Bergen (1997) model Quick Response (QR), a movement in the fashion apparel industry, in a manufacturer-retailer supply chain. Achieved by any of a number of process improvements, QR is simplified to mean lead time reduction, i.e. a delay of the point at which the supply chain makes its quantity commitment. The retailer benefits since its orders may then be placed under an improved state of information (as modeled within a Bayesian updating framework), yet the manufacturer can actually be made worse off. Specifically, since this manufacturer is assumed to operate in pure make-to-order mode, its payoff is determined once the retailer orders, regardless of how the uncertain market demand eventually resolves. Unaffected by any overage risk, the manufacturer naturally prefers high retailer orders, even if this includes large amounts of safety stock that never get sold. Thus the manufacturer may resist efforts that will reduce lead-times precisely because improved forecasts enable the retailer to reduce its safety stock position. According to the authors, this may explain the various side-agreements which have been observed to accompany QR efforts, such as commitments to higher service levels to the end customer, higher wholesale prices, or volume commitments across multiple products. These mechanisms all work by forcing the retailer to buy and/or pay more than it would with QR alone, enough so that the manufacturer's original profit is preserved. The key conclusion is that Pareto-improving contractual combinations do exist whereby the implementation of lead time reduction can proceed with the blessings of both parties.

Grout and Christy (1993) discuss the incentives faced by a supplier in quoting a delivery time to a buyer, and the implications for the likelihood of on-time delivery. During the contracting process associated with the one-time purchase of some item, the buyer offers a lump-sum bonus of B for on-time delivery, and the supplier in turn specifies a delivery time A. The supplier's risk is due to uncertainty in the production time. If the supplier completes production prior to time A, he collects B but incurs a cost of α per unit of time for holding the item until delivery. If instead a production delay causes tardiness, the supplier incurs a cost of β per unit of time late. Given this structure, the supplier's optimal A turns out to be a critical fractile of the distribution of the random production time, reflecting the relative values of α, β and B. Anticipating this, the buyer specifies B so as to minimize his own expected shortage cost (incurred at a rate of δ per unit of time the delivery is late) and expected bonus payment. The analysis focuses on the role of the bonus by comparing the case of $B = 0$ to a contract in which the buyer sets the bonus with channel coordination in mind. Indeed, $B = 0$ leads to inefficient performance since the supplier's decision does not take the buyer's shortage cost into account. A B which achieves the first-best outcome is shown to exist, which leads to a recommendation regarding the make-or-buy decision: compare the expected bonus paid to an independent supplier under the first-best B against the cost of vertical integration, and choose the cheaper option.

Moinzadeh and Ingene (1993) call attention to the supplier's perspective on the design of what amounts to a dual lead-time supply arrangement. The supplier carries two different products which are partially substitutable. Good 1 is held in inventory, hence offers immediate delivery if in stock. Good 2 is available by special order only, imposing a one-period delay on customers. Each product is the first choice of some segment of the population, as characterized by two stationary Poisson processes. (While the discussion is placed in a context of a population of individual consumers, the overall demand pattern could potentially describe the purchase preferences of a single downstream organization.) The net demand for each item is computed by examining the customer's 3 possible reactions to unavailability of the stocked good: (i) "walk", meaning that the sale is lost altogether, (ii) "wait", which imposes a backorder cost, or (iii) "switch" by special-ordering good 2 instead. The rate of occurrence of each of these is known, with an interpretation based on consumer utility. The analysis considers the supplier's problem in maximizing its long-run profit (discounted over an infinite horizon), assuming that good 1 has fixed price and is managed according to a base-stock inventory policy. The decisions are the base-stock level of good 1 (R) and the markup on good 2 (m_2). For a given m_2, the optimal R is shown to have newsvendor-style structure, which balances the margin on good 1 against the marginal cost of being out of stock on good 1. The latter includes lost sales and backorders as in a traditional model, but also the possible benefit that accrues when customers switch to good 2. Naturally, the relative magnitudes of these factors depend on m_2. In fact, if a sufficient fraction of good 1 customers show a tendency to either switch or wait (which can be induced by the choice of m_2), the supplier may prefer not to stock good 1 at all. More generally, the profit maximizing strategy can involve setting a price that encourages switching in order to reduce holding costs for the zero lead-time item. A numerical method for setting m_2 under additional structural assumptions is presented, as is some discussion of how this parameter might be used to compensate customers for the dissatisfaction that results from their not obtaining their first-choice item.

10.4.8 Quality

The papers discussed thus far are concerned primarily with the timing and quantity of material flows and the associated financial transfers. However, any supply relationship is premised on the quality of the delivered product. This may be formalized by conditions of the contract.

The economics literature has an extensive history of modeling supply settings in which product quality is a management choice. The representation of quality is relatively abstract, treating it as a product attribute which has a positive effect on both sales volume and production cost. This is typically encoded in a deterministic demand curve that is downward-sloping in price and shifts upward with quality, in conjunction with a production cost function that increases in both volume and quality. Attention then turns to characterizing the decisions that are optimal for a given market structure, and commenting on the consequences for profits and social welfare. Early examples of this approach include Spence (1975) and Dixit (1979). Similar efforts appeared later in the marketing literature, one example being the treatment of non-price variables by Jeuland and Shugan (1983). Because of the generality of this structure, virtually identical models have also been observed with "service" or "advertising" taking the place of the quality parameter in the formulation.

A great deal of insight into supplier and buyer behavior results from relaxing the assumption that product quality is common knowledge to all parties. In many models in economics, the buyer must attempt to infer in advance the true quality of the product. This may be based on signals conveyed by other terms of the contract, such as the selling price or the supplier's willingness to offer a warranty. Or, if transactions recur over time, the buyer may rely on previous experiences with that supplier's products. In turn, the supplier may incur the cost of initially offering high quality in order to establish a reputation that will lead to repeat business. See Chapter 2 of Tirole (1988) for a textbook treatment of this body of literature, which relies heavily on game theoretic constructs.

Quality has also long been of concern to the inventory management community, and, by extension, the SCM community. The models in this area examine the production process in more detail and therefore tend to have a much more concrete notion of quality, operationalizing it primarily in one of two ways: (1) as a probability that a particular item is defective or non-conforming, or (2) as a yield rate (either deterministic or stochastic). Many inventory models treat the process quality capability as an exogenous variable and then determine the appropriate lot-sizing behavior (see Yano and Lee 1995 for an extensive review). Others consider the choice of the quality level, but more from the vantage point of a single organization contemplating how to design its internal practices in light of its own costs of quality. For example, Porteus (1986) modifies a basic EOQ setting to consider the manufacturer's option to invest in the improvement of process quality (defined as a probability of going "out of control") as a way to manage the cost of reworking defective lots. Starbird (1994) examines how a risk-averse supplier's choice of quality level depends on the acceptance sampling method used by the buyer, and Starbird (1997a) performs similar analysis for an expected-cost-minimizing supplier with EOQ-style setup and holding cost concerns.

Models which explore the negotiation for quality are much less common in the SCM literature, and many of the phenomena described earlier (e.g. signaling behavior) are usually beyond the scope of the analysis. This is because, as noted, the relative emphasis on operational-level details obstructs the in-depth consideration of issues such as information asymmetry. Below we describe some efforts to examine the motivations that determine the quality terms in supply chain relationships.

Reyniers and Tapiero (1995) use a simple game-theoretic formulation of a supplier-producer channel to examine the impact of contract structure on the supplier's quality and the producer's inspection practices, and the implications for the quality of the end product. The supplier chooses one of two production methods, indexed by i, which differ in output quality (modeled as the probability of defect, p_i); the production cost T_i is higher for the process with higher quality. The producer may choose to perfectly inspect the incoming item at cost of m, or incorporate it directly into the end product. The supply contract stipulates that if the inspection reveals a defect, the supplier pays the producer $(C + \Delta\pi)$, where C covers the cost of repair and $\Delta\pi$ represents any additional rebate. If a defective input reaches the end customer, a failure cost of R is incurred with certainty. A fraction α of this is paid by the supplier, the rest by the producer. All parties are risk-neutral, all parameters are common knowledge, and the game is played only once. This results in a simple 2x2 matrix representation of the payoffs, which can be analyzed for Nash equilibria in the standard way. The game has an equilibrium in mixed strategies, whose structure depends on the value of $\Delta T/\Delta p$ $(= (T_1 - T_2)/(p_1 - p_2))$ and the other cost parameters in the following intuitive ways: (i) the probability that the producer uses inspection is increasing in $\Delta T/\Delta p$, (ii) the probability that the supplier provides low quality is increasing in m, and (iii) the quality of the end product is decreasing in α and m, and increasing in $\Delta T/\Delta p$. The "value of cooperation", which is the value of moving from independent to joint decision-making given the contract structure, is shown to be decreasing in $\Delta\pi$ for both parties, and decreasing in α for the producer.

Tagaras and Lee (1996) focus on a manufacturer who has the option to increase the quality of an input material by paying a higher unit procurement cost. (The menu of cost-quality combinations may come from multiple differentiated vendors, or could also be different offerings from a single vendor.) The input item is defective with probability p, and the manufacturer's production process fails with probability q. Costs are assigned to the various root causes which might lead to a defective final product: r_1 if the input is defective, r_2 if the manufacturing process is at fault, and r_{12} if both apply. The manufacturer's expected cost per unit processed is then $\phi(p) = p(1-q)r_1 + q(1-p)r_2 + pqr_{12}$. This is compared against $C(p)$, the unit cost charged by the vendor for a defect rate of p ($C(p) = c(1-p)$ and $C(p) = c(1-p)^2$ are considered). Incoming inspection is also a possibility (at a unit cost of a, with any necessary rework costing r_i). The analysis reveals that the manufacturer's proper choice of vendor quality depends not only on the vendor's price, but also on the capability of the process using the item as an input (the structure of the dependence reflects the relative magnitudes of r_1, r_2, r_{12}). Contrary to a view that is popular in the modern quality movement, under some circumstances the manufacturer is better off buying lower quality inputs at a lower cost, because the value of high quality inputs is negated by internal process problems.

Starbird (1997b) examines supplier buyer behavior in a model which features a careful accounting of the quality-related costs categorized by Joseph Juran (cf. Juran and Gryna 1988). Prevention and internal failure (e.g. scrap, rework) costs are incurred by the supplier, and appraisal (inspection) and external failure costs (warranty, replacement costs, etc.) are paid by the buyer. The buyer, who faces a deterministic market demand, procures from the supplier in lot sizes (L) that are economically chosen to minimize the expected cost of ordering, holding, purchasing, inspection, and external failure. The supplier minimizes the sum of expected setup, holding, manufacturing, prevention, and scrap costs by choosing a production lot size (Q) and a quality level (ϕ, the probability that an individual item is defective). The probability that a procurement lot will be accepted by the buyer is determined by the acceptance sampling rule, and has the binomial form $P_A(\phi) = \sum_{d=0}^{c} \binom{n}{d} (1-\phi)^d \phi^{n-d}$ where n is the sample size and c is the acceptance number ($c \leq 1$ simplifies the analysis). The author characterizes the resulting non-cooperative Nash equilibrium outcome, which demonstrates that the independence of the parties may lead the supplier to choose a quality level that is either higher or lower than the quality level that would arise under cooperation.

10.5 CONCLUSION

In this paper we have reviewed some recent efforts to study how the design of contracts affects supply chain behavior and performance. We believe this to be a very important and challenging field of research. The scope of issues that this literature has addressed thus far has been limited, as the variety and complexity of contracts used in practice do not lend themselves easily to mathematical modeling. This is one reason why so many of the papers reviewed here are done in the single-period "newsvendor" setting.

Our review suggests several opportunities for future research. Following the evolution of inventory theory, these analyses could naturally benefit from extension to production systems of greater structural complexity. This could include consideration of multiple planning periods, a larger number of products, or multi-layered and branching supply "networks" in which each party might have contracts with several others. Other issues worthy of further attention derive from the multi-party aspect of real systems, a few of which are described below.

Despite recent advances in information technology and trends towards sharing information with supply chain partners (cf. Kumar 1996, Verity 1996, Lee et al 1997), information asymmetry remains a key feature of real supply relationships. However, virtually all multi-player models in this review rely at some level on common knowledge of all parameters. One of the difficulties of including information asymmetry is that the analysis must then consider the multiple points of view. Since the probability structure of an uncertain event may be perceived differently by the parties to the contract, there may be disagreement in the calculation of expected profits. Hence, the notions of efficiency and optimality are not clearly defined. Addressing this issue might require a substantially more complex informational structure. This can include, for example, conditions regarding the buyer's beliefs about the supplier's beliefs, and vice versa.

A related concern is the assumption of risk neutrality, as the notion of efficiency again becomes unclear once the various players are allowed to have different objective functions. Further, decisions made by individuals are often motivated by a variety of complex incentive and compensation schemes instituted by their own firms. Not only are such schemes difficult to formulate analytically in the kind of models described here, they are often unknown to the other contracting parties. Nevertheless, contracts may be a way to reconcile the differing preferences the parties may have towards the uncertainty in the outcomes.

Another deficiency in the current literature is the lack of attention to competition, either between multiple buyers or multiple suppliers. Buyers that share a common supplier and compete in the same consumer market might behave in a way that obstructs their competitors' access to suppliers. In turn, the supplier might consider playing the buyers off one another to obtain price or purchase commitments. Multiple suppliers to a common buyer might need to alter their price, service, lead time, or flexibility offerings in light of the competitive environment. The kinds of contracts discussed here could play a role in structuring such relationships so as to improve efficiency and/or reallocate the risks.

330

While the stylized models we have reviewed offer many insights into supply chain behavior, they fail to address a variety of issues that become relevant to actual implementation. For example, as mentioned earlier, these models ignore many of the legal, public policy and social issues associated with contracts. Further, contractual arrangements can substantially affect the roles of particular individuals within any organization. If the affected parties feel threatened, or if their incentive systems are not appropriately modified, implementation of such contracts could be far from successful. Another practical concern that is unattended to by the existing SCM contract literature is that of prescribing how the benefits from coordination ought to be divided among the parties, a decision which might require extensive bargaining and negotiations. Clearly, there are opportunities to integrate the existing literature with the substantial body of knowledge from the field of Game Theory.

We believe that these issues offer a rich set of possibilities for future research on contracts in supply chains, and look forward to the interesting work in this area that the current literature will spawn.

Acknowledgments: The authors would like to thank the following individuals for helpful comments: Karen Donohue, Steve Gilbert, Warren Hausman, Ananth Iyer, Marty Lariviere, Chung-Lun Li, Kamran Moinzadeh, Dave Pyke, Sridhar Tayur, and Jin Whang.

References

Agrawal, N. and M.A. Cohen, "Optimal Material Control in an Assembly Environment with Component Commonality," Working Paper, Leavey School of Business, Santa Clara University, 1997.

Agrawal, N., S.A. Smith, and A.A. Tsay, "Multi-Vendor Sourcing in a Retail Supply Chain," Working Paper, Leavey School of Business, Santa Clara University, 1998.

Agrawal, N. and A.A. Tsay, "The Impact of Intrafirm Performance Incentives on Supply Chain Contracts," Working Paper, Leavey School of Business, Santa Clara University, 1998.

Anupindi, R. and R. Akella, "An Inventory Model with Commitments," Working Paper, Northwestern University, 1993.

Anupindi, R. and Y. Bassok, "Analysis of Supply Contracts with Total Minimum Commitment and Flexibility," *Proceedings of the 2nd International Symposium in Logistics*, University of Nottingham, U.K., 1995.

Anupindi, R. and Y. Bassok, "Approximations for Multiproduct Contracts with Stochastic Demands and Business Volume Discounts: Single Supplier Case," Working Paper, Northwestern University, 1997. To appear in *IIE Transactions*.

Baker, K.R., "Safety Stocks and Component Commonality," *Journal of Operations Management*, **6**, 1 (1985), 13-22.

Bannerjee, A., "A Joint Economic-Lot-Size Model for Purchaser and Vendor," *Decision Science*, **17** (1986), 292-311.

Barnes-Schuster, D., Y. Bassok, and R. Anupindi, "The Effect of Delivery Lead Time in a Two-Stage Decentralized System," Working Paper, Graduate School of Business, University of Chicago, 1997.

Barnes-Schuster, D., Y. Bassok, and R. Anupindi, "Supply Contracts with Options: Flexibility, Information and Coordination," Working Paper, Graduate School of Business, University of Chicago, 1998.

Bassok, Y. and R. Anupindi, "Analysis of Supply Contracts with Forecasts and Flexibility," Working Paper, Northwestern University, 1995.

332

Bassok, Y. and R. Anupindi, "Analysis of Supply Contracts with Total Minimum Commitment," *IIE Transactions*, **29**, 5 (1997a), 373-381.

Bassok, Y. and R. Anupindi, "Analysis of Supply Contracts with Commitments and Flexibility," Working Paper, Northwestern University, 1997b.

Bessler, S.A. and A.F. Veinott, "Optimal Policy for a Dynamic Multi-Echelon Inventory Model," *Naval Research Logistics Quarterly*, **13**, 4 (1966), 355-389.

Blair, B.F. and T.R. Lewis, "Optimal Retail Contracts with Asymmetric Information and Moral Hazard," *Rand Journal of Economics*, **25** (1994), 284-296.

Cachon, G.P. and M.A. Lariviere, "Capacity Choice and Allocation: Strategic Behavior and Supply Chain Performance," Working Paper, The Fuqua School of Business, Duke University, 1996.

Cachon, G.P. and M.A. Lariviere, "Contracting to Assure Supply or What Did the Supplier Know and When Did He Know It?", Working Paper, The Fuqua School of Business, Duke University, 1997a.

Cachon, G.P. and M.A. Lariviere, "Capacity Allocation Using Past Sales: When to Turn-and-Earn," Working Paper, The Fuqua School of Business, Duke University, 1997b.

Chen, F., "Decentralized Supply Chains Subject to Information Delays," Working Paper, Graduate School of Business, Columbia University, 1997.

Clark, A.J, "An Informal Survey of Multi-Echelon Inventory Theory," *Naval Research Logistics Quarterly*, **19** (1972), 621-650.

Clark, A.J. and Scarf, H., "Optimal Policies for a Multiechelon Inventory Problem," *Management Science*, **6** (1960), 475-490.

Cohen, M.A and N. Agrawal, "An Analytical Comparison of Long and Short Term Contracts," Working Paper, Department of OPIM, The Wharton School, University of Pennsylvania, 1998.

Cohen, M.A., P.R. Kleindorfer, and H.L. Lee, "Service Constrained (s,S) Inventory Systems with Priority Demand Classes and Lost Sales," *Management Science*, **34**, 4 (1988), 482-499.

Corbett, C.J. and X. de Groote, "Integrated Supply-Chain Lot Sizing Under Asymmetric Information," Working Paper, UCLA, 1997.

Daniel, K.H., "A Delivery-Lag Inventory Model With Emergency," in H.E. Scarf, D.M. Gilford and M.W. Shelly (Eds.), *Multistage Inventory Models and Techniques*, Stanford University Press, Stanford, CA, 1963.

Dixit, A.K., "Quality and Quantity Competition," *Review of Economic Studies*, **46** (1979), 587-599.

Donohue, K.L, "Supply Contracts for Fashion Goods: Optimizing Channel Profit," Working Paper, Department of OPIM, The Wharton School, University of Pennsylvania, 1996.

Emmons, H. and S.M. Gilbert, "Note: The Role of Returns Policies in Pricing and Inventory Decisions for Catalogue Goods, " *Management Science*, **44**, 2 (1998), 276-283.

Eppen, G.D., "Effects of Centralization on Expected Costs in a Multi-Location Newsboy Problem," *Management Science*, **25**, 5 (1979), 498-501.

Eppen, G.D. and A.V. Iyer, "Backup Agreements in Fashion Buying-The Value of Upstream Flexibility," *Management Science*, **43** (1997), 1469-1484.

Eppen, G.D. and L. Schrage, "Centralized Ordering Policies in a Multi-Warehouse System with Lead Times and Random Demand," In L.B. Schwarz, (Ed.), *Multi-Level Production/Inventory Control Systems: Theory and Practice*, Netherlands: North-Holland Publishing Co., 1981, 51-67.

Federgruen, A., "Centralized Planning Models for Multi-Echelon Inventory Systems Under Uncertainty," in S.C. Graves, A.H.G. Rinnooy Kan and P.H. Zipkin (Eds.), *Handbooks in Operations Research and Management Science, Vol. 4 (Logistics of Production and Inventory)*, Elsevier Science Publishing Company B.V., Amsterdam, The Netherlands, 1993, 133-173.

Fukuda, Y., "Optimal Policies for the Inventory Problem with Negotiable Lead-time," *Management Science*, **10** (1964), 690-708.

Gerchak, Y. and M. Henig, "Component Commonality in Assemble-To-Order Systems: Models and Properties," *Naval Research Logistics Quarterly*, **36** (1989), 61-68.

Gilbert, S.M. and Z.K. Weng, "Incentive Effects Favor Non-Consolidating Queues in a Service System: The Principal-Agent Perspective," Working Paper, Weatherhead School of Management, Case Western Reserve University, 1997.

Grout, J.R. and D.P. Christy, "An Inventory Model of Incentives for On-Time Delivery in Just-in-Time Purchasing Contracts," *Naval Research Logistics*, **40** (1993), 863-877.

Ha, A.Y., "Supply Contract for a Short-Life-Cycle Product with Demand Uncertainty and Asymmetric Cost Information," Working Paper, Yale School of Management, 1997a.

Ha, A.Y., "Inventory Rationing in a Make-to-Stock Production System with Several Demand Classes and Lost Sales," *Management Science*, **43**, 8 (1997b), 1093-1103.

Hammond, J.H., "Barilla SpA (A)," HBS Case 9-694-046, Harvard Business School Publishing, Boston, MA, 1994.

Hausman, W.H., H.L. Lee, and L. Zhang, "Optimal Ordering for an Inventory System with Dual Lead Times," Working Paper, Department of Industrial Engineering-Engineering Management, Stanford University, 1994.

Harris, F.W., *Operations And Cost (Factory Management Series)*, Shaw, Chicago, 1915.

Iyer, A. and M.E. Bergen, "Quick Response in Manufacturer-Retailer Channels," *Management Science*, **43**, 4 (1997), 559-570.

Jackson, P.L., "Stock Allocation in a Two-Echelon Distribution System or 'What to Do Until Your Ship Comes In'," *Management Science*, **34**, 7 (1988), 880-895.

Jeuland, A.P. and S.M. Shugan, "Managing Channel Profits," *Marketing Science*, **2** (1983), 239-272.

Jonsson, H. and E.A. Silver, "Overview of a Stock Allocation Model for a Two-Echelon Push System Having Identical Units at the Lower Echelon," In S. Axsater, C. Schneeweiss and E.A. Silver, (Eds.), *Multi-Stage Production*

Planning and Inventory Control: Lecture Notes in Economics and Mathematical Systems, 266, Berlin: Springer-Verlag, 1981, 44-49.

Juran, J.M. and F.M. Gryna, *Juran's Quality Control Handbook*, 4th Edition, McGraw-Hill, New York, 1988.

Kandel, E., "The Right to Return," *Journal of Law and Economics*, **39** (1996), 329-356.

Katz, M.L., "Vertical Contractual Relations," in R. Schmalensee and R.D. Willig (Eds.), *Handbook of Industrial Organization: Volume I*, Elsevier Science Publishers B.V., New York, NY, 1989.

Kumar, A., P.S. Giridharan, and R. Akella, "Control Structures, Performance and Incentives: An Operations Management Perspective," Working Paper, Department of Engineering Economic Systems, Stanford University, 1996.

Kumar, N., "The Power of Trust in Manufacturer-Retailer Relationships," *Harvard Business Review*, **74** (1996), 92-106.

Lawson, D.G. and E.L. Porteus, "Dynamic Lead Time Management," Research Paper #1409, Graduate School of Business, Stanford University, 1996.

Lee, H.L., "Effective Inventory and Service Management Through Product and Process Redesign," *Operations Research*, **44** (1996), 151-159.

Lee, H.L., C. Billington, and B. Carter, "Hewlett-Packard Gains Control of Inventory and Service Through Design for Localization," *Interfaces*, **23** (1993), 1-11.

Lee, H.L., V. Padmanabhan, and S. Whang, "Information Distortion in a Supply Chain: The Bullwhip Effect," *Management Science*, **43** (1997), 546-558.

Lee, H.L. and M.J. Rosenblatt, "A Generalized Quantity Discount Pricing Model to Increase Supplier's Profits," *Management Science*, **30** (1986), 1179-1187.

Lee, H.L. and S. Whang, "Decentralized Multi-Echelon Supply Chains: Incentives and Information," Working Paper, Graduate School of Business, Stanford University, 1997.

Li, C. and P. Kouvelis, "Flexible and Risk-Sharing Supply Contracts Under Price Uncertainty," Working Paper, Olin School of Business, Washington University, 1997.

Milner, J.M. and M.J. Rosenblatt, "Two-Period Supply Contracts: Order Adjustments and Penalties," Working Paper, Olin School of Business, Washington University, 1997.

Moinzadeh, K. and C. Ingene, "An Inventory Model of Immediate and Delayed Delivery," *Management Science*, **39**, 5 (1993), 536-548.

Moinzadeh, K. and S. Nahmias, "Adjustment Strategies for a Fixed Delivery Contract," Working Paper, University of Washington, 1997.

Monahan, J.P., "A Quantitative Discount Pricing Model to Increase Vendor Profits," *Management Science*, **30** (1984), 720-726.

Moorthy, K.S., "Managing Channel Profits: Comment," *Marketing Science*, **6**, 4 (1987), 375-379.

Nahmias, S., *Production and Operations Analysis*, Irwin, Homewood, IL, 1997.

Nahmias, S. and W.S. Demmy, "Operating Characteristics of an Inventory System with Rationing," *Management Science*, **27** (1981), 1236-1244.

Narayanan, V.G. and A. Raman, "Assignment of Stocking Decision Rights under Incomplete Contracting," Working Paper, Harvard Business School, 1997.

Pasternack, B.A., "Optimal Pricing and Returns Policies for Perishable Commodities," *Marketing Science*, **4** (1985), 166-176.

Porteus, E.L., "Stochastic Inventory Theory," in D.P. Heyman and M.J. Sobel (Eds.), *Handbook in Operations Research and Management Science, Volume 2 (Stochastic Models)*, Elsevier Science Publishers B.V., 1990, 605-652.

Porteus, E.L. and S. Whang, "On Manufacturing/Marketing Incentives," *Management Science*, **37**, 9 (1991), 1166-1181.

Reyniers, D.J. and C.S. Tapiero, "The Delivery and Control of Quality in Supplier-Producer Contracts," *Management Science*, **41**, 10 (1995), 1581-1589.

Rosenblatt, M.J. and H.L. Lee, "Improving Profitability with Quantity Discounts under Fixed Demand," *IIE Transactions*, **17**, 4 (1985), 388-395.

Rosling, K., "Optimal Inventory Policies for Assembly Systems Under Random Demands," *Operations Research*, **37** (1989), 565-579.

Schwarz, L.B., *Multi-Level Production/Inventory Control Systems: Theory and Practice*, North-Holland Publishing Company, New York, NY, 1981.

Spence, A.M., "Monopoly, Quality, and Regulation," *Bell Journal of Economics*, **6** (1975), 417-429.

Spengler, J.J, "Vertical Restraints and Antitrust Policy," *Journal of Political Economy*, **58** (1950), 347-352.

Starbird, S.A., "The Effect of Acceptance Sampling and Risk Aversion on the Quality Delivered by Suppliers," *Journal of the Operational Research Society*, **45**, 3 (1994), 309-320.

Starbird, S.A., "Acceptance Sampling, Imperfect Production, and the Optimality of Zero Defects," *Naval Research Logistics*, **44** (1997a), 515-530.

Starbird, S.A., "Supply Chain Management, Joint Cost Optimization, and Quality," Working Paper, Leavey School of Business, Santa Clara University, 1997b.

Tagaras, G. and H.L. Lee, "Economic Models for Vendor Evaluation with Quality Cost Analysis," *Management Science*, **42**, 11 (1996), 1531-1543.

Tirole, J., *The Theory of Industrial Organization*, The MIT Press, Cambridge, MA, 1988.

Tsay, A.A, "The Quantity Flexibility Contract and Supplier-Customer Incentives," Working Paper, Leavey School of Business, Santa Clara University, 1996.

Tsay, A.A. and W.S. Lovejoy, "Quantity Flexibility Contracts and Supply Chain Performance," Working Paper, Leavey School of Business, Santa Clara University, 1998. To appear in *Manufacturing & Service Operations Management*.

336

Verity, J.W., "Clearing the Cobwebs from the Stockroom," *Business Week*, October 21, 1996, p.140.

Weng, Z.K., "Channel Coordination and Quantity Discounts," *Management Science*, **41** (1995), 1509-1522.

Weng, Z.K., "Pricing and Ordering Strategies in Manufacturing and Distribution Alliances," *IIE Transactions*, **29** (1997a), 681-692.

Weng, Z.K., "Coordination Strategies for Aligning Divergent Interests in a Manufacturing and Distribution Supply Chain," Working Paper, School of Business, University of Wisconsin-Madison, 1997b.

Weng, Z.K., "A Decision Framework for the Manufacturer to Coordinate with the Buyer," Working Paper, School of Business, University of Wisconsin-Madison, 1997c.

Whang, S., "Coordination in Operations: A Taxonomy," *Journal of Operations Management*, **12** (1995), 413-422.

Yano, C.A. and H.L. Lee, "Lot Sizing with Random Yields: A Review," *Operations Research*, **43**, 2 (1995), 311-334.

11 MODELING THE IMPACT OF INFORMATION ON INVENTORIES

Ananth. V. Iyer

Krannert School of Management
1310 Krannert Building, Room # 541
Purdue University
West Lafayette, IN 47907-1310
(765)-494-4514

aiyer@mgmt.purdue.edu

In this chapter, we study models of supply chains that focus on the impact of demand information on demand uncertainty and the consequent impact on the inventory levels required to maximize expected profit. We will also focus on the different impacts of information on the manufacturer and the buyer expected profits. This permits us to study the effect of contractual agreements between the buyer and the supplier that may be required to share the benefits of information on a supply chain. The bulk of the material in this chapter is derived from Eppen and Iyer (1997a), (1997b) and Iyer and Bergen (1997).

We will first explore the impact of information on manufacturer and retailer profits in the absence of any constraints on the supply that is possible *after* the retailer receives current information about demand. We will then explore a specific contractual arrangement between a manufacturer and a retailer that sets constraints on the amount of supply that will be provided after information is received by the retailer. We conclude with a literature review and conclusions.

11.1 MODEL STRUCTURE

Consider a system in which orders are placed at time 0 by the retailer. These orders are delivered L units of time later to the retailer for sale in the season beginning at time L. We have two levels of demand uncertainty. The *first* source of uncertainty concerns inherent demand uncertainty about the product i.e., even if we know the mean demand θ, the demand during the season is Normally distributed with mean θ and variance σ^2 i.e., $f(x \mid \theta) \sim N(\theta, \sigma^2)$. This model captures uncertainty regarding the number of people who come to the store, their propensity to buy, etc. The *second* source of uncertainty models uncertainty regarding θ at time 0. Information regarding θ at time 0 is modeled as a Normal distribution with mean μ and variance τ^2 i.e., $g(\theta) \sim N(\mu, \tau^2)$. Thus, at time 0, our model of demand is a Normal distribution with mean μ and variance $\sigma^2 + \tau^2$ (Berger, 1985) i.e., $m(x) \sim N(\mu, \sigma^2 + \tau^2)$.

The retailer faces the following costs: a cost of c per unit to purchase the product from the manufacturer, a goodwill cost of π per unit of demand not satisfied during the season, a holding cost of h per unit of product left over at the end of the season. The manufacturer experiences a cost per unit of w to produce a unit of product. Each unit of product sold by the retailer to customers generates a revenue of r per unit.

We will refer to the system where the retailer has to make decisions at time 0 as the *old system*. The retailer chooses an inventory level, Q, at time 0, to maximize the following expected profit which is

$$\int_{-\infty}^{Q} rxm(x)dx + \int_{Q}^{\infty} rQm(x)dx - h\int_{-\infty}^{Q}(Q-x)m(x)dx$$
$$-\pi\int_{Q}^{\infty}(x-Q)m(x)dx - cQ$$

Since this is the standard Newsboy model, it can be verified that this function is concave in Q and the optimum initial inventory is

$$I_{old} = \mu + Z(s)\sqrt{\sigma^2 + \tau^2} \tag{11.1}$$

where the optimal *service level* (s) is defined as $\frac{r+\pi-c}{r+\pi+h}$ and $Z(s)$ is the Z value of a standard Normal distribution that generates a cumulative probability of s. The associated maximum expected profit in the Old System is

$$EP_{old-ret} = (r-c)\mu - \{(c+h)Z(s) + (r+h+\pi)b_r(Z(s))\}\sqrt{\sigma^2+\tau^2} \quad (11.2)$$

where $b_r(Z(s))$ is the right linear loss function of a standard Normal distribution at $Z(s)$. The expected quantity *sold* during the season in the old system ($I_{old-sold}$) is $I_{old-sold} = \int_{-\infty}^{I_{old}} xm(x)dx + \int_{I_{old}}^{\infty} I_{old}m(x)dx$ which simplifies to $I_{old} = \mu - b_r(Z(s))\sqrt{\sigma^2+\tau^2}$. The expected quantity *left* over at the end of the second period in the old system is $I_{old-left} = \int_{-\infty}^{I_{old}}(I_{old}-x)m(x)dx = (Z(s)+b_r(Z(s)))\sqrt{\sigma^2+\tau^2}$

The *manufacturer* expected profit under this system is obtained using equation (1) as

$$EP_{old-mfr} = (c-w)\{\mu + Z(s)\sqrt{\sigma^2+\tau^2}\} \quad (11.3)$$

11.1.1 The impact of Information on Demand

Suppose that at time 0 the retailer starts collecting information regarding sales of related products. At $L_1(\leq L)$ the retailer places an order with the manufacturer. The manufacturer produces the order for delivery $L_2 = L - L_1$ units of time later for sale during the season beginning at time L. The values of L are reported to range from 5 to 8 months. The values of L_1 range from 2 to 5 months. We will refer to this system as the *info* system.

The lower lead time (L_1) enables data collected during L_1 regarding sales of related items to be used to decrease forecast error for the item being ordered[1]. Data collected during L_1 is converted by the buyer into an estimate of the demand for the item under consideration i.e., d_1[2]. This d_1 can be used to generate a posterior distribution for demand. Thus, given d_1, the conditional distribution of demand during the season is $g(\theta \mid d_1) \sim N(\mu(d_1), \frac{1}{\rho})$, where $\rho = \frac{1}{\sigma^2} + \frac{1}{\tau^2}$ and $\mu(d_1) = \frac{\sigma^2\mu}{\sigma^2+\tau^2} + \frac{\tau^2 d_1}{\sigma^2+\tau^2}$. This implies that $m(x \mid d_1) \sim N(\mu(d_1), \sigma^2 + \frac{1}{\rho})$ (Berger, 1985).

In the presence of information, the retailer's inventory choice involves two steps:

1. Observe the demand for related items during L_1 and use it to generate an estimate for the item under consideration i.e., d_1 . Use that demand to generate the posterior distribution.

2. Choose the optimal inventory to maximize expected profits given the posterior distribution. This would imply different inventory levels depending on d_1. Intuitively if the retailer estimates a low d_1 then he would be more likely to choose low inventory levels because he expects a low probability of high demand in the season.

It can be verified that the expected quantity ordered by the retailer, the corresponding expected profit, the expected quantity sold and the expected quantity left over are as follows:

$$EI_{info} = \mu + Z(s)\sqrt{\sigma^2 + \frac{1}{\rho}} \tag{11.4}$$

$$EP_{info} = (r - c)\mu - \{(c + h)Z(s) + (r + h + \pi)b_r(Z(s))\}\sqrt{\sigma^2 + \frac{1}{\rho}} \tag{11.5}$$

$$EI_{info-sold} = \mu - b_r(Z(s))\sqrt{\sigma^2 + \frac{1}{\rho}}$$

$$EI_{info-left} = \{Z(s) + b_r(Z(s))\}\sqrt{\sigma^2 + \frac{1}{\rho}}$$

We note that $EI_{info-sold} \geq EI_{old-sold}$ and $EI_{info-left} \leq EI_{old-left}$. Thus current demand information enables the retailer to decrease left over inventory and yet increase the customer fill rate.

The expected profit for the *manufacturer under* QR is obtained using equation as

$$EP_{info-mfr} = (c - w)\{\mu + Z(s)\sqrt{\sigma^2 + \frac{1}{\rho}}\} \tag{11.6}$$

Lemma 1 *If* $s \geq \frac{1}{2}$, *then information gathering results in a decrease in expected manufacturer profit but an increase in expected retailer profits i.e.,* $EP_{info-mfr} \leq EP_{old-mfr}$ *and* $EP_{info} \geq EP_{old}$.

Proof: If $s \geq \frac{1}{2}$ then $Z(s) \geq 0$. Verify from equations (11.3) and (11.6) that $EP_{info-mfr} \leq EP_{old-mfr}$. •.

In Lemma 1, we show that when the service level for products is higher than $\frac{1}{2}$, then information gathering and lower lead times will not be Pareto improving[3] without some additional action. Specifically, it is not profitable for the manufacturer.

11.2 COMMITMENTS USING SERVICE LEVEL

We begin with a discussion of the role of service level commitments in making the use of information Pareto improving. A service level commitment at time 0 implies that a higher service level (s', with $s' \geq s$) is offered by the retailer, with the actual quantities ordered conditional on the estimated demand d_1.

A commitment is not considered credible unless it will be in the incentives of the retailer to take those actions at a later time. We thus model the change in service level by changing the goodwill cost π.[4] Commitments to variables that change goodwill costs to the retailer make the commitment to a higher service level credible from the manufacturer's perspective. Although negotiated changes in goodwill costs are not traditional in the production literature, it is often possible to influence the goodwill cost component π_N either contractually, or by making the in stock position of the product more salient to the consumer. Formally, we show that there *always* exists a service level arrangement that makes retailer use of information Pareto improving.

Theorem 1 *If* $s \geq \frac{1}{2}$, $d = \frac{\sigma}{\tau}$ *and* $y = \sqrt{\frac{1 + \frac{1}{1+d^2}}{1 + \frac{1}{d^2}}}$ *then increasing* π *to* π', *where* $\pi' = \frac{s'(r+h)+c-r}{1-s'}$ *and* s' *such that* $Z(s') = \frac{Z(s)}{y}$ *makes the use of information Pareto improving.*

Proof: See Appendix.

Note, however, that the effect of this service level arrangement is to make the retailer place the same expected order size as in the old system. The following lemma shows that even though the expected order size is the same as in the old system, the expected quantity sold to the customer increases and the expected quantity left over decreases, making the retailer better off.

Lemma 2 *If* $s \geq \frac{1}{2}$, $d = \frac{\sigma}{\tau}$ *and* $y = \sqrt{\frac{1 + \frac{1}{1+d^2}}{1 + \frac{1}{d^2}}}$ *then increasing* π *to* π', *where* $\pi' = \frac{s'(r+h)+c-r}{1-s'}$ *and* s' *such that* $Z(s') = \frac{Z(s)}{y}$, *results in* $EI_{info-sold}(s') \geq I_{old-sold}$ *and* $EI_{info-left}(s') \leq I_{old-left}$

Proof: See Appendix.

At this point in the chapter, we have shown that without any agreements, for reasonable values of service level ($\geq$ 50%) the use of information by the retailer and lower lead times for the manufacturer generates a system that is not Pareto improving. We have also shown that there exists a Pareto improving service level agreement. Under this agreement, the retailer purchases the same expected quantity as in the old system but has a higher expected profit. We will next explore an agreement where the retailer's supply (after current information is available) is constrained.

11.3 MODELING THE IMPACT OF INFORMATION ON A CONSTRAINED SUPPLY AGREEMENT

We specifically focus on backup agreements. In a backup agreement the customer, (e.g. a catalog company) commits to a certain number of units (y) as much as ten months before the catalog is mailed. The supplier agrees to hold a certain percentage (say ρ) of these y units in reserve and delivers the remaining $(1 - \rho)y$ units before the catalogue is mailed. After some specified and typically brief period of time after the catalog is mailed (typically about two weeks), the catalog company has the opportunity to buy any or all of the items that were on backup, at the original purchase cost, but incurs a penalty cost of b for each unit not taken from backup. For example, Anne Klein, Finity, and DKNY all offer Catco backup agreements with $\rho = 0.2$ and $b = 0$. Andrea Jovine sets $\rho = 0.33$ and $b = 0$, whereas Liz Claiborne sets $\rho = 0.25$ and $b = 0.2$ in agreements with Catco. Backup agreements are common in the multibillion dollar catalog business. Each customer must decide on the commitment quantity, y, and then how many units, if any, to take from the backup in the face of substantial uncertainty and contract parameters that vary from vendor to vendor.

We will create a stochastic dynamic programming formulation of the backup agreement problem and use this model to derive the optimal inventory policy, i.e. the optimal commitment y and the optimal amount to take from backup as a function of the observed demand. The model includes a Bayesian updating structure that uses early sales data to update the demand distribution.

11.4 THE MODEL

11.4.1 The Process Flow

Let y be the commitment for the season. At the beginning of the first period, $y(1 - \rho)$ units are delivered to the catalog company at a cost of c per unit. The other $y\rho$ units are held by the manufacturer in backup. Product is offered for sale to customers in the catalogue at a price of r per unit, and the random demand in the first period, ξ_1, then occurs. The catalog company pays a per unit holding cost of h_1 for each unit on hand at the end of this period; and π per unit of unsatisfied demand in the (unlikely) event that a stockout should occur in the first period. We consider the no backlog case since the standard policy is to cancel orders if no inventory is in stock. The first period is typically about two weeks long which represents the point in time where about 10 % of the demand for the season has been received. Clearly h_1 will be small. It could, of course, be set to zero with no impact on the general results.

The catalog company thus starts period 2 with either:(1) $y(1 - \rho) - \xi_1$ units on hand if it does not run out in period 1 or (2) 0 units on hand if it does run out. Since the product sells at a price of r per unit, the catalog company receives net revenue of $r\xi_1$ (if $\xi_1 < y(1 - \rho)$) or $ry(1 - \rho)$ (if $\xi_1 \geq y(1 - \rho)$).

The catalog company has the option to buy as many of the $y\rho$ units from backup as it wants for c each. It also pays b for each unit held in backup that it does not buy. Deciding how many units to buy from backup at the beginning of period 2 is essentially a Newsboy problem with the complication of returns, the restriction on the backup quantity and the backup penalty b for items not taken. The demand ξ_2 is a random drawing from a probability mass function that is conditional on ξ_1. The holding cost per unit at the end of period 2 is h_2, the penalty cost is still π, and the salvage value is s where $s < c$. One could equivalently use a holding cost h_2' where $h_2' = h_2 - s$.

11.4.2 The Demand Process

We assume that demand during the season is generated by one of several pure demand process. By a pure demand process we mean a process that provides a probability distribution of demand for period 1 and period 2 as well as for the season. Another level of uncertainty is provided by the fact that the buyer is not sure which of the several pure processes will actually generate the demands. She thus must assign prior probabilities to these processes. Defining the demand model is a two step process: (1) Specify a set (three or four) of pure demand processes. Let $\Phi_1^i(x), \Phi_2^i(x)$ and $\Phi_{12}^i(x)$ be the cumulative demand distributions (CDF) for the first period, the second period and across both periods respectively for pure process i $(i = 1, 2, \ldots, N)$. Also, let $\phi_1^i(x), \phi_2^i(x)$ and $\phi_{12}^i(x)$ be the respective probability density functions for these CDFs. Our approach restricts attention to processes that satisfy three properties: family consistency, stochastic dominance and a monotone likelihood ratio property. Three common distributions: (i) Normal distributions with a common variance, (ii) Poisson distributions and (iii) Negative Binomial distributions with a common n satisfy these conditions. For a more complete discussion see Eppen and Iyer (1997a). (2) Select a set of prior probabilities at the start of the first period that each of these pure demand processes will be the process that actually produces the demand (Let P_{1i} be the prior probability for pure demand process i).

11.4.3 The Dynamic Programming Model

Let $f_1(0)$ be the optimal expected profit for the two period problem assuming that 0 items are on hand at the beginning of period 1. Then $f_1(0) =$ Maximize$_{y \geq 0} G_1(y)$ where $G_1(y) = -cy(1-\rho) + \int_0^\infty f_2(y, \xi_1)\phi_1(\xi_1)d\xi_1$ In this expression, $f_2(y, \xi_1)$ is the maximum expected profit in period 2 if y items were committed and ξ_1 was the demand in period 1 and $\phi_1(\xi_1)$ is the density function of demand in the first period. In this formulation, all of the costs and revenues (except the purchase cost) that actually occur in the first period are accounted for in the second period. This unorthodox formulation enables us to handle the lost sales case in a convenient manner. The maximization problem for $f_2(y, \xi_1)$ is as follows:

$$f_2(y, \xi_1) = \begin{cases} Max_{0 \leq y_2 \leq y\rho}\{ry(1-\rho) - cy_2 + \\ \quad \pi(y(1-\rho) - \xi_1) - \\ \quad b(\rho y - y_2) + G_2(y_2, \xi_1)\} & \text{if } y(1-\rho) \leq \xi_1 \\ Max_{y(1-\rho)-\xi_1 \leq y_2 \leq y-\xi_1} r\xi_1 - c(y_2 - y(1-\rho) + \xi_1) \\ \quad -h_1(y(1-\rho) - \xi_1) - \\ \quad b\{\rho y - (y_2 - y(1-\rho) + \xi_1)\} + G_2(y_2, \xi_1) & \text{otherwise.} \end{cases}$$

where $G_2(y_2, \xi_1) = \int_0^{y_2} r\xi_2\phi_2(\xi_2 \mid \xi_1)d\xi_2$
$+ \int_{y_2}^\infty ry_2\phi_2(\xi_2 \mid \xi_1)d\xi_2 - (h_2 - s)\int_0^{y_2}(y_2 - \xi_2)\phi_2(\xi_2 \mid \xi_1)d\xi_2$
$-\pi \int_{y_2}^\infty(\xi_2 - y_2)\phi_2(\xi_2 \mid \xi_1)d\xi_2$

where $\phi_2(\xi_2 \mid \xi_1)$ is the conditional density for ξ_2, the demand in period 2 given ξ_1.

11.4.4 Results

We first use the model to derive the form of the optimal policy. These results are summarized in the following theorem.

Theorem 2 *The function $G_1(y)$ is concave in y and $G_2(y_2, \xi_1)$ is concave in y_2. Thus the optimal policy in period 1 is an order-up-to policy defined by y^* and the optimal policy in period 2 is an order-up-to policy defined by $y_2^*(\xi_1)$.*

Proof: The proof is provided in the Appendix. $\square$

It is intuitively appealing that increasing values of b decrease the value of backup agreements. More precisely, it seems that if $b \geq c + h_2 - s$ then Catco should take all of the backup independent of the value of ξ_1 since it is cheaper to buy an item and have it on hand then it is to pay the penalty for not taking it from backup. In this case there is no value to the information included in the demand in period 1 and thus no economic advantage to a backup agreement (other than the savings due to h_1). To support these ideas we first prove the following Lemma.

Lemma 3 *If $b \geq c + h_2 - s$ then the optimal policy is to take all of the backup for all values of ξ_1.*

Proof: The proof is provided in the appendix.

Figure 11.1 shows the impact of a backup and a no backup agreement on the inventory in the second period for a given order commitment y. The x-axis provides different values of ξ_1, the first period demand. The solid line provides the value of $y_2^*(\xi_1)$ – the optimal unconstrained inventory for the second period (from theorem 2 in the Appendix). For the no backup case, the available inventory in the second period is provided by a line with slope -1 for $\xi_1 \leq y$ i.e., when we do not run out in the first period. If we run out in the first period, i.e., $\xi_1 \geq y$, then the inventory available in the second period is 0. The available inventory levels for the no backup case are indicated by the dashed line.

The value of ξ_1^I in Figure 11.1 is the value of ξ_1 below which it is optimal to take no items from backup. Similarly, ξ_1^{II} is the value of ξ_1 above which it is optimal to take all units from backup. The arrows show the available inventory in period two under a backup agreement. The parameter ρ affects the gap between ξ_1^I and ξ_1^{II} and thus affects how closely the catalog company can duplicate the optimal unconstrained starting inventory for period two.

We will use Figure 11.1 to motivate the proof of a theorem that states that a backup agreement should motivate a catalog company to increase the commitment. For a given value of b, let $Y^*(\rho)$ be the optimal order commitment if a backup contract defined by ρ is available and $Y^*(0)$ be the optimal order commitment if no backup contract is available. Theorem 3 (below) thus states that if it is extremely unlikely that we run out in the first period using the commitment that is optimal when no backup is available i.e., $\Phi_1((1-\rho)Y^*(0)) \approx 1$, then the optimal commitment with backup (for any ρ) is greater than the optimal commitment when no backup is offered. We remark that the condition that $\Phi_1((1-\rho)Y^*(0)) \approx 1$ is a mild condition for this problem, since it only requires that sufficient inventory be available initially so that catalog company does not run out in two weeks (out of a six month season).

Theorem 3 *If* $\Phi_1((1-\rho)Y^*(0)) \approx 1$, *then* $Y^*(\rho) \geq Y^*(0)$.

Proof: The proof is provided in the Appendix. □

Note that in the presence of supply constraints such as a backup agreement, it may be optimal for the buyer (the catalog company) to increase the committed quantity over that in a no backup agreement system. Such systems might, in some cases, even generate pareto improving outcomes for both the supplier and the buyer (Eppen and Iyer(1997b)). We have thus shown the interaction between information, demand, optimal inventory levels and a supply contract.

11.5 LITERATURE REVIEW

There is an enormous literature that deals with stochastic inventory models and it is not our intention to review it here. Useful review articles include Graves (1981), Silver (1981), Veinott (1966) and Wagner (1981). Few stochastic inventory control models include the option of dumping items. Exceptions include Fukuda (1961) and the cash management models of Eppen and Fama (1968,1969) as well as the work of Neave (1970) on the same topic.

We are aware of eight papers that consider inventory problems in the fashion goods industry, but do not include a Bayesian approach. The first three consider production problems in a fashion goods environment: Matsuo (1990) considers the issue of common setups in a family of products; Crowston (1973) and Hausman (1972) consider producing in advance for a short sales season when the information about demand changes over time. Papers by Chang (1971), Hausman (1969), and Hertz (1960) concentrate on forecasting. Ravindran (1972) investigates a problem in which the length of the selling season is a decision variable. Finally, Hausman (1973) uses data from the sale of women's sportswear to suggest a model for generating demand data as well as a decision making framework. Several papers consider the explicit use of observed data in the inventory control decision, but are not based on a Bayesian approach. Examples are papers by Blinder (1982), Pindyck (1982), Harpaz, Lee, and Winkler(1982) and Miller (1986).

Finally there is a series of papers that are closely related to this work in the sense that they use Bayesian updating. The papers are Scarf (1959), Scarf (1960a), Iglehart (1964), Murray and Silver (1966), Azoury (1984), Azoury (1985) and Lovejoy (1990). Each of these papers focuses on a specific class of demand functions: the exponential class Scarf (1958), the exponential and the range class Iglehart (1964) and the Gamma family Scarf (1960a). In the latter case the prior distribution must also be of a specific form. The two papers by Azoury (1984) and (1985) as well as Lovejoy (1990) assume that if ξ_t is the demand in period and $D(\xi_t \leq A)$ is the probability that ξ_t is less than or equal to A then $D(\xi_t \leq A) = F_t(\frac{A}{q_t(S)})$ where $F_t(\bullet)$ is a known cumulative density function, S is a sufficient statistic for previous demand data and $q_t(\bullet)$ is a known function. Murray and Silver(1966) assume a Beta prior and a binomial process in each period.

Various buyer-supplier arrangements are the focus of a number of papers in the marketing and management science literature. See Stern and El-Ansary (1988), Anupindi and Akella (1993) as well as Kumar, Akella and Cornuejols (1992). Bassok and Anupindi (1994) consider minimum order commitment based contracts while Tsay and Lovejoy (1995) consider flexible contracts with a constraint on the possible change in forecast values. Iyer and Bergen (1994) describe contracts in the context of Quick Response while Moses and Seshadri (1995) consider negotiations between the manufacturer and the retailer to manage inventory and service levels. A recent paper by Fisher and Raman (1996) incorporates the impact of early demand information.

11.6   RESEARCH QUESTIONS AND CONCLUSIONS

We have focused in this chapter on a Bayesian model of information and its impact on demand uncertainty. We have also considered the impact of information on the profits for the buyer and the supplier. Note that in the absence of any supply constraints, the expected quantity purchased from a supplier decreases as the buyer has access to more information about demand. Intuitively the increased information enables a reduction in the safety stocks that the retailer would have had to carry in the absence of curremt demand information. However, as Lemma 1 shows, such a system is not Pareto improving. Thus, in order to make the gains from information Pareto improving, we need to have contractual agreements such as a service level agreement in Theorem 1. Alternately, we see the impact of backup agreements as contributing to an increase in the committed quantity (theorem 3). Thus we suggest that the interaction between information and demand has to be examined from the point of view of the individual parties to the supply and demand of the product.

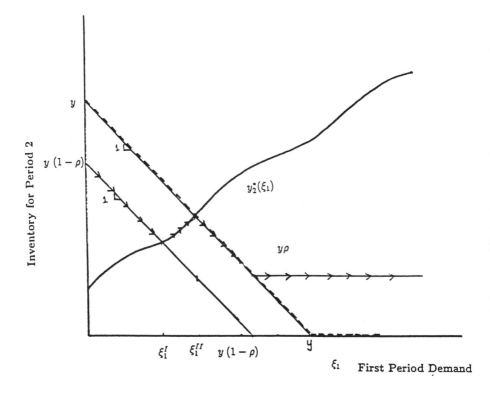

Figure 11.1 The inventory in period two for the backup and no backup case after optimal control decisions are made. The dashed lines refer to the second period inventory for the no backup case. The arrows refer to the second period inventory for the backup case.

Notes

1. Consider the practices of retailer A in forecasting demand for women's blouses for the Spring 1996 season. Retailer A stores extensive Point Of Sale data that can be analyzed along any one of over 1000 possible attributes, such as color, silhouette, pattern, fabric, buttons or zippers to name a few. Thus, during the summer and fall seasons in 1995, a fashion buyer can access information along any particular attribute under consideration, for instance the color red. The buyer examines data regarding sales of other items that share the same color red over the last one to two months and can compare that to recent sales in the previous few months, or compare them to sales in the previous year. These comparisons enable the buyer to assess whether the color red exhibits a fashion trend for the upcoming Spring 1996 season. Similar analyses are undertaken for other attributes, including patterns, silhouettes, and styles. This example is representative of buying for fashion items across retailers.

2. We note, that because the data is generated from similar (but not the same items) the remaining uncertainty, σ^2 reflects uncertainty both due to inherent demand uncertainty as well as due to the use of similar but not the same items for error reduction before the start of the season.

3. A scheme is Pareto improving if neither of the parties involved is worse off and atleast one of them is strictly better off

4. We will consider π used by the retailer as made up of a combination of two factors, one exogenous (π_0) and one negotiated between the manufacturer and the retailer(π_N). The exogenous portion of goodwill costs (π_0) cannot be negotiated or influenced by the manufacturer and includes industry standards, customer expectations from previous buying experiences and general customer buying characteristics in that area. Negotiations between manufacturer and retailer generates π_N. These components of π are combined in a non-decreasing function (γ) of π_0 and π_N. The simplest form of this function is as the sum of these two terms i.e., $\pi = \gamma(\pi_0, \pi_N) = \pi_0 + \pi_N$. Our analysis would extend to more complicated forms of the goodwill cost.

References

Agrawal, N, and Smith, S.C., "Estimation Methods for Retail Inventory Management with Unobservable Lost Sales", Working Paper, Santa Clara University, 1994.

Anupindi, R., and Akella, R., "An Inventory Model with Commitments", Working Paper, Carnegie Mellon University, GSIA, 1993.

Arrow, K., Harris, T., and Marschak, J., "Optimal Inventory Policy", *Econometrica*, Vol. 9, No. 3, July 1951, pg 250-272.

Azoury, K., "Bayes Solution to Dynamic Inventory Models under Unknown Demand Distribution", *Management Science*, Vol. 31 No. 9, September 1985, pp. 1150 - 1160.

Azoury, K. S., and Miller, B. L., "A Comparison of the Optimal Ordering Levels of Bayesian and Non-Bayesian Inventory Models", *Management Science*, Vol. 30, No. 8, August 1984, pp. 993 - 1003.

Bassok, Y, and Anupindi, R., "Analysis of Supply Contracts with Total Minimum Commitment", *IIE Transactions*, 1994.

Berger, James O., *Statistical Decision Theory and Bayesian Analysis*, Springer-Verlag, Second edition, New York, 1985.

Blackburn, J.D., *Time-Based Competition*, Business One Irwin, Homewood, IL, 1991, pg 246-269.

Blinder, A., "Inventories and Sticky Prices: More on the Microfoundations of Macroeconomics", *American Economic Review*, Vol. 72, 1982, pg 332-348.

Chang, S.H. and Fyffe, D. E., "Estimation of Forecast Errors for Seasonal-Style-Goods Sales", *Management Science*, Vol 18, No.2, October 1971, pp B-89-B-96.

Chung, K.L., *A First Course in Probability Theory*, Harcourt, Brace and World, New York, 1968.

Crowston,W.B., Hausman , W. H., and Kampe, W.R., "Multistage Production for Stochastic Seasonal Demand", *Management Science*, Vol 19, No. 8,April 1973.

Eppen, G.D., and Fama, E.F., "Solutions for Cash Balance and Simple Dynamic Portfolio Problems", *Journal of Business*, January 1968.

Eppen, G.D., and Fama, E.F., "Cash Balance and Simple Dynamic Portfolio Problems with Proportional Costs", *International Economic Review*, Vol. 10, No. 2, June 1969, pg 119-133.

Eppen, G.D., and Iyer, A.V.,"Improved Fashion Buying Using Bayesian Updates", *Operations Research*, November–December 1997

Eppen, G.D., and Iyer,A.V.,"Backup Agreements in Fashion Buying – The Value of Upstream Flexibility",*Management Science*, November 1997.

Fisher, M.A, and Raman, A., "Reducing the Cost of Demand Uncertainty Through Accurate Response to Early Sales", *Operations Research*, Vol. 44, No. 1, January-February 1996.

Fukuda, Y., "Optimal Disposal Policies", *Naval Research Logistics Quarterly*, Vol. 8, No. 3, September 1961, Pg 221-227.

Harpaz, G., Lee, W.Y. and Winkler, R.L., "Optimal Output Decisions of a Competitive Firm",*Management Science*, Vol. 28, 1982, pg 589-602.

Hausman, W. H., "Sequential Decision Problems: A Model to Exploit Existing Forecasters",*Management Science*, Vol. 16. No.2, October 1969, pp 93-111.

Hausman, W. H. and Peterson, R., "Multiproduct Production Scheduling for Style Goods with Limited Capacity, Forecast Revisions and Terminal Delivery", *Management Science*, Vol. 18. No. 7, March 1972, pp 370-383.

Hausman, W.H., and Sides, R., "Mail-Order Demands for Style Goods: Theory and Data Analysis", *Management Science*, Vol. 20. No. 2, October 1973, pp 191-202.

Hertz, D. B. and Schaffir, K. H.,"A Forecasting Method for Management of Seasonal Style-Goods Inventories",*Operations Research*, Vol. 8, 1960, pp45-52.

Iglehart, D., "The Dynamic Inventory Problem with Unknown Demand Distribution",*Management Science*, Vol. 10, 1964, pp. 429-440.

Iyer, A.V., and Bergen, M., "Quick Response in Manufacturer–Retailer Channels", *Management Science*, April 1997.

Kumar, A., Akella, R., and Cornuejols, G., "Supply Contracts Under Bounded Order Quantities", Working Paper, Carnegie Mellon University, GSIA, 1992.

Lovejoy, W. S., "Myopic Policies for Some Inventory MOdels with Uncertain Demand Distributions",*Management Science*, Vol. 36, No. 6, June 1990, pp. 724 - 738.

Matsuo, H.,"A Stochastic Sequencing Problem for Style Goods with Forecast Revisions and Hierarchical Structure",*Management Science*, Vol. 36. N0. 3, March 1990, pp. 332 - 347.

Miller, B., "Scarf's State Reduction Method, Flexibility and a Dependent Demand Inventory Model",*Operations Research*, Vol. 34, No. 1, Jan-Feb 1986, pp. 83-90.

Morrison, D.G., and Schmittlein, D.C., "Generalizing the NBD Model for Customer Purchases: What are the Implications and Is it Worth the Effort", *Journal of Business and Economic Statistics*, Vol. 6, No. 2, April 1988, pp. 145-159.

Moses, M., and Seshadri, S., "Policy Mechanisms for Supply Chain Co-ordination", Working Paper, Stern School of Business, New York University, 1995.

Murray, G. R., Jr. and Silver, E.A.,"A Bayesian Analysis of the Style Goods Inventory Problem",*Management Science*, Vol. 12, No. 11, July 1966, pp. 785-797.

Neave, E.H.,"The Stochastic Cash Balance Problem with Fixed Costs for Increases and Decreases",*Management Science*, Vol. 16, March 1970, pg 472-490.

Pindycj, R., "Adjustment Costs, Uncertainty and the Behavior of the Firm",*American Economic Review*, Vol. 72, 1982, pg 415-427.

Ravindran, A.,"Management of Seasonal style-Goods Inventories", *Operations Research*, Vol. 20, No. 2, March-April 1972, pp. 265-275.

Scarf, H., "Some Remarks on Bayes Solutions to the Inventory Problem",*Naval Research Logistics Quarterly*, Vol. 7, 1958, pg 591-596.

Scarf, H., "Bayes Solutions of the Statistical Inventory Problem",*Annals of Mathematical Statistics*, Vol. 30, 1959, pg 490-508.

352

Stern, L.W., and El-Ansary, A.I., *Marketing Channels*, Prentice Hall, Englewood Cliffs, N.J. 07632, 1988.

Tsay, A., and Lovejoy, W.S., "Supply Chain Control with Quantity Flexibility", Working Paper, Santa Clara University, Santa Clara, CA. 95053, 1995.

Veinott, A.F. Jr., "The Status of Mathematical Inventory Theory", *Management Science*, Vol. 12, No. 11, July 1966, pg 745-777.

11.7 APPENDIX

Theorem 1 If $s \geq \frac{1}{2}$, $d = \frac{\sigma}{\tau}$ and $y = \sqrt{\frac{1+\frac{1}{1+d^2}}{1+\frac{1}{d^2}}}$ then increasing π to π', where

$\pi' = \frac{s'(r+h)+c-r}{1-s'}$ and s' such that $Z(s') = \frac{Z(s)}{y}$ makes the use of information Pareto improving.

Proof: To maintain manufacturer expected profits, we require that $EP_{old-mfr} = EP_{info-mfr}$. We guarantee this relation by changing the service level from s to s' under *info* such that $Z(s') = \frac{Z(s)}{y}$, where $s' = \Phi(Z(s'))$, $\Phi(Z(s'))$ is the cumulative density function of a standard Normal distribution, $y = \sqrt{\frac{1+\frac{1}{1+d^2}}{1+\frac{1}{d^2}}}$, $d = \frac{\sigma}{\tau}$ (Note that $y \leq 1$).

We will implement this new service level at the retailer by changing π to π' such that $s' = \frac{r+\pi'-c}{r+\pi'+h}$. We have to verify whether the retailer expected profits under this service level s' and information use is greater than in the old system. Verify that from equations (11.2) and (11.4), we have to show that $\frac{b_r(Z(s))}{b_r(\frac{Z(s)}{y})} \geq \frac{(r+h+\pi')y}{r+h+\pi}$ which simplifies to showing that $v(Z(s)) \geq v(\frac{Z(s)}{y})$ where $v(Z) = \frac{\phi(Z)}{Z(1-\Phi(Z))}$, where $\phi(Z)$ is the probability density function of a standard Normal distribution.

Note that it is sufficient to show that $\frac{dv(Z)}{dZ} \leq 0$ for all $Z \geq 0$ (because $y \leq 1$). Verify that $\frac{dv(Z)}{dZ} = Zv(Z)(v(Z) - \frac{1+Z^2}{Z^2})$. Also, from Chung (1968), we know that $1 - \Phi(Z) \geq \frac{Z\phi(Z)}{1+Z^2}$ for all $Z > 0$. Thus we have $v(Z) \leq \frac{1+Z^2}{Z^2}$ and $\frac{dv(Z)}{dZ} \leq 0$ for all $Z \geq 0$. The retailer expected profits under *info* and the service level s' are greater than or equal to the retailer profits under the old system. Hence the use of information is Pareto improving with this change in service level from s to s'. ●.

Lemma 2: If $s \geq \frac{1}{2}$, $d = \frac{\sigma}{\tau}$ and $y = \sqrt{\frac{1+\frac{1}{1+d^2}}{1+\frac{1}{d^2}}}$ then increasing π to π', where $\pi' = \frac{s'(r+h)+c-r}{1-s'}$ and s' such that $Z(s') = \frac{Z(s)}{y}$, results in $EI_{info-sold}(s') \geq I_{old-sold}$ and $EI_{info-left}(s') \leq I_{old-left}$

Proof: To show that $EI_{info-sold}(s') \geq I_{old-sold}$, we have to show that $\mu - b_r(\frac{Z(s)}{y})\sqrt{\sigma^2 + \frac{1}{\rho}} \geq \mu - b_r(Z(s))\sqrt{\sigma^2 + \tau^2}$. This implies that $\frac{b_r(Z(s))}{b_r(\frac{Z(s)}{y})} \geq y$ This is clearly true from theorem 1 (note that $\pi' \geq \pi$). Hence $EI_{info-sold}(s') \geq I_{old-sold}$. Since we know that $EI_{info}(s') = I_{old}$, we see that $EI_{info-left}(s') \leq I_{old-left}$. ●.

Theorem 2 The function $G_1(y)$ is concave in y and $G_2(y_2, \xi_1)$ is concave in y_2. Thus the optimal policy in period 1 is an order-up-to policy defined by y^* and the optimal policy in period 2 is an order-up-to policy defined by $y_2^*(\xi_1)$.

Proof:

This proof proceeds in a manner that is somewhat different than that used for the typical dynamic inventory control problem. The fact that the amount committed in the first period determines an upper bound on what we can buy in the second period implies that knowing: (1) the inventory on hand after demand occurs in period 1 and (2) the demand in period 1 is not sufficient to determine the optimal policy in period 2. The proof is established in four steps: *Step 1*: $G_2(y_2, \xi_1)$ is defined in section 11.4.3, we first show that this function, plus two linear terms, is concave in y_2 for every ξ_1, *Step 2*: We use this fact to derive the optimal policy for period 2, *Step 3*: We then use the form of the optimal policy to prove that $f_2(y, \xi_1)$ is concave in y for all ξ_1, (Note that this y is the amount committed at the beginning of period 1), *Step 4*: From this fact, the expression for $G_1(y)$ in section 11.4.3, and the fact that a convex combination of concave functions is concave, we conclude that $G_1(y)$ is concave. The form of the optimal policy follows directly.

Step 1:

$-cy_2 + by_2 + G_2(y_2, \xi_1) =$

$-cy_2 + by_2 + \int_0^{y_2} r\xi_2 \phi_2(\xi_2 \mid \xi_1)d\xi_2 + \int_{y_2}^{\infty} ry_2 \phi_2(\xi_2 \mid \xi_1)d\xi_2$

$-(h_2 - s)\int_0^{y_2}(y_2 - \xi_2)\phi_2(\xi_2 \mid \xi_1)d\xi_2$

$-\pi \int_{y_2}^{\infty}(\xi_2 - y_2)\phi_2(\xi_2 \mid \xi_1)d\xi_2$

The derivative with respect to y_2 yields

$-c + b - \Phi_2(y_2)\{r + (h_2 - s) + \pi\} + \{r + \pi\}$

The second derivative with respect to y_2 yields

$-\phi_2\{r + (h_2 - s) + \pi\}$

which is clearly ≤ 0 indicating that the function is concave for all y_2.

The optimal $y_2^*(\xi_1)$ is obtained by setting the first derivative equal to zero to yield

$$\Phi_2(y_2^*(\xi_1) \mid \xi_1) = \frac{\{r + \pi + ((b - c))\}}{\{r + (h_2 - s) + \pi\}} \quad (1)$$

Step 2:

The optimal policy then is an order-up-to policy defined by $y_2^*(\xi_1)$. Specifically, the amount taken from backup given an observed first period demand of ξ_1 is termed $BU(\xi_1)$ and is determined as follows: (i) If $y(1 - \rho) \geq \xi_1$, and $y(1 - \rho) - \xi_1 \geq y_2^*(\xi_1)$, then $BU(\xi_1) = 0$. (ii) If $y(1 - \rho) \geq \xi_1$, and $y(1-\rho) - \xi_1 < y_2^*(\xi_1) \leq y - \xi_1$, then $BU(\xi_1) = y_2^*(\xi_1) - [y(1-\rho) - \xi_1]$. (iii) If $y(1-\rho) \geq \xi_1$, and $y_2^*(\xi_1) > y - \xi_1$, then $BU(\xi_1) = \rho y$. (iv) If $y(1-\rho) < \xi_1$, and $0 < y_2^*(\xi_1) \leq y\rho$, then $BU(\xi_1) = y_2^*(\xi_1)$. (vi) If $y(1 - \rho) < \xi_1$, and $y_2^*(\xi_1) > y\rho$, then $BU(\xi_1) = \rho y$.

Step 3:

To prove that $f_2(y, \xi_1)$ is concave in y for every ξ_1, we first pick a value of ξ_1 and then vary y. Note that choosing ξ_1, determines $y_2^*(\xi_1)$. This is important since the value of $f_2(y, \xi_1)$ depends on the location of y with respect to the parameters A, B and C where $A = \frac{y_2^*(\xi_1) + \xi_1}{1 - \rho}$, $B = y_2^*(\xi_1) + \xi_1$, $C = \frac{\xi_1}{1 - \rho}$

The proof is tedious because these parameters can assume three configurations: (1) $C \leq B \leq A$, (2) $B \leq C \leq A$ and (3) $B \leq A \leq C$. We will provide details for the first case. This case applies for all ξ_1 such that $y_2^*(\xi_1) \geq \xi_1 \frac{\rho}{1-\rho}$. All other cases follow similarly.

There are four possible regions for y: (1) $y \leq C$, (2) $C \leq y \leq B$, (3) $B \leq y \leq A$, (4) $A \leq y$. We consider them in order.

Case 1: $C \leq B \leq A$

Under this condition, there are four possible regions for y and the associated $f_2(y, \xi_1)$ values are as follows:

$y \leq C$: This is the region where $y(1-\rho) < \xi_1$ and we cannot reach $y_2^*(\xi_1)$ in the second period because $y_2^*(\xi_1) > y - \xi_1$. i.e., $y \leq B$. Thus we set $y_2 = \rho y$ and the value of $f_2(y, \xi_1)$ is as follows:

$$= ry(1-\rho) - c\rho y + \pi(y(1-\rho) - \xi_1) + G_2(\rho y, \xi_1)$$

The slope of $f_2(y, \xi_1)$ with respect to y is thus $r(1-\rho) - c\rho + \pi(1-\rho) + (\frac{dG_2(\rho y, \xi_1)}{dy})\rho$

Given that $G_2(y_2, \xi_1)$ is concave in y_2, $f_2(y, \xi_1)$ is concave in y for $y < C$.

$C < y \leq B$: In this region, we have $y(1-\rho) \geq \xi_1$ but we cannot reach $y_2^*(\xi_1)$ in the second period because $y_2^*(\xi_1) > y - \xi_1$ i.e., $y \leq B$. Thus we set $y_2 = y - \xi_1$ and the value of $f_2(y, \xi_1)$ is as follows:

$$= r\xi_1 - c\rho y - h_1(y(1-\rho) - \xi_1) + G_2(y - \xi_1, \xi_1)$$

The slope of $f_2(y, \xi_1)$ with respect to y is thus $-c\rho - h_1(1-\rho) + (\frac{dG_2(y-\xi_1, \xi_1)}{dy})$

As before, since $G_2(y_2, \xi_1)$ is concave in y_2, $f_2(y, \xi_1)$ is concave in y for $C < y \leq B$. Note that at $y = C$, we have to show that the slope to the left of C is greater than or equal to the slope to the right of C. This is verified by setting $y = \frac{\xi_1}{1-\rho}$ in the slopes for the two cases we have seen. The resulting condition that $\frac{dG_2(\rho y, \xi_1)}{dy} \leq r + \pi + h_1$ is verified from the fact that $\frac{dG_2(y_2, \xi_1)}{dy_2} \leq r + \pi$ from Step 1.

$B < y \leq A$: In this region, we have $y(1-\rho) \geq \xi_1$ and we can reach $y_2^*(\xi_1)$. Thus we have $y_2 = y_2^*(\xi_1)$. The value of $f_2(y, \xi_1)$ in this region is as follows:

$$= r\xi_1 - c(y_2^*(\xi_1) - y(1-\rho) + \xi_1) - h_1(y(1-\rho) - \xi_1) - b\{\rho y - (y_2^*(\xi_1) - y(1-\rho) + \xi_1\} + G_2(y_2^*(\xi_1), \xi_1)$$

The slope of $f_2(y, \xi_1)$ with respect to y is thus $-h_1(1-\rho) + c(1-\rho) - b$

Since this slope is constant $f_2(y, \xi_1)$ is concave for $B < y \leq A$. It can be verified that the slope to the left of B is $\geq$ the slope to the right of B.

$y \geq A$: In this region, we have $y(1-\rho) \geq \xi_1$, however, taking no units from backup is the optimal strategy i.e., $y_2 = y(1-\rho) - \xi_1$. The value of $f_2(y, \xi_1)$ in this region is as follows:

$$= r\xi_1 - h_1(y(1-\rho) - \xi_1) - b\rho y + G_2(y(1-\rho) - \xi_1, \xi_1)$$

The slope of $f_2(y, \xi_1)$ with respect to y is thus

$-h_1(1-\rho) - b\rho + (\frac{dG_2(y(1-\rho) - \xi_1, \xi_1)}{dy})(1-\rho)$

Again, since $G_2(y_2, \xi_1)$ is concave in y_2, $f_2(y, \xi_1)$ is concave for $y > A$. It can be verified that the slope to the left of A is $\geq$ the slope to the right of A. We have thus shown that $f_2(y, \xi_1)$ is concave in y for Case 1. Similar analysis can be used to show that as long as $r \geq c$, $f_2(y, \xi_1)$ is concave in y for every ξ_1 for the other two cases when $B \leq C \leq A$ and when $B \leq A \leq C$.

Step 4:

We have shown that $f_2(y, \xi_1)$ is concave in y. Using the definition of $G_1(y)$ and the fact that a convex combination of concave functions is concave we conclude that $G_1(y)$ is concave in y. Hence the order up to policy follows. $\square$

Lemma 3: If $b \geq c + h_2 - s$ then the optimal policy is to take all of the backup for all values of ξ_1.

Proof: Note that if we set $b = c + h_2 - s$, then the right hand side of equation 1 is equal to 1 so that $y_2^*(\xi_1)$ is equal to ∞. This implies, from Step 2 in the proof of theorem 1, that it is optimal to take all items from backup independent of the observed value of ξ_1.

Theorem 3 Let $Y^*(0)$ be the optimal order commitment from theorem 1 for $\rho = 0$. Also, let $Y^*(\rho)$ be the optimal order commitment from theorem 1 for a given ρ. If $\Phi_1((1 - \rho)Y^*(0)) \approx 1$, then $Y^*(\rho) \geq Y^*(0)$.

Proof: We will prove this result by showing that $\frac{dG_1(y)}{dy}|_{Y^*(0)} \geq 0$. In order to prove this, we will evaluate the value of $-c(1 - \rho) + \frac{df_2(y, \xi_1)}{dy}|_{Y^*(0)}$ for each possible value of ξ_1, and integrate across values of ξ_1 to get the result. The condition $\Phi_1(Y^*(0)(1 - \rho)) \approx 1$ implies that we need to consider only values of $\xi_1 \leq Y^*(0)(1 - \rho)$.

Define ξ_1^I and ξ_1^{II} as follows: $(1 - \rho)Y^*(0) - \xi_1^I = y_2^*(\xi_1^I)$, and $Y^*(0) - \xi_1^{II} = y_2^*(\xi_1^{II})$. Assume that $\xi_1^I \geq 0$ and $\xi_1^{II} \geq 0$. Otherwise set these values equal to zero to get the correct expressions. There are two cases, when $Y^*(0)(1 - \rho) \geq \xi_1^I$ and when $Y^*(0)(1 - \rho) \leq \xi_1^I$. We will present results for the first case. The second case follows similarly. Figure 11.1 and the discussion in section 11.4.4 provides the intuition behind this proof.

For the situation with a backup parameter ρ, the values of ξ_1 fall into three regions: (i) $\xi_1 \leq \xi_1^I$, (ii) $\xi_1^I \leq \xi_1 \leq \xi_1^{II}$, (iii) $\xi_1^{II} \leq \xi_1 \leq y(1 - \rho)$. For the region $0 \leq \xi_1 \leq \xi_1^I$, the inventory on hand at the beginning of period 2 is larger than $y_2^*(\xi_1)$ so we take no items from backup. For the region $\xi_1^I \leq \xi_1 \leq \xi_1^{II}$, the inventory is smaller than $y_2^*(\xi_1)$ and we can take items from backup to reach $y_2^*(\xi_1)$ so we order up to $y_2^*(\xi_1)$. For the region $\xi_1^{II} \leq \xi_1 \leq (1 - \rho)y$, we cannot order up to $y_2^*(\xi_1)$ so we take all the units from backup. This is also the region where the inventory for the backup and no backup agreements are the same. Finally, when $\xi_1 \geq y(1 - \rho)$, we run out in the first period and the inventory in the second period is ρy. The inventory after choosing the optimal policy is indicated by the arrows in Figure 11.1. The other case will change the location of $y(1 - \rho)$ with respect to ξ_1^I and ξ_1^{II}.

Since by assumption, $\Phi_1(Y^*(0)(1 - \rho)) \approx 1$, $\frac{dG_1(y)}{dy}|_{Y^*(0)} = \int_0^{y(1-\rho)} \{-c(1 - \rho) + \frac{df_2(y, \xi_1)}{dy}|_{Y^*(0)}\}\phi_1(\xi_1)d\xi_1$. We can use the proof of theorem 2 to read off the value of $\frac{df_2(y, \xi_1)}{dy}$ for each of the three regions for ξ_1 i.e.,

$$\frac{dG_1(y)}{dy}|_{Y^*(0)} = \int_0^{\xi_1^I}(-c(1 - \rho) - h_1(1 - \rho) - b\rho + (\frac{dG_2(y(1-\rho)-\xi_1,\xi_1)}{dy})|_{Y^*(0)}$$
$$)(1 - \rho))\phi_1(\xi_1)d\xi_1$$
$$+ \int_{\xi_1^I}^{\xi_1^{II}}(-b - h_1(1 - \rho))\phi_1(\xi_1)d\xi_1 + \int_{\xi_1^{II}}^{y(1-\rho)}(-c - h_1(1 - \rho) + (\frac{dG_2(y-\xi_1,\xi_1)}{dy})|_{Y^*(0)}$$
$$)\phi_1(\xi_1)d\xi_1.$$

Note that the slope $\frac{dG_1(y)}{dy}\mid_{\rho=0}$ (which is the slope of $G_1(y)$ evaluated at $y = Y^*(0)$) is obtained as $\frac{dG_1(y)}{dy}\mid_{\rho=0,Y^*(0)} = \int_0^{Y^*(0)(1-\rho)}(-h_1-c+(\frac{dG_2(y-\xi_1,\xi_1)}{dy}\mid_{Y^*(0)}))\phi_1(\xi_1)d\xi_1$. If we examine the difference between the terms for ξ_1 between 0 and ξ_1^I in the two equations presented just above, we get

$$\rho c + h_1\rho - b\rho + (1-\rho)(\frac{dG_2(y(1-\rho)-\xi_1,\xi_1)}{dy}\mid_{Y^*(0)}) - (\frac{dG_2(y-\xi_1,\xi_1)}{dy}\mid_{Y^*(0)})$$

$$= \rho h_1 + \rho\{(c-b) - \frac{dG_2(y(1-\rho)-\xi_1,\xi_1)}{dy}\mid_{Y^*(0)}\}$$

$$+\{\frac{dG_2(y(1-\rho)-\xi_1,\xi_1)}{dy}\mid_{Y^*(0)} - \frac{dG_2(y-\xi_1,\xi_1)}{dy}\mid_{Y^*(0)}\}$$

It is clear that $\rho h_1 \geq 0$. The second term in $\{\bullet\}$ is ≥ 0 because $y^*(0)(1-\rho) - \xi_1 \geq y_2^*(\xi_1)$ and hence the slope of $G_2(\bullet,\xi_1)$ is $\leq c-b$. Also the last term in $\{\bullet\}$ is ≥ 0 by concavity of $G_2(y_2,\xi_1)$.

If we examine the difference between terms for ξ_1 between ξ_1^I and ξ_1^{II} in the equations, we get $h_1\rho - \{-c+b+\frac{dG_2(y-\xi_1,\xi_1)}{dy}\mid_{Y^*(0)}\}$. This is ≥ 0 as the second term in $\{\bullet\}$ is ≤ 0 because $Y^*(0) - \xi_1 \geq y_2^*(\xi_1)$.

For $\xi_1 \geq \xi_1^{II}$, the terms in the equations differ only by $h_1\rho$ which is ≥ 0. Hence we have shown that $\frac{dG_1(y)}{dy}\mid_{Y^*(0)} \geq \frac{dG_1(y)}{dy}\mid_{\rho=0,Y^*(0)} = 0$. Thus, using theorem 1 we conclude that $Y^*(\rho) \geq Y^*(0)$. $\square$

12

MODELING IMPACTS OF ELECTRONIC DATA INTERCHANGE TECHNOLOGY

Sunder Kekre, Tridas Mukhopadhyay and Kannan Srinivasan

Graduate School of Industrial Administration
Carnegie Mellon University
Pittsburgh, PA 15213

12.1 Introduction

Supply chains today are increasingly depending on effective and efficient information exchanges between the value chain partners. Over the last decade, practices such as just in time management, quick response manufacturing and lean production systems have required coordinated and reliable information exchanges between trading partners (HBS 1991). Massive investments in information technology (IT) have been made by manufacturers, suppliers and logistics providers with the hope of achieving successful JIT implementation in their supply chains (Kekre et al., 1992; Schonberger 1986). From a managerial standpoint, these companies face the following questions:

- What are the operational gains accrued by the manufacturer and their suppliers?
- Does the technology payoff?

Empirical researchers seeking answers to these questions, in turn, have faced the following challenges:

- How does one integrate the conceptual frameworks from both IT and supply management literature to measure the impact of technology?
- What are the key methodological issues the researcher should be aware of?

In our field studies, we chose to investigate the role of Electronic Data Interchange (EDI) in JIT supply chains. EDI provides electronic exchange of standard business documents between trading partners. It has often been considered as an enabler of JIT systems. The increased investments in this technology and its usage has paralleled the changes in the logistics and product delivery functions. This provided us with a natural setting to assess the impact of IT drivers as changes with JIT procurement took hold.

We adopted the research paradigm shown in Figure 1 as we sought answer to the role of EDI in different supply chain settings. This chapter reports three major field studies that were undertaken whose common goal was to identify the *drivers* of performance in vertically information integrated supply chains (Srinivasan et al., 1994, Mukhopadhyay et al., 1995, Mukhopadhyay et al., 1998) The application of the research paradigm will be illustrated as we present the individual field studies in subsequent sections. A brief summary of the three field studies follows.

We begin in Section 2 by examining the operational gains from EDI from a manufacturer's perspective. The first set of models examined the effects of EDI technology on shipment performance at Chrysler's assembly centers. The shipment discrepancy analysis used a logit model that captured both material flow and information exchange characteristics. From a modeling standpoint, the control for heterogeneity across suppliers significantly enhanced the explanatory power of our models.

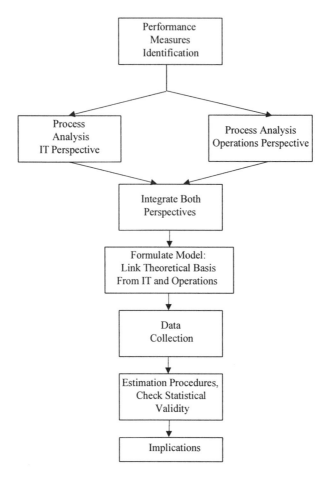

Figure 1: Research Paradigm for EDI Impact on Supply Chains

In Section 3, we present the second set of models that take a vendor's perspective of the role of EDI in supply chains, and examine the gains accrued to Kennametal, a distributor of tooling systems. We adopted an integrative framework that ties together transaction cost economics, information economics, and operation management principles. We developed a probit model to quantify the impact of EDI use on the timeliness of payments received by Kennametal, and highlighted the role of information exchange in the order payment cycle.

Section 4 discusses models to answer the critical question: What is the value created by EDI in JIT supply chains? It provides a careful analysis of the performance data of Chrysler's assembly centers over the past decade. Cross-sectional and time series data were combined to estimate the impact of varying levels of EDI penetration on JIT cost pools. Taken together, all three sets of models confirm the significance of integrated information exchanges and highlight the drivers of successful JIT supply chains from both material and information flow perspectives.

We conclude in Section 5 with a discussion of some of the methodological issues that researcher face in undertaking such field research. The next section presents the first study focusing on the analysis of shipment discrepancies at Chrysler.

12.2 Operational Gains from EDI: Manufacturer's Perspective

In JIT supply chains, the coordination of material flows through information exchange is becoming increasingly important. With technologies such as Electronic Data Interchange with suppliers, the rapid transmission of information with trading partners allows manufacturers to efficiently manage logistics activities. From a research stand point, the question, therefore, is to identify the drivers of performance of JIT supply chains, specifically the role of technology in interorganizational information exchange.

We seek answers to the following two key questions:
- What are the determinants of shipment performance?
- What is the impact of the vertical information integration with vendors in the supply chain?

12.2.1 Research Site

The assembly centers of Chrysler provided a fascinating example of the role of EDI technology to synchronize material and information flows. We conducted an ex-post analysis of the JIT programs that Chrysler had initiated in the mid-eighties. This program required all JIT vendors to adopt EDI technology. The analysis, however, became complicated because of the differences in the logistics complexities and information exchange patterns across the vendor base. From a material flow stand point, the complexity arose due to differences in factors such as the number of parts supplied, the number of shipping destinations, etc. Some suppliers also specialized in the auto components industry, and had developed special capability to adhere to auto industry standards.

From an information exchange stand point, some vendors had invested heavily to achieve full levels of information integration. Others had developed only the minimal capability to meet the minimum JIT information exchange requirements mandated by Chrysler. Thus, the level of information integration differed across the vendor base with some suppliers having achieved full integration capabilities by mapping the incoming information directly into their internal systems. Chrysler, in turn, provided some vendors with JIT schedules thereby facilitating synchronization of their internal operations with Chrysler assembly centers.

364

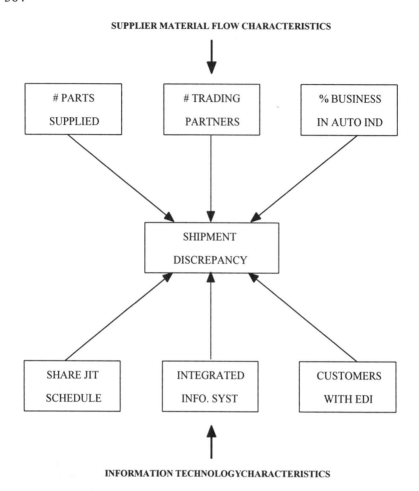

Figure 2: Drivers of Shipment Discrepancy

The performance metric we examined was the shipment discrepancy (D) which is a binary measure. A discrepancy is said to occur when the shipment arrives outside the window of JIT delivery as indicated in the advanced shipping notice, or if the material is delivered with errors in either destination, quantity, or shipping documents. Our model attempted to tease out the effect of the supplier material flow and information flow characteristics as shown in Figure 2. These characteristics (independent variables) together with the dependent variable were defined as follows.

Independent Variables:
A. Information Exchange

SCHEDULE	= 1, if vendor received JIT schedule,
	= 0, otherwise
INTEGRATION	= 1, if vendor had integrated electronic link with Chrysler
	= 0, otherwise
ELINKS	= Fraction of vendor's trading partners with EDI links

B. Material Flow Complexity

PARTS	= Number of distinct part numbers shipped
PARTNERS	= Number of vendor's trading partners
AUTO	= Fraction of vendor's business with the auto industry

Dependent Variable:

D	= 1 if the shipment has a discrepancy
	= 0 otherwise.

12.2.2 Model

Our dependent variable (D) was a binary variable and hence is bounded between 0 and 1. Therefore the OLS procedure leads to inefficient estimates (Kmenta 1986). We adopted the logit model for our analysis, and used the maximum likelihood estimation procedure (Judge et al., 1985). Therefore, the probability of any shipment being defective was

$$\text{Prob}(D=1) = \frac{e^{\alpha+\Sigma\beta x}}{1 + e^{\alpha+\Sigma\beta x}} \tag{1}$$

where

$$
\begin{aligned}
\alpha + \sum \beta x = \; & \alpha + \beta_1 SCHEDULE + \beta_2 INTEGRATION \\
& + \beta_3 ELINKS + \beta_4 PARTS \\
& + \beta_5 PARTNERS + \beta_6 AUTO
\end{aligned} \tag{2}
$$

Combining the two equations, we obtained the following likelihood function for the kth shipment being defective as

$$L = \prod_{k=1}^{K} \left(\frac{e^{\alpha+\Sigma\beta x_k}}{1 + e^{\alpha+\Sigma\beta x_k}} \right)^{D_k} \left(\frac{1}{1 + e^{\alpha+\Sigma\beta x_k}} \right)^{(1-D_k)} \tag{3}$$

where $D_k = 1$ if the k-th shipment is defective. Note that x_k is the vector of explanatory variables for the k-th shipment.

The baseline model was enriched to capture the unobserved heterogeneity across the supplier base such as management skills and other idiosyncratic factors (Chamberlain 1980). In particular, our model allowed the intercept term in equation (2) to vary across suppliers. Hence we defined $\alpha_n = \alpha + v_n$ where v_n is assumed to be normally distributed with a mean of 0 and a standard deviation of σ.

Another source of heterogeneity we observed concerned some suppliers who consistently had no shipment discrepancy during the study period. We adopted the stayer-mover structure (Gonul and Srinivasan 1993), where we defined S as the probability of a consistent supplier. Therefore the likelihood function with a random intercept term and stayer-mover structure was

$$
\begin{aligned}
L = \quad & \frac{1}{\sqrt{2\pi}\sigma} \prod_{n=1}^{N} \left[\left\{ \left(\int_{-\infty}^{\infty} \prod_{j=1}^{J_n} \left(\frac{e^{\alpha+v_n\Sigma\beta x_j}}{1 + e^{\alpha+v_n\Sigma\beta x_j}} \right)^{D_j} \right. \right. \right. \\
& \times \left. \left(\frac{1}{1 + e^{\alpha+v_n+\Sigma\beta x_j}} \right)^{(1-D_j)} e^{-(\alpha v_n^2/2\sigma^2)} dv_n \right) (1-S) \right\}^{I_n} \\
& \times \left\{ \left(\int_{-\infty}^{\infty} \prod_{l=1}^{I_n} \left(\frac{e^{\alpha+v_n+\Sigma\beta x_j}}{1 + e^{\alpha+v_n+\Sigma\beta x_j}} \right)^{D_j} \right. \right. \\
& \times \left. \left(\frac{1}{1 + e^{\alpha+v_n+\Sigma\beta x_j}} \right)^{(1-D_j)} e^{-(\alpha v_n^2/2\sigma^2)} dv_n \right) \\
& \times (1-S) + S \}^{(1-I_n)} \bigg].
\end{aligned}
$$
(4)

12.2.3 Results and Discussion

The estimates of the model are displayed in Table 1. We conducted a likelihood ratio test to ascertain the efficacy of the extended model (4) relative to the baseline model (3). The χ^2 test rejected the base line model in favor of the extended model (p<.0001) underscoring the importance for controlling for heterogeneity in our sample. Note that the U^2 value is analogous to R^2 in OLS and is often less than 50% even when the model has very good fit (Hauser 1978). Our results suggested that the extended model improved the U^2 by 100%.

The extended model shed light on the significant impact of information exchanges in inter-organizational systems. For instance, sharing of JIT schedules with suppliers had led to reduced shipment discrepancies. Furthermore, suppliers who had invested in integrated information systems reaped higher benefits as revealed by the coefficient of INTEGRATION. As expected, material flow complexity adversely affected performance. Notice the positive coefficients of PARTS and PARTNERS. Hence our model provided a quantitative measure of the benefits of electronic data interchange for varying levels of logistics complexity and investments in EDI.

Table 1. Estimation Results

Coefficient	Variable	0 Baseline	Extended
		1 Logit	
β_0	Intercept	-2.917***	-2.227***
β_1	SCHEDULE	-0.306*	-0.410***
β_2	INTEGRATION	-0.054	-0.910***
β_3	E-LINKS	-0.411*	-0.225***
β_4	PARTS	0.376***	0.370***
β_5	PARTNERS	0.071***	0.079***
β_6	AUTO	0.398	-0.205
	S	—	0.113***
	σ	—	1.315***
	Log Likelihood	-772	-682
	U^2	0.136	0.24

***Significant at 1% Level.

** Significant at 5% Level.

* Significant at 10% Level.

12.2.4 Commentary on the Modeling Framework

The estimation of the model as shown in Table 1 was based on a parametric specification (4) with heterogeneity across suppliers modeled with a normally distributed random intercept term and consistent suppliers captured by a stayer-mover structure. The assumption of normal distribution for the random intercept term was subsequently relaxed in favor of a non-parametric specification (Heckman and Singer 1984). We found the results of the two approaches to be highly consistent.

While this study was an attempt to quantify the role of information coordination in JIT systems, the results raised questions with respect to the gains achieved by the vendors in the supply chain. While the manufacturer (Chrysler) had benefited from lower shipment discrepancies at its assembly plants, did the supplier also benefit in terms of superior operational performance? We modeled this issue as described in the next section.

12.3 Operational Gains from EDI: Supplier's Perspective

In the previous section, we examined how effectively the suppliers of Chrysler met the shipment requirements of the assembly centers for a JIT environment. EDI technology was found to facilitate the delivery of products through reduced errors. We have also studied the flip side of this relationship by modeling the operational gains accrued by the supplier. One of the key benefits suppliers could enjoy from EDI technology is timely payments for the services provided. We model this issue in the context of JIT distribution systems for tools, tooling systems, and services.

We sought answers to the following key questions.
- Did EDI facilitate timely payment by manufacturers to the supplier?
- Did process improvement related to the order completion cycle significantly impact payments?

12.3.1 Research Site

Our research site was Kennametal, a Fortune 500 distributor of tooling systems. It had well over 300 customers with varying levels of electronic integration. While some customers were capable of advanced EDI transaction sets (e.g., electronic invoicing), others conduct business using paper based purchased orders. This setting allowed us to examine the differential gains from using EDI technology as reflected in the timely payment cycles. Similar to the Chrysler study above, we had to control for the order complexity across the customer base.

We collected data for fiscal year 1994 for a sample of over 300 customers of Kennametal. Half of these customers had electronic linkages with Kennametal (though to different degrees); the other half were non-EDI customers, and were selected so as to match their EDI counterparts in terms of annual sales and SIC codes. This matching was done to allow us to tease out the benefits of electronic information exchange. Since the matching is based on exogenous characteristics, the empirical analysis does not change from that of a random sample (Hausman and Wise 1981).

The dependent variable in this case was the probability of a delayed payment for a customer, and was measured as the fraction of the orders with delayed payment for a specific customer. We randomly selected 238 customers (half EDI and their non-EDI counter parts) for model estimation, and used the remaining data to test the predictive ability of our estimation results.

The explanatory variables in our model belonged to two categories as shown in Figure 3. The timeliness of payment depended upon the order complexity of customer accounts as well as the nature of information exchange to process the payment. This gave rise to two sets of variables as defined below.

Independent Variables:
A. *Information Exchange*

EORDER	= 1 if the customer sends order electronically
	= 0 otherwise
EINVOICE	= 1 if the customer sends invoice electronically
	= 0 otherwise

B. Order Complexity

    BLANKET   = Fraction of orders associated with blanket agreements
    STANDARD = Fraction of orders associated with standard items
    ITEMS   = Average number of line items per order

Dependent Variable:

    P   = Proportion of orders with delayed payment for a customer

INFORMATION EXCHANGE CHARACTERISTICS

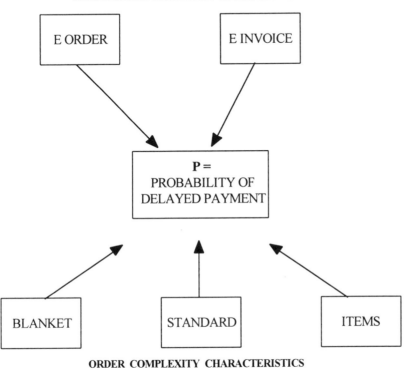

ORDER COMPLEXITY CHARACTERISTICS

Figure 3: Drivers of Delayed Payment of Orders

12.3.2 Model

We used the standard Probit model to estimate the probability of delayed payment Pindyck et al., 1981). For each customer,

$$P = F(\alpha + \beta_1 EORDER + \beta_2 EINVOICE + \beta_3 BLANKET$$
$$+ \beta_4 STANDARD + \beta_5 ITEMS)$$
$$= F(z) \tag{5}$$

where

$$F(Z) = \frac{1}{\sqrt{2\pi}} \int_{-\infty}^{Z_i} e^{-s^2/2} ds \qquad (6)$$

This model was estimated using the weighted least square estimator (Amemiya 1985), and exhibited excellent explanatory power as shown by $R^2 = .597$ in Table 2. The standard tests for heteroskedasticity were also performed. We also tested the predictive ability of the model. We used the average value of P from the main sample as an estimate for the holdout observations and calculated the sum of squares residuals. A χ^2 test based on these values rejected the null hypothesis that all the model coefficients are zero at 99% significance level (Greene 1990).

Table 2. Estimation of Delayed Payment Model

Variable Name	Coefficient Value	Significance Level
Constant	0.362	0.000
EORDER	-0.120	.354
EINVOICE	-0.351	.003
BLANKET	0.208	.043
STANDARD	-0.615	.001
ITEMS	0.384	.000

12.3.3 Results and Discussion

The signs of the estimated coefficients of both information and order complexity variables were as expected. From an information exchange standpoint, our results showed that the key driver for timely payment is electronic invoicing, and not electronic orders. At first glance this appeared counter intuitive. However, a close examination of the process revealed that electronic invoicing did make all the difference. It made invoicing information available to the customer earlier than manual invoicing. Furthermore, the information content of the invoice relative to the manual mode is of much higher accuracy. The customer's accounting department was thus able to process the payment authorization much earlier resulting in an efficient payment matching process. With electronic orders but in the absence of electronic invoicing, the processing of payments gets bogged down due to manual processing of invoices.

Similar to the Chrysler study, the order complexity variables were found to have significant impact on timely payments. The number of items per order (ITEMS) stretched the payment cycle as the matching process became more complicated. On the other hand, we found significant beneficial effect of standard orders. A surprising finding was that the blanket orders extended the payment cycle. This is contrary to JIT philosophy which has all along promoted the benefits of long term blanket agreements. Further scrutiny revealed the true cause of this apparent anomaly. The customers who had blanket agreements with Kennametal did not indicate this in their orders. As a result, such orders were treated as ordinary orders in the first round of processing, and were subsequently delayed till the price discrepancy was resolved.

12.3.4 Commentary on the Modeling Framework

The benefits of the EDI technology to a supplier are also accrued in other order processing activities. In the Kennametal case, we also found significant gains in terms of number of orders requiring reprocessing. In addition, we also analyzed the impact of advanced EDI capabilities on sales volume for various accounts. Our results indicate that a supplier can derive significant benefits through increased sales awarded by customers when the supplier enhances the electronic link with the customer. For brevity we do not discuss these models here (for details, refer to Mukhopadhyay et al.1998).

While increased business, smoother internal order processing and faster payments were enjoyed by Kennametal as a result of effective EDI investments and use, a question remained if EDI translated to bottom line benefits. We wanted to ascertain if the financial impact of inter-organizational information exchange was reflected in accounting numbers. This was the focus of the next study . We went back to Chrysler and reexamined if the assembly plants were able to generate cost savings as a result of using EDI in their JIT programs.

12.4 Business Value of EDI

Despite the operational gains quantified in the previous sections, we wanted to resolve the controversy on the financial impact of EDI. Some studies in the IT area have reported little or no IT impact with some researchers suggesting that the marginal dollar would best have been sent on non-IT inputs. Other studies have reported more promising results (Barua et al., 1995). There seems to be little consensus about the impact of IT for two reasons (Brynjolfsson 1993). First, the researchers have to contend with data that is sometimes hard to get. Second, to tease out the IT effects, one has to account for other non-IT related inputs.

The Chrysler study provided us with a unique opportunity to ascertain if the EDI technology in the JIT supply chains had an impact on major cost categories. By tracking the cost performance of the assembly centers over time, we sought answers to the following key questions relating to the business value of EDI at Chrysler:

- What is the locus of impact of increased EDI penetration in JIT supply chains?
- How does the impact vary across different cost categories and across plants?

372

12.4.1 Research Site

As part of its corporate goal to reduce vehicle costs by 30%, the procurement, manufacturing and logistics areas of Chrysler were asked to coordinate and improve the materials management process. EDI was harnessed to change materials flows and implement JIT practices. Each assembly center, as a part of the JIT program, had to track for each model year metrics such as inventories, obsolete inventories, transportation and premium freight. Likewise, the IT group over the years recorded the penetration of EDI each year as more suppliers came on board their JIT programs.

Having demonstrated the improved operational performance from reduced shipment discrepancies at the assembly centers (Section 1), we took the next step of probing if the measures tracked by both the IT and logistics groups could establish the business value of EDI in the Chrysler logistics network. Cross-sectional and time series data from the IT and logistics groups were thus combined to assess the impact of EDI. The summary statistics for the logistics network are summarized in Table 3. It underscores the magnitude and complexity of this vast network.

Table 3. Key Data

Time Period	1981-1990
Number of Assembly Plants	9
Total Transportation Costs	Over $300 million
Number of Supply Lines	7000
Number of Daily Truck Deliveries (June 1991)	1483
Obsolete Inventory Cost (1990)	$5.9 million
Gross Productive Inventory (1990)	$328.7 million

The dependent variables in this study were the major cost categories related to materials and logistics functions as defined below:

Dependent Variables:

INVT = Inventory Turnover
OBS = Annual material dollars written off at a plant

The independent variables, as before, are divided into two categories related to information exchange and material flow complexity.

Independent Variables:

A. *Information Exchange*
PROGRAM = 1, after launch of EDI program
= 0, otherwise
EDIP = Percentage of materials dollars under EDI program
B. *Material Complexity*
VOLUME = Annual vehicle production
PARTS = 1, if parts variety exceeds the average level

MINOR   = 0, otherwise
                    = 1, if 5-15% of the parts were new in the model year
                    = 0, otherwise

MAJOR   = 1, if more than 15% of the parts were new in the model year
                    = 0, otherwise

12.4.2 Model

Our conceptual model is shown in Figure 4. It captures the impact of both information exchange characteristics and material complexity variables as well as their interactions. For the sake of brevity, our discussion of the model is limited to the analysis of inventory and obsolescence costs. The combined cross-sectional and time series data on these two cost categories consisted of a panel data set. We expected serial correlation, and corrected for autocorrelation using a non-iterative procedure for each assembly plant. While the Breusch-Pagan (1979) test rejected the null hypothesis of homoskedastic error for INVT, the Goldfeld-Quandt (1972) rejected the same for OBS. Therefore we corrected for heteroskedasticity using White's procedure (White 1980). We also checked for multicollinearity (Belsley et al., 1980). Finally, we performed the Spencer and Berk (1981) test to determine if any dependent variable specification should contain the other dependent variable.

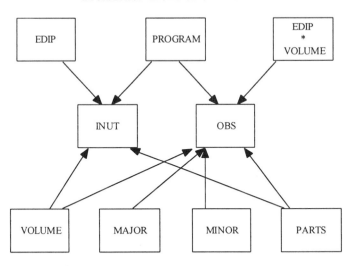

INFORMATION EXCHANGE CHARACTERISTICS

MATERIAL COMPLEXITY CHARACTERISTICS

Figure 4: Drivers of Inventory and Material Obsolescence

From a modeling standpoint, we recognized the contemporaneous correlation between the residuals across the plants due to common management policies and procedures. Thus, if contemporaneous correlation were present, the estimates were likely to be unbiased and consistent, but inefficient. To overcome this problem, Seemingly Unrelated Regressions (SUR) could be used. However, the gains form SUR estimates were likely to be small for three reasons: (1) the two equations contained similar explanatory variables, (2) the estimation was based on trended data, (3) the correlation coefficients among the explanatory variables, by year, were greater than 0.5. Moreover, in this case the correlation between the residuals was found to be small indicating little gains in efficiency if SUR methods were deployed. Nonetheless, we estimated both equations using SUR estimates as well as OLS, after correcting for autocorrelation and heteroskedasticity. The results for both dependent variables are displayed in Table 4.

Table 4. Estimation Results

Variable Name	Inventory Turnover (INV)	Obsolete Inventory (OBS)
Constant	42.81^{***}	253^{*}
PROGRAM	6.80^{**}	-168^{***}
EDIP	0.219^{***}	
VOLUME	$-.074^{**}$	5.633^{*}
EDIP*VOLUME		-0.038^{**}
PARTS	-9.94^{*}	94.7
MINOR		48.384^{**}
MAJOR		237^{***}
N	90	90
F-Stat	28.9	22.4
Adj-R^2	0.87	0.74

$^{*},^{**},^{***}$ indicate α = .1, .05, and .01 respectively.

12.4.3 Results and Discussion

Consider the impact of information exchange on both cost categories. The impact of PROGRAM was favorable on both inventory turnover (INVT) as well as obsolescence cost (OBS). As shown in Table 4, INVT increased by 6.8 with the adoption of the EDI program, and OBS decreased considerably. With increased penetration (EDIP), the inventory turnover increased showing that the assembly centers were able to make their operations leaner with higher level of electronic exchange. Similar positive impacts were seen with the reduction of OBS with higher level of electronic integration. From a material flow standpoint, higher level of PARTS reduced inventory turns and increased OBS. The impact of EDI on inventory turnover (INVT) was found to be dependent on the size of the plant as indicated by the negative coefficient of VOLUME in Table 4. The end of the year obsolescence cost increased with higher proportion of new parts in the model year as confirmed by the coefficients of MINOR and MAJOR.

The value created from the reduction in inventories and reduced write-offs at the end of the model year are displayed in Figures 5 and 6. The trajectory of inventory showed a marked improvement at the start of the program, and was dependent on the size of the plant. The savings were greater from plants with lower initial turns (larger production volume). In other words, plants with higher initial turns had less to gain from improved information exchange. The rate of savings, as expected diminished with increased EDI penetration.

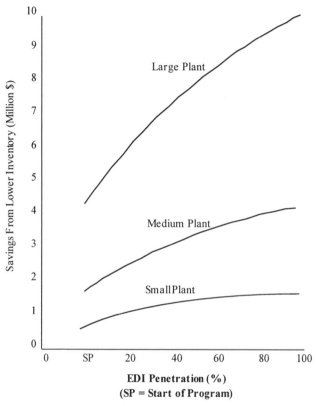

Plant size: Large = 300,000 Vehicles/Year
Medium = 200,000 Vehicles/Year
Small = 100,000 Vehicles/Year

Source: MIS Quarterly/June 1995

Figure 5: Savings From Lower Inventory

The cost reduction in material obsolescence was also felt as soon as the JIT program was adopted. The mangers attributed this gain to the inhouse data cleaning, system and operating policy changes that had to be undertaken before initiating the JIT program The accrued benefits were higher with more EDI penetration. On an average, the initiation of the EDI program reduced materials write-offs by 0.17

million dollars per plant. The extent of savings from EDI penetration increased with plant size as shown in the Figure.

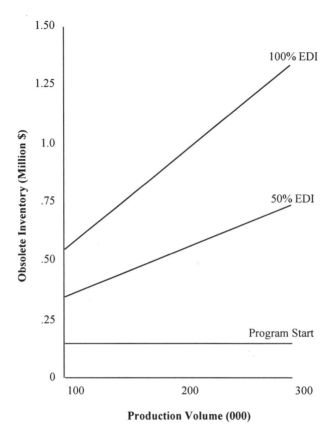

Figure 6: Obsolete Inventory Cost Savings

12.4.4 Commentary on the Modeling Framework

The analysis of the ten years of longitudinal data on cost trends and EDI penetration demonstrated vividly the financial impact at Chrysler. The four cost categories (inventory, obsolete materials, transportation and premium freight) related to the JIT program were significantly affected by the new mode of information exchange with the first tier suppliers (Mukhopadhyay et al., 1995). The total expected savings from these cost categories were about $60 per vehicle for an assembly center producing 200,000 vehicles per year and at 90% EDI penetration. Including the additional savings from EDI document preparation and transmission, the total benefits of EDI system wide were computed to be $220 million translating to $100 per vehicle.

From a managerial standpoint, the business value analysis underscores the gains from EDI as a dominant method of communication between buyers and suppliers in JIT supply chain. Second, the material complexity variables highlight the importance of managing logistics complexity. The three studies taken together illustrated the efficacy of the research paradigm discussed in Section 1 to tease out the interplay of logistics management and information flows. While this paradigm has a great deal of promise for future research in supply chain management, care must be taken in three major areas to ensure the validity and proper interpretation. These methodological issues are discussed next.

5. Methodological Issues

Our experience in the research projects on supply chain management and EDI has taught us some key lessons on methodological issues. We pool them into three categories: (1) Method Related, (2) Measurement Related, and (3) Data Related concerns.

Method Related: In all three cases, the conceptual framework evolved as a result of blending the IT perspective and the materials management perspective. Our research team, being interdisciplinary, spent considerable effort mapping the process from both an information flow stand point as well as materials flow perspective. It was during these mappings that the conceptual model was developed based on the theoretical underpinnings from both the functional areas. Also, most empirical studies are plagued with several confounding factors. To mitigate this problem, we had to investigate alternate specifications and employ multiple estimation procedures to ensure the robustness of the results and derive consistent management implications.

Measurement Related: The researcher is often faced with the choice of the appropriate level of aggregation to capture the true drivers. In the Chrysler business value study (Section 4), we measured the use of EDI penetration at an aggregate level in terms of materials dollars being procured under the EDI program. In contrast, the analysis of shipment discrepancy at Chrysler assembly centers (Section 2) required a finer measure of EDI usage.

Data Related: The researcher has to also take extreme care while analyzing both cross-sectional and time series data. The longitudinal data set in Chrysler business value study (Section 4) provided us with sufficiently long time window to track the impact of increased EDI usage over time. On the other hand, we had to control for other confounding factors such as changes in parts complexity, models, and program volume that could corrupt the performance measures being tracked. Obtaining quality data on these exogenous factors became a serious challenge. The cross sectional data sets (Kennametal order payment study in Section 3 and Chrysler study in Section 2) did not provide us longitudinal data over an extended period of time. We exploited the cross sectional variation to obtain insights into the efficacy of supply chain management policies.

Future Challenges: Supply chain management is critical is achieving cost control and reduction in a number of industries. Firms aggressively invest in information technologies (the explosive growth in enterprise integration is a case in point) but often have little or no direct validation of the impact on efficient material flows. The field studies reported here present a systematic analysis of the effectiveness of supply chain management. In spite of the data collection and estimation challenges, replication of our analysis across a number of firms is necessary to generalize the findings reported here.

References

Amemiya, T. 1985. *Advanced Econometrics*, Cambridge, MA: Harvard University Press.

Barua, A., C. H. Kriebel and T. Mukhopadhyay (March 1995). "Information Technologies and Business Value: An Analytic and Empirical Investigation." *Information Systems Research* 6(1), 1-24.

Belsley, D. A., E. Kuh and R. E. Welsch (1980). *Regression Diagnostics*, New York: Wiley.

Breusch, T. S. and A. R. Pagan (September 1979). "A Simple Test for Heteroskedasticity and Random Coefficient Variation." *Econometrica* 47(9), 1287-1294.

Brynjolfsson, E. (December 1993). "The Productivity Paradox of Information Technology." *Communications of the ACM* 36(12), 66-77.

Chamberlain, G. (1980). "Analysis of Co-variance with Qualitative Data." *Review of Economic Studies* 47, 225-238.

Goldfeld, S. M. and R. E. Quandt (1972). *Nonlinear Methods in Econometrics*, Amsterdam: North-Holland.

Gonul, F. and K. Srinivasan (Summer 1993). "Modeling Multiple Sources of Heterogeneity in Multinomial Logit Models: Methodological and Managerial Issues." *Market Sci.* 12, 213-229.

Greene, W. H. (1990). *Econometric Analysis*, New York: MacMillan.

Hauser, J. R. (May 1978). "Testing the Accuracy, Usefulness and Significance of Probabilistic Choice Models: An Information Theoretic Approach." *Oper. Res.* 26, 406-421.

Hausman, J. and D. Wise (1981). "Stratification on Endogenous Variables and Estimation: The Gary Income Maintenance Experiment." *Structural Analysis of Discrete Data with Econometric Applications*, Cambridge, MA: MIT Press.

HBS (1991). "Chrysler Corporation: JIT and EDI (A)." Harvard Business School Case Study, 9-191-146, MA: Cambridge, MA.

Heckman, J. J. and B. Singer (1984). "A Method for Minimizing the Impact of Distributional Assumptions in Econometric Data for Duration Data." *Econometrica* 52, 271-320.

Judge, George G., W. E. Griffiths, R. Carter Hill, N. Lütkepohl and T. C. Lee (1985). *Theory and Practice of Econometrics*, New York: John Wiley.

Kekre, S. and T. Mukhopadhyay (December 1992). "Impact of Electronic Data Interchange Technology on Quality Improvement and Inventory Reduction Programs: A Field Study." *International Journal of Production Economics* 28(3), 265-282.

Kmenta, Jan (1986). *Elements of Econometrics*, New York: MacMillan.

Mukhopadhyay, T, S. Kekre and S. Kalathur (June 1995). "Business Value of Information Technology: A Study of EDI." *MIS Quarterly*, 137-156.

Mukhopadhyay, T., S. Kekre and T. Pokorney (1998). "Strategic and Operational Benefits of Electronic Data Interchange." GSIA Working Paper, Carnegie Mellon University, Pittsburgh, PA.

Pindyck, R. S. and D. L. Rubinfeld (1981). *Econometric Models and Economic Forecasts*, New York: McGraw-Hill.

Schonberger, R. J. (1986). *World Class Manufacturing: The Lessons of Simplicity Applied*, New York: The Free Press.

Spencer, D. E. and K. N. Berk (July 1981). "A Limited Information Specification Test." *Econometrica* 49(7), 1079-1085.

Srinivasan, K., Sunder Kekre and Tridas Mukhopadhyay (1994). "Impact of Electronic Data Interchange Technology Support on JIT Shipments." *Management Science* 40(10), 1291-1304.

White, H. (April 1980). "A Heteroskedasticity-consistent Covariance Matrix Estimator and a Direct Test For Heteroskedasticity." *Econometrica* 48(4), 817-838.

13 BUSINESS CYCLES AND PRODUCTIVITY IN CAPITAL EQUIPMENT SUPPLY CHAINS

Edward G. Anderson Jr. and Charles H. Fine

Department of Management
University of Texas
Austin, Texas 78733

Sloan School of Management
Massachusetts Institute of Technology
Cambridge, Massachusetts 02139

13.1   Introduction[1]

Cyclicality is a commonly observed phenomenon in market economies. Less well understood, however, is the amplification of cyclicality as one progresses up the supply chain from original equipment manufacturer (OEM) to first-, second-, and third-tier suppliers. Recent studies have focused on management techniques to minimize inventory costs when faced with amplification in product distribution chains (Baganha and Cohen 1996; Lee, Padmanabhan, and Whang 1997; Sterman 1989a).[2] This paper instead examines long-term supplier productivity as influenced by amplification in capital goods supply chains.

This paper will focus on the machine tool industry because it exhibits three characteristics common to many capital suppliers: high leverage, extraordinarily volatile business cycles, and a highly trained workforce. It should be noted that, while this paper specifically examines the machine tool industry, the results of this study may be generalizable to suppliers from many other industries such as electronics and semiconductors.

The machine tool industry is a high-leverage industry because, although relatively small, it may exert significant influence on a large fraction of the economy. While U.S. annual sales in the machine tool industry have never exceeded $10 billion, business leaders and government officials from many nations focus on the machine tool industry as a source of competitive advantage for the nation's industrial infrastructure (Dertouzos, Lester, and Solow 1989). Summing up the reasons for this concern at a national level, Dertouzos and his co-authors offer the following explanation:

> The machine tool industry stands at the heart of the nation's manufacturing infrastructure, and it is far more important than its relatively small size might suggest. All industries depend on machine tools to cut and shape parts. The entire industrial economy suffers if a nation's machine tools are too slow, cannot hold tight tolerances, break down often, or cost too much. If American manufacturers must turn to foreign sources for machine tools (or for other basic processing systems...), they can hardly hope to be leaders in their industries, because overseas competitors will often get the latest [production] advances sooner.[3]

This is not only a macro-level concern. Many oft-cited examples of efficient firms such as Honda and Toshiba believe that producing some of their own machine tools within the firm is a source of competitive advantage (Fine and Whitney 1995).

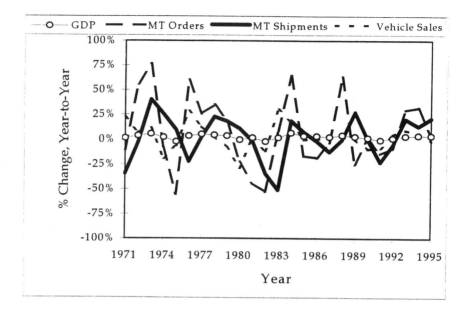

Figure 1: U.S. Machine Tool Order Volatility, Year-over-Year Percent Change

		Data Sources:
GDP change	σ = 2%	GDP: 1997 Federal Budget; Machine Tools: The
MT Orders change	σ = 37%	Economic Handbook of the Mach. Tool Industry;
MT Shipments change	σ = 22%	Vehicles: Bureau of the Census (SIC 3711).
Vehicle Sales change	σ = 13%	

Another characteristic of the industry is the extreme volatility of orders for machine tools. Figure 1 compares the volatility of changes in the U.S. gross domestic product, U.S. automotive sales, and U.S. machine tool orders and shipments. The automobile industry is often considered the prototype of a "cyclical" industry. However, the standard deviation of year-to-year changes in machine tool shipments is roughly *twice* that of automotive sales and *ten* times that of GDP. Moreover, machine tool orders are over half again more volatile than shipments. Adjusting production to accommodate this volatile order stream has been quite difficult for the machine-tool industry. Up until the mid-1970's, the industry used large backlogs to smooth production. With the appearance of foreign competitors promising rapid delivery, American machine tool manufacturers lost this option. Since that time, there has been a shake-out in the industry with the disappearance of over 40% of the 500 active U.S. firms through acquisition or dissolution (March 1988).

Finally, the machine tool industry is characterized by a highly trained workforce. In general, the production employees are highly skilled, requiring years of training to obtain any degree of proficiency in their profession. Firms which lay off trained production employees during an economic downturn find it very difficult to re-hire them during the next upturn. This scarcity is due to a chronic shortage of skilled trades in the United States coupled with their usefulness to less volatile industries, ironically including many machine tool customers (Parker 1996). The long lead times associated with training employees seriously complicates the task of accommodating the industry's order volatility.

To inform our analysis, this paper offers a simple model of the machine tool industry that uses dynamic programming and control theory concepts to model the exceptional volatility seen in the industry and develop a strategy which suppliers might pursue to minimize employee costs. In developing an understanding of these phenomena, this paper has built upon the work of Fine (1998); the analysis in Anderson (1997a); and the simulation work of Anderson, Fine, and Parker (1996), and Kallenberg (1994); and finally Sterman's empirical analysis of managers' stock-replenishment behavior during the MIT "Beer Game." We have also benefited from the work of Lee, Padmanabhan, and Whang (1997) who developed necessary conditions for volatility amplification in supply chains. We are also indebted to the work of Graves (1997) on simple models of supply chain volatility characterized by non-stationary demand as well as Baganha and Cohen's (1996) paper on using inventories to dampen volatility amplification in supply chains. A final, vital influence is the ground-breaking work on supply chains by Forrester (1958).

The remainder of the paper is organized as follows. Section 13.2 illustrates the mechanism behind volatility amplification in capital supply chains and places it in the context of the so-called "Bullwhip Effect" and "Beer Game" literature. Section 13.3 models a simple capital supply chain with one customer and one supplier. Section 4 develops a minimum mean-square-error predictor based on this model. Section 5 obtains an approximately optimal policy for managing supplier employee levels based on the predictor's forecast of incoming orders. Section 6 develops analytic measures for machine order volatility, employee productivity, and workforce shortfalls under conditions of varying volatility, machine lifetimes, customer order policies, and employee training times. Section 7 offers a numerical example to illustrate the efficacy of this approach. Finally, Section 8 recapitulates the paper's results and proposes future lines of research inquiry.

13.2 Mechanism of Capital Supply Chain Volatility

Lee et al. (1997) developed a theoretical model to show that volatility amplification—often referred to as the beer-game or bullwhip effect—will be absent in distribution chains only if all of the following conditions hold: (1) demand is stationary with its distribution known to all members, (2) all orders are delivered on time in requested quantity, (3) inventory levels are monitored every period and replenishing orders are issued immediately, and (4) the price at each node remains the same across all periods. Needless to say, these conditions rarely hold as a group. They also considered four sources of volatility amplification resulting from an absence of each condition, viz. demand signal processing, rationing games, order batching, and price variations. Sterman (1989b) treats the first two sources in depth. This paper will be even more restrictive, focusing on demand signal processing by separating it into two components: information delay and target stock readjustment. (Economists often refer to target stock readjustment of capital goods as the investment accelerator.) Unlike the distribution problem, we assert that these two factors alone explain much of the volatility in the machine tool industry.

In fact, the core difference between the two chains lies in the parameters characterizing the stock readjustment mechanism. The residence time of an item in finished goods inventory is rather short, typically on the order of sixty days or less. In contrast, the residence time of a capital stock item is its lifetime, typically on the order of ten *years*. This makes the average ratio of desired stock to order rate in the two chains differ by at least an order of magnitude. Hence, any change in desired stock resulting from a change in product demand will have a much greater effect upon orders in a capital goods chain than in a distribution chain.

To help visualize this difference, consider a stylized two-period numerical example with representative parameters. The example will describe a firm's components supply chain—similar in mechanics to a distribution chain—and its capital goods supply chain. Assume that a firm which has traditionally seen a demand of 100 products per quarter receives a 5% increase in demand in Quarter 2. Assume that the firm instantaneously adjusts all its inventories and ordering rates to the new demand level. In this case, it will need to adjust its component replacement rate from 100 to 105 components per quarter to balance the new demand. If the firm holds two quarters of component inventory, it will also need to order enough components to adjust the inventory upwards from 200 to 210 components. Thus, the firm will need to order a total of 115 components in Quarter 2, a 15% increase in orders from Quarter 1. Hence, the initial change in product demand amplifies the change in component orders by a factor of 3.

Now consider the firm's stock of machine tools. Assume the firm has a stock of 100 machine tools (each machine tool can produce 1 product per quarter). Thus, the number of machine tools in the machine stock is the same as the number of components in component inventory. However, the machine tools have an average lifetime of 50 quarters (12.5 years). Traditionally, the firm orders 2 machine tools per quarter, but in Quarter 2 it will need to adjust the stock from 100 to 105 machines. Hence, it will need to order $2 + 5 = 7$ machines, an increase of 250% from Quarter 1. The original product demand change of 5% has been amplified 50 times. This amplification is *over 16 times greater* than that in the components chain. A moment's reflection will reveal that the volatility amplification in this simple model is roughly proportionate to the lifetime of the stock of inventory under consideration. While this illustration is much less complex than the real world, it does show how the long lifetime of capital stock creates the dramatic volatility seen in capital supply chains.

13.3   Methodology and Model Description

13.3.1 Base Model Description

In order to examine the volatility present in the machine tool industry, this paper builds a simple analytic model of a typical machine tool order supply chain. The model is based on the simulation work of Anderson, Fine, and Parker (1996), which was developed from interviews with machine tool buyers and suppliers. That simulation work is simplified to be amenable to dynamic programming and control theory analyses, which are particularly suitable to modelling volatility issues resulting from delays and feedback loops (Ogata 1970). This simplified model will not seek to capture the broad and rich dynamics inherent in the machine tool industry but rather demonstrate the essential dynamics relevant to capacity management. As such, some explanatory accuracy will be traded for ease of explication. In particular, we will use continuous representations for both time and the involved stocks. While discrete formulations might be more realistic, the benefits would be slight in comparison with the overhead involved. For example, assuming that the firms involved make monthly ordering and hiring decisions, the decision horizons described in the model below would contain in excess of 30 ordering periods. Each period would represent only 3% of the total horizon. Hence, a continuous representation of time should not be too distorting. Similarly with the employee stock, the typical large machine tool supplier has at least 500 employees. Each discrete employee will only represent 0.2% of the total number. Finally, due to the vast proliferation of machine tool types (it is not a great exaggeration to say that each machine tool ordered is unique), most records of industry and firm operational performance are kept in dollars rather than units (Association for Manufacturing Technology 1970-94). As the average machine tool costs thousands of dollars, a continuous representation seems most appropirate here also.

The industry structure modelled is of two simple interacting firms: a machine tool customer and a machine tool supplier. Figure 2 presents an overview of the model. The demand for product $g(t)$ is modelled in (1) as a Wiener process (or Brownian motion), which is the continuous-time version of the random walk. The non-stationary assumption for demand requires some explanation, as many operations management models assume a stationary mean. While a stationary approximation is appropriate for the short-term models typical of many operations problems, a great deal of econometric evidence exists that even de-trended aggregate demand over the long term is non-stationary in the mean (Hamilton 1989; Nelson and Plosser 1982). And, most assuredly for individual firms, long-term market share is non-stationary. Hence, a non-stationary demand assumption seems reasonable for purposes of long-term staffing plans that characterize this model. The Wiener process assumption is the simplest of non-stationary assumptions, requiring only that the change in demand $dg(t)/dt$ be normally distributed and completely uncorrelated with its past and future values.

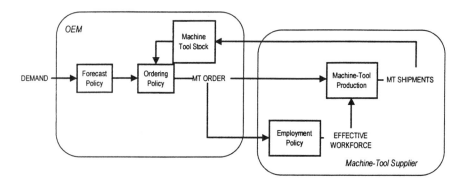

Figure 2: Overview of MT Model

All the random disturbances, such as *v(t)*, in the model are assumed to be normally distributed with zero mean and constant variance. (For notational convenience, Roman letters will represent time-dependent variables, Greek letters will represent time-invariant parameters.) Hence, changes in product demand *g(t)* resulting from changes in technology are not explicitly represented in the model.

$$dg(t)/dt = v(t) \sim N(0, \sigma_v^2) \qquad (1)$$

$$g(0) = g_0 \geq 0 \qquad (2)$$

We will refer to the machine-tool customer as the OEM. The OEM4 must make some estimate of future product demand in order to plan its capacity requirements. In the model it does this by exponentially smoothing demand over the recent past. (The circumflex operator indicates an estimate.) We choose this particular estimation method because of its simplicity and utility. Makridakis et al. (1993) showed that exponential smoothing of past realizations is an excellent forecast methodology of volatile time series. Of course, the most accurate forecast of a random walk is the most recent period's realization. However, gathering the appropriate information together is typically a time-consuming task. Thus, most companies update their forecasts on a less than instantaneous basis. In the model, the estimate, $\hat{g}(t)$ represents the OEM's best guess of what future demand will be. This estimate is updated by the actual demand *g(t)* at a fractional rate per year of ϕ.

$$d\hat{g}/dt = \phi(g - \hat{g}) \tag{3}$$

$$\hat{g} \geq 0 \tag{4}$$

At this point, the OEM's forecast is converted into the number of machines needed to meet the forecast demand by dividing the estimated future product demand by the factor productivity of machine tools κ,

$$m^* = \kappa^{-1}\hat{g} \tag{5}$$

This number $m^*(t)$ is compared with the actual machines in stock m. Any gap between the two is corrected at a constant fractional rate γ. Simultaneously, the machine tool stock is deteriorating at a constant fractional rate of ε. The gap correction is summed together with deterioration replacement to produce an order stream for new machine tools $r(t)$.

$$r = \max(0, \gamma(m^* - m) + \varepsilon m) \tag{6}$$

Note, that under typical real-world operating parameters such as those discussed later, $r(t)$ will rarely if ever reach zero.

The backlog of orders at the machine tool supplier accumulates the difference between OEM machine tool orders $r(t)$ and machine tool shipments $p(t)$.

$$db(t)/dt = r - p\left(\frac{b}{w}\right) \tag{7}$$

The function p represents the capacity constraint of the machine tool supplier. Typically, employees in the machine tool industry enter the industry capable of little or no productivity and gain a reasonably high level of proficiency only after five years.[5] Due to these long lead times, matching capital to employees in the machine tool firm is fairly simple, and so we neglect capital considerations in (7). Instead $p(t)$, the rate of machine tool shipments, is typically an increasing, concave function of the workforce utilization, that is the machine order backlog $b(t)$ divided by the effective workforce $w(t)$.

The effective workforce $w(t)$ is equal to the number of experienced employees plus a discounted number of new employees. The discounting reflects the fact that the new employees are less productive than their experienced counterparts.

$$w = e + \delta n \tag{8}$$

The target machine tool supplier workforce is set equal to a constant fraction of the order rate, where ζ is the factor productivity of labor.

$$w^* = \zeta^{-1}r \qquad (9)$$

The level of experienced employees $e(t)$ can change in one of two ways. (Retirements will be ignored as their effect on the results of this paper would be relatively minor in relation to the additional complexity entailed in modelling them.) Experienced employees can increase as new workers $n(t)$ graduate from their training after being employed for τ years. During downturns, however, the machine tool supplier may need to lay off employees. New employees are laid off first. If all the new employees have been laid off, that is $n(t) = 0$, further layoffs will come from the experienced employees. Equation 11 represents this restriction where $l(t)$ is the rate of experienced employee layoffs. The variables n and l are also restricted to be nonnegative.

$$de(t)/dt = \frac{n}{\tau} - l \qquad (10)$$

$$n(t)l(t) = 0 \qquad (11)$$

$$n, l \geq 0 \qquad (12)$$

Finally, the machine tool stock at the OEM increases with shipments from the machine tool supplier and decreases with deterioration.

$$dm(t)/dt = p(b, w) - \varepsilon m \qquad (13)$$

13.3.2 Forecast Model

The nonlinearity inherent in $p(t)$ embedded in the feedback loops of our model creates two analytical problems. One is its concavity. This could, in principle, be dealt with by linearization. However, $p(t)$ is also a function of workforce utilitzation, which is the ratio of two time-dependent variables. This seriously complicates finding an exact closed-form solution for volatility amplification using an optimal employee scheduling policy. Instead, we will adopt a hierarchical approach to simplify the issue in which we ignore the effect of the staffing constraint upon the machine stock. Such approaches have had good success in managing other manufacturing planning and control problems (Bitran 1993).

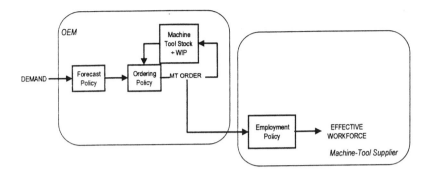

Figure 3: Hierarchical View of MT Model

Figure 3 shows a hierarchical model in which we determine machine tool orders without regard to the supplier's capacity constraints. This has intuitive appeal because the effect of staffing decisions today should affect future machine tool order rates only after several years. Under this approach we first perform an open-loop forecast of future machine tool orders and then use this forecast in a dynamic program to determine an appropriate employment policy. We then use the policy to adjust the employee workforce as required for the current period. Then in the next period, we will update the forecast using new information and repeat the process. This is a common approach to dynamic programs with uncertainty (Bertsekas 1987).

To further simplify order forecasting, we now allow $m(t)$ to represent both the machine tool stock in the OEM and the backlog of new machine tool orders at the supplier. The time constant ε now represents the average length of time from the order of a particular machine tool to its retirement. The parameter κ is also adjusted to reflect the augmentation of the stock m for purposes of determining orders. Finally, the random disturbance to the order rate and the nonnegativity constraint on g is suppressed for analytical purposes. Our new model then becomes:

$$d\hat{g}(t) / dt = \phi(g - \hat{g}) \tag{14}$$

$$dm(t) / dt = r - \varepsilon m \tag{15}$$

$$r = \gamma(\kappa^{-1}\hat{g} - m) + \varepsilon m \tag{16}$$

Is this open-loop model sufficiently accurate for our purposes? Historically, the delivery delay in the industry varies between 5 and 18 months (Association for Manufacturing Technology 1970-94). This is much shorter than the typical machine tool lifetime of ten years or longer. Hence, it seems reasonable to approximate ε by adding together the machine tool lifetime and the mean delivery delay. This will allow us to create a two-state, linear, and time-invariant approximation of the ordering process. Fitting[6] this model using the U.S. Durables Industrial Production Index[7] (U.S. Federal Reserve Board 1996) from 1960 to 1993 for product demand and the U.S. Machine Tool Order Rate (Association for Manufacturing Technology 1970-94) for machine tool orders yields the parameter estimates shown in Table 1. The estimated parameters appear reasonable given our field work. A comparison of resulting simulated monthly machine tool order rate time-series with its historical counterpart is presented in Figure 4.

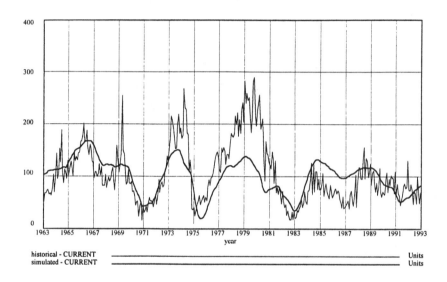

Figure 4: Simulated and Historical Monthly U.S. Machine Tool Orders (1963-1993)

Note that the historical data has been geometrically detrended and normalized so that the mean order rate is 100.

Table 1: Parameters for Historical Fit

Machine Replacement Rate	ε	0.121	years^{-1}
Factor Productivity of MT's	κ	0.121	(prod. orders/year)/machine
Forecast Rate	ϕ	1.33	years^{-1}
Correction Rate	γ	1.40	years^{-1}

The typical goodness-of-fit statistic for forecasting Mean Absolute Percent Error (MAPE) measures point-to-point divergence between simulated and actual series. In this case, the MAPE at 37% is reasonably good given the high coefficient of variation of the historical series. In particular, the prediction of turning points throughout the simulation is quite good. The amplitude prediction, however, is somewhat low during months 1973 through 1982. However, the econometric analysis in Anderson (1997a) suggests that very little of this error is attributable to either the lack of a supplier capacity constrain or to disaggregated OEM and WIP machine tool stocks. Rather, the bulk of this unexplained variation seemed to be explained by changes in real interest rates, the effects of which are outside the scope of this paper.

For the remaining sections of the paper, we will need the transfer function from $dg(t)/dt$ to $dr(t)/dt$. Using the Laplacian operator s for differentiation and rearranging Equations 14, 15, and 16 yields the following:

$$\frac{sR(s)}{sG(s)} = \frac{\kappa^{-1}\gamma\phi(s+\varepsilon)}{(s+\gamma)(s+\phi)} \tag{17}$$

We capitalize r and g and use s as their arguments to indicate that the functions being considered are the Laplacian transforms of $r(t)$ and $g(t)$.

13.4 Order Predictor

The model assumes that only the order rate is known to the machine tool supplier, as is typically the case in most real capital supply chains. From the order rate, the supplier can use the simplified model just described to construct state estimates and forecast the future stream of machine tool orders. For the current purposes, we use a Wiener predictor of the model, which minimizes the mean squared error of forecasts.

Proposition 1

For the forecast model described above, the transfer function of the minimum mean square error predictor $\hat{r}(t,T)$ is:

$$\hat{R}_{t,T}(s,T) = \left(C_0 + C_1 e^{-\phi T} + C_2 e^{-\gamma T}\right)R(s) \tag{18}$$

where,

$$C_0 = \left\{\varepsilon(\phi-\gamma)s^2 + \varepsilon(\phi^2-\gamma^2)s + \varepsilon\phi\gamma(\phi-\gamma)\right\}/D$$
$$C_1 = \left\{\gamma(\varepsilon-\phi)s^2 + \gamma^2(\varepsilon-\phi)s\right\}/D$$
$$C_2 = \left\{\phi(\gamma-\varepsilon)s^2 + \phi^2(\gamma-\varepsilon)s\right\}/D$$
$$D = \phi\gamma(\phi-\gamma)(s+\varepsilon)$$

An outline of the proof is included in the Appendix. ∎[8]

The transfer function from Proposition 1 is the sum of two exponentially decaying terms and a constant. Essentially, the forecast begins at the present time being equal to the current order rate $r(t)$, but experiences some transients that with typical real-world parameters die out in less than four years. At that point, the forecast reaches a steady state, which we will term the long-term forecast $\hat{r}_{LRF}(t,T) = C_0 r(t)$.

Intuition concerning Proposition 1's result is difficult. Fortunately, however, the long-term forecast is much easier to interpret and is the primary driver of the control law derived later in this paper. If we let the OEM forecast update and error correction rates be equal ($\phi = \gamma$), and look only at the long term forecast, we obtain:

$$\hat{r}_{LRF}(t,T) \approx r(t) + \frac{\varepsilon}{\phi^2}\left(\frac{dr(t)}{dt}\right) \qquad (19)$$

Thus, the long-term forecast is roughly determined by the current machine tool order stream and the order stream's rate of change. However, the trend is discounted to the extent that the machine lifetime ε^{-1} is long or the reaction rate ϕ of the OEM to product demand changes is rapid. This discounting comes about because large values of either parameter tend to increase the volatility amplification of the system, making it more "nervous". Equation 19 discounts recent trends to account for this volatility. Hence, a supplier using this forecast policy will be less likely to mistake system "nervousness" for evidence of a real shift in product demand.

13.5   Control Law

To obtain a reasonable employment policy, we look forward from the present time t_0 using the Wiener predictor's forecast of future orders seeking to balance capacity shortfalls against excess employee costs. Let $\tilde{t} = t - t_0$ represent time units from the present. We form the following dynamic program.

$$\min_{u(\tilde{t})} \int_0^\infty \left\{\alpha^2\left[w*(\tilde{t}) - e(\tilde{t})\right]^2 + \beta^2\left[\tau u(\tilde{t})\right]^2\right\}d\tilde{t} \qquad (20)$$

such that

$$de(\tilde{t})/d\tilde{t} = u(\tilde{t})$$
$$w*(\tilde{t}) = \zeta^{-1}\hat{r}(t_0,\tilde{t})$$
$$u(\tilde{t}) = \frac{n(\tilde{t})}{\tau} - l(\tilde{t})$$
$$n(\tilde{t})l(\tilde{t}) = 0$$
$$e(\tilde{t}), n(\tilde{t}), l(\tilde{t}), \alpha, \beta, \lambda, \tau \geq 0$$

The objective function places a quadratic weight on the difference between target workers and the effective employee workforce. For purposes of presentation, new employees are assumed to have zero productivity. (Conceptually, this presents little problem because employees usually begin with negative productivity when one nets in their required mentoring time from experienced employees. However, new employees become gradually more productive, becoming net productivity contributors, until eventually they themselves are fully trained and become mentors themselves. This allows us to adjust the time required to train employees so that the average new employee has zero productivity.) For the same reasons, costs are not discounted with respect to time. (Given real world parameters, the transients in the forecast will die out within five years. Thus, discounting using typical costs of capital would not greatly change the results. Also, the "costs" in this model include items such as quality or lost market share that the supplier may be unwilling to discount in the foreseeable future.)

The objective function also places a quadratic weight on the rate of change in experienced employees. When the change is positive, $\tau u(\cdot)$ represents the number of new employees required in the training pipeline to produce the desired increase in the number of experienced employees. When the change is negative, it still seems reasonable that the cost of laying off experienced employees will go up with the employees' skill level, proxied here by training time. The quadratic weights reflect the escalating costs associated with these variables. For instance, a small shortfall of employees can be compensated for with overtime. However, employees can only work so many hours per week, so a large employee shortfall will force the supplier to either lengthen delivery delays or turn down possible business. Similarly, training a large fraction of new employees is much more difficult than training a small fraction, because the training has to be performed by experienced employees taken away from production.

The cost function integrates these two costs from the present to infinitely far in the future. To ease computation, we will assume, without loss of generality, that alpha is equal to unity. The second constraint represents that the number of target workers at time $\tilde{t}$ in the future is proportional to the forecast order rate. The third constraint interprets $u(\cdot)$ as either training new employees or laying off experienced ones. The fourth constraint prevents firing experienced employees while training new ones. The final constraint prohibits negative costs, layoffs, or employees.

Proposition 2

If $w*(\tilde{t})$ is of the form $W_0 + W_1 e^{-\phi \tilde{t}} + W_2 e^{-\gamma \tilde{t}}$ in the dynamic program in (20), then the optimal rate of change of experienced employees, $u(t) = u(\tilde{t} = 0)$, is:

$$u(t) = \frac{W_0}{\beta\tau} + \frac{W_1}{\beta\tau(\gamma\beta\tau + 1)} + \frac{W_2}{\beta\tau(\phi\beta\tau + 1)} - \frac{e(t)}{\beta\tau} \tag{21}$$

The proof is in the Appendix. ∎

We will also need the transfer function for the complete control policy, including the Wiener predictor, in order to analyze the volatility of the system.

Corollary 1

The transfer function of the complete control policy is:

$$U(s) = R(s)\zeta^{-1}s\frac{\beta\tau(\varepsilon\beta\tau+1)s^2+(\varepsilon\beta\tau+1)(\phi\beta\tau+\gamma\beta\tau+1)s+\varepsilon(\phi\beta\tau+1)(\gamma\beta\tau+1)}{(s+\varepsilon)(s\beta\tau+1)(\phi\beta\tau+1)(\gamma\beta\tau+1)} \quad (22)$$

The Proof is included in the Appendix. ∎

Unfortunately, Equation 22 is intuitively difficult, but if we again let $\phi = \gamma$ and combine the results of Propositions 1 and 2, we obtain a more intuitive form in the time domain:

$$u(t) = \frac{1}{\beta\tau}\left\{\left[\dot{w}_{LRF}(t,T)-e(t)\right]+\frac{\dot{w}^*(t)-\dot{w}_{LRF}(t,T)}{\phi\beta\tau+1}\right\} \quad (23)$$

where $\dot{w}_{LRF}(t,T)$ is the size of the workforce required to produce the long-term forecast of machine tool orders $\hat{r}_{LRF}(t,T)$. Given real world parameters, the bulk of the behavior of $u(t)$ is determined by the left-hand term in Equation 23. This term drives the current number of experienced employees $e(t)$ towards that required to produce the long-term forecast of machine tool orders. The rate of change is slowed, however, to the extent that the cost of new employees is high or the time required to train them is long. On the other hand, the right-hand term will adjust the first term's correction somewhat towards the current target workforce if the base rate at which the OEM can change its machine tool order rate, ϕ, is much more sluggish than the rate at which the supplier can change its employee stock (governed by $[\beta\tau]^{-1}$). Given real-world parameters, however, this is typically a second-order effect.

13.6 The Effects of Volatility Amplification

13.6.1 Effect on Capital Equipment Orders

In order to evaluate the effects of the employment policy, we will need an expression for the volatility amplification along the supply chain from changes in product demand to changes machine tool orders. In general, the time-averaged variance along a given random process's sample path is a complicated expression dependent on the length of the averaging interval. However, because the covariance function for machine orders is square-summable, the time-averaged variance for a given sample path will be covariance-ergodic (Papoulis 1991). This allows us to use the easy-to-derive ensemble average—that is the variance of machine orders over all possible sample paths—as an approximation for the variance of machine orders along any given path. Hence, to study volatility amplification along a supply chain, we need only study the variance of each variable of interest.

Recall from Equation 17 the transfer function from product demand to capital equipment orders:

$$\frac{sR(s)}{sG(s)} = \frac{\kappa^{-1}\gamma\phi(s+\varepsilon)}{(s+\gamma)(s+\phi)} \qquad (17)$$

We could measure the variance amplification directly from this function. However, as a change of 1 order per month is more important to a firm with a typical monthly demand of 100 orders than one with 1000 orders, it would be fruitful to express the volatility amplification as a ratio of normalized expressions for the volatility of product demand and machine orders. If we normalize (17) by the expected ratio of product demand to machine orders, we will obtain $L(s)$, which is—roughly speaking—a ratio of the fractional rate of change of machine orders to that of product demand.9 The result in Equation 24[10] allows us to prove Proposition 3:

$$L(s) = \frac{\gamma\phi(s+\varepsilon)}{\varepsilon(s+\gamma)(s+\phi)} \qquad (24)$$

Proposition 3

The normalized ratio of changes to machine tool orders to changes in product demand, that is the volatility amplification of orders, can be approximated by:

$$L_{tot} = \left[\frac{\phi\gamma\left(\varepsilon^2 + \phi\gamma\right)}{2\varepsilon^2\left(\phi + \gamma\right)}\right]^{1/2} \qquad (25)$$

The proof is included in the Appendix. ∎

Note from the above that order amplification, as expected from our example earlier, will increase in machine lifetime ε^1. It will also increase in the OEM forecast demand update rate, and, OEM machine-stock correction rate. Thus, the more quickly the OEM reacts to changes in market demand, the more volatile the resulting machine tool order stream will be.

13.6.2 Effect on Employment

It is most important not only to understand the effect of volatility amplification along the supply chain from product demand to capital equipment orders, but also the volatility amplification from product demand to the capital equipment supplier's workforce. Using arguments similar to those in the previous section obtains the following result.

Proposition 4

Under the near-optimal policy described in Section 5, the amplification of the fractional change in orders from product demand g(t) to that of experienced employees e(t), which will be termed M_{tot}, can be approximated by:

$$M_{tot}^2 = \phi\gamma \frac{\left[\begin{array}{l}(\phi\gamma\varepsilon)^2(\phi+\gamma)\beta^5\tau^5 + 3\phi\gamma\varepsilon^2(\phi+\gamma)^2\beta^4\tau^4 + \\ \varepsilon(\phi+\gamma)\big[\varepsilon(\phi^2+7\phi\gamma+\gamma^2)+2\phi\gamma(\phi+\gamma)\big]\beta^3\tau^3 \\ +\big[\varepsilon^2(3\phi^2+7\phi\gamma+3\gamma^2)+4\varepsilon\phi\gamma(\phi+\gamma)+\phi\gamma(\phi^2+\gamma^2)+3(\phi\gamma)^2\big]\beta^2\tau^2 \\ +3(\phi+\gamma)(\varepsilon+\phi\gamma)\beta\tau+(\varepsilon^2+\phi\gamma)\end{array}\right]}{2\varepsilon^2(\phi+\gamma)(\phi\beta\tau+1)^3(\gamma\beta\tau+1)^3} \quad (26)$$

An outline of the Proof is included in the Appendix. ∎

This leads to the following important corollary which will be used later to examine employee productivity.

Corollary 2

The amplification of volatility from changes in product demand to changes in the machine supplier workforce increases with machine lifetime. Furthermore, under typical conditions (the fractional rates of forecast updating ϕ and stock correction γ are twice that of machine deterioration ε,), volatility amplification will increase with training time and training/firing costs. That is, V_{tot} decreases in ε, and, if ϕ, $\gamma \geq 2\varepsilon$, then V_{tot} increases in τ and β.
The proof is shown in the Appendix. ∎

It is also important to explore the expected effects of increasing volatility transfer on new employee levels and overall productivity in the machine tool firm. The non-stationarity of g(t) and consequently e(t) presents a conceptual difficulty because, in a strict sense, these processes have no mean. However, for purposes of convenience, we will assume that we are looking at a finite time interval, in which case we can use the initial value of employees in the interval to stand in for the expected value.

Corollary 3

The approximate number of new employees will be:

$$\mu_n = \left(\frac{\sigma_v^2 \tau^2 \varepsilon^2 \kappa^{-2} \zeta^{-2}}{2\pi} \right)^{\frac{1}{2}} M_{tot} \tag{27}$$

The proof is included in the Appendix. ∎

13.6.3 Effect on Employee Workforce Shortfalls

Another measure of how well the control law matches the supplier workforce to its target levels is the average workforce shortfall. Let $d(t)$ represent the shortfall from the supplier's target workforce.

13.6.3.1

Proposition 5

The approximate value for the average shortfall is:

$$\mu_d = \kappa^{-1} \zeta^{-1} \phi \gamma \tau^2 \beta^2 \left[\frac{\begin{pmatrix} \phi \gamma \varepsilon^2 \tau^3 \beta^3 (\phi + \gamma) + \phi \gamma \tau^2 \beta^2 (\phi \gamma + 2\phi \varepsilon + 2\gamma \varepsilon + \varepsilon^2) + \\ 3\phi \gamma \tau \beta (\phi + \gamma) + 3\phi \gamma - 2\varepsilon(\phi + \gamma) + \phi^2 + \gamma^2 + \varepsilon^2 \end{pmatrix}}{4\pi(\phi + \gamma)(\phi \beta \tau + 1)^3 (\gamma \beta \tau + 1)^3} \right]^{\frac{1}{2}} \tag{28}$$

The proof is included in the Appendix. ∎

13.6.4 Effect on Productivity

From the propositions above, we can begin to develop some intuition for the effect of parameter changes upon employee productivity. First, we need to develop a definition for productivity. The non-stationarity of $g(t)$ and consequently $e(t)$ presents a conceptual difficulty because, in a strict sense, these processes have no mean. However, for purposes of convenience, we will assume that we are looking at a finite time interval, in which case we can use the initial value of employees in the interval to stand in for the expected value.

Accordingly, let $\bar{n}$ represent the expected mean number of new employees over a given time interval from t_0 to t_f, and $\bar{e}$ represent the expected mean number of experienced employees. Assume further, that $e(t)$ and $g(t)$ remains positive over the interval, and that the initial value of product demand g_0 is known. Let

$R(\beta, \varepsilon, \phi, \gamma, \tau)$ be the new employee ratio defined by $R = \bar{n} / (\bar{e} + \bar{n})$ relating the number of expected new employees to the number of expected total employees, i.e. $\bar{e} + \bar{n}$. Let $P(\beta, \varepsilon, \phi, \gamma, \tau)$ represent the expected productivity as defined by $P = \bar{e} / (\bar{e} + \bar{n})$ for the same time period. (Note, however, that any positive values for g_0 will enable the proof of the proposition below.)

Proposition 6

The fraction of new employees in the supplier workforce will increase with the amplification of volatility from changes in product demand to changes in experienced employees, and the productivity of the workforce will decrease. (These statements will not, however, necessarily hold for changes in volatility induced by training time.) That is, let ψ stand for any parameter (other than τ) in R and P. If the amplification of volatility from v(t) to u(t), $V_{tot}(\psi)$ increases (decreases) in ψ, then so too will R(ψ) increase (decrease) in ψ. Furthermore, under the same conditions P(ψ) will decrease (increase).

An outline of the proof is contained within the Appendix. ■

Combined with the Corollary 2, Proposition 6 suggests three lessons. First, the productivity of a supplier using the approximately optimal policy drops *ceteris paribus* if the lifetime of machine tools increases. Thus, because capital equipment produced for higher-tech industries such as the electronics industry has a shorter lifetime than machine tools, one might expect that they are more immune to the vitiating effects of supply chain volatility amplification on productivity. Secondly, the productivity of a firm improves as the cost of training or firing experienced employees increases. This seems to indicate that a supplier should avoid layoffs of experienced employees that are expensive to fire or replace. On the other hand, *such a workforce should also not be expanded too rapidly during upturns*. Finally, under the assumptions in Proposition 3, increases in the rate at which the OEM updates its forecasts, γ, or corrects its machine stock, ϕ, will diminish its capital supplier's productivity. This final result is particularly interesting, because it implies that under many real world conditions, the OEM may be "shooting itself in the foot" if it changes its capital order stream too rapidly.

Unfortunately, equivalent results for Proposition 6 cannot be made for training times. For low training times, the extra τ in (A20) implies that the increase in the size of new employee stock for each month's additional training time may dominate the employment policy's increasing tendency to minimize experienced employee volatility. However, we will demonstrate in the next section that, in many real world cases, the effect of training time can be quite important.

With the above results, we can now proceed to developing some numerical intuition for the difficulty in managing a capital equipment supplier.

13.7 Numerical Example

To illustrate the impact of these policies on a typical machine tool firm, we offer a numerical example. The change in product demand is assumed to be a white-noise input with product demand set initially at 100,000 products per year. We assume that the machine tool firm begins with 1000 employees, each employee having a productivity of 100 machine tools per year. Further, each production employee costs, with full fringe benefits, $100,000 per year. We also assume that the machine tool firm's production employee costs represent half its revenue in the base case. Finally, interviews with Cincinnati Milacron, among others, indicate that machine tool firms typically maintain more employees than required to meet expected demand (Parker 1996). Cincinnati Milacron estimates that they maintain about 20% more employees than they would need in a less volatile environment. While costly, machine tool firms feel they need these "excess" employees as buffer capacity to avoid the penalties associated with insufficient capacity during unexpected upturns. We will capture this by making the annual workforce shortfall cost per employee be twice the cost of an overage. In other words, each unit of employee shortfall will cost $200,000 per year. The remaining stylized parameters are presented in Table 2 and are derived either from the estimation in Section 3 or from discussions with machine tool suppliers. Note that the parameters marked by an asterisk are arbitrary and may be changed without loss of generality.

Table 2: Stylized Representative Parameters for Numerical Analysis

Machine Replacement Rate	ε	0.125	years^{-1}
OEM Forecast Rate	ϕ	1.50	years^{-1}
OEM Correction Rate	γ	1.50	years^{-1}
MT Factor Productivity*	κ	0.125	(prod. orders/yr)/machine
Exp. Employee Productivity*	ζ	100	(machines/yr)/employee
Hiring/Firing Penalty	β^2	0.5^2	$/(yr*person)^2
Training Time	τ	2.5	years

Note that the parameters marked by an asterisk are arbitrary with respect to the results of our analysis.

The results are presented in Table 3. The first column represents the standard deviation of changes in the product order rate normalized by the expected mean of the order rate over a finite interval. The second column represents the resulting normalized standard deviation of the machine tool order rate, which is obtained from Proposition 3. The total number of employees in Column 3 is determined from the initial conditions and Proposition 4. The new employee percentage in Column 4 results from the number of new employees (Proposition 4) divided by Column 3. Column 5, the machine tool maker's annual costs result from our assumptions in the previous paragraph combined with the total number of production employees resulting from Column 3. Workforce productivity in Column 6 is calculated as the ratio of the total number of employees in Column 3 over the total number of employees required in the case of a constant product demand. Finally, Column 7, the capacity shortfall, is a result of Proposition 5.

Table 3: Effects of Increasing Volatility on MT Supplier

Product Order Rate Volatility (norm. σ)	MT Order Rate Volatility (norm. σ)	MT Employ- ees (mean)	New Employee Percentage (mean)	MT Maker Annual Costs ($M/yr)	Work- force Prod'y. vs. Max	Mean % Exper. Employee Shortfall
0%	0%	1,000	0%	200	100%	0%
5%	37%	1,100	8.8%	223	91%	6.6%
10%	74%	1,190	16%	246	84%	13.2%

Using the above assumptions in Table 3, we see that increases in product demand volatility are amplified roughly seven-fold in the machine tool order rate. Volatility increases also drive up machine tool employees, which in turn increases costs by roughly a million dollars per percentage point of employment. At higher order volatilities, these effects are more pronounced due to the adjustment of the target employment levels to avoid large capacity shortfalls. This increase in costs can quite significantly impact profits, which for a manufacturing firm are typically on the order of five percent of revenue. And the calculated personnel costs are a conservative measure of the costs arising from an increasing new employee percentage. For instance, someone has to train all these new employees. Since the trainers are experienced employees who would otherwise be making machines, productivity will drop even further than indicated by the above calculations. Additionally, it is quite likely that a less experienced workforce will create more quality problems, which will show up in rework, product defects, and perhaps ultimately lost sales. These back of the envelope calculations also do not account for the increasing overhead probably needed to maintain more employees. Thus, increasing product demand volatility can deleteriously affect a machine tool firm's profitability and quality.

Table 4: Effects of Smoothing Ordering Policies (σ_v =10%)

Product Order Smoothing Rate (γ,ϕ=yrs^{-1})	MT Order Rate Volatility (norm. σ)	MT Employees (mean)	New Employee Percentage (mean)	MT Maker Annual Costs ($M/yr)	Work-force Prod'y. vs. Max	Mean % Exper. Employee Shortfall
0.5	0%	1,100	9.0%	216	91%	3.0%
0.75	26%	1,140	12%	225	88%	5.7%
Base 1.5	74%	1,190	16%	246	84%	13.2%

From the previous discussion, one might suspect that an OEM policy which smoothes machine tool orders should benefit the supplier. We can test this by reducing the fractional rate of forecast updating ϕ and machine stock correction γ from their base values as shown in Column 1. (The parameter σ_v represents the normalized product order standard deviation. A value of 10% is typical for the automotive industry.) This will cause the OEM to react more slowly to any changes in product demand. The operating results shown in Table 4 confirm our intuition. By smoothing the rate of machine tool orders, the OEM can save the machine tool supplier 8% of its workforce and at least 12% of its cost structure. Furthermore, these savings of do not account for the reductions in other costs such as quality, overhead, delivery delays etc. described earlier that are associated with increasing volatility.

Table 5: Effects of Employee Training Time (σ_v =10%)

Employee Training Time (years)	MT Order Rate Volatility (norm. σ)	MT Employ- ees (mean)	New Employee Percentage (mean)	MT Maker Annual Costs ($M/yr)	Work- force Prod'y. vs. Max	Mean % Exper. Employee Shortfall
0.25	74%	1,070	6.3%	209	94%	0.1%
1	74%	1,170	14%	230	86%	6.5%
Base 2.5	74%	1,190	16%	246	84%	13.2%

Table 5 displays how shorter employee training times affect the machine tool supplier. We accomplish this by varying τ, the employee training time, in Column 1. While this may not be generally true of all capital equipment industries, reducing the training time for machine tool employees can have significant effects on the supplier's operating characteristics. However, it must be borne in mind that reducing the training time does *not* mean reducing the learning. It merely means reducing the time it takes for a new employee to achieve full productivity.

Table 6: Effects of Machine Tool Lifetime on MT Supplier (σ_v =10%)

Machine Tool Lifetime (years)	MT Order Rate Volatility (norm. σ)	MT Employ- ees (mean)	New Employee Percentage (mean)	MT Maker Annual Costs ($M/yr)	Work- force Prod'y. vs. Max	Mean % Exper. Employee Shortfall
4	37%	1,110	10%	225	90%	6.8%
8	74%	1,190	16%	246	84%	13.2%
12%	110%*	1,280	22%	267	78%	20%

Finally, Table 6 displays how increasing machine tool lifetime affects order rate volatility and machine tool supplier operating variables. We accomplish this by varying ε^1, the machine tool lifetime, in Column 1, and increasing the productivity of machines κ to compensate. Increasing lifetimes from four to twelve years dramatically increases the new employee fraction and reduces productivity. Costs also increase by 19%. While it must be said that the model's assumption of linearity becomes somewhat suspect in the twelve-year case, it does seem likely that different capital lifetimes significantly affect volatility and operating measures.

13.8   Discussion

Using information delays and the investment accelerator in a simple model, this paper has replicated much of the dynamic supply chain behavior seen in the industrial world in which machine tool companies operate. We have also derived a reasonable policy for managing supplier employees in the face of this volatility, and arrived at several conclusions about system behavior—albeit in the restricted model we have constructed. Several results stand out. One result is that even though machine tool suppliers can do little to reduce the order volatility they are faced with, they can minimize the impact of that volatility on their operational efficiency by using a policy such as the one described in this paper. A second more important result is that machine tool customers can significantly reduce the volatility for their suppliers through their choice of ordering policies. Companies that more slowly update their demand forecasts and correct their machine tool stock more slowly impose less of their volatility on the capital supply base. While there is a natural tendency to cut capital equipment orders at the first sign of a downturn in order to smooth the stream of OEM profits (Anderson 1996b), this model provides evidence that such smoothing may be costly. By reducing order volatility, the OEM reduces long-run supplier costs, which should in turn reduce machine tool prices. Of course, this must be balanced against giving up some of the OEM's ability to match capacity to demand. However, by constraining shipment rates, supplier capacity limitations will always limit the practical results of any ordering flexibility. Thus, there is probably some room in which to create a win-win situation for the OEM and the supplier by smoothing capital orders. Hence, we propose capital order smoothing as one element of a strategy of proactive supply chain design and management.

The smoothing rule may be even more beneficial than the preceding analysis suggests. The production function used in this model could perhaps represent the output of knowledge workers, such as engineers producing a given rate of product improvements per year. If this is reasonable, then the same improvements seen in production efficiency could be translated into improved technical progress at a lower cost per machine tool engineer. This would give a further impetus for a machine tool customer to smooth its order rate.

In an ideal environment, machine tool customers would take the lead in supporting suppliers by flattening out the worst effects of cyclicality and encouraging innovation and continuous process improvements. However, many customers do not take this leadership role, perhaps under-estimating the effect they have on the stability of the technology suppliers upon which they rely. One implication of this is that machine tool makers may benefit from taking the lead in creating new relationships with their customers aimed at reducing volatility. Cincinnati Milacron is one example of a company which is building such a relationship through their partnership with Eaton (Fine and Parker 1993-1994). In another example, the head of a purchasing group responsible for several hundred million dollars of machine tool orders per year stated that his division was able to save between 20 and 40% on equipment costs through the use of multi-year contracts (Anderson 1997b).

The solution described above is of course only one way to improve the efficiency of the supply chain. Another simple (and direct) way to insulate capital equipment firms from the effects of volatility is to purchase them outright. Since these companies are typically so much smaller than their customers, supporting the suppliers during dry periods is not much of a burden. Japanese firms in particular are notorious for keeping their capital equipment capabilities in house. Toshiba has developed an interesting variation of this strategy to deal with capital equipment demand volatility. This highly diversified manufacturing company maintains a central division which supplies manufacturing equipment to the various divisions: metal machining, semiconductor wafer processing, circuit board assembly, etc. When demand from one division slows, personnel are re-assigned to work on equipment for other divisions (Fine 1994). No doubt employees who normally work on semiconductor equipment are less effective when assigned to machine tool building, but there are some skills which should transfer relatively easily, such as controls programming. This practice also encourages expertise from one area to be applied to another.

A much lesser commitment to in-house capital equipment expertise can also be beneficial. In the manufacturing systems business, standard machine tools are often customized with the addition of applications engineering and systems integration. Many German and Japanese manufacturing firms consider applications engineering and system integration to be core capabilities which must be done in-house. This view has changed somewhat in recent years with the dramatic economic downturn in those countries, but it remains a general observation. Completely addressing the benefits and costs of this practice lies outside the scope of this paper, but one result of higher customer capability is that machine tool builders are able to stock more standardized machines, which the customers can then customize to their own requirements.[11] This implies that production smoothing is more feasible for those companies who have customers with highly capable internal engineering staffs.[12]

At an even simpler level, there may some benefit to enforcing more standardization in transfer line design. Product design has benefited enormously in recent years from parts commonization efforts. It has also profited from the platform concept, which spins many designs off from a common platform. Both policies require, however, a level of discipline greater than that often pursued in large U.S. manufacturing firms. The prime benefit, however, would be reduction of purchased machine types. This would again allow the supplier to stock some machines in advance, and allow the customer to pre-purchase machines more easily a hedge against future shortages.

A final policy solution to the problem is for public policy to encourage more apprenticeship programs like those found in Germany. The apprenticeship programs there provide a ready supply of skilled workers with which to stock their highly developed capital goods industry. As the new employees are already skilled when they join a firm, many of the problems associated with managing skilled employees in the U.S. recede because they need much less training than their American counterparts. Such programs have been pursued on a private basis in the United States by such organizations as the Sacred Heart Seminary in Detroit (Anderson 1996a). Their primary interest is to help provide good jobs for the unemployed in Michigan, but a secondary benefit is to help maintain Detroit as a manufacturing center despite what many firms consider an unfriendly business environment. Recently, the state of Wisconsin has begun a similar program based explicitly on the German model (Economist 1997). If these programs are adopted, then the need to coordinate capital supply chain purchasing and production policies becomes less problematic.

The model also suggests another path of empirical analysis. In the model, volatility amplification along the capital supply chain is affected by several difficult-to-change industrial factors; among them are capital lifetime and employee training time. This suggests that industries characterized by shorter capital lifetimes or shorter training times should have *ceteris paribus* more productive and efficient suppliers. Fine's (1996) work on technology supply chains stresses the importance of these parameters which he considers elements in his concept of "clockspeed". Higher clockspeed firms are characterized by more rapid industrial changes including technological progress. If one assumes that capital replacement rates climb with an industry's rate of technological progress, the suppliers of high-tech industries may benefit from the high clockspeeds in their industries. This suggests an empirical comparison of a high-technology capital supply industry, such as the photolithography industry, with the machine tool industry. The expectation is that photolithography firms, which supply the critical equipment needed to manufacture semiconductors, may be operationally healthier than their machine tool counterparts. Similarly, a comparison of two high technology industries characterized by differing capital-order rate volatilities may be fruitful. For instance, technological change in the printed-circuit board assembly equipment industry tends to be "lumpier" than that seen in the photolithography industry, even though both are certainly "high-tech" (Anderson and Joglekar 1994). In this case, one might expect that the operational efficiency of PCB equipment industry to be less than that of the photolithography industry. Also, from an industrial structure point of view, OEMs in "lumpier" or lower-tech industries may derive quite an advantage from minimizing the volatility of their suppliers. If the technology is critical enough, OEMs should want to maximize their share of a capital-supplier's business over time in order to reduce volatility. One way they can do this through a combination of long-term contracts combined with single- or dual-sourcing. Or if the OEM's business is large enough, an exclusive relationship may be indicated. This may explain why so many high-performing firms seem to keep at least a capital "customization" capability in house (Fine and Whitney 1995). Since such customization tends towards the more knowledge-intensive end, i.e. requiring more training and experience, of capital equipment production, it is probably the most important part of the industry to buffer from volatility. A useful contribution would therefore be a comparative empirical study examining efficiency and structure in industries characterized by differing clockspeed parameters.

---

[1] This paper is adapted in part from work in Anderson (1997a; 1997c), and Fine (1998).

[2] (Sterman 1989a) examines supply chain volatility empirically, and briefly describes capital equipment demand amplification in his survey of stock management systems.

[3] Of course this statement may not apply so well to newer industries such as the electronics and semiconductor industries. However, they have their own critical capital suppliers which have problems parallel to the machine tool industry's.

[4] The abbreviation OEM stands for "original equipment manufacturer", which encompasses such firms as General Motors, Boeing, or IBM. We will use this term to denote the machine-tool firm's customer in the model, although strictly speaking, the customer could be a components supplier to an OEM such as Eaton which makes automotive components for Ford.

[5] This is probably a conservative consumption. Cincinnati Milacron indicated that new employees are of negative productivity because they require mentoring time from experienced employees (Parker 1996).

[6] The estimation algorithm used was Powell's (1972) optimization algorithm

[7] Given the time parameters in this study, the short-term differences between production and consumption rates of durables is negligible.

[8] Under testing, $\hat{r}_{LRF}(t)$ never fell below zero while using real-world parameters and data. However, while we will not explicitly deal with this corner condition in this paper, it would seem prudent in implementation to place a floor on $\hat{r}_{LRF}(t)$ at zero to avoid an absurd result.

[9] This will obtain as long as the coefficient of variation of product demand per stochastic cycle of machine orders is relatively small compared with that of machine orders.

[10] The expected ratio of r(t) to g(t) is found from the expression L(s) when s is set equal to zero.

[11] In general, there is anecdotal evidence which suggests that performing applications engineering and system integration "in house" makes a company better able to pursue successful concurrent engineering projects and to undertake continuous improvement projects. The trade-off for these capabilities is the higher fixed costs of the personnel necessary to perform these functions (Fine and Whitney, 1996).

[12] See AMT (1995) for an article discussing the migration of U.S. customers toward more standard machines, with a concurrent demand for more applications engineering support from machine-tool builders.

13.9   Appendix

13.9.1   Outline of Proof for Proposition 1

The following derivation uses the Bode-Shannon approach to Wiener prediction (Shanmugan and Breipohl 1988).

Let $H_r(s)$ represents the real system while $H(s)$ represents the transfer function of the open-loop forecast model developed previously in (17).

By breaking up $H(s)$ into partial fractions and multiplying by e^{sT}, we obtain:

$$H(s)e^{sT} = e^{sT}\left(\frac{\varepsilon\kappa^{-1}}{s} + \frac{\kappa^{-1}\phi(\varepsilon-\gamma)}{(\gamma-\phi)(s+\gamma)} + \frac{\kappa^{-1}\gamma(\varepsilon-\phi)}{(\phi-\gamma)(s+\phi)}\right) \tag{A1}$$

Transforming (18) term-by-term back into the time-domain yields,

$$H(t+T) = \left(\varepsilon\kappa^{-1} + \frac{\kappa^{-1}\phi(\varepsilon-\gamma)}{(\gamma-\phi)}e^{-\gamma(t+T)} + \frac{\kappa^{-1}\gamma(\varepsilon-\phi)}{(\phi-\gamma)}e^{-\phi(t+T)}\right)U(t+T) \tag{A2}$$

where $U(t)$ is the Heaviside unit step function, which takes a value of unity for nonnegative arguments, zero otherwise. Taking the non-negative time portion of (19) yields:

$$\{H(t+T)\}_{t\geq 0} = \left(\varepsilon\kappa^{-1} + \frac{\kappa^{-1}\phi(\varepsilon-\gamma)}{(\gamma-\phi)}e^{-\gamma T}e^{-\gamma t} + \frac{\kappa^{-1}\gamma(\varepsilon-\phi)}{(\phi-\gamma)}e^{-\phi T}e^{-\phi t}\right)U(t) \tag{A3}$$

Transforming back into the s-domain.

$$\{H(s)e^{sT}\}_{t\geq 0} = \frac{\varepsilon\kappa^{-1}}{s} + \frac{\kappa^{-1}\phi(\varepsilon-\gamma)}{(\gamma-\phi)(s+\gamma)}e^{-\gamma T} + \frac{\kappa^{-1}\gamma(\varepsilon-\phi)}{(\phi-\gamma)(s+\phi)}e^{-\phi T} \tag{A4}$$

Finally, taking the product of (21) and $H^{-1}(s)$ and simplifying yields the required transfer function. ■

13.9.2 Proof for Proposition 2 and Corollary 1

Using the cost function and first constraint of (20), and the fact from Proposition 1 that $w*(\tilde{t})$ is of the form $W_0 + W_1 e^{-\phi\tilde{t}} + W_2 e^{-\gamma\tilde{t}}$ (where W_0, W_1, and W_2 are constants), we obtain the following necessary conditions for an optimal control policy. (As the objective function's integrand and the state constraint are both jointly convex in e and u, the following conditions also guarantee sufficiency.)

$$2\beta^2\tau^2 e'(\tilde{t}) + \lambda(\tilde{t}) = 0 \tag{A5}$$

$$\lambda'(\tilde{t}) = 2[W_0 + W_1 e^{-\phi\tilde{t}} + W_2 e^{-\gamma\tilde{t}} - e(t)] \tag{A6}$$

$$\lambda(\infty) = 0 \tag{A7}$$

$$e(t_0) = E_0 \tag{A8}$$

Equation A5 represents the optimality condition of the Hamiltonian with respect to $u(\cdot)$ where $\lambda(\cdot)$ represents the Lagrangian multiplier associated with the employee state constraint. Equation A6 is the adjoint equation. Equation A7 is the transversality condition on the end state of $e(\tilde{t})$. Equation A8 reflects the fact that experienced employees cannot be adjusted instantaneously from their initial level.

Solving for $u(\tilde{t})$ yields the following expression:

$$\frac{(W_0 - E_0)\left[\beta^2\tau^2(\phi^2\gamma^2\beta^2\tau^2 - \phi^2 - \gamma^2) + 1\right] + W_1(1 - \gamma^2\beta^2\tau^2) + W_2(1 - \phi^2\beta^2\tau^2)}{\beta\tau(\phi\beta\tau + 1)(\phi\beta\tau - 1)(\gamma\beta\tau + 1)(\gamma\beta\tau - 1)}e^{-\frac{\tilde{\iota}}{\beta\tau}} + \quad (A9)$$

$$\frac{W_1\phi e^{-\phi\tilde{\iota}}(\gamma^2\beta^2\tau^2 - 1) + W_2\gamma e^{-\gamma\tilde{\iota}}(\phi^2\beta^2\tau^2 - 1)}{\beta\tau(\phi\beta\tau + 1)(\phi\beta\tau - 1)(\gamma\beta\tau + 1)(\gamma\beta\tau - 1)}$$

Setting $\tilde{\iota}$ equal to zero, substituting $E_0=e(t)$, and simplifying yields the control law for this instant required for Proposition 2.

Let K_0, K_1, K_2, K_e respectively represent the coefficients of W_0, W_1, W_2, E_0 in (30). Combining this with the equation for the Wiener predictor yields the following (where ζ is the productivity of the supplier's employees),

$$U(s) = \zeta^{-1}\left(K_0 C_0 + K_1 C_1 + K_2 C_2\right)\frac{R(s)}{1 - K_e/s} \quad (A10)$$

(The third factor in the above equation represents the effect of feeding back the current employee state into the control law.) Substituting in for the K's and C's yields the required result for Corollary 1. ∎

13.9.3 Outline of Proof for Proposition 3

A well-known theorem from signal processing states that a linear system's amplification of a white noise process can be found directly from the system's transfer function (Shanmugan and Breipohl 1988). This is done by substituting in $s = 2\pi f\iota$ (where f represents the frequency and $\iota = \sqrt{-1}$), taking the square of the absolute value of the transform, and integrating it with respect to f. Performing this procedure with $L(s)$ yields the required result. ∎

13.9.4 Proof for Proposition 4

To obtain an approximate transfer function from changes in product demand to changes in the number of experienced employees, we integrate (22) and combine it with the approximate product demand-to-machine tool order transfer function from (17).

$$\frac{U(s)}{V(s)} = \phi\gamma\kappa^{-1}\zeta^{-1}\frac{\beta\tau(\varepsilon\beta\tau + 1)s^2 + (\varepsilon\beta\tau + 1)(\phi\beta\tau + \gamma\beta\tau + 1)s + \varepsilon(\phi\beta\tau + 1)(\gamma\beta\tau + 1)}{(s + \phi)(s + \gamma)(s\beta\tau + 1)(\phi\beta\tau + 1)(\gamma\beta\tau + 1)} \quad (A11)$$

In order to understand the transfer of volatility through the system, we again need a function to represent the percentage changes in the two variables per period, which we will denote M_{tot}. Repeating the process used to normalize machine orders in Equation 24, we obtain:

$$M_{tot}(s) = \phi\gamma \frac{\beta\tau(\varepsilon\beta\tau+1)s^2 + (\varepsilon\beta\tau+1)(\phi\beta\tau+\gamma\beta\tau+1)s + \varepsilon(\phi\beta\tau+1)(\gamma\beta\tau+1)}{\varepsilon(s+\phi)(s+\gamma)(s\beta\tau+1)(\phi\beta\tau+1)(\gamma\beta\tau+1)} \quad (A12)$$

Repeating the procedure used in Proposition 3 to extract a variance from a transfer function yields the required result. ■

13.9.5 Outline of Proof for Corollary 2

Clearly, if the derivative of the integrand of $M_{tot}^2 = \int_{-\infty}^{\infty} |M(2\pi f\iota)| df$, where f represents frequency and $\iota = \sqrt{-1}$, under a set of given conditions, so too will M_{tot}.

Let an asterisk indicate the complex conjugate of a given value. Then, let

$$N(f) = |M(2\pi f\iota)|^2 = M(2\pi f\iota)M*(2\pi f\iota) \quad (A13)$$

Substituting (A12) into (A13),

$$N(f) = \phi^2\gamma^2 \frac{\left\{\begin{array}{l}(2\pi f)^4\,\beta\tau(\varepsilon\beta\tau+1)^2 + \varepsilon^2(1+\phi\beta\tau)(1+\gamma\beta\tau) \\ (2\pi f)^2(\varepsilon\beta\tau+1)\left[\varepsilon\beta^3\tau^3(\phi^2+\gamma^2)+(\beta\tau)^2(\phi+\gamma)^2+2\beta\tau(\phi+\gamma)-\varepsilon\beta\tau+1\right]\end{array}\right\}}{\varepsilon^2(4\pi^2 f^2+\phi^2)(4\pi^2 f^2+\gamma)(4\pi^2 f^2\beta^2\tau^2+1)(\phi\beta\tau+\gamma\beta\tau+\phi\gamma\beta^2\tau^2+1)^2} \quad (A14)$$

Finding the partial derivative of N with respect to ε,

$$\frac{\partial N(f)}{\partial\varepsilon} = \frac{-8\phi^2\gamma^2\pi^2 f^2\left\{\begin{array}{l}(2\pi f)^2(\beta^2\tau^2)(\beta\tau+1)+\varepsilon(\beta^3\tau^3)(\phi^2+\phi\gamma+\gamma^2)+ \\ \varepsilon(\beta^2\tau^2)(\phi+\gamma)+2\phi\gamma\beta^2\tau^2+2\beta\tau(\phi+\gamma)+\beta^2\tau^2(\phi^2+\gamma^2)+1\end{array}\right\}}{\varepsilon^3(4\pi^2 f^2+\phi^2)(4\pi^2 f^2+\gamma)(4\pi^2 f^2\beta^2\tau^2+1)(\phi\beta\tau+\gamma\beta\tau+\phi\gamma\beta^2\tau^2+1)^2} \quad (A15)$$

The right-hand side of Equation A15 must always be negative because of the nonnegativity constraints imposed upon the parameters. Hence, $N(f)$ decreases in ε.

Taking the partial derivative of N with respect to τ yields the following proportion,

$$\frac{\partial N(f)}{\partial\tau} = -\frac{\begin{array}{l}16\pi^4 f^4\left[\beta^4\tau^4(2\phi\gamma-\varepsilon^2)+\beta^3\tau^3(\phi+\gamma-\varepsilon)\right]+ \\ 4\pi^2 f^2\beta^4\tau^4\left[2\phi\gamma(\phi^2+\gamma^2-\varepsilon^2)-(4\phi^2\gamma^2-\phi^2\varepsilon^2-\gamma^2\varepsilon^2)\right]+ \\ 8\pi^2 f^2\beta^3\tau^3\left[\phi(4\phi^2\gamma-\varepsilon^2)+\gamma(4\phi\gamma^2-\varepsilon^2)\right]-8\pi^2 f^2\beta^2\tau^2\left[5\phi\gamma-\varepsilon^2\right]+ \\ 4\pi^2 f^2\beta\tau\left[3(\phi+\gamma-\varepsilon)+\beta\tau\phi(3\phi-\varepsilon)+\beta\tau\gamma(3\gamma-\varepsilon)\right]+ \\ \beta^2\tau^2\left[\phi^3(\gamma-\varepsilon)+\gamma^3(\phi-\varepsilon)\right]-3\beta\tau\left[\phi^2(\gamma-\varepsilon)+\gamma^2(\phi-\varepsilon)\right]+ \\ 2\left[\phi\gamma-(\phi+\gamma)\varepsilon\right]+o.p.t.\end{array}}{o.p.t} \quad (A16)$$

where *o.p.t.* stands in for other positive terms. Clearly, a sufficient condition for (A16) to be negative is that ε is less than ϕ, γ, and $\frac{\phi\gamma}{\phi+\gamma}$. Hence, if 2ε is less than ϕ and γ, negativity will be guaranteed. Accordingly, $N(f)$ will also decrease in τ.

Note that β is symmetric with τ in $M(s)$. Thus, the result for τ will also hold for β. ∎

13.9.6 Proof for Corollary 3

The expectation of the positive half of a Gaussian process $x(t)$ is related to its variance σ_x^2 as follows,

$$\mu_x = \frac{1}{2}\bullet 0 + \int_0^\infty \frac{\tilde{x}}{\sqrt{2\pi\sigma_x^2}}\exp\left(-\frac{\tilde{x}^2}{2\sigma_x^2}\right)d\tilde{x} = \left(\frac{\sigma_x^2}{2\pi}\right)^{\frac{1}{2}} \tag{A13}$$

Thus, following the proof for Proposition 4, the number of new employees will be:

$$\mu_n^2 = \frac{\sigma_v^2}{2\pi}\int_{-\infty}^\infty\left|\left(\frac{\tau}{s}\bullet\frac{U(s)}{G(s)}\right)_{s=2\pi f i}\right|^2 df = \frac{\sigma_v^2\tau^2\varepsilon^2\kappa^{-2}\zeta^{-2}}{2\pi}\int_{-\infty}^\infty|M(2\pi f i)|^2 df = \frac{\sigma_v^2\tau^2\varepsilon^2\kappa^{-2}\zeta^{-2}}{2\pi}M_{tot}^2 \tag{A14}$$

Taking the square root of this expression yields the required result. ∎

13.9.7 Proof for Proposition 5

To find a value for the average shortfall, first find an expression for the workforce gap.

$$C(s) = W*(s) - E(s) = \zeta^{-1}R(s) - E(s) \tag{A15}$$

Substituting from Corollary 1 for *e(s)*,

$$C(s) = \zeta^{-1}R(s)\left[1 - \frac{\beta\tau(\varepsilon\beta\tau+1)s^2 + (\varepsilon\beta\tau+1)(\phi\beta\tau+\gamma\beta\tau+1)s + \varepsilon(\phi\beta\tau+1)(\gamma\beta\tau+1)}{(s+\varepsilon)(s\beta\tau+1)(\phi\beta\tau+1)(\gamma\beta\tau+1)}\right] \tag{A16}$$

Substituting in from Equation 17 to eliminate *R(s)*,

$$\frac{C(s)}{G(s)} = \kappa^{-1}\zeta^{-1}\phi\gamma\beta^2\tau^2 s\frac{(\beta\tau\phi\gamma+\phi+\gamma-\varepsilon)s+\varepsilon\beta\tau\phi\gamma+\phi\gamma}{(s+\phi)(s+\gamma)(s\beta\tau+1)(\phi\beta\tau+1)(\gamma\beta\tau+1)} \tag{A17}$$

Then to find the variance of *c(t)*, differentiate the above expression and integrate over the frequency domain,

$$\text{var}(c(t)) = \int_{-\infty}^{\infty} \sigma_v^2 \left| \left[\frac{C(s)}{sG(s)} \right]_{s=2\pi fi} \right|^2 df \tag{A18}$$

Evaluating the integral yields,

$$\text{var}(c(t)) = \kappa^{-2} \zeta^{-2} \phi^2 \gamma^2 \tau^4 \beta^4 \frac{\begin{pmatrix} \phi\gamma\varepsilon^2 \tau^3 \beta^3 (\phi+\gamma) + \phi\gamma\tau^2 \beta^2 (\phi\gamma + 2\phi\varepsilon + 2\gamma\varepsilon + \varepsilon^2) + \\ 3\phi\gamma\tau\beta(\phi+\gamma) + 3\phi\gamma - 2\varepsilon(\phi+\gamma) + \phi^2 + \gamma^2 + \varepsilon^2 \end{pmatrix}}{2(\phi+\gamma)(\phi\beta\tau+1)^3(\gamma\beta\tau+1)^3} \tag{A19}$$

Since $c(t)$ is Gaussian, $d(t)$ is the magnitude of the negative half of a Gaussian process. Thus, the expected value of $d(t)$, μ_d is the square root of var(c(t))/2π. This yields the required result. ∎

13.9.8 Proof for Proposition 6

Using Equation 27 to estimate P by exploiting the known expected ratio of $e(t)$ to $g(t)$,

$$P = \frac{\bar{e}}{\bar{e} + \bar{n}} = \frac{\bar{e} \cdot \bar{g}}{\bar{e} \cdot \bar{g} + \bar{n} \cdot \bar{g}} = \frac{\varepsilon\kappa^{-1}\zeta^{-1}}{\varepsilon\kappa^{-1}\zeta^{-1} + \tau\varepsilon\kappa^{-1}\zeta^{-1}g_0^{-1}\sqrt{(2\pi)^{-1}}\sigma_v^2 M_{tot}^2} = \frac{1}{1 + \tau g_0^{-1} M_{tot}\sqrt{(2\pi)^{-1}}\sigma_v^2} \tag{A20}$$

Clearly, if M_{tot} increases (decreases) in ψ, P will decrease (increase). And because $R = 1 - P$, if V increases (decreases) in ψ, R will increase (decrease). ∎

13.10 References

Anderson, Edward G. (1996a). Interview with Roger DePlaunty of Metro Metals Inc.

Anderson, Edward G. (1996b). Interview with William Colwell of Ford Motor Company.

Anderson, Edward G. (1997a). *The Effects of Business Cycles upon Capital Supplier Productivity and Technological Capability*. Ph.D., Sloan School of Management, Massachusetts Institute of Technology, Cambrige, MA.

Anderson, Edward G. (1997b). Interview with Head of Big 3 Purchasing Group. Cambridge, MA: Massachusetts Institute of Technology.

Anderson, Edward G. (1997c). Managing Skilled-Employee Productivity in Capital Equipment Supply Chains. Austin, TX: University of Texas.

Anderson, Edward G., Charles H. Fine, and Geoffrey G. Parker (1996). "Upstream Volatility in the Supply Chain: The Machine Tool Industry as a Case Study." Paper read at INFORMS MSOM Conference, June, 1996.

Anderson, Edward G., and Nitin R. Joglekar (1994). Visits to Delco and Ford Electronics Division.

Association for Manufacturing Technology (1970-94). *The Economic Handbook of the Machine Tool Industry*. McLean, VA.

Association for Manufacturing Technology (AMT) (1995). "Building Value into Machine Tools." *American Machinist* 139 (10):A2.

Baganha, Manuel P., and Morris A. Cohen (1996). "The Stabilizing Effects of Inventory in Supply Chains." Paper read at INFORMS MSOM Conference, June 1996, at Dartmouth College.

Bertsekas, Dimitri P. (1987). *Dynamic Programming: Deterministic and Stochastic Models.* Englewood Cliffs, NJ: Prentice-Hall.

Bitran, Gabriel R. (1993). "Hierarchical Production Planning." In *Logistics of Production and Inventory,* edited by S. C. Graves, A. R. Kan and P. Zipkin. Amsterdam: Elsevier Science.

Dertouzos, Michael L., Richard K. Lester, and Robert M. Solow (1989). *Made in America.* New York: Harper Perennial.

Economist (1997). "The Heartland's German Model." *The Economist,* February 15, 1997.

Fine, Charles H. (1994). Interview at Toshiba Manufacturing Center in Osaka, Japan: Massachusetts Institute of Technology.

Fine, Charles (1996) "Industry Clockspeed and Competency Chain Design." *Proceedings of the 1996 Manufacturing and Service Operations Management Conference.* Dartmouth College, Hanover, New Hampshire June 24-25, 1996, pp. 140-143. (Downloadable from http://web.mit.edu/afs/athena/org/c/ctpid/www/people/FinePapers.html%20)

Fine, Charles H. (1998). *Clockspeed: Winning Industry Control in the Age of Temporary Advantage.* Reading, MA: Perseus Books, Fall 1998.

Fine, Charles H., and Geoffrey G. Parker (1993-1994). Field Visits to Cincinnati Milacron, Eaton, and Cincinnati Milacron/Eaton Meetings.

Fine, Charles H., and Daniel Whitney (1995). Is the Make/Buy Decision a Core Competence? Cambridge, MA: Massachusetts Institute of Technology.

Forrester, Jay W. (1958). "Industrial Dynamics: A Major Breakthrough for Decision Makers." *Harvard Business Review* 36 (4):37-66.

Graves, Stephen C. (1997). A Single-Item Inventory Model for a Non-Stationary Demand Process. Cambridge, Mass.: Massachusetts Institute of Technology.

Hamilton, James D. (1989). "A New Approach to the Economic Analysis of Nonstationary Time Series and the Business Cycle." *Econometrica* 57:357-384.

Kallenberg, Robert (1994). *Analysis of Business Cycles in the U.S. Machine Tool Industry Using the System Dynamics Method.* Masters, Rheinische-Westfaelische Techische Hochschule Aachen, Aachen, Germany.

Lee, Hau, P. Padmanabhan, and S. Whang (1997). "Information Distortion in a Supply Chain: The Bullwhip Effect." *Management Science* 43 (4):516-558.

Makridakis, Spyros, C. Chatfield, M. Hibon, M. Lawrence, T. Mills, K. Ord, and L. F. Simmons (1993). "The M2-Competition: A Real-Time Judgmentally Based Forecasting Study." *International Journal of Forecasting* 9 (1):5-22.

March, Artemis (1988). The US Machine Tool Industry and its Foreign Competitors," MIT Commission on Industrial Productivity Working Paper - background for Made In America. Cambridge, MA: MIT Commission on Industrial Productivity.

Nelson, Charles R., and Charles I. Plosser (1982). "Trends and Random Walks in Macroeconomic Time Series: Some Evidence and Implications." *Journal of Monetary Economics* 10:139-162.

Ogata, Katsuhiko (1970). *Modern Control Engineering.* Englewood Cliffs, NJ: Prentice-Hall.

Papoulis, Athanasios (1991). *Probability, Random Variables, and Stochastic Processes.* New York: McGraw Hill.

Parker, Geoffrey G. (1996). Interviews with Cincinnati Milacron. Cambridge, Massachusetts: MIT Technology Supply Chain Research Project.

Powell, M. J. D. (1972). "Problems Related to Unconstrained Optimization." In *Numerical Methods for Unconstrained Optimization*, edited by W. Murray. New York: Academic Press.

Shanmugan, K. Sam, and A. M. Breipohl (1988). *Random Signals: Detection, Estimation, and Data Analysis*. New York: Wiley.

Sterman, John D. (1989a). "Misperceptions of Feedback in Dynamic Decision Making." *Organizational Behavior and Human Decision Processes* 43:301-335.

Sterman, John D. (1989b). "Modeling Managerial Behavior: Misperceptions of Feedback in a Dynamic Decision Making Experiment." *Management Science* 35 (3):321-339.

U.S. Federal Reserve Board (1996). *Industrial Production Indices*. Washington, DC: http://www.bog.frb.fed.us/commerce.htm.

14 THE BULLWHIP EFFECT: MANAGERIAL INSIGHTS ON THE IMPACT OF FORECASTING AND INFORMATION ON VARIABILITY IN A SUPPLY CHAIN

Frank Chen[1], Zvi Drezner[2], Jennifer K. Ryan[3]
and David Simchi-Levi[4]

[1] Department of Decision Sciences,
National University of Singapore

[2] Department of Management Science/Information Systems,
California State University, Fullerton

[3] School of Industrial Engineering,
Purdue University

[4] Department of Industrial Engineering and Management Sciences,
Northwestern University

14.1 INTRODUCTION

An important observation in supply chain management, popularly known as the **bull-whip effect**, suggests that demand variability increases as one moves up a supply chain. For example, empirical evidence suggests that the orders placed by a retailer tend to be much more variable than the customer demand seen by that retailer. This increase in variability propagates up the supply chain, distorting the pattern of orders received by distributors, manufacturers and suppliers.

As pointed out by Lee, Padmanabhan and Whang (1997 a, b), the term "bullwhip effect" was coined by executives of Procter and Gamble (P&G), the company which manufacturers the Pampers brand of diapers. These exectutives observed that while the consumer demand for Pamper's diapers was fairly constant over time, the orders for diapers placed by retailers to their wholesalers or distributors were quite variable, i.e., exhibited significant fluctuations over time. In addition, even larger variations in order quantitities were observed in the orders that P&G received from its wholesalers. This increase in the variability of the orders seen by each stage in a supply chain was called the "bullwhip effect". For a further discussion of this phenomenon, see, for example, Baganha and Cohen (1995), Caplin (1985), Kahn (1987), Kaminsky and Simchi-Levi (1996) and Sterman (1989).

The bullwhip effect is a major concern for many manufacturers, distributors and retailers because the increased variability in the order process (i) requires each facility to increase its safety stock in order to maintain a given service level, (ii) leads to increased costs due to overstocking throughout the system, and (iii) can lead to an inefficient use of resources, such as labor and transportation, due to the fact that it is not clear whether resources should be planned based on the average order received by the facility or based on the maximum order.

To better understand the impact of the bullwhip effect on an entire supply chain, consider the case of a simple, two stage supply chain consisting of a single retailer and a single manufacturer. The retailer observes customer demand and places orders to the manufacturer. To determine how much to order from the manufacturer, the retailer must **forecast** customer demand. Generally, the retailer will use the observed customer demand data and some standard forecasting technique to calculate these forecasts.

Next, consider the second stage in our supply chain, the manufacturer. The manufacturer observes the retailer's demand and places orders to his supplier. To determine these order quantities, the manufacturer must forecast the retailer's demand. In many supply chains, the manufacturer does not have access to the actual customer demand data. Therefore, he must use the orders placed by the retailer to perform his forecasting. If, as the bullwhip effect implies, the orders placed by the retailer are significantly more variable than the customer demand observed by the retailer, then the manufacturer's forecasting and inventory control problem will be much more difficult than the retailer's forecasting and inventory control problem. In addition, the increased variability will force the manufacturer to carry more safety stock or to maintain higher capacity than the retailer in order to meet the same service level as the retailer.

In this paper we discuss the causes of the bullwhip effect, as well as methods for reducing its impact. The analysis is based on the recent work by Chen, Drezner, Ryan and Simchi-Levi (1998), Chen, Ryan and Simchi-Levi (1998) and Lee, Padmanabhan and Whang (1997 a, b). We start, in the following section, with a discussion of the five main causes of the bullwhip effect. We then present, in the second section, some new results which enable us to quantify the increase in variability due to two of these causes: demand forecasting and lead times. These results provide some useful managerial insights on controlling the bullwhip effect, which are discussed in the third section. We then extend these results, in the fourth section, to consider the impact of centralized demand information on the bullwhip effect. Finally, we conclude with a discussion of some methods for reducing the impact of the bullwhip effect.

14.2 IDENTIFYING THE CAUSES OF THE BULLWHIP EFFECT

It is clear that the bullwhip effect can lead to significant increases in costs and inventory levels throughout the supply chain. In order to control or eliminate the bullwhip effect we must first understand its causes. Lee, Padmanabhan and Whang (1997 a, b) identify five main causes of the bullwhip effect: demand forecasting, lead times, batch ordering, supply shortages and price variations. Caplin (1985) considers the impact of batch ordering on the bullwhip effect. Chen, Drezner, Ryan and Simchi-Levi (1998) and Chen, Ryan and Simchi-Levi (1998) consider the impact of demand forecasting and lead times on the bullwhip effect. In this section, we will briefly discuss how each these factors can cause the bullwhip effect.

14.2.1 Demand Forecasting

Consider our example of a simple two stage supply chain in which each stage uses some form of demand forecasting to determine its desired inventory level and order quantity. Recall that a simple **order-up-to inventory policy** requires each stage of the supply chain to raise its inventory level up to a given target level in each period. One common form of this policy is to set the target inventory level in period t, y_t, equal to

$$y_t = \hat{\mu}_t^L + z\hat{\sigma}_t^L, \tag{14.1}$$

where $\hat{\mu}_t^L$ is an estimate of the mean lead time demand, $\hat{\sigma}_t^L$ is an estimate of the standard deviation of the forecast errors over the lead time, and the parameter z is chosen to meet a desired service level.

In order to determine its target inventory level, each stage of the supply chain must forecast both the expected demand and the standard deviation of demand. This forecasting can be done using any of a number of forecasting techniques, for example, moving average or exponential smoothing. In the moving average forecast, the forecast average demand per period is simply the average of the demands observed over some fixed number of periods, say p periods. In the exponential smoothing forecast, the forecast average demand per period is a weighted average of *all* of the previous demand observations, where the weight placed on each observation decreases with the age of the observation.

Any forecasting technique can cause the bullwhip effect. To understand this, note that one property of most standard forecasting methods is that the forecast is updated each time a new demand is observed. Therefore, at the end of each period, the retailer will observe the most recent demand, update her demand forecast based on this demand, and then use this updated forecast to update her target inventory level. It is this updating of the forecast and order-up-to point in each period that results in increased variability in the orders placed by the retailer.

Finally, note that if the retailer follows a simple order-up-to inventory policy in which she does not update the desired inventory level in each period, then she would not see the bullwhip effect. In other words, if in every period the retailer places an order to raise the on-hand inventory to the same **fixed** level, then the orders seen by the manufacturer would be exactly equal to the customer demand seen by the retailer. Therefore, the variability in the orders seen by the manufacturer would be exactly equal to the variability of the customer demand, and there would be no bullwhip effect.

14.2.2 Lead Times

The lead time is defined as the time it takes an order placed by the retailer to be received at that retailer. Lead times can add to the bullwhip effect by magnifying the increase in variability due to the demand forecasting. To understand this, note that lead times increase the target inventory level, i.e., the longer the lead time, the larger the inventory level required. In addition, if as discussed above, the retailer updates her target inventory level in each period (using demand forecasting), then longer lead times will lead to larger changes in the target inventory level, and thus more variability in the orders placed by the retailer.

For example, if the demands seen by the retailer are independent and identically distributed (i.i.d.) from a normal distribution with mean μ and variance σ^2, then an approximately optimal order-up-to inventory level in period t is

$$ y_t = L\hat{\mu}_t + z\sqrt{L}\hat{\sigma}_t, $$

where L is the lead time plus 1, $\hat{\mu}_t$ is an estimate of μ, and $\hat{\sigma}_t$ is an estimate of σ. In this case, we clearly see that if, from time t to $t+1$, our estimate of μ changes by Δ, then our order-up-to level will change by $L\Delta$, where $L \geq 1$. In other words, any changes in our estimates of the parameters of the demand process will be magnified by the lead time.

14.2.3 Batch Ordering

Another major cause of the bullwhip effect is the "batching" of orders. The impact of batch ordering is quite simple to understand: if the retailer uses batch ordering, then the manufacturer will observe a very larger order, followed by several periods of no orders, followed by another large order, and so on. Thus the manufacturer sees a distorted and highly variable pattern of orders.

Caplin (1985) considers the impact of batch ordering on the bullwhip effect. He considers a retailer who follows a continuous review (s, S) inventory policy in which the retailer continuously monitors the inventory level, and when the inventory level drops to s, places an order to raise the inventory level to S. Note that for an inventory policy of this form, when the retailer places an order, the size of that order, $Q = S - s$, is fixed and known. In this case, since the size of the order is fixed, the variability of the orders placed by the retailer is due only to the variability in the time between orders, i.e., the variability in the time it takes for the inventory level to fall from S to s. Caplin proves that, if the demands faced by the retailer are i.i.d., then the variance of the orders placed by the retailer is greater than the variance of the customer demand observed by the retailer, and that the variance of the orders increases linearly in the size of the orders, i.e., the variability will increase linearly in Q.

14.2.4 Supply Shortages

A fourth cause of the bullwhip effect is associated with anticipated supply shortages. If a retailer anticipates that a particular item will be in short supply, he may place an inflated, unusually large, order with his supplier, because experience suggests that the supplier will be rationing the product among his customers based on the size of their orders. This "rationing" or "gaming" distorts the true demand pattern and gives the manufacturer a false impression of the market demand for the product. Lee, Padmanabhan and Whang (1997 a, b) discuss in detail the impact of this type of gaming on the variability in a supply chain.

14.2.5 Price Variations

A final cause of the bullwhip effect is the frequent price variations seen throughout a supply chain. For example, many retailers will offer products at a regular retail price with periodic price promotions and clearance sales. Clearly, when the price for an item changes, the customer demand for that item will also change. For example, during a promotion the retailer will see higher than usual demand, while after the promotion, the retailer may observe unusually low demand. Therefore, periodic price promotions can cause distorted demand patterns and increased variability in demand.

This phenomenon is observed at other stages of the supply chain as well. For example, when a manufacturer offers a trade promotion, retailers may place unusually large orders and stockpile inventory, and may not order again for several periods. Again, this causes distorted demand patterns and increases the variability in demand.

14.3 QUANTIFYING THE BULLWHIP EFFECT

So far, we have discussed a number of the causes and effects of the bullwhip effect. In order to better understand and control the bullwhip effect, it would also be useful to **quantify** the bullwhip effect, i.e., quantify the increase in variability that occurs at every stage of the supply chain. This would be useful not only to demonstrate the magnitude of the increase in variability, but also to show the relationship between the demand process, the forecasting technique, the lead time and the increase in variability.

To quantify the increase in variability for a simple supply chain, consider a two stage supply chain with a retailer who observes customer demand in time t, D_t, and places an order, q_t, to the manufacturer. Assume the customer demands seen by the retailer are random variables of the form

$$D_t = \mu + \rho D_{t-1} + \epsilon_t, \qquad (14.3)$$

where μ is a non-negative constant, ρ is a correlation parameter with $|\rho| < 1$, and the error terms, ϵ_t, are i.i.d. from a symmetric distribution with mean 0 and variance σ^2.

Suppose that the retailer faces a fixed lead time, L, such that an order placed by the retailer at the end of period t is received at the start of period $t + L$. Also, suppose the retailer follows a simple order-up-to inventory policy in which, in every period, she places an order to bring her inventory level up to a target level. That target inventory level is as given in equation (14.1).

14.3.1 Moving Average Forecasts

In order to implement this inventory policy, the retailer must estimate the mean and standard deviation of demand based on her observed customer demand. Suppose the retailer uses one of the simplest forecasting techniques: the moving average. In other words, in each period the retailer estimates the mean leadtime demand as $\hat{\mu}_t^L = L\hat{\mu}_t$ where $\hat{\mu}_t$ is an average of the previous p observations of demand, i.e.,

$$\hat{\mu}_t = \frac{\sum_{i=1}^{p} D_{t-i}}{p}.$$

She estimates the standard deviation of the L period forecast errors using the sample standard deviation of the single period forecast errors, $e_t = D_t - \hat{\mu}_t$.

Our objective is to quantify the increase in variability. To do this, we must determine the variance of q_t relative to the variance of D_t, i.e., the variance of the orders placed by the retailer to the manufacturer relative to the variance of the demand faced by the retailer. For this purpose, we write q_t as

$$q_t = y_t - y_{t-1} + D_{t-1}.$$

Observe that q_t may be negative, in which case we assume, similarly to Kahn (1987) and Lee, Padmanabhan and Whang (1997b), that this *excess inventory is returned without cost*. This assumption is not a standard one in inventory models, but it is necessary to obtain analytical results. In addition, simulation results indicate that this assumption has little impact on the results presented in this paper. For a discussion of these simulation results, see Chen, Drezner, Ryan and Simchi-Levi (1998). For further discussion of this assumption see Lee, Padmanabhan and Whang (1997b)

Given the estimates of the mean and standard deviation of the lead time demand, we can write the order quantity q_t as

$$\begin{aligned} q_t &= \hat{\mu}_t^L - \hat{\mu}_{t-1}^L + z(\hat{\sigma}_t^L - \hat{\sigma}_{t-1}^L) + D_{t-1} \\ &= (1 + L/p)D_{t-1} - (L/p)D_{t-p-1} + z(\hat{\sigma}_t^L - \hat{\sigma}_{t-1}^L). \end{aligned}$$

And taking the variance of q_t we get

$$\begin{aligned} Var(q_t) &= (1 + L/p)^2 Var(D_{t-1}) + (L/p)^2 Var(D_{t-p-1}) \\ &\quad -2(L/p)(1 + L/p)Cov(D_{t-1}, D_{t-p-1}) \\ &\quad +2z(1 + 2L/p)Cov(D_{t-1}, \hat{\sigma}_t^L) \\ &\quad +z^2 Var(\hat{\sigma}_t^L - \hat{\sigma}_{t-1}^L) \\ &= \left[1 + \left(\frac{2L}{p} + \frac{2L^2}{p^2}\right)(1 - \rho^p)\right] Var(D) \end{aligned}$$

$$+2z(1 + 2L/p)Cov(D_{t-1}, \hat{\sigma}_t^L)$$
$$+z^2 Var(\hat{\sigma}_t^L - \hat{\sigma}_{t-1}^L),$$

where the second equation follows from $Var(D) = \frac{\sigma^2}{1-\rho^2}$ and $Cov(D_{t-1}, D_{t-p-1}) = \frac{\rho^p}{(1-\rho^2)}\sigma^2$.

To further evaluate $Var(q_t)$ we need the following lemma:

Lemma 14.3.1 *If the retailer uses a simple moving average forecast with p demand observations, and if the demand process satisfies (14.3), then*

$$Cov(D_{t-i}, \hat{\sigma}_t^L) = 0 \, for \, all \, i = 1, \ldots, p.$$

We therefore have the following lower bound on the increase in variability from the retailer to the manufacturer:

Theorem 14.3.2 *If the retailer uses a simple moving average forecast with p demand observations, an order-up-to inventory policy defined in (14.1), and if the demand process satisfies (14.3), then the variance of the orders, q^{MA}, placed by the retailer to the manufacturer, satisfies*

$$\frac{Var(q^{MA})}{Var(D)} \geq 1 + \left(\frac{2L}{p} + \frac{2L^2}{p^2}\right)(1 - \rho^p), \qquad (14.7)$$

where $Var(D)$ is the variance of customer demand. The bound is tight when $z = 0$.

Several observations can be made from this relationship. First, we notice that the increase in variability from the retailer to the manufacturer is a function of three parameters, (1) p, the number of observations used in the moving average, (2) L, the leadtime between the retailer and the manufacturer, and (3) ρ, the correlation parameter. The impact of the number of observations and the lead time is intuitive. The increase in variability is a decreasing function of p, the number of observations used in the moving average, and an increasing function of L, the lead time.

The correlation parameter, ρ, can have a significant impact on the increase in variability. First, if $\rho = 0$, i.e., if demands are independent and identically distributed (i.i.d.), then

$$\frac{Var(q^{MA})}{Var(D)} \geq 1 + \frac{2L}{p} + \frac{2L^2}{p^2}. \qquad (14.8)$$

Second, if $\rho > 0$, i.e., if the demands are positively correlated, then the larger ρ, the smaller the increase in variability. In addition, we see that positively correlated demands lead to less variability than i.i.d. demands. On the other hand, if $\rho < 0$, i.e., if the demands are negatively correlated, then we see some strange behavior. For even values of p, $(1 - \rho^p) < 1$, while for odd values of p, $(1 - \rho^p) > 1$. Therefore, for $\rho < 0$, the lower bound on the increase in variability will be larger for odd values of p than for even values of p.

Finally, we note that the lower bound is tight in the case of $z = 0$, i.e., if the retailer's order-up-to level is just a multiple of the mean demand. We will discuss inventory policies of this form later in this paper. In addition, simulation estimates indicate that this the lower bound is quite good for other values of z. In other words, the lower bound given in (14.7) accurately describes the behavior of the system even in cases where $z \neq 0$.

Figure 1 shows the lower bound (solid lines) on the increase in variability for $L = 1$ and $z = 2$ for various values of ρ and p. It also presents simulation estimates of the exact value of $Var(q_t)/\sigma^2$ (dotted lines). We see that the increase in variability is larger for small values of the correlation than for high values of the correlation. Figure 2 shows the lower bound (solid lines) on the variability amplification as a function of p for various values of the lead time, L, for $\rho = 0$ and $z = 2$. It also presents simulation estimates of the exact value of $Var(q_t)/\sigma^2$ (dotted lines). For additional simulation results, see Chen, Drezner, Ryan and Simchi-Levi (1998).

14.3.2 Exponential Smoothing Forecasts

We can perform a similar analysis for any forecasting method. Consider, for example, exponential smoothing, one of the most common forecasting techniques in practice. In this case, the retailer estimates the mean leadtime demand as $\hat{\mu}_t^L = L\hat{\mu}_t$ with

$$\hat{\mu}_t = \alpha D_{t-1} + (1 - \alpha)\hat{\mu}_{t-1},$$

where $0 < \alpha \leq 1$ is the smoothing constant. In other words, the current forecast, $\hat{\mu}_t$, is just the weighted average of the previous period's demand and the previous period's forecast demand, where α is the relative weight to be placed on the current period's demand. We note that, in general, $0 < \alpha \leq 0.30$. We assume that the retailer estimates the standard deviation of the forecast errors using the standard estimate based on the mean absolute deviation of the forecast errors. For more information on exponential smoothing forecasts, see Hax and Candea (1984, Chap. 4).

Given these estimates of the mean and standard deviation, the retailer can form an order-up-to inventory policy as defined in equation (14.1), and we can repeat the above analysis. In this case, if the demands seen by the retailer satisfy (14.3), we have the following lower bound on the increase in variability from the retailer to the manufacturer.

Theorem 14.3.3 *If the retailer uses a simple exponential smoothing forecast with smoothing parameter α, $0 < \alpha \leq 1$, and an order-up-to inventory policy defined in (14.1), then the variance of the orders, q^{EX}, placed by the retailer to the manufacturer, satisfies*

$$\frac{Var(q^{EX})}{Var(D)} \geq 1 + \left(2L\alpha + \frac{2L^2\alpha^2}{2 - \alpha}\right)\left(\frac{1 - \rho}{1 - \beta\rho}\right), \qquad (14.10)$$

where $Var(D)$ is the variance of customer demand. The bound is tight when $z = 0$.

Again, several important observations can be made from this relationship. First, we notice that the increase in variability from the retailer to the manufacturer is a function of three parameters, (1) α, the smoothing parameter used in the exponential smoothing, (2) L, the leadtime between the retailer and the manufacturer, and (3) ρ, the correlation parameter.

The impact of the smoothing parameter and the lead time is quite intuitive. The increase in variability is an increasing function of α, the smoothing parameter. This can be explained as follows. Recall that α is the weight placed on the most recent observation of demand in the exponential smoothing forecast. Therefore, the more weight we place on a single observation, the greater the increase in variability. Also notice that, as with the moving average, the increase in variability is an increasing function of L, the lead time.

The correlation parameter, ρ, can also have a significant impact on the increase in variability. First, if $\rho = 0$, i.e., if demands are independent and identically distributed, then

$$\frac{Var(q^{EX})}{Var(D)} \geq 1 + 2L\alpha + \frac{2L^2\alpha^2}{2 - \alpha}. \tag{14.12}$$

Next, notice that if $0 \leq \rho < 1$,

$$\frac{1 - \rho}{1 - \beta\rho} \leq 1,$$

while if $-1 < \rho \leq 0$,

$$\frac{1 - \rho}{1 - \beta\rho} \geq 1.$$

Therefore, for **positively correlated** demands, the increase in variability will be **less** than for i.i.d. demands ($\rho = 0$). On the other hand, for **negatively correlated** demands, the increase in variability will be **greater** than for i.i.d. demands.

Finally, we again note that the lower bound is tight in the case of $z = 0$, and that simulation estimates indicate that this the lower bound is quite good for other values of z. In other words, the lower bound given in (14.3.3) accurately describes the behavior of the system even in cases where $z \neq 0$.

Figure 3 presents the increase in variability for a retailer using an exponential smoothing forecast when $L = 1$ and $z = 0$ for various values of the correlation parameter, ρ. The figure demonstrates that negatively correlated demands can lead to significantly higher variability than positively correlated demands. For additional results see Chen, Ryan and Simchi-Levi (1998).

14.3.3 Multiple Retailers and a Single Manufacturer

The model presented above clearly does not capture many of the complexities involved in real world supply chains. For example, we have not considered a multi-stage system with multiple retailers and manufacturers. Extending our results to the multiple retailer case when demand between retailers may be correlated is straightforward.

Assume that there are n retailers and one manufacturer. Retailer k observes customer demand in time $t - 1$, D_{t-1}^k, uses a moving average with p observations to forecast the mean, $\hat{\mu}_t^k$, and standard deviation, $\hat{\sigma}_t^{L,k}$, of the lead time demand, calculates the order-up-to point, y_t^k, and places an order, q_t^k, to the manufacturer. Then, as in the previous section, assuming that excess inventory is returned at no cost, we have,

$$q_t^k = y_t^k - y_{t-1}^k + D_{t-1}^k$$

$$= (1 + L/p)D_{t-1}^k - L/pD_{t-p-1}^k + z(\hat{\sigma}_t^{L,k} - \hat{\sigma}_{t-1}^{L,k}).$$

Here we have assumed that every retailer faces the same lead time, L. The results presented in this section can be modified to include different lead times, L_k, for each retailer.

Given the orders q_t^k placed by each retailer in period t, the manufacturer sees a total order

$$Q_t = \sum_{k=1}^n q_t^k = (1 + L/p)\sum_{k=1}^n D_{t-1}^k - L/p\sum_{k=1}^n D_{t-p-1}^k + z\sum_{k=1}^n(\hat{\sigma}_t^{L,k} - \hat{\sigma}_{t-1}^{L,k}).$$

To quantify the bullwhip effect in this case, we need to determine the variance of the total order seen by the manufacturer relative to the variance of the total demand seen by the retailers. We assume that the demands observed by the retailers are i.i.d. over time, but may be correlated across the retailers. Then, under certain conditions on the symmetry of the joint distribution of demand across retailers, or if $z = 0$, we can state the following lower bound on the variance of the total order seen by the manufacturer relative to the variance of the total customer demand.

Theorem 14.3.4 *If each retailer uses a simple moving average forecast with p demand observations, and an order-up-to inventory policy defined in (14.1), then the variance of the total order, Q, placed by the retailers to the manufacturer, satisfies*

$$\frac{Var(Q}{Var(\sum_{k=1}^n D^k)} \geq 1 + \frac{2L}{p} + \frac{2L^2}{p^2}. \tag{14.14}$$

The bound is tight when $z = 0$.

It is now clear that the lower bound on the relative increase in variability seen by the manufacturer in the case of multiple retailers is identical to the lower bound on the relative increase in variability seen by the manufacturer in the case of a single retailer, given in (14.8). For a more detailed discussion of the multiple retailer case, see Ryan (1997).

14.4 MANAGERIAL INSIGHTS ON THE BULLWHIP EFFECT

In this section we present some important insights regarding the bullwhip effect that can be derived from the expressions given above for the variance of the orders placed by the retailer relative to the variance of customer demand. In addition, we also present some important insights into the impact of different forecasting methods and different types of demand patterns, e.g., demands with an increasing mean, on the bullwhip effect.

14.4.1 The bullwhip effect is due, in part, to the need to forecast demand.

As discussed above, the bullwhip effect is caused, in part, by the need to forecast the mean and standard deviation of demand. In fact, if the retailer follows a simple order-up-to inventory policy in which she does not update the forecast of the mean and standard deviation of demand in each period, then the target inventory level will remain the same in each period, and there will be no increase in variability, i.e., there will be no bullwhip effect. The results presented in the previous section clearly demonstrate the impact of demand forecasting on the bullwhip effect.

Although this insight seems to suggest that we should eliminate demand forecasting, this clearly is not the answer. Forecasting, of course, provides valuable information to each stage of the supply chain. Without demand forecasting, we would be forced to simply "guess" the expected demand in each period.

14.4.2 Smoother demand forecasts can reduce the bullwhip effect.

As discussed above, if a retailer follows a simple order-up-to inventory policy, such as (14.1), then any forecasting technique which updates the mean and standard deviation of demand in each period will cause the bullwhip effect. We can reduce this effect by using a forecasting technique that produces smooth forecasts, that is, a forecasting technique in which the estimates of the mean and standard deviation of demand do not change dramatically from period to period.

Consider, for example, the moving average forecast, in which we estimate the mean demand as the average of the previous p demand observations. Clearly, when p is small, the estimate of the mean will change more rapidly than when p is large. In other words, larger values of p create smoother forecasts. This implies that the larger p, i.e., the more demand information used to construct the forecast, the lower the bullwhip effect. This is exactly the effect we saw in the previous section. Similarly, for exponential smoothing, small values of α produce smoother forecasts. Therefore, as seen in the previous section, small values of α mean less variaibility.

14.4.3 With long lead times we must use more demand data to reduce the bullwhip effect.

We have already seen that lead times can add to the bullwhip effect by magnifying the increase in variability due to the demand forecasting. To understand the relationship between the lead time and forecasting technique more fully, note that the expression for the variance of the orders placed by the retailer given in equation (14.7) is an increasing function of the ratio of the lead time to the number of demand observations. In other words, the variability of the orders placed by the retailer increases in the ratio L/p. Similarly, equation (14.10) is an increasing function of $L\alpha$. This fact leads to the following conclusion: a retailer faced with long lead times must use more demand data in her forecasts in order to reduce the bullwhip effect. In other words, when the lead time increases, the retailer must use smoother demand forecasts in order to prevent a further increase in variability.

14.4.4 The magnitude of the increase in variability depends on the forecasting method.

It is now clear that demand forecasting can cause the bullwhip effect. We can also ask whether the **magnitude** of the increase in variability depends on the forecasting method. To answer this question, we compare the magnitude of the increase in variability for the two forecasting methods considered above, moving average and exponential smoothing, for i.i.d. demands and $z = 0$. To do this, we design the forecasts, i.e., choose the parameters p and α, such that both forecasts have the same variance of the forecast errors. It is easily verified, see Nahmias (1993), that if the demands seen by the retailer are i.i.d., then the standard deviation of the forecast errors for exponential smoothing forecasts with smoothing parameter α, is given by

$$\sigma_e = \sigma \sqrt{\frac{2}{2-\alpha}},$$

while for moving average forecasts, the standard deviation of the forecast errors is

$$\sigma_e = \sigma \sqrt{\frac{p+1}{p}},$$

where p is the number of observations used in the moving average. If we equate these standard deviations, we obtain

$$\alpha = \frac{2}{p+1}.$$

Using this relationship between α and p, by equation (14.12), we have

$$\frac{Var(q^{EX})}{\sigma^2} = 1 + \frac{4L}{p+1} + \frac{4L^2}{p(p+1)}.$$

On the other hand, when the retailer uses a simple moving average with p observations, we have (14.8). Since

$$\frac{4L}{p+1} + \frac{4L^2}{p(p+1)} = \left(\frac{2L}{p} + \frac{2L^2}{p^2}\right)\frac{2p}{p+1} > \frac{2L}{p} + \frac{2L^2}{p^2},$$

for $p \geq 1$, we see that the increase in variability for simple exponential smoothing is greater than the increase in variability for the simple moving average. In other words, if demands are i.i.d. and $z = 0$, the bullwhip effect will be larger for a retailer using an exponential smoothing forecast than for a retailer using a moving average forecast.

14.4.5 The magnitude of the increase in variability depends on the demand process.

We have already seen that the increase in variabilty due to demand forecasting depends on the exact nature of the customer demand process. For example, we can see from (14.7) and (14.10) that the magnitude of the increase in variablity depends on ρ, the correlation parameter. In particular, we've seen that negatively correlated demands can lead to higher variability than positively correlated demands.

So far we have considered forecasting only the mean and standard deviation of demand under the assumption that the mean demand remains constant over time. Of course, it is frequently the case that the average customer demand changes over time. For example, the average demand may be steadily increasing over time, as in the case of a new product whose market share is growing steadily over time. On the other hand, the average demand for an item may fluctuate over time, as in the case of a seasonal product whose demand may be higher in the summer and lower in the winter. We can also look at the impact of these different types of demand patterns on the bullwhip effect.

As an example, consider the simple supply chain model discussed in the previous section, but now assume that customer demands are independent random variables of the form

$$D_t = \mu + bt + \epsilon_t, \tag{14.17}$$

where μ represents the constant demand level at time $t = 0$, b denotes a linear trend factor, and ϵ_t is the random error term. Assume that the error terms are i.i.d. from a symmetric distribution with mean 0 and variance σ^2. This implies that the D_t are independent with mean $\mu + bt$ and variance σ^2. In this case, we can use the standard two parameter exponential smoothing method to forecast customer demand, as described in Hax and Candea (1984, p. 160). As in exponential smoothing with no trend, the estimate of the constant level, μ, is

$$\hat{\mu}_t = \alpha_1 D_{t-1} + (1 - \alpha_1)\hat{\mu}_{t-1}.$$

Similarly, the estimate of the trend parameter, b, is

$$\hat{b}_t = \alpha_2(\hat{\mu}_t - \hat{\mu}_{t-1}) + (1 - \alpha_2)\hat{b}_{t-1},$$

where $0 < \alpha_1 \leq 1$ and $0 < \alpha_2 \leq 1$ are smoothing constants. Given these estimates, the forecast L period demand is

$$\hat{\mu}_t^L = L\hat{\mu}_t + \left(\frac{L}{\alpha_1} + \frac{L(L-1)}{2}\right)\hat{b}_t.$$

We can repeat the analysis above to obtain the following bound on the variance of the orders placed by a retailer using exponential smoothing to forecast a demand process with a trend, $Var(q_t^{EXT})$:

$$\frac{Var(q_t^{EXT})}{\sigma^2} \geq 1 + 2L\alpha_1 + \frac{2L^2\alpha_1^2}{2 - \alpha_1}$$
$$+ [2L\alpha_2 + L(L-1)\alpha_1\alpha_2] \times$$
$$[1 + \frac{L\alpha_1}{2 - \alpha_1} + \frac{(\alpha_1 + \alpha_2)(L\alpha_1 + L\alpha_2 + \frac{L(L-1)}{2}\alpha_1\alpha_2)}{(2 - \alpha_1)(\alpha_1 + \alpha_2 - \alpha_1\alpha_2)}].$$

From this expression, we can see that, for a retailer using exponential smoothing with $\alpha = \alpha_1$, any choice of α_2, and $z = 0$, the increase in variability for the linear trend demand model is greater than the increase in variability, given by (14.12), for the i.i.d. demand model.

We also note that the increase in variability does not depend on the true value of the trend parameter, b, but rather on the smoothing parameter for the linear trend, α_2. In other words, the additional increase in variability is not due to the magnitude of the linear trend, but to the need to estimate an additional parameter, b, when forecasting a demand process with a linear trend. In particular, consider a retailer who faces i.i.d. demands, but who believes she is facing a demand process with a linear trend, and therefore uses an exponential smoothing forecast with a linear trend. The orders placed by this retailer will be more variable than those placed by the same retailer using a simple (without trend) exponential smoothing forecast.

14.5 THE IMPACT OF CENTRALIZED INFORMATION ON THE BULLWHIP EFFECT

One of the most frequent suggestions for reducing the bullwhip effect is to centralize demand information within a supply chain, that is, to provide each stage of the supply chain with complete information on the actual customer demand. For example, Lee, Padmanabhan, and Whang (1997a) suggest that "one remedy is to make demand data at a downstream site available to the upstream site." To understand why centralized demand information can reduce the bullwhip effect, note that if demand information is centralized, each stage of the supply chain can use the actual customer demand data to create more acurate (less variable) forecasts, rather than relying on the orders received from the previous stage, which, as we have seen, can be significantly more variable than the actual customer demand.

In this section we consider the value of sharing customer demand information within a supply chain. To determine the impact of centralized demand information on the bullwhip effect, we will consider two types of supply chain, with and without centralized demand information, both of which are described below.

14.5.1 Centralized Demand Information

Consider a multi-stage supply chain in which the first stage (i.e., the retailer) shares all demand information with each of the subsequent stages. In other words, in each period the retailer provides every stage of the supply chain with complete information on customer demand. Assume that all stages of the supply chain use a moving average forecast with p observations to estimate the mean demand, and that the demands seen by the retailer are i.i.d.. Therefore, since each stage has complete information on customer demand, each stage will use the same estimate of the mean demand, $\hat{\mu}_t = \sum_{i=1}^{p} D_{t-i}/p$.

Finally, assume that each stage follows an order-up-to inventory policy in which the order-up-to point in each period is just the forecast demand over a fixed lead time. That is, assume that each stage, k, follows an order-up-to policy where the order-up-to point is of the form

$$y_t^k = L_k \hat{\mu}_t.$$

Note that this inventory policy is just a special case of the policy defined in (14.1) with $z = 0$. Frequently, when a policy of this form is used in practice, an inflated value of the lead time L_k is used, with the excess inventory representing safety stock.

For example, a retailer facing an order lead time of three weeks may choose to keep inventory equal to four weeks of forecast demand, with the extra week of inventory representing his safety stock. In our experience, policies of this form are often used in practice. Indeed, we have recently collaborated with a major U.S. retail company that uses a policy of this form. See also Johnson *et al.* (1995).

In this case, if we perform an analysis similar to that presented above, we have the following expression for the variance of the orders placed by stage k, q^k, relative to the variance of customer demand.

$$\frac{Var(q^k)}{Var(D)} = 1 + \frac{2(\sum_{i=1}^{k} L_i)}{p} + \frac{2(\sum_{i=1}^{k} L_i)^2}{p^2} \qquad \forall\, k. \qquad (14.22)$$

Notice that this expression is quite similar to the single stage bound presented in (14.8), but with the single stage lead time, L, replaced by the echelon lead time, $\sum_{i=1}^{k} L_i$. This result also provides a lower bound on the increase in variability when the retailer uses an inventory policy in which z is not equal to zero, as defined in (14.1).

In many supply chains where customer demand information is shared between stages an echelon inventory policy is used. Consider an echelon inventory policy where the order-up-to point is of the form

$$y_t^k = \left(\sum_{i=1}^{k} L_i \right) \hat{\mu}_t,$$

where $\sum_{i=1}^{k} L_i$ is the echelon lead time. The expression given in (14.22) for the variance of the orders placed by each stage of the supply chain also holds in this case. In other words, if demand information is centralized, then the increase in variability seen by each stage of the supply chain is the same whether the supply chain follows an echelon inventory policy or not.

Finally, notice that (14.22) demonstrates that even when (i) all demand information is centralized, (ii) every stage of the supply chain uses the same forecasting technique, and (iii) every stage of the supply chain uses an echelon inventory policy, we will still see an increase in variability at each stage of the supply chain. In other words, by centralizing customer demand information and coordinating inventory control we have not completely eliminated the bullwhip effect.

14.5.2 *Decentralized Demand Information*

Next, consider a supply chain similar to the one just analyzed, but without centralized customer demand information. In this case, the retailer does not provide the upstream stages with any customer demand information. Therefore, each stage determines its forecast demand based on the orders placed by the previous stage, not based on actual customer demand. Assume that each stage of the supply chain follows an order-up-to policy of the form

$$y_t^k = L_k \hat{\mu}_t^{(k)},$$

where L_k is the lead time between stages k and $k+1$,

$$\hat{\mu}_t^{(1)} = \frac{\sum_{i=1}^{p} D_{t-i}}{p},$$

and

$$\hat{\mu}_t^{(k)} = \frac{\sum_{j=0}^{p-1} q_{t-j}^{k-1}}{p} \qquad \forall\, k \geq 2,$$

where q_t^k is the order placed by stage k in period t.

In this case, we have the following lower bound on the variance of the orders placed by each stage of the supply chain:

$$\frac{Var(q^k)}{Var(D)} \geq \prod_{i=1}^{k} \left(1 + \frac{2L_i}{p} + \frac{2L_i^2}{p^2} \right) \qquad \forall\, k. \tag{14.23}$$

Note that this expression for the variance of the orders placed by the kth stage of the supply chain is similar to the expression for the variance of the orders placed by the retailer given in (14.8), but now the variance increases multiplicatively at each stage of the supply chain. Also, again note that the variance of the orders becomes larger as we move up the supply chain, so that the orders placed by the manufacturer are more variable than the orders placed by the retailer.

14.5.3 Managerial Insights on the Value of Centralized Information

We are interested in determining the impact of centralized customer demand information on the bullwhip effect. To do this, we must compare the increase in variability at each stage of the supply chain for the centralized and decentralized systems. We have already seen that for either type of supply chain, i.e., centralized or decentralized, the variance of the order quantities becomes larger as we move up the supply chain, so that the orders placed by the manufacturer are more variable than the orders placed by the retailer, and so on. The difference in the two types of supply chains is in terms of how much the variability grows as we move from stage to stage.

The results above indicate that, for supply chains with centralized information, the increase in variability at each stage is an additive function of the lead time and the lead time squared, while for supply chains without centralized information, the lower bound on the increase in variability at each stage is multiplicative. In other words, a decentralized supply chain, in which only the retailer knows the customer demand, can lead to significantly higher variability than a centralized supply chain, in which customer demand information is available at each stage of the supply chain, particularly when lead times are large. So we conclude that **centralizing customer demand information can significantly reduce the bullwhip effect.**

It is clear that by sharing demand information with each stage of the supply chain we can significantly reduce the bullwhip effect. This reduction in variability is due to the fact that when demand information is centralized, each stage of the supply chain can use the actual customer demand data to estimate the average demand. On the other hand, when demand information is not shared, each stage must use the orders placed by the previous stage to estimate the average demand. As we have already seen, these orders are more variable than the actual customer demand data, and thus the forecasts created using these orders are more variable, leading to even more variable orders.

Figure 4 presents the variability in orders placed by the manufacturer (i.e., stage 2 in the supply chain) for the case of centralized (dashed lines) and decentralized (dotted lines) demand information when $L = 1$. In each case, the figure presents the increase in variability for $z = 0$ (from (14.22) in the centralized case and (14.23) in the decentralized case) and for $z = 2$ (based on simulation results). It also presents, as a reference, the increase in variability for the first stage (solid lines) for both $z = 0$ (from (14.8)) and $z = 2$ (simulation results).

Figure 5 presents the variability of the orders placed by stages 1, 2 and 3 of the supply chain, when $z = 0$ and $L_1 = L_2 = L_3 = 3$, for centralized (dashed lines) and decentralized supply chains (dotted lines). This figure, when compared with Figure 4, indicates that in multi-stage supply chains, the lead time can have a significant impact on the increase in variability. In addition, this figure indicates that the **difference** between the variance in the centralized and decentralized supply chains **increases** as we move up the supply chain, i.e., as we move from the second to the third stage of the supply chain.

Finally, both Figures 4 and 5 demonstrate that centralizing customer demand information can reduce variability. In addition, we see that for *both* the centralized and the decentralized systems there is an increase in variability when we move from the first stage to the second stage to the third stage. This figure reinforces an important insight: **the bullwhip effect exists even when demand information is completely centralized and all stages of the supply chain use the same forecasting technique and inventory policy**. So we conclude that centralizing customer demand information can significantly reduce, but will not eliminate, the bullwhip effect.

14.6 METHODS FOR REDUCING THE IMPACT OF THE BULLWHIP EFFECT

The research on identifying and quantifying the causes of the bullwhip effect has led to a number of suggestions for reducing the bullwhip effect or for eliminating its impact. These suggestions, as discussed in Lee, Padmanabhan and Whang (1997 a, b), include reducing uncertainty, reducing the variability of the customer demand process, reducing lead times and engaging in strategic partnerships. These suggestions, and how they relate to the new insights presented above, are discussed briefly below.

14.6.1 Reducing Uncertainty

As mentioned above, one of the most frequent suggestions for reducing or eliminating the bullwhip effect is to reduce uncertainty throughout the supply chain by centralizing customer demand information. The results presented in the previous section demonstrate that centralizing demand information can in fact reduce (but will not eliminate) the bullwhip effect.

Lee, Padmanabhan and Whang (1997a) also point out, however, that even if each stage uses the same demand data, they may use different forecasting methods and different buying practices or ordering policies, both of which may contribute to the bullwhip effect. The results presented in the previous section indicate that even when each stage uses the same demand data, the same forecasting method and the same ordering policy, the bullwhip effect will continue to exist.

14.6.2 Reducing Variability

We can also reduce the bullwhip effect by reducing the variability inherent in the customer demand process. For example, if we can reduce the variability of the customer demand seen by the retailer, then even if the bullwhip effect occurs, the variability of the demand seen by the manufacturer will also be reduced.

We can reduce the variability of customer demand through, for example, the use of an "every day low pricing" strategy, or EDLP. When a retailer uses EDLP, he offers a product at a single consistent price, rather than offering a regular price with periodic price promotions. By eliminating price promotions, a retailer can eliminate many of the dramatic shifts in demand that occur along with these promotions. Therefore, every day low pricing strategies can lead to much more stable, i.e., less variable, customer demand patterns.

14.6.3 Leadtime Reduction

The results presented in this paper clearly indicate that lead times serve to magnify the increase in variability due to demand forecasting and demonstrate the dramatic effect that lead times can have on the variability at each stage of the supply chain. Therefore, it is clear that lead time reduction can significantly reduce the bullwhip effect throughout a supply chain.

The results presented here also demonstrate the relationship between the lead time and the forecasting technique. In particular, we have demonstrated that with longer lead times a retailer must use more demand data (smoother forecasts) in order to reduce the bullwhip effect.

14.6.4 Strategic Partnerships

Finally, we can also eliminate the bullwhip effect by engaging in any of a number of strategic partnerships. These strategic partnerships change the way information is shared and inventory is managed within the supply chain, and can therefore reduce or eliminate the impact of the bullwhip effect. For example, in vendor managed inventory (VMI), the manufacturer manages the inventory of his product at the retailer, determining for himself how much inventory to keep on hand and how much to ship to the retailer in every period. Therefore, the manufacturer does not rely on the orders placed by the retailer, and thus avoids the bullwhip effect entirely.

Other types of partnerships can be designed to reduce the bullwhip effect. For example, the results presented in this paper indicate that centralizing demand information can dramamtically reduce the variability seen by the upstream stages in a supply chain. Therefore, it is clear that these upstream stages would benefit from a strategic partnership which provides an incentive for the retailer to make customer demand data available to the rest of the supply chain.

436

References

Baganha, M. and M. Cohen, "The Stabilizing Effect of Inventory in Supply Chains," To appear in *Operations Research*, 1995.

Caplin, A.S., "The Variability of Aggregate Demand with (S,s) Inventory Policies," *Econometrica*, **53** (1985), 1396–1409.

Chen, F., J. K. Ryan and D. Simchi-Levi, "The Impact of Exponential Smoothing Forecasts on the Bullwhip Effect," Working Paper, Northwestern University, 1998.

Chen, F., Z. Drezner, J. K. Ryan, and D. Simchi-Levi, "Quantifying the Bullwhip Effect in a Simple Supply Chain: The Impact of Forecasting, Leadtimes and Information," Working Paper, Northwestern University, 1998.

Johnson, M.E., H.L. Lee, T. Davis and R. Hall (1995), Expressions for Item Fill Rates in Periodic Inventory Systems. *Naval Research Logistics*, **42**, pp. 39–56.

Kahn, J., "Inventories and the Volatility of Production," *The American Economic Review*, **77** (1987), 667–679.

Kaminsky, P. and D. Simchi-Levi, A New Computerized Beer Game: Teaching the Value of Integrated Supply Chain Management. To appear in the book *Supply Chain and Technology Management*. Hau Lee and Shu Ming Ng, eds., the Production and Operations Management Society, (1996).

Lee, H., P. Padmanabhan and S. Whang, "The Bullwhip Effect in Supply Chains," *Sloan Management Review*, **38** (1997a), 93–102.

Lee, H., P. Padmanabhan and S. Whang, "Information Distortion in a Supply Chain: The Bullwhip Effect," *Management Science*, **43** (1997b), 546–58.

Ryan, J. K., "Analysis of Inventory Models with Limited Demand Information," Ph.D. Dissertation, Department of Industrial Engineering and Management Sciences, Northwestern University, 1997.

Sterman, J. D., "Modeling Managerial Behavior: Misperceptions of Feedback in a Dynamic Decision Making Experiment," *Management Science*, **35** (1989), 321–339.

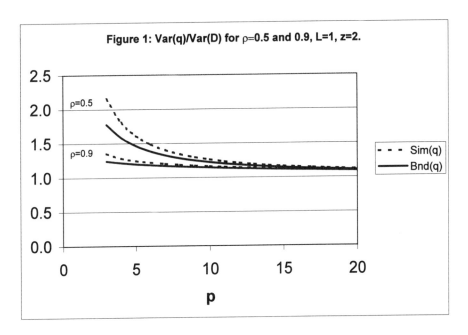

Figure 1: Var(q)/Var(D) for ρ=0.5 and 0.9, L=1, z=2.

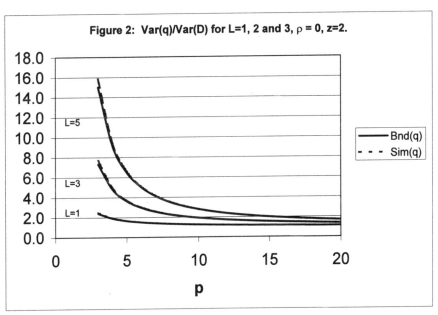

Figure 2: Var(q)/Var(D) for L=1, 2 and 3, ρ = 0, z=2.

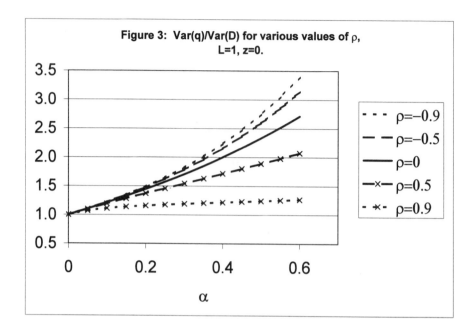

Figure 3: Var(q)/Var(D) for various values of ρ, L=1, z=0.

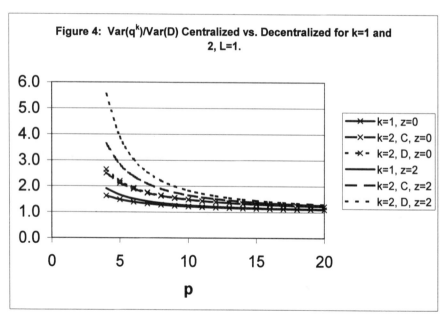

Figure 4: Var(q^k)/Var(D) Centralized vs. Decentralized for k=1 and 2, L=1.

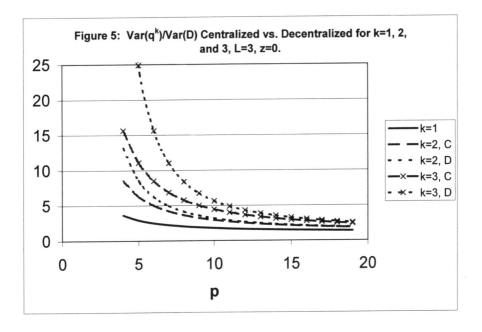

Figure 5: Var(q^k)/Var(D) Centralized vs. Decentralized for k=1, 2, and 3, L=3, z=0.

15 VALUE OF INFORMATION SHARING AND COMPARISON WITH DELAYED DIFFERENTIATION

Srinagesh Gavirneni

Schlumberger
Austin, TX 78726

Sridhar Tayur

Graduate School of Industrial Administration
Carnegie Mellon University
Pittsburgh, PA 15213

15.1 INTRODUCTION

The industrial supplier-customer relations have undergone radical changes in recent years as the philosophy behind managing manufacturing systems continues to be influenced by several Japanese manufacturing practices. As more organizations realize that successful in-house implementation of Just-In-Time alone will have limited effect, they are seeking other members of their supply chain to change their operations. This has resulted in a certain level of co-operation, mainly in the areas of *supply contracts* and *information sharing*, that was lacking before. This is especially true when dealing with customized products, and is most commonly seen between suppliers and their larger customers.

The motivation to study the benefits of information arose because of differing reactions to Electronic Data Interchange (EDI) benefits from industrial sources: while some were very happy with improved information, others were disappointed at the benefits (see [1] and [16]). Thus, while information is always beneficial, we would like to know when it is most beneficial and when it is only marginally useful. In the latter case, some other characteristic of the system, such as end-item demand variance or supplier capacity may have to be improved before expecting significant benefits from information. Thus, our computational efforts will be directed towards understanding some of these issues.

In section 1 we incorporate information flow between a supplier and a customer in a two-echelon model that captures the capacitated setting of a typical supply chain. We consider two situations: (1) the supplier has the information of the (s, S) policy used by the customer as well as the end-product demand distribution; and (2) the supplier has full information about the state of the customer. We show that order up-to policies continue to be optimal and develop solution procedures to compute the optimal parameters. Study of these models enables us to understand the relationships between capacity, inventory and information at the supplier level and how they are affected by customer $S - s$ values and the end-item demand distribution. We estimate the savings at the supplier due to information flow and study when information is most beneficial.

In section 2 we extend the models in section 1 to incorporate two customers, each using an (s, S) policy, ordering two different products or the same product. The customers may or may not be providing information about their inventories to the supplier. This gives rise to four models as depicted in Figure 15.8. These four models are connected through delayed differentiation and information sharing. The goal of this section is to study (from the supplier's perspective), in a simple yet representative setting, when one strategy would be preferred over the other and how one can enhance the other if applied simultaneously.

Majority of the previous research in this area, i.e. incorporating information flow into inventory control and supply chains, has assumed the presence of infinite capacity which is not the case in this chapter. Zheng and Zipkin

[17] showed that using the information about the outstanding orders of the products resulted in improvement of system performance in a two product setting. Hariharan and Zipkin [10] incorporated information about order arrivals, termed demand lead times, into inventory control and studied its effect on system performance. They observed that demand lead times behave in a fashion that is exactly the opposite of supply lead times. An increase in demand lead time improves the system performance exactly like a reduction in supply lead time. Chen [4] studied the benefit of information flow by observing that echelon base-stock policies and installation base-stock policies in a multi-echelon environment differ in their informational requirements. He also studied the effect of parameters such as number of stages, lead times, demand variance, and customer service on the benefit of information flow. Cachon and Fisher [3] studied the benefits of information flow in a one-warehouse multi-retailer system.

Our framework is that of discrete time production-inventory models. The models described in this chapter lead to capacitated inventory control models that face a non-stationary demand. Capacitated inventory control models have been the focus of many research articles. Federgruen and Zipkin [5, 6] established that a modified order up-to policy is optimal for the stationary demand case. Capacitated non-stationary inventory control models have recently received some attention. Aviv and Federgruen [2] and Kapuscinski and Tayur [11] studied the periodic demand case. The models in this chapter generalize the models in Kapuscinski and Tayur [11]. Though they are not as general as the models in Scheller-Wolf and Tayur [15], these models have special structure that we use to establish some monotonicities which we exploit in our computational procedures. Glasserman and Tayur [8, 9] studied multi-echelon systems operated via a base stock policy. They developed and validated solution procedures based on Infinitesimal Perturbation Analysis (IPA) to compute the optimal base stock levels. We apply this technique in our computational study.

15.2 SINGLE CUSTOMER USING AN (s, S) POLICY

We consider a periodic review inventory control problem (at the supplier level) with linear holding and penalty costs (h and b per unit), finite capacity (C), and no purchase or salvage costs. We do not allow for any disposal of inventory except at end of horizon. The unit variable purchase cost does not affect our results. The customer is following a (s, S) policy. The (s, S) policy was shown to be optimal for infinite horizon inventory problem with periodic review, fixed ordering cost, linear holding and penalty costs, constant lead time, facing stationary demands and full backlogging by Scarf [14]. This is the setting of the customer. Let us first consider Model 1 (see Figure 15.1) where the supplier knows that the customer is using a (s, S) policy and is also aware of the end-item demand distribution. The sequence of events in every period is as follows. The supplier decides on his production quantity for the period. The customer realizes the end-item demand for the period and after satisfying the demand, if her inventory level is below s, she places an order with the supplier

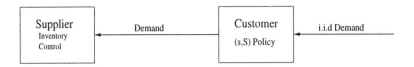

Figure 15.1 Model 1 - Supply chain with some information flow.

to bring her inventory level to S. If the supplier cannot satisfy the full order of the customer, we assume that the customer acquires the missing part of the order elsewhere (or via expediting) and brings her inventory level to S. All this happens with no lead time. A fixed non-zero delivery lead time does not change the analysis. The supplier pays holding cost if there is inventory at his level at the end of the period or pays penalty cost if a portion of the demand could not be satisfied from inventory. The objective is to find a control policy for the supplier that minimizes the total holding and penalty costs.

At the supplier level, incorporation of this information into inventory models results in a non-stationary demand process although the end-item demand is stationary. From the suppliers point of view, every period is in one of k possible 'states' (see remark below). *State* is defined as the number of periods since the last demand. Let p_i be the probability that a demand to the supplier will be realized in state i. If the demand is realized in state i, it is from a cumulative distribution function (cdf) $\Phi_i(.)$ (and probability distribution function, pdf, $\phi_i(.)$) with mean μ_i and the next period will be in state 1. If the demand is not realized, then the next period will be in state $i + 1$. We further assume that $p_i \leq p_{i+1} \forall i < k, p_k = 1$ and $\Phi_i(.) \leq_{st} \Phi_{i+1}(.)$ i.e. the chance and quantity of realization of demand increases in time since last demand ('monotonicity'). If a period reaches state k, a demand will be realized in that period from distribution $\Phi_k(.)$. This monotonicity assumption is satisfied if the end-item demand to the customer has an Increasing Failure Rate (IFR) distribution (such as Uniform, Normal or Erlang distributions); see [7] for a proof. We further assume that the end-item demand has a finite mean, and so $\mu_i < \infty$ for all i.

Remarks. We have made two assumptions above that simplify our analysis. (1) We assume a finite k; this is satisfied if the end-item demand is non-zero in every period, bounded below by some $\epsilon > 0$. Then we can choose $k = \lfloor \frac{\Delta}{\epsilon} \rfloor + 1$. (2) We allow the customer to obtain all the requirements not met from this supplier at a higher price from elsewhere (that includes expediting from this supplier or from an emergency location), and pass on the cost difference to this supplier. This cost differential is what b represents. If the customer at the time of placing her order could not be guaranteed of the full amount demanded, then she will not follow a simple stationary (s,S) policy.

In Model 2 (see Figure 15.2), the *state* captures the inventory level at the customer exactly. As we shall see in section 2.2, Model 2 is structurally equivalent to Model 1 (although definition of states is very different), and some of the demand assumptions made above can be relaxed. In what follows, we will deal with Model 1 and provide a summary of results for Model 2 in Section 15.2.2.

Figure 15.2   Model 2 - Supply chain with full information flow.

Note that at the supplier-customer interface there is an implicit fixed order-ing cost in our model, and hence the customer uses a (s, S) policy (with optimal parameters based on her holding and penalty costs). Once the supplier uses the optimal policy which we derive for models 1 and 2, and if the holding cost at the supplier is lower than the holding cost at the customer, then the *entire supply chain* is 'optimal'. This is a pleasing property obtained due to the simplicity of our setting.

15.2.1   Model 1

We briefly note some results for Model 1. As defined in [5, 6], a modified order up-to policy with level z is one where if the inventory level is less than z, we raise it to z; if this level cannot be reached, we exhaust the available capacity; if the inventory level is above z, we produce nothing.

Property 1.   For finite horizon and infinite horizon (discounted and average cost) an order up-to policy is optimal.

Proof.   The proof follows standard steps as in [5, 6] and [11]. See [7].   □

In the infinite horizon the optimal order up-to level in a period depends only on the state of the system in that period. Let z_i be the infinite horizon optimal order up-to level in state i. In the finite horizon, the optimal order up-to level also depends on the horizon length. Let y_i^n be the optimal order up-to level in state i for an n-period problem.

Property 2.   Optimal order up-to levels are ordered in both finite horizon and infinite horizon (discounted and average cost), i.e. $y_i^n \leq y_{i+1}^n$ and $z_i \leq z_{i+1}$.

Proof.   The proof follows from the monotonicity of p_i and $\Phi_i(.)$. For details see [7].   □

We compute the optimal order up-to levels via simulation based optimization (using infinitesimal perturbation analysis (IPA)). As in [8] and [11] we write the simulation recursions and differentiate them. Validation of the procedure follows standard steps for finite horizon and infinite horizon. We exploit the monotonicity of order up-to levels to show that regeneration occurs if $z_{i+1} \leq \max\{z_i + C, [z_k - \Delta]^+ + C\}$ for $i < k$ where $z_0 = 0$ and $x^+ = \max(0, x)$. See [7].

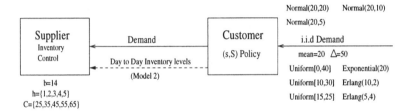

Figure 15.3 The experimental setup.

15.2.2 Model 2

In Model 2, as shown in Figure 15.2, the supplier has information about the day-to-day inventory levels at the customer as well as the (s, S) values and the end-item demand distribution. The state of the system is no longer captured fully by the number of periods since the last demand. So, we redefine the state space to incorporate the information about the total demand seen so far by the customer. Thus, his inventory model corresponding to Model 2 has the following characteristics. *State* is defined as the total demand seen by the customer since receiving her last order from the supplier. Every period is in one of possible Δ states $\{0, 1, 2, \ldots, \Delta - 1\}$ where $\Delta = S - s$. As before, p_i is the probability that a demand to the supplier will be realized in state i. If the demand to the supplier (exactly equal to i plus the end-item demand at the customer for one period) is realized in state i, it is from a distribution with cdf $\Phi_i(.)$ with mean μ_i and the next period will be in state 0. If the demand is not realized, then the next period will be in state $j > i$ with probability $\pi_{(j-i)}$ which is exactly the probability that the end-item demand in a single period is $(j - i)$. (In Model 1, $j = i + 1$.) It is easily verified that $p_i \leq p_{i+1} \forall i < \Delta - 1, p_{\Delta-1} = 1$ and $\Phi_i(.) \leq_{st} \Phi_{i+1}(.)$. i.e. the chance and quantity of realization of demand increases in time. Therefore we do not need the IFR assumption on the end-item demand distribution for monotonicity, nor a lower bound on the end-item demand for finiteness of states.

All of the properties of Model 1 also hold for Model 2. Briefly: (1) the cost function is convex and the optimal policy is modified order up-to; (2) the order up-to levels are ordered; and (3) IPA can be used to find the optimal order up-to levels.

15.2.3 Computational Results

In this section we implement the IPA solution procedure both the models. Our goal is to understand the trade-offs between inventories, capacities and information. We vary C, b, Δ, and variance of end-item demand.

The experimental design is shown in Figure 15.3. The values of b and mean demand were held constant at 14 and 20 respectively. The holding cost h was varied from 1 to 5 in increments of 1, while the capacity was varied from 25 to 65 in increments of 10. Δ was allowed to take values from 50 to 300 in increments of

50. Three distributions (Uniform, Erlang, and Normal) were used for the end-item demand seen by the customer. For each distribution, three values of the parameters were chosen: Uniform[0,40], Uniform[10,30], and Uniform[15,25]; Exponential(20), Erlang(10,2), and Erlang(5,4) (mean 20, standard deviation of 10); Normal(20,20), Normal(20,10), and Normal(20,5) where the first number is the mean and the second number is the standard deviation. It should be noted that while generating demands using the Normal distribution, the non-positive demands were discarded and so the non-truncated mean was selected so that the modified demands had a mean of 20. The standard deviations of the observed demands for these three cases were 13.5, 9.5, and 5 and these values were used while analyzing the results.

15.2.3.1 Costs and Savings. We observed that for all the experiments the total cost increased with increase in holding cost, decrease in capacity, or increase in variance. These results are expected and so we will not elaborate on them here. We will look more closely at the savings realized due to information flow and their change due to changes in capacity, holding cost and variance. The percentage savings from Model 1 to Model 2 were estimated using the formula:

$$\% \text{ savings} \quad = \quad \frac{\text{Model 1 cost} - \text{Model 2 cost}}{\text{Model 1 cost}}.$$

We want to understand how these savings are affected by parameters such as capacity, the ratio of holding cost to penalty cost, the variance of the end-item demand (as seen by the customer), and Δ. Notice that the cost for Model 2 was always smaller than the cost for Model 1. This would imply that information flow will always result in savings. This leads us to conclude that *information is always beneficial.* This is not a surprising result as information only presents the supplier with more options.

The plots of these savings with respect to capacity, penalty cost, variance, and Delta values are given in Figures 15.4, 15.5, 15.6, and 15.7. These savings have a wide range, varying from 1% to 35%. Details of the savings due to changes in capacity, penalty cost, variance, and Δ are described below.

1. EFFECT OF CAPACITY. For every value of capacity and each demand distribution we averaged the percentage savings over all the holding costs. A representative sample of these results are presented in Figure 15.4. It is easily noticed that as the capacity was increased, the percentage savings also increased. When the capacity is 25, since the end-item mean demand is 20, one does not have much choice other than to produce in every period ($S - s$ value does not really affect this) to meet the demand. However when the capacity is high, the supplier has flexibility and thus can use the information to delay production or produce a larger quantity in a given period (if necessary). We believe this ability to react to information is the main reason that *the benefit of information flow is higher at higher capacities.*

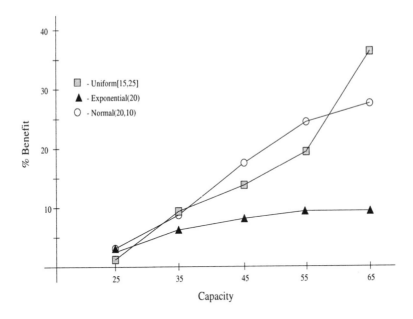

Figure 15.4   Plot of % savings comparing Model 1 to Model 2 versus capacity.

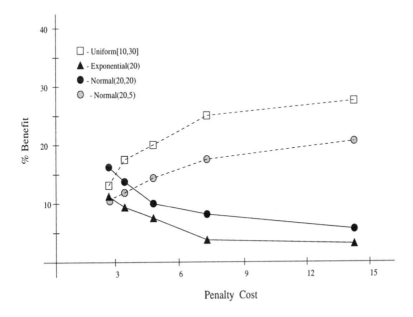

Figure 15.5   Plot of % savings comparing Model 1 to Model 2 versus penalty cost.

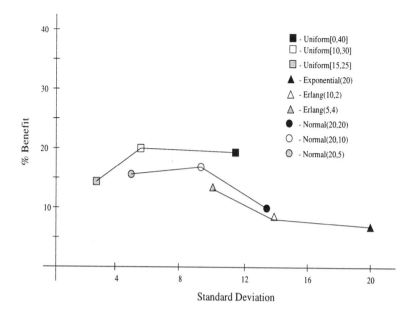

Figure 15.6 Plot of % savings from Model 1 to Model 2 versus variance.

2. EFFECT OF PENALTY COST. Figure 15.5 shows the information benefits as function of penalty cost. For small variances, the information value increases as the penalty cost increases, while for medium to high variances, the benefits decrease. This is because, at higher variances, there remains significant uncertainty on the total demand quantity to the supplier (although the uncertainty of timing may have reduced). In the extreme case when the end-item demand has an Exponential distribution, the excess over Δ has the same distribution, independent of the state at the end of the previous period. Under such large uncertainty in demand, as penalty cost increases the system is more difficult to manage and the information is relatively less beneficial. Similarly at low demand variances, when the penalty cost decreases the system becomes very easy to manage and the information does not play an important role in reducing costs. So we conclude that *if the variance of the demand seen by the customer is small (high), we can expect the benefit of information flow to increase (decrease) with increase in penalty cost.*

3. EFFECT OF END-ITEM DEMAND DISTRIBUTION AND VARIANCE. Figure 15.6 contains the plot of percentage savings versus standard deviation for each of the demand distributions. For the uniform distribution as the standard deviation was decreased from 11.55 to 5.78 to 2.89 the percentage savings decreased while for the Erlang distributions as the variance was decreased from 20 to 14.14 to 10 the percentage savings increased. For the normal distribution as the standard deviation was changed from

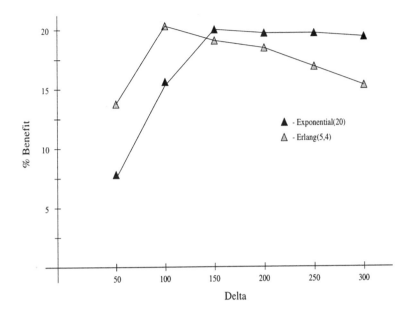

Figure 15.7 Plot of % savings from Model 1 to Model 2 versus Δ.

13.5 to 9.5 to 5.0 the percentage savings initially increased from 7.55% to 9.37% and then dropped to 7.37%. The plot seems to indicate a bell shaped relation between variance and percentage savings. This could be due to the fact that, as in item 2 above, when the variance is very high the uncertainty in quantity to the supplier is not reduced significantly. In other words the information available relative to the uncertainty is small and thus does not reduce costs as effectively. When the variance is very low the system almost acts like a deterministic system and the information is not really beneficial. *Information is most beneficial at moderate values of variance.*

4. EFFECT OF Δ. The percentage savings between models 1 and 2 as a function of Δ are given in Figure 15.7 for two distributions, exp(20) and Erlang(5,4). For both the distributions, as the value of Δ is increased these savings increase initially and then start to decrease. This behavior can be explained as follows. The value of Δ determines the size of the order. Due to finite capacity, in anticipation of a large order we start building up inventory over a horizon which we call "production horizon". When Δ is large (compared to capacity and mean end-item demand), the production horizon is so large that (1) holding costs accrue and (2) Central Limit Theorem applies reducing the coefficient of variation of quantity demanded. On the other hand, when Δ is small (relative to mean end-item demand), the end-item demand is passed through to the supplier almost every period and there is no significant difference between

Models 1 and 2. So we conclude that *information is less beneficial at extreme values of* Δ.

15.2.4 Conclusions

In this section, we incorporated information flow into inventory control models. This gave rise to interesting non-stationary demand processes. Optimal policy structures are order up-to policies. An extensive computational study provided us with insights on the savings and relative benefits due to information flow.

15.3 DELAYED DIFFERENTIATION VERSUS INFORMATION SHARING

In this section we extend the models in section 2 to incorporate the presence of two customers. Our first two models assume that the two customers, each using (s,S) policy and facing end-item demands, are ordering two different products (one each) from a single capacitated supplier: (1) Model 1, where the supplier only knows that customers are using (s, S) policies and is aware of the values of s, S, and the demand distributions of the end-items, and (2) Model 2, where the customers provide also information about their day-to-day inventory levels. The supplier can use this information, available in Model 2, to better predict the time and size of the next demand and thus better manage his/her inventories. From the solutions of these two models we compute the *benefit of information* in the presence of two customers and two products.

Delayed product differentiation is a strategy that has recently become popular for managing product variety (see Lee and Tang [12]). The main idea behind this strategy is to redesign the product and/or the production process so that the point of differentiation (i.e., the stage after which the products assume their unique identities) is delayed as much as possible. This provides the manager with the ability to not commit the work-in-process inventory to a particular finished product until later in the production process. Thus the production process is more flexible and requires lower inventories for the same customer service levels. Hewlett-Packard used this strategy to considerably reduce its inventories of desk-jet printers (see Lee et al. [13]). Hewlett-Packard achieved this by delaying the localization process (adding components that dictate the final destination of the product). To study the benefits of delayed product differentiation, we consider two more models (Model 3 and Model 4) that naturally flow from Models 1 and 2. Since the most we can delay the differentiation is to the point of shipment to the customer, we consider the case when the products have been redesigned and appear to be the same product from the suppliers point of view. This gives us an upper bound on the benefits that can be achieved by the strategy of delayed product differentiation. This collapse of two products to one results in models in which there are two customers using (s, S) policies and ordering the same product: Model 3 falls out from Model 1 and Model 4 from Model 2 The four models are depicted in Figure 15.8.

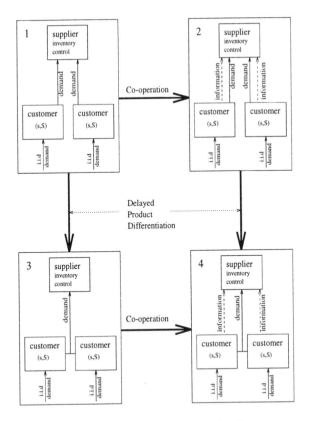

Figure 15.8 Two customers using (s, S) policies

A study of these four models will help us compare and contrast the relative benefits of information sharing and delayed product differentiation. Comparing Model 3 with Model 1 and comparing Model 4 with Model 2 provides us with the effect of delayed differentiation. Comparing Model 2 with Model 1 and comparing Model 4 with Model 3 shows the benefit of improved information. Note that we are studying the benefits of the two strategies at the supplier level. Our computational study will be aimed at computing the relative benefits of these two strategies and studying how they are affected by various system parameters such as supplier capacity, holding cost (of inventory) to penalty cost (of expediting) ratio at the supplier level, end-item demand variance (at the customer level), and relative sizes of the customers.

15.3.1 Inventory Models

We use discrete time models (periodic review). In each period, end-item demands occur at the customers. In all four models, the two customers each follow a (possibly different) (s,S) policy and each face an end-item demand process that is independent (not necessarily identical) from each other. At any customer, the demand is independent and identical from period to period. If the inventory at a customer falls below her s, they place an order to the supplier, an amount that will bring their level to her S.

At the supplier level, the holding costs are linear at h dollars per unit per period. There is finite capacity, C units per period, for production at the supplier. The lead time between the supplier and customers is deterministic, and we assume for convenience that it is zero. In models 1 and 2, we further assume that the supplier has pre-allocated capacity to each customer; so C is split a priori to some C_1, C_2 with $C_1 + C_2 = C$. Thus at the supplier level we have two separate channels serving the two customers and the capacity for one product can not be utilized to produce the other. In models 3 and 4, the two capacities are pooled (as a single product is being produced for both customers). Thus, we will compute the maximum possible benefit from delayed product differentiation. With this assumption, the analysis of models 1 and 2 can be deduced from the models in the previous section and so we skip their details here. We will detail the analysis of Model 3 and mention briefly how it can be adapted to Model 4. As was the case in section 2, we assume that the customer demands that can not be satisfied from supplier's inventory are to be obtained from another source at an additional cost of b dollars per unit, paid by the supplier. This b is thus a 'penalty cost' at the supplier.

15.3.2 Delayed Differentiation without Information Sharing: Model 3

Recall that a (s, S) policy is one in which the customer orders up-to S when her inventory falls below s. Thus the customer does not order every period and when she orders, the order quantity will be at least $\Delta (= S - s)$ units. The demand process originating from such a customer was formulated in Section 15.2. If i periods have passed since a demand was seen from the customer, then the customer is said to be in state i in this period. let p_i^k be the probability that the customer $k \in \{1, 2\}$ places an order in state i. If the demand is realized in state i, it will be from a cumulative distribution function (cdf) $\Phi_i^k(.)$ (and probability distribution function, pdf, $\phi_i^k(.)$) and the customer will be in state 1 in the next period. If the demand is not realized, then customer will be in state $i + 1$ in the next period. It was established in Section 15.2 that under fairly general conditions, for a customer k, $p_i^k \le p_{i+1}^k$ and there exists finite K such that $p_K^k = 1$. Further, the demands are also monotonic in the sense that $\Phi_i^k \le_{st} \Phi_{i+1}^k$. That is, it is more likely to get a demand if the time since the last demand is longer, and the quantity demanded also is higher as i increases.

If customer 1 is in state i and customer 2 is in state j, then the current state of the system is represented by (i, j). Since there are two customers who may or may not place an order, in every period there exist four possible transitions for the state space. From a state of (i, j), the state in the next period is

$(i+1, j+1)$	If neither customer places an order	with prob. $(1 - p_i^1)(1 - p_j^2)$
$(1, j+1)$	If only customer 1 places an order	with prob. $p_i^1(1 - p_j^2)$
$(i+1, 1)$	If only customer 2 places an order	with prob. $(1 - p_i^1)p_j^2$
$(1, 1)$	If both customers place orders	with prob. $p_i^1 p_j^2$

Property 3. For finite horizon and infinite horizon (discounted and average cost) an order up-to policy is optimal.

Proof. The proof follows standard steps as in [5, 6] and [11]. See [7]. □
In the infinite horizon the optimal order up-to level in a period depends only on the state of the system in that period. Let z_{ij} be the infinite horizon optimal order up-to level in state (i, j). In the finite horizon, the optimal order up-to level also depends on the horizon length. Let y_{ij}^n be the optimal order up-to level in state (i, j) for an n-period problem.

Property 4. Optimal order up-to levels are ordered in both finite horizon and infinite horizon (discounted and average cost), i.e. $y_{ij}^n \leq y_{i(j+1)}^n$ and $y_{ij}^n \leq y_{(i+1)j}^n$ and $z_{ij} \leq z_{i(j+1)}$ and $z_{ij} \leq z_{(i+1)j}$.

Proof. The proof follows from the monotonicity of p_i and $\Phi_i(.)$. For details see [7]. □
We use Infinitesimal Perturbation Analysis (IPA) to compute these optimal order up-to levels. Similar to our approach in the previous section, we write the simulation recursions and differentiate them. Validation of the procedure follows standard steps for finite horizon and infinite horizon. We exploit the monotonicity of order up-to levels to establish existence of regeneration.

15.3.3 Delayed Differentiation and Information Sharing: Model 4

In Model 4 the supplier has information about the day-to-day inventory levels at the customers as well as the (s, S) values and the end-item demand distributions. A customer is said to be in state i if the total demand seen by him since his last order is i units. The system is said to be in state (i, j) when customer 1 is in state i and customer 2 is in state j. As before, p_i^k represents the probability that an order will be placed by customer k in state i. If an order is placed by the customer in state i, it will be from the cumulative distribution $\Phi_i^k(\cdot)$ and the customer will be in state 0 in the next period. If a demand is not realized in state i then the customer will be in some state $\hat{i}(= i + \text{end-item demand in this period})$ in the next period. These probabilities and the demand distributions satisfy the monotonicity properties $p_i^k \leq p_{i+1}^k$ and $\Phi_i^k \leq_{st} \Phi_{i+1}^k$. The only difference is in the transitions between the states. From state (i, j), the following are the only possible transitions (where $\hat{i} > i, \hat{j} > j$):

to $(\hat{i},\hat{j})$	if neither customer places an order	with prob. $(1-p_i^1)(1-p_j^2)$;
to $(0,\hat{j})$	if only customer 1 places an order	with prob. $p_i^1(1-p_j^2)$;
to $(\hat{i},0)$	if only customer 2 places an order	with prob. $(1-p_i^1)p_j^2$;
to $(0,0)$	if both customers place orders	with prob. $p_i^1 p_j^2$.

With minor adjustments the properties presented for Model 3 apply to this model as well. Briefly: (1) the cost function is convex and the optimal policy is modified order up-to; (2) the optimal policy exists and has finite cost for the discounted cost criterion and the average cost criterion; (3) the infinite horizon order up-to levels are ordered; (4) for the uncapacitated system, a recursive procedure can be used to find the optimal order up-to levels; and (5) when the capacity is finite, IPA can be used to find the optimal order up-to levels.

15.3.4 Computational Results

We estimate the benefits of co-operation and delayed differentiation and study when one strategy outperforms another and how the to strategies complement each other. Some of the issues we address are: (1) Is delayed differentiation more beneficial than information sharing at higher holding costs or lower holding costs? (2) How do capacity restrictions affect the relative performances of delayed differentiation and information sharing? (3) What is the effect of end-item demand variance on the performance of these two strategies? (4) How does the relative sizes of the two customers affect the answers to the first three questions?

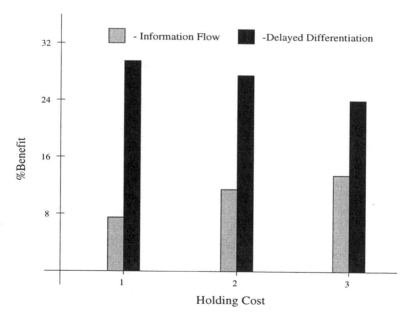

Figure 15.9 Benefits as a function of holding cost for symmetric customers

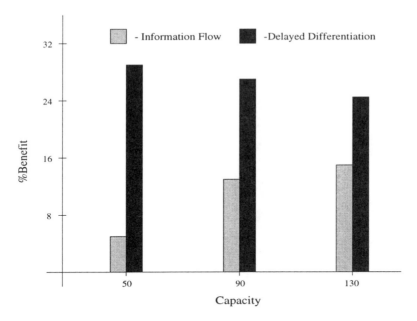

Figure 15.10 Benefits as a function of capacity for symmetric customers

The details of our computational study are as follows. The penalty cost was set equal to 14 while the holding cost took values 1, 3, and 5. The capacity was values used are 50, 90, and 130. The total mean demand of the two customers was set to 40. Thus in the symmetric case we had two customers with a mean of 20 each. The distributions used for the end-item demand were Exponential(20) and Erlang(5,4). In this case the Δ value for each customer was set to 50. For the case of asymmetric customers we assumed that the larger customer had a mean of 35 and the smaller customer had a mean of 5. Thus the larger customers end-item demand distribution was either Exponential(35) or Erlang(8.75,4) while the smaller customers end-item demand distribution was either Exponential(5) or Erlang(1.25,4). Also for this scenario the Δ values for the two customers were set equal to 87.5 and 12.5 respectively. Thus there are $3(3)(2) = 18$ instances (by varying holding cost, capacity and end-item demand) of symmetric customers and 18 instances of asymmetric customers. In each of the 36 instances, we solved for (cptimal levels and) cost for each of the four models.

Remark. We have further assumed that this pre-allocation of capacity in models 1 and 2 is proportional to the mean demands. We note that other allocations are possible and may in fact outperform this allocation in costs. However, the insights gained about the relative benefits of information sharing and delayed product differentiation do not change at other allocations.

First we report the results for the case of symmetric customers.

458

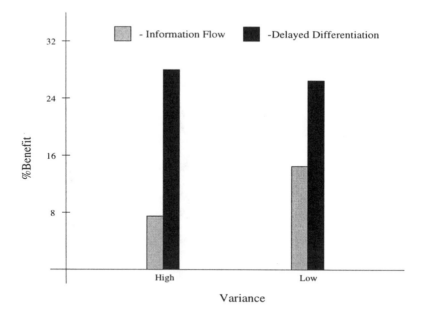

Figure 15.11   Benefits as a function of variance for symmetric customers

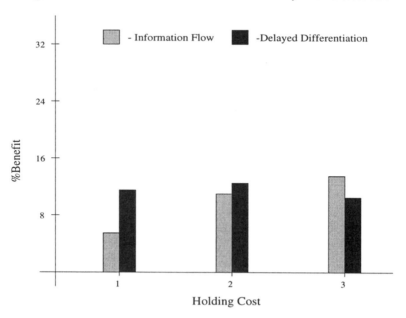

Figure 15.12   Benefits as a function of holding cost for asymmetric customers

15.3.4.1   Symmetric Customers.   We observed that for these experiments the benefit of information flow varied between 2.03% and 13.92% for the Exponential(20) distribution. This range was between 4.26% and 25.86% for the

Erlang(5,4) distribution. The benefit of delayed differentiation varied between 20.33% and 33.99% for the Exponential(20) distribution while it varied between 17.89% and 34.18% for the Erlang(5,4) distribution. From this we observe that *for the case of symmetric customers, delayed differentiation seems to be more beneficial than information flow.* This is because when the customers are symmetric, implementing delayed differentiation results in a significant drop in the co-efficient of variation – the ratio of standard deviation to the mean – of demand (at the supplier). This leads to lower inventories and lower costs. Note that delayed differentiation does *not* affect the timing of the demands from the customers.

We now study how various system parameters such as holding cost, capacity, and demand variance affect these benefits.

1. EFFECT OF HOLDING COST. Figure 15.9 contains the plot of benefits of both these strategies as a function of holding cost. To obtain this plot we averaged the percentage benefit of a strategy over all the experiments that had the same holding cost. For example, to obtain the benefit of information flow for holding cost of 1, we compared Model 3 with Model 1 and compared Model 4 with Model 2 for each of the 6 instances (obtained by varying capacity and end-item demand). Observe that the percentage benefit of information flow ranged from 7.30% to 13.68% and generally seemed to increase with holding cost. This is because when the holding costs are low (penalty/holding ratios are high), the system holds larger inventories (at optimality). Due to finite capacity, the build up of inventory starts well before the demands occur. The system is not able to respond significantly to the new information, and thus the resulting benefits are lower. On the other hand, the percentage benefit of delayed differentiation varied between 23.81% and 29.51% and generally seemed to decrease with increase in holding cost. This is because, since delayed differentiation reduces the overall demand co-efficient of variation, it is able to result in lesser expediting and payment of b. This results in higher savings at higher penalty/holding ratios. This leads us to conclude that *delayed differentiation is a dominant strategy when the holding costs are low and its dominance decreases with increase in holding cost* (if b is kept fixed).

2. EFFECT OF CAPACITY. Figure 15.10 illustrates the benefits of the two strategies as a function of capacity. To obtain this plot, as in the case of the holding cost, we averaged the percentage benefit of a strategy over all the experiments that had the same capacity. Notice that the percentage benefit of information flow ranged between 4.98% and 14.86% and generally seemed to increase with capacity while the percentage benefit of delayed differentiation ranged between 24.41% and 29.07% and generally seemed to decrease with increase in capacity. First, at higher capacities the system is no longer struggling to meet the required inventory levels and so does not start production early. As capacity increases the system

is able to respond more effectively to the information flow – as the ability to predict the timing of demand can be effectively utilized – and thus benefits of information flow increase with capacity. At higher capacities, the negative effect of demand (quantity) variance is lower, and so the benefit of delayed differentiation is not that significant. Thus we conclude that *at lower capacities delayed differentiation results in higher benefits* while the *benefit of information increases with increased capacity.* (Note that capacity being high or low is relative to the customer S-s value. Also, if the customer S-s value is small relative to end-item demand mean – or zero, if a base stock policy is being used – then the information is of no use as there is demand at the supplier almost every period. Similarly, if the S-s is very large compared to supplier capacity, the supplier has to start production early, and there is low value of information. So the benefit of information is maximum at moderate values of S-s.)

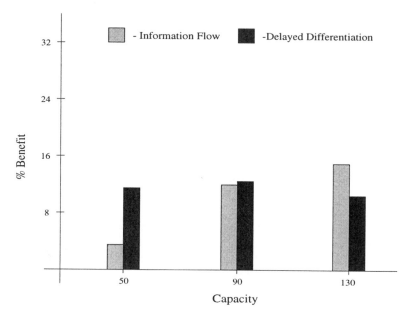

Figure 15.13   Benefits as a function of capacity for asymmetric customers

3. EFFECT OF VARIANCE.   The plot of benefits of the two strategies as a function of end-item demand variance is given in Figure 15.11.   To obtain this plot we used the strategy similar to the two plots above. Observe that when the variance dropped the percentage benefit of information increased from 7.37% to 14.48% while the percentage benefit of delayed differentiation dropped from 27.71% to 26.34%. When the demand is highly variable, the information available reduces the (timing and quantity) uncertainty by a small fraction and thus results in lower benefits. However, at higher variances, delayed differentiation is effective

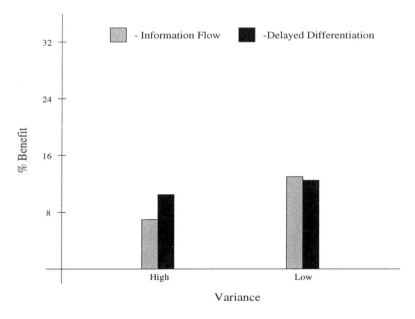

Figure 15.14 Benefits as a function of variance for asymmetric customers

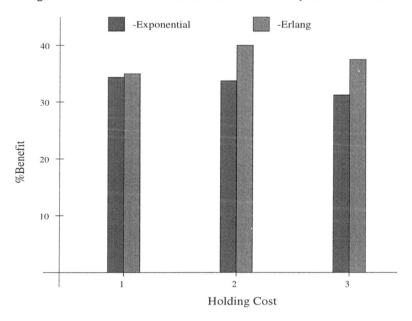

Figure 15.15 Combined benefits as a function of holding cost for asymmetric customers

most since it is able to significantly reduce the variance. Thus, we conclude that information sharing becomes increasingly more beneficial as the variance drops. However, at very low variances, neither information

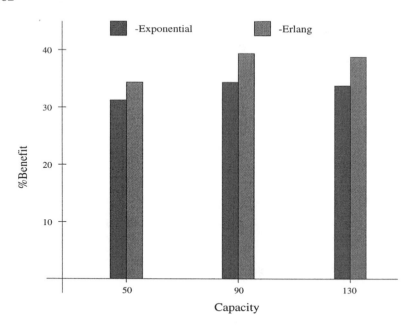

Figure 15.16 Combined benefits as a function of capacity for symmetric customers

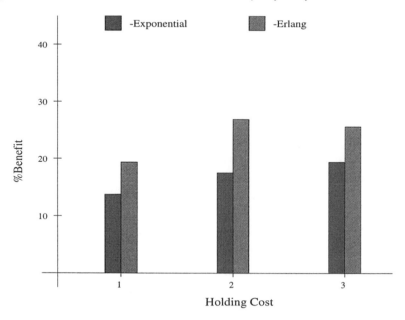

Figure 15.17 Combined benefits as a function of holding cost for asymmetric customers

flow nor delayed differentiation are required as the supply chain becomes deterministic. So, *at moderate values of variability is when information is most useful.*

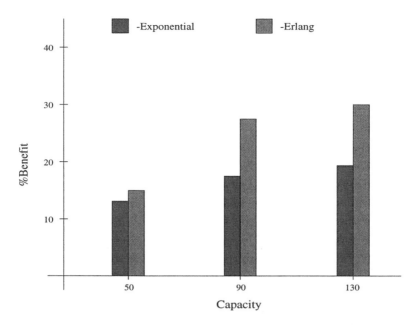

Figure 15.18 Combined benefits as a function of capacity for asymmetric customers

In general, it appears that the system parameters affect the benefit of information flow more dramatically than they affect the benefit of delayed differentiation. In other words, delayed differentiation is a beneficial strategy under a wide variety of conditions while information flow is beneficial only under a smaller set of conditions such as high holding cost, high capacity, and low variance. While comparing the relative benefits of these two strategies we ignored the costs of implementing these two strategies. These costs will ultimately play a critical role in deciding a suitable strategy. In general, implementation of delayed differentiation requires a redesign of the product and/or the production process. Such a redesign could be much more expensive than establishing an information system required to share information. Also the benefit of delayed differentiation computed here is the maximum possible which can not be achieved in most of the situations. So after considering these two observations, even though in terms of the percentage benefit, information flow does not dominate delayed differentiation in any of the the plots above, it may be a preferred strategy in many cases after accounting for the costs of implementation.

15.3.4.2 Asymmetric customers. In this section we report the results for the case of asymmetric customers. We assumed that one customer had a mean of 35 while the second customer had a mean of 5. For this situation the percentage benefit of information flow ranged from 1.19% to 14.35% for the Exponential distributions while it ranged from 1.46% to 25.75% for the Erlang distributions. The percentage benefit of delayed differentiation varied between 8.15% and 13.66% for the Exponential distribution while it varied between

9.34% and 16.04% for the Erlang distributions. Observe that when compared to the case of symmetric customers the percentage benefit of information did not change significantly. In contrast, percentage benefit due to delayed differentiation recorded a significant drop as compared to the case of symmetric customers. This drop in benefits due to delayed differentiation is best understood by considering the extreme case of one customer having mean demand of 40 units while the other customer having a mean demand of zero units. Clearly in such a situation delayed differentiation has no benefits. In any asymmetric case, it can easily be shown that improvement in variance is bounded by that in the symmetric case. Thus *when the customers are of unequal size, delayed differentiation results in lower benefits.*

Figures 15.12, 15.13, and 15.14 contain the plots of benefits of these two strategies as functions of holding cost, capacity, and variance respectively for the case of asymmetric customers. Observe that in these graphs at higher holding costs, higher capacities, and lower variances the strategy of information has higher benefits than the strategy of delayed differentiation (mainly because the latter drops off and not because the former increases). Thus we conclude that *in the presence unequal customers, information flow dominates delayed differentiation at high holding costs, high capacities and low variances.* It also appears that the presence of asymmetric customers hastened the trends observed in section 4.1.

15.3.4.3 Combined Benefits. In this section, we report the results of implementing these two strategies simultaneously. The combined benefit is computed as the percentage difference between the cost of Model 1 and the cost Model 4. Figures 15.15, 15.16 contain the plots of these benefits as functions of holding cost and capacity for symmetric customers. Similarly figures 15.17, 15.18 contain the plots asymmetric customers. ¿From these plots we can make the following observations.

First let us study the effects of relative sizes of customers. The benefits for the symmetric customers ranged from 31.29% to 39.64% while the benefits for the case of asymmetric customers ranged from 13.36% to 29.71%. This is because as observed in the previous section when the customers are asymmetric the benefits due to information sharing do not change much while the benefits due to delayed differentiation drop significantly. Thus *the benefits of using these strategies simultaneously are higher when the customers are of equal size.*

Next we study the effect of end-item demand variance on these combined benefits. For the exponential distribution these benefits ranged from 13.36% to 34.68% while they ranged from 14.85% to 39.64% for the Erlang distribution. From the figures 15.15, 15.16, 15.17, and 15.18 it is clearly seen that for every situation the benefits for the case of Erlang distribution were higher than the benefits for the exponential distribution. As we discussed before, there is no benefit at zero variance. Thus we conclude that *the combined benefits are highest at moderate variances.*

With respect to holding cost and capacity the figures do not show consistent trends. For example with respect to holding cost observe that the benefits for the case of exponential distribution increase in figure 15.17 while they decrease in figure 15.15. Similarly with respect to capacity, for the case of Erlang distribution the benefits in figure 15.16 are largest at medium capacity while the benefits in figure 15.18 are increasing with increase in capacity. This behavior can be explained by our observations in the previous section. There we observed that the benefits due to information sharing and delayed differentiation have opposing trends with changes in capacity or holding cost. Due to these opposing trends, under different conditions the maximum combined benefit occurs at different values of holding cost and capacity.

15.3.5 Conclusions

In this section, we analyzed four inventory control models to study the benefits of information flow and delayed differentiation. We used a discrete time framework and found that the optimal policy structure is a state dependent order up-to policy for each model. We provided procedures to compute the optimal parameters. A computational study of the optimal policies of these four models suggests that information flow should be preferred over delayed differentiation when there are high holding costs (low penalty costs), high capacities, moderate variances, or unequal customers (and vice-versa). This choice is more beneficial when two or more of these conditions exist simultaneously. We also notice that these two strategies complement each other well when used simultaneously.

References

[1] Armistead, C.G. and Mapes, J., "The impact of supply chain integration on operating performance," *Logistics Information Management,* 6 (1993), 9-14.

[2] Aviv, Y. and Federgruen, A., "Stochastic Inventory Models with Limited Production Capacity and Periodically Varying Parameters," *Probability in The Engineering and Informational Science,* 11 (1997), 107-135.

[3] Cachon, G., "Value of Information in a One Warehouse Multi Retailer System," Fuqua School of Business, Duke University, 1996.

[4] Chen, F., "Echelon Reorder Points, Installation Reorder Points, and the Value of Centralized Demand Information," Graduate School of Business, Columbia University, 1995.

[5] Federgruen, A. and P. Zipkin, "An inventory model with limited production capacity and uncertain demands I : The average-cost criterion," *Mathematics of Operations Research,* 11 (1986), 193-207.

[6] Federgruen, A. and P. Zipkin, "An inventory model with limited production capacity and uncertain demands II : The discounted-cost criterion," *Mathematics of Operations Research,* 11 (1986), 208-215.

[7] Gavirneni, S. "Inventories in Supply Chains under Co-operation," *GSIA Doctoral Dissertation,* Carnegie Mellon University, Pittsburgh, PA, August 1997.

[8] Glasserman, P. and S. Tayur, "The stability of a capacitated, multi-echelon production-inventory system under a base-stock policy," *Operations Research,* 42 (1994), 913-925.

[9] Glasserman, P. and S. Tayur, "Sensitivity analysis for base-stock levels in multi-echelon production-inventory systems," *Management Science,* 42 (1995), 263-281.

[10] Hariharan, R. and P. Zipkin, "Customer-order Information, Lead times, and Inventories," *Management Science,* 41 (1995), 1599-1607.

[11] Kapuscinski, R. and S. Tayur, "A Capacitated Production-Inventory Model with Periodic Demand," *GSIA working paper,* Carnegie Mellon University, Pittsburgh, PA, November 1995; revised January 1996, to appear in *Operations Research.*

[12] Lee, H. and C. S. Tang, "Modeling the Costs and Benefits of Delayed Product Differentiation," *Management Science,*43 (1997), 40-53.

[13] Lee, H. et al., "Hewlett-Packard Gains Control of Inventory and Service through Design for Localization," *Interfaces,*23 (1993), 1-11.

[14] Scarf, H.E., " The Optimality of (s, S) policies in the dynamic inventory problem," in K.J. Arrow, S. Karlin, and P. Suppes (editors), *Mathematical Methods in Social Sciences,* Stanford University Press, 1962.

[15] Scheller-Wolf, A. and S. Tayur, "Reducing International Risk through Quantity Contracts," *GSIA working paper,* Carnegie Mellon University, Pittsburgh, PA, April 1997.

[16] Takac, P.F., " Electronic data interchange (EDI): an avenue to better performance and the improvement of trading relationships?," *International Journal of Computer Applications in Technology,* 5 (1992), 22-36.

[17] Zheng, Y. and P. Zipkin, "A Queuing Model to Analyze the Value of Centralized Inventory Information," *Operations Research,* 38 (1990), 296-307.

16 MANAGING PRODUCT VARIETY: AN OPERATIONS PERSPECTIVE

Amit Garg

IBM, T. J. Watson Research Center,
Yorktown Heights, NY 10598

Hau L. Lee

Department of Industrial Engineering-Engineering Management,
Stanford University, Stanford CA 94305

16.1 INTRODUCTION

Researchers from diverse fields such as economics, marketing, and operations management have studied product variety. Economists have studied product variety from two perspectives. First, they have studied the impact of a welfare criterion on socially optimal degree of product variety. And second, they have studied various market equilibria resulting from different levels of competitive relationships, product differentiation, and product variety in each firm in the market. Lancaster [17] presents a survey of this line of research on product variety.

Roberts and Lilien [26] and Menon and Kahn [23] and Dodd et al [5] focus on the marketing perspective of product variety. They have studied the reasons why consumers seek variety and how companies can exploit consumers' behavior to increase sales.

As the title of this chapter suggests, we will focus on the operations perspective of product variety. Researchers in the area of operations management define product variety as the breadth and the depth of product lines (MacDuffie et al [22]). This definition is similar to that of researchers in economics and marketing. However, research in operations management differs from that in economics and marketing largely due to its inward-looking focus on cost of variety within a company. Research in economics and marketing on the other hand, has focused on the competitive and consumer choice aspects of product variety.

Several empirical studies (Child et al [4], Yeh et al [30], and MacDuffie et al [22]) of product variety support the view that greater product variety increases cost by making the business more complex. MacDuffie et al [22] define five types of complexities to capture different aspects of product variety. The end-result of complexity resulting from increased product variety is a proliferation of products, parts, and suppliers, and a multiplicity of processes performed within a company. The big problem with all the this complexity is the uncertainty that plagues it. Uncertainties in demand realization, delays in deliveries, machine breakdowns, etc., lead to either excess inventories, or stocks that are insufficient to meet customer demands. Therefore, not surprisingly, quantitative models developed by researchers in the area of operations management have tried to reduce the impact of product variety on inventory costs.

Research in the area of product variety with an operations perspective can be categorized into two broad classes, strategies for reducing the level of complexity, and models for reducing the impact of uncertainties by reducing cycle times. We call them

- Non-lead time reduction strategies, and

- Lead time reduction strategies

respectively.

Non-lead Time Reduction Strategies:. The main objective of this class of strategies is to reduce the complexity of the system by reducing the number of parts and processes, and by mitigating the effect of uncertainties on total costs in the system. Part commonality, postponement, and process sequencing are some of non-lead time reduction strategies.

Lead Time Reduction Strategies:. In general, short-term forecasts for an item are more accurate than long-term ones. Lead time reduction is very useful in mitigating the effect of uncertainties because it also reduces the length of the forecasting horizon. Therefore, error between the forecast and the actual realization of demand is much smaller, resulting in lower safety-stock levels. Production line structuring and Quick Response (QR) systems are two effective means of reducing total lead times.

This chapter is organized as follows. Section 16.2 reviews work done in the area of non-lead time reduction strategies. Section 16.3 reviews papers on lead time reduction techniques. Finally, we conclude in Section 16.4 by summarizing the current state-of-the-art in this area and outline directions for future research.

16.2 NON-LEAD TIME REDUCTION STRATEGIES

Early research on managing product variety studied standardization of components (Collier [3], Baker et al [1], and Gerchak et al [15]), one of the simplest ways of reducing variety. These researchers show that commonality results in lower safety stocks of common components. Lower safety stocks of common components are largely due to the risk-pooling effect from demands of different end-products.

Lately researchers have started taking a more holistic viewpoint by trying to reduce costs in a supply chain instead of just in manufacturing. Postponement[1] and process sequencing are two non-lead time reduction strategies that have been studied by researchers. The term postponement refers to delaying differentiation of products until as late as is cost-effective. Postponement implies both component-level and process-level standardizations and is strong way of gaining through economies of scale and scope. Process sequencing is a useful in two ways. First, it is a means of effecting postponement. Second, it is means of reducing the cost during value creation in the supply chain. Basically, benefits due to process sequencing can flow from adding most value to the products as close as possible to the point of realization of their demands. This research was first motivated by applications in industry and is an area with a rich set of open problems. In this section we first describe some of the papers studying various aspects of postponement and then review papers on process sequencing.

---

[1]Other terms used to describe this concept include late customization and delayed differentiation.

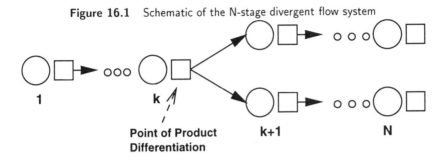

Figure 16.1 Schematic of the N-stage divergent flow system

16.2.1 Research on Postponement

As you may recall, under the postponement strategy, products within a family share common parts and processes until their point of differentiation. At the point of differentiation a "fan-out" occurs because end-products require some special components and/or processes (Figure 16.1). This naturally leads to a divergent flow network resembling a distribution network that has been studied by many researchers (Eppen and Schrage [6], Federgruen and Zipkin [7], Schwarz [27], and Garg and Lee [9]) in the past. Therefore, most of the models studying the impact of postponement have drawn upon this body of research.

Research on modeling postponement can be sliced and diced in several different ways. One could classify models on postponement based on:

■ the means of effecting postponement (standardization and process sequencing),

■ the operating mode (make-to-stock and make-to-order), and

■ the supply chain control mechanism (centralized and decentralized control).

We shall follow the first criterion in describing research on postponement in this section. We first review research on effecting postponement through standardization of parts and/or components, then some research on effecting postponement through process sequencing, followed by papers on effecting postponement through standardization and process sequencing, and finally papers that study other aspects of postponement.

16.2.1.1 Postponement through Standardization. Postponement can be effected by standardizing parts and/or processes. Lee et al [18], Lee [19], and Garg and Tang [12] are some of the papers that study postponement through standardization.

Lee et al [18] is one of the first papers on postponement and is based on work done at Hewlett-Packard's Deskjet-Plus printer line. At the end of manufacturing, generic printers are "localized" by adding appropriate power supply modules (with the correct voltage and plugs) and the appropriate manual with the printer. Previously, the factory manufactured printers destined for all countries and localized them before shipping them to the distribution centers (DCs) covering those countries. They call this the *factory-localization* strategy. The paper examines the cost-effectiveness of adopting a *DC-localization* strategy wherein the localization process is performed at the DCs instead of at the factory. The authors developed a single-site, single-item, inventory model (Nahmias [24]) to estimate the savings in system-wide inventories if the DC-localization strategy were adopted.

The basics of their model are as follows. Let μ_i and σ_i^2 be the mean and the variance respectively, of demand at the DCs from country i. Let S_i be the target base-stock level for printers destined for country i, l the supplier lead time (a random variable), and r the review period length. The variance of demand for printer during its replenishment lead time (i.e., over its lead time and review period) from country i, X_i, is given by

$$\mathsf{Var}\,(X_i) = \sigma_i^2\,(\mathsf{E}(l) + r) + \mathsf{Var}(l)\mu_i^2 \tag{16.1}$$

The base-stock level, S_i, is given by $S_i = \mu_i(\mathsf{E}(l)+r) + k\sqrt{\mathsf{Var}(X_i)}$, where k is the safety-stock factor corresponding to given fill rate. $ss_i = k\sqrt{\mathsf{Var}(X_i)}$ is the safety stock for printers destined for country i. Average on-hand inventory of printers for country i at the DC is given by $ss_i + \mu_i/2$.

In the DC-localization case, the DC stocks generic printers and localization material separately and customizes the printers on demand. Assume the DC serves N countries and demands for each country are *iid*, therefore, the total demand for generic printers has a mean and a variance of $\sum_i^N \mu_i$ and $\sum_i^N \sigma_i^2$ respectively. One can now derive an expression analogous to equation (16.1) for the variance of demand for the generic printer over the replenishment lead time. Inventory savings between the factory- and DC-localization cases is the difference in total safety stocks of generic printers between the two strategies and is given by

$$ss_{factory} - ss_{DC} = k \sum_{i=1}^{N} \left\{ (\mathsf{E}(l) + r)\,\sigma_i^2 + \mathsf{Var}(l)\mu_i^2 \right\}^{1/2}$$
$$-k \left\{ (\mathsf{E}(l) + r) \sum_{i=1}^{N} \sigma_i^2 + \mathsf{Var}(l) \sum_{i=1}^{N} \mu_i^2 \right\}^{1/2}, \tag{16.2}$$

where $ss_{factory}$ and ss_{DC} denote safety-stock levels in the factory- and DC-localization alternatives respectively. We can see that $ss_{factory} - ss_{DC} > 0$. However, the system-wide inventory of localization material is greater in the DC-localization case. Since the unit cost of localization materials was much lower than that of the generic printer, the DC-localization alternative resulted in lower system-wide inventories.

Lee [19] is among the first theoretical analyses of postponement. He studies the benefits of postponement in a make-to-order (MTO) and a make-to-stock (MTS) production system. In the MTO production system, an intermediate stage of the product is built to stock and is customized into various end-products on demand. In the production system operating in an MTO mode, it takes t time units to manufacture the generic intermediate product and $T - t > 0$ time units to customize the intermediate product into end-products. Therefore, we can view time t as the point of product differentiation.

Model for the MTO case uses response time to customer orders (denoted by Y) as the performance measure. In particular, the specific performance measure could be the expected response time, $\mathsf{E}Y$, or the response time reliability, i.e., the probability of the response time being less than a target response time, R, $\mathsf{P}(Y \le R)$. Let $X(r)$ be the demand in r time units, and $F(x|r) = \mathsf{P}(X(r) \le x)$. Response time to a customer order can be expressed as $Y = T - t + W$, where W $(0 \le W \le t)$ is the waiting time if the intermediate stockpile does not have sufficient inventory to satisfy a customer's order.

Since a base-stock policy is being used to manage the intermediate product stockpile, the two measures of response times can be written as follows.

$$\mathsf{E}Y = T - \int_0^t F(S|\tau)d\tau, \tag{16.3}$$

$$\mathsf{P}(Y \le R) = F(S|T - r). \tag{16.4}$$

Given a target $\mathsf{E}Y$ or the target response time reliability, one can compute the base-stock level for the intermediate product. Effecting postponement entails redesigning products and/or processes so that the intermediate stockpile is held at t' $(T > t' > t)$ time units after starting production.

In order to evaluate the cost-effectiveness of postponement, let $g(t)$ denote the unit inventory holding cost rate for intermediate inventory is held at t time units after starting production. $g(t)$ is assumed to be non-decreasing in t since value is added progressively in production. Also define $H(t)$ to be the expected holding cost per unit time for the intermediate product. Since the total average work-in-process is invariant whether or not we delay differentiation, it is sufficient to focus only on the inventory level at the stockpile of intermediate product. The holding cost rate for the intermediate product is

$$H(t) = g(t) \int_0^S F(x|t)dx \tag{16.5}$$

For postponement to be beneficial, we want $\partial H/\partial t \le 0$. On differentiating equation (16.5) w.r.t. t, we get

$$\frac{\partial H}{\partial t} = g(t)\left[F(S|t)\frac{\partial S}{\partial t} + \int_0^S \frac{\partial F(x|t)}{\partial t}dx\right] + g'(t)\int_0^S F(x|t)dx \tag{16.6}$$

In the above equation, $\partial F(x|t)/\partial t \le 0$ and $\partial S/\partial t \le 0$. The latter condition can be verified by differentiating equation (16.3) w.r.t. t, and setting the derivative to 0. However, it is still not clear if postponement is beneficial because $g'(t) \ge 0$. Postponement would be beneficial if $g'(t)$ is small and sufficiently close to

0. Therefore, the value added (or cost added) to the intermediate product determines the cost-effectiveness of postponement.

In the MTS case, the paper uses a model based on the work of Eppen and Schrage [6]. The system under MTS mode operates under centralized control and stocks only finished products. As in the MTO case, the total lead time to manufacture each end-product is T time units. The products are in a generic form for the first t time units of production after which they are customized (or "allocated") into various end-products. Assume the demand for end-product i is normally distributed with μ_i and σ_i^2 as its mean and variance respectively. End-product demands are independent across time but may be correlated within a time unit. Let ρ_{ik} denote the coefficient of correlation between the demands for end-products i and k in a time unit. Under this scenario, the mean and variance of the inventory level for end-product i, I_i is given by

$$\mathsf{E}I_i = A_i(S_i) - R_i T \sum_j \mu_j, \tag{16.7}$$

$$\mathsf{Var}I_i = t R_i^2 \left(\sum_j \sigma_j^2 + \sum_{j \neq k} \rho_{jk} \sigma_j \sigma_k \right) + (T-t)\sigma_i^2, \tag{16.8}$$

where $A_i(S_i)$ is a function independent of t, and $R_i = \sigma_i / \sum_j \sigma_j$. Given the target service level for the products, and mean and variance of I_i, we can determine the base-stock level, S_i, for product i. Clearly, smaller the mean and variance of I_i, smaller the base-stock level, S_i, for a given service-level requirement. Note that $\mathsf{E}I_i$ is independent of t, therefore, delaying product differentiation will affect the base-stock level, S_i, only through the variance of I_i. Therefore, $\partial S_i / \partial t = \partial \mathsf{Var}(I_i) / \partial t$, which is given by

$$\frac{\partial \mathsf{Var}(I_i)}{\partial t} = R_i^2 \left[\sum_j \sigma_j^2 - \sum_{j \neq k} \rho_{jk} \sigma_j \sigma_k \right] - \sigma_i^2. \tag{16.9}$$

Since $\rho_{jk} \leq 1$, for all j and k, we can see that $\partial S_i / \partial t \leq 0$. Therefore, delaying product differentiation will always result in lower inventories of finished products.

In the models described above, the product line has only one point of differentiation, or has only one differentiating feature. In reality, most product lines have many differentiating features. Garg and Tang [12] study postponement strategies for product lines with multiple differentiating features. In order to illustrate their results, they take a two differentiating-feature case depicted in Figure 16.2. This figure depicts the lead times at each stage of this three-stage divergent system, and the two types of differentiation points: the *family differentiation point* and the *product differentiation point*. At the family differentiation point, the product line gets differentiated into various product families,

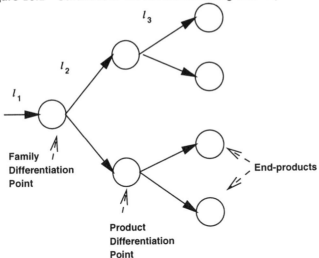

Figure 16.2 Schematic of the two differentiating-feature product line

which get further differentiated into end-products at the product differentiation point. Therefore, each type of differentiation point corresponds to one of the features of the product line.

The authors call delaying differentiation at the family differentiation point *early postponement,* and delaying differentiation at the product differentiation point *late postponement.* A stage-dependent lead-time vector can be used to characterize the two types of postponement. The lead-time vector when there is no postponement is (l_1, l_2, l_3), with early postponement it becomes $(l_1 + 1, l_2 - 1, l_3)$, while with late postponement it becomes $(l_1, l_2 + 1, l_3 - 1)$. The main question this paper addresses is the set of drivers that determine the postponement strategy yielding greatest marginal benefits. The paper addresses this question for systems operating under centralized and decentralized control.

The product structure can be defined as follows. Let I be the index for product families, R_I be the set of products belonging to family I. Let i be the index for products. Demand for product i is normally distributed with a mean of μ_i, and a variance of σ_i^2. Let ρ_{ij} denote the coefficient of correlation in demands of products i and j. Also define $\sigma_I^2 = \sum_{i,j \in R_I}(\sigma_i^2 + \sum_{j \neq i} \rho_{ij}\sigma_i\sigma_j)$, $\mu_I = \sum_{i \in R_I} \mu_i$, $\bar{\sigma}_I = \sum_{i \in R_I} \sigma_i$, $\sigma^2 = \sum_I (\sigma_I^2 + \sum_{J, J \neq I} \sum_{i \in R_I} \sum_{j \in R_J} \rho_{ij}\sigma_i\sigma_j)$, $\mu = \sum_I \mu_I$, and $\tilde{\sigma} = \sum_I \sigma_I$. Note that $\sigma \leq \tilde{\sigma} \leq \sum_I \bar{\sigma}_I$.

In the case of centralized control, the authors extend the work of Eppen and Schrage [6]. Their model of a three-echelon divergent flow network (Figure 16.2) requires an additional assumption, the *correlation assumption,* i.e. $\bar{\sigma}_I/\sigma_I = \beta$, for all I, where β is a constant. This constant is an aggregate measure of the correlations between demands of products within a product family. The authors show that both early and late postponements result in inventory savings. However, late postponement yields greater marginal benefits if $2 - \sigma^2/\tilde{\sigma}^2 \leq \beta^2$. In a centralized control system, the main determinants of postponement strategy are relative market sizes and the structure of demand correlations.

The authors illustrate their results through three special cases. They show that if the demand correlation structure remains unchanged, savings due to postponement increase as relative market shares of products become balanced. For example, the savings due to early postponement are greater if the two product families (Figure 16.2) command a market share of 50% each than if the two families have market shares of 10% and 90% respectively. This result is intuitive because in the equal market share case, the risk-pooling effect due to postponement is greater. Similarly, they show that benefits of postponement along an feature increase as the correlations between demands from various choices of that feature decrease. Lower correlations between demands of choices within a feature increase the risk-pooling effect, increasing savings in inventories.

The drivers of postponement strategy in the case of decentralized systems are different from those in centralized systems. In decentralized systems, relative lead times between echelons play an important role in determining the postponement strategy. In fact, it is not even necessary for postponement to result in inventory savings. In order to illustrate this result, let us analyze the impact of early postponement in a decentralized control system. As a result of early postponement, the lead-time vector for the divergent network will change from (l_1, l_2, l_3) to $(l_1 + 1, l_2 - 1, l_3)$. The average total inventory in the decentralized system, $\mathcal{I}$, before and after early postponement respectively can be written as

$$\mathcal{I}(l_1, l_2, l_3) = \frac{3}{2}\mu + z\sigma\sqrt{l_1 + 1} + z\tilde{\sigma}\sqrt{l_2 + 1} + z\sqrt{l_3 + 1}\sum_I \overline{\sigma}_I, \quad (16.10)$$

$$\mathcal{I}(l_1 + 1, l_2 - 1, l_3) = \frac{3}{2}\mu + z\sigma\sqrt{l_1 + 2} + z\tilde{\sigma}\sqrt{l_2} + z\sqrt{l_3 + 1}\sum_I \overline{\sigma}_I, \quad (16.11)$$

where μ, σ, $\tilde{\sigma}$, and $\overline{\sigma}_I$ are as defined earlier, and z is the safety-stock factor. We can express the marginal saving due to early postponement, $\Delta\mathcal{I}^e = \mathcal{I}(l_1 + 1, l_2 - 1, l_3) - \mathcal{I}(l_1, l_2, l_3)$, as

$$\Delta\mathcal{I}^e = z\sigma\left(\sqrt{l_1 + 2} - \sqrt{l_1 + 1}\right) + z\tilde{\sigma}\left(\sqrt{l_2} - \sqrt{l_2 + 1}\right) \quad (16.12)$$

Note that $\sigma < \tilde{\sigma}$ and $\sqrt{x+1} - \sqrt{x}$ is decreasing in x. Therefore, we can see from the equation (16.12) that relative values of lead times, l_1 and l_2, can determine if early postponement will lower inventories. The analysis of late postponement is similar.

One may wonder why relative lead time between echelons do not affect the postponement strategy in the centralized control system. This is primarily because the centralized system modeled by the authors does not carry inventory at intermediate stages. It carries inventories only at the finished items level.

16.2.1.2 Postponement through Process Sequencing. Garg [10] and Lee and Tang [21] study postponement through process sequencing. Garg [10] is described in Section 16.2.2. Like Garg and Tang [12], Lee and Tang [21] also consider a product line with two differentiating features. However, Lee and Tang focus on the sequence in which the features should be installed in order to minimize inventory costs. The authors model a three-echelon divergent network operating with decentralized control, stocking inventory at each stage. Each stockpile in the network is managed by a periodic-review, base-stock inventory policy. Let the two features be denoted by A and B, and let X_{ij} denote the demand for choice i of feature A and choice j of feature B, $i, j = 1, 2$.

Since the model considered in the Lee and Tang paper has two features, each having two choices, we have product line with four end-products. The paper assumes demand for the product line to be N, a random variable with mean and variance denoted by μ and σ^2 respectively. Demands for the end-products, denoted by $(X_{11}, X_{12}, X_{21}, X_{22})$, are assumed to be multinomially-distributed random variables with parameters $(N; \theta_{11}, \theta_{12}, \theta_{21}, \theta_{22})$. θ_{ij} represents the probability that the customer will purchase (or alternatively, the market share of) an end-product with choice i of feature A and choice j of feature B. In this demand model, $\mathsf{E}(X_{ij}|N) = N\theta_{ij}$, $\mathsf{Var}(X_{ij}|N) = N\theta_{ij}(1 - \theta_{ij})$, and $\mathsf{Cov}(X_{ij}, X_{mn}|N) = -N\theta_{ij}\theta_{mn}$, for $ij \neq mn$. Note that all end-product demands are negatively correlated.

The market shares of each end-product, θ_{ij}, can be expressed as functions of the following conditional probabilities. Let p denote the probability a customer will purchase a product with choice 1 of feature A, given that he/she will purchase a product. Hence $1 - p$ is the conditional probability of a customer buying the product with choice 2 of feature A. Define $f_i(p) = \mathsf{P}(B = 1|A = i)$ to be the conditional probability the customer buys the product with choice 1 of feature B, given that the customer has decided to purchase the product with choice i of feature A, for $i = 1, 2$. Probabilities $f_i(p)$ can be used to model interaction between features and choices. Given this probability structure, market shares of each end-product can be expressed as: $\theta_{11} = pf_1(p)$, $\theta_{12} = p(1 - f_1(p))$, $\theta_{21} = (1 - p)f_2(p)$, and $\theta_{22} = (1 - p)(1 - f_2(p))$.

The authors use this demand model to analyze the impact of installing feature A before feature B, sequence $A - B$, and sequence $B - A$ on inventories in the supply chain. Specifically, the authors try to minimize demand variability in the supply chain because demand variability drives inventories. For the three-echelon system, the authors assume lead times and the value-added by a feature at each stage to be identical. Under these assumptions, changing the sequence from $A - B$ to $B - A$ will not affect the costs of end-product inventories. Similarly, change in the operation sequence will not affect the mean and the variance of demands for the generic product. The only stage at which operation sequence will affect variability is the intermediate stage. Hence, the authors use variability at the intermediate stage as their performance metric.

By applying the conditional variance formula, the authors derive the total variance at the intermediate stage for sequence $A - B$ to be $(\mu - \sigma^2)p(1 - p) + \sigma^2$, and that for sequence $B - A$ to be $(\mu - \sigma^2)[pf_1(p) + (1-p)f_2(p)]\{1 - [pf_1(p) + (1-p)f_2(p)]\} + \sigma^2$. Now one can see that sequence $A - B$ will have a smaller total variance than $B - A$ if $C(p) < 0$, where

$$
\begin{aligned}
C(p) &= (\mu - \sigma^2)\{p(1-p) - [pf_1(p) + (1-p)f_2(p)] \times \\
&\quad \{1 - [pf_1(p) + (1-p)f_2(p)]\}\}.
\end{aligned}
\tag{16.13}
$$

The authors construct special cases under three conditions to derive insights into operations sequencing decisions:

- Independent feature choices, i.e., $f_i(p) = q$, for $i = 1, 2$, where q is the probability of a customer buying an end-product with choice 1 of feature B, identical lead times at each stage, and value added by each feature is identical.

- Feature-dependent lead times and value-added, and independent feature choices, and

- Choice interactions.

We will now summarize the main results in the three cases mentioned above. From equation (16.13) we can see that the optimal sequence will change depending upon the sign of $(\mu - \sigma^2)$. If $\mu > \sigma^2$, the authors call that condition to be the "stable" demand case, otherwise it is deemed the "unstable" demand case. Table 16.1 summarizes results for the case of independent feature choices with stable demands. The optimal sequence would be reversed if the demand were unstable, or the factors were not true.

Table 16.1 Summary of the results for independent choice probabilities and stable demands.

Condition	Factor	Result
2 Features	Imbalance between market shares of choices within a feature	Process the feature with greater imbalance in market share of its choices FIRST.
2 Features & equal probability of a choice within a feature	Number of choices within a feature	Process the feature with fewer choices FIRST.
3 or more features, 2 choices/feature	Probability of choice 1 in feature i, p_i	Process features in a ascending sequence of $p_i(1 - p_i)$

When lead times and the value added are feature-dependent, variability at the intermediate stage is not a sufficient to compare sequences $A - B$ and

$B - A$. The variability measure has to account for the total inventory cost in the network in order to compare the the two sequences. Again, the assumption of independent feature choices holds. Therefore, sequence $A - B$ has a lower costs if

$$\left(\mu - \sigma^2\right) \left(l_A - l_B\right) \mathcal{V}\mathcal{P} + 2\left[2\left(\mu - \sigma^2\right) p(1 - p) + \sigma^2\right]$$
$$\times \left(l_A v_A - l_B v_B\right) > 0, \quad (16.14)$$

where l_A and l_B lead times for installing features A and B respectively, v, v_A, and v_B are the value added by the generic product, feature A, and feature B, respectively, $\mathcal{V} = (v + v_A + v_B)$, and $\mathcal{P} = \left\{1 - \left[p^2 + (1 - p)^2\right]^2 - 2p(1 - p)\right\}$. Now if demand is stable, i.e., $\mu > \sigma^2$, the sufficient conditions for equation (16.14) to hold are $l_A > l_B$ and $v_B/l_B > v_A/l_A$. This is a rather intuitive result and is depicted in Figure 16.4. It implies if feature A takes longer to process and if the value-added per unit processing time for feature A is smaller than that of feature B, process feature A first. This sequence would reduce the cost of work-in-process inventory.

Kapuscinski and Tayur [16] also study variability reduction through operations reversal. In the case of a product line with two features, each having two choices, they show that sequence $A - B$ would result in lower variance and hence lower inventories if one of the following is true

$$\mathsf{Cov}\left(X_{11} + X_{12}, X_{21} + X_{22}\right) > \mathsf{Cov}\left(X_{11} + X_{21}, X_{12} + X_{22}\right) \quad (16.15)$$
$$\mathsf{Cov}\left(X_{11}, X_{12}\right) + \mathsf{Cov}\left(X_{21}, X_{22}\right) < \mathsf{Cov}\left(X_{11}, X_{21}\right) + \mathsf{Cov}\left(X_{12}, X_{22}\right) \quad (16.16)$$

If one were to consider reduction of system-wide standard deviation, in the case of independent feature choices, i.e. $f_i(p) = q, \forall i$, the authors show that sequence $A - B$ is preferable when $|0.5 - p| > |0.5 - q|$. This result implies that one would want to sequence the feature with fewer choices first irrespective of the standard deviations.

16.2.1.3 Postponement through Standardization and Process Sequencing.

The models described until now researchers have focused on using only one of the techniques for effecting postponement. Lee and Tang [20], Garg and Lee [11], and Swaminathan and Tayur [29] are some of the papers applying both standardization and process sequencing to effect postponement. In this section we summarize results from Lee and Tang [20] and Garg and Lee [11]. We refer the reader to Chapter 19 of this volume for more details on Swaminathan and Tayur [29].

Most of the quantitative models have focused only on inventories as the metric to evaluate the cost-effectiveness of postponement. Postponement usually changes fixed and variable costs in the system. Inventory costs are one of the variable costs. Another variable cost that may change is the unit processing cost. And clearly, the redesign effort requires some fixed investment as well. In order to truly evaluate the cost-effectiveness of postponement, one needs to consider all relevant costs. Lee and Tang [20] and Garg and Lee [11] are two papers that consider these costs.

Table 16.2 Implications of the basic approaches of delaying differentiation.

Basic Approach	Redesign Process	Conditions for Effectiveness
Standardization	Design a common part	– Investment cost is low – Incremental processing cost is low
Modular Design	Divide part into 2 modules: common & custom modules. Delay assembly of custom module.	– Incremental lead time is low – Incremental processing cost is low – Unit inventory holding cost is low
Process Sequencing: Postpone an Operation.	Divide an operation into 2 steps: common & special operations. Postpone the special operation.	– Lead time of common operation significantly longer than that of the special operation. – Special operation is high value-add activity.
Process Sequencing: Operation Reversal	Reverse the order of 2 operations. As a result, the first operation is common to all products.	– Deferring high value-add operation.

Lee and Tang [20] model a manufacturing system with two end-products in which each product requires N stages of processing. Each stage has a buffer to store its work-in-process. The first k stages in this system are common to both end-products and after this stage the products get differentiated (see Figure 16.1). Therefore, k is the point of differentiation. They develop a discrete-time model of a decentralized control system in which the demand for each end-product is *iid* and normally distributed with $ED_i(t) = \mu_i$, $\text{Var}(D_i(t)) = \sigma_i^2$, and $\text{Cov}(D_1(t), D_2(t)) = \rho \sigma_1 \sigma_2$, for $i = 1, 2$.

The authors evaluate the benefits of delaying product differentiation through a total relevant cost function. This cost function considers the average investment cost per period for redesigning the products and/or processes, the change in unit processing cost when products and/or processes are redesigned, and inventory holding costs. They use this cost function to evaluate the benefits of postponement through standardization of parts, modular design, and process re-sequencing by constructing simplified cases of the total relevant cost function. The main conclusions of this paper are summarized in Table 16.2.

Garg and Lee [11] study the postponement through standardization and process sequencing by using a model of divergent flow system with more than two echelons, stage-dependent lead times, and buffer stocks at the end of each intermediate stage and two end-products (see Figure 16.3). Mean and covariance of demands for the end-products in period t, $D(t)$, are $ED_i(t) = \mu_i$, $\text{Var}D_i(t) = \sigma_i^2$, for $i = 1, 2$, and $\text{Cov}(D_1(t), D_2(t)) = \rho \sigma_1 \sigma_2$.

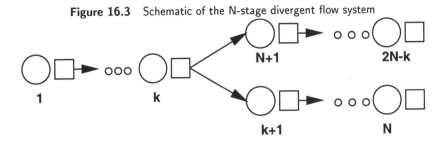

Figure 16.3 Schematic of the N-stage divergent flow system

Each stage of this divergent flow system has a target inventory position. Every period, an order is placed to bring the inventory position at each stage up to its target level. This replenishment policy is similar to the periodic-review, base-stock policy. The difference between the traditional base-stock policy and the policy used in this paper is the way in which order quantities are determined. The authors use a linear production control rule to determine the order each stage places with the upstream stage. If $Y_i(t)$ denotes the inventory position at stage i at the beginning of period t, T_i the target stock level at stage i, l_i the lead time at stage i, and $X_i(t)$ the order that arrives at stage i at the beginning of period t. Then the production control rule that determines the order quantity, $X_i(t)$, is given by

$$X_i(t + l_i) = \mu_1 + \mu_2 + \sum_{j=i}^{2N-k} (T_j - Y_j(t)) \, r_j, \quad \text{for } i = 1,\dots,k, \quad (16.17)$$

$$X_i(t + l_i) = \mu_1 + \sum_{j=i}^{N} (T_j - Y_j(t)) \, r_j, \quad \text{for } i = k+1,\dots,N, \quad (16.18)$$

$$X_i(t + l_i) = \mu_2 + \sum_{j=i}^{2N-k} (T_j - Y_j(t)) \, r_j, \quad \text{for } i = N+1,\dots,2N-k. \,(16.19)$$

In the production rule presented above, the first term on the right hand side represents the average level of production required to meet expected demand, while the second term represents the adjustments made to this average production level to compensate for the deviations of the downstream inventory positions from their target levels. r_i is the *restoration* coefficient that smooths changes in order quantities from period to period.

The authors use this modeling framework to derive the "steady-state" (defined as asymptotic covariance stationarity) operating characteristics of the divergent network. These characteristics are then used to analyze postponement decisions. They first show that postponement through standardization will result in inventory savings. They then develop a total relevant cost function that includes inventory holding costs, fixed investment for redesigning products and/or processes, and the change in unit processing cost due to re-designing components. In order to derive insights into postponement decisions, the authors construct special cases of cost functions.

The authors also study the joint standardization and process sequencing decisions. They define three types of standardizations in a supply chain:

- Component standardization (products at that stage may be differentiated).

- Process standardization (products at that stage may be differentiated).

- Product standardization.

Note that product standardization is a much stronger condition than the other two types of standardizations. For example, even if the components or the processes at a stage are standardized, products flowing through a stage may be different. However, if products at a stage are standardized, components and processes at that stage have to be standardized.

The system considered in their paper consists of two products and N operations (or processes). These operations need to be sequenced. Since there are N operations, there have to be N stages (or positions) in the sequence. The two products may be common for the first k stages and then may get customized into end-products (Figure 16.1). The joint process sequencing and delayed differentiation problem results in an optimization problem that can be formulated as follows. Let s_{ij} represent the cost of performing operation i at stage j of the sequence. Let α_{ij} be the net savings resulting from standardizing components and processes used in operation i at stage j of the sequence. These savings reflect benefits due to risk-pooling in component inventories, effect of quantity discounts in purchasing, economies of scope, minus redesign costs. Note that α_{ij} can be negative. Also let h_{i1} be the inventory holding cost of items in the buffer of operation i at a stage in which the product is standardized, and h_{i2} the inventory holding cost of items in the buffer of operation i at a stage in which the product is not standardized. These inventory holding costs are obtained using the operating characteristics of the divergent flow network. Also define the following indicator variables:

$$z_{ij} = \begin{cases} 1 & \text{if operation } i \text{ is performed at stage } j \\ 0 & \text{otherwise} \end{cases}$$

$$u_j = \begin{cases} 1 & \text{if products are standardized at stage } j \\ 0 & \text{otherwise} \end{cases}$$

The objective function for this problem can be written as

$$TC = \sum_i \sum_j \left(s_{ij} z_{ij} + (1 - u_j) h_{i2} - u_j z_{ij} h_{i1} + \alpha_{ij} u_j z_{ij} \right) \qquad (16.20)$$

The optimization problem can now be written as

$$\min TC$$

s. t.

$$\sum_i z_{ij} = 1, \quad \forall j \tag{16.21}$$

$$\sum_j z_{ij} = 1, \quad \forall i \tag{16.22}$$

$$u_j \geq u_{j+1}, \quad j \in \{1, \ldots, N-1\} \tag{16.23}$$

$$u_j, z_{ij} \in \{0, 1\}, \forall i, j$$

Constraints (16.21) and (16.22) of this optimization problem are the assignment constraints. Constraint (16.23) ensures that products can be standardized at stage j only if they are also standardized for all stages 1 through $j-1$, $u_N = 0$ and $u_0 = 1$. The optimal solution to this problem can be obtained by solving N linear assignment problems. Note that the point of product differentiation can vary between 0 and $N-1$. The point of product differentiation determines the vector of u_js. Once this vector is determined, the resulting problem is a linear assignment problem that can be solved easily.

In this formulation, the authors have not considered operation precedence explicitly. Operation precedence can be incorporated implicitly in the cost coefficients s_{ij}. If one were to include precedence constraints explicitly in the formulation, it may not be easy to obtain an optimal solution.

Swaminathan and Tayur [29] also study the joint delayed differentiation and process sequencing problem. Their model is more comprehensive than that of others. They consider redesign costs and inventory holding costs, capacity constraints, and precedence constraints on the operations. As a result of all the factors they have considered in their model, it is not possible to obtain analytical results. In order to solve their problem, they develop heuristics involving stochastic integer program with recourse.

16.2.1.4   Assortment Problem or the Vanilla Box Problem. Swaminathan and Tayur [28] model a product line with multiple end-products. Their objective is to determine the set of semi-finished products (or vanilla boxes) that must be stocked. These vanilla boxes are then customized into end-products. This idea is similar to the probabilistic assortment problem studied by Pentico [25]. Their model and solution approach is similar to that for Swaminathan and Tayur [29] described in the previous section.

16.2.1.5 Other Aspects of Postponement. Gavirneni and Tayur [14] study postponement and information sharing. They focus on relative benefits of information sharing and delayed product differentiation. They conduct numerical tests to study the impact of holding costs, capacity, and imbalance in the market shares of different end-products on benefits of postponement and information sharing. We refer the reader to Chapter 15 of this volume where this work is described in greater detail.

Garg et al [13] study the impact of various operational parameters, some of which were influenced by the terms and conditions of supply agreements, on postponement. This work is based on work done for one of IBM's product divisions. They model a divergent flow system operating under a decentralized control policy. The operational parameters they consider include relative lead times between echelons, demand variability, value added at each stage, and the level of part commonality. The supply agreements the division has with its customers and vendors, affect lead times and variance of demands at different stages of the network. They identify relative lead times between stages, variance of demand at each stage, value-added at each stage, and the level of part commonality, as the major drivers of cost-effectiveness of postponement.

16.2.2 Process Sequencing

Researchers such as Lee and Tang [21], Garg [10], and Swaminathan and Tayur [28] have studied processing sequencing in the past. Most of this work has focused on process sequencing as a means of effecting postponement. However, process sequencing is a powerful means of reducing cost in the supply chain. We use Figure 16.4 to provide intuition on how process sequencing can reduce costs in the supply chain. This figure depicts two process sequences, A and B. On the x-axis is the cumulative lead time since the beginning of production, while the y-axis shows the cumulative value of the product at a given stage of its processing. Assuming the inventory costs are proportional to the value of the product at that stage, the area beneath each curve represents the total inventory carrying costs for that sequence. The shaded area represents the difference in inventory costs between sequences A and B. Therefore, the process sequence that minimizes adds more value to the product closer to the end of its manufacturing.

Among the research on process sequencing, Lee and Tang study feature sequencing in a multi-feature, product line as a means of reducing variability in the system. Swaminathan and Tayur focus on assembly sequence as a means of delaying differentiation. Garg [10] studies process sequencing as means of delaying differentiation and as a means of reducing costs in the supply chain. He studies a system with N stages with stage k ($0 \leq k < N$) as the point of differentiation.

In order to outline his approach, we define the following notation. Let $Pos(j)$ denote the position of process j in the sequence of stages, and let the ith position in the sequence be denoted by $\lfloor i \rfloor$. $Pos(j) = \lfloor i \rfloor$ implies that process j is in the ith position in the sequence. Note we assume that $\lfloor k \rfloor$ is the point

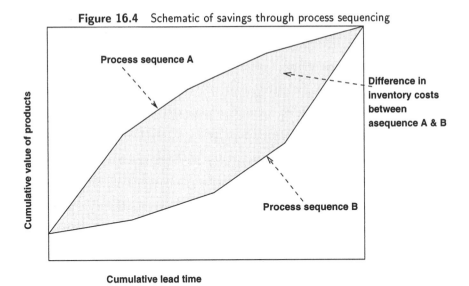

Figure 16.4 Schematic of savings through process sequencing

of differentiation. He studies the incremental cost of *pre-performing* a stage. This operation can generate all possible sequences of stages, given N stages. Define ψ_{ij}, the pre-performing operator, to be the operator that generates a new sequence by moving stage $\lfloor j \rfloor$ from its current position and performing it immediately *before* stage $\lfloor i \rfloor$, where $\lfloor i \rfloor < \lfloor j \rfloor$. Therefore, if $\lfloor i \rfloor \leq \lfloor k \rfloor < \lfloor j \rfloor$, we are effecting postponement.

The incremental cost of pre-performing a stage includes the savings in inventory holding costs, product and process redesign cost, and change in unit process costs. Inventory costs for a sequence are obtained from the model of a divergent flow system that is similar to that used in Garg and Lee [11]. This model is outlined in Section 16.2.1.3. He then constructs several special cases of cost function and value-add function. In one of the special cases, he shows that the minimum inventory cost process sequence would sequence the processes in the ascending order of their value-added per unit lead time, a result analogous to the one depicted in Figure 16.4.

16.3 LEAD TIME REDUCTION STRATEGIES

Industries such as apparel and electronics manufacturing are characterized by a large number of end-products with very short life-times, and with the supply chains with long lead times. One may need to start manufacturing some of the components more than a year in advance. Therefore, production targets are often based on long-range forecasts. These long-range forecasts tend to be inaccurate for two main reasons. First, in these industries changes in consumer tastes and/or technology are quite rapid. Second, because of sheer variety of end-products, it is very difficult to forecast demands for each end-product accurately. As a result, manufacturers often have either excess inventories, or have insufficient amount of each end-product to satisfy consumer demands.

Lead time reduction strategies take product variety as a fact of life and try to alleviate problems stemming from variety by reducing lead times. Quick Response (QR) and production line structuring are the two strategies of reducing inventories by cutting lead times in the supply chain.

QR is a strategy of reducing inventories in the supply chain by cutting lead times in production and improving coordination between different stages of the supply chain. QR systems have become very popular in apparel and fashion goods industries. Fisher et al. [8] describe one such application at a fashion products supplier. As a part of implementing QR, the fashion products supplier reduced manufacturing lead time. Empirical analysis by the authors showed that early season demand of an item is a better predictor of its demand for the rest of the season than the forecast generated before the season. Smaller manufacturing lead times enabled the manufacturer to adjust its production of various products based on their demand during the first few weeks of the season.

Production line structuring (or modularization) is another means of reducing the total lead time. This concept relates to structuring the products and the production line so that many end-products can be assembled from relatively independent modules and auxiliary systems. The benefit of this approach is that these modules can be manufactured concurrently instead of serially, thereby reducing the total manufacturing lead time. Lee and Tang [20] address study the impact of modularization and delayed differentiation.

Besides lead time reduction, production line structuring also reduces the complexity in the system because one can manufacture many different end-products by using modules in various combination with one another. In addition to these benefits, production line structuring also increases flexibility in field repair, ease of upgrade, etc. Barkan [2] discusses the benefits of this approach and cites examples of companies like General Electric, Allen-Bradley, and Chicago Pneumatic that have successfully employed production line structuring.

16.4 CONCLUSIONS AND FUTURE RESEARCH

We have reviewed some research on various strategies of managing product variety such as component commonality, postponement, process sequencing, quick response systems, and production line structuring. Research in this area is motivated by practices in industry, and hence, quite a few of the papers are based on work done at various companies. Implementation of these strategies often requires product and process redesign. As a result, both fixed and variable costs for that manufacturing that product may change. Most of the papers in this area have focused on one of the variable costs, inventory costs. This is probably because inventories are relatively easy to quantify. Few of the papers that do consider other costs have been able to provide insights only in some special cases.

Researchers have used different approaches and operating conditions to study the impact of postponement. Based on work that has been done in this area, we identify some of the drivers of inventory in postponement:

- Demand variability and correlations.

- Relative market shares (or balance) between the demands of end-products.

- Relative lead times between echelons.

- Value-added to the product at each echelon of the supply chain.

- Capacity.

Most of the papers reviewed in this chapter show how some of these drivers affect postponement decisions. Models that consider all these (and perhaps more) would be useful in providing greater insight into the impact of postponement.

Similarly, process sequencing and production line structuring have been studied largely as a means of effecting postponement. Both these strategies can be useful for managing product variety even if they do not delay product differentiation. This is another area for further research.

In general, the superiority of one strategy over the others is very hard to characterize. One would expect each strategy to exhibit decreasing marginal benefits. Barkan [2] details the pitfalls of following one product and process design strategy too single-mindedly. Therefore, perhaps one can obtain the best results by applying these strategies in combinations with one another. Models that enable one to apply different strategies and to compare their benefits would also very useful.

References

[1] Baker, K. R., M. J. Magazine, and H. Nuttle, The Effect of Commonality on Safety Stock in a Simple Inventory Model, *Management Science,* **32**, 8, 1986, 982-988.

[2] Barkan, P., Strategic and Tactical Benefits of Simultaneous Engineering, *Design Management Journal,* Spring 1991, 39-42.

[3] Collier, D. A., Aggregate Safety Stock Levels and Component Part Commonality, *Management Science,* **28**, 11, 1982, 1296-1303.

[4] Child, P. R. Diederichs, F-H Sanders, S. Wisniowski, SMR Forum: The Management of Complexity, *Sloan Management Review,* Fall 1991, 73-80.

[5] Dodd, T. H., B. E. Pinkleton, A. W. Gustafson, External Information Sources of Product Enthusiasts: Differences between Variety Seekers, Variety Neutrals, and Variety Avoiders, *Psychology & Marketing,* **13**, 3, 1996, 291-304.

[6] Eppen, G., and L. Schrage, Centralized Ordering Policies in a Multi-Warehouse System with Lead Times and Random Demand, in: L. B. Schwarz (ed.), *Multi-level Production/Inventory Control Systems: Theory and Practice,* North-Holland, Amsterdam, 1981, 51-68.

[7] Federgruen, A., and P. Zipkin, Approximations of Dynamic Multilocation Production and Inventory Problems, *Management Science,* **30**, 1, 1984, 69-84.

[8] Fisher, M. L., and A. Raman, Reducing the Cost of Demand Uncertainty through Accurate Response to Early Sales, *Operations Research,* **44**, 1, 1996, 87-99.

[9] Garg, A., and H. L. Lee, A Divergent Production Control Model with Stage-Dependent Lead Times, *Proceedings of the Third Industrial Engineering Research Conference,* Atlanta, GA, May 1994.

[10] Garg, A., Product and Process Design Strategies for Effective Supply Chain Management, *Unpublished PhD Dissertation,* Stanford University, Stanford, CA 94305, 1995.

[11] Garg, A., and H. L. Lee, Effecting Postponement through Standardization and Process Sequencing, *IBM Research Report RC 20726,* IBM, T. J. Watson Research Center, Yorktown Heights, NY 10598, 1996.

[12] Garg, A. and C. S. Tang, On Postponement Strategies for Product Families with Multiple Points of Differentiation, *IIE Transactions,* **29**, 8, 1997, 641-650.

[13] Garg, A., N. Nayak, R. Jayaraman, Is Postponement Always Beneficial?, *IBM Research Report RC 21150,* IBM T. J. Watson Research Center, Yorktown Heights, NY 10598, 1998.

[14] Gavirneni, S. and S. Tayur, Delayed Product Differentiation and Information Sharing, *Working Paper,* GSIA, Carnegie-Mellon University, Pittsburgh, PA, 1997.

[15] Gerchak, Y., M. J. Magazine, and A. B. Gamble, Component Commonality with Service Level Requirements, *Management Science,* **34**, 6, 1988, 753-760.

[16] Kapuscinski, R. and S. Tayur, Variance vs. Standard Deviation — Note on: Variability Reduction through Operations Reversal in Supply Chain Re-engineering, *Working Paper,* GSIA, Carnegie-Mellon University, Pittsburgh, PA, 1997.

[17] Lancaster, K., The Economics of Product Variety: A Survey, *Marketing Science,* **9**, 3, 1990, 189-206.

[18] Lee, H. L., C. A. Billington, and B. Carter, Hewlett-Packard Gains Control of Inventory and Service through Design for Localization, *Interfaces,* **23**, 4, 1993, 1-11.

[19] Lee, H. L., Effective Inventory and Service Management through Product and Process Redesign, *Operations Research,* **44**, 1, 1996, 151-159.

[20] Lee, H. L., and C. S. Tang, Modeling the Costs and Benefits of Delayed Product Differentiation, *Management Science,* **43**, 1, 1997, 40-53.

[21] Lee, H. L., and C. S. Tang, Variability Reduction through Operations Reversal, *Management Science,* **2**, 1998, 162-172.

[22] MacDuffie, J. P., K. Sethuraman, M. L. Fisher, Product Variety and Manufacturing Performance: Evidence from the International Automotive Assembly Plant Study, *Management Science,* **42**, 3, 1996, 350-369.

[23] Menon, S., and B. Kahn, The Impact of Context on Variety Seeking in Product Choices, *Journal of Consumer Research,* **22**, 3, 1995, 285-295.

[24] Nahmias, S., Production and Operations Analysis, Second Edition, *Richard D. Irwin,* Boston, MA, 1993.

[25] Pentico, D. W., The Assortment Problem with Probabilistic Demands, *Management Science,* **21**, 3, 1974, 286-290.

[26] Roberts, J. H. and G. L. Lilien, Explanatory and Predictive Models of Consumer Behavior, in *Marketing—Handbooks in Operations Research and Management Science,* J. Eliashberg and G. L. Lilien (eds.), Elsevier Science Publishers, Amsterdam, The Netherlands, 1993.

[27] Schwarz, L. B., Model for Assessing the Value of Warehouse Risk-pooling over Outside Supplier Lead Times, *Management Science,* **35**, 7, 1989, 828-842.

[28] Swaminathan, J. and S. Tayur, Managing Broader Product Lines through Delayed Differentiation using Vanilla Boxes, *Working Paper*, GSIA, Carnegie-Mellon University, Pittsburgh, PA, 1996.

[29] Swaminathan, J. and S. Tayur, Managing Design of Assembly Sequences for Product Lines that Delay Product Differentiation, *Working Paper,* GSIA, Carnegie-Mellon University, Pittsburgh, PA, 1997.

[30] Yeh, K.-H., C.-H. Chu, Adaptive Strategies for Coping with Product Variety Decisions, *International Journal of Operations and Production Management,* **11**, 8, 1991, 35-47.

[31] Zinn, W., and D. J. Bowersox, Planning Physical Distribution with the Principle of Postponement, *Journal of Business Logistics,* **9**, 1988, 117-136.

17 RETAIL INVENTORIES AND CONSUMER CHOICE

Siddharth Mahajan

The Wharton School
University of Pennsylvania

Garrett J. van Ryzin

Graduate School of Business
Columbia University

17.1 INTRODUCTION

In this chapter, we examine retail inventory management under consumer choice. Research on this topic is motivated by the simple recognition that consumers are frequently willing to buy a different color, size or brand within a product category if their preferred variant is either not offered or is offered but temporarily out of stock. In other words, faced with limited choices, consumers are often willing to *substitute* rather than go home empty handed. Intuitively, a retailer's decisions about the level of variety they offer in an assortment and the quantity of inventory they stock ought to account for such behavior. Understanding precisely how choice behavior can be incorporated in inventory models and what impact it has on assortment decisions and operating performance are the main research challenges in this area.

This problem is a real one. Retailers are aware of the importance of coordinating assortment and inventory management decisions, and a growing number are pursuing so-called *efficient assortment* and *category management* programs [54]. These programs seek to rationalize assortment and inventory decisions at the category level based on detailed cost-benefit analysis. Given the widespread penetration of supply chain management thinking, this trend should not come as a surprise. For a retailer who is taking a holistic view of the various activities of their entire distribution channel, it is only logical to question whether the inventory costs of added variety on their retail shelves justifies the value it provides to end consumers.

Yet until recently, inventory management and consumer choice behavior have been largely isolated areas of investigation within academic literature. Inventory theorists typically have assumed that demands for different variants are mutually independent and that product variety decisions have been determined exogenously. (See Graves et al. [26] for a comprehensive collection of survey articles on production and inventory theory.) At the same time, most marketing models of retail assortment decisions either ignore inventory costs or, at best, have a rudimentary model of inventory cost and performance. (See, for example, Anderson and Amato [4], Bultez and Naert [14], Corstjens and Doyle [17] and Hansen and Heisenbroek [30].)

Therefore, there is both a practical and theoretical need to better understand the joint interaction of assortment variety, inventory policies and consumer choice behavior. We believe the research surveyed in this chapter is an important step in this direction.

That said, we have deliberately limited the scope of this chapter. There is extensive literature on the economics of product variety, operations/marketing product line selection, manufacturer-controlled substitution and shelf-space allocation, that addresses closely related problems. (In the next section, we provide a brief literature survey of this work.) However, our main focus is on work that directly addresses the interaction of inventory management and consumer choice. In particular, the papers of Agrawal [2], Anupindi, Dada and Gupta [6], Noonan [47], [48], Mahajan and van Ryzin [41], Smith and Agrawal [58] and van Ryzin and Mahajan [66].

Each of these papers uses a combination of both inventory and consumer choice models to investigate retail variety and inventory management, either from a computational or structural standpoint or both. We note that, at this time, all these papers have yet to appear in journals, though several are accepted or in advanced stages of review. This is a reflection of the fact that the topic is quite new. Indeed, all these papers were originally drafted only in the last five years. It is, therefore, very much a topic that is still in its early stages of development. It is our hope that this chapter serves to unify and disseminate results in the area and thereby stimulate further research developments.

The remainder of the chapter is organized as follows: In Section 17.2, we survey related literature outside the main topic of this chapter. Then, in Section 17.3, we provide a tutorial on consumer choice models. We do this in order to make the chapter self-contained and easily accessible to readers in the operations management community. Readers familiar with consumer choice theory can skip this section, although we have given careful thought in organizing this material to illustrate important connections of choice theory to operations management questions.

The core material of the chapter is contained in Sections 17.4 and 17.5. In Section 17.4 we discuss the work of Smith and Agrawal [58] and van Ryzin and Mahajan [66], both of which analyze what we call *static choice* models, in which consumer choice decisions depend on which variants are selected for the assortment but not on their immediate inventory status. Practical examples of such situations include mail-order catalogues and products in which the use of "floor models" is common. In both cases, the status of the on-hand inventory is not readily discernable. In Section 17.5, we discuss the work of Noonan [47] and Mahajan and van Ryzin [41] on *dynamic choice* models, in which consumer choice decisions are based on the actual on-hand inventory at each point in time.

Finally, in Section 17.6 we briefly discuss estimation of the parameters of consumer choice models, in particular maximum likelihood estimators of random utility models and the recent results of Anupindi, Dada and Gupta [6] on estimation of substitution probabilities based on retail transaction data. Our conclusions and thoughts on future research in this area are given in Section 17.7.

17.2 RELATED LITERATURE

17.2.1 Economics of Product Variety

The economics literture on product variety (See Hotelling [32], Chamberlin [15], Kaldor [34], Lancaster [35], Dixit and Stiglitz [20], Spence [57] and Perloff and Salop [53].) is broadly related to our topic in that it addresses the costs and benefits of product variety. However, the main focus is a market-level analysis of product variety. Single product firms produce goods which are not identical, so the level of product variety is determined by the number of firms in the market. Typical questions of interest include characterizing the number of firms in equilibrium and determining whether the level of variety offered under equilibrium is socially optimal. See Anderson, De Palma and Thisse [5] and Lancaster [35] for surveys.

In this literature, considerable attention is devoted to understanding how demand for differentiated products is modeled. However, the models of operating cost are much less detailed. Usually, a firm is assumed to have a fixed cost for operating and a marginal cost for producing an item. More complex inventory or production cost structures are not considered. Moreover, because the models are aggregate, market-level models, they typically do not consider uncertainty in demand or transient phenomenon, such as inventory stock-outs, both of which are key considerations in retail inventory management.

17.2.2   Joint Operations/Marketing Product Line Design

There has been a growing body of research on coordinating operating and marketing product line decisions, for example de Groote [19], Chen, Eliashberg and Zipkin [16], Dobson and Kalish [21] and Shugan [56]. This work recognizes that customers are heterogenous in taste, and that they are often willing to pay a higher price for products with attributes closer to their desired attributes. Thus, a firm has an incentive to offer a broad variety of products to better cover the possible range of consumer tastes. However, offering a broad product line adds to manufacturing cost due to increased setup for the machines, increased raw material or work-in-process holding costs, or investment in more sophisticated technology. Thus, a trade-off exits.

de Groote [19] considers a monopolist with products differentiated by a single attribute. Consumers tastes for the single attribute are uniformly distributed, and Hotelling's model is used to derive demand for each product. The marketing decisions are the number and position of products in attribute space and the price of each product. The multiproduct economic order quantity model (EOQ) is used to model manufacturing costs, where the flexibility of the production process is characterized by the setup cost. A key result of de Groote is that iteratively moving between the two decisions of choosing setup cost and then the number of products may produce suboptimal sets of decisions.

In Chen, Eliashberg and Zipkin [16], a similar problem of pricing and positioning of a product line is considered. The authors use a more general version of Hotelling's model, where the distance function representing the penalty of not meeting the consumer's ideal attribute is modeled as a convex function. Consumers differ both in their reservation prices and in their ideal attribute choice. Different manufacturing cost structures corresponding to either deterministic or stochastic demand are allowed. The authors develop structural properties of optimal product lines and exploit these properties to develop dynamic programming algorithms. They show numerically that more products are offered if consumers are more heterogeneous in taste, or more rigid in their preferences, or if setup cost is lowered.

Dobson and Kalish [21], formulate the product line design and pricing problem as an uncapacitated plant location problem. The market facing the firm consists of multiple consumer segments. The raw data for the problem consists of the utilities associated by each consumer segment for each product and the fixed and variable costs incurred by the firm for offering each product. The decision corresponding to the products to offer is modeled using appropriate binary variables in the formulation. The authors formulate two versions of the problem based on a welfare criterion and a profit maximizing criterion, develop heuristics and use them on test problems. It is suggested that the data corresponding to the utilities associated by each customer segment for each product be estimated via conjoint analysis techniques developed in the marketing literature.

Shugan [56] analyzes a product line problem motivated by the ice cream industry. Citing examples of Haagen-Daz, Baskin-Robbins and Bresler, the author notes that super-premium ice creams tend to be offered in fewer flavors than economy ice creams. A model with three producers is formulated to help explain this observation. The level of variety corresponds to the different flavors of ice cream offered by each producer. The quantity produced by each producer as a function of the price and the level of variety of the others is specified using a econometric model in which there is a unit cost for production and a manufacturing cost proportional to the number of flavors offered. He shows that a Nash equilibrium exists in this three producer market and identifies higher variable cost for super-premium ice cream as well as a smaller market size as possible explanations for the narrower product lines observed in premium ice cream lines.

While the general thrust of this product line research is quite similar to the retail assortment problem, there are some important differences. A distinguishing feature of retail variety decisions, as opposed to production variety decisions, is that there are few *direct* costs (set-ups, change-overs, etc.) of variety for a retailer. [1] (We note that Chen, Eliashberg and Zipkin [16] do consider stochastic inventory models without fixed costs.) Rather, it is the *indirect* costs of stockouts and overstocking that impose an implicit cost on variety. Retail buyers typically discuss this trade-off in terms of "breadth versus depth" of assortment. Operating cost structures need to model these important phenomenon of retail inventories. In addition, retailers rarely make product *design* and positioning decisions; rather, they typically make only product *selection* decisions. [2] This in turn has important implications for how one models consumer choice. We discuss this point in more detail in Section 17.3.

---

[1] There are some direct transactions and information processing cost that are driven by the number of SKUs a retailer carries; however, in most cases these are relatively small costs compared to the cost of goods and other operating expenses.

[2] Again, this distinction is not strict. In particular, retailers who offer private label merchandise must design and position their products relative to competing products just as a manufacturer would.

17.2.3 Manufacturer-Controlled Substitution Models

The general idea of using inventory of one type of product to satisfy demand for another product is explored in the literature on manufacturer-controlled substitution. In these models, unsatisfied demand for an item can be satisfied by another item, usually with a certain penalty cost. For example, a high-speed CPU could substitute for a low-speed CPU, or a long, high tensile strength beam could be used to satisfy demand for shorter, medium-strength beams. The manufacturer must decide in advance how much inventory of each product to stock and then how best to divert demands when faced with a stockout. Demand for items could be deterministic or stochastic.

Substitution problems with deterministic demands have been considered by Sadowski [55], Wolfson [67], Tryfos [63] and Pentico [52]. The analysis depends on the nature of the costs considered, on whether the products are substituted on a single dimension, as in the chip example, or in two dimensions, as in the case of steel beams, and on whether demand is discrete or continuous. Typical costs include a fixed cost for keeping the product in stock, an inventory holding cost which could be linear or concave, and a substitution cost which is additive and linear.

Substitution problems with stochastic demand have been considered by Pasternack and Drezner [51] and Bassok, Anupindi and Akella [9]. Pasternack and Drezner [51], consider a two item single period problem with item specific holding, shortage and salvage costs. They distinguish between the revenue received from an item depending on whether the item satisfied original demand or substitute demand. The authors prove concavity of the profit function and provide sensitivity results.

Bassok, Anupindi and Akella [9], consider a single period, N product inventory model with item-specific holding and shortage costs. Only downward substitution is allowed, i.e. a demand for an item can be satisfied only by items which are better than it based on a given attribute. Every time a substitution occurs a fixed penalty cost is incurred. The authors formulate the problem as a two stage stochastic program and show concavity of the resulting profit function. Using a numerical study with two products, they consider the benefits of incorporating substitution effects in ordering decisions under different cost and demand scenarios.

While this body of work does address substitutable inventory management, consumer substitution is fundamentally different. Consumer substitution decisions are not *controlled* by the retailer; rather, they are made by a large number of independently-minded (and self-interested) consumers. A retailer can only *indirectly* affect consumers decisions through his/her inventory policy. Nevertheless, the inventory processes under these two types of substitution share certain structural properties.

17.2.4 Shelfspace Allocation

In the marketing literature, there is a body of research that addresses how shelf space should be allocated. Some work representative of this line of research is Anderson and Amato [4], Bultez and Naert [14], Corstjens and Doyle [17] and Hansen and Heisenbroek [30]. Space allocation can be at an aggregate level, among categories in a store, or at a category level, among variants within a given category. These models typically assume deterministic demand that is elastic with space, i.e. if more shelf space is allocated to a variant in a category, more demand will be generated for it. Substitution effects are incorporated indirectly through the use of cross-elasticities for space, i.e. increasing the space allocated to a variant in a category, decreases the demand for other variants in the category.

Generally, inventory cost considerations in this literature are ignored or treated in a rudimentary fashion, incorporating only a marginal cost for stocking an item or for using space. A significant exception is the recent work of Fruend and Matsuo [23]. They consider a shelfspace allocation problem under Poisson demand and use a detailed representation of inventory cost. The inventory control system for each item is of a periodic-review-order-up-to type with zero lead time. The problem is formulated as a constrained optimization problem with constraints on shelf space and on service level which is measured as the probability of stockout. The decision variables include the review period for each item and the amount of shelf space assigned to it, where shelf space is determined in terms of number of facings.

However, Fruend and Matsuo [23], do not consider consumer choice effects, which are a primary focus of the work surveyed in this chapter. While they propose different assumptions on demand elasticity with space, demand for a given product is a function only of the space allocated to it. This assumption is reasonable if one is considering allocating space to dissimilar products (e.g. corn chips and tooth paste), but is less attractive for competing products in the same category (corn chips and potato chips).

However, many of the basic questions Fruend and Matsuo [23] address are very similar in spirit to those covered in this chapter. For example, using a disguised data set from a convenience store, the authors show that when inventory costs are included in the shelf space allocation problem, assortments with lower variety are more profitable than the store's current assortments.

17.2.5 Retail Variety Costs and Benefits

Two papers that analyze the level of variety in a retail assortment are Baumol and Ide [10], who consider the problem of store-level variety, and Anupindi, Gupta, Venkataraman [7], who model assortment decisions at the category level. They differ from the work on shelf space allocation considered above in that they do not model space elasticity of demand. Neither paper combines both choice and inventory models; however, they do address the general question of the costs and benefits of assortment variety.

In Baumol and Ide [10], variety is measured by a scalar variable (the number of categories). Consumers visit the store with a higher probability, if more variety is offered. Consumers also incur a cost for visiting and a search cost for shopping within the store, which is an increasing function of the number of categories offered. Based on these costs, the authors find a region around the store beyond which consumers will not shop at the store. Making assumptions on consumer density within this area, they find expressions for store sales, as a function of the number of categories stocked. The inventory costs for each category are modeled as an EOQ cost. However, in their work Baumol and Ide [10] do not model substitution effects; rather, in their model, offering more categories increases the number of categories that the consumer can purchase from for a given fixed cost of visiting a store. The negative effect of excess variety is an increased within-store search cost. Thus, while the work is quite interesting and relevant, it does not address the precise topic of this chapter.

Anupindi, Gupta and Venkataraman [7], analyze assortment stocking decisions for a single category. The authors model a market with multiple customer segments, with each segment having a deterministic utility for the set of variants in the assortment. Consumers choose the variant with the highest utility. This effectively models static choice with deterministic demand. On the cost side, there is a fixed cost for including a product in the assortment and a variable cost for each item sold.

The assortment optimization problem is formulated as an uncapacitated plant location problem, where the decision to include a variant in the assortment is analogous to opening a plant. The objective function is a convex combination of profit and a consumer disutility measure. Estimation methods based on the multinomial logit model are proposed. The optimization model is fit and tested on scanner data from a margarine category.

While the theme of managing assortment variety explored in Anupindi, Gupta and Venkataraman [7] is very closely linked to the topic of this chapter, the authors do not consider inventory dynamics and costs, nor do they account for phenomenon such as demand uncertainty or temporary stock-outs. For this reason, we do not include it in the body of the chapter.

17.3 CHOICE PROCESSES

This section provides a self-contained survey of the essential ideas from the theory of discrete choice. It is provided as a service to those readers who are unfamiliar with consumer choice models and theory. We describe the key models and discuss their merits for use in analyzing operations management problems.

This section is organized as follows: We first define preference relations and present an important theorem linking preference relations to utility maximization. In Section 17.3.1 we describe various deterministic models for constructing preference relations and utilities. We then consider probabilistic versions of these models, which can be used to represent heterogeneity of preference among a population of consumers. In Section 17.3.4, we consider an alternative approach to modeling probabilistic choice due to Luce [39], which is motivated by the empirical finding that consumers do not always behave as if they have well-defined preference relations among alternatives. The primitives in Luce's model, therefore, are choice probabilities not preference relations.

We caution the reader that the authors consider themselves, at best, merely "consumers" of choice theory. Our main concern is not the development of choice theory itself, but its use in operations management applications. For a more in-depth treatment of choice theory, the reader is encouraged to consult standard references by Anderson, dePalma and Thisse [5], Ben-Akiva and Lerman [12] and Manski and McFadden [43].

17.3.1 Choice Axiomatized by Preference Relations

Given two alternatives, a *choice* corresponds to an expression of preference of one alternative over another. Similarly, given n alternatives, choice can be defined in terms of the preferences expressed for all pair-wise comparisons between the n alternatives.

The mathematical construct that formalizes this notion of choice is a *preference relation.* It is defined as follows: Let X be a set of alternatives and let $\prec$ be a binary relation on the set X. (We shall consider X to be finite, but this restriction can be relaxed.) For alternatives $x, y \in X$, $x \prec y$, denotes that y is strictly preferred to x. Consider the following properties of the binary relation $\prec$:

Asymmetry   If x is strictly preferred to y, then y is not strictly preferred to x.

Negative Transitivity   If x is not strictly preferred to y and y is not strictly preferred to z, then x is not strictly preferred to z.

Asymmetry and negative transitivity can be considered as "minimal consistency properties" for a expression of preference among a set of alternatives. Hence,

Definition 1 *A binary relation $\prec$ on a set X is called a preference relation, if it is asymmetric and negatively transitive.*

A remarkable fact is that such preference relations are intimately related to utility maximization. Indeed, (See Kreps [33] for a proof.):

Theorem 1 *If X is a finite set, a binary relation $\prec$ is a preference relation if and only if there exists a function $u : X \to \Re$ (called a utility function) such that*

$$x \prec y \quad iff \quad u(x) < u(y) \qquad (17.1)$$

From Theorem 1, it follows that when a preference relation exists, making a choice among the alternatives in any set S is equivalent to choosing the alternative with the maximum utility for some appropriately defined utility function. As shown next, this result has important implications for constructing choice models.

17.3.2 Mechanisms for Generating Preference Relations

Theorem 1 suggests two generic approaches for modeling choice: 1) construct preference relations directly, or 2) construct utilities and then apply utility maximization. While Theorem 1 shows that both approaches are essentially equivalent, there are some important qualitative differences in actual applications. Below we present examples of both approaches and discuss their relative merits for operations management research.

17.3.2.1 Attribute Models. One approach to modeling choice is to construct preference relations directly. This typically consists of modeling attributes of each alternative and then specifying a decision rule for ranking alternatives based on their attributes.

One key advantage of attribute models of choice is that consumer preferences can be linked directly to (potentially controllable) attributes of a firm's products. As a result, this approach is well suited to operations management problems involving product design or product positioning, in which a firm can be reasonably be viewed as having control over the design features of its products. When product design is not controllable, there is a less compelling reason to use attribute models. We discuss two such attribute models, lexicographic and address models.

Lexicographic models. In a lexicographic rule, a product is made up of binary attributes. For example, a tennis racquet may or may not have a wide head, it may or may not use graphite material, etc.. The consumer decision rule is to rank all attributes and then eliminate alternatives which do not possess the most important attribute. If more than one alternative remains, the next most important attribute is chosen as a criterion for elimination of alternatives, and so on. A lexicographic rule, therefore, generates a preference relation among the alternatives, and we can directly link these preferences to the existence of various attributes in each alternative. Note, however, that this model implies attributes strictly dominate each other. That is, a consumer always prefers a racquet with a wide head over one with a standard head, regardless of the other attributes each racquet posseses.

To illustrate the result of Theorem 1, we show how to construct a utility function that generates the same choices as the lexicographic model. Suppose there are n alternatives that have m attributes. Let the attributes be ordered so that 1 represents the highest-valued attribute and m the lowest. Let $a_{jk}, k = 1, ..., m$, be binary digits representing whether alternative j, possesses attribute k. Then a utility satisfying Theorem 1 is the binary number,

$$U_j = a_{j1}a_{j2}...a_{jm}.$$

Maximizing over these utilities leads to the same decision as the lexicographic model.

Address models. Address models (See Hotelling [32] and Lancaster [35].) link attributes to preference without imposing the restriction that some attributes strictly dominate others. We have n alternatives, each of which has m attributes that take on real values. Therefore, alternatives can be represented as n points, $z_1, .., z_n$, in $\Re^m$, which is called *attribute space*.

Each consumer has an ideal point (address) $y \in \Re^m$, reflecting his/her most preferred combination of attributes. A consumer chooses the product closest to his/her ideal point in attribute space, where distance is defined by a metric ρ on $\Re^m \times \Re^m$ (e.g. Euclidean distance). These distances define a preference relation, i.e. $i \prec j$ if and only if $\rho(z_i, y) > \rho(z_j, y)$.

Again, we point out that Theorem 1 guarantees that an equivalent utility maximization model exits which generates the same choices. In this case, it is easy to see that for consumer y the utilities

$$U_j(y) = c - \rho(z_j, y),$$

where c is an arbitrary constant, produce the same decision rule as the address model.

As with the lexicographic model, the address model is best suited to problems of product design and positioning. For example, these models are used in the work of Chen, Eliashberg and Zipkin [16] and de Groote [19] on joint marketing-operations product line design.

17.3.2.2 Utility Models. If one does not need to model product design decisions, then one can simply specify the utility values of each alternative, $U_1, .., U_n$, directly. It is also possible to specify different market segments (e.g. different customer types) and associate a different utility vector with each segment. As compared to attribute models, utility models are more naturally suited to problems of product *selection*. For example, in Sections 17.4 and 17.5 we show how this approach is used to model selection of products to include in a retail assortment. Also, the work of Anupindi, Gupta and Venkataraman [7] and Dobson and Kalish [21] are examples of deterministic utility models of this type applied to assortment selection.

We note, however, that the distinction between attribute and utility models is not entirely sharp. Indeed, when using utility models it is not uncommon to specify a function relating attribute values to utilities directly. A specific example that is used frequently in transportation choice models is the *linear in attributes* utility model (See Ben-Akiva and Lerman [12].), in which the utility is expressed as a linear function of an alternative's attributes. Estimation of this type of model is discussed in Section 17.6.

17.3.3 Probabilistic Mechanisms for Generating Preference Relations

When modeling the choice decisions of a population of consumers, it is often desirable to have a probabilistic model of choice. Probabilistic models are useful for several reasons: First, not all consumers have the same preference relations, so probabilistic models can be used to represent unobservable heterogeneity of preference among a population of consumers. In addition, uncertain choice behavior can arise due to the inability of a decision maker to observe all the relevant variables affecting a given consumer's choice. Finally, consumers may exhibit "variety seeking" behavior, and deliberately change their preferences over time (movie or meal choice, for example).

For all these reasons, it is often reasonable to assume that a decision maker has only incomplete information on the preference relation of any given consumer. This uncertainty can be modeled using natural randomizations of the choice mechanisms considered in Section 17.3.2 above.

Alternatively, probabilistic choice can be associated with consumers whose behavior is inherently unpredictable. That is, consumers may not behave consistent with a well-defined preference relation, and, at best, they may exhibit only some probabilistic tendency to prefer one alternative over another. Luce [39] developed a model of this type based purely on a set of axioms on the choice probabilities. His model is discussed in Section 17.3.4.

The distinction between models based on randomized preferences and those based on random choice behavior is important primarily to behavioral theorists. (A seminal work in this area is Block and Marschak [13].) However, for most operations management problems, what matters most is the demand process produced by a given model. As shown below, in some important cases both approaches lead ultimately to the same probabilistic model of demand.

17.3.3.1 Attribute Mechanisms with Randomization. Parelleling our discussion in Section 17.3.2, we next consider randomized versions of the lexicographic and address models.

Lexicographic: The Tversky Model. Tversky's [65] work was motivated by the desire to model probabilistic choice behavior. However, one can also view his model as a randomization of the deterministic lexicographic rule. For purposes of illustration, this is the approach we take here.

The Tversky model assumes that each alternative consists of several binary attributes. The decision rule is a probabilistic version of the lexicographic rule discussed previously. It proceeds as follows:

1. Attributes common to all alternatives are deleted.

2. One of the remaining attributes is selected with probability proportional to a specified (deterministic) utility value of the attribute.

3. Alternatives not having the selected attribute are eliminated.

4. If a single alternative remains, it is chosen. If several alternatives remain, a new attribute is determined and more alternatives are eliminated by repeating the above process. If all remaining alternatives have the same attributes, then one is chosen with equal probability.

Since different sequences of elimination are possible under this model, the probability of choosing a particular alternative, is the sum of probabilities of all sequences of choices which lead up to this alternative. Because the sequence of attributes chosen as a criterion for elimination of alternatives is random, Tversky's model can be considered as a randomization of the lexicographic rule.

Address Models. As above, each of the n alternatives has m real-valued attributes and can be represented as elements $z_1, .., z_n$, of $\Re^m$. Similarly, each individual has an ideal point (address) $y \in \Re^m$. The only difference is that we now associate a probability density with the location of consumer ideal point, y, in attribute space. Let $g(y), y \in \Re^m$, denote the density function for these ideal points. Then we can associate a utility with each alternative j of the form

$$U_j(y) = c - \tau \parallel z_j - y \parallel, \qquad (17.2)$$

where c is the utility of the customer's ideal point y and τ represents the disutility due to deviations from this ideal point. The market space of variant j is given by

$$M_j = \{y \in \Re^m : U_j(y) \ge U_i(y), i = 1, ..., n\} \qquad (17.3)$$

The probability that a consumer buys variant j is then given by,

$$P_j = \int_{M_j} g(y) dy \qquad (17.4)$$

17.3.3.2 Random Utility Models .

It is also possible to randomize utility models of choice. This leads to so-called *random utility* models. These models originated with the early work of the mathematical psychologist Thurston [60] and were later formalized by economists, most notably Manski [42] and McFadden [45]. (See Manski and McFadden [43].)

Random utility models are defined as follows: Let the n alternatives be denoted $j = 1, ..., n$. A consumer associates a utility for alternative j, denoted U_j. This utility is decomposed into two parts, a *representative* component u_j that is deterministic and a *random* component, ϵ_j, with mean zero. (The assumption of zero mean is without loss of generality.) Therefore,

$$U_j = u_j + \epsilon_j, \qquad (17.5)$$

and the probability that an individual selects alternative j is

$$q_j = P(U_j = \max_{i=1,...,n} U_i).$$

This probability depends, of course, on the joint distribution of the random components ϵ_j. We discuss the most common versions next.

Binary Probit. If there are only two alternatives and the error terms $\epsilon_j, j = 1, 2$ are independent, normally distributed random variables with mean zero and identical variances σ^2, then the probability that Variant 1 is chosen is given by

$$q_1 \;=\; P(\epsilon_2 - \epsilon_1 \leq u_1 - u_2) \;=\; \Phi(\frac{u_1 - u_2}{\sqrt{2}\sigma}) \tag{17.6}$$

where $\Phi(.)$ denotes the standardized cumulative normal distribution. This model is known as the *binary probit* model. While the normal distribution is an appealing model of disturbances in utility (i.e. it can be viewed as the sum of a large number of random disturbances), the resulting probabilities do not have a closed form solution. This has lead researchers to seek other more analytically tractable models.

Binary Logit. The binary logit model leads to choice probabilities similar to those given by the binary probit model, but is much simpler to analyze. The assumption made here is that the error term $\epsilon = \epsilon_1 - \epsilon_2$ has a *logistic* cumulative distribution, i.e.

$$F(\epsilon) = \frac{1}{1 + e^{-\frac{\epsilon}{\mu}}} \tag{17.7}$$

where $\mu > 0$ is a scale parameter and $-\infty < \epsilon < \infty$. Here ϵ has a mean zero and variance $\frac{\mu^2 \pi^2}{3}$. The logistic distribution provides a good approximation to the normal distribution, though it has "fatter tails". The probability that variant 1 is chosen is given by

$$\begin{aligned} q_1 \;=\;& P(\epsilon_2 - \epsilon_1 \leq u_1 - u_2) \\ =\;& \frac{e^{\frac{u_1}{\mu}}}{e^{\frac{u_1}{\mu}} + e^{\frac{u_2}{\mu}}} \end{aligned}$$

$$\tag{17.8}$$

Multinomial Logit. The multinomial logit model (MNL) is a generalization of the binary logit model to n alternatives. It is derived by assuming that the ϵ_j are i.i.d. random variables with a Gumbel (or double exponential) distribution [29], with cumulative density function

$$F(x) = P(\epsilon_j \leq x) = e^{-e^{-(\frac{x}{\mu} + \gamma)}}$$

where γ is Euler's constant $(= 0.5772...)$ and μ is a scale parameter. The mean and variance are

$$E[\epsilon_j] = 0, \qquad Var[\epsilon_j] = \frac{\mu^2 \pi^2}{6} \tag{17.9}$$

The Gumbel distribution has some useful analytical properties, the most important of which is that it is closed under maximization. That is, the distribution of the maximum of n independent Gumbel random variables with the same scale parameter μ is also a Gumbel random variable. If two random variables, ϵ_1 and ϵ_2 are Gumbel distributed with mean 0 and scale parameter μ, then $\epsilon = \epsilon_1 - \epsilon_2$ has a logistic distribution with mean 0 and variance, $\frac{\mu^2 \pi^2}{3}$; hence, the connection to the binary logit.

For the MNL model, the probability that an alternative j is chosen from a set $S \subset \{1, 2, ..., n\}$ that contains j, denoted $P_S(j)$, is given by

$$P_S(j) = \frac{e^{\frac{u_j}{\mu}}}{\sum_{i \in S} e^{\frac{u_i}{\mu}}}. \tag{17.10}$$

(See Anderson, de Palma and Thisse [5] for a proof.) If $\{u_j : j \in S\}$ has a unique maximum and $\mu \to 0$, the variance of the $\epsilon_j, j = 1, ..., n$ tends to zero, and the MNL reduces to a deterministic model, viz

$$\lim_{\mu \to 0} P_S(j) = \begin{cases} 1 & \text{if } u_j = \max_{i \in S} u_i \\ 0 & \text{otherwise} \end{cases} \tag{17.11}$$

Conversely, if $\mu \to \infty$, the variance of the $\epsilon_j, j = 1, ..., n$ tends to infinity and the systematic component of utility, u_j, becomes negligible. Specifically,

$$\lim_{\mu \to \infty} P_S(j) = \frac{1}{|S|}, \quad j \in S. \tag{17.12}$$

The MNL has been widely used as a model of choice; however, it possesses a somewhat restrictive property known as the *independence from irrelevant alternatives* (IIA) property:

Definition 2 (IIA Property) *For all $S \subset N$, and $T \subset N$ such that $S \subset T$ and for all $i, j \in S$,*

$$\frac{P_S(i)}{P_S(j)} = \frac{P_T(i)}{P_T(j)} \tag{17.13}$$

Equation (17.13) says that the ratio of choice probabilities for i and j is independent of the choice set containing these alternatives. This property is not realistic if the choice set contains alternatives that can be grouped such that alternatives within a group are more similar than alternatives outside the group, because adding a new alternative reduces the probability of choosing similar alternatives more than dissimilar alternatives.

A famous example illustrating this point is the "blue bus/red bus paradox," due to Debrue [18]. In it, an individual has to travel and can use one of two modes of transportation: a car or a bus. Suppose the individual selects them with equal probability. Let the set $S = \{$ car, bus $\}$. Then

$$P_S(\text{car}) = P_S(\text{bus}) = \frac{1}{2} \tag{17.14}$$

Suppose now that another bus is introduced, which is identical to the current bus in all respects except color: one is blue and one is red. Let the set T denote

{ car, blue bus, red bus }. Then it can be shown that the MNL predicts

$$P_T(\text{car}) = P_T(\text{ blue bus}) = P_T(\text{red bus}) = \frac{1}{3}. \quad (17.15)$$

However, it is clearly more realistic to imagine that, just as before, buses are chosen with the same probability as cars, so a more natural outcome is

$$P_T(\text{car}) = \frac{1}{2}$$
$$P_T(\text{blue bus}) = P_T(\text{red bus}) = \frac{1}{4}$$

$$(17.16)$$

These probabilities violate IIA.

As a result of IIA, the MNL model must be used with caution. It should be restricted to choice sets which contain alternatives that are, in some sense, "equally dissimilar." Despite this deficiency, however, the MNL model is widely used. For example, see Guadagni and Little's [28] work on determining brand share in the presence of marketing variables such as advertising and promotion. It has also seen considerable application in estimating travel demand, e.g. see Ben-Akiva and Lerman [12]. The popularity of MNL is due to the fact that it is analytically tractable, relatively accurate (if applied correctly) and can be estimated easily using standard statistical techniques (See Section 17.6.)

Variations of the MNL have been introduced to avoid the IIA problem, the most prevalent of which is the nested MNL of Ben-Akiva [11]. We also note that a key contribution of Tversky's model, discussed in section 17.3.3.1, is that it avoids the IIA property. Alternatives with different degrees of similarity are allowed and according to Tversky, "the introduction of an additional alternative hurts similar alternatives more than dissimilar ones" [65].

17.3.3.3 Equivalence Between Address and Random Utility Models.

As suggested by Theorem 1, there is a strong connection between preference relations based on attributes and utility maximization. Therefore, it is not too surprising that there exists an equivalence between randomized attribute models and random utility models. Indeed, one can show that, under fairly general conditions, given an address space model it is possible to construct a random utility model that produces statistically identical choices, and vice versa. (See Section 4.3 of Anderson, dePalma and Thisse [5] for a proof.) Again, this result reinforces the theme that these approaches are in many ways equivalent, and which modeling approach one chooses is a matter of analytical convenience and research objectives.

17.3.4 *Choice Axiomatized by Probabilities: Luce Model*

As described above, psychologists and economists alike have frequently questioned the assumption that people have strict preference relations among alternatives. To these behavioral theorists, a model that allows for imperfect discriminatory power and limited information processing ability (i.e. bounded rationality) is more realistic.

The most famous model of this type is due to Luce [39]. Luce proposed a set of axioms on choice probabilities that play a role analogous to the properties of asymmetry and negative transitivity that define a deterministic preference relation. Let N, denote the set of all alternatives. As before, given $S \subset N$ and $j \in S$, $P_S(j)$, denotes the probability that an individual chooses the alternative j from the set S. Let $S \subset N$ and $T \subset N$ be any two sets such that $S \subset T$. The axioms proposed by Luce are:

i) *If for given $j \in S$, $P_{\{i,j\}}(j) \neq 0, 1$ for all $i \in T$, then*

$$P_T(j) = P_T(S)P_S(j)$$

ii) *If $P_{\{i,j\}}(j) = 0$ for some $i, j \in T$, then for all $S \subset T$*

$$P_T(S) = P_{T-\{j\}}(S - \{j\})$$

Luce's main result is the following theorem:

Theorem 2 *The choice axioms are satisfied if and only if there exists a positive real valued function u defined on S, unique up to multiplication by a constant, such that*

$$P_S(j) = \frac{u(j)}{\sum_{i \in S} u(i)} \qquad (17.17)$$

The value $u(i)$ is interpreted by Luce as the deterministic utility of alternative i, and (17.17) suggests that individuals choose based on utility, but do so imperfectly, in the sense that their probability of choice is proportional to utility but their actual choice is uncertain.

Note that these probabilities are identical in form to the MNL model probabilities (17.10) if we equate $u(j)$ with $e^{\frac{u_j}{\mu}}$. Consequently, the Luce choice probabilities have the IIA property. Indeed, the Luce model produces the same probabilistic demand process as the MNL, despite the fact that the two models are derived using very different approaches.

17.4 INVENTORY MANAGEMENT UNDER STATIC CHOICE

We now turn to our main subject: inventory management under consumer choice. We consider a retail assortment consisting of substitutable product variants. The retailer's decision problem is to determine which subset, S, from a universe N of possible variants should be included in the assortment and how much to stock of each variant. Ideally, these decisions should account for the effects of customer choice on total demand and on the inventory dynamics of each variant. Intuitively, each variant added to the assortment enlarges the choice set and therefore increases the likelihood that a customer purchases something from the assortment. (Not purchasing is included as an option.) However, since items are substitutes, including more variants reduces the likelihood that any particular variant is chosen, reducing the volume of demand for each variant. This thinning - or fragmenting - of demand increases demand variability, which in turn tends to drive up inventory costs. This trade-off between the benefits and costs of increased assortment variety lies at the heart of the problem of inventory management under choice.

In this section, we consider a class of models we term *static choice* models. In these models, customer choices do not depend on the transient inventory status of the variants in the assortment. Rather, their choices depend only on the subset of variants, S, selected for the assortment. This is a restrictive assumption; however, it is a reasonable approximation in certain cases and has the advantage of considerably simplifying the inventory analysis. Models of *dynamic choice*, in which consumers choose from the available inventory at each point in time, are discussed in Section 17.5.

More precisely, in static choice models we make the following two assumptions:

(A1) The initial choice of a variant is independent of the inventory status of the variants in S.

(A2) If a customer selects a variant in S and the store does not have it in stock, the customer does not undertake a second choice, and the sale is lost.

Assumption A1 corresponds to a situation where customers choose without knowing the inventory status of the variants. In several retail settings this assumption is quite reasonable. For example, customers of a catalogue retailer do not know the inventory status of variants prior to ordering from the store. Only after attempting to place an order do they find out whether their chosen variant is in stock. Another case is when customers choose based on inspection of "floor models," and the status of the on-hand inventory is not readily discernable. A shoe store is an example. Customers make a decision based on the display of shoes in the store, only later learning whether the shoe they selected is available in their particular size. Similar buying processes are found at many consumer electronics and appliance retailers. Finally, as pointed out by Smith and Agrawal [58], the opportunity cost of shortages in retailing are typically quite high, and, as a result, optimal inventory decisions hedge strongly against stockouts; hence it can be argued that most customers are indeed choosing from the full assortment.

Assumption A2 is, admittedly, harder to defend and is required primarily for analytical tractability. One possible defense of A2 is to view customers as relatively uninformed *prior* to visiting the store. Upon inspecting the choice set S (e.g. the floor models), a customer becomes informed and identifies a variant $j \in S$ that they like. If the store then turns out to be out of stock, the newly-informed customer elects to go elsewhere to obtain j rather than settling for another alternative. Thus, the act of inspecting S changes the information a customer has about the possible choices, making the first and second choice decisions fundamentally different.

Another setting in which both Assumptions A1 and A2 are plausible is when S is viewed as a store's strategic variety decision, and customers, rather than viewed as being in the store, are viewed as making a choice whether or not to visit a store based on their knowledge of S gained from past shopping experience. For example, S may be the collection of brands a store carries, and brand-loyal customers may or may not visit a store based on their knowledge of which brands the store carries. Since customer's are not physically in the store when they make their store-visit decision, they are unaware of the inventory status of individual brands. Upon visiting the store, they learn if their preferred brand is in stock, and, if it is not in stock, they may seek their preferred brand elsewhere.

We next discuss two papers that address different versions of this static choice model. Smith and Agrawal [58] consider a model based on a probabilistic choice mechanism. van Ryzin and Mahajan [66] model consumer choice using the MNL random utility model. In both models, the supply process of the retailer is modeled as a newsboy problem (single-period, lost sales model).

Before proceeding, we define some notation common to both models:

N The set $\{1, 2, ..., n\}$ of all variants available in the market.

S The subset of variants stocked by the retailer.

x_j Decision variable representing the initial (start-of-period) inventory of variant j.

x Inventory level vector, $\{x_j : j \in N\}$.

Y Random variable representing the total number of customers making choice decisions (store traffic).

λ The mean of Y.

Y_j Random variable representing the demand for the jth variant (a function of S).

$q_j(S)$ Probability that variant j is chosen by an arriving customer.

17.4.1 The Smith and Agrawal Model

In the Smith and Agrawal [58] model, the optimal assortment S and the inventory levels x are found by formulating a mathematical programming problem which maximizes expected profit per period subject to constraints on floor space or inventory investments. We first describe their demand model and then describe the resulting optimization problem.

17.4.1.1 Demand Model. Smith and Agrawal's choice model is based on a probabilistic decision rule as follows: each customer has an initial preference for variant $j \in N$ with probability f_j. If $j \in S$, the customer chooses to purchase variant j. If variant $j \notin S$, the customer will substitute with another variant $i \in N$ with a probability s_{ji}. A customer chooses not to substitute with probability $L_j = 1 - \sum_{i \in N} s_{ji}$. Smith and Agrawal also consider a variation of this model in which an unlimited number of independent substitution attempts are made, according to the probabilities s_{ji}, until either an item in S is selected or the no-purchase option is selected.

Given this choice model, the probability that a variant j is chosen from a set S under the static choice assumptions A1 and A2 is given by

$$q_j(S) = f_j + \sum_{i \notin S, i \neq j} f_i s_{ij}.$$

In the case of unlimited substitution attempts, this probability is obtained by solving the system of equations,

$$q_j(S) = f_j + \sum_{i \notin S, i \neq j} q_i(S) s_{ij}, \quad j \in S.$$

Smith and Agrawal also propose related expressions using the end-of-period stock out probability that approximately correct for dynamic choice effects.

In the Smith and Agrawal model, total store traffic, Y, is modeled as a negative binomial random variable. This modeling choice is motivated by earlier theoretical and empirical work by Agrawal and Smith [3] showing that the negative binomial is a good model of retail demand. Let ψ represent the probability mass function for aggregate demand and $\psi_j, j \in N$ represent the probability mass function of demand for variant j. Then,

$$\psi(y) = \binom{k+y-1}{k-1} t^k (1-t)^y$$

where $y = 0, 1, 2, ...,$, k and t are parameters of the negative binomial distribution, and

$$\lambda = \frac{k(1-t)}{t}.$$

Given the total demand $Y = y$ and the subset of variants S, the number of customers who prefer variant j, $Y_j, j \in S$ has a binomial distribution with parameters y and $q_j(S)$. By unconditioning over the aggregate demand, it can be shown that the number of customers who prefer variant j, also has a negative binomial distribution. Let $\psi_j(y_j, S)$ represent the probability mass function for the preference for each variant j given S. Then,

$$\psi_j(y_j, S) = \binom{k+y_j-1}{k-1} r_j^k (1-r_j)^{y_j} \tag{17.18}$$

where

$$r_j = \frac{t}{t + q_j(S)(1 - t)} \tag{17.19}$$

Let $\Psi_j(y_j, S) = \sum_{d=0}^{y_j} \psi_j(d, S)$ denote the cumulative distribution function of Y_j.

With this model, different probabilistic mechanisms can be considered by altering the structure of the substitution probabilities s_{ij}. Smith and Agrawal consider three specific examples. For the case $n = 4$ and $L_j = L$ for all $j \in N$, these are

Random Substitution

$$\begin{bmatrix} 0 & \frac{1-L}{3} & \frac{1-L}{3} & \frac{1-L}{3} \\ \frac{1-L}{3} & 0 & \frac{1-L}{3} & \frac{1-L}{3} \\ \frac{1-L}{3} & \frac{1-L}{3} & 0 & \frac{1-L}{3} \\ \frac{1-L}{3} & \frac{1-L}{3} & \frac{1-L}{3} & 0 \end{bmatrix}$$

Adjacent Substitution

$$\begin{bmatrix} 0 & 1-L & 0 & 0 \\ \frac{1-L}{2} & 0 & \frac{1-L}{2} & 0 \\ 0 & \frac{1-L}{2} & 0 & \frac{1-L}{2} \\ 0 & 0 & 1-L & 0 \end{bmatrix}$$

One Variant Substitution

$$\begin{bmatrix} 0 & 1-L & 0 & 0 \\ 0 & 1-L & 0 & 0 \\ 0 & 1-L & 0 & 0 \\ 0 & 1-L & 0 & 0 \end{bmatrix}.$$

In Random Substitution, consumers will substitute any of the other variants with equal probability. If the probabilities of the initial preferences are identical, i.e. $f_1 = f_2 = f_3$, then this substitution structure is equivalent to a MNL model with symmetric utilities. In Adjacent Substitution, consumers substitute only variants which are adjacent to their initial preference. For example, they are willing to substitute up or down at most one size. In One Variant Substitution, one variant serves as a universally acceptable substitute for all customers. For example, vanilla ice cream might serve as an acceptable substitute for consumers who have an initial preference for a wide range of more exotic flavors. This flexibility in modeling substitution structure is an important feature of this model.

17.4.1.2 Profit Function. Smith and Agrawal use a single-period, lost sales model (newsboy model) of inventory. The total profit function includes the following components:

k_j The fixed cost to stock variant j.

m_j The unit profit margin obtained from selling variant j.

c_{oj} The overage cost for variant j.

c_{uj} The underage cost for variant j.

The profit obtained from stocking x_j units of variant j, denoted $\pi_j(S, x_j)$, is

$$\pi_j(S, x_j) = m_j E[Y_j] - c_{oj} \sum_{d=0}^{x_j} (x_j - d)\psi_j(d, S) - c_{uj} \sum_{d=0}^{x_j} (d - x_j)\psi_j(d, S) - k_j,$$

where $\psi_j(\cdot)$ is give by (17.18). The total profit, denoted $\pi(S, x)$, is then simply

$$\pi(S, x) = \sum_{j \in S} \pi_j(S, x_j) \tag{17.20}$$

17.4.1.3 Assortment Optimization. Smith and Agrawal propose a nonlinear integer programming formulation to solve for the optimal subset S^* and the optimal stocking quantities x^*. Note that from the static choice assumptions A1 and A2, given S the optimal stocking decision for each variant, x_j^*, is determined by solving a simple, single-item newsboy problem, viz

$$\Psi_j(x_j^* - 1, S) < \frac{c_{uj}}{c_{oj} + c_{uj}} \leq \Psi_j(x_j^*, S).$$

This property is the main analytical advantage to the static choice model assumptions. Substituting these values into the profit function (17.20) yields a discrete optimization problem with a nonlinear objective function.

Smith and Agrawal formulate this problem using binary variable z_j indicating whether a item j is included in S ($z_j = 1$ if $j \in S$). We omit the exact formulation, since it follows in a straightforward way given (17.18) and (17.20). Within this formulation, they consider a variety of constraints on x and S. One is a display space constraint,

$$\sum_{j \in S} t_j z_j x_j \leq T \tag{17.21}$$

where T is the total display space and t_j is the display space occupied per unit of variant j. A simpler constraint is a limit on the total number of variants due to limited availability of fixtures,

$$\sum_{j \in S} z_j \leq T \tag{17.22}$$

where T is the total number of fixtures available for the merchandise category.

Because there are no good, general purpose methods for solving nonlinear integer programs, this exact formulation is difficult to solve, and, in general, one must resort to complete enumeration. However, Smith and Agrawal do propose a quadratic 0-1 programming approximation for the problem. This quadratic 0-1 program can be reduced to a linear integer program by adding variables, so one can exploit standard integer programming codes to solve the problem. This approach works with the total fixture constraint (17.22) but does not extend to the space constraint (17.21).

17.4.1.4 Numerical Results. Smith and Agrawal applied their model to the $n = 5$ case described above and solved it using enumeration. They investigated how profits vary with factors such as margins and the initial preference and substitution probabilities.

Their experiments are based on the $n = 5$ case under the three substitution mechanisms outlined above. A base case is also considered in which no substitution occurs. In each case, the authors first solve for the optimal assortment fixing the number of variants in S. They then vary the size of S from 1 to 5 variants to determine which level of variety generates the maximum profit. The sensitivity of the assortment decisions to various parameters of the problem was investigated.

The numerical results reveal some interesting findings. First, the optimal profit for the No Substitution model is always lower than the substitution model at any given level of variety, so ignoring substitution effects may result in underestimating the profitability of an assortment. Second, the optimal assortments for the three different substitution mechanisms are quite different. Generally, Random Substitution requires the largest optimal assortment, Adjacent Substitution a somewhat smaller optimal assortment and One Variant Substitution is often maximized with only one variant (the "universal substitute"). An interesting result of the One Variant Substitution case is that it is often better to stock every customer's "second favorite" variant rather than trying to cover each customer's first preferences.

Smith and Agrawal also study the sensitivity of the assortment to the probability of not substituting, L. They conclude that as L increases, the number of variants in the optimal assortment increases. This follows, because for large L, substitution is relatively rare event and the variants become essentially independent and profits become purely additive in the number of variants. If L is low, customers are very willing to substitute, and hence it may not be necessary (e.g. profitable) to offer all possible variants. While these results are obtained in the presence of fixed costs per variant stocked, similar results are reported in the absence of fixed costs, though fixed costs clearly tend to reduce the optimal number of variants in an assortment. These result are also consistent with the results of van Ryzin and Mahajan in Theorem 4 b) discussed below.

17.4.1.5 Extensions. Agrawal [2] extends the above model to allow for joint pricing and assortment decisions. The assumptions and method of analysis follows closely that of Smith and Agrawal [58]. To model the dependence of demand for each variant on price, a demand function is defined. This function is used to parameterize how the distribution of demand for each variant changes with price.

By fixing the decision on which variants to include in the assortment, sensitivity results are derived for how the optimal price and profit change with the level of substitution, in the choice process used. A nonlinear integer programming formulation is developed to jointly determine prices and assortment decisions.

17.4.2 The van Ryzin and Mahajan Model

van Ryzin and Mahajan [66] propose and analyze an alternative static choice model for assortments. A significant difference is that the van Ryzin and Mahajan model uses a MNL random utility model of consumer choice. Other assumptions of the model are similar but somewhat more stylized than those in the Smith and Agrawal model. In particular, the model requires uniform prices and costs for all variants, does not incorporate fixed costs or constraints on the assortment and uses a normal rather than negative binomial model of store traffic. However, the resulting model is amenable to analysis, so it provides useful theoretical insights into the structure of optimal assortments and the factors that affect optimal assortment variety.

17.4.2.1 Prices and Costs. In the van Ryzin and Mahajan model, each variant has an identical retail price, p, and has an identical unit cost, c. This is a restriction that is needed primarily for analytical tractability. Examples where this assumption is a reasonable approximation include different titles in an assortment of music CD's or paperback books; casual trousers, where variants are different color-size combinations; men's ties, where each variant is distinguished by pattern and color; canned soup and ice cream, where variants are different flavors, etc. Fixed costs are not explicitly considered, though such costs are not difficult to add and do not fundamentally alter the structure of the problem.

17.4.2.2 The Choice Model. Customer choice is modeled using the MNL discussed in Section 17.3.3.2. Each customer associates a random utility U_j with the variants $j \in S$. In addition, a no purchase option, denoted $j = 0$, with associated utility U_0 is introduced, which represents the alternative of not buying anything or perhaps buying the same product from another retailer. Customers choose the variant with the highest utility among the set of available choices, $\{U_j : j \in S \cup \{0\}\}$. Note that increasing the choice set S makes it more likely that a given consumer chooses to purchase, because a consumer chooses to purchase only if $\max_{j \in S}\{U_j\} > U_0$. On the other hand, if the set S is enlarged by adding a variant i, a customer will switch from i to j if $U_i > U_j$.

Given S, the probability that variant j is selected, $q_j(S)$, is given by a slight variation of (17.10), in which the no-purchase alternative ($j = 0$) is added. In particular,

$$q_j(S) \; = \; P\left(U_j = max\{U_i : i \in S \cup \{0\}\}\right)$$

$$= \; \frac{v_j}{\sum_{j \in S} v_j + v_0} \qquad (17.23)$$

where,

$$v_j = \begin{cases} e^{\frac{u_j}{\mu}} & j \in N \\ e^{\frac{u_0}{\mu}} & j = 0. \end{cases} \qquad (17.24)$$

Recall, u_j is the systematic component of utility and μ is the scale parameter of the random component of utility. The quantities v_j are called *preferences*, because these values are increasing in the systematic component of utility, u_j. For purposes of characterizing the choice process, the vector of preferences $v = \{v_j : j \in S\}$ completely specifies a category.

17.4.2.3   Aggregate Demand Models.   The above description details how an individual customer makes his or her decision. Total demand, however, is determined by the aggregate decisions of all customers who visit the store, and this depends on how individual customer decisions are related to one another. van Ryzin and Mahajan consider the following two models of aggregate demand:

The Independent Population Model.   In this model, each customer assigns utilities to the variants in the subset S based on *independent* samples of the MNL model. The utility U_{ij} that customer i assigns to variant j is given by

$$U_{ij} = u_j + \xi_{ij}$$

where the $\{\xi_{ij}; i \geq 1\}$ are iid random variables. Hence, customers make i.i.d. choices among the variants offered by the retailer according to the MNL probabilities $q_j(S)$. Since customers are heterogeneous and independent, the observation of one customer's choice reveals no additional information about the choice of subsequent customers.

The mean number of customers arriving during the season is denoted λ. The number selecting variant j is denoted by Y_j, and by assumptions A1 and A2 above, $EY_j = \lambda q_j(S)$. Y_j is assumed to be normally distributed with a standard deviation that is a power function of the mean; that is, the standard deviation is $\sigma(\lambda q_j(S))^\beta$, where $\sigma > 0$ and $0 \leq \beta < 1$. [3] A natural special case of this model is when the total volume of customers visiting the store is Poisson with mean λ. Then the total demand for variant j, Y_j, is also Poisson with mean $\lambda q_j(S)$, because the MNL results in a thinning

---

[3]The assumption that $\sigma > 0$ and $0 \leq \beta < 1$ implies that the coefficient of variation is decreasing in λ. If the coefficient of variation is constant, the assortment profit is simply proportional to λ. The case where the coefficient of variation is increasing in volume does not seem to be a very realistic case and is omitted.

of the aggregate Poisson demand. In this case, a normal approximation to the Poisson distribution yields $\sigma = 1$ and $\beta = 1/2$. (This approximation is justified if $EY_j \gg 1$.)

The independent purchase model is useful for basic product categories, in which aggregate consumer preference is relatively stable and the primary uncertainty is over attributes such as color, size or flavor that inherently vary from one individual to the next.

The Trend-Following Population Model. The trend-following population model assumes an opposite extreme. A fixed number λ of customers visits the store. Each customer has *identical* valuations of the utilities for the variants, and these utilities are determined by a *single* sample of the MNL model. Hence, $\{\xi_{ij} = \xi_{1j}, \forall i > 1\}$. As a result, each customer makes the same choice, and once the outcome of one customer's choice is observed the decision of subsequent customers are perfectly predictable. However, it is assumed that these common utility values are not observable to the retailer prior to making assortment decisions. The retailer therefore has an incentive to hedge against this uncertainty by stocking more than one variant.

Under this model, demand for variant j, denoted Y_j, is (a scaled) Bernoulli random variable with probability mass function,

$$P(Y_j = y) = \begin{cases} q_j(S) & y = \lambda \\ 1 - q_j(S) & y = 0 \\ 0 & \text{otherwise} \end{cases} \tag{17.25}$$

This demand process can be thought of as a stylized model of trend-following behavior among customers, as one might find, for example, in color or style variety in high-fashion apparel.

17.4.2.4 Inventory Models. van Ryzin and Mahajan consider two closely-related inventory models. One is the single-period, lost sales (newsboy) model which is the same as that used in the Smith and Agrawal model. This is called the *seasonal merchandise* model. Note that one can also interpret this as the single-period cost in a periodic review system with lost sales as in Smith and Agrawal [58]. The other inventory model is an infinite-horizon base-stock model, appropriate for replenishable merchandise with backordering.

Both inventory models are analyzed under the independent and trend-following demand models, though the case of replenishable merchandise with trend-following demand is omitted because it is somewhat logically inconsistent. [4] All cases yield qualitatively similar profit functions, but provide different insights into the assortment decision.

---

[4]The replenishable supply model is an infinite horizon model in which it is logical to assume that a long history of purchase choices is available. In such a case, the choice of the trend-following population could be perfectly predictable and there would be no need to stock more than one variant.

Seasonal Merchandise with an Independent Population. Without loss of generality, we assume the sales season lasts one unit of time. The number of customers selecting variant j, Y_j, is normally distributed with mean $\lambda q_j(S)$ and standard deviation $\sigma(\lambda q_j(S))^\beta$ as described above. For each unit not sold, the loss is the cost c. (There is no salvage value for unsold units, though this assumption can easily be relaxed.) For each unit short, the opportunity cost is the loss in margin, $p - c$.

The maximum expected profit given S and v, denoted $\pi_I(S, v)$, is

$$\pi_I(S, v) = \max_{x \geq 0} \sum_{j \in S} E[p \min\{x_j, Y_j\} - c x_j]. \tag{17.26}$$

The optimal profit depends on S and v through their effect on the choice probabilities $q_j(S)$ in (17.23), which in turn determine the demand Y_j.

Under the assumption of a normal distribution, the optimal stocking level of variant j, denoted x_j^*, is given by

$$x_j^* = \lambda q_j(S) + z\sigma(\lambda q_j(S))^\beta, \tag{17.27}$$

where

$$z = \Phi^{-1}(1 - \frac{c}{p}), \tag{17.28}$$

and $\Phi(z)$ denotes the c.d.f. of a standard normal random variable. (We let $\phi(z) = \frac{1}{\sqrt{2\pi}} e^{-\frac{z^2}{2}}$ denote the standard normal density function.) Substituting this expression for x_j^*, in (17.26) we obtain

$$\pi_I(S, v) = (p - c)\lambda \sum_{j \in S} q_j(S) - \frac{p\sigma\lambda^\beta e^{-\frac{z^2}{2}}}{\sqrt{2\pi}} \sum_{j \in S} q_j^\beta(S), \tag{17.29}$$

where we have used the fact that for a standard normal random variable Z, $E(Z - z)^+ = \phi(z) - z(1 - \Phi(z))$.

Seasonal Merchandise with a Trend-Following Population. In this case, a constant number λ of customers visit the store and each makes identical utility valuations. The valuations are unknown to the retailer at the time the assortment decisions are made, and the distribution of demand for variant j, Y_j, is given by (17.25), as described above.

The expected profit from variant j given x_j is again $E[p \min\{Y_j, x_j\} - c x_j]$. The optimal expected profit given S and v, denoted $\pi_T(S, v)$, is

$$\pi_T(S, v) = \sum_{j \in S} (p q_j(S) - c)^+ \lambda, \tag{17.30}$$

using the fact that the optimal procurement level is

$$x_j^* = \begin{cases} 0 & \text{if } p q_j(S) < c \\ \lambda & \text{if } p q_j(S) \geq c \end{cases}$$

Replenishable Merchandise with an Independent Population. For replenishable merchandise, consider an infinite horizon of sales and the independent population model of aggregate demand. If a customer's first choice is not in stock it is assumed that they are willing to tolerate a back order and the sale is not lost. However, the store incurs costs at rate b per unit time for each unit back ordered and a holding cost h per unit time for each unit held in stock. In this case, the static choice assumption A2 is slightly modified as follows:

(A2') If a customer selects a variant in S and the store does not have it in stock, the customer is willing to accept a backorder and does not undertake a second choice.

From a behavioral standpoint, this assumption is, in some ways, more palatable than the original assumption A2. That is, it is easier to imagine that customers might be willing to accept a backorder on a product rather than settle for a less desirable substitute (e.g. catalogue apparel purchases). However, even if backorders are an option, some consumers would no doubt opt to substitute rather than wait to receive their first choice.

The lead time for replenishment is a constant l units and the store uses a one-for-one inventory replenishment policy (*base-stock* policy; see [37]). The base-stock level, x_j, is the stocking decision variable for the store in the replenishable merchandise case. The lead-time demand for variant j, $D_j(t-l, t]$, is assumed to be normally distributed with mean $\lambda l q_j(S)$ and standard deviation $\sigma(\lambda l q_j(S))^\beta$. One can then show that the optimal base-stock level, x_j^*, is $x_j^* = \lambda l q_j(S) + z_R \sigma(\lambda l q_j(S))^\beta$ where $z_R = \Phi^{-1}(\frac{b}{b+h})$ and, as before, $\Phi(z)$ is the c.d.f. of a standard normal random variable. This yields the profit function

$$\pi_R(S, v) = (p - c)\lambda \sum_{j \in S} q_j(S) - \frac{(b + h)\sigma(\lambda l)^\beta e^{-\frac{z_R^2}{2}}}{\sqrt{2\pi}} \sum_{j \in S} q_j^\beta(S), \qquad (17.31)$$

Note that this profit is nearly identical in form to the profit in the seasonal case (See (17.29)), with the exception that the lead-time only enters in the second term.

17.4.2.5 Assortment Optimization. With these different profit functions at hand, one can formulate the optimal assortment selection problem by solving

$$\max_{S \subseteq N} \pi(S, v).$$

where the appropriate profit function, (17.29), (17.30) or (17.31), is inserted above.

Let S^* denote an optimal solution to this problem. Then the corresponding pair (S^*, x^*) defines an *optimal assortment* for the store. Note that there are no constraints on the number of units stocked. Such constraints would certainly affect the structure of a store's assortment in the short run. On the other hand, over a long time horizon such constraints can be relaxed at a cost, either by reallocating store space among categories or by expanding the selling space of the store. Under such conditions, one could potentially include the shadow price of selling space or appropriately amortized expansion costs in the unit cost c.

17.4.2.6 The Structure of the Optimal Assortment. Under this model, one can analyze the structure of the optimal assortment (S^*, x^*). To do so, let the variants be ordered according to decreasing values of v_j, so that $v_1 \geq v_2 \geq \ldots \geq v_n$. The following theorem shows that the optimal set S^* is structurally quite simple.

Theorem 3 *Let $A_i = \{1, \ldots, i\}$ for $1 \leq i \leq n$. Then for each of the assortment problems defined above, there exists an $S^* \in \{A_1, \ldots, A_n\}$ that maximizes store profits.*

In words, the optimal assortment can be restricted to one of n possible types. One simply choose the best i variants, where $1 \leq i \leq n$. While this result may seem intuitively obvious, it is more subtle than it appears. For example, it is *not* true in general that the best subset of i variants consists of the best i variants; That is, the set $\{1, 8\}$ may be *more* profitable than the set $\{1, 2\}$. However, the theorem says that the most profitable subset does consist of the most popular i variants for some value $i = 1, \ldots, n$.

The fact that retail assortments will be constructed using the most popular variants first is certainly intuitive. Yet this result would appear to exclude the possibility of "niche" retailers with specialized and/or unique merchandise. It is important to recognize, however, that choice models describe a population of customers that visit a given store - and not all stores share the same population (segment) of customers. Factors such as the store image, layout, location, advertising - and past assortment decisions themselves - affect the type of customers a store attracts. Theorem 3 simply says that the merchandise assortment should match the profile of customers a store is attracting. Conversely, a store may be constrained to offer only those assortments that match the profile of the customers it is able to attract.

Finally, we note that this result stands in contrast to the results of Smith and Agrawal [58], wherein it was shown that it can be optimal to stock a single variant that is not the most popular first preference, provided it is an acceptable substitute for a wide range of consumers. This difference stems fundamentally from differences in the two choice models. Under the MNL, the IIA property implies that the ratio of choice probabilities is independent of the set in which they are considered. Thus, customers cannot have a weak initial preference for a variant j as a first choice and then switch and have an overwhelmingly strong preference for j as a second choice. However, because the Smith and Agrawal choice model allows for a dependence between a customer's first and second preference, customers can have strong preference for a variant as a second choice even though they may not rank it highly as a first choice.

17.4.2.7 **Factors Affecting the Optimal Level of Variety.** The simple structure of the optimal assortment makes it easy to define the level of variety a store offers, because more variety corresponds directly to a higher index i among the possible subsets $\{A_1, ..., A_n\}$. This raises interesting questions about what affects the level of variety offered in an optimal assortment. The answers, which are straight forward to show given Theorem 3 and the various profit functions, are summarized in the following theorem:

Theorem 4 *For all $n > i \geq 1$,*

  a) $\pi(A_{i+1}, v) > \pi(A_i, v)$ for sufficiently high selling price p,
  b) $\pi(A_{i+1}, v) < \pi(A_i, v)$ for sufficiently low no-purchase preference v_0.
  c) $\pi(A_{i+1}, v) > \pi(A_i, v)$ (replenishable case only) for sufficiently low lead-time l.
  d) $\pi(A_{i+1}, v) > \pi(A_i, v)$ (independent population cases only) for sufficiently high store volume λ.

These results have intuitively satisfying implications. Part (a) states that high selling prices create an incentive to stock higher levels of variety. This occurs because as margins increase the risk of lost sales dominates the risk of overstocking. Therefore, a wide variety is offered in order to minimize the likelihood of customers not purchasing. The result suggests that as market prices rise in category, not only will new firms be enticed into entering the market, but existing retailers will have an incentive to broaden the range of merchandise they carry. [5]

Part (b) considers the no-purchase utility as a driver of variety. A high no-purchase utility could correspond to a product category that is somewhat frivolous (e.g. souvenirs, toys or jewelry), in which not purchasing is a common outcome. Alternatively, a high no-purchase utility can represent the existence of many attractive outside alternatives, including other stores in close proximity carrying similar merchandise. In either case, as the no purchase utility declines, the prospect of losing a purchase to an external option decreases while the threat of within-assortment cannibalization increases. Therefore, it is in the retailer's interest to decrease the breadth of the assortment. For example, this result predicts that stores in less competitive retail environments will tend to offer lower variety than similar stores in more competitive retail environments.

Part (c) provides an interesting insight in the replenishable case. It predicts that improved supplier delivery performance (lower lead-time) induces retailers to carry a wider range of a supplier's products. This occurs because lower lead-times reduce inventory risk and make even relatively less popular variants profitable to carry. This appears consistent with empirical and case evidence on

---

[5]We note that Shugan [56] shows that it is possible for lower levels of variety to be optimal for higher priced categories when demand is highly price sensitive. The reason is that the reduced purchase volume from high prices can be enough to offset the positive effect of higher margins. Part b) assumes store volume and purchase probabilities are fixed.

supplier improvement programs (see Abernathy et al. [1]), where it is has been observed that improved delivery performance by a supplier results not only in higher sales volumes but also a wider range of products sold to retailers.

Part (d) implies that, in the independent demand case, as the volume of business increases, high variety becomes increasingly more profitable, and for a sufficiently high volume a store will carry all variants. That is, there are *scale economies* in offering variety, and as traffic grows, a store not only stocks more units of each variant, but also tends to stock more variants.

The reason higher variety becomes more desirable in the independent population case is due to the risk pooling inherent in a large number of independent purchase decisions. As the volume of purchase decisions goes up, the relative amount of overstocking and understocking error goes down and the cost of having fragmented purchase decisions become relatively smaller. Hence, more variety becomes profitable. Super-stores (Borders, Home Depot, Staples, Toys-R-Us, etc.) are plausible examples of this sort of scale effect. Indeed, the results suggests a reason why the super-store format is economically viable. There may be a natural, positive feedback in this format; the large variety attracts a high volume of traffic which in turn allows the super-store to profitably offer a large variety of merchandise.

But this scale economy only exists for the independent purchase case. In the trend-following case, (17.30) shows that profit is directly proportional to λ. That is, while increased store volume makes higher variety more profitable in the independent population case, in the trend-following population case, volume has no effect on the relative profitability of different levels of assortment variety. Hence, there are no scale economies; a large store simply places bigger bets and ends up making proportionately bigger losses.

While this is clearly a stylized model, the general conclusions appear consistent with actual retailing practices. For example, consider the categories of merchandise offered by the super stores mentioned above. They tend to have merchandise categories that experience highly fragmented but largely independent purchases, e.g. Borders (books, magazines), Home Depot (hardware, home remodeling), Staples (office supplies) and Toys-R-Us (children's toys). In contrast small stores, independent boutique clothing stores for example, tend to thrive with merchandise categories that have strong trend-following customers. Here, store scale is not an impediment to profitability. Indeed, in the trend-following case, competitive advantage lies primarily in getting better information on the likely choice the population will make. This information is often best obtained by having intimate knowledge of local customers, which is precisely the advantage small, independent stores have.

17.4.2.8 Defining Fashion Based on Majorization Ordering. An important benefit of the van Ryzin and Mahajan model is that it can be used to formally compare merchandise categories. Specifically, the profit of a category is shown to be closely linked to the "evenness" of the preference vector v, and this notion of "evenness" can be made precise using the theory of majorization [44].

We first review some basic concepts on majorization ordering. For a vector $x \in \Re^n$, let $[i]$ denote a permutation of the indices $\{1, 2, \ldots, n\}$ satisfying $x_{[1]} \geq x_{[2]} \geq \ldots \geq x_{[n]}$. Then,

Definition 3 For $x, y \in \Re^n$, x is said to be majorized by y, $x \prec y$ (y majorizes x), if $\sum_{i=1}^n x_{[i]} = \sum_{i=1}^n y_{[i]}$ and

$$\sum_{i=1}^k x_{[i]} \leq \sum_{i=1}^k y_{[i]}, \quad k = 1, \ldots, n-1$$

Intuitively, a nonnegative vector y that majorizes x tends to have more of its "mass" concentrated in a few components. van Ryzin and Mahajan argue that majorization provides the right measure for the degree of fragmentation in consumer preference within a given category of merchandise and propose the following definition:

Definition 4 A merchandise category v is said to be more _fashionable_ than w if $v \prec w$. If $v \prec w$, we refer to w as the _basic_ category and v as the _fashion_ category.

This definition says that, for fashion categories, the preferences are more evenly spread out across variants. Alternatively, in the trend-following demand case, the degree of fragmentation of the retailer's prior information of consumer preferences for variants is higher for the fashion category than for the basic category. Finally, we note that one can interpret v and w either as two categories at the same point in time or one category observed at two different points in time. In the latter case, one can then meaningfully talk of a category becoming "more fashionable" over time. Pashigian [50] studies bed sheets (white vs. fancy) as an example of such a category.

To apply this definition, we must make the assumption that both categories have the same number of variants n. However, this is not restrictive, since one can always add an arbitrary number of variants j with preference values $v_j = 0$ without altering the problem. The assumption that the preference values sum to the same quantity is also not restrictive, since we can scale all values of v (and the no-purchase preference, v_0) by an arbitrary multiplier and not affect the resulting choice probabilities.

The following intermediate result shows that the optimal profit obtained from two merchandise categories under similar demand and cost assumptions is intimately related to the majorization ordering.

Theorem 5 Consider two merchandise categories, v and w, that have identical cost structures (p and c), demand volumes, λ and equal no purchase utilities ($v_0 = w_0$). Define r and l such that

$$\pi(A_r, v) = \max_{S \subseteq N} \pi(S, v)$$

524

and

$$\pi(A_l, w) = \max_{S \subseteq N} \pi(S, w),$$

where $\pi(\cdot)$ can be any of the assortment profit functions, (17.29), (17.30) or (17.31), Then if $v \prec w$, (category v is more fashionable than category w), $\pi(A_r, v) \leq \pi(A_l, w)$.

In contrast to the previous assumptions on the vectors v and w, the condition $v_0 = w_0$ is restrictive. Roughly, it corresponds to assuming that both categories as a whole are equally attractive relative to not purchasing.

The theorem says if one merchandise category is more fashionable than another, then, all other things being equal, the optimal profit of the fashion category will be lower than that of the more basic category. The reason for this is that the risk of inventory overage and underage is higher due to the higher fragmentation of consumer purchase decisions in the fashion category. Thus, even under optimal variety and stocking decisions, the category is less profitable at a given price.

However, an equilibrium argument suggests that the market will compensate retailers for the added costs of fashion categories by allowing higher market prices. That is, retailers recognizing the higher profitability of the basic categories will tend to enter the market (i.e. add these categories to their store). The resulting increase in supply will, in turn, tend to reduce market prices until the basic categories become comparable in profitability to the fashion categories and further new entrants are discouraged. A corollary of Theorem 5 formalizes this reasoning.

Corollary 1 *Assume $v \prec w$, all other parameters except price, p, are identical and that prices satisfy the equilibrium profit condition,*

$$\pi(A_r, v) = \pi(A_l, w) = \pi_e,$$

where $\pi(\cdot)$ is any of the assortment profit functions (17.29), (17.30) or (17.31) and π_e is an arbitrary, nonnegative equilibrium profit level. Let the equilibrium prices be denoted p_v and p_w, respectively, for categories v and w. Then

$$p_v \geq p_w$$

Of course, this equilibrium profit hypothesis represents a highly simplistic view of retail markets. In reality, other cost and competitive factors will not doubt produce variances in profitability from one category to another. Nevertheless, it provides a simple and intuitively appealing explanation for the fact that fashion goods tend to have higher margins than basic goods (See Lazear [36] and Pashigian [50] for an alternative theory for this effect and some empirical evidence.) In short, Corollary 1 simply says that higher mark-ons serve to compensate retailers for the increased inventory risks induced by the highly fragmented purchase choices of the fashion category.

17.4.2.9 Supply Chain Implications. The equilibrium price results of Corollary 1 and the scale economy results of Theorem 4 have interesting implications for a retailer's supply chain strategy. Corollary 1 states that equilibrium margins are higher for fashion categories. These high margins, in turn, may justify changes in the way goods are supplied to stores. In particular, a viable strategy to manage a fashion category with dependent (trend-following) purchase behavior may be to use fast - and potentially expensive - logistics processes (e.g. air freight) to replenish stocks of popular variants in season, rather than relying on stocking to forecasts. However, from Theorem 4, we see that for the independent (non-trend-following) population case there are scale economies to offering variety. Therefore, such a replenishment strategy may not be warranted in this case, since large-scale retailers can mitigate inventory risks without resorting to expensive logistics options.

Dvorak and Paasschen [22], writing in *McKinsey Quarterly*, describe several retailer's logistical strategies that are consistent with these general conclusions. The authors describe the operations of one (anonymous) "fast-to-market" retailer as follows:

> High fashion is a high-risk business. Trend-setters cannot simply react to consumer demand. They have to amaze customers with original collections, but need to know their innovations will sell...
>
> ... Retailers usually air-ship only those items that have unexpectedly run out. But this retailer air-ships all items that have sold better than expected during tests... None of this speed is cheap, but the expense is more than covered by higher sales and fewer mark-downs... Careful market research spots possible sales; then fast, focused logistics deliver the goods to make sure each potential sale is captured.

For a retailer of casual cotton clothes, which correspond more closely to the independent (non-trend-following) case, they describe a very different strategy:

> What is important is to make sure stores are always stocked with the right color, size and design... Production lead times are long, because goods are manufactured overseas to keep costs to a minimum. Price constraints also rule out shipping goods by air ... The thrust of the logistics strategy is therefore to achieve not speed, but a smooth, seamless transition from one wave of goods to the next.. ... to cope with variations in demand - for specific colors, sizes, or designs - a second strand of the retailer's strategy introduces flexibility. Regional warehouses, located close enough to stores to allow cost-effective, frequent replenishment, provide buffer stocks ready to fill gaps on the store's shelves.

This description is consistent with the lower margin, scale sensitive characteristics of the non-trendy case.

17.5 INVENTORY MANAGEMENT UNDER DYNAMIC CHOICE

The static choice assumptions A1 and A2 simplify the analysis of inventory models under choice, but they are somewhat unsatisfying, especially for categories such as cigarettes, grocery items, soft drinks etc in which consumers readily substitute when products are out of stock. In this section, we consider models that allow for dynamic substitution behavior based on the inventory status at the time a customer makes his/her choice. The price one pays for these more realistic models is that they are less tractable, and one must rely more heavily on computational studies to understand the problem.

Some questions the models address are: Does the inventory cost posses any intuitively appealing and/or simplifying structural properties? How can assortments be optimized efficiently given the complexities of dynamic choice behavior? If assortments can be optimized efficiently, then how much difference does it make in gross profits? Finally,what distortions are introduced in a retailer's assortment decisions if one ignores substitution effects?

The models we review below are those of Noonan [48] and Mahajan and van Ryzin [41]. However, there are two papers from the production literature that address related topics which deserve mention.

The first is Parlar and Goyal, [49], who consider a two item single period problem, with identical costs for all items. Out of stock items are substituted for in-stock items with a certain probability, and one can view this as a type of consumer choice decision. They establish concavity of the profit function and find approximate order quantities for each item.

The second paper is McGillivray and Silver [46]. They analyze an n item problem, with an inventory control policy of (R, Q) type for each item, with identical holding and penalty costs for all items. If an item is out of stock, there is a fixed probability of the customer substituting another item for it. Again, this can be viewed as a form of customer choice. For the two item case they develop a heuristic approach to find the order-upto levels. They find that if the probability of substitution is sufficiently high, fewer items of each type need to be ordered. This result is consistent with those of Mahajan and van Ryzin [41] below.

We also note that Lariviere and Porteus [38] discussed, but did not publish, some interesting work on dynamic choice models of inventory that was based on a Markov chain model of the inventory dynamics. Indeed, to the best of our knowledge, their work and the work of Noonan [48] were the first in this area. Smith and Agrawal [58] also formulate a Markov chain model of the problem but do not analyze it. While the Markov chain approach is a natural one to try, it is limited by the high dimensionality of the state space required.

Lastly, we note that there is one conceptual difference between the static and dynamic choice cases that is important to bear in mind. In the static choice case, the decision of which set of variants to select S and the stocking levels x were viewed as distinct decisions. In the dynamic choice case, this is not necessary. The vector of initial stocking levels, x, completely specifies the retailer's decision. That is, if variant j is not included in the assortment, it is equivalent to starting out with $x_j = 0$. Thus, the dynamic choice problem, on the surface, more closely resembles a traditional inventory stocking problem, though it implicitly includes a assortment variety decision as well.

17.5.1 The Noonan Model

Noonan [48] considers a single period inventory model of a merchandise category made up of multiple product variants, where each variant has a different unit selling price and unit procurement cost. He assumes customers have a first choice and a second choice and that demand is generated in two stages. In the first stage, primary (first-choice) demand is realized and satisfied as much as possible with available inventory. Then, in the second stage, unfilled primary demand is converted to secondary (second-choice) demand for products based on *deterministic* proportions. The resulting total demand for each product is then analyzed in terms of multidimensional integrals over the space of initial demand realizations.

Specifically, as before x is the vector of stock levels x_j, $j \in N$, and $c_j, p_j, j \in N$, denote, respectively, the unit procurement cost and the unit selling price for variant j. Let s_{ij} be the fixed proportion of unfilled demand for variant i that is transferred to variant j after stockouts. The substitution mechanism Noonan proposes takes place in two stages:

Stage 1: Original Demand Customer demand for each variant is realized. This demand is filled to the extent possible from available stock. Unfilled demand for a variant in Stage 1 is converted into substitution demand for another variant in Stage 2.

Stage 2: Substitution Demand Let $U \subset N$ denote the variants with unfilled demand after Stage 1 and $O = N/U$ denote those with remaining stock. Of the unfilled demand for variant $i \in U$, a deterministic proportion s_{ij} is converted into substitution demand for variant $j \in O$. Total substitution demand for variants in the set O is filled to the extent possible. If total substitution demand for a particular variant $j \in O$ exceeds the available stock of variant j, then the substitution demand from each variant i is partially filled in equal proportions.

Note that under this model, the substitution demand and total inventory remaining are deterministic functions of the original demand, so demand realizations can be mapped deterministically into total sales and profit outcomes.

Noonan shows how the newsboy logic for finding the critical fractile by marginal analysis can be extended, at least conceptually, to the substitution problem. To simplify the presentation, we consider the case $n = 2$. The two dimensional space of all possible demand realizations is divided into the following mutually exclusive and collectively exhaustive regions that are functions of the stocking levels x_1 and x_2:

R_ϕ Region of demand where neither variant stocks out.

$R_\forall$ Region of demand where both variants stock out on original demand.

R_1 Variant 1 stocks out on original demand, but the substitution demand is satisfied using Variant 2.

R_1^2 Variant 1 stocks out on original demand and its substitution demand is large enough to stock out Variant 2.

R_2, R_2^1 Defined symmetrically to R_1 and R_1^2.

Let P_A denote the probability that the demand vector lies in region R_A. The expression for expected profit can be written by integrating over the different regions above. Differentiating then leads to the following first-order necessary conditions:

$$c_1 = p_1[P_V + P_1^2 + P_2^1 + P_1] - s_{12}p_2P_1 \tag{17.32}$$

and symmetrically,

$$c_2 = p_2[P_V + P_1^2 + P_2^1 + P_2] - s_{21}p_1P_2 \tag{17.33}$$

Under reasonable assumptions on the parameters, second-order sufficiency of these conditions can be established.

Equation (17.32) can be explained as follows. The probability of Variant 1 stocking out is $P_V + P_1^2 + P_1$. The probability of Variant 2 stocking out and then using up all of Variant 1's remaining stock is P_2^1. Therefore the gain from an additional unit of Variant 1 is $p_1[P_V + P_1^2 + P_2^1 + P_1]$. However, in region R_1, this incremental unit would fill a demand of Variant 1 that would otherwise have been filled as substitute demand for Variant 2. So, we subtract $s_{12}p_2P_1$ to arrive at the RHS of equation (17.32), which represents the gain from stocking an additional unit of Variant 1. The cost from stocking an additional unit of Variant 1, is c_1. Equation (17.33) can be explained similarly.

It is interesting to compare (17.32) to the traditional newsboy model conditions. Note in (17.32) that there are two extra terms: One is $p_1R_2^1$, which represents the extra marginal contribution to sales from demand substituting into Variant 1 from Variant 2. The other extra term is $-s_{12}p_2P_1$, which one can interpret as a reduction in underage cost due to the willingness of customers to substitute Variant 2 for Variant 1. Compared to the traditonal newsboy solution, the first term tends to increase the optimal stocking level while the second tends to decrease it. Which effect dominates depends on the particular parameter values.

This line of analysis can be extended to the n product case, by partitioning the multidimensional space of demand realizations into different regions and using marginal analysis, but the number of outcome types grows exponentially in n. A further difficulty in using these conditions is that they only indirectly specify the inventory levels. That is, one must pick x_1 and x_2, define the regions based on x_1 and x_2, and then check to see if (17.32) and (17.33) are satisfied. Based on this idea, Noonan proposes the following iterative algorithm to find the inventory level vector x (See Noonan [48] for a complete specification):

1. Start with an initial inventory level vector x

2. Sample demand realizations to compute estimates of the probabilities that a demand vector realization would lie in one of the different partitions of the n dimensional space. (Monte Carlo integration)

3. Use the inverse of the marginal probability distribution for each variant to find a new inventory level vector x.

4. GOTO Step 2.

This algorithm is not guaranteed to converge, but Noonan reports that it is robust across a wide range of parameter values on small problems. However, the complexity increases rapidly with n because of the number of samples required to get good estimates in the Monte Carlo integration step.

17.5.2 The Mahajan and van Ryzin Model

Mahajan and van Ryzin [41], analyze a dynamic choice version of the assortment problem using a general random utility model. During the sales period, a sequence of heterogeneous customers choose from among the in-stock variants (or they choose not to purchase) based on a utility maximization criterion. A sample path analysis is used, so minimal distributional assumptions are made on both the statistics of customer arrivals and the utilities they assign to variants.

17.5.2.1 Model Formulation . The set of natural numbers (non-negative integers) is denoted by Z. All vectors are in $\Re^n$ unless otherwise specified. The notation y^T stands for the transpose of a vector y. Where possible, components of vectors are now denoted by superscripts while subscripts denote elements of a sequence. For example, x_t^j denotes the j-th component of a vector x_t in a sequence $\{x_t : t \geq 1\}$. $\mathbf{I}$ denotes the identity matrix and $e^i \in Z^n, 1 \leq i \leq n$ denotes the j-th unit vector; that is, a column vector with a 1 in the i-th position and a 0 elsewhere; we also extend this definition and let e^0 denote a column of all zeros. We use $\mathbf{1}$ to denote the column vector with a 1 as every element; a.s. means *almost surely* and c.d.f. is short for *cumulative distribution function*.

As before, the merchandise category consists of n substitutable variants, with selling prices p^j, procurement cost of c^j and a single-period (newsboy-like) inventory model in which the retailer's only decision is the vector of initial inventory levels x for each of the variants. Without loss of generality, there is no salvage value for any of the variants.

Mahajan and van Ryzin use a sample path analysis. Let T denote the number of customers on a sample path. Each customer $t = 1, ..., T$ chooses from the variants that are in-stock when he/she arrives. Let $x_t = (x_t^1, ..., x_t^n)$ denote the vector of inventory levels observed by customer t, and note that $x_1 = x$, where x is the initial stocking decision mentioned above.

For any real inventory vector y, let $S(y) = \{j \bigcup \{0\} : y_j > 0\}$ denote the set of variants with positive inventory (the set of *in-stock* variants) together with the *no-purchase* option, denoted by the element 0. Customer t can only make a choice $j \in S(x_t)$. Because inventory levels are nonincreasing over time, we have that $S(x_{t+1}) \subseteq S(x_t)$.

A customer's choice is based on utility maximization: Each customer t assigns a utility U_t^j to the variants $j \in N$ and to the no-purchase option, U_t^0. Let $U_t = (U_t^0, U_t^1, \ldots, U_t^n)$ denote the vector of utilities assigned by customer t. Based on the inventory level x_t and utility vector U_t, customer t makes the choice, $d(x_t, U_t)$, that maximizes his/her utility:

$$d(x_t, U_t) = \arg \max_{j \in S(x_t)} \{U_t^j\}. \tag{17.34}$$

The resulting decision could be either to buy a variant or to not purchase at all, depending on whether $d(x_t, U_t) = j$ for some $j \in S(x_t)$, $j > 0$ or whether $d(x_t, U_t) = 0$, respectively.

Let $\omega = \{U_t : t = 1, \ldots, T\}$ denote a sample path from some probability space $(\Omega, \mathcal{F}, P)$. The only assumptions made on this space are that the sequence is bounded w.p.1, i.e. $P(T \leq C) = 1$ for some finite C, and that each customer t makes a unique choice w.p.1.; that is

$$P(U_t^i \neq U_t^k) = 1 \text{ for all } i \neq k, \ i, k \in N \bigcup \{0\} \ .$$

This latter assumption is satisfied, for example, if the utilities have continuous distributions.

The retailer does not know the particular realization ω but does know the probability measure P, so we think of P as characterizing the retailer's knowledge of future demand. The retailer's objective is to choose initial inventory levels x that maximize total expected profit.

To express this total profit concisely, let $\eta^j(x, \omega)$ denote the number of sales of variant j made on the sample path ω given initial inventory levels x. Let $\eta(x, \omega) = (\eta^1(x, \omega), \ldots, \eta^n(x, \omega))$. Then the sample path profit, denoted $\pi(x, \omega)$, is given by

$$\pi(x, \omega) = p^T \eta(x, \omega) - c^T x, \tag{17.35}$$

where p and c are the vectors of prices and costs, respectively. Therefore, the retailer's objective is to solve

$$\max_{x \geq 0} E[\pi(x, \omega)]. \tag{17.36}$$

Since the only complicated quantities in (17.35) are the functions $\eta^j(x, \omega)$, we focus on understanding their properties.

To do so, it is convenient to consider a recursive formulation of the problem. First, define the *system function*

$$f(x_t, U_t) = x_t - e^{d(x_t, U_t)}, \tag{17.37}$$

where $d(x_t, U_t)$ is, as defined above, the decision made by customer t and $e^j, j = 1, .., n$ are the n unit vectors and e^0 denotes the zero vector. Then $x_{t+1} = f(x_t, U_t)$, so that $f(\cdot)$ describes how the inventory evolves over time.

Next, define a sequence of *sales-to-go* functions, $\eta_t^j(x_t, \omega)$ for $t = 1, \ldots, T$, via the recursion,

$$\eta_t^j(x_t, \omega) = x_t^j - f^j(x_t, U_t) + \eta_{t+1}^j(x_{t+1}, \omega) \tag{17.38}$$
$$x_{t+1} = f(x_t, U_t),$$

with initial conditions

$$\eta_{T+1}^j(x_{T+1}, \omega) = 0, \quad j \in N$$
$$x_1 = x.$$

Note $\eta_t^j(x_t, \omega)$ gives the number of sales of variant j on sample path ω from customer t onward (the "sales-to-go"), and (17.38) decomposes this total sales-to-go as the sum of the sales of variant j resulting from customer t and the total sales-to-go of variant j for the remaining customers $t+1, t+2, ..., T$. Of course, the total sales of variant j are simply the total sales-to-go for customers $1, ..., T$, so $\eta^j(x, \omega) = \eta_1^j(x_1, \omega)$. One can therefore use this recursion to investigate properties of the sales functions $\eta^j(x, \omega)$.

17.5.2.2 Structural Result. The properties of $\eta^j(x, \omega)$ require some background material on parametric monotonocity. The following definitions and results can be found in Sundaram [59] (See also Topkis [62].): Let S and Θ be subsets of $\Re^n$ and $\Re^l$ respectively.

Definition 5 *A function $h : S \times \Theta \to \Re$ satisfies decreasing differences in (z, θ), if for all $z' \geq z$ and $\theta' \geq \theta$*

$$h(z', \theta) - h(z, \theta) \geq h(z', \theta') - h(z, \theta').$$

Suppose that S denotes a space of feasible actions and Θ denotes a parameter space. Consider the following optimization problem, for fixed $\theta \in \Theta$,

$$\max\{h(z, \theta) : z \in S\},$$

and define the *optimal action correspondence* (point-to-set mapping)

$$D^*(\theta) = \{z^* : h(z^*, \theta) \geq h(z, \theta) \ \forall z \in S\}$$

Parametric monotonicity means that the optimal action correspondence $D^*(\cdot)$ is monotone in θ. Let

$$z^*(\theta) = \max\{z : z \in D^*(\theta)\}. \tag{17.39}$$

The following result shows that the property of decreasing differences implies parametric monotonicity.

Lemma 1 *Suppose that the optimization problem given by*

$$\max\{h(z, \theta) : z \in S\}$$

has at least one solution for each $\theta \in \Theta$, $S \subset \Re$ and that h satisfies decreasing differences in θ. Then $z^*(\theta)$ is monotone decreasing in the parameter θ.

To connect these parametric monotonicity results to the problem at hand, let z be a scalar, $\theta \in \Re^n$ and define

$$h^j(z, \theta) = \eta^j(ze^j + \theta, \omega).$$

Thus, z changes the inventory level of variant j while θ changes the inventory levels of any variant. The main structural result of Mahajan and van Ryzin [41] is the following theorem:

Theorem 6
i) The function $h^j(z, \theta)$ is componentwise concave in z for all ω.
ii) The function $h^j(z, \theta)$ satisfies the property of decreasing differences in (z, θ) for all sample paths ω.

These properties are quite intuitive. Componentwise concavity implies that adding more stock of any particular variant produces decreasing marginal benefits. Therefore, if one were myopically optimizing the level of one particular variant only, a critical stocking level (base-stock level) would be optimal. On the other hand, the decreasing differences property implies that the marginal benefit of an additional unit of inventory of any particular variant decreases as the inventory levels of the other variants are increased. That is, the component-wise optimal stocking level of a particular variant is decreasing in the inventory of the other variants. These results are quite natural properties for the profit function to posses and reflect the fact that variants are indeed substitutes.

Unfortunately, these same intuitive properties do not hold for the total profit function. Indeed, Mahajan and van Ryzin [41] give counter-examples showing that the total assortment profit is not component-wise concave and does not posses the decreasing differences property. This is strong evidence that the profit functions of dynamic substitution problems are, in general, quite complex.

17.5.2.3   A Fluid Model Relaxation. Having a continuous version of the dynamic choice problem is desirable for two reasons: 1) to address theoretical questions of quasi-concavity, and 2) to facilitate computational approaches using gradient-based algorithms. Mahajan and van Ryzin [41] develop a natural continuous relaxation of the problem which has the integer problem as a special case.

To define the continuous problem, inventory is viewed as a fluid. Each customer t requires a certain quantity of fluid, Q_t, which could vary from customer-to-customer and could be non-integral. Different fluids are ordered based on the customer's preference. The customer drains the inventory of the most preferred fluid. If this fluid runs out, the customer drains the inventory of the second most preferred fluid and so on. This process continues until either the customer's requirement is met or the inventory of all fluids valued higher than the no-purchase utility is exhausted. Note that if each customer requires exactly one unit of fluid and if the initial fluid levels are integral, then this model is equivalent to the discrete one considered earlier.

For the relaxed problem, the sample path is $\omega = \{(U_1, Q_1), (U_2, Q_2), ..., (U_T, Q_T)\}$ and the system function is modified is a straight-forward way (see [41]). With such a continuous model, one can investigate joint concavity properties of the profit function. The main result here is a negative one. Namely, in general, the sample path profit function is not even jointly quasi-concave in the starting inventory level vector:

Theorem 7 *There exist sample paths w and starting inventory level vectors x on which the sample path profit function for the relaxed problem, $\pi(x, \omega)$ is not quasi-concave.*

The significance of this result is that without quasi-concavity, we cannot preclude the possibility that there may be local optima in the expected profit maximization problem (17.36).

17.5.2.4 Optimizing Assortment Inventories. To solve this problem algorithmically, Mahajan and van Ryzin [41] develop a sample path gradient algorithm. Note that in view of Theorem 7, finding globally convergent algorithms for the relaxed problem is likely to be difficult. Therefore, the best one can hope for is algorithms that converge to stationary points of the expected profit function. That is, points x satisfying

$$E[p^T \nabla \eta^T(x, \omega) - c] = 0. \tag{17.40}$$

This can be accomplished by using sample path gradient $\nabla \eta(x, \omega)$.

17.5.2.5 Computing $\nabla \eta(x, \omega)$. Using the knowledge of the choice decision made by each customer, one can easily calculate the inventory level vector observed by the subsequent customer on the sample path. This is done in a *forward pass*. After computing the inventory level vectors, $\{x_t : t = 1, ..., T\}$, the sample path gradient $\nabla \eta(x, \omega)$ is found using a *backward pass* on the sample path.

Specifically, these steps are:

Forward Pass

$$x_1 = x$$
$$x_{t+1} = f(x_t, U_t, Q_t) \quad \forall t = 1, ..., T$$

Backward Pass

$$\nabla \eta_T(x_T, \omega) = \mathbf{I} - \nabla f(x_T, U_T, Q_T) \tag{17.41}$$
$$\nabla \eta_t(x_t, \omega) = \mathbf{I} - \nabla f(x_t, U_t, Q_t) \tag{17.42}$$
$$+ \nabla f(x_t, U_t, Q_t)[\nabla \eta_{t+1}(f(x_t, U_t, Q_t), \omega)]$$

for $t = 1, ..., T - 1$, where the gradient $\nabla f(x_t, U_t, Q_t)$ is calculated using the fluid relaxation of equation (17.37). Then $\nabla \eta(x, \omega) = \nabla \eta_1(x_1, \omega)$.

17.5.2.6 The Sample Path Gradient Algorithm. The algorithm requires an initial starting inventory level y and a sequence of step sizes, $\{a_k\}$, with the following properties.

$$\sum_{k=0}^{\infty} a_k = \infty \quad \text{and} \quad \sum_{k=0}^{\infty} a_k^2 < \infty.$$

(For example, $a_k = 1/k$.)

It proceeds as follows:

1. Initialize: $k = 0$ and $y_0 = y$

2. At iteration k

    i) Generate a new sample path ω_k

    ii) Calculate sample path gradient $\nabla \eta(y_k, \omega_k)$ and step size a_k.

    iii) Update the starting inventory level for the next iteration, using the equation

$$y_{k+1} = y_k + a_k \nabla \pi(y_k, \omega_k).$$

3. $k := k + 1$ and GOTO Step 2.

Under relatively mild assumptions, Mahajan and van Ryzin [41] show that the sample path gradient algorithm converges to an inventory level vector in the set of stationary points of the expected profit function.

17.5.2.7 Numerical Experiments and Comparisons with Heuristic Policies. We next examine the results of some numerical experiments using the sample path gradient algorithm. The assortments produced by the algorithm are compared to those produced by several other policies which do not take into account dynamic choice effects. We first describe the assumptions behind the numerical experiments and then present the results.

In the examples, the MNL model described in Section 17.4.2 was used to generate the sequence of utilities $\{U_t\}$. The systematic component of utility is broken down as

$$u_j = y + a_j - p_j \tag{17.43}$$

and

$$u_0 = y + a_0,$$

where y stands for consumer income, a_j is a quality index and p_j is the price for variant j.

The examples have $n = 10$ variants, with linearly decreasing quality indices

$$a_j = 12.25 - 0.5(j - 1) \quad \forall j = 1, ..., 10, \tag{17.44}$$

and $a_0 = 4.0$. Consistent with the MNL, the error terms ξ_t^j are i.i.d., Gumbel distributed with parameter $\mu = 1.5$, so the variance of ξ_t^j is 1.18.

The procurement cost for a unit is set at $c_j = 3.0$ for all j, and so is the price,

$$p_j = p, \quad j = 1, ..., 10,$$

where p takes values in the range $3 - 9$. This simplification facilitates comparison with other heuristic policies and also makes it easier to investigate how different performance measures vary with price. The number of customers in the sequence (total store traffic), T, is a Poisson random variable with mean 30 and each customer demands exactly one unit of product ($Q_t = 1$ for all t).

In addition to tracking profit on each sample path, the following measures are computed:

Consumer Surplus.

$$CS(x,\omega) = \sum_{t=1}^{T}\sum_{j=0}^{n} U_t^j 1_{\{d(x_t,U_t)=j\}}. \tag{17.45}$$

Social Welfare.

$$SW(x,\omega) = \pi(x,\omega) + CS(x,\omega). \tag{17.46}$$

Consumer surplus measures the total benefit generated for consumers on the sample path. Social welfare is the total benefit generated for both the retailer and the consumer on the sample path.

17.5.2.8 Heuristic Policies. To gain some insight into the relative importance of dynamic choice behavior, Mahajan and van Ryzin [41] also analyzed several heuristic policies. These policies are simple approaches that ignore dynamic choice effects. As such, they are meant to mimic decisions a retailer might take in practice.

The policies are:

Independent Newsboy. The Independent Newsboy policy is essentially equivalent to the static choice model discussed in Section 17.4.2. Thus, the optimal Independent Newsboy procurement level, denoted x_I^j, is

$$x_I^j = \lambda q_j(S) + z(\sqrt{\lambda q_j(S)}), \quad j \in N,$$

where $q_j(S)$ are given by (17.23), λ is the mean of the Poisson distribution, and $z = \Phi^{-1}(1 - \frac{c}{p})$ where $\Phi(z)$ denotes the c.d.f. of a standard normal random variable.

Pooled Newsboy. In this policy, the entire category is aggregated and viewed as a single variant. An aggregate newsboy inventory level is then calculated for the category based on the probability the category will be chosen over the no-purchase option. This aggregate inventory is then allocated proportional to the preference values v_j in (17.24). Specifically, let

$$v_L = \sum_{j \in N} v_j,$$

Then the probability that the category will be chosen over the no-purchase option is given by

$$q_L = \frac{\sum_{j \in N} v_j}{\sum_{j \in N} v_j + v_0}.$$

Under the normal approximation to the Poisson distribution, the optimal aggregate procurement level for this variant, denoted x_L, is given $x_L = \lambda q_L +$

$z(\sqrt{\lambda q_L})$, where z is the newsboy fractile defined as before in (17.28). The Pooled Newsboy inventory for variant j, denoted x_P^j, is then determined by allocating the aggregate inventory proportional to v^j:

$$x_P^j = x_L \frac{v_j}{\sum_{j \in N} v_j + v_0}, \quad j \in N.$$

Naive Gradient. In this method, inventory levels are calculated similarly to the sample path gradient algorithm. The only difference is that we approximate the random gradient at each iteration of the algorithm naively. Specifically, we decrease the inventory level of all variants that are left in stock at the end of the sample path and increase the inventory level of all variants that are out of stock at the end of the sample path. The actual magnitude of these changes depends on the procurement cost, selling price and step size, as shown below. This mimics a simple feedback reaction to stockouts.

Let $\nabla \kappa(x, \omega)$ denote the naive gradient. As before, let T be the number of customers observed on the sample path ω. Then x_{T+1} denotes the inventory level vector at the end of the sample path. Define,

$$\nabla \kappa^j(x, \omega) = \begin{cases} p - c & x_{T+1}^j = 0 \\ -c & x_{T+1}^j > 0. \end{cases}$$

A sample path gradient method is used in exactly the same way as before, except that the sample path gradient $\nabla \pi(y_k, \omega_k)$ at each iteration k of the algorithm is replaced with the naive gradient $\nabla \kappa(y_k, \omega_k)$.

Note that this algorithm can be viewed as an adaptive version of the independent newsboy model, because the naive gradient for each variant is calculated based on its inventory level at the end of the sample path irrespective of the inventory levels of other variants. One can show easily that Naive Gradient policy finds the optimal newsboy quantity if demand for each item is truly independent.

17.5.2.9 Performance Comparison. The profit comparison for the four methods is given in Table 17.1. The results clearly show that the sample path gradient method outperforms the other policies. The relative magnitude of the profit improvement ranges from a low of 1.0% over Pooled Newsboy at $p = 5$ to a high of 12.6% over Naive Gradient at $p = 5$. The improvement over the Independent Newsboy is approximately 3% at both price levels. While small in absolute terms, these differences are nevertheless significant, considering they are differences in gross profits.

We next examine the stocking levels obtained under each heuristic policy relative to those obtained by the sample path gradient algorithm.

Independent Newsboy. The total inventory under this policy is underestimated at price 5.0 and overestimated at price 8.0 relative to the sample path gradient method. The inventory levels are systematically lower for the first

Table 17.1 Gross Profit of Dynamic Choice Policies

Price	5.0	8.0
Sample Path Gradient	47.82	98.15
Independent Newsboy	46.31	95.40
Pooled Newsboy	47.35	96.92
Naive Gradient	42.48	92.74

two variants and systematically higher for the last eight variants at both price levels. That is, inventory is more evenly spread across variants.

The intuition for this behavior is essentially the same as that for Noonan's condition (17.32). Namely, dynamic choice behavior has two effects on the standard newsboy logic. First, the demand for a given variant will be higher than the independent newsboy predicts because each variant receives some additional substitute demand from other variants that are out of stock. This effect increases the level of demand, which provides an incentive to *increase* inventory. On the other hand, the unit underage cost is lower than the independent newsboy predicts because an underage in one variant does not always result in a lost sale; customers may substitute rather than not purchase. This reduces the effective underage cost, which creates an incentive to *decrease* inventory. Which of these two opposing effects dominates depends on the variant, with popular variants having lots of additional substitution demand and fewer chances for "back-up" alternatives, while less popular variants receive little substitution demand and have many attractive back-up alternatives if they are out of stock. The net result is that that the independent newsboy is biased; it stocks too little of the popular variants and too much of the less popular variants. Results are shown in the Figure 17.1 for $p = 8$.

Pooled Newsboy. Figure 17.1 shows the allocation under the Pooled Newsboy policy for $p = 8$. The total inventory of the pooled newsboy is remarkably close to that of the sample path gradient algorithm at both price levels. However, inventory is also more evenly spread across variants than is optimal, as in the Independent Newsboy policy. This difference in the allocation of inventory among variants does matter. As seen in Table 17.1, the simple allocation of the Pooled Newsboy policy results in a reduction of roughly 1% in gross profits. Even bigger losses can occur under other allocation schemes. For example, if we allocate total inventory evenly, which is clearly not optimal, then at a price of 5.0 the profit drops to 41.87 from its previous value of 47.35, a loss of 11.6%.

Naive Gradient. The inventory level for each variant under the naive gradient is systematically higher than the sample path gradient method. (The inventory level vectors were generated using the same sample paths in all cases.)

538

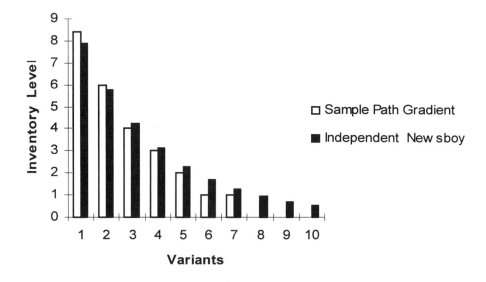

Figure 17.1   Inventory levels of Independent Newsboy policy

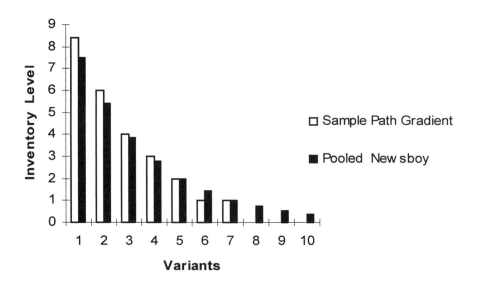

Figure 17.2   Inventory levels of Pooled Newsboy policy

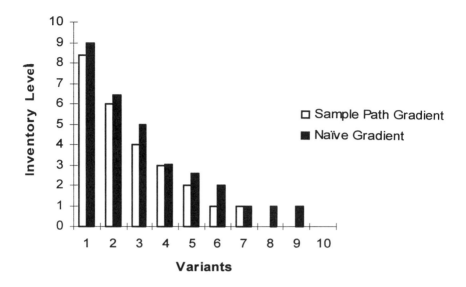

Figure 17.3 Inventory levels of Naive Gradient policy

This can be explained as follows: when inventory of a particular variant is left over at the end of a sample path, both the naive gradient and the sample path gradient method choose a gradient direction which decreases the inventory level of that variant. When a variant is sold out at the end of the sample path, the naive gradient method always increases the inventory level of that variant for the next iteration. However the sample path gradient method does not always increase the inventory level, because an incremental unit may only result in a substitution rather than an additional sale. Therefore, the naive gradient method overestimates the inventory level when compared with the sample path gradient method. Results are shown in the Figure 17.3 for $p = 8$.

17.5.2.10   Price and Scale Effects Under Dynamic Choice.   We next examine some numerical results based on the dynamic choice model. Under a static choice model, we saw in Section 17.4.2 that variety increases if the store volume (number of arriving customers) increases. We also saw that there was a strong connection between the level of fashion in a category and equilibrium profit margins. The numerical results in this section indicate that these insights hold under a dynamic choice model as well.

By varying the price term in the random utility model and examining the consumer surplus measure, one can characterize the total benefit an assortment provides to consumers.

As in the numerical examples above, we assume the price of all variants in the category is the same. Using the specific form of the utility given (17.43), we see that the quality indices $\{a_j : j \in N\}$ and the price p completely characterize the vector $v = \{v_0, v_1 ..., v_n\}$, where v_j is defined by (17.24). (The income term y has no affect on the choice probabilities.)

Let $\pi(x, p, v, \omega)$ denote the profit for a category v priced at p on a sample path ω given a starting inventory level x. Mahajan and van Ryzin consider three prices for a given category v:

Equilibrium Price. An equilibrium price, denoted p_e^v, is the price at which the maximum expected profit achieves an equilibrium profit π_e:

$$\pi_e = max_{x \geq 0} E[\pi(x, p_e^v, v, \omega)].$$

(If there are multiple solutions, then we take the lowest equilibrium price, since this is the only one that is "stable".)

Monopoly Price. The monopoly price, p_m^v, is the price which maximizes retailer profits. That is,

$$p_m^v = \arg \max_{p \geq 0} \max_{x \geq 0} E[\pi(x, p, v, \omega)].$$

Social Welfare Maximizing Price. The social welfare generated by a category is given by (17.46). Analogous to profit, we define $\mathcal{SW}(x, p, v, \omega)$ as the social welfare on the sample path ω for category v given price p and starting inventory level x. Let p_s^v denote the price which maximizes expected social welfare for category v. Then

$$p_s^v = \arg \max_{p \geq 0} \max_{x \geq 0} E[\mathcal{SW}(x, p, v, \omega)].$$

For a grid of 25 price levels ranging from $p = 3.2, 3.4, ..., 8$, Mahajan and van Ryzin computed starting inventory levels x using the sample path gradient algorithm. For each starting inventory level at its corresponding price value a simulation was run to study the performance of the inventory level-price combination. The parameters for the experiment are as described earlier. The basic category w used is the same as in equation (17.44). The fashion category v is chosen so that $v \prec w$. Specifically,

$$a_j = 10.86 \quad \forall j = 1, ..., 10.$$

The numerical results can be summarized as follows:

Observation 1
a) If $v \prec w$, then $p_e^v \geq p_e^w$
b) $p_m^v > p_s^v$.
c) *Higher variety is offered as store traffic λ increases.*

Part a) of Observation 1, is consistent with the results in Section 17.4.2 from the static choice model. Namely, fashion categories have higher mark-ons in

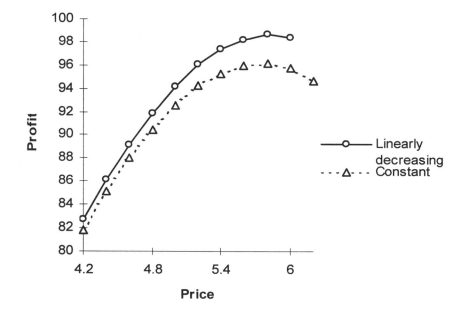

Figure 17.4 Profit as a function of price

equilibrium than basic categories. The reasoning is essentially the same; in fashion categories, there is increased inventory costs due to highly fragmented purchase choices. The higher mark-ons in equilibrium compensate retailers for these added costs. Figure 17.4 shows how the expected profit varies with price for the fashion and basic categories. We find that profit for the basic category is higher at all price levels, which in turn implies the equilibrium price is lower for the basic category.

Part b) is supported by the data in Table 17.2, which shows performance measures for the monopoly and social welfare maximizing price. The social welfare is maximized at a price of 5.0 and the profit maximizing monopoly price is 8.2. The numerical results showed that as price increases, the benefit obtained by the retailer as measured by expected profit increases, and the benefit obtained by the consumer as measured by consumer surplus decreases. A similar disparity in these prices occurs for the fashion category.

If one equates consumer surplus to the value customers derive as a result of their shopping experience, these examples suggest that pricing and inventory strategies that deliver long-term value may be quite different from those that maximize short-run profits.

Table 17.2 Maximized Profit and Welfare Measures

Price	Profit	Cons. Surplus	Welfare	$\dfrac{\text{Cons. Surplus}}{\text{Welfare}}$ (%)
5.0	47.8	200.2	247.6	80.7
8.2	98.6	88.7	187.3	47.4

Table 17.3 Optimal Variety at Different Traffic Volumes

Mean # arrivals	# Variants stocked
10	3
30	5
100	9
1000	10

Part c) implies that there are scale economies to offering variety. This conclusion is supported by the data in Table 17.3. Again, this finding is consistent with the results of Section 17.4.2 for the static choice model and the basic intuition behind it is the same.

17.6 ESTIMATION OF DEMAND UNDER CONSUMER CHOICE

Ultimately, to understand consumer behavior empirically or to operationalize choice-based inventory models, one must be able to estimate a choice model. In this section, we briefly discuss two such estimation methodologies, one for the MNL and another proposed by Anupindi, Dada and Gupta [6] for a probabilistic model of choice involving two variants. Both are based on maximum likelihood estimation (MLE) methods. The discussion in this section is not meant to be exhaustive; rather, we simply want to give the reader some flavor for the available estimation techniques and illustrate that it is indeed possible to estimate choice models in practice. An excellent reference on estimation of discrete choice models is the book by Ben-Akiva and Lerman [12].

17.6.1 Maximum Likelihood Estimation of the MNL

We first discuss how one can estimate a MNL dynamic choice model, such as was used in the numerical example of the Mahajan and van Ryzin model described in Section 17.5.2. To do so, we must estimate the systematic components of utility u_j required for computing the choice probabilities of the MNL in (17.10). Because multiplying all utilities by a scalar constant does not alter the choice probabilities, we can, without loss of generality, assume utilities are scaled so that $\mu = 1$ in (17.10). Hence, we do not need to estimate μ.

In applications, we often want to specify a functional form for the systematic component of utilities, u_j, based on observable attributes of each variant. Examples of such attributes could include price, brand, size, color, etc. (binary attributes are allowed). The most convenient functional form in practice is the *linear in attributes* model, defined as follows: Let each variant $j = 1, ..., n$ be associated with a m-vector of attribute values, $y_j = (y_{1j}, .., y_{mj})$. In addition let $\beta = (\beta_1, ..., \beta_m)$ be a vector of co-efficients, which determine how attribute values are weighted to obtain the systematic component of utility. Then,

$$u_j = \beta^T y_j, \quad j = 1, ..., n. \tag{17.47}$$

If we know the vector β, we can determine u_j and then the choice probabilities from equation (17.10).

To estimate β based on observed sales and inventory data, we can use maximum likelihood estimation. We shall assume we have data describing T customer choice decisions. Let S_t denote the set of alternatives available to Customer t and define Customer t's decision by the values

$$z_{jt} = \begin{cases} 1 & \text{if alternative } j \text{ is chosen by Customer } t \\ 0 & \text{otherwise} \end{cases} \tag{17.48}$$

We assume the data consists of the values of $\{z_{jt} : t = 1, ..., T, j = 1, ..., n\}$ together with the choice sets $\{S_t : t = 1, ..., T\}$. These could be determined, for example, by tracking the on-hand inventory status prior to each purchase. However, we are assuming that we have data on the inventory status prior to every choice decision as well as all choice outcomes, including no-purchase outcomes. As shown below, Anupindi, Dada and Gupta [6] consider an estimation problem with only partial information on the inventory status and also lack of information on no-purchase outcomes.

The log-likelihood function, $\mathcal{L}(\beta)$, is then (see Ben-Akiva and Lerman [12]):

$$\mathcal{L}(\beta) = \sum_{t=1}^{T} \sum_{j \in S_t} z_{jt} \left(\beta^T y_j - \ln \sum_{i \in S_t} e^{\beta^T y_i} \right) \tag{17.49}$$

To estimate β, one simply maximizes this function over β; the optimal value is the maximum likelihood estimate. McFadden [45] shows that the log-likelihood function of the MNL is globally concave. Moreover, there are closed-form expressions for the Hessian of $\mathcal{L}(\beta)$. Therefore, it is easy to maximize $\mathcal{L}(\beta)$ using Newton's method or other standard nonlinear programming methods.

Under relatively general conditions, the MLE estimator of β is consistent, asymptotically efficient and asymptotically normally distributed. As a result, a variety of useful test statistics and goodness-of-fit measures can be easily computed. Again, we refer the reader to Ben-Akiva and Lerman [12] for a thorough treatment of these issues.

For all these reasons, we conclude that estimation of the MNL is certainly possible in practice given the appropriate data. Indeed, ease of estimation is an important reason behind the popularity of the MNL model.

17.6.2 *Estimation of Two-Product Probabilistic Substitution*

Another example of choice estimation is the recent work of Anupindi, Dada and Gupta [6], who develop a maximum likelihood method for estimating substitution probabilities for a two-product, dynamic choice system. As mentioned, two important issues they address are estimation when inventory levels are unobservable (in particular when we only have data on the starting and ending inventory in a period) and estimation when the no-purchase outcome cannot be observed.

Anupindi, Dada and Gupta [6] consider a category of two variants, which are restocked after every T time units. Two information systems are considered. In the periodic review system, inventory is reviewed only at the end of the period. In the transactions reporting system, information on each sale is available.

Customers arrive according to a Poisson process with a mean arrival rate λ per unit time. Customers can choose to purchase Variant 1 or Variant 2 or choose not to purchase. If both variants are in-stock, the probability of choosing variant j is given by $p_j, j = 1, 2$ and the probability of not purchasing is $1 - p_1 - p_2$. However, if only Variant 1 is in stock, the customer purchases Variant 1 with a probability $p_1^+ > p_1$ and chooses to not purchase with probability $1 - p_1^+$. Similarly, if only Variant 2 is in stock, the probability of choice is $p_2^+ > p_2$. Define $\lambda_1 = p_1 \lambda$. Similarly, define λ_2, λ_1^+ and λ_2^+. These have the interpretation of being the effective arrival rates for Variants 1 and 2 in the absence or presence of stockouts, respectively.

17.6.2.1 Estimation. The customer arrival rates λ_1, λ_2, λ_1^+ and λ_2^+ are unknown and need to be estimated. We first describe the estimation methodology for the case of the periodic review system. Let x_1 and x_2 be the stocking quantities of Variants 1 and 2. Let $P_{y,z}$ be the probability that the sales of Variant 1 equal y and the sales of Variant 2 equal z. At the end of the interval, one of four mutually exclusive, collectively exhaustive events can happen: i) only Variant 1 stocks out, ii) only Variant 2 stocks out, iii) both variants stock out, or iv) neither variant stocks out. Let $y < x_1$ and $z < x_2$. Under the Poisson model, Anupindi, Dada and Gupta develop closed form expressions for $P_{y,z}$, $P_{x_1,z}$, P_{y,x_2} and P_{x_1,x_2} in terms of the arrival rates λ_1, λ_2, λ_1^+ and λ_2^+.

The data consist of M periods each of length T, with potentially different initial inventories at the beginning of each time period. At the end of the time period, the final inventory levels and stockout events are recorded. Let I, J, K and L be the number of observations in which events i), ii), iii) and iv) defined above occur, respectively. Then

$$I + J + K + L = M \qquad (17.50)$$

We can then find the values of the purchase probabilities that maximize the likelihood of observing I, J, K and L observations of each type.

This MLE proceeds as follows: Let x_1^i (resp., x_1^k) denote the starting inventory levels for Variant 1 for Event i) (resp., Event iii)). Similarly, let x_2^j (resp., x_2^k) denote the starting inventory levels for Variant 2 for Event ii) (resp., Event iii)). Finally, let y^j (resp., y^l) denote the number of sales of Variant 1 under Event ii) (resp.,Event iv)). and let z^i (resp., z^l) denote the number of sales of Variant 2 under Event i) (resp., Event iv)). Then the purchase probabilities are estimated by maximizing the log-likelihood

$$\mathcal{L} = \sum_{i=1}^{I} log P_{x_1^i, z^i} + \sum_{j=1}^{J} log P_{y^j, x_2^j} + \sum_{k=1}^{K} log P_{x_1^k, x_2^k} + \sum_{l=1}^{L} log P_{y^l, z^l} \quad (17.51)$$

This log-likelihood function may not be concave. The first order conditions to maximize the likelihood function above result in a set of simultaneous nonlinear equations that are solved using iterative approaches developed in nonlinear programming.

For the case of the transactions reporting system, all purchase transaction information is available. However, under the Poisson model it is sufficient to know only the timing of the stockout events. This has two ramifications for the estimation methodology described above. Firstly, we now have five types of observations, because we now need to know the sequence of stockout events. Secondly, the derivations of the probabilities of the five events changes. Otherwise, conceptually, the MLE proceeds similarly to the periodic case.

Finally, Anupindi, Dada and Gupta show that when the choice probabilities follow the IIA property, it is possible to provide an estimator for the rate of no-purchase, even when no-purchase outcomes are unobservable. This is an important consideration in practice because one typically has data only on sales transactions.

For the transactions reporting system, the extension of the above estimation methodology to more than 2 variants is straightforward. For the periodic system, the computational burden may be prohibitive, since evaluation of multidimensional integrals is necessary to derive the probabilities of each event types.

17.6.2.2 Empirical Test.

Anupindi, Dada and Gupta apply their estimation model to an interesting experiment involving a Virtual Vending Machine (VVM) on the World Wide Web. The VVM stocks two beverages, Diet Coke and Diet Pepsi. A sample of 1000 respondents were sent email messages to visit the VVM and told that they would receive an incentive for doing so. On visiting the VVM, the respondents were offered 60 cents, which is the price of the beverage. They had the choice to spend it and physically receive the beverage later or to not purchase and receive the money. The respondents were able to view whether the VVM stocked out, and those respondents who faced a stockout were asked what their first preference was. Information on every transaction was available from the server, so this is a transactions reporting system. [6]

---

[6] We note in passing that in this type of web-based commerce, it is indeed possible to observe no-purchase behavior, because one can easily tack visits to a site ("hits").

The experiment was conducted for 20 periods, i.e. $M = 20$, and 359 respondents visited the VVM in all. Of the 20 periods, in 8 one variant stocked out, in 8 both variants stocked out and in 4 no variant stocked out. So observations in all the four categories were observed. The Poisson distribution provided a good fit to the arrival data. The purchase probabilities were computed based on the estimation method described earlier. The data showed that there was significant substitution in the presence of stockouts. Further, from the purchase probabilities it was concluded that 60% of Diet Coke drinkers substitute Diet Pepsi, while only 30% of Diet Pepsi drinkers substitute Diet Coke. These numbers were found to agree closely with the survey responses of those respondents who faced a stockout.

17.7 CONCLUSIONS

While research on inventory management under consumer choice is new and still evolving, several themes clearly emerge from the literature to date.

First, the phenomenon of consumer choice does affect inventory management, as show by the empirical work of Anupindi, Dada and Gupta [6], and choice behavior has a significant impact on both stocking decisions and profits. In particular, under dynamic substitution one needs to stock relatively more of popular variants and relatively less of unpopular variants than a traditional newsboy analysis indicates, due to excess substitution demand combined with reduced underage costs from having substitute variants as backups. However, as demonstrated by the static substitution models of Smith and Agrawal [58] and van Ryzin and Mahajan [66], the particular type of choice behavior (e.g. MNL, Adjacent Substitution, One Variant Substitution) strongly impacts which variants are most profitable to stock. Finally, both static and dynamic substitution models suggest narrower assortments are optimal if there is a higher level of substitution among variants in a category.

A second theme of this literature is that while inventory dynamics and cost functions under choice can be complex, they are nevertheless amenable to both structural and computational analysis. In the dynamic substitution model of Mahajan and van Ryzin [41], which is the most general of the models analyzed, the marginal profit function has the decreasing difference property, but the total profit function does not and moreover is not even quasi-concave. However, good computational approaches have been developed that allow one to analyze the problem numerically under very general assumptions and also find good inventory policies. In terms of estimation, classical MLE of the MNL can be applied if full inventory and choice data are available, and the work of Anupindi, Dada and Gupta [6] provides approaches for the case of partial information. The use of POS terminals in supermarkets and the advent of web-based commerce will lead to an increasing ability to utilize such estimation approaches. Thus, despite the apparent difficulties, operationalizing dynamic substitution models, both from an estimation and an optimization standpoint, seems possible.

Finally, a third theme in this work is that choice-based inventory models provide useful insights into what factors determine optimal variety in a retail category. The static substitution models, in particular, show how margins, store volume, outside alternatives, lead-time, substitution structure and level of fashion affect optimal assortment variety. They also lead to a definition of fashion as a measure of the level of fragmentation in consumer preference, and highlight a distinction between fashion and trend-following consumer behavior that has interesting consequences for supply chain strategy. Finally, they show that fashion categories have higher equilibrium prices than basic categories. While all these results are only shown rigorously for the static choice case, they appear to hold in the dynamic choice case as well.

While this literature has made substantial progress, there are certainly many open areas of possible investigation. One is to consider the impact of using alternative choice and inventory models. It would be interesting to learn to what extent the conclusions reached above are sensitive to the particular structure of the choice and inventory models. It would also be desirable to endogenize the customer's store visit decision and account for within-store search costs, as is done in a deterministic setting in the work of Baumol and Ide [10]. Endogenizing the price decision would also be a worthwhile line of future research.

It would also be interesting to study assortment selection from a broader point of view, taking into account the possibility of complimentary merchandise categories. For example, one category might be carried as a "loss leader" to generate volume for other complimentary categories. Alternatively, categories might be negatively correlated, allowing the store to hedge the risks of over and under stocking.

Under dynamic substitution, a relevant question is how to approximate complex choice effects heuristically. For example, the simple Pooled Newsboy heuristic performs remarkably well, even in the presence of complicated consumer choice effects. Perhaps treating an entire category as if it were a single variant and then performing a simple allocation of the aggregate inventory is a reasonable way to manage assortments in practice. This raises interesting issues about how a merchandise category should be aggregated for such purposes (e.g. all dress shirts, all size M shirts, all blue shirts, etc.).

Finally estimation methodologies for choice models under incomplete information merit more attention, and there is a need for empirical work to confirm or refute the phenomenon identified by the theoretical models.

References

[1] Abernathy, F.H., J.T. Dunlop, J.H. Hammond and D. Weil (1995). "The Information-Integrated Channel: A Study of the U.S. Apparel Industry in Transition," Brookings Papers on Microeconomic Activity, Harvard Business School, Cambridge, MA.

[2] Agrawal, N. (1997). "Joint Pricing and Inventory Planning in Retail Inventory Systems with Demand Substitution," Working Paper, Decision and Information Sciences, Santa Clara University.

[3] Agrawal, N. and S.A. Smith (1996). "Estimating Negative Binomial Demand for Retail Inventory Management with Unobservable Lost Sales," *Naval Research Logistics*, **43**, 839-861.

[4] Anderson, E.E. and H.N. Amato (1974). "A Mathematical Model for Simultaneously Determining the Optimal Brand-Collection and Display-Area Allocation," *Operations Research*, **22**, 13-21.

[5] Anderson, S.P., A. de Palma, and J.F. Thisse. *Discrete Choice Theory of Product Differentiation*, The MIT Press, Cambridge, MA, 1992.

[6] Anupindi,R., M. Dada and S. Gupta (1997). "A Dynamic Model of Consumer Demand with Stock-Out Based Substitution," Working Paper, Kellogg School of Management, Northwestern University.

[7] Anupindi, R., S. Gupta and N. Venkataraman (1997). "Managing Variety on the Retail Shelf: Using Scanner Data to Rationalize Assortments," Working Paper, Kellogg Graduate School of Management, Northwestern University.

[8] Banks, J., J.S. Carson and B.L. Nelson (1996). *Discrete Event System Simulation*, Prentice Hall, NJ.

[9] Bassok, Y., R. Anupindi and R. Akella (1997). "Single Period Multi-Product Inventory Models with Substitution," Working Paper, Kellogg School of Management, Northwestern University.

[10] Baumol, W.J. and E.A. Ide (1956). "Variety in Retailing," *Management Science*, **3**, 93-101.

[11] Ben-Akiva, M. (1973). "Structure of Passenger Travel Demand Models," Ph.D. Dissertation, Department of Civil Engineering, MIT, Cambridge, Massachusetts.

[12] Ben-Akiva, M. and S.R. Lerman, *Discrete Choice Analysis*, The MIT Press, Cambridge, Massachusettes.

[13] Block, H.D. and J. Marschak (1960). "Random Orderings and Stochastic Theories of Responses," in I. Olkin (ed.), *Contributions to Probability and Statistics*, Stanford University Press, Stanford.

[14] Bultez, A. and P. Naert (1988). "SH.A.R.P.: Shelf Allocation for Retailer's Profit," *Marketing Science*, **7**, 211-231.

[15] Chamberlin, E.H. (1933)'. *The Theory of Monopolistic Competition*, Harvard University Press, Boston.

[16] Chen, F., J. Eliashberg and P.H. Zipkin, (1997). "Customer Preferences, Supply-Chain Costs and Product-Line Design," Working Paper, Columbia Business School.

[17] Corstjens, M. and P. Doyle (1981). "A Model for Optimizing Retail Space Allocations," *Management Science*, **27**, 822-833.

[18] Debreu, G. (1960). "Review of R.D. Luce, Individual Choice Behavior: A Theoretical Analysis," *American Economic Review*, **50**, 186-188.

[19] de Groote, X. (1994). "Flexibility and marketing/manufacturing coordination," *International Journal of Production Economics*, **36**, 153-167.

[20] Dixit, A.K. and J.E. Stiglitz (1977). "Monopolistic Competition and Economic Product Diversity," *American Economic Review*, **67**, 297-308.

[21] Dobson, G. and S. Kalish (1993). "Heuristics for Pricing and Positioning a Product Line," *Management Science*, **39**, 160-175 .

[22] Dvorak, R.E. and F. van Paasschen (1996). "Retail Logistics: One Size Doesn't Fit All," *The McKinsey Quarterly*, **2**, 120-129.

[23] Fruend, R.B. and H. Matsuo (1997). "Retail Inventory and Shelf Space Allocation," Working Paper, Univ. of Texas at Austin.

[24] Garcia, C.B. and W.I. Zangwill (1981). *Pathways to Solutions, Fixed Points and Equilibria*, Pretice-Hall, NJ.

[25] Glasserman, P. (1994). "Perturbation Analysis of Production Networks," Chapter 6 in Yao, D. (ed.), *Stochastic Modeling and Analysis of Manufacturing Systems*, Springer-Verlag, New York, 1994.

[26] Graves, S.C., A.H.G. Rinnooy Kan and P.H. Zipkin (1993). *Logistics of Production and Inventory*, Vol. 4 in *Handbooks in Operations Research and Management Science*, Elzevier, Amsterdam.

[27] Green, P. and A. Krieger (1985). "Models and Heuristics for Product Line Selections," *Marketing Science*, **4**, 1-19.

[28] Guadagni, P.M. and J.D.C. Little (1983). "A Logit Model of Brand Choice Calibrated on Scanner Data," *Marketing Science*, **3**, 203-238.

[29] Gumbel, E.J. (1958). *Statistics of Extremes*, Columbia University Press, New York.

[30] Hansen, P. and H. Heinsbroek (1979). "Product Selection and Space Allocation in Supermarkets," *European Journal of Operational Research*, **3**, 58-63.

[31] Hanson W. and K. Martin (1996). "Optimizing Multinomial Logit Profit Functions," *Management Science*, 42, 992-1003.

[32] Hotelling, H. (1929). "Stability in Competition," *Economic Journal*, **39**, 41.

[33] Kreps, D.M. (1988). *Notes on the Theory of Choice*, Westview Press, London.

[34] Kaldor, N. (1935). "Market Imperfection and Excess Capacity," *Economica*, **2**, 35-50.

[35] Lancaster, K. (1990). "The Economics of Product Variety: A Survey," *Marketing Science*, **9**, 189-210.

[36] Lazear, E.P. (1986). "Retail Pricing and Clearance Sales," *American Economic Review*, **76**, 14-32.

[37] Lee, H.L. and S.Namhias (1993). "Single-Product, Single-Location Models," in Graves, S.C., A.H.G. Rinnooy Kan and P.H. Zipkin (ed.) (1993). *The Logistics of Production and Inventory*, Handbooks in Operations Research and Management Science Vol. 4, Elsevier, Amsterdam.

550

[38] Lariviere, M. and E.L. Porteus (1996). "The Ice Cream Stocking Problem," presented in Session SC06, INFORMS, Washington D.C., May 5, 1996.

[39] Luce, R. (1959). *Individual Choice Behavior: A Theoretical Analysis*, Wiley, New York.

[40] Luce, R. and P. Suppes (1965). "Preference, Utility and Subjective Probability," in *Handbook of Mathematical Psychology. Vol. 3*, R. Luce, R. Bush, and E. Galanter, eds., Wiley, New York.

[41] Mahajan, S. and G. van Ryzin (1998). "Stocking Retail Assortments Under Dynamic Consumer Substitution," Working Paper, Columbia Business School, NY.

[42] Manski, C. (1977). "The Structure of Random Utility Models," *Theory and Decisions*, **8**, 229-254.

[43] Manski, C.F. and D. McFadden (eds.) (1981). *Structural Analysis of Discrete Data with Econometric Applications*, MIT Press, Cambridge, Massachusetts.

[44] Marshall, A.W. and, I. Olkin (1979). *Inequalities: Theory of Majorization and its Application*, Academic Press, New York.

[45] McFadden, D. (1980). "Econometric Models of Probabilistic Choice Among Products," *Journal of Business*, **53**, 513-529.

[46] McGillivray, A. and E.A. Silver (1978). "Some Concepts for Inventory Control Under Stochastic Demand," *INFOR*, **16**, 47-63.

[47] Noonan, P.S. (1993). "A Decision Analysis of Single-Period Stocking Problems with Consumer-Driven Substitution," Ph.D. Dissertation, Harvard University, Cambridge, Massachusetts.

[48] Noonan, P.S. (1995). "When Consumers Choose: A Multi-Product, Multi-Location Newsboy Model with Substitution," Goizueta Business School Working Paper, Emory University, Atlanta, Georgia.

[49] Parlar, M. and S. Goyal (1984). "Optimal Ordering Decisions for Two Substitutible Products with Stochastic Demands," *OPSEARCH*, **21**, 1-15.

[50] Pashigian, P.B. (1988). "Demand Uncertainty and Sales: A Study of Fashion and Markdown Pricing," *American Economic Review*, **78**, 936-953.

[51] Pasternack, B. and Z. Drezner (1991). "Optimal Inventory Policies for Substitutable Commodities with Stochastic Demands," *Naval Research Logistics*, **38**, 221-240.

[52] Pentico, W. (1988). "The Discrete Two-Dimensional Assortment Problem," *Operations Research*, **36**, 324-332.

[53] Perloff, J. and S.C. Salop (1985). "Equilibrium with Product Differentiation," *Review of Economic Studies*, **52**, 107-120.

[54] Progressive Grocer (1993). "Category Dynamics," Executive Report, September 1993, 1-46.

[55] Sadowski, W. (1959). "A Few Remarks on the Assortment Problem," *Management Science*, **6**, 13-24.

[56] Shugan, S.M. (1989). "Product Assortment in Triopoly," *Management Science*, **35**, 304-320.

[57] Spence, A.M. (1976). "Product Selection, Fixed Costs and Monopolistic Competition," *Review of Economic Studies* , **43**, 217-235.

[58] Smith, S.A. and N. Agrawal (1996). "Management of Multi-Item Retail Inventory Systems with Demand Substitution," Working Paper, Decision and Information Sciences, Santa Clara University, Santa Clara, CA.

[59] R.K. Sundaram (1996). *A First Course in Optimization Theory*, Cambridge University Press.

[60] Thurstone, L.L. (1927a). "Psychological Analysis," *American Journal of Psychology*, **38**, 368-389.

[61] Thurstone, L.L. (1927a). "A Law of Comparative Judgement," *Psychology Review*, **34**, 273-286.

[62] Topkis, D.M. (1978). "Minimizing a Submodular Function on a Lattice," *Operations Research*, **26**, 305-321.

[63] Tryfos, P. (1985). "On the Optimal Choice of Sizes," *Operations Research*, **33** , 678-684.

[64] Tversky, A. (1969). "Intransitivity of Preferences," *Psychological Review*, **76**, 31-48.

[65] Tversky, A. (1972). "Choice by Elimination," *Journal of Mathematical Psychology* , **9** , 341-367 .

[66] van Ryzin, G. and S. Mahajan (1996). "On the Relationship Between Variety Benefits and Inventory Costs in Retail Assortments," Working Paper, Columbia Business School, NY.

[67] Wolfson, M.L. (1965). "Selecting the Best Length to Stock," *Operations Research*, **13**, 570-585.

18 THE BENEFITS OF DESIGN FOR POSTPONEMENT

Yossi Aviv

Olin School of Business
Washington University
St. Louis, MO 63130

Awi Federgruen

Graduate School of Business
Columbia University
New York, NY 10027

18.1 INTRODUCTION

Delayed product differentiation and *Quick Response* rank among the most beneficial strategic mechanisms to manage the risks associated with product variety and uncertain sales.

The product portfolio offered by a company often consists of families of closely related products which differ from each other in terms of a limited number of differentiating features only. Consider for example the apparel industry. A given design or style is usually offered in many distinct sizes and colors. Grocery and dry food products are typically sold in several package sizes, with a proliferation of differentiating features, e.g. different fragrances added to detergents. Automobile manufacturers offer a virtually endless variety of model configurations within a few basic product lines, while a given computer or printer model is distributed with a variety of accessories (e.g., power supply modules, key pads or manuals written in different languages). Delayed differentiation or *postponement* strategies attempt to reduce the risks associated with this product variety, by exploiting the commonality between items and by designing the production and distribution processes so as to delay the point(s) of differentiation. In the same vein, coordinating several geographically dispersed sales outlets via a regional distribution center, delays the point of differentiation to the final points of sale.

In particular under delayed product differentiation strategies, one finds that the production and distribution process consists of several stages, each with a significant leadtime. The Quick Response concept consists of introducing systematic reductions in the average value and variability of these leadtimes. Such reductions can be achieved by setup time reductions, the adoption of faster and more reliable production technology, electronic submission of purchase orders, and contractual agreements with suppliers stipulating quick and reliable delivery times, to mention but a few possibilities. Kurt and Salomon Associates, Inc. (1993) estimate that a Quick Response system should result in a 7-12% increase in retailers' returns on assets and a 20-40% increase in retail sales, see Kurt and Salomon Associates, Inc. (1993) and Hammond and Kelly (1990).

In this chapter, we provide a survey of analytical models which can be used to assess the benefits and costs associated with delayed product differentiation in a large variety of settings. The concept of delayed product differentiation was first introduced in the marketing literature. Introduced by Alderson (1950), it was subsequently developed in qualitative terms by Cox and Goodman (1956), the textbook of Vaile et al. (1952) and Bucklin (1965). The above mentioned analytical models include Eppen and Schrage (1981), Federgruen and Zipkin (1984a,b,c), Schwarz (1989) and Lee (1993, 1996), Lee and Tang (1997), Aviv and Federgruen (1997b,c), Swaminathan and Tayur (1996) and Gavirneni and Tayur (1997) among others.

In Section 18.2, we give a classification of the different ways in which product differentiation can be delayed, and review several documented implementations and case studies. In reviewing analytical models, we first consider those relating to a single point of differentiation (§18.3-§18.5).

In Section 18.3, we first discuss a simple model due to Lee (1993, 1996) which is applicable when only the intermediate product is made to stock. This is followed by an extension, based on Lee and Tang (1997), in which each of the two stages is modeled in more detail, as consisting of a number of discrete stages, such that each stage maintains inventories, both those before and those after the point of differentiation. The first model can be analyzed via a single location, single item model. The second model assumes that each stage is governed by a local order-up-to rule and it evaluates each stage as if its preceding stage never runs out of stock. Under these simplifying assumptions, the second model decomposes straightforwardly into a number of simple single item systems, one for each stage.

In Section 18.4, we consider models for settings where inventories are kept for all end products, possibly in combination with inventories for the intermediate product. A distinguishing feature of the models in this system is that it considers centralized strategies which optimize system-wide performance. In this section, we also consider settings where observed sales data can be used to ascertain improved forecasts for future demand distributions, either because demands in consecutive periods are correlated or because some of the parameters of the demand distributions are unknown. We show that such settings gave rise to increased benefits for delayed product differentiation beyond those observed in standard models without this "learning effect". In Section 18.5, we discuss a few models that characterize the benefits of capacity sharing as enabled by delayed differentiation. Section 18.6 concludes this chapter with an overview of models addressing multiple points of differentiation. We particularly focus on a model designed to analyze the optimal sequencing of differentiating operations, one of the types of postponement strategies described in Section 18.2.

18.2 DIFFERENT TYPES OF DELAYED DIFFERENTIATION STRATEGIES: EXAMPLES

Delayed product differentiation can be achieved via a variety of design changes in the production and distribution processes. These include:

(I) Standardization of components and subassemblies: often, related items in a family start out being manufactured from distinct, initial subassemblies and components. Standardization is achieved by substituting these by common ones. These substitutions may invoke (re)engineering changes. An example given in Lee and Tang (1997) is a company offering a mono (black ink) and a color (multicolor ink) printer. The manufacturing process consists of three stages: (i) assembly of printed circuit boards, (ii) final assembly and testing, and (iii) customization into mono or color printers. Standardization of the first stage may be achieved by designing or adopting a common key component, the so-called head driver board, for both product types. Similarly, standardization of the final assembly step requires the use of a common key component of that stage, i.e., the print mechanism interface.

(II) <u>Modular design</u>: sometimes it is impossible to substitute a given part or subassembly, for which a different choice is made in the design of each end item, by a common one. On the other hand, this part could be redesigned as an assembly of two or more modules such that one or several of the modules can be chosen uniformly across all end items and such that the assembly of the differentiating module(s) can be postponed till a later stage in the process. As an example, Lee and Tang (1997) cite the manufacturing and distribution of dishwashers of a variety of colors. The end products are differentiated from each other, in terms of the metal frames which in the traditional design of dishwashers are assembled in a second manufacturing stage referred to as "integration and shipping", preceding distribution to the various distribution centers in the company's network. No single, common, metal frame could be used for all end items, i.e., standardization of the metal frames is out of the question. On the other hand, the metal frame could be replaced by a combination of a uniform metal frame and a plastic frame specific to each end item, i.e., color, assembled on top of the metal frame. Assembly of the relatively light plastic frames (of different colors) can be postponed until the end of the distribution process. The original metal frame is thus modularized into two modules and assembly of the differentiating module is postponed to a later point in the process.

(III) <u>Postponement of operations</u>: sometimes, one or more differentiating operations can be postponed till a later stage in the process, prolonging the initial phase which precedes the point of differentiation. Returning to the example of families of printers, often the same basic printer is sold in many different countries where it becomes differentiated in terms of the required power supply cords, keypads and manuals. In the initial design of the production and distribution process, these accessories may be packaged with the printer at the plant, i.e., prior to distribution to a regional distribution center. Product differentiation may be delayed by shipping generic printers from the plant to the regional centers and by adding the differentiating accessories at the very end of the distribution process. Van Doremalen and Fleuren (1991) give similar examples where a generic electronic device is differentiated into a list of distinct end items by offering a variety of compatible monitors.

(IV) <u>Resequencing of operations</u>: sometimes, product differentiation occurs in several stages of the process, each differentiating operation being associated with a separate attribute. To the extent that the sequences in which these operations are performed can be chosen (subject to technological precedence constraints), it is intuitive that those operations associated with more "variable" attributes be postponed to the later stages to achieve operational benefits. Until recently, it has been an open question whether this intuition can be substantiated and if so, how different attributes should be compared in terms of their degree of variability. Examples of reversible operations arise in the context of apparel manufacturing where a family of garments is differentiated by color (attribute A) and style/size combination (attribute B). The dyeing operation determines the color and a knitting operation the style/size combination. While traditionally yarns are dyed first and dyed yarns subsequently knitted into different styles and sizes, companies like Benetton have experienced major performance improvements reversing the operations for some of their garment families, see HBS case (1986), Bruce (1987) and Dapiran (1992). Other examples arise in the area of manufacturing hard drives or printers.

The above classification of different types of postponement strategies is due to Lee and Tang (1997) and is most useful for analysis purposes. Zinn and Bowersox (1988) provide a different taxonomy based on the type of manufacturing operation which is postponed. The four types of postponement are: labeling postponement, packaging postponement, assembly postponement and manufacturing postponement. Alternative names for the design for differentiation concept, are *design for postponement* and *design for customization*. If the differentiation involves making the product ready for a specific market (country or global region), the term *design for localization* is often used. See, e.g., the above example of localizing printers by providing country- (and language-) specific manuals, keypads and power supply modules.

We now review some successful implementations of postponement strategies. One of the best documented implementations relates to Hewlett Packard's printer business. Design for postponement, via postponement of operations, was achieved by delaying certain assembly operations, see (III) above. Major efficiency improvements were accomplished vis-a-vis the European market; these are described and analyzed in Lee et al. (1993) and, as a teaching case, by Kopczak and Lee (1993). (The latter is reprinted in Flaherty (1996).) As far as the case is concerned, the analysis of the benefits, provided in the teaching note by Flaherty et al. (1996) is based on a sample of sales data for six Deskjet printers over 12 consecutive months. A more refined analysis, considering capacity constraints and seasonal variations was provided in Aviv and Federgruen (1997b). Lee et al. (1993) estimated that for Hewlett Packard, design for differentiation would lead to a reduction by no less than 18% of the total inventory investment. Hewlett Packard also reaped the benefits of postponement by standardizing several initial components in the manufacturing process; see (I) above. See also Lee and Billington (1994).

As mentioned above, Benetton achieved major efficiency improvements by reversing the sequence of certain manufacturing operations, in particular the dyeing and knitting operations, type (IV) of the four postponement types. See Signorelli and Heskett (1989) for a detailed case description of this implementation, and Hamond and Kelly (1991) for a general review of Quick Response and postponement strategies in the apparel industry. In the food industry, Gingrich and Metz (1990) report how Kellogg achieved comparative advantage in its cereal foods line by postponing packaging operations to a later stage and by storing the product at the "in process" stage. Low cost color analyzer' systems allow the more advanced producers in the paint industry to offer a virtual infinite variety of paints, matching any customer's desired color by last minute, electronically driven mixing of a few basic color paints. Ealey and Mercer (1992) describe how the main advantage of the advent of active suspensions in the automobile industry results from the ability to standardize. A single type of suspension system is kept in stock, with last minute tuning for each type of car, by reprogramming the computers driving the hydraulic rams. As an example of design for localization, General Motors announced a redesign of its distribution network supporting the Cadillac product line, by the opening of regional distribution centers maintaining inventory instead of the dealers. Other examples include Compaq (see New York Times (1997)), Sun Microsystems (see Mrena (1997)) and Toyota, see Gonsalvez (1991) and Federgruen (1993).

18.3   INDEPENDENT INVENTORY SYSTEMS

In this section, we review several models in which the system behaves as a unique single-item inventory system or as several independent single-item systems. The simplest such model arises when only the common intermediate product in Figure 18.1 is made-to-stock, while all final products are built-to-order from the intermediate product. Lee (1996) considers this setting with the following additional assumptions: periodically, inventories are reviewed and decisions are taken. At the onset of each period, unit size customer orders arrive for each of the final products. The inventory of the intermediate product is used to satisfy all customer orders, i.e., to customize the intermediate product into the different end items. If the aggregate demand exceeds the inventory level, the latter is randomly assigned to the competing customer orders and the excess is backlogged. Next, we assess the remaining inventory position of the intermediate product (= inventory level + work in process) and raise it to a given target level S by placing an order for a new batch. Inventory carrying costs are incurred at a unit rate $h(L)$ which may depend on the point of differentiation, L. Presumably, $h(L)$ is increasing in L, as it increases with the total value added to the (intermediate) product at the time of differentiation.

Given the restriction to order-up-to policies, only a single parameter (S) needs to be determined. Lee (1996) suggests doing this by trading off long-run average holding costs, with some constraints on the steady-state response time Y (e.g. an upper bound on the expected response time or the probability that it exceeds a given value x). Here, Y is defined as the response time to all customer orders arriving in an arbitrary period. Alternatively, we may wish to impose backlogging costs, as in most traditional inventory models. As long as the backlogging cost parameters are identical for all end items, only the aggregate demand distribution in each period needs to be known, i.e., the performance of the system is invariant to J, the number of end items, the relative sales volume of each, or the individual standard deviations and correlation matrix. This is in sharp contrast to the models described in the next section. Similarly, given

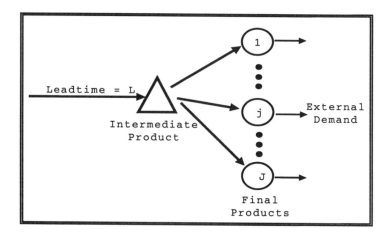

Figure 18.1   Two-stage inventory system

the definition of the response time Y, only the aggregate demand distribution is relevant in this variant of the model as well.

Thus, let $\{D_1, D_2, \ldots\}$ denote the sequence of aggregate demands, assumed to be independent and distributed with cdf $F(\cdot)$. Let $F(\cdot|t)$ denote the cdf of the t-fold convolution of F, $t \geq 1$. Using standard arguments, the traditional cost minimization problem with backlogging cost rate p reduces to:

$$\min_{S}\{ h(L) \int_0^S (S - w)dF(w|L) + p \int_S^\infty (w - S)dF(w|L) \} \qquad (18.1)$$

The optimal order-up-to value S^* satisfies the critical fractile equation:

$$F(S^*|L) = \frac{p}{p + h(L)} \qquad (18.2)$$

Assume that the sequence of cdfs $\{F(\cdot|t)\}_{t=1}^\infty$ can be embedded in a family of cdfs with <u>continuous</u> parameter t, and that $F(x|t)$ is continuously differentiable with respect to t (as well as x). (For example, if $F(\cdot)$ is a Gamma distribution with scale parameter μ, and shape parameter α, $F(\cdot|t)$ is Gamma with scale parameter $t\mu$ and scale parameter α.) Likewise, assume $h(L)$ is continuously differentiable in L. Applying the implicit function theorem to (18.2) to obtain $\frac{\partial S^*}{\partial L}$, one easily verifies:

Proposition 18.1 *Under the cost minimization problem (18.1), delayed product differentiation always results in an increase of system-wide costs.*

The above analysis ignores the fact that, as the point of differentiation is postponed, the customization lead time (l) is reduced, resulting in a reduction of the customers' response time. A more appropriate model arises, therefore, when minimizing expected holding costs subject to a bound on the expected response time $E[Y]$. Fixing the total lead time τ in the system, note that $Y = \tau - L + W$, where W is the waiting time if the inventory is insufficient to satisfy the customer orders. The distribution of W was derived by Chen and Zheng (1992): fix $x > 0$ and let W_t denote the waiting time experienced by customers arriving in period t, $IL(t+x)$ the inventory level at time $(t+x)$ and $D(t, s]$ the aggregate demand in periods $t + 1, \ldots, s$. Then,

$$
\begin{aligned}
Pr(W_t > x) &= Pr(IL(t + x) < -D(t, t + x]) \\
&= Pr(S - D(t + x - L, t + x] < -D(t, t + x]) \\
&= Pr(D(t + x - L, t] > S) = 1 - F(S|L - x). \quad (18.3)
\end{aligned}
$$

Thus, minimizing expected holding costs subject to the constraint $E[Y] \leq Y^0$ reduces to:

$$
(P) \qquad C^* = \min \; h(L) \int_0^S F(u|L)\,du \qquad (18.4)
$$

$$
\text{s.t.} \quad \tau - \sum_{r=1}^L F(S|r) \leq Y^0 \qquad (18.5)
$$

The optimal order-up-to level S^* satisfies (18.5) as an equality. Once again, embedding the cdfs $F(\cdot|t)$ in a family with continuous parameter t, and replacing the summation in (18.5) by an integral, we obtain:

$$
\int_0^L F(S^*|w)\,dw = \tau - Y^0 \qquad (18.6)
$$

It follows from the implicit function theorem, that $\frac{\partial S^*}{\partial L} \leq 0$ while $\frac{\partial C^*}{\partial L} = h'(L) \int_0^{S^*} F(u)\,du + h(L)F(S^*|L)\frac{\partial S^*}{\partial L}$. We conclude

Proposition 18.2 *Under model (P), delayed product differentiation always reduces system-wide costs, if $h(L)$ is constant or $h'(L)$ sufficiently close to zero.*

On the other hand, delayed product differentiation may increase costs if the holding cost rate increases sufficiently fast with L. Most generally, there may be an optimal value for L in the interior of the total lead time interval $[0, \tau]$. The above analysis may be extended to allow for stochastic lead times. We note also that delayed differentiation is never beneficial if the constraint $E[Y] \leq Y^0$ is replaced by a constraint of the type $Pr(Y \leq Y^0) \leq \alpha$. (In the latter case, $\frac{\partial S^*}{\partial L} = 0$.)

We now turn to a more refined model, due to Lee and Tang (1997), in which each of the two phases is modeled as consisting of a number of discrete operations and corresponding stages, with inventories kept at each of them. Thus, assume the total manufacturing/distribution lead time consists of τ stages, with stage k the last common operation. Let $n_i(k)$ denote the lead time of operation i, $p_i(k)$ the processing cost rate at stage i and $h_i(k)$ the inventory carrying cost for every unit kept at or in process towards stage i, all with k

the last common operation. Finally, S_i denotes the amortized investment cost (per period) if operation i is turned into a common operation. Lee and Tang demonstrate how different types of delayed product differentiation (see (I)-(IV) in section 18.2) can be modeled by appropriate choices of the $n_i(\cdot)$, $p_i(\cdot)$ and $h_i(\cdot)$ functions.

Let d_{jt} denote the demand for item j in period t ($j = 1, \ldots, J; t = 1, 2, \ldots$). For each $j = 1, \ldots, J$ the variables $\{d_{jt} : t = 1, 2, \ldots\}$ are assumed to be i.i.d. with mean μ_j and standard deviation σ_j. $D_t = \sum_j d_{jt}$ denotes the aggregate demand in period t and has mean $\mu = \sum_j \mu_j$ and standard deviation $\sigma \leq \sum_j \sigma_j$. (For example, when demands are independent across items, $\sigma = \sqrt{\sum_j \sigma_j^2}$.)

The system may be represented as a network comprised of $k + J(\tau - k)$ inventory nodes. Nodes $i = 1, \ldots, k$ represent the common, undifferentiated product at the completion of operation i; nodes (i, j) with $i = k+1, \ldots, \tau$ and $j = 1, \ldots, J$ represent item j at the completion of operation i, see Figure 18.2. The model further assumes that each node in the network is governed by a simple order-up-to policy . The sequence of events is as follows: at the beginning of period t, the demands for the end items $\{d_{jt}\}$ get realized. These demands immediately trigger orders to inventory nodes $\{(\tau - 1, j) : j = 1, \ldots, J\}$, which in turn trigger orders to the next echelon of nodes $\{(\tau - 2, j) : j = 1, \ldots, J\}$, et cetera. Since each inventory node is governed by an order-up-to policy (with respect to the order inventory position), each node (i, j) with $i = k+1, \ldots, \tau$ and $j = 1, \ldots, J$ experiences a demand of size d_{jt} in period t, and all nodes at the stages before the point of product differentiation, i.e., nodes $i = 1, \ldots, k$ experience a demand of size D_t.

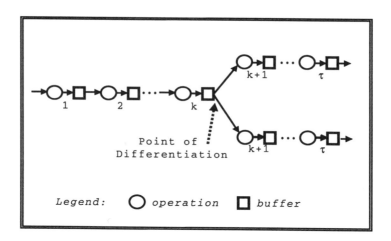

Figure 18.2 Two-stage inventory systems with discrete operations in each stage

In selecting the order-up-to levels for the different inventory nodes, Lee and Tang (1997) adopt the following heuristic: the order-up-to level is set at the mean of lead time demand plus a safety stock, given by a predetermined number (z) of standard deviations of leadtime demand. Here, the leadtime is taken as the normal leadtime, ignoring waiting time because of stockouts at the supplying stage. Finally, in evaluating the cost performance of this strategy, the same assumption of zero waiting times because of stockouts is made. Under this assumption, each of the inventory nodes can be evaluated independently as a single location inventory system. The resulting system-wide cost function is

$$
\begin{aligned}
C^* \;=\; & \sum_{i=1}^{k} S_i + \mu \sum_{j=1}^{J} p_i(k) + \mu \sum_{i=k+1}^{\tau} h_i(k) n_i(k) \\
+ \; & \sum_{i=1}^{k} h_i(k) [\tfrac{1}{2}\mu + z\sigma\sqrt{n_i(k)+1}] \\
+ \; & \sum_{i=k+1}^{\tau} h_i(k) [\tfrac{1}{2}\mu + z(\sum_j \sigma_j)\sqrt{n_i(k)+1}] \qquad (18.7)
\end{aligned}
$$

This cost model may present us with a true trade-off in selecting the optimal point of differentiation. The last two terms in (18.6), representing carrying costs for inventories held at the different stages, are clearly minimized by increasing k (since $\sigma \le \sum_j \sigma_j$) provided $h_i(k)$ does not increase too fast with k. On the other hand, the first term may offset the cost savings due to delayed differentiation, and the same may apply to the second and third terms.

We conclude with a specific example from Lee and Tang (1997), i.e., the above example of standardizing one or several stages in the manufacturing processes of mono and color printers. The process consists of three stages: printed circuit board assembly (PCA), final assembly and testing (FA&T) and final customization. Thus, $\tau = 3$ and $k = 0, 1, 2$ are possible points of differentiation. The choice $k = 1$ and $k = 2$ require standardizing either PCA by itself, or PCA and FA&T together. Both standardizations involve significant investments, more expensive (common) components and additional processing cost rates. This can be modeled by specifying: $S_1, S_2 > 0$, $p_i(k) = p_i + \beta$, if $i \le k$ and $p_i(k) = p_i$ if $i > k$; and $h_i(k) = h_i + \delta_1 + \delta_2 + \cdots + \delta_{\min(i,k)}$. The δ-components reflect additional value added at each standardized stage due to additional processing costs and more expensive components. As the authors explain, the optimal point of differentiation depends on the relative magnitude of various cost components.

We refer to Chapter 19 for a discussion of stochastic programming models demonstrating the benefits of production and storage of general use intermediate products (*vanilla boxes*), see also Swaminathan and Tayur (1996) and the references therein.

18.4   MULTI-ITEM SYSTEMS

In this section we reconsider the two-stage production system of the previous section (Figure 18.1), now assuming that inventories are kept of all final items, possibly in combination with inventories for the intermediate product or blank. Moreover, to incorporate the impact of capacity restrictions, we impose a capacity limit on each period's order for the intermediate product. In the eighties, several papers developed analytical models to explain and quantify the operational benefits of delayed differentiation, see Eppen and Schrage (1981), Federgruen and Zipkin (1984a,b,c) and Schwarz (1989). These models, in the general tradition of inventory theory, assume that while demands in each period are random, they are <u>independent</u> across time and their distribution is <u>perfectly</u> known, i.e., sales forecasts do not need to be updated as time progresses. Under this assumption, the benefits of delayed differentiation, in terms of inventory related performance measures, are restricted to <u>two</u> factors:

(i) (*Statistical economies of scale*) Assume that τ denotes the total production or replenishment leadtime. Product differentiation occurs after a first stage of $L \leq \tau$ periods which terminates with a common intermediate product. The delayed differentiation permits one to specify only *aggregate* orders at the beginning of the replenishment process and to commit resources to individual products only at the end of the first stage of L periods. It also allows one to observe the inventory status of the products at the end of the first stage and thus to make better informed allocations to the individual products. Eppen and Schrage (1981) coined the phrase "statistical economies of scale" for this effect. Others refer to it as *"risk pooling"*.

(ii) (*Risk pooling via a common buffer*) To the extent inventories of the common intermediate product are stocked, these may serve as a <u>common</u> buffer from which all products can draw relatively quickly in case of need. The common buffer further reduces the magnitude of system-wide safety stocks to guarantee given service levels.

The key assumption of demand distributions being perfectly known from the outset is, in general, rather restrictive, with few exceptions such as staple goods facing mature markets. Many products face a short life cycle or are subject to dynamic and competitive market forces. Thus, even the most basic characteristics of the demand distributions (e.g., their means) may not be known with sufficient accuracy. However, estimates can be significantly improved on the basis of observed sales data. The fashion and high technology industries represent extreme examples of this phenomenon. Fisher and Raman (1996) document dramatic improvement in forecast accuracy which can be achieved after observing only 20% of initial sales in a sales season for fashion items. Delayed differentiation allows one to use observed sales data during the first common phase (of L periods), not just to get updated information about the products' inventory status (see (i) above), but also to exploit a third factor, i.e.:

(iii) (*The learning effect*) The generation of significantly more accurate forecasts of future demand distributions, further improving allocation to the individual products at the completion of the first phase. The learning effect arises even in settings where the demand distributions are known with accuracy but consecutive demands are correlated. Here too, observed sales data during the first common phase can be exploited to revise forecasts for future demands affecting the allocations to individual products. Clearly, both types of learning effects often prevail simultaneously and their benefits compound on each other.

18.4.1 The multi-item model

We initially assume that no inventories of blanks are maintained. Assessing the benefits of delayed differentiation under the restriction of zero intermediate stock, results in a <u>lower</u> bound on the full benefits achievable. In §18.4.2 we extend our discussion to the case in which intermediate stock is allowed.

Unsatisfied demand is backlogged. Production costs in both phases are proportional with the production volumes. All other cost components incurred for a product (in particular holding and shortage costs) can be expressed as a function of this product's inventory position (= inventory on hand + blanks being transformed into units of this final product − backlogs). The capacity limit, all cost parameters and functions, and the demand distributions follow a periodic pattern, with periodicity K. The objective is to minimize expected system-wide discounted costs over a finite or infinite horizon or their long-run average value.

At the beginning of each period $n = 1, 2, \ldots$ a decision is made whether to order a batch of blanks, and if so, of what size. When the batch is completed at the beginning of period $n + L$, it is allocated to the J final products; the blanks allocated to product j $(j = 1, \ldots, J)$ become available as units of the final product j at the beginning of period $n + L + l_j$, possibly after a second manufacturing stage. Demands in the k-th period of any cycle of K periods (alternatively called "a period of type k") are identically distributed as the vector $d^k = (d_1^k, \ldots, d_J^k)$. Let:

x_j = the inventory position of item j at the beginning of a period, <u>before</u> allocation of this period's production batch of blanks,

y_j = the inventory position of item j at the beginning of a period, <u>after</u> allocation of this period's production batch of blanks,

w = the size of the batch of blanks ordered at the beginning of a period.

w^r = the size of the batch of blanks ordered r periods before the beginning of a given period $(r = 1, \ldots, L)$,

b^k = the capacity in periods of type k $(k = 0, \ldots, K - 1)$,

γ^k = the variable (first stage) production cost rate for blanks, in periods of type k $(k = 0, \ldots, K - 1)$,

c_j^k = the variable second stage production cost rate for item j, in periods of type k $(j = 1, \ldots, J; \ k = 0, \ldots, K - 1)$.

Future costs are discounted with a factor $\alpha \leq 1$. We assume that the expected value of all other cost components for item j $(j = 1, \ldots, J)$ that are charged to a period of type k $(k = 0, \ldots, K - 1)$ can be expressed as a function $G_j^k(y_j)$. Assume, for example, that the carrying (backlogging) cost incurred for an inventory (backlog) of x_j^+ (x_j^-) units at the end of a period of type k is given by a function $h_j^k(x_j^+)$ $(p_j^k(x_j^-))$. Using a standard accounting device, we charge to each period the expected discounted holding and backlogging costs incurred one lead-time later, i.e. $G_j^k(y_j) = \tilde{G}_j^k(y_j; l_j)$ where

$$
\begin{aligned}
\tilde{G}_j^k(y_j; l) &= \alpha^l \mathsf{E}\{h_j^{k\oplus l}([y_j - d_j^k - d_j^{k\oplus 1} \cdots - d_j^{k\oplus l}]^+) \\
&+ p_j^{k\oplus l}([d_j^k + d_j^{k\oplus 1} \cdots + d_j^{k\oplus l} - y_j]^+)\},
\end{aligned}
\tag{18.8}
$$

and where $a \oplus b \doteq (a + b) \bmod K$.

We first specify the model for the case where $L = \gamma^k = 0$, i.e. where production occurs in a single stage. The model can be formulated as a Markov Decision Process (MDP) with countable state space $S = \{(x, k) : x \text{ is integer}, k = 1, \ldots, K\}$ and (finite) action sets $A(x, k) = \{y : x \leq y \text{ and } \sum_{j=1}^J y_j \leq \sum_{j=1}^J x_j + b^k\}$. In other words, the state of the system is given by the prevailing vector of inventory positions and the period type k. Let $v_t^*(x, k)$ denote the minimal expected discounted cost over a horizon of t periods when starting in state $(x, k) \in S$. The functions v_t^* satisfy $v_0^* \equiv 0$ and

$$
v_t^*(x, k) = -\sum_{j=1}^J c_j^k x_j + \min_{y \in A(x,k)} H_t(y, k); \quad t \geq 1,
\tag{18.9}
$$

where $H_t(y, k) = \sum_{j=1}^J [c_j^k y_j + G_j^k(y_j)] + \alpha \mathsf{E}[v_{t-1}^*(y - d^k, k \oplus 1)]$.

The recursion in (18.9) can be simplified by the following transformation of functions: let $\bar{v}_t(x, k) \doteq v_t^*(x, k) + \sum_{j=1}^J c_j^k x_j$ and $\bar{G}_j^k(y_j) \doteq G_j^k(y_j) + (c_j^k - \alpha c_j^{k\oplus 1})y_j + \alpha c_j^{k\oplus 1}\mathsf{E}[d_j^{k\oplus 1}]$. By simple algebra, $\bar{v}_t(x, k) = \min_{y \in A(x,k)}\{\sum_{j=1}^J \bar{G}_j^k(y_j) + \alpha \mathsf{E}[\bar{v}_{t-1}(y - d^k, k \oplus 1)]\}$.

Since the state space is of dimension $J + 1$, it is impractical, in general, to compute an optimal policy. However, a lower bound approximation[1] can be obtained for this problem by relaxing the action sets $A(x, k)$ to sets $\tilde{A}(x, k) = \{y : y \text{ is integer and } \sum_{j=1}^J x_j \leq \sum_{j=1}^J y_j \leq \sum_{j=1}^J x_j + b^k\}$; in other words, by replacing the individual lower bounds $y \geq x$ by the aggregate constraint $\sum_{j=1}^J y_j \geq \sum_{j=1}^J x_j$. It can be shown that the value functions in this relaxed model depend on the vector of inventory positions x only via its aggregate sum $X = \sum_{j=1}^J x_j$, and are given by $V_t : \mathbf{Z} \times \{1, \ldots, K\} \to \mathbf{R}$, defined recursively via $V_0 \equiv 0$ and:

$$
V_t(X, k) = \min_{X \leq Y \leq X + b^k}\{R^k(Y) + \alpha \mathsf{E}[V_{t-1}(Y - \sum_{j=1}^J d_j^k, k \oplus 1)]\},
\tag{18.10}
$$

where

$$
R^k(Y) = \min\{\sum_{j=1}^J \bar{G}_j^k(y_j) : \sum_{j=1}^J y_j = Y\},
\tag{18.11}
$$

We observe that $V_t(X, k)$ represents the minimum expected cost over a horizon of t periods in a *single-item* capacitated model with periodic parameters[2].

It has $D^k = \sum_{j=1}^{J} d_j^k$, $R^k(\cdot)$ and b^k as the demand, one-step expected cost function and capacity limit in periods of type k. Under mild convexity and regularity assumptions, Aviv and Federgruen (1997b) show that a modified base-stock policy is optimal for the relaxed single-item model (possibly with a different base-stock level for each period type $k = 0, \ldots, K - 1$) for finite and infinite horizons and whether considering total discounted- or average-costs. Furthermore, in each case, the optimal policy can be computed via a value-iteration scheme.

When $L > 0$ (i.e. when production occurs in <u>two</u> phases), it is easily verified that the single-item, single-stage lower-bound model, described by (18.10), continues to apply, merely replacing $R^k(\cdot)$ by $\tilde{R}^k(Y) \doteq \mathsf{E}\{R^k[Y - (D^k + D^{k+1} + \cdots + D^{k+L})]\}$; Y now denotes the system-wide echelon inventory position of blanks, i.e., all blanks being manufactured, transformed into final products or as part of final products' inventories. The approximate lower bound model suggests several heuristics strategies. See Aviv and Federgruen (1997b) for a detailed description.

Analysis of the approximate model reveals that delayed product differentiation, i.e., an increase of L for a fixed total leadtime τ, is <u>always</u> beneficial, as are <u>reductions</u> in the individual leadtimes of the two phases, L and l_j's. Recall that the same qualitative conclusions could be drawn in some but not all of the models discussed in Section 18.3. The extent of the benefits of delayed product differentiation depends on the number of end-items, the pattern of correlations between the items' demands among other factors. We defer further discussion to §18.4.4.

Aviv and Federgruen (1997b) conducted an extensive numerical study, exhibiting very small gaps (on average less than 0.45%) between the cost values of the proposed strategies and the lower bound. This implies that the lower bound is accurate and that the proposed heuristic strategies are close-to-optimal. As part of their experimental study, Aviv and Federgruen (1997b) investigated a set of instances generated from a well-known case study of Hewlett Packard's European distribution process of deskjet printers, see e.g. Flaherty (1996) or Lee et al. (1993). Their analysis shows that capacity constraints, as well as seasonality factors have a significant impact on both the cost estimates and the optimal strategies.

An important variant of this model arises when fixed setup costs are incurred whenever a new production batch of blanks is initiated. In the absence of capacity limits, a similar analysis results again in a lower bound approximate model which may be interpreted as the dynamic program of a single-item inventory system, see Federgruen and Zipkin (1984a,c). The optimal ordering policy, in this case, is of the (s, S)-type, i.e., the inventory position is increased to a level S whenever it is found to be at or below s. (The parameters s and S may vary by period.)

18.4.2 Systems with intermediate inventory

In this section, we give a brief outline of how the above results can be extended to allow for intermediate inventories. An exact representation of the problem requires a dynamic program with a $(L + J + 1)$-dimensional state space. In addition to the vector x of the final items' inventory positions, the state description includes the L-vector of outstanding orders for blanks $(w^1, \ldots, w^L)$ as well as

$$
\begin{aligned}
v^d &= \text{the echelon inventory of blanks before arrival of an order} \\
&= \text{inventory of blanks} + X.
\end{aligned}
$$

For the sake of brevity, we confine ourselves to the finished goods' holding and backlogging cost structure used in (18.8). Let h_0^k denote the cost of carrying a blank unit in inventory in a period of type k, and reinterpret h_j^k as the incremental cost of holding a unit of item j in inventory during a period of type k (beyond h_0^k). Without loss of generality, $h_j^k \geq 0$ for all j, k. We account for all expected inventory carrying costs, by charging in periods of type k, the basic rate h_0^k for each unit in the echelon inventory and for all $j = 1, \ldots, J$, $\alpha^l h_j^{k+l}$, the incremental holding cost for every expected unit of inventory of finished good j, l periods later. The actions at the start of every period consist of (i) $w = $ the number of new blanks to be ordered; (ii) $z = $ the amount to be withdrawn from the inventory of blanks to start the second manufacturing stage , and (iii) the vector $(z_1, \ldots, z_J)$ with z_j the part of z allocated to final item j.

This intractable dynamic program can again be closely approximated, this time by a pair of interdependent dynamic programs for single item inventory systems. The approximation starts, again, with a relaxation of the nonnegativity constraints $z_j \geq 0$. Under this relaxation, the current choice of $(z_1, \ldots, z_J)$ in no way affects the future achievable values of the items' inventory positions, or their distributions. We may thus choose the vector $(z_1, \ldots, z_J)$ to minimize immediate costs only, i.e., to solve problem (18.11) in periods of type k.

We refer to the pair of single-dimensional dynamic programs as the "finished goods model" and the "blanks model". The finished goods model is defined by the following recursion, which resembles (18.10):

$$
V_t(X, k) = \min_{z \geq 0} \{R^k(X + z) + \alpha \mathsf{E}[V_{t-1}(X + z - \sum_{j=1}^{J} d_j^k, k \oplus 1)]\}. \quad (18.12)
$$

If $R^k(\cdot)$ is convex, and orders in this model unbounded, it follows from Zipkin (1989) that a simple (periodic) base-stock policy solves the problem and that the function $V_t(\cdot, k)$ is convex and has a finite minimizer $X_{t,k}^*$, for all t and k. The optimal base-stock policy in (18.12) fails, in and of itself, to be an optimal allocation policy, simply because its recommended withdrawal quantity may be infeasible given a limited inventory of blanks. However, similar to the Clark and Scarf (1960) -model, the functions $V_t(\cdot, k)$ are used to arrive at an optimal strategy for the lower bound model.

It is easily verified that if the available echelon inventory Y, at the beginning of period t, is in excess of $X_{t,k}^*$, the optimal withdrawal quantity z^* in (18.12) is in fact feasible (and optimal). If $Y < X_{t,k}^*$, the withdrawal quantity is limited to $(Y - X_{t,k})$ and the expected finished goods-related costs correspondingly higher. We thus define for all t and k, induced penalty functions $P_t(Y, k)$ by

$$
P_t(Y, k) = \begin{cases} 0 & \text{if } Y \geq X_{t,k}^* \\ R^k(Y) - R^k(X_{t,k}^*) + \alpha E[V_{t-1}(Y - \sum_{j=1}^{J} d_j^k, k \oplus 1)] & \\ \quad -\alpha E[V_{t-1}(X_{t,k}^* - \sum_{j=1}^{J} d_j^k, k \oplus 1)] & \text{if } Y < X_{t,k}^* \end{cases}
$$

which are easily verified to be convex and non-negative. These functions are used to specify the second dynamic problem, i.e., the "blanks model", which is defined by the recursion:

$$
W_t(w^1, \ldots, w^L; v^d, k) = \min_{0 \leq w \leq b^k} \{\gamma^k w + h_0^k(v^d + w^L) + P_t(v^d + w^L, k)
$$
$$
+ \quad \alpha E[W_{t-1}(w, w^1, \ldots, w^{L-1}; v^d + w^L - \sum_j d_j^k, k \oplus 1)]\}. \quad (18.13)
$$

It can be shown that:

(a) Except for certain constant terms, and given the above cost accounting schemes, the expected total cost with t periods to go and given the current period is of type k, is represented by $V_t(X, k) + W_t(w^1, \ldots, w^L; v^d, k)$.

(b) Under certain (mild) convexity and regularity assumptions, a modified base-stock policy with period-dependent base-stock levels optimizes (18.12) and can be found by a simple value-iteration method. This policy is an optimal order policy for the lower bound model. For this lower bound model, it is further optimal to set the withdrawal quantity as close as feasible to $\{X_{t,k}^*\}$ and to allocate it so as to solve the optimization problem in (18.11).

The lower bound cost model thus provides an easily computable cost estimate for the system. The policy optimizing the model can again be transformed into an effective heuristic strategy for the original system, see Aviv and Federgruen (1997b). As with the model in §18.4.1, fixed costs for initiating the production of a batch of blanks may be added, in the absence of capacity constraints. The optimal policy for the "blank model" is then of an (s, S)-type, see Federgruen and Zipkin (1984b).

18.4.3 The learning effect in design for postponement

As mentioned in the introduction of this section, the key assumption of demand distributions being perfectly known from the outset is, in general, rather restrictive. In order to characterize the benefits of delayed differentiation in the presence of imperfect knowledge of the demand distributions, we make the following modification to the problem defined in §18.4.1: the planning horizon consists of N ($\leq \infty$) periods. For all $j = 1, \ldots, J$, we assume that the d_{jn} variables are Normally distributed with mean μ_j ($j = 1, \ldots, J$) and standard deviation σ_j ($j = 1, \ldots, J$), for all $n = 1, \ldots, N$; i.e., the distributions are identical across time. However, contrary to standard inventory models, we assume that while the parameters $\{\sigma_j : j = 1, \ldots, J\}$ are known, the means $\{\mu_j : j = 1, \ldots, J\}$ are not. More specifically,

$$d_{jn} = \mu_j + \epsilon_{jn}, \qquad n = 1, 2, \ldots, \quad \text{and} \quad j = 1, \ldots, J \qquad (18.14)$$

where the ϵ_{jn}-variables have a known Normal distribution, and our initial uncertainty about the mean demand μ_j is characterized by a prior distribution, itself assumed to be Normal with mean μ_{0j} and standard deviation $\sigma_{0j} = \eta \sigma_j$ for some $\eta \geq 0$ ($j = 1, \ldots, J$). The assumed distributional form gives rise to a so-called conjugate pair. Posterior distributions for the demands in any given period, conditioned upon observed demands, are again shown to be Normal. In particular, we assume that estimates of the parameters of the demand distributions are revised on the basis of observed sales data in a Bayesian framework.

Note, that while the conditional distributions $(d_{jn}|\mu_j)$ are independent across time, the unconditional distributions are not. In other words, uncertainty about some of the parameters in the distributions generates correlation across time.

It is well known that the mean and standard deviation of the effective lead-time demands (EM, ESD) are the prime cost determinants for single item inventory models. For example, under perfect knowledge regarding the demand distributions, the minimum achievable long-run average cost is directly proportional with this effective standard deviation, when the ordering cost function is linear and the demand distributions Normal, see Scarf (1958) and Gallego and Moon (1993). Similarly, when the ordering cost functions contain fixed components and (s, S)-type policies are optimal, it is known from Ehrhardt (1979) that the optimal safety stock is roughly proportional to $\text{ESD}^{1.706}$. In our approximate model, EM and ESD equal

$$\text{EM} = (L + l + 1) \frac{n\eta^2 \bar{D}_n + \sum_j \mu_{0j}}{n\eta^2 + 1}, \qquad \text{and}$$

$$\text{ESD}^2 = \left(\frac{(l+1)^2 \eta^2}{(n+L)\eta^2 + 1} + (l+1) \right) \cdot \left(\sum_{j=1}^{J} \sigma_j \right)^2$$

$$+ \left[\frac{(l+1)\eta^2}{(n+L)\eta^2 + 1} + 1 \right]^2 \left(\frac{L^2 \eta^2}{n\eta^2 + 1} + L \right) \cdot \left(\sum_{j=1}^{J} \sigma_j^2 \right) \quad (18.15)$$

We now investigate how the effective standard deviation depends on (i) the relative leadtimes l and L, in particular how it changes when the point of differentiation is postponed; (ii) the initial degree of uncertainty regarding the mean demands which is characterized by η; (iii) n, the number of periods in which demands have been observed; (iv) the degree of demand dispersion over a variety of finished items, characterized by J, and (v) the standard deviations of the one-period demands, as characterized by $\{\sigma_j\}$. We first recall from Aviv and Federgruen (1997c):

Corollary 18.1 *ESD is increasing in l, η and in each of the standard deviations σ_j, $(j = 1, \ldots, J)$. ESD is decreasing in n.*

Perhaps surprisingly, ESD may fail to increase when L, the leadtime of the first manufacturing stage increases. Note that the first term in (18.14) is decreasing in L and this term may dominate, in particular when $(\sum_j \sigma_j)^2$ is much larger than $(\sum_j \sigma_j^2)$. (For example, when all $\sigma_j = \sigma$, $(\sum_j \sigma_j)^2 = J(\sum_j \sigma_j^2)$ so that the first term may dominate the second term in (18.14) by an arbitrarily large factor as $J \to \infty$.) Intuitively, this phenomenon may arise because an extension of L, the first manufacturing stage of the common product, permits one to observe additional demand values prior to making detailed allocations to the individual final products. The potential for this phenomenon decreases as time progresses and clearly increases with η, the degree of uncertainty surrounding the mean demand values, vanishing in the traditional model where $\eta = 0$.

Observe also that ESD grows linearly with the manufacturing leadtime τ. This phenomenon is in sharp contrast to standard models where ESD grows as the square root of the leadtime when, as in our case, the period-by-period deviations from the means are independent of each other.

To characterize the benefits of delayed differentiation, let $\tau \doteq L + l$, be the total manufacturing leadtime. Note that the mean of the effective leadtime demand in the approximate model only depends on the total manufacturing leadtime τ, i.e., it is independent of the point of differentiation. The square of the effective standard deviation, as a function of τ and L equals

$$\text{ESD}^2 = \frac{[(n + \tau + 1)\eta^2 + 1]^2}{\eta^2[(n + L)\eta^2 + 1]} \left[(\sum_{j=1}^{J} \sigma_j)^2 - (\sum_{j=1}^{J} \sigma_j^2) \right]$$
$$- \frac{(n + \tau + 1)\eta^2 + 1}{\eta^2} \left[(\sum_{j=1}^{J} \sigma_j)^2 - \frac{(n + \tau + 1)\eta^2 + 1}{n\eta^2 + 1} (\sum_{j=1}^{J} \sigma_j^2) \right]. \quad (18.16)$$

Note that, given a fixed value of τ, the second term in (18.16) is independent of L. Aviv and Federgruen (1997c) conclude:

Corollary 18.2 *Assume that $\tau \doteq L + l$ is fixed. ESD is monotonically decreasing with L. The magnitude of this decrease is by itself a decreasing function of the number of periods of observed demands, n, and an increasing function of η.*

In other words, Corollary 18.2 shows that delayed differentiation always reduces ESD and that the magnitude of the reduction is especially large at the

572

beginning of the planning horizon, and monotonically increasing with the degree of uncertainty in our a priori knowledge regarding the mean demand values. Thus, the more differentiation is postponed, the more we can benefit from statistical economies of scale as well as the pooling benefits associated with the learning effect; moreover, the magnitude of this benefit of postponed differentiation is all the larger as we face increased uncertainty about the mean demand values $\{\mu_j\}$, either because of limited historical data (n) or because of large initial uncertainty in our a priori assessments of these means (η). Indeed, when $\eta \to 0$ or $n \to \infty$, the benefit associated with the learning effect disappears and the benefit of a postponement of the point of differentiation by a single period is given by $(\sum_j \sigma_j)^2 - (\sum_j \sigma_j^2) > 0$, as in the model with perfect knowledge, see Federgruen and Zipkin (1984a).

We demonstrate the dependence of ESD on its determining factors, via a series of figures. All consider instances with $\sigma_j = \sigma$ for all $j = 1, \ldots, J$.

Figure 18.3 shows how ESD decreases as time progresses and more and more demand realization are observed; five curves are displayed for different values of η, i.e., representing different degrees of uncertainty about the mean demands. Observe that the magnitude of the reduction of ESD greatly varies with η: when $\sigma_0 = \sigma$, i.e., when the degree of uncertainty surrounding the mean demands is identical to that of the period-by-period deviations from the mean, ESD decreases by 30.6% after 10 observations, compared to its value after the first period.

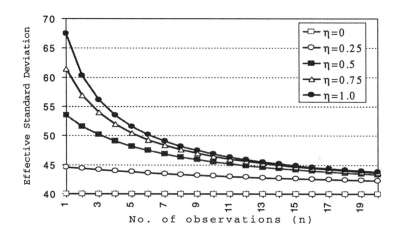

Figure 18.3 Effective Standard Deviation (ESD) as a function of the number periods of observed demands n. ($J = 2$ identical items, $L = l = 2$ and $\sigma_j = 10$.)

Figures 18.4 and 18.5 exhibit the dependence of ESD on the two leadtimes, L and l; each figure consists of 4 curves, for different values of J ($J = 2, 5, 10$ and 100). We express ESD as a percentage of its value in the (base) case where $l = L = 2$. As Corollary 18.1 indicates, and Figure 18.4 confirms, ESD is always increasing in l and the rate of increase is increasing in the number of final products J. Figure 18.5 shows, as discussed above, that the dependency on L is more complex. For $J = 2$ items, in this example, an increase in the first manufacturing leadtime consistently results in an increase of ESD. On the other hand, with $J = 100$ final items, ESD consistently decreases with L over the range $L = 0, 1, \ldots, 6$, where for the two intermediate cases, with $J = 5$ and $J = 10$ items, ESD first decreases and then increases.

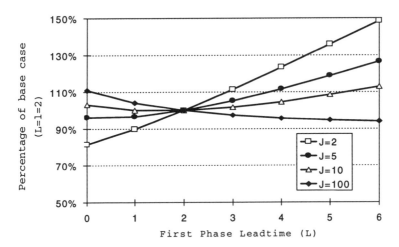

Figure 18.4 ESD, as a percentage of base case ($L = l = 2$), as a function of the first phase leadtime L. ($n = 2$, $\eta = 1$ and $\sigma_j = 10$.)

Figure 18.6 illustrates Corollary 18.2. For a fixed value of $\tau = 4$, we display the reduction in the ESD value when postponing the point of differentiation from $L = 0$ to 4. Once again, we display 5 curves for 5 values of η. Postponement is always beneficial (see Corollary 18.2) but the benefit of a 50% postponement ($L = l = 2$) is 2.7 times larger when $\eta = 1$ as when $\eta = 0$ (i.e., when the mean demands are perfectly known) and the benefit of <u>maximal</u> postponement ($L = 4$) is 1.9 times as large, comparing the same pair of η-values.

Aviv and Federgruen (1997c) show that the learning effect always results in <u>increased</u> benefits of delayed differentiation and leadtime reductions (through Quick Response programs), and that the incremental benefits can be very significant, indeed. In addition to analyzing the benefits of delayed differentiation, Aviv and Federgruen (1997c) characterize the structure of optimal ordering policies (for the common intermediate product) in the approximate single-item model (similar to those described in §18.4.1 and §18.4.2). As in standard inventory models, the structure depends heavily on the form of the order cost functions, but now, the optimal order-up-to levels or (s, S)-parameters depend on the number of sales observations and the observed mean aggregate sales value (a sufficient statistic of the observed sales history).

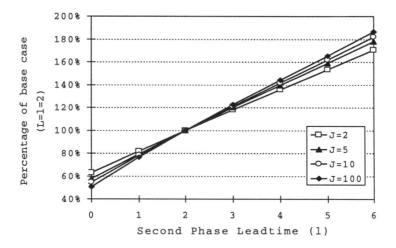

Figure 18.5 ESD, as a percentage of base case ($L = l = 2$), as a function of the second phase leadtime l. ($n = 2$, $\eta = 1$ and $\sigma_j = 10$.)

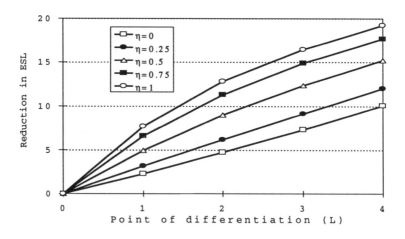

Figure 18.6 Reduction in ESD as a function of the point of differentiation L. ($J = 2$ identical items, $n = 2$, $\tau = L + l = 4$ and $\sigma_j = 10$.)

18.4.4 Inter-temporal correlation

The case where demand distributions are unknown and a Bayesian approach is taken represents one setting in which consecutive demands are correlated. Intertemporal correlation may however arise even when the joint distribution of demands is perfectly known, e.g., when demands follow an autoregressive time-series model, a common assumption in many statistical forecasting systems (e.g., exponential smoothing, Box-Jenkins processes)[3].

To investigate the impact of intertemporal correlations, Aviv and Federgruen (1997c) first consider settings where the period-by-period variations from the means (i.e., the $\{\epsilon_{jn}\}$-variables) follow an autoregressive time series model of first order. More specifically,

(i) $\epsilon_{j0} \sim N(0, \sigma_j)$,

(ii) $\epsilon_{jn} = \theta\epsilon_{j,n-1} + \delta_{jn}$, $n = 1, 2, \ldots$, with $\{\delta_{jn}\}_{n=1}^{\infty}$ i.i.d. and independent of ϵ_{j0}; $\theta < 1$ and

(iii) $\delta_{j,n} \sim N(0, \sqrt{1 - \theta^2}\sigma_j)$.

As with the model in §18.4.3, the ϵ_{jn} variables are independent across items and their distributions have <u>known</u> parameters σ_j $(j = 1, \ldots, J)$ and θ. Assuming perfect knowledge regarding the demand distributions, Erkip *et al.* (1990) address a (slight) variant of this first order autoregressive pattern, in which demands for the individual items are equal to a fixed deterministic percentage of their aggregate. These authors restrict themselves upfront to base-stock policies, combined with myopic allocations, and a <u>fixed</u> base-stock level. Aviv and Federgruen (1997c) show that under linear or convex order costs, the order policies, in the approximate model, should be selected from the class of base-stock policies with base-stock levels dependent on the last observed demand. In their setting, the base-stock levels should be dependent on <u>this</u> as well as a <u>second</u> sufficient statistic of the observed history of demands.

18.5 BENEFITS DUE TO CAPACITY SHARING

Additional benefits of delayed product differentiation, beyond statistical economies of scale, risk pooling via common buffers and the "learning effect" (cf. Section 18.4) may arise due to the ability to pool <u>capacities</u>. For example, if initially a subassembly or operation, used in the manufacturing process of a family of J end-items, was differentiated requiring a dedicated capacity of b_j units for item $j = 1, \ldots, J$, standardization may allow for a single capacity pool of size $B \le \sum_j b_j$. The models in Section 18.4 are well suited to quantify the benefits of *capacity pooling*. They can be used to compute the operational savings or service improvements if $B = \sum_j b_j$, i.e., pooling does not result in a capacity reduction. Alternatively, given targets for operational costs and service levels, the models may be used to assess what value of $B \le \sum_j b_j$ is required, i.e., how much capacity can be saved. The benefits can indeed be substantial, in particular under high utilization rates and significant uncertainties, see Aviv and Federgruen (1997b).

Gavirneni and Tayur (1997) consider a decentralized system with 2 end-items each of which is governed by an (s, S) policy and supplied by an outside manufacturer. The model focuses on the cost performance of the manufacturer under linear holding and backlogging costs. If the 2 items become differentiated at the manufacturer, each is produced in a separate facility (or production line), with b_1 and b_2, the capacity per period. If differentiation can be postponed until the distribution stage, a single capacity pool of size $B = b_1 + b_2$ can be used. The model assesses the manufacturer's cost performance under alternative information structures: (i) the manufacturer knows the order policies of the two items and is instantaneously informed about each period's demands; (ii) the manufacturer only knows the ordering policies, and (iii) the manufacturer only observes his orders without knowing by what policies they are triggered.

Without standardization, it is optimal for the manufacturer to adopt a separate modified order-up-to policy for each of the two items. The order-up-to level varies with the cumulative amount of demand [number of periods elapsed] since the last order, under information base (i) [(ii)], and is constant under (iii). Under standardization, a single modified order-up-to policy suffices for the unique common product, where the order-up-to level now depends on the pair of cumulative demands [number of periods elapsed] since the last pair of orders, under (i) [(ii)], and is again constant under (iii). We refer to Chapter 2 for more details.

18.6 MULTIPLE POINTS OF DIFFERENTIATION

In the previous section, we have focused on planning models for processes with a single point of differentiation. In this section, we discuss several extensions suitable for settings where differentiation occurs at multiple points in the production/distribution process.

Reconsider the network in Figure 18.1 and assume now that the manufacturing/distribution process consists of <u>three</u> phases with the items at the bottom echelon of Figure 18.1 being further differentiated in a third and final stage. This gives rise to an expanded network with three echelons. Let K denote the (maximum) number of variants into which each of the items $j = 1, \ldots, J$ in the second echelon is differentiated. The end-item (j, k) with $j = 1, \ldots, J$ and $k = 1, \ldots, K$ denotes the k-th variant of (intermediate) item j. The models described in Section 18.4 can all be extended to this three echelon network, as long as the last manufacturing operation has the same characteristics as the second phase operation, i.e., no capacity limits or fixed costs prevail for this operation. Let d_{jkt} denote the demand for end-item (j, k) in period t $(j = 1, \ldots, J; k = 1, \ldots, K$ and $t = 1, 2, \ldots)$.

We outline the approach for the most basic of models, i.e., where no inventories are kept of any of the intermediate products at the first or second echelons and where the demands $\{d_{jkt} : t = 1, 2, \ldots\}$ of each of the end-items (j, k) are assumed to be i.i.d. with a cdf $F_{jk}(\cdot)$ with known parameters, in particular with known mean μ_{jk} and standard deviation σ_{jk}. As with the models in Section 18.4, the demands of the different end-items within a given period t may be correlated and follow a general multivariate distribution. Let $\sigma_j \leq \sum_{k=1}^{K} \sigma_{jk}$ denote the standard deviation of the aggregate demand for all items emanating from the intermediate product j $(j = 1, \ldots, J)$ and let $\sigma \leq \sum_{j=1}^{J} \sigma_j$ denote the standard deviation of system-wide aggregate demands in any given period. Finally, let l_{jk} denote the leadtime required to transform the intermediate product j into the final item (j,k).

The approximation method developed in Section 18.4 may be adapted to generate an approximate lower bound model which can be interpreted as the dynamic program of a single location inventory system. First, by relaxing for all $j = 1, \ldots, J$, non-negativity constraints with respect to the decision variables describing allocations of batches of intermediate product j to each of the end-items $\{(j,k) : k = 1, \ldots, K\}$, one can show that the exact dynamic programming model can be replaced by a two-echelon model. In this approximating model, each of the subtree emanating from one of the intermediate products, is replaced by a single node or item. The subtree emanating from intermediate product j is represented by a single item with

$$M_j = \sum_{k=1}^{K} (l_j + l_{jk} + 1)\mu_{jk} \qquad (18.17)$$

as the mean leadtime demand, but with effective standard deviation of leadtime demand (ESD) given by:

$$\text{ESD}_j = [l_j \sigma_j^2 + (\sum_{k=1}^{K} \sqrt{l_{jk} + 1} \sigma_{jk})^2]^{\frac{1}{2}} ; \qquad j = 1, \ldots, J \qquad (18.18)$$

We are now left with a two-echelon model like the one discussed in the previous section. Expressions (18.17) and (18.18) give the first two moments of leadtime demand for the j-th "end-item" in the approximating two-echelon system. If all end-items $\{(j,k) : k = 1, \ldots, K\}$ share the same holding, backlogging and processing cost rates, these apply to the j-th "end-item" in the approximating model as well; otherwise, appropriate weighted averages of these parameters need to be constructed, see Section 18.4. Thus, applying the methods of Section 18.4 to the approximate two-echelon system, we obtain a final approximation in terms of a single item model with mean leadtime demand given by

$$\text{EM} = \sum_{j=1}^{J} \sum_{k=1}^{K} (L + l_j + 1)\mu_{jk} \qquad (18.19)$$

and standard deviation of leadtime demand

$$\text{ESD} = [L\sigma^2 + (\sum_{j=1}^{J} \text{ESD}_j)^2]^{\frac{1}{2}} \qquad (18.20)$$

Substituting (18.18) into (18.20) we conclude that

$$\text{ESD}^2 = L\sigma^2 + (\sum_{j=1}^{J} \sqrt{l_j \sigma_j^2 + (\sum_{k=1}^{K} \sqrt{l_{jk} + 1} \sigma_{jk})^2})^2 \qquad (18.21)$$

This expression again reveals that delayed product differentiation (whether an extension of L, followed by an equal reduction of the leadtime in the second or third phase, or an extension of one of the second phase leadtime l_j, followed by an equal reduction of the leadtimes l_{jk}, $k = 1, \ldots, K$) always results in a reduction of inventory related costs.

The above analysis is based on the last part of Federgruen and Zipkin (1984a). Apparently unaware of the latter, Garg and Tang (1997) developed a heuristic treatment for this three echelon model which results in a different approximation. According to this approximation, the system is equivalent to one in which all items emanating from the intermediate product j are perfectly substitutable so that their inventories can be pooled and only their aggregate demand (across all variance of the intermediate product j) matters. It is well known that this representation results in significant distortions. The above analysis can easily be extended to more complex settings. First, the above described sequential approximation method can be applied to an arbitrary network with an arbitrary number of echelons. Second, similar analyses can be construed for settings where inventories are kept for the common first stage product and for the intermediate second phase products, or where the demands for the end-items are correlated across time, or some of the parameters in their distributions are unknown, and need to be updated by Bayesian analysis.

We now turn to the model by Lee and Tang (1998) which is well suited to analyze the optimal sequence of differentiating operations. Lee and Tang focus on a special case of the three echelon network discussed above. Let $L = 0$, so that the first phase is vacuous and we are left with two operations, the first of which (I) generates J types, differentiating according to attribute A, and the second of which (II) generates K types, differentiating according to attribute B. See Section 18.2 for a discussion of various examples where an optimal sequence of operations may be selected.

As in the model above, Lee and Tang (1998) assume that aggregate demands in consecutive periods are independent as in the model above, and identically distributed with mean μ and standard deviation σ, and that each demanded unit is with probability θ_{jk} for the finished good (j, k), i.e., for the j-th choice of attribute A and the k-th choice of attribute B $(j = 1, \ldots, J, k = 1, \ldots, K)$. The attribute preferences of the customers are independent of each other, and customers are not willing to adopt a "second" choice if their preferred item is out-of-stock. Operations I and II have deterministic leadtimes L_A and L_B. Contrary to the centralized systems described above, Lee and Tang (1998) assume that the two stage system operates as a pull system in which each of the finished goods as well as the intermediate items are governed by an (independent) uncapacitated base-stock policy.

Lee and Tang analyze the sequencing question under the assumption that stockouts of intermediate products are so rare (perhaps because of high service level requirements) as to have negligible impact on the performance of the inventory systems of the final products. Under this assumption, and if the leadtimes for the two operations are identical, the performance of the final products' inventory systems is identical under both sequences of manufacturing operations (I-II or II-I). Thus, only the costs of the inventory systems of the intermediate products depend on this sequence choice.

The authors argue that the total expected cost incurred for the intermediate products is proportional with the sum of the <u>variances</u> of demands in a period, for each of these intermediate products. This results in a number of unintuitive conclusions. For example, in the special case where $J = K = 2$ (i.e., two options for each attribute), the authors show that the sequence A-B is to be preferred over the sequence B-A, if

$$(\mu - \sigma^2)[p(1 - p) - q(1 - q)] < 0 \quad \text{where} \quad p = \theta_{11} \quad \text{and} \quad q = \theta_{21} \quad (18.22)$$

Thus, if the variance-to-mean ratio σ^2/μ is less than one, operation I is to precede operation II if and only if the corresponding attribute A is less variable than attribute B ($|0.5 - p| > |0.5 - q|$), thus confirming our intuition. However, the <u>opposite</u> is true if $\sigma^2 > \mu$. Similarly, in the special case where each of the $N = JK$ final products is equally likely to be chosen, i.e., $\theta_{jk} = \frac{1}{N}$ for all (j, k), operation I is to be sequenced before operation II, if and only if

$$(\mu - \sigma^2)\left(\frac{1}{J} - \frac{1}{K}\right) < 0 \quad (18.23)$$

Once again, we would expect that the attribute with the fewer number of options should be processed first. Our intuition is confirmed by (18.23) but only in case the variance-to-mean ratio of aggregate demands is less than one, while Lee and Tang conclude that the more variable attribute should be processed first if $\sigma^2 > \mu$. The authors state their surprise: *"this is a surprising result. It states that the nature of total demand uncertainties is very critical in determining whether operations reversal is an effective means to to reengineer the supply chain."* In their conclusion section, the authors list five qualitative properties which hold when the total demand for all options is fairly stable (i.e., $\sigma^2 < \mu$), but they conclude that *"when the total demand is highly variable, then, interestingly, the reverse of the above properties is true."*

Kapuscinski and Tayur (1998) critique Lee and Tang, arguing that when $L_A = L_B$, the appropriate measure to be optimized is a weighted average or sum of the <u>standard deviations</u> of the demands for the intermediate products. They demonstrate that under this objective, for the two special cases discussed above (the case $J = K = 2$ and the case where all end-item choices are equally likely), the operation associated with the <u>less variable</u> attribute is <u>always</u> to be scheduled first, regardless of any other parameters.

Staying with the case $L_A = L_B$, Federgruen (1998) gives a precise definition of when attribute A is more <u>variable</u> than attribute B in terms of a general condition for the variety matrix $[\theta_{jk}]$. He then shows that operation I (II) should precede operation II (I) if attribute A (B) is less variable than attribute B (A), irrespective of the entire distribution of aggregate demands in general, and μ and σ in particular. If this general condition for the variety matrix θ fails to be satisfied, i.e., if attribute A is neither more variable than attribute

B, nor attribute B more variable than attribute A, or if the leadtimes of the operations are of unequal length, a different closed form inequality needs to be verified, in which μ and σ, and possibly the leadtimes themselves, may play a decisive role.

To introduce the variability condition, assume that $J = K$. If $J < K$, it is possible to add $(K - J)$ choices for attribute A, each with zero probability for being selected, i.e., $\theta_{jk} = 0$ for all $j = J + 1, \ldots, K$ and $k = 1, \ldots, K$. If $J > K$, $(J - K)$ zero demand options are added to attribute B. Let $\theta_j = \sum_{k=1}^{K} \theta_{jk}$ for all $j = 1, \ldots, J$ and $\theta^k = \sum_{j=1}^{J} \theta_{jk}$ $(k = 1, \ldots, K)$ denote the row and column sums of the Θ-matrix and $\Theta(I) = (\theta_1, \ldots, \theta_n)$ and $\Theta(II) = (\theta^1, \ldots, \theta^n)$ the corresponding vectors. For any pair of n-dimensional vectors $x, y \in R^n$ with $\sum_{i=1}^{n} x_i = \sum_{i=1}^{n} y_i$ we say that x majorizes y, written $x \succeq y$ if all of the partial sums of the ordered components of x are larger than or equal to the corresponding partial sums of the ordered components of y. More formally, let $x_{[i]}$ and $y_{[i]}$ denote the i-th largest component of the vectors x and y, respectively. $x \preceq y$ if $\sum_{i=1}^{k} x_{[i]} \leq \sum_{i=1}^{k} y_{[i]}$ for all $k = 1, \ldots, n - 1$ and $\sum_{i=1}^{n} x_i = \sum_{i=1}^{n} y_i$ Attribute A is more [less] variable than attribute B if

$$\Theta(I) \preceq \Theta(II) \quad [\Theta(I) \succeq \Theta(II)]$$

Theorem 18.1 *Assume the Scarf or Normal approximation is applied to assess the optimal cost value of all inventory systems for the intermediate products, and assume that all intermediate products have identical cost parameters and/or service levels under both sequences of operations. If attribute A is less [more] variable than attribute B, i.e., $\Theta(I) \preceq \Theta(II)$ $[\Theta(I) \succeq \Theta(II)]$, the sequence I-II [II-I] is to be preferred over the sequence II-I [I-II].*

See Federgruen (1998) for a specification of the Scarf and the Normal approximation, and for a discussion of the variability ranking condition and its generalization to an arbitrary number of attributes and corresponding differentiating operations. (With more than two attributes, the condition is especially easy to use and to interpret if attribute choices are made independently of each other.)

Notes

1. See Aviv and Federgruen (1997b). This type of approach is similar to that employed in Eppen and Schrage (1981), Federgruen and Zipkin (1984a,b) and Chen and Zheng (1994) for uncapacitated models with stationary data (but possibly more complex production costs).

2. See, e.g., Aviv and Federgruen (1997a) and Kapuscinski and Tayur (1995)

3. For a variety of such single-item autoregressive models, assuming that ordering and inventory costs are linear, see e.g., Veinott (1965), Johnson and Thompson (1975), Sobel (1988), Miller (1986) and Lovejoy (1990)

References

Alderson, W., "Marketing Efficiency and the Principle of Postponement," *Cost and Profit Outlook*, September (1950) 3.

Aviv, Y. & A. Federgruen, "Stochastic Inventory Models With Limited Production Capacity and Periodically Varying Parameters," *Probability in the Engineering and Informational Science* 11 (1997a) 107-135.

Aviv, Y. & A. Federgruen, "Capacitated Multi-Item Inventory Systems with Random and Seasonally Fluctuating Demands," Working Paper (1997b), *Graduate School of Business, Columbia University, New York, NY*

Aviv, Y. & A. Federgruen, "Design for Postponement: A Comprehensive Characterization of its Benefits Under Unknown Demand Distributions," Working Paper (1997c), *Graduate School of Business, Columbia University, New York, NY*

Bruce, L., "The Bright New Worlds of Benetton," *International Management*, November (1987) 24-35.

Bucklin, L.P., "Postponement, Speculation and the Structure of Distribution Channels," *Journal of Marketing Research*, Volume 2, February (1965) 26-31.

Chen, F. & Y.S. Zheng, "Waiting Time Distribution in (T, S) Inventory Systems," *Operations Research Letters* **12** (1992) 145-151.

Chen, F. & Y.S. Zheng, "Lower Bounds for Multi-Echelon Stochastic Inventory systems," *Management Science* **40** (1994) 1426-1443.

Clark, A.J. & H. Scarf, "Optimal Policies for a Multi-Echelon Inventory Problem," *Management Science* **6** (1960) 475-490.

Cox, R. & C.S. Goodman, "Marketing of Housebuilding Materials," *The Journal of Marketing* **21** (1956) 55-56.

Dapiran, P., "Benetton – Global Logistics in Action," *Asian Pacific International J. Business Logistics*, (1992) 7-11.

Ealey, L. & G. Mercer, "The Showroom as Assembly Line," *The McKinsey Quarterly* **3** (1992) 103-120.

Eppen, G. & L. Schrage, "Centralized Ordering Policies in a Multiwarehouse System with Leadtimes and Random Demands," in *Multi-Level Produc-*

tion/Inventory Control Systems: Theory and Practice, (L. Schwarz, Ed.), North-Holland, Amsterdam, (1981) 51-69.

Erhardt, R., "The Power Approximation for Computing (s, S) Inventory Policies," *Management Science* **25** (1979) 777-786.

Erkip, N., W.H. Hausman & S. Nahmias, "Optimal Centralized Ordering Policies in Multi-Echelon Inventory Systems with Correlated Demands," *Management Science* **36** (1990) 381-392.

Federgruen, A., "Recent Advances in Production and Distribution Management," in *Perspectives in Operations Management*, (R. Sarin, Ed.), Kluwer Academic Publishers, Norwell, Mass. (1993)

Federgruen, A., "Comments on: Variability Reduction through Operations Reversal in Supply Chain Re-Engineering," *Columbia University* (1998).

Federgruen, A. & P. Zipkin, "Approximations of Dynamic, Multilocation Production and Inventory Problems," *Management Science* **30** (1984a) 69-84.

Federgruen, A. & P. Zipkin, "Allocation Policies and Cost Approximations for Multilocation Inventory Systems," *Naval Research Logistics Quarterly* **31** (1984b) 97-129.

Federgruen, A. & P. Zipkin, "Computational Issues in an Infinite Horizon Multi-Echelon Inventory Model," *Operations Research* **32** (1984c) 818-832.

Fisher, M.L. & A. Raman, "Reducing The Cost of Demand Uncertainty Through Accurate Response to Early Sales," *Operations Research* **44** (1996) 87-99.

Flaherty, M.T., *Global Operations Management*, McGraw-Hill (1996).

Flaherty, M.T., M. Cohen, L. Kopczak, H. Lee & D. Pyke, "Teaching Note for Hewlett Packard: DeskJet Printer Supply Chain (A)," *Wharton School, University of Pennsylvania, Philadelphia, PA* (1996).

Gallego, G. & I. Moon, "The Distribution Free Newsboy Problem: Review and Extensions," *Journal of Operational Research Society* **44** (1993) 825-834.

Garg A. & C.S. Tang, "On Postponement Strategies for Product Families with Multiple Points of Differentiation," *IIE Transaction* **29** (1997) 641-650.

Gavirneni S. & S. Tayur, "Delayed Product Differentiation and Information sharing," Working Paper, *GSIA, Carnegie Mellon University* (1997).

Gingrich J. & H. Metz, "Conquering the Cost of Complexity," *Business Horizons* (1990) 64-71.

Gonsalvez, D., "Integrating Sales and Manufacturing Order Management Processes, Applied Scheduling : Integration of Disciplines," *Columbia University*, October 24-25 (1991), New-York, NY.

Hammond J. & M. Kelly, *"Quick Response in the Apparel Industry,"* Harvard Teaching Note N9-690-038 (1990).

Harvard Business School, *"Benetton (A) and (B),"* Harvard Teaching Case 9-685-014 (1986).

Johnson, G. & H. Thompson, "Optimality of Myopic Inventory Policies for Certain Dependent Demand Processes," *Management Science* **21** (1975) 1303-1307.

Kapuscinski, R. & S. Tayur, "A Capacitated Production-Inventory Model with Periodic Demand," Working Paper, *GSIA, Carnegie Mellon University, Pittsburgh, PA* (1995).

Kapuscinski, R. & S. Tayur, "Variance vs. Standard Deviation – Note on: Variability Reduction Through Operations Reversal in Supply Chain Re-Engineering," Working Paper, *University of Michigan and Carnegie Mellon University* (1998).

Kopczak, L. & H. Lee, *Hewlett Packard: DeskJet Printer Supply Chain (A)*, Stanford University, Stanford, CA (1993).

Kurt Salomon Associates Inc., *"Efficient Consumer Response: Enhancing Consumer Value in The Grocery Industry,"* Food Marketing Institute, Washington, D.C. (1993).

Lee, H., "Design for Supply Chain Management: Concepts and Examples," in *Perspectives in Operations Management*, (R. Sarin Ed.), Kluwer Academic Publishers, Norwell, Mass. (1993).

Lee, H., "Effective Management of Inventory and Service Through Product and Process Redesign," *Operations Research* **44** (1996) 151-159.

Lee, H., & C. Billington, "Designing Products and Processes for Postponement," in *Management of Design: Engineering and Management Perspectives*, (S. Dasu and C. Eastman Eds.), Kluwer Academic Publishers, Norwell, Mass. (1994), 105-122.

Lee, H., C. Billington & B. Carter, "Hewlett-Packard Gains Control of Inventory and Service through Design for Localization," *Interfaces*, August (1993) 1-11.

Lee, H., & C.S. Tang, "Modeling the Costs and Benefits of Delayed Product Differentiation," *Management Science* **43** (1997).

Lee, H., & C.S. Tang, "Variability Reduction Through Operations Reversal," *Management Science* **44** (1998) 162-173.

Lovejoy, W.S., "Myopic Policies for Some Inventory Models with Uncertain Demand," *Management Science* **36** (1990).

Miller, B., "Scarf's State Reduction Method, Flexibility, and a Dependent Demand Inventory Model," *Operations Research* **34** (1986) 83-90.

Mrena, C., "Supply Chain Strategies at Sun Microsystems," Presentation at the *Supply Chain Management Conference*, Santa Clara University, California, May 15 (1997).

"Compaq Plans A Big Change in its PC Sales," *New York Times*, May 5 (1996).

Scarf, H., "A min-max solution of an inventory problem," Chapter 12 in *Studies in the Mathematical Theory of Inventory and Production*, Stanford University Press, (1958).

Schwarz, L.B., "Model for Assessing the Value of Warehouse Risk-Pooling: Risk Pooling over Outside-Supplier Leadtimes," *Management Science* 35 (1989) 828-842.

Signorelli, S. & J.L. Heskett, "Benetton (A)," *Harvard Business School*, (1989).

Sobel, M., *Dynamic Affine Logistics Models*. Report. SUNY at Stony Brook (1988).

Swaminathan, J. & S. Tayur, "Managing Broader Product Lines Through Delayed Differentiation Using Vanilla Boxes," Working Paper, *GSIA, Carnegie Mellon University* (1996).

Vaile R.S., E.T. Grether & R. Cox, *Marketing in the American Economy*, New York, Ronald Press (1952) 149-150.

Van Doremalen, J. & H. Fleurn, "A Quantitative Model for the Analysis of Distribution Network Scenarios," in *Modern Production Concepts: Theory and Applications*, Fandel and Zapfel, Springer Verlag, Berlin, (1991) 660-673.

Veinott, A. Jr., "Optimal Policy for a Multi-Product, Dynamic Non-Stationary Inventory Problem," *Management Science* **12** (1965) 206-222.

"GM Expands Plan to Speed Cars to Buyers," *Wall Street Journal,* October 21 (1996).

Zinn W. & Bowersox, D.J., "Planning Physical Distribution with the Principle of Postponement," *J. Business Logistics,* **9** (1988) 117-136.

Zipkin, P., "Critical Number Policies for Inventory Model with Periodic Data," *Management Science* **35** (1989) 71-80.

19

STOCHASTIC PROGRAMMING MODELS FOR MANAGING PRODUCT VARIETY

Jayashankar M. Swaminathan[1]

Sridhar R. Tayur[2]

[1]Walter A. Haas School of Business
University of California
Berkeley, CA-94720
msj@haas.berkeley.edu

[2]Graduate School of Industrial Administration
Carnegie Mellon University
Pittsburgh, PA-15213
stayur@grobner.gsia.cmu.edu

19.1 INTRODUCTION

In the last decade, there has been an increasing emphasis on producing specialized products for different segments in the market as a means for providing better service to customers. Manufacturers have been adopting strategies including (but not limited to) (i) integration of marketing, manufacturing and distribution activities within the supply chain (see Cohen and Lee 1988 and Lee and Billington 1993), (ii) accurate response to changes in market needs (see Fisher *et al.* 1994), (iii) flexibility in manufacturing process using concepts such as delayed differentiation (see Lee 1996 and Lee and Tang 1998), (iv) incorporation of commonality at the product design stage (see Hayes, Wheelright and Clark 1988 and Ulrich and Eppinger 1995), (v) adoption of flexible manufacturing technologies and (vi) implementation of just-in-time practices such as set-up reduction (see Hall 1983), in order to reduce costs and improve customer satisfaction. Kekre and Srinivasan (1990) suggest that such strategies may mitigate any possible adverse effect of product proliferation.

In this paper, we present models based on stochastic programming for large scale industry problems for three different aspects related to managing product variety at the final assembly stage, namely, utilizing component commonality, delaying differentiation (postponement) using vanilla boxes, and integrating assembly task design and operations to enable postponement strategies. We also present efficient solution methodologies and computational insights where appropriate. The models and results presented in this paper are based on earlier research papers on each of the individual topics (see Tayur 1994, Swaminathan and Tayur 1995 and Swaminathan and Tayur 1997). We model the final assembly stage of production as a two stage process where either components are directly assembled into final products (component commonality) or semi-finished sub-assemblies (vanilla boxes) are customized to create finished products after demand is realized. A two stage stochastic program with recourse captures the dynamics of such a system very well (see Wets 1982 and Kall and Wallace 1995 for a detailed description of stochastic programs). The first stage decisions (inventory of components in the component commonality case or the choice of components in the semi-finished sub-assemblies and their inventory in the vanilla box case) are taken before the random event (demand for product) occurs and the second stage decisions (the assembly of final products from components and vanilla boxes) are taken as recourse action after demand gets realized in order to minimize the expected holding and shortage costs. All the models developed are motivated by real applications at firms including IBM (commonality and vanilla box) and USFilter (assembly task design).

Component commonality problem has been studied by many researchers in the past (see Collier 1982, Sauer 1985, Gerchak and Henig 1986, Baker *et al.* 1986 and Gerchak *et al.* 1988). Most of their analysis has been driven with an aim to obtain useful analytical insights on the problem utilizing simple models with one or two components and a limited number of products. Our approach on the other hand, has been to generate effective solution procedures for large scale (industry size) problems and provide computational insights. The vanilla box approach takes the postponement strategies studied by earlier researchers in a supply chain setting (see Lee and Billington 1993 and Garg and Lee 1996) and tailors it to a final assembly setting. The model presented in this paper was one of the earliest models to recognize the vanilla box approach and the computational insights from the research enabled IBM managers to take important decisions regarding the assembly of their RS6000 product line. Assembly task design problem has mostly been studied from an engineering perspective (see Nevins and Whitney 1989 and Stadzisz and Henrioud 1995) where the objective has been to find algorithms and heuristics to generate assembly sequence for a given product or product family without much considerations about operations. Qualitative research in operations has explained practices in industry and provided insights on the link between assembly design, product architecture and operations (see Pine 1993, Boothroyd *et al.* 1994, Ulrich and Eppinger 1995, and Ulrich *et al.* 1997). Gupta and Krishnan (1996) and Dell (1996) are examples of work in this area which have developed quantitative models for decision support in this area. Once again, our approach has been to address the large scale problem and provide computational insights.

The rest of the paper is as follows. In section 2, we discuss the component commonality problem. In section 3, we discuss the vanilla box problem and in section 4, we discuss the assembly task design problem. We conclude in section 5.

19.2 THE COMPONENT COMMONALITY PROBLEM

Our original motivation to study this problem came from a need to analyze IBM's PS/2 line of business which has the following characteristics. The lead times for components are long and the product demands are highly variable and correlated. The end-products may share several common components in different ratios. In this model we assume that the assembly capacity is unconstrained and determine the optimal inventory of components to store in each period under a myopic policy (the general case is a complex stochastic program). We model the finite horizon problem as a multi-period stochastic program.

19.2.1 A Discrete-time Model

We consider a finite horizon (indexed by t, $1 \leq t \leq T$) model with stochastic demand for the multiple end-products (indexed by j, $1 \leq j \leq m$) that use (perhaps) common multiple components (indexed by i, $1 \leq i \leq n$). We need to compute quantities q_{it}, representing the amount of type i component to be delivered in period t. One unit of end-product j needs u_{ij} units of component i. Demands for the end-products are continuous, inventories are reviewed periodically, and unfilled orders are backlogged. We assume that $F_{1,...,m;1,...,T}(\{\psi_{jt}\})$, the joint distribution of end-product demands is known (as is the density $f_{1,...,m;1,...,T}$), and that random variates $\{\psi_{jt}^r\}$, $1 \leq r \leq R$ can be generated easily. The holding cost for each unit of component i is h_i and the penalty cost for each unit of end-product j backorder is w_j. We detail the operation of the first period problem, then explain how the T-period system works.

Consider the first period. In this period, events occur in the following order: First, components arrive for this period. Denote them by $q_{i1}, i = 1, \ldots, n$. Second, demands of the end-products occur. Denote these demands by ψ_{j1}. Third, using the components available, the production level s_{j1} $(j = 1, \ldots, m)$ for product j is decided. The end-product demands are filled (or a backlog occurs) using s_{j1}. The holding cost is on the inventory of components left-over at the end of the period (h_i per unit for component i); the penalty cost is on backorder of end-product demand (w_j per unit for end-product j). We never assemble more products than we can sell in any period because the assembly capacity is assumed unlimited.

Let $\mathbf{h} = (h_1, \ldots, h_n)$, $\mathbf{w} = (w_1, \ldots, w_m)$, $\mathbf{q_1} = (q_{11}, \ldots, q_{n1})$ and $\mathbf{s_1} = (s_{11}, \ldots, s_{m1})$. Let $\mathbf{U}$ denote a matrix of elements u_{ij}. As it will be obvious from the context, we will not explicitly state whether a vector is a row or a column vector, nor will we display its dimension.

We have

$$\text{(P)} \qquad \min_{\mathbf{q_1}} \mathbf{h} \cdot \mathbf{q_1} + \mathbf{E}\mathbf{Q}(\mathbf{q_1}, \psi_1)$$

$$\text{where} \quad \mathbf{Q}(\mathbf{q_1}, \psi_1) = -\max\left[(\mathbf{h} \cdot \mathbf{U} + \mathbf{w}) \cdot \mathbf{s_1} - \mathbf{w} \cdot \psi_1\right]$$

$$\text{such that} \qquad \mathbf{U} \cdot \mathbf{s_1} \leq \mathbf{q_1}$$

$$0 \leq \mathbf{s_1} \leq \psi_1.$$

The stochastic program (P) is to be interpreted as follows. A decision $\mathbf{q_1}$ is made (components are ordered), then a demand ψ_1 (of end-products) occurs, to which we respond by making $\mathbf{s_1}$ (of end-products) that optimizes the imbedded linear program (denoted by $P(\mathbf{q_1}, \psi_1)$). We are to find a $\mathbf{q_1}$ that minimizes the total expected cost of holding the left-over components as well as the penalty for unsatisfied end-product demands. In a multi-period setting this is a myopic approach since we select $\mathbf{s_1}$ without considering the future periods. We will follow a similar approach in the tth period problem as well.

In periods $t \geq 2$, both the left-over components at the end of period $t - 1$ (if any) and the backorders at the end of period $t - 1$ (if any) carry over to period t. Two features thus distinguish the multi-period system from a single period system: (1) Components scheduled to arrive at the beginning of period t, namely q_{it}, are added to those left over from the previous period. We denote available quantity of component type i in period t by $\tilde{q}_{it}$. (2) The total number of end-products that need to be produced now equal the sum of the new demand ψ_{jt} and the backlog from the previous period. We denote this quantity by $\tilde{\psi}_{jt}$.

The t-th period problem then is as follows.

$$\min_{\mathbf{q_t}} \mathbf{h} \cdot \tilde{\mathbf{q}}_t + \mathbf{E}\mathbf{Q}(\tilde{\mathbf{q}}_t, \tilde{\psi}_t)$$

$$\text{where} \quad \mathbf{Q}(\tilde{\mathbf{q}}_t, \tilde{\psi}_t) = -\max\left[(\mathbf{h} \cdot \mathbf{U} + \mathbf{w}) \cdot \mathbf{s}_t - \mathbf{w} \cdot \tilde{\psi}_t\right]$$

$$\text{such that} \quad \mathbf{U} \cdot \mathbf{s}_t \leq \tilde{\mathbf{q}}_t$$

$$0 \leq \mathbf{s}_t \leq \tilde{\psi}_t$$

The T-period problem (*not* to be confused with the T-th period problem), then, is the following:

$$(PT) \qquad \min_{\mathbf{q}_1,\dots,\mathbf{q}_T} \sum_{t=1}^{T}(\mathbf{h} \cdot \tilde{\mathbf{q}}_t + \mathbf{EQ}(\tilde{\mathbf{q}}_t, \tilde{\psi}_t))$$

$$\text{where} \quad \mathbf{Q}(\tilde{\mathbf{q}}_t, \tilde{\psi}_t) = -\max\left[(\mathbf{h} \cdot \mathbf{U} + \mathbf{w}) \cdot \mathbf{s}_t - \mathbf{w} \cdot \tilde{\psi}_t\right] \quad 1 \leq t \leq T$$

$$\text{such that} \quad \mathbf{U} \cdot \mathbf{s}_t \leq \tilde{\mathbf{q}}_t \qquad 1 \leq t \leq T$$

$$0 \leq \mathbf{s}_t \leq \tilde{\psi}_t \qquad 1 \leq t \leq T$$

$$\text{and} \quad \tilde{\mathbf{q}}_t = \tilde{\mathbf{q}}_{t-1} + \mathbf{q}_t - \mathbf{U} \cdot \mathbf{s}_{t-1} \qquad 1 \leq t \leq T$$

$$\tilde{\psi}_t = \tilde{\psi}_{t-1} - \mathbf{s}_{t-1} + \psi_t \qquad 1 \leq t \leq T$$

with $\tilde{\mathbf{q}}_0 = \tilde{\psi}_0 = \mathbf{s}_0 = 0$.

Unlike the first-period problem (P), the t-period problem (just the same as a T-period problem but we replace T by t) has the *balance* equations: the equations that link components and backorders carried over from one period to the next. (The base plans $\mathbf{q}_t, 1 \leq t \leq T$ are provided to the supplier before actually seeing any demand.) Note that the production decisions in any period are myopic and do not consider information about future periods. We can recast this T-period problem as a nested set of stochastic programs with recourse with a structure similar to the first period case as follows.

Define, for $1 \leq t \leq T$,

$$\hat{\mathbf{q}}_t = \sum_{l=1}^{t} \mathbf{q}_l$$

that denotes the total number of components that will arrive in periods $1, 2 \dots, t$. Also, define

$$\hat{\mathbf{s}}_t = \sum_{l=1}^{t} \mathbf{s}_l$$
$$\hat{\psi}_t = \sum_{l=1}^{t} \psi_l$$

to denote the cumulative production and demands that takes place in periods $1 \leq l \leq t$. Rewriting all the tth-period problems in terms of the new (cumulative) variables defined above, we obtain the following set of stochastic programs, one for each $1 \leq t \leq T$, representing *up to* period t-problem:

$$(\hat{P}_t) \qquad \min_{\hat{\mathbf{q}}_t} \mathbf{h} \cdot \hat{\mathbf{q}}_t + \mathbf{EQ}(\hat{\mathbf{q}}_t, \hat{\psi}_t)$$

$$\text{where} \quad Q(\hat{\mathbf{q}}_t, \hat{\psi}_t) = -\max(\mathbf{h} \cdot \mathbf{U} + \mathbf{w}) \cdot \hat{\mathbf{s}}_t + \mathbf{w} \cdot \hat{\psi}_t$$

$$\text{such that} \quad \mathbf{U} \cdot \hat{\mathbf{s}}_t \leq \hat{\mathbf{q}}_t$$

$$0 \leq \hat{\mathbf{s}}_t \leq \hat{\psi}_t.$$

Each $(\hat{P}_t)$ can be thought of as a one period problem with demands having a distribution of $\hat{\psi}_t$. (Notice that for $t = 1$, we obtain the first-period problem.)

We are to find components given by $\hat{q}_t$; the production quantities for any realized demand are given by $\hat{s}_t$. Observe that solving the above T problems will provide us with optimal values of $\hat{q}_t$, from which we can obtain the optimal values of q_t. Since every $q_t \geq 0$, we need to ensure that $\hat{q}_t \geq \hat{q}_{t-1}$ for all t. We don't put these constraints in the above problems as they can be easily incorporated while updating q_t^{iter} in our gradient based algorithm. Similarly, we ensure that $\hat{s}_t \geq \hat{s}_{t-1}$ on any sample path .

Very simply, our original T-period problem is the sum of $(\hat{P}_t), 1 \leq t \leq T$ with the added constraints $\hat{q}_t \geq \hat{q}_{t-1}$ and $\hat{s}_t \geq \hat{s}_{t-1}$. Solving $(\hat{P}_t)$ starting from $t = 1$ to $t = T$ sequentially, provides the solution for (PT).

Proposition 2.1. If the demands for all the products are non-negative with finite means and variances, then the objective function of (P) is convex with respect to q_1.

Proposition 2.2. If the demands for all the products in all the time periods are non-negative with finite means and variances, then (PT) is a convex program with respect to $\hat{q}_t$ for $t = 1, \ldots, T$.

19.2.2 Algorithm

Our solution procedure determines the optimal procurement of components utilizing a sub-gradient based algorithm. We show that averaging the derivatives, whenever they exist, to estimate the derivative of the expected value is valid in the sense that the sample-path derivatives they generate are unbiased estimators of derivatives of expectations. We use the simple fact that for a linear program, the dual value of a constraint is the derivative of the objective function with respect to the right hand side of that constraint. If the derivative does not exist at a point, then our result holds for the right-derivative (as against the left-derivative) at this point.

Proposition 2.3. $(EQ(q_1, \psi_1))' = E(Q(q_1, \psi_1)')$.

Let us consider (PT). First consider the derivative with respect to $\hat{q}_{it}$ keeping other $\hat{q}$ fixed: This is just a one-period problem that affects $(\hat{P}_t)$ only. Next, consider the derivative with respect to some q_{kt} keeping all other q's fixed. When we increase q_{kt} by ϵ (keeping all other q's fixed), we do not affect $(\hat{P}_s)$ for any $s < t$, and we affect all $(\hat{P}_s)$ for $s \geq t$. In fact, for each of $(\hat{P}_s)$ with $s \geq t$, $\hat{q}_{ks}$ increases by ϵ. Thus, we can obtain the derivative of the objective function of (PT) with respect to q_{kt} by summing over all $s \geq t$ the derivatives of the objective functions of $(\hat{P}_s)$ with respect to $\hat{q}_{ks}$. We thus have:

Proposition 2.4.

1. For any $1 \leq s \leq T$, $\frac{dEQ(\hat{q}_s, \hat{\psi}_s)}{d\hat{q}_{ks}} = E(\frac{dQ(\hat{q}_s, \hat{\psi}_s)}{d\hat{q}_{ks}})$.

2. Furthermore, the derivative of the expected T-period cost with respect to q_{kt} equals $(T - t + 1)h_k - \sum_{s \geq t} E(\frac{dQ(\hat{q}_s, \hat{\psi}_s)}{d\hat{q}_{ks}})$.

There are several ways of dealing with continuous demands in a multi-period situation. We adopt the following approach: we sample from the distribution of ψ_t (say R samples of each), and obtain samples for $\hat{\psi}_t$ by simply adding up corresponding samples from ψ_t. Furthermore, on every sample path, we ensure that $\hat{s}_t \geq \hat{s}_{t-1}$. This does not affect convexity of feasible region because these constraints are linear. This does not affect feasibility on any sample path because the optimal solution to any imbedded LP $P(\hat{q}_t, \hat{\psi}_t)$ is feasible for an imbedded LP $P(\hat{q}_{t+1}, \hat{\psi}_{t+1})$. We are *simulating* the multi-period setting R times and averaging the results. (Exchanging the derivative and expectation when F is not discrete is essentially Infinitesimal Perturbation Analysis, IPA). Note that any discretization procedure (so have p_r for sample r) that has the property that it approaches the true distribution as it is made finer, will work in theory. Other details about the model and the algorithm can be found in Tayur(1994).

19.2.3 Computational Results

We tested the algorithm of Berger, Mulvey and Ruszczyinski (1994) (DQA, a state-of-the-art general purpose algorithm for multi-period stochastic programming) on our problem. The testing was done on a 3-period problem on a R4000 (Personal Iris machine) chip using sequential DQA. For $m = n = 2$ with 100 (10 samples of period 1 demand, 5 samples of period 2 demand, 2 samples of period 3 demand), 200 (multiply 10,10,2) and 400 (multiply 10,10,4) scenarios, the times are 20, 38 and 131 seconds on a R4000 chip; for $m = 20$, $n = 10$, with a total of 100, 200 and 400 scenarios the times were 354, 786, 1791 seconds respectively. The corresponding times for our gradient procedure on a SPARC 20 (about 20 % faster than R4000 chip) are 15 seconds (10 samples in first period, 50 samples in second period, 100 samples in third period), 35 seconds (10 samples in first period, 100 samples in second period, 200 samples in third period), and 63 seconds (10 samples in first period, 100 samples in second period, 400 samples in problem would take significantly longer on a parallel machine (with 4 processors) with DQA than the time our gradient approach would take on an ordinary workstation (sequentially): see Figure 8 of Berger, Mulvey and Ruszczynski (1994) for low accuracy DQA. For example with 3 periods, $m = n = 50$ with 100 scenarios per period, DQA will take *at least* 3 hours with 4 processors on Silicon Graphics machine. Our gradient based method takes less than 2 hours on a serial implementation. Furthermore, as the number of time periods increase, the DQA grows exponentially in time. These results indicate that our method is useful for moderate and larger problems, while for smaller problems the sequential DQA has a better time performance.

Our basic conclusions are the following.

1. More samples (replications) are needed as the variance of demand increases. The number of replications depend more on the variance of demand rather than m or n. For example, if u_{ij} is selected from U[0,4], 200 replications were sufficient when demand was less variable than U(40, 85), and 300 were sufficient for all demands with variance less than or equal

to U(5, 120); see Table 1 for $m = n = 2$. This remained true for m and n values are 5 (see Table 2), 10 and 20 (not shown).

2. It is better to start with a small number of replications, and after finding the optimum with this number of replications, use it as a starting point to optimize with larger number of samples. (For a very elegant algorithm that adds one sample at a time, see Higle and Sen, 1991). For example, if 400 samples are required, we start with 100 samples and find a good solution, then move to 200 samples starting from this solution and then finally to 400 samples.

So to solve large problems, our suggestion is the following. Start with small number of replications (say from 100 replications and increase them in stages, 100 at a time), use a standard gradient step that is initially quite large (like (components)*(1% of right hand side) for example) and is then gradually reduced as the gradient oscillates, and stop when either the gradient is very small or the moving average of cost does not change significantly. Line search is not required.

Table 19.1 An example with $m = n = 2$. A stands for demands U(5, 120), B for demands U(25, 100) and C for demands U(50, 75). All times are in seconds on a SPARC 2.

Demand		A		B			C		
Samples	Levels(q)	Cost	Time	Levels	Cost	Time	Levels	Cost	Time
100	147, 184	7.047	53.2	161, 207	4.597	35.6	178, 235	1.52	13.3
200	142, 175	6.603	120.5	158, 201	4.299	80.5	177, 233	1.426	30.9
300	136, 165	6.602	205.7	154, 195	4.302	134.8	176, 231	1.431	47.3
400	136, 165	6.57	273.3	154, 194	4.282	181.8	176, 231	1.427	62.7
1000	137, 166	6.46	671.1	155, 196	4.215	462.2	177, 232	1.405	158.6

We found the following observations with regard to the solution.

1. Increasing the variance increases cost (for example, see Table 1).

2. Increasing variance either monotonically decreases stocking values or monotonically increases them; the direction depends whether we are stocking less than the mean values (holding is expensive), or higher than the mean values (penalty cost is high). In Table 1, the mean values are 187.5 and 250, and so decreasing variance is increasing stocking levels.

3. Even for fixed m, n, the total time to compute the optimal can vary significantly (this is somewhat disconcerting, but has been observed in other algorithms also; see Higle and Sen, 1991). Note that total time = (number of iterations)*(time per iteration). See Table 5 for some values. Not apparent from the table, but we noticed that the number of iterations required varies depending on $F, \mathbf{h}, \mathbf{w}$ and the re-optimization time varies depending on $\mathbf{U}, \mathbf{h}, \mathbf{w}$.

Table 19.2 EFFECT OF U ON R. An example with $m = n = 5$, showing the effect of sample size and commonality on optimal solution $(q_1, \ldots, q_5)$. (A) corresponds to u_{ij} from U[0,4] and (B) corresponds to u_{ij} from U[1,4]. Note that in (A) 300 samples suffice; but in (B), 500 samples are required. For (B) with 10000 samples, the optimal levels were (425, 481, 652, 780, 771). The demands are from U(40, 85).

Samples	50	100	200	300	500	1000
Levels	311	321	318	321	323	320
(A)	243	247	247	250	250	250
	583	592	588	598	600	596
	536	547	544	548	551	546
	753	767	769	778	780	774
Samples	100	200	300	500	1000	5000
Levels	424	425	431	429	429	427
(B)	478	478	484	485	483	481
	649	649	660	659	656	661
	780	783	791	788	788	783
	766	770	780	778	778	772

Table 19.3 Times, in minutes, for problems with 200 samples on a HP 720 Workstation for a one-period problem.

(m, n)	(20, 20)	(30, 30)	(40, 40)	(50, 50)
Time	16.75	51.7	105.1	135.1

Table 19.4 Times, in minutes, for problems with 100 samples on a HP 720 Workstation. $m = 50$. All solutions are within 1% of the ultimate value found by increasing R and having tighter termination criteria.

n	$T = 1$	$T = 2$	$T = 3$
20	14.3	21.5	34.7
30	16.6	18.5	35.1
40	40.1	49.2	83.5
50	66.3	70.5	116.3

Table 19.5 Showing the wide variance in times per iteration as well as iterations as $U, \mathbf{h}, \mathbf{w}$ and F are varied. $m = 50$, $R = 100$ and $T = 1$ in all cases. 5 problems were solved at each n.

n	time/iteration ratio of max/min	iterations (min, max)
20	1.52	(300, 404)
30	2.35	(66, 238)
40	2.16	(199, 452)
50	2.24	(72, 436)

4. Typically the constraints $\hat{q}_t \geq \hat{q}_{t-1}$ are not active at optimality.

5. The cases when some $\hat{q}_{it} \geq \hat{q}_{it-1}$ is active have the following characteristics: there is a product (say j_1) with low w that uses component i more than other products do; product j_1 shares a component k (with a high h) with another product j_2 with high w; the demand for j_2 has low variance until period $t-1$ and has high variance in period t while the demand for j_1 has high variance until period $t-1$ but low variance in period t. In this situation, component i will be used less in period t because component k is very likely to be a tight resource, and so if we solve $\hat{P}_t$ and $\hat{P}_{t-1}$ *separately* (or independently) we notice that $\hat{q}_{it}$ will be lower than $\hat{q}_{it-1}$ at optimality. In all of the cases when this happened, $\hat{q}_{it}$ did not differ from $\hat{q}_{it-1}$ significantly, explaining our next observation.

Several other insights about the commonality problem are discussed in Tayur (1994).

19.3 THE VANILLA BOX PROBLEM

The vanilla box strategy takes the concept of delayed differentiation or postponement to the final assembly stage. This research was motivated by the the vanilla box strategy piloted in one of the assembly plants for the RS6000 server business at IBM in order to quickly respond to demands while customizing it to the greatest extent. The strategy was to store inventory of semi-finished products called *vanilla boxes* and to assemble final products from the vanilla boxes after a customer order was realized. Thus, a vanilla assembly process delays differentiation of final products by keeping inventory of semi-finished products called *vanilla boxes*.

Consider a simple example of a product line shown in Figure 1a. It consists of products $P1 \ldots P4$ each of which has up to four *features* ($a \ldots d$). The manufacturer offers two *options* in features a and d, namely, $a1, a2$ and $d1, d2$ respectively. Let us hypothetically assume that each of the products is a computer and say feature a is main memory then options $a1$ and $a2$ could correspond

PRODUCT LINE X1

a. Bill of Material

	a1	a2	b1	c1	d1	d2
P1	1	0	1	1	0	0
P2	1	0	1	1	0	1
P3	0	1	1	1	0	1
P4	1	0	1	1	1	0

b. Assembly Sequences

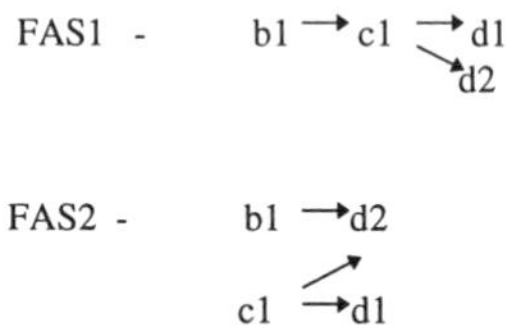

FAS1 - b1 → c1 → d1
                  ↘ d2

FAS2 - b1 → d2
          ↗
        c1 → d1

c. Reduced Bill of Material

	a	b	c	d
P1	1	1	1	0
P2	1	1	1	1
P3	1	1	1	1
P4	1	1	1	1

Figure 19.1 Product Structure, Vanilla Boxes and Assembly Sequences.

to memory sizes of 16MB and 32MB respectively. In that case $P1, P2$ and $P4$ are computer models with 16MB memory whereas $P3$ is a model with 32MB memory. A vanilla box $V1 = (b1, c1)$ supports assembly of all the products and helps in providing quick response to customer demands because when a product such as $P1$ is assembled from a vanilla box $V1$ rather than starting assembly from raw components, customers experience only the time it takes to assemble $a1$ to the vanilla box. Thus a vanilla assembly process enabled assembly of customized products within much shorter lead times, though the manufacturer had to carry additional inventory of vanilla boxes. The vanilla assembly process also captures the traditional modes of operation: make-to-stock and assemble-to-order. If we restrict the vanilla boxes to be finished products (rather than semi-finished products) then the vanilla assembly process represents the make-to-stock environment. On the other hand, if we decide that all vanilla boxes are restricted to be single components then we get an assemble-to-order environment.

19.3.1 Model

In the basic model, we consider a discrete time single period model with a finite capacity C for (final) assembly. C is the capacity available (after the realization of demand and within the customer response time window) to assemble additional components to the vanilla boxes already present or assemble products starting from components. All vanilla boxes are produced before the beginning of the period. Demands are realized at the beginning of the period before decisions need to be made regarding the assembly of the final products. Every product i $(1 \ldots N)$ may either be assembled directly from its components, or from any vanilla box whose component set is a subset of those required by i. As a result, we do not allow redundant components in a product. Demands are random but follow one of L given scenarios, each with a likelihood. Holding costs are incurred on unused vanilla boxes at a box type specific rate, and shortage costs for unsatisfied demand, at a product specific rate. Without loss of generality, we assume that the bill of material in terms of the components is binary.

The notation used in our formulation is as follows. We will index the products by $i(1 \ldots N)$, the components by $j(1 \ldots n)$ and the vanilla boxes by $k(1 \ldots K)$.

- C: capacity available for assembly;

- π_i: per unit per period stock-out cost for the product i;

- h_k: per unit per period holding cost for a vanilla box k;

- $\mathbf{q} = (q_1 .. q_K)$: vector of starting inventory levels of vanilla boxes k $(1 \ldots K)$.

- t_{i0}: per unit assembly time for product i from components;

- t_{ik}: per unit assembly time for product i from vanilla box k ($t_{ik} = \infty$ if product i cannot be made from vanilla box k).

- ξ_l: a realization $(\xi_{1l} \ldots \xi_{Nl})$ of product demands in scenario l where $\xi_1 \ldots \xi_N$ have a joint distribution F.

- B: set containing boolean values - $0, 1$;

- $\mathbf{E}$: expectation operator with respect to F;

- $Q(\mathbf{q}, \mathbf{U}, \xi)$: single period cost function when demand is ξ;

- $\mathbf{U}$: 0-1 matrix that represents the bill of material of the K vanilla boxes in terms of components u_{kj} ($\mathbf{U_k}$ refers to constitution of the kth vanilla box);

- s_{ikl}: quantity of product i that were made in the period using vanilla box k in scenario l ($k = 0$ implies that product i is assembled directly from components);

Our objective is to minimize the expected cost which consists of the stock-out cost (when demand is not met) and the cost of holding vanilla boxes. The number of vanilla boxes K is considered fixed at this stage. This problem is formulated as (P).

$$(P): \quad \min_{\mathbf{q},\mathbf{U}} \mathbf{E}_l Q(\mathbf{q}, \mathbf{U}, \xi_l)$$

where

$$Q(\mathbf{q}, \mathbf{U}, \xi_l) = \min \sum_{i=1}^{N} (\pi_i(\xi_{il} - \sum_{k=0}^{K} s_{ikl})) + \sum_{k=1}^{K} (h_k(q_k - \sum_{i=1}^{N} s_{ikl}))$$

s.t.

$$\sum_{i=1}^{N} \sum_{k=0}^{K} t_{ik}(\mathbf{U}) s_{ikl} \leq C \quad \forall l \tag{1}$$

$$\sum_{i=1}^{N} s_{ikl} \leq q_k \quad \forall k \geq 1, \forall l \tag{2}$$

$$\sum_{k=0}^{K} s_{ikl} \leq \xi_{il} \quad \forall i, l \tag{3}$$

$$u_{kj} \in B \tag{4}$$

$$s_{ikl}, q_j \in R_+ \tag{5}$$

The above formulation is a stochastic program with recourse with mixed first stage variables and continuous second stage variables. The first stage decision variables (which are decided before demand occurs) are - (i) vanilla box configuration $\mathbf{U}$ (indicates which components are present in the K vanilla boxes), (ii) inventory of the K vanilla boxes $\mathbf{q}$. The second stage decision variables (which are determined after demand occurs) are (i) assembly of products from vanilla boxes or components in each scenario (s_{ikl}). Earlier work in two stage stochastic programming with binary first stage variables (see Wollmer 1980 and Laporte and Louveaux 1993) has mainly dealt with problems where the first stage binary variables do not affect the constraint matrix of the second stage problem. In our problem, the first stage binary variables $\mathbf{U}$ affect the constraint matrix at the second stage since t_{ik} depends on the constitution $\mathbf{U_k}$ of vanilla box k. As a result, the first stage variables $\mathbf{U}$ are forming a non-linear constraint with the second stage variables s_{ikl} in constraints (1). For a given $\mathbf{U}$ the problem can be represented as a two stage stochastic program with recourse having only continuous variables given by $P1(\mathbf{U})$.

$$P1(\mathbf{U}): \quad \min_{\mathbf{q}} \mathbf{E}_1 Q(\mathbf{q}, \mathbf{U}, \xi_1)$$

where

$$Q(\mathbf{q}, \mathbf{U}, \xi_1) = \min_{S_1} \sum_{i=1}^{N} (\pi_i(\xi_{il} - \sum_{k=0}^{K} s_{ikl})) + \sum_{k=1}^{K} (h_k(q_k - \sum_{i=1}^{N} s_{ikl}))$$

s.t.

$$\sum_{i=1}^{N}\sum_{k=0}^{K} t_{ik} s_{ikl} \leq C \quad \forall l \qquad (1)$$

$$\sum_{i=1}^{N} s_{ikl} \leq q_k \quad \forall k \geq 1, \forall l \quad (2)$$

$$\sum_{k=0}^{K} s_{ikl} \leq \xi_{il} \quad \forall i, l \qquad (3)$$

$$s_{ikl}, q_j \in R_+ \qquad (4)$$

Note that $P1(\mathbf{U})$ is a stochastic program which is convex (based on Proposition 2.1) and as a result we utilize a sub-gradient based algorithm to find the optimal inventory levels.

The basic model above can be expanded to settings where the assembly capacity is used to produce vanilla boxes as well as final products, and even to multiperiod settings under the following two assumptions: (1) Demands in consecutive periods are independent; (2) A base stock policy is adopted for managing the inventory of vanilla boxes so that each period starts with the same number of vanilla boxes for each type $k = 1 \ldots K$.

Under these two assumptions, each period's planning problem is entirely independent of that of the other periods. We assign overtime costs for bringing the inventory level to the target level if it is lower at the end of a period. Here are some additional notations- (1) $\mathbf{q_0}$: Vector of target inventory levels of vanilla boxes; (2) q_{1kl}: production of vanilla box k in scenario l; (3) r_k: Additional cost incurred (overtime charges) per unit of vanilla type k assembled at the end of the period; (4) v_k: Amount of assembly capacity utilized per unit of vanilla type k assembled during the period. We have $P2(\mathbf{U})$ similar to $P1(\mathbf{U})$ in the basic model.

$P2(\mathbf{U})$: $\quad\min_{\mathbf{q_0}} \mathbf{E}_1 Q(\mathbf{q_0}, \mathbf{U}, \xi_1)$

where $\quad Q(\mathbf{q_0}, \mathbf{U}, \xi) = \min \sum_{k=1}^{K} r_k \left(\sum_{i=1}^{N} s_{ikl} - q_{1kl} \right) + \sum_{i=1}^{N} \left(\pi_i \left(\xi_{il} - \sum_{k=0}^{K} s_{ikl} \right) \right)$

$$+ \sum_{k=1}^{K} \left(h_k \left(q_{0k} - \sum_{i=1}^{N} s_{ikl} \right) \right)$$

s.t.

$$\sum_{i=1}^{N} \sum_{k=0}^{K} t_{ik} s_{ikl} + \sum_{k=1}^{K} (v_k q_{1kl}) \;\leq\; C \qquad \forall l \qquad (1)$$

$$\sum_{i=1}^{N} s_{ikl} \;\leq\; q_{0k} \qquad \forall k \geq 1, \forall l \quad (2)$$

$$\sum_{k=0}^{K} s_{ikl} \;\leq\; \xi_{il} \qquad \forall i, l \qquad (3)$$

$$q_{1kl} \;\leq\; \sum_{i=1}^{N} s_{ikl} \quad \forall k \geq 1, \forall l \quad (4)$$

$$s_{ikl}, q_{0k}, q_{1kl} \;\in\; R_+ \qquad\qquad (5)$$

19.3.2 Algorithm

We present a solution methodology for solving the vanilla box problem by structurally decomposing the problem and fixing the binary variables (corresponding to the configuration of the vanilla boxes) using the knowledge about the bill of material (BOM) matrix. Note that this approach works for both P1 and P2 but we will use P1 as an illustrative example.

Our proposed planning approach starts with the enumeration of all possible vanilla box configurations using at most K types. For a given choice for the configuration of the vanilla boxes, the remaining problem consists of determining optimal inventory levels of the boxes used, and for each scenario, an optimal assembly plan. We obtain that by solving $P1(\mathbf{U})$ to optimality utilizing a sub-gradient based algorithm.

Definition 3.1: A vanilla box is *maximal* if addition of any component to it reduces the number of products that can be assembled using that vanilla box.

For products $P1 \ldots P4$ in figure 1a the vanilla box $(b1, c1)$ is maximal. Note that we do not allow redundant components in products so addition of any component to $(b1, c1)$ reduces the number of products which can be assembled from it.

Proposition 3.1. If holding cost of all vanilla boxes are identical then an optimal vanilla box has to be *maximal*.

Proof. See Swaminathan and Tayur(1995) for details.

$$
\begin{pmatrix}
 & a & b & c & d & e \\
P1 & 1 & 0 & 0 & 1 & 1 \\
P2 & 0 & 1 & 0 & 1 & 1 \\
P3 & 1 & 1 & 0 & 1 & 1 \\
P4 & 1 & 1 & 1 & 1 & 0 \\
P5 & 1 & 0 & 1 & 1 & 1
\end{pmatrix}
$$

Figure 19.2 Bill of Material for X1.

19.3.3   Computational Insights

In this section, we present our computational insights.

19.3.3.1   Pilot Study.   In order to explain the results of our pilot study clearly, we utilize the results from one problem X1 and explain it in detail. In the process, we also explain our experiment generation which remains the same throughout our computational study unless specified otherwise.

Problem X1:

- Bill of Material: X1 had five products ($P1$ to $P5$) and five components (a to e). The bill of material for X1 is given in figure 2. This indicates $P1$ contains (a, d, e), $P2$ contains (b, d, e) and so on.

- Speed up in Assembly: We assume in our computations that it takes 1 unit of time to assemble a component. As a result, the time to assemble any product from raw components is equal to the number of components in the product. For $P1$ it is 3 units whereas for $P4$ it is 4 units. Now if a vanilla box (a, d) is used to assemble $P1$, the time for assembly reduces to 1 unit and for $P4$ it reduces to 2 units.

- Capacity: Capacity is computed based on the mean demand for products. So, if the mean demands were assumed to be 50 in each period then a 40 % capacity would correspond to $0.4 * (150 + 150 + 200 + 200 + 200) = 360$ time units. 90% capacity would correspond to $0.9 * (900) = 810$ time units.

- Product Demands: The demand for products are generated from a multivariate normal distribution using a standard procedure. All demand scenarios have equal probability (which can be easily changed if required). In each scenario demand is represented by a N-vector of real numbers.

Here are our main results from the pilot study. It is noted to be that the results that we highlight here for X1 were found to be true across all the 40 problems considered in the pilot study. Tables 6 and 7 represent the cost incurred by the vanilla process ($K = 1, 2$), make-to-stock(MTS) and assemble-to-order(ATO) under different conditions related to capacity, correlation and variance in product demands for the problem X1. Table 8 shows the changes in vanilla configuration and inventory with an increase in capacity for problem X1. All insights discussed in this section (3.3.1) correspond to the vanilla assembly process.

■ **Variance and Correlation of Product Demands.** (1) We find that an increase in variance increases the cost incurred (Tables 6 and 7). (2) We find that the vanilla process incurs lower cost under negative correlation as compared to the case when product demands are positively correlated (Tables 6 and 7). Under negative correlation in demand, while the demands for individual products can be very variable, the total demand across all products is nearly constant. As a result, the demand for common components across the product line is fairly stable. The demand for vanilla boxes made out of these common components can be predicted accurately. So, one would expect a vanilla process to perform better under negative correlation.

Table 19.6 Effect of changes in capacity and variance of demand on the cost incurred under negative correlation ($\rho = -0.25$).

| | | High Variance ($cv = 0.2$) | | | | Low Variance ($cv = 0.1$) | | |
Capacity	K=1	K=2	ATO	MTS	K=1	K=2	ATO	MTS
Low - 35%	11532.80	1844.47	29722.71	1278.61	7193.37	332.05	23878.92	276.39
Med. - 65%	1793.79	297.10	16967.93	575.36	92.05	0.00	12256.44	0.85
High - 95%	202.27	10.98	6227.49	154.22	0.00	0.00	1033.46	0.00

Table 19.7 Effect of changes in capacity and variance of demand on the cost incurred under positive correlation ($\rho = 0.5$).

| | | High Variance ($cv = 0.2$) | | | | Low Variance ($cv = 0.1$) | | |
Capacity	K=1	K=2	ATO	MTS	K=1	K=2	ATO	MTS
Low - 35%	23709.84	9314.11	52359.06	1839.39	14166.75	3655.16	37137.78	655.44
Med. - 65%	12497.41	2506.60	39588.44	1151.24	4391.04	489.28	24996.42	290.30
High - 95%	5461.47	1062.58	28733.18	798.70	490.30	121.29	14646.09	77.08

■ **Capacity Restrictions.** (1) An increase in capacity reduces the total cost incurred (Tables 6 and 7). Since capacity appears on the right hand side of the recourse step of the stochastic program, we can show that the total cost function is convex and decreasing with respect to capacity. (2) In order to understand the effect of capacity on the configuration of vanilla boxes, we varied it in steps. We observe the following stages as capacity is increased (Table 8): (a) the inventory of the vanilla boxes increase, but the configuration does not change; (b) the configuration changes, with the vanilla box typically having less number of components; (c) the inventory level reduces but the configuration does not change. Our intuition is as follows. Under very tight capacity restrictions vanilla boxes even if available in plenty cannot be utilized because there is not enough capacity to assemble products even from vanilla boxes, and as a result, the optimal inventory is low. When the capacity restrictions are medium then vanilla boxes are more useful, and as a result, there is an increase in optimal inventory. Finally, when the capacity is high, many vanilla boxes may not be required since products can be assembled from components, and as a result, the optimal inventory decreases.

- **Number of Vanilla Types (K).** (1) An increase in the number of vanilla boxes leads to a decrease in the cost incurred (Tables 6 and 7).

Table 19.8 Influence of capacity on optimal vanilla configurations and inventory ($K = 2$).

Capacity	15%	35%	65%	95 %
Vanilla	((a d e), (a b c d))	((a d e), (a b c d))	((a d e), (b d))	((a d e), (b d))
Inventory	(535.19, 299.21)	(700.64, 299.21)	(585.66, 521.73)	(501.88, 448.70)

19.3.3.2 Make-to-Stock and Assemble-to-Order.
A make-to-stock and an assemble-to-order process are special cases of the vanilla assembly where the configuration of vanilla boxes corresponds to the set of final products and when $K = 0$ respectively. It is to be noted that while computing the optimal inventory levels for the different products in MTS, we take into account the limited assembly capacity. In this section, we compare the performance of the vanilla assembly process to assemble-to-order (ATO) and make-to-stock (MTS) processes. We utilized the same 40 problems used in the pilot study.

Table 9 displays the comparison between vanilla assembly process and MTS under negative correlation for the 40 problems used in the pilot study. $(A - B - C)$ in this table indicates that out of the 40 problems, MTS did better in A, MTS and vanilla were equal in B and vanilla assembly was better in C.

- **Demand Correlation.** (1) We find that the vanilla process (with $K = 2$) performs significantly better than MTS under negative correlation and medium and large capacity (Table 9). (2) MTS performs better than vanilla assembly process ($K = 1, 2$) in all cases under positive demand

Table 19.9 Comparison between MTS and Vanilla under Negative Correlation where $(A - B - C)$ in this table indicates that out of the 40 problems, MTS did better in A, MTS and vanilla were equal in B and vanilla assembly was better in C.

		Low Capacity		Medium Capacity		High Capacity	
		K=1	K=2	K=1	K=2	K=1	K=2
Low Var.		(40-0-0)	(31-0-9)	(36-2-2)	(2-8-30)	(5-28-7)	(0-30-10)
High Var.		(40-0-0)	(39-0-1)	(40-0-0)	(14-0-26)	(22-2-16)	(4-4-32)

correlation. It is to be noted that a vanilla assembly process with $K = N$ will outperform the MTS since MTS is a special case of that vanilla assembly system. (3) The performance of vanilla assembly process and MTS performance is better than or equal to ATO in all the cases under both positive and negative correlation (Tables 6 and 7).

- **Capacity Restrictions.** (1) Under medium and high capacity, vanilla assembly process with $K = 2$ performs better than MTS under negative correlation (Tables 6,7 and 9). (2) We determined the relative performance of MTS and the vanilla process $(K = 1, 2)$ respectively by computing the difference in expected cost incurred as compared to ATO as a percentage of ATO. For example, in Table 1, the relative performance of MTS under low capacity and high variance is given by (29722.71-1278.61)/29722.71. We find that under all conditions, the relative performance of MTS and the vanilla process improves with increase in capacity (Table 6,7). (3) ATO with additional capacity could outperform both vanilla and MTS assembly processes. Providing additional capacity to an assemble-to-order system is a viable alternative to a vanilla assembly process depending on the additional capacity required.

- **Variance.** (1) The costs incurred by ATO, MTS and the vanilla assembly process are smaller under lower variance as compared to higher variance (Tables 6 and 7). (2) Both MTS and vanilla process perform better compared to ATO under lower variance. An intuitive explanation being that both vanilla assembly and MTS tend to incur higher costs (either due to holding or stock-out) when there is greater variance in demand. The performance of ATO is not affected that much by variance in a relative sense (Tables 6 and 7).

- **Number of Vanilla Boxes.** (1) We find that the number of vanilla boxes improves the performance of the vanilla assembly process (Tables 6,7,9). Swaminathan and Tayur (1995) note that the decrease in expected cost incurred with increase in K appears to have a convex structure. In addition they note that when $K = N$, an optimal vanilla process could be better and different from MTS which stocks the N final products. For example, if the products are (a, b, c), (b, c, d) and (a, c, d) then the best three vanilla boxes could be (b, c), (a, c) and (c, d).

Swaminathan and Tayur (1995) provide examples with 9 and 10 products where the performance of vanilla assembly with two vanilla boxes is much superior to the make-to-stock (MTS) environment.

19.3.3.3 Application to Industry Size Problems. Our research site was an assembly plant of an IBM product line. There were 50 main products in the product line and each one of them was assembled from 10 major components (features). The ten major components (subassemblies) including - memory, hardfile, powerunit, processor, disk drive, floppy drive, graphic cards and communication cards constituted 80-90% of the cost of a product. The management was interested in efficiently managing the assembly of these components. Typical steps in final assembly involved getting components together (kitting), putting them in the right place (assembly), testing, loading software (preloading) and packing the final product. The existing mode of operations was an assemble-to-order (ATO) process. A make-to-stock environment was not being used because these products were customized from a set of components and it was a daunting task to keep track of all the possible choices. The 50 products considered in our study constituted more than 90% of the demand for the product line. In addition, demands for products were random and correlated. For example, sometimes there was demand for only 20% of the above 50 products and at other times the remaining 80% were in great demand. In such an environment, an assemble to order system provided flexibility to change production based on demand realized. The main problem associated with that approach related to meeting the customer due dates. Orders were going late and many times customer orders were lost due to long lead times quoted.

The management decided to experiment with delayed differentiation using vanilla boxes at the final assembly stage. The assembly plant had a satellite plant where vanilla boxes could be made and shipped. The management was interested in answering the following questions - (1) if a vanilla assembly process would be better than their existing process; if so, (2) how many and what type of vanilla boxes to keep; (3) how should the inventory of vanilla boxes be set; and (4) how to allocate vanilla boxes to final products. The management restricted the number of types of vanilla boxes to utmost 3 in order to make it easier to pilot the process. In addition, they wanted to gradually introduce the different types one at a time and stop when they were either satisfied with the performance or if they found that the vanilla assembly process was too difficult to implement.

Since the problem size was large for our algorithm to handle, we developed a *sequential heuristic* for the problem. Our heuristic is a greedy heuristic which at each stage n selects the vanilla box which provides the maximum cost reduction when used with the previous $n-1$ selected vanilla boxes. So, at any stage n, we fix the configuration of the previous $n-1$ optimal vanilla boxes and optimize the configuration of the nth vanilla box and the inventory of all the n vanilla boxes. Swaminathan and Tayur (1995) report that the above heuristic provides optimal solutions for 87% of the sample set of 80 problems with number of vanilla box types K equal to five.

Table 19.10   Performance of Sequential Heuristic on Industry Size Problems.

	Time Taken	Vanilla Process			ATO	MTS
		K=1	K=2	K=3		
P11	8.51	109525	59114.33	26958.32	180485.5	29461.15
P12	4.68	76214.60	23962.81	2041.42	144485.53	23938.83
P13	6.06	54279.21	6591.45	114.19	137682.34	35230.33
P14	5.15	88486.10	28215.86	3562.71	171986.05	19893.32
P15	6.13	191343.25	134830.52	98155.55	274825.20	48301.84

We obtained the bill of material data for the product line as well as the percentage of component ratios in the product line. Using that data, we created 5 problems: P11, P12 (50 products and 9 components) and, P13, P14 and P15 (50 products and 10 components) and ran our heuristic. The problems P11-P15 do not directly correspond to any IBM problem though they are of the same size. In each of these problems we considered 100 scenarios to capture the demand process. The number of vanilla configurations that were evaluated were in the range of 1000 to 1500 (i.e., 400 to 500 choices out of 1023 were found to be feasible choices for any vanilla type). Table 10 displays the costs of the vanilla, ATO and MTS processes as well as the time required (in hours) to compute these numbers. We find that a greedy sequential heuristic provides effective solutions and can solve the industry size problem in reasonable time. In addition, we also find that the vanilla process significantly outperforms the current assemble-to-order process in all examples and the MTS in all but P15. It is to be noted that the vanilla boxes found by our heuristic may not be optimal and as a result, the performance of the optimal vanilla assembly could be even better.

More details about the vanilla box problem are available in Swaminathan and Tayur (1995).

19.4   ASSEMBLY TASK DESIGN PROBLEM

In the previous models, we have only considered the operations aspect of managing product variety. In this model, we consider the design problem related to finding an efficient sequence for assembly tasks for a product line that is delaying the differentiation of products. Vanilla boxes help in providing quick response to customers for a broad product line while customizing their products by exploiting the inherent commonality in the product family. Consider the product line introduced in section 3 (refer figure 1a). Assume that all options corresponding to a feature (like $a1, a2$) are assembled at the same position in the assembly sequence and let the existing sequence of assembly for the product line $X1$ be given by $Y1 = (a \prec b \prec c \prec d)$. The assembly sequence $Y1$ indicates that features $a \ldots d$ are put in the following sequence first a then b then c followed by d. In such a situation the vanilla box $V1 = (b1, c1)$ is not feasible because it can not be assembled using the assembly sequence since feature a has to be included before b or c can be added to the assembly. On the other hand, if the assembly sequence is $Y2 = (b \prec c \prec a \prec d)$ then the assembly of vanilla box $V1$ is possible. However, additional design changes may to be made to the components to enable the assembly sequence $Y2$. For example, Hewlett Packard had to redesign the power supply unit for their printers in order to provide an ability to insert a common power supply unit at local distribution centers (Lee 1996). These changes might increase the design costs and so a model that can quantify the operational benefit is required for decision making.

19.4.1 Model

In this section we describe our two models (Model-VA and Model-AV) which differ in the sequence in which the decisions about assembly sequence design and operational decisions in terms of vanilla boxes are made. If the manufacturer decides to utilize vanilla boxes during the design phase for components then it could design products to enable delayed differentiation. We model that situation through Model-VA where first the best vanilla configuration is found for the product line and then the best assembly sequence is designed which can facilitate the assembly of set of products and vanilla boxes. On the other hand, often manufacturers make the decision about utilizing concepts such as delayed differentiation after the products have been designed. We model such situations through Model-AV where first the assembly sequence for a given product line is generated and only those vanilla boxes are chosen which can be assembled with that sequence. Both Model-AV and Model-VA have two submodels (ASDP and VCP). The assembly sequence design model (ASDP) determines the optimal assembly sequence for a given product line and vanilla boxes. The objective function measures the cost that needs to be incurred while designing the components in order to enable such an assembly sequence. The vanilla configuration model (VCP) determines the optimal vanilla boxes and inventory. This model is the same as P2 discussed in section 3.1. The assembly sequence design model measures a one time design cost, whereas the operational model measures the expected costs per period during the life cycle of the product.

19.4.1.1 Assembly Sequence Design Model. Consider the example given in Figure 2a. We have four products ($P1 \ldots P4$) and six components ($a1, a2, b1,$

$c1, d1, d2$). The product line is defined by the bill of materials shown in the figure. It indicates that $P1 = (a1, b1, c1)$, $P2 = (a1, b1, c1, d2)$, $P3 = (a2, b1, c1, d2)$ and $P4 = (a1, b1, c1, d1)$. We define a feasible assembly sequence to be one that enables assembly of all the products in the product line. An assembly sequence is defined by a set of precedence constraints on components in the product line. Let us assume that there is no restriction that all the options corresponding to a feature have to occur at the same position in assembly. Then FAS1 in Figure 2b represents a feasible assembly sequence for the product line which is defined by the set of precedence constraints between different components as given below: $(b1 \prec c1; b1 \prec d1; b1 \prec d2; c1 \prec d1; c1 \prec d2)$. In the above assembly sequence $a1$ and $a2$ are unrelated (so are $a1, b1$; $a2, b1$; $a1, c1$ etc.). This implies that they could be assembled in any order. We call such an arrangement an independent relation between the pair of components. On the other hand, the relationship between say, $b1$ and $c1$ is very well defined in that $b1$ has to precede $c1$ in the assembly sequence. Suppose, if we add another product $P5 = (b1, c1, d1)$ to the product line, FAS1 is still a feasible assembly sequence because one could assemble this product P5. However, if we added $P5 = (a1, c1, d1)$ then FAS1 is no longer a feasible assembly sequence because in order to be able to assemble $c1$ we need to have $b1$ in the assembly but P5 does not have $b1$. In that case, a new assembly sequence such as FAS2 is feasible. FAS2 is defined by the following set of precedence constraints: $(b1 \prec d2; c1 \prec d1; c1 \prec d2)$. Next, if we operate this product line with a vanilla box $V1 = (b1, c1)$ then the assembly sequence FAS2 remains feasible. On the other hand, if we had a vanilla box $V1 = (b1, d1)$ then FAS2 is no longer valid because $c1$ should precede assembly of $d1$, and, as a result vanilla box $V1$ can not be assembled.

Our assumptions are as follows.

- (1) We define a feasible assembly sequence to be one that enables assembly of all the products and vanilla boxes. This is in line with the efforts at many manufacturers including Toyota where they are trying to aggregate similar products into a product family and generate assembly sequences for them.

- (2) We model the cost of designing flexibility in components based on precedence constraints on each pair of components. For every pair of components (a, b), there is a cost of assembling one before the other or keeping them independent. One way of determining these costs is to determine what additional design changes have to be made to b if a is assembled before b. If a and b are left independent then the design costs would consists of costs of design changes to both a and b so that a could occur before b and b could occur before a. These design costs may not be readily available and may require additional effort on the part of design engineers. However, utilizing this information and integrating it with the operations cost to design an efficient assembly sequence could significantly improve the overall performance of the system.

- (3) We assume that the design cost for making a pair of components independent is greater than the design cost for facilitating a particular precedence relationship between the components. Our assumption is based on the grounds that independence among components may require design changes to both the components and as a result will cost at least as much as the cost incurred to change one component.

The notation used in our formulation is as follows. We will index the products by i, the components by j and the vanilla boxes by k.

- N: number of products being assembled $(i = 1 \ldots N)$;

- n: number of components $(j = 1 \ldots n)$;

- K: number of different types of vanilla boxes being used $(k = 1 \ldots K)$;

- u_{kj}: contents of the kth vanilla box in terms of the components;

- $\mathbf{U}$: vanilla configuration (matrix of u_{kj}). $\mathbf{U_k}$ refers to kth vanilla box;

- a_{ij}: bill of material (BOM) for the products in terms of the components;

- g_{pq}: cost of assembling component p before component q;

- e_{pq}: cost of allowing complete independence between component p and component q.

- y_{pq}: boolean variable set to 1 if component p is chosen to be always assembled before component q else is set to 0;

- $\mathbf{Y}$: an assembly sequence defined by the set of y_{pq}'s;

- B: set containing boolean values - $0, 1$;

An assembly sequence is defined through Y above, where $y_{pq} = 1$ implies that $p \prec q$ or in other words, component p has to be assembled before component q. We assign design costs for enabling such a precedence relationship. When two components are independent (also called parallel in other parts of this paper) in that either one of them can be assembled before the other then we assume that more cost has to be incurred in designing the components (e_{pq}) as compared to the case when one component necessarily comes before the other (g_{pq}). The difference in these above costs is denoted by $c_{pq} = g_{pq} - e_{pq}$. Our objective is to minimize the total cost incurred (or maximize the benefit). The formulation below corresponds to a particular vanilla configuration U hence, it is called ASDP(U).

$$ASDP(U): \quad \min_{\mathbf{Y}} \sum_{p=1}^{n} \sum_{q=1}^{n} c_{pq} y_{pq}$$

s.t.

$$
\begin{array}{rcll}
u_{kp}(1 - u_{kq}) & \leq & 1 - y_{qp} & \forall p, q, k \quad (1) \\
a_{ip}(1 - a_{iq}) & \leq & 1 - y_{qp} & \forall i, p, q \quad (2) \\
y_{pq} + y_{qp} & \leq & 1 & \forall p, q \quad (3) \\
y_{pq} + y_{qr} & \leq & 1 + y_{pr} & \forall p, q, r \quad (4) \\
y_{pp} & = & 0 & \forall p \quad (5) \\
y_{pq} & \in & B & \forall p, q \quad (6)
\end{array}
$$

In the above formulation, constraint (1) represents that if a component q is not present in a vanilla box then it cannot be a predecessor of any component p in that vanilla box. In other words, if $u_{kp} = 1$ and $u_{kq} = 0$ then y_{qp} must be 0. These constraints make sure that all vanilla boxes can be assembled using the assembly sequence Y. Constraint (2) represents that if a component q is not present in a product then it cannot be a predecessor of any component p in that product. In other words, if $a_{ip} = 1$ and $a_{iq} = 0$ then y_{qp} must be 0. These constraints make sure that all products can be assembled using the assembly sequence Y. Constraint (3) indicates that two components are either unordered in the assembly sequence ($y_{pq} = y_{qp} = 0$) or there exists a unique ordering of these components in the assembly sequence ($y_{pq} = 1$ or $y_{qp} = 1$ but not both). Constraint (4) maintains the transitivity relationship between components. Constraint (5) indicates that all components of the same type are at the same level in the assembly sequence. Constraint (6) is an integrality (B) constraint.

Lemma 4.1. A maximal vanilla does not introduce any new constraints of type (1) in the formulation for ASDP(U).

Proof. Recall the definition of maximal vanilla boxes from section 3.2. Let there be a *maximal* vanilla box V that supports a set of products given by P_V. For every component j not present in V constraints of type (1) are introduced in the formulation which prevent j from being a predecessor to components in the vanilla box. Component j does not occur in at least one product say, $P_j \in P_V$ because if it occurred in all products in P_V then j could be added to V without reducing the number of products it supports and as a result, our initial assumption that V is *maximal* will not be valid. Since P_j has to be assembled using the same assembly sequence, there already exist constraints of type (2) which prevent j (a component that is absent in the product) from being a predecessor to components present in the product. Since the components in V are a subset of components in P_j all the type (1) constraints are already present in the form of type (2) constraints. Thus, no new constraints of type (1) are added on addition of a maximal vanilla box $\square$.

Proposition 4.1. If all the vanilla boxes being used during assembly are maximal then delaying the differentiation does not increase the design cost.

- Initialize *Top-Vanil-List* $= \phi$. Enumerate all feasible vanilla configurations.

- For each configuration **U**

  - Find the optimal inventory and expected cost by solving P2(**U**).
  - If **U** is in the top Z vanilla configurations in terms of expected cost then insert it into Top-Vanil-List.

- For each configuration **U** in Top-Vanil-List

  - Find the optimal assembly sequence and the design cost by solving ASDP(**U**).
  - If **U** has the minimum total cost then store **U**, the inventory levels and the assembly sequence associated with it as Min-Cost-Vanil.

- Output: Min-Cost-Vanil.

Figure 19.3 Our Heuristics for Model-VA.

Proof. Based on Lemma 4.1 we know that no additional constraints of type (1) are added to ASDP when maximal vanilla boxes are used. Thus, the optimal assembly sequence is the same as the optimal assembly sequence for assembling the products alone. As a result, delaying differentiation using vanilla boxes does not increase the design cost □.

The above result has a very important and significant managerial insight. It tells us that additional flexibility could be gained in operations at no additional cost during the design stage if maximal vanilla boxes are used in final assembly.

19.4.2 Algorithm

In this section, we describe our solution methodology for Model-VA and Model-AV.

For Model-VA, we first generate the optimal vanilla configuration for a given product line utilizing the algorithm from Swaminathan and Tayur (1995) which solves a stochastic integer program with recourse (see Figure 3). While solving the above problem, we assume the assembly time for both products and vanilla boxes is equal to the number of components present in them. So, for a product $P1 = (a, b, c, d)$ the per unit assembly time is 4. For a vanilla box $V1 = (b, c)$ the per unit assembly time is 2. If the product $P1$ is assembled from the vanilla box $V1$ then the remaining assembly time is 2 units.

The top Z vanilla configurations (each with K vanilla boxes) along with the inventory levels and the expected cost incurred are stored. For each of these configuration U we find the optimal assembly sequence by solving the assembly design sequence problem ASDP(U) using CPLEX mixed-integer programming routines. It is to be noted that ASDP(U) problem is very similar to the linear ordering problem which has been well studied and is known to be a NP-hard problem. The constraints of type (1) and type (2) are not present in the linear ordering problem. However, Grotschel et al. (1984) have shown that the linear ordering problem can be solved quite easily in many instances. Our experience has been quite similar with the ASDP(U) problem in that fairly large instances can be solved very quickly using CPLEX. We pick the vanilla configuration and the corresponding optimal design cost which incurs the minimum total cost consisting of one time design cost and number of period times the operations cost incurred per period. The performance of our heuristic depends on the value of Z that is chosen. A smaller Z makes it faster but the solution obtained may be far away from the optimal. In our computational study, we first decide on a suitable value for Z through experimentation and then use the same Z across all other experiments.

For Model AV, we first generate the top A assembly sequences for the given product line (see Figure 4). We find the best assembly sequence by solving ASDP (using CPLEX) without the vanilla box constraints (1) in the formulation. The next assembly sequence is generated by introducing cuts in the formulation that remove the earlier optimal solution from the feasible region. This is continued to obtain the top A sequences. For each of these assembly sequence Y we find the optimal vanilla configuration for a given product line using the algorithm in Swaminathan and Tayur (1995) with the following modifications. While solving the problem we enumerate only those vanilla configurations which are feasible under the given assembly sequence Y. We generate the speed-up due to usage of vanilla boxes by taking into account both set-up times and processing times. We assume that independent operations can be done in the same set-up. (These assumptions on setups are based on our observations at US Filter, and we believe that it is a representative application.) The processing times are assumed to be equal to the number of components in the vanilla box or product. For example, if we consider FAS1 (refer Figure 1b), the number of set-ups required to assemble product (a1,b1,c1,d1) is 3. If we assume a set-up fraction of 0.5 then the total set-up time for this product is 1.5 and processing time 4. Thus total time for assembly is 5.5. Consider a vanilla box (b1,c1), its total assembly time is 3. If the vanilla box is used to assemble the product the remaining time for assembly is 2.5 unit.

Once the best vanilla configurations have been found for all the top A assembly sequences we pick the vanilla configuration and the corresponding optimal design cost which incurs the minimum total cost consisting of one time design cost and number of period times the operations cost incurred per period. The performance of our heuristic depends on the value of A that is chosen. In our computational study, we first decide on a suitable value for A through experimentation and then use the same A across all other experiments.

- Initialize *Top-Assembly-List* = ϕ. Generate A top assembly sequences one at a time.

- For each assembly sequence Y in *Top-Assembly-List*

  - Enumerate all feasible vanilla configurations which can be assembled using Y.
  - For each configuration **U**
    * Generate speed-ups for the vanilla boxes taking set-up times into consideration.
    * Find the optimal inventory and expected cost by solving P2(**U**).
  - Find the optimal vanilla configuration **U**. If **U** has the minimum total cost then store **U**, the inventory and the assembly sequence Y as Min-Cost-Vanil.

- Output: Min-Cost-Vanil.

Figure 19.4   Our Heuristic for Model-AV.

19.4.3   Computational Insights

In this section we describe our computational experience in detail. In the following passages we describe generation of problems for the computational study in this paper.

Product Line: We define a product line by a vector such as $(n_1, n_2, \ldots n_p)$ which indicates that there are p features in the product line and each feature i has n_i options in it. We assume that any product has a subset of features present in the product line and contains at most one option of a feature. In addition we assume that there is a one-to-one mapping between options and components in the product line. Thus, the total number of components in the product line is $\sum_{i=1}^{p} n_i$. For example, table 12 shows the results for a set of products where the number of features each providing two options increase from zero to five. The number of products in the product line is assumed to be given and the 0-1 bill of materials is generated using a parameter controlling the sparcity of the bill of materials matrix.

Assembly Time: We assume that it takes one unit of assembly time to add a component to a product when set-up time is not considered (unless specified otherwise). In experiments where set-up time is considered then it is defined as a fraction of the processing time. For example, if the set-up fraction is 50% then it takes 0.50 unit of time to set-up and one unit of time for processing so the total assembly time for adding that component is 1.5 units. In addition, we generate the assembly time for products and vanilla boxes based on the assembly sequence in Model-AV since the vanilla boxes are determined after the assembly sequence is determined. Here our assumption is that if two components are independent then they can be performed in the same set-up (see section 4.2 for an example).

Assembly Capacity: We assume that the total capacity available (variable C in VCP) for production is 60% of the capacity required to satisfy the mean demand for products starting from raw components.

Product Demands: In our experiments we generate product demands from multivariate normal distributions. We generate fifty equally probable scenarios to capture the demand process. We assume that the mean demand lies between a minimum and a maximum value that are specified for the product line. We assume a 10% increase in mean demand per feature offered in the product and a 5% increase in mean demand for each option within a feature. For example, if a product has 4 features and two of the feature have two options each then the increase in demand is 60%. This implies that if the minimum mean demand is 100 and maximum mean demand is 200 then the mean demand for this product is 160. In addition, we maintain the same standard coefficient of variation for all the products in the product line. This is in line with intuitive arguments that an increase in features and options leads to increased market coverage and as a result, should lead to more demand. Marketing literature indicates that it is possible to model markets in terms of ideal points of customer segments which in turn could be used to generate demand for a product using customer preferences and utility functions.

Holding Cost: Holding cost for vanilla boxes are assumed to be equal to the sum of holding cost of components in them. In addition, we assumed the same holding cost for all components. The stock-out to holding cost ratio was kept around 10:1.

Design Costs: The pair-wise design costs for different components is generated from a normal distribution. A random positive number is added to the maximum of the pair-wise sequential costs between every two components in order to obtain the cost for keeping those components independent.

Component Types: We also consider three types of components namely *pseudo simple, pseudo complex and variant* discussed in Stadzisz and Henrioud (1995). Pseudo components (simple or complex) are positioned at the same place in the assembly sequence irrespective of the particular option present in a product. As a result, they do not add any additional constraints to the formulation ASDP. The features that are provided through simple pseudo components have the same pair-wise design costs independent of the number of options because providing options does not pose any additional difficulties in assembly. For complex pseudo components, we increase the pair-wise design costs based on the number of options provided in that feature. For example, if a feature has two options then the pair-wise design costs between this feature and other features is increased by a factor of two. Variant components do not require that options in the component occur in the same place in assembly.

19.4.3.1 Choice of Z in Model-VA. The ability of our algorithm to find the optimal solution depends on the value of Z. From Proposition 4.1 we know that if the optimal vanilla configuration is maximal then the corresponding assembly sequence is optimal. We generated 100 random problems consisting of 3 to 12 products and 4 to 8 components. Note in these experiments each feature had only one option, so the number of features was equal to the number of components.

Table 19.11 Differences in Operations and Design Costs Rankings $(K = 1)$. Rank indicates the position of the vanilla configuration in terms of operations cost.

		Rank	Vanilla Box	Op. Cost	Des. Cost
P1		1	(a,c)	460.6	209154
		2	(a,b,c,e)	643.1	203944

We restricted the number of types of vanilla boxes (K) to one or two. We applied our algorithm to each of these problems with Z equal to n, $2n$, $3n$, $4n$ and 2^n where n is the number of components. To our surprise, we found that in all but one problem (P1 in Table 11), the optimal vanilla configuration was maximal. Thus, except for this problem, there was no effect of increasing Z on the quality of solution obtained. In problem P1, with $K = 1$, the design cost corresponding to vanilla box (a,b,c,e) is lower than the design cost for (non-maximal) vanilla box (a,c). Since (a,b,c,e) is maximal there is no need to consider other vanilla boxes. In all our subsequent experiments, we decided to use $Z = 10$ in that we store the top ten vanilla configurations and generate the assembly sequences for them.

19.4.3.2 Options Provided in Features. In this section we studied the effect the number of features and options on expected costs incurred. As discussed in Ulrich *et al.* (1997), companies differentiate themselves strategically by providing a large number of options on a limited number of features. They indicate that such a phenomenon might occur as a result of core competencies of the firm. Our results provide an operational reasoning for the same phenomenon.

In our experiments we kept the number of features constant at 5 and increased the options on features. While generating the product structure we made sure that only one option of a feature occurs in any product. In the following tables, a feature mix like (12221) means that there are two options in the 2nd, 3rd and 4th features. We find the following.

Number of Features that Provide Options: In this part, we increase the number of features that provide options while keeping everything else constant.

Table 12 shows our results for a typical problem like P10 which has eight products. (1) We find that the operations cost increases with the number of features that provide options while using variant as well as pseudo variant components. An intuitive reason for this trend is that the mean demand for products go up as we increase the number of features that offer options. However, the assembly capacity remains the same, as a result the operations cost goes up due to shortages. (2) The design costs increase for both complex pseudo and variant components. One naturally expects the design cost to go up in both complex pseudo and variant components because they add more complexity to the assembly sequence. We also find that the increase in costs is much greater for complex pseudo components. (3) The operations cost is same across the three types of components because the bill of material remains the same and the optimal vanilla configuration was maximal in each case.

Table 19.12 Number of Features Providing Options.

		Pseudo Complex	Pseudo Simple	Variant
Features	Op. Cost	Design Cost	Design Cost	Design Cost
(11111)	3373.9	20062.8	20062.8	20062.8
(21111)	6883.9	28745.1	20062.8	28684.7
(22111)	9297.5	38701.3	20062.8	38882.2
(22211)	13761.9	50928.0	20062.8	49546.55
(22221)	17905.2	65645.7	20062.8	63288.6
(22222)	21923.1	80791.1	20062.8	80528.8

Number of Options in Features: In this part, we keep the number of features times options at a constant and change the distribution of options across features. (1) We find that in most of our problems the operations cost reduces when we aggregate options together within a feature (see Table 13). An intuitive reason is that as the number of options in one feature increases the options in the others decrease, as a result, a vanilla box formed without the components which offer options is likely to suit delayed differentiation very well. (2) We find that the design cost goes down when the options are aggregated in the case of variant components (see Table 14). The assembly sequence that is generated tends to place the feature with more options at the end of the sequence. As a result, it provides an ideal sequence for delaying differentiation where a feature that has maximum variability is placed at the end of the sequence. For example, Figure 5 considers a problem P13 with a product line having a feature mix of (51111). This implies that components $a - e$ are options on the first feature.

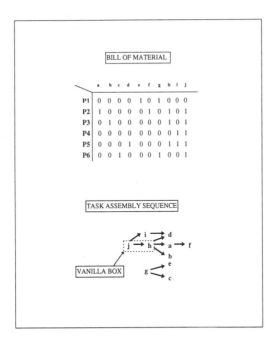

Figure 19.5 Task Sequencing for Delayed Differentiation.

The optimal vanilla box (h, j) and the final assembly sequence obtained show that components with higher variability $a - e$ occur towards the end of the assembly process. These observations (1) and (2) re-enforce arguments presented in Lee and Tang (1998) where they show that features with higher variability in the product line should be added towards the end of the assembly process in order to reduce cost.

Table 19.13 Operations Cost as Number of Options Increase.

	(222211)	(322111)	(421111)	(511111)
P11	32334.4	27214.1	26037.3	20268.3
P12	52830.2	45399.3	44057.0	40590.3

Table 19.14 Design Cost as Number of Options Increase.

		(222211)	(322111)	(421111)	(511111)
	P11	85925.9	84503.4	82673.4	81410.7
	P12	87064.8	85387.5	83940.6	82932.3

19.4.3.3 Set-up Considerations. This subsection provides insights on the model with set-ups (Model-AV). We varied *setup-fraction* – which determines time for each set-up as a fraction of processing time – from 0.2 to 1 in increments of 0.2 while keeping other factors constant. Recall from section 4.2 how we use setup and processing time information to determine the speed-up. We find the following results:

Operations Cost: We find that the operations cost increases with increase in set-up fraction (see Table 15). An intuitive explanation is that increase in set-up increases the total processing time for products and vanilla boxes which leads to an increase in cost when capacity is kept constant.

Vanilla Configuration: The optimal vanilla configuration may change with an increase in set-up fraction (see Table 16). This is mainly due to the fact that change in set-up fraction changes the processing times and speed-ups of vanilla boxes thus making an alternative vanilla configuration more suitable for the product line.

Assembly Sequence: An increase in set-up fraction changes the assembly sequence favoring a sequence that has more parallel operations (see Table 16). An intuitive explanation is that if operations are parallel then the effect of increase in set-up fraction does not get reflected in the processing times for products or vanilla boxes. An interesting example P16 is presented in Table 7 where the vanilla box remains the same but it is produced from a more parallel assembly sequence when set-up fraction increases.

Design Cost: As indicated in the above result, an increase in set-up fraction may increase the number of parallel operations thereby, increasing the design cost.

Other computational insights and details about the model can be found in Swaminathan and Tayur (1997).

19.5 CONCLUSIONS

In this paper, we addressed important issues related to commonality, postponement and operations resequencing within the context of managing product variety. We presented three different models that utilize stochastic programming approach and provided efficient algorithms to solve those problems. In an en-

Table 19.15 Effect of Set-up Fraction on Costs.

		Set-up Fraction		
		0.2	0.6	1.0
P15	Design Cost	296841.0	299599.1	299599.1
	Operations Cost	8753.15	10074.35	10847.9

Table 19.16 Effect of Set-up Fraction on Vanilla Configuration and Assembly Sequence.

		Set-up Fraction		
		0.2	0.6	1.0
P15	Vanilla Configuration	(b,c,e,f)	(c,f)	(c,f)
	Assembly Sequence	$[c \prec e]$	$[\phi]$	$[\phi]$
P16	Vanilla Configuration	(b,e,f)	(b,e,f)	(b,e,f)
	Assembly Sequence	$[b \prec d, b \prec f, e \prec f]$	$[b \prec f, e \prec f]$	$[b \prec f, e \prec f]$

vironment where demands are random (and correlated) our models provide an ability to capture large scale realistic situations. We utilize the inherent structure of the problems to develop computationally efficient algorithms based on sub-gradient methods.

References

BAKER K.R., M.J. MAGAZINE AND H.L.W. NUTTLE, The Effect of Commonality on Safety Stocks in a Simple Inventory Model, *Management Science*, 32, 982-988, 1986.

BERGER, A., J. MULVEY AND A. RUSZCZYINSKI, *An Extension of the DQA Algorithm to Convex Stochastic Programs*, SIAM J. Optimization, 4(4), 735-753, November 1994.

BOOTHROYD G., P. DEWHURST AND W.A. KNIGHT, *Product Design for Manufacturing*, Marcel Dekker, New York, 1994.

COHEN M.A. AND H.L. LEE, Strategic Analysis of Integrated Production Distribution Systems, *Journal of Operations Research*, 36(2), 216-228, 1988.

COLLIER D.A., Aggregate Safety Stock Levels and Component Part Commonality, *Management Science*, 28(11), 1296-1303, 1982.

DEFAZIO T.L. AND D.E. WHITNEY, Simplified Generation of Mechanical Assembly Sequences, *IEEE Journal of Robotics and Automation*, 3(6), 640-656, 1987.

DELL J.S. A Methodology for Product Architecture Decisions Based on Market Segmentation, *Unpublished Masters Thesis*, MIT, 1996.

FISHER M.L., J.H. HAMMOND, W.R. OBERMEYER AND A. RAMAN, Making Supply Meet in an Uncertain World, *Harvard Business Review*, 72(3), 83-89, 1994.

GARG A. AND H.L. LEE, Effecting Postponement through Standardization and Process Sequencing, *Working Paper*, Department of IEEM, Stanford University, 1996.

GERCHAK Y. AND M. HENIG, An Inventory Model with Component Commonality, *Oper. Res. Lett.*, 5(3), 157-160, 1986.

GERCHAK Y., M.J. MAGAZINE AND A.B. GAMBLE, Component Commonality with Service Level Requirements, *Management Science*, 34(6), 753-760, 1988.

GROTSCHEL M., M. JUNGER AND G. REINELT, A Cutting Plane Algorithm for the Linear Ordering Problem, *Journal of Operations Research*, 32960, 1195-1220, 1984.

GUPTA S. AND V. KRISHNAN, Product Family-Based Assembly Sequence Design to Advance The Responsiveness-Customization Frontier, *Working Paper*, Department of Management, University of Texas at Austin, 1996 (forthcoming in *IIE Transactions*).

HALL R., *Zero Inventories*, Dow-Jones Irwin, Homewood IL, 1983.

HAYES R., S. WHEELRIGHT AND K.B. CLARK, *Dynamic Manufacturing*, The Free Press, New York, 1988.

HIGLE, JULIA AND SURVAJEET SEN, *Stochastic Decomposition: An Algorithm for Two-Stage Linear Programs With Recourse*, Math. Oper. Res., 16(3), 650-669, August 1991.

KALL, P. AND S.W. WALLACE, *Stochastic Programming*, John Wiley and Sons, New York, 1995.

KEKRE, S. AND K. SRINIVASAN, Broader Product Line: A Necessity to Achieve Success ?, *Management Science*, 36(10), 1216-1231, 1990.

LAPORTE, G. AND F.V. LOUVEAUX, The Integer L-shaped Method for Stochastic Integer Problems with Complete Recourse, *Operation Research Letters*, 13(1993), 133-142, 1993.

LEE H.L., Effective Inventory and Service Management through Product and Process Redesign, *Journal of Operations Research*, 44(1), 151-159, 1996.

LEE, H.L. AND C. BILLINGTON, Material Management in Decentralized Supply Chains, *Journal of Operations Research*, 41(5), 835-847, 1993.

LEE, H.L. AND C.S. TANG, Variability Reduction through Operations Reversal, *Management Science*, 44(2), 162-172, 1998.

NEVINS J.L. AND D.E. WHITNEY, *Concurrent Design of Products and Processes: A Strategy for the Next Generation in Manufacturing*, McGraw-Hill Publishing Company, New York, 1989.

PINE J.B., *Mass Customization: A New Frontier in Business Competition*, Harvard Business School Press, Boston, 1993.

SAUER, G.L., Commonality in the Multi-Part and Product Newsboy Problem, *Technical Report*, Tuck School of Business, Dartmouth College, 1985.

SRINIVASAN, R., R. JAYARAMAN, R. ROUNDY AND S. TAYUR, Procurement of Common Components in Presence of Uncertainty, *IBM Technical Report*, 1992.

STADZISZ P.C. AND J.M. HENRIOUD, Integrated Design of Product Families and Assembly Systems, *Proceedings. IEEE International Conference on Robotics and Automation*, IEEE Vol. 2, 1290-1295, 1995.

SWAMINATHAN J.M. AND S.R. TAYUR, Managing Broader Product Lines through Delayed Differentiation using Vanilla Boxes, *Working Paper*, GSIA, Carnegie Mellon University, 1995 (last revised 1998) (forthcoming in *Management Science*).

SWAMINATHAN J.M. AND S. TAYUR, Managing Design of Assembly Sequences Product Lines that Delay Product Differentiation, *Working Paper*, GSIA, Carnegie Mellon University, 1997 (revised 1998).

TAYUR, S., Computing Optimal Stock Levels for Common Components in an Assembly System, *Working Paper*, GSIA, Carnegie Mellon University, 1994 (revised 1995;1997).

ULRICH K.T. AND S.D. EPPINGER, *Product Design and Development*, McGraw-Hill, NY, 1995.

ULRICH K., M.L. FISHER M.L. AND D. REIBSTEIN, Managing Product Variety: A Study of the Bicycle Industry, *Product Variety Conference*, UCLA, 1997.

WETS, R., Stochastic Programming: Solution Techniques and Approximation Schemes, *Mathematical Programming: The State of the Art*, Springer, Berlin, 566-603, 1982.

WETS, R., *Handbooks in Operations Research and Management Science, Vol. I*, North Holland, 573-630, 1989.

WOLLMER, R.D., Two Stage Linear Programming under Uncertainty with 0-1 First Stage Variables, *Mathematical Programming*, 19(1980), 279-289, 1980.

20 GLOBAL SOURCING STRATEGIES UNDER EXCHANGE RATE UNCERTAINTY

Panos Kouvelis

Olin School of Business
Washington University
One Brookings Drive
Campus Box 1133
St. Louis, MO 63130

20.1 INTRODUCTION

Firms competing in the "global" marketplace have to take into account the comparative advantages of various countries in forming their manufacturing and sourcing strategies. Developing countries, for instance, have advantages over the developed nations in terms of low labor and raw material costs. Such cost advantages provide strong motivation for multinational firms to seek offshore sourcing arrangements in developing countries. In other cases, critical technological components and/or process equipment are available only from few foreign sources, which are in many cases located in technologically advanced countries other than the firm's home country. In such cases the firm has no choice but to look at these foreign suppliers for its sourcing needs. Because of these reasons, global sourcing is increasingly emerging as a key strategy for companies seeking competitive advantages, and it represents a sizable amount of the economic activities of multinational firms. Nearly 10 percent of Chrysler's $8.6 billion outsourcing budget is spent on global purchases. Westinghouse spent more than 7 percent of its total purchasing dollars on international procurement.

In this paper, we focus on an in-depth presentation and analysis of global sourcing strategies as operational hedging mechanisms for responding to fluctuating exchange rates. The multiple managerial challenges, as well as benefits, arising from global sourcing arrangements have been listed in the management literature (see an excellent reference by Carter and Narasimhan (1990)). The main risk in global sourcing contracts is that, depending on the direction of the exchange rate movement and the time the actual payment is made, a buyer can end up paying substantially more or less than the original contract price. A study by McKinsey & Co. (Hertzell and Casper 1988) found that the unpredictability of currency fluctuations makes strategies based on single supplier sourcing, even with supplier selection accounting for currency swings through forecasts, ineffective. On the other hand, volatile exchange rates provide the buyer with an opportunity to exploit favorable price changes, or avoid unfavorable price differentials, through supplier selection in multi-sourcing arrangements, adjustments in procured quantities and the careful timing of purchases. Because the impact of these currency fluctuations is so large (fluctuations of 1% in a day of 20% in a year are not unheard of), the authors of the above study suggest that companies should consider selecting suppliers on the basis of contribution to balancing currency flows, rather than price and performance. Thus, firms can operationally hedge against currency risk by establishing a portfolio of suppliers in different countries and exploiting its sourcing flexibility.

In this paper we explore the advantages of operational hedging through global supplier selection within the following stylized modeling framework: A firm needs to procure a product from various sources (vendors), some of whom are located in foreign countries. The firm has to decide which vendors to choose, and the quantity that should be sourced from each vendor. The supplier choice and sourced quantity can be adjusted dynamically at the expense of switchover costs, or within the constraints of a given supply contract, in response to fluctuating exchange rates. The firm wants to develop a global sourcing strategy that will minimize its total expected purchasing cost, which might also include any relevant inventory holding costs, in the presence of exchange rate uncertainty. This problem is henceforth referred to as the *global sourcing problem*.

There has been considerable interest in the global sourcing problem. We can classify the literature on the problem in three general categories: vendor selection, international purchasing, and the recent contingent claims ("real options") literature with applications in global operations management issues. On the international purchasing side, Carter and Narasimhan (1990) discuss the emergence of global purchasing as a strategic weapon for U.S. firms and provide guidelines for managers engaged in global sourcing. Carter and Vickery (1988, 1989) illustrate, through a series of numerical examples, the effectiveness of various strategies, mostly financial ones, for managing volatile exchange rates in global sourcing.

Even though in most cases not explicitly addressing issues in an international setting, the vendor selection literature provides a framework for approaching the global sourcing problem. Current and Weber (1994) formulate the vendor selection problem as a variant of the facility location problem. When the vendors have no capacity constraints, the authors show that the problem can be modeled as a simple plant location problem. Weber and Current (1993) solve the multiobjective vendor problem as a mixed integer linear program. Rosenthal et. al (1995) study the vendor selection problem with bundling, where multiple products are to be purchased, and vendors offer product bundles at discounted prices. They propose a mixed integer linear program for the buyer that minimizes the total purchasing cost. While all of the above literature does not deal with uncertain exchange rates, it sets up a plant location framework for the sourcing problem. This is later expanded by Gutierrez and Kouvelis (1993) to capture exchange rate uncertainties. Gutierrez and Kouvelis (1993) use the "robust" framework to model the global supplier selection problem. Their approach develops the supplier network in a way that adequately hedges the firm's performance against the worst contingency in terms of realizable real exchange rate shocks over a planning horizon. Their framework of the problem is a min-max variant of the single plant location problem. However, all of the above papers fail to account for dynamic responses of the firm to fluctuating exchange rates by appropriately exploiting its sourcing flexibilities (e.g., adjusting quantities sourced from various suppliers over time and/or completely switching its sourcing among competing suppliers).

Our approach in structuring the global sourcing problem is influenced by the recent literature on the use of "real options" in evaluating strategic investment opportunities (some of the prominent references in the literature are Dixit (1989a,b), Dixit and Pindyck (1994), Hodder and Triantis (1993), McDonald and Siegel (1986), Trigeorgis (1993)). In the global sourcing problem the firm has the real option to switch among alternative suppliers as the exchange rates change over time. The firm's ability to adapt its sourcing strategy in response to altered macroeconomic conditions increases the firms' value by improving its upside potential while limiting downside losses. For example, if a firm sources from two foreign suppliers, and the cost of one of them increases due to its currency appreciation relative to the home currency, then the firm has the option to switch some or all of the sourced quantity, to the other source if its cost structure is more favorable. Switching of suppliers results in switchover costs, which are incurred because of termination penalties, loss of transportation economies due to lower sourced volume from a specific country, cost of finding new sources, cost of developing needed infrastructure such as a local freight forwarders and other administrative personnel for handling sourcing details, local office overhead, etc. The irreversible nature of supplier switching due to switchover costs, and the future uncertainty in exchange rates, make the choice of timing in switching of suppliers a critical issue, which is best captured through a real options valuation procedure.

The economics and international operations management literature have studied variants of the global sourcing problem in the presence of uncertain exchange rates (see Baldwin and Krugman (1989), Brennan and Schwartz (1995), Dixit (1989a and b) for a sample in the economics literature, and Huchzermeier and Cohen (1996), Kogut and Kulatilaka (1994), Dasu and Li (1997) and Scheller-Wolf and Tayur (1997) for a sample of the international operations management literature). The economics literature has placed emphasis on the problem of firms exiting and entering foreign markets when the exchange rate between the two countries are uncertain. In the international sourcing context, the entry and exit from countries can be thought of as switching between two suppliers in different countries with one of them being the home country (market where the product is sold). The insights obtained for such decisions (through the use of a real options approach) are illuminating. The option to enter a foreign market (i.e., source from a foreign supplier) is similar to a call option where the exercise price is the cost of the investment required to enter the market (i.e., set up the foreign supplying arrangement). The option to enter (to source from foreign supplier) is not exercised unless expected operating profits (purchasing cost savings) are at least sufficient to provide the required rate of return on committed capital. The option to exit (use home country supplier) is not exercised until operating profits (purchasing cost savings) are insufficient to provide the required rate of return on the salvageable capital. The gap between the expected profits that trigger entry and exit induces a **hysteresis** (inertia, inaction) in the firm's investment (foreign sourcing) behavior. An increase in exchange rate volatility widens the gap, thus increasing hysteresis.

There are important differences between the entry/exit models (e.g., Dixit (19989a)) and the global sourcing problem studied here. In the entry/exit models the firm chooses only between two alternatives and there is a single uncertainty factor (i.e., always one supplier is from the home country and the uncertain factor is the exchange rate between the two countries). We allow the firm to choose among more than two alternatives (consideration of the timing and sourced quantity decision simultaneously can lead to more than two alternatives even in the case of two suppliers). We also allow all suppliers to be in countries other than the home country and as a result the uncertain factors, i.e., exchange rates, are multiple and potentially correlated. While the entry/exit literature emphasizes constant switchover costs, we also deal with the case where the magnitude of the switchover cost is linear function of the adjustments in procured quantities from the suppliers. The issues of supplier selection and capacity allocation between various suppliers in uncertain environments have not been explored in the literature so far, and our paper provides some interesting insights on them.

The global operations management literature dealt with the issue of valuing the flexibility and the operational hedging advantages of global multiplant networks. Kogut and Kulatilaka (1994) explicitly value the option of shifting production between two plants located in different countries as exchange rate fluctuates. In any period only one plant produces the required quantities to meet demand. Within a cost minimization framework, where the exchange rate influences the unit production costs, the authors determine when the firm should exercise the option to switch between plants. The optimal production switching behavior exhibits *hysteresis* due to switchover costs associated with shutdowns and startups, labor contracting and managerial time commitment. These results can be interpreted in a similar way to the previous entry/exit results and the differences from our work are exactly the same as for the entry/exit models.

Huchzermeier and Cohen (1996) develop a stochastic dynamic programming formulation for the valuation of global supply chain network options with switching costs. A numerical evaluation scheme (a multinomial approximation of correlated exchange rate processes) is proposed for the valuation of a global supply chain network for a discrete set of alternative network options with costly switching between them. This work is the first to numerically evaluate the option value of operational flexibility of a network of factories and demonstrates some of the benefits of operational hedging in environments of volatile exchange rates. However, the work maintains a numerical focus and does not explicitly deal with the structure of the optimal operating policies. The work of Dasu and Li (1997) is along the lines of Kogut and Kulatilaka (1994), with the additional feature of capacity constrained plants. Scheller-Wolf and Tayur (1997) study a global sourcing problem taking the perspective of a small supplier. All transactions are conducted in the supplier's currency. The paper analyzes a minimum order quantity contract in the presence of international exchange rates. Such contracts are beneficial to the supplier as they transfer some of the uncertainty to the buyer in exchange for a reduction in the unit price charged by the supplier. In addition, such contracts can be used as aids in financial hedging by reducing the variance that needs to be hedged against. Finally, Kouvelis, Axarloglou, and Sinha (1997) study the effect of real exchange rates on the long term production strategies of firms entering foreign markets within a profit maximization framework (i.e., exporting, joint ventures and wholly owned production facilities) using the real options methodology.

Thus the main contribution of our paper are: (i) the presentation of a formal modeling framework for the evaluation of multi-supplier sourcing in volatile exchange rate environments; (ii) the development of insightful structural results on the form of optimal sourcing strategies in the case of two suppliers; and (iii) the presentation of useful managerial insights on the effects of the variance and the correlation of the exchange rate processes of the supplier currencies on the cost of the sourcing strategy and the allocation of sourced quantities among suppliers.

Also in this paper, we develop a new modeling approach in studying supplier selection and switching of supplier strategies for environments of deterministic demand but with uncertain prices for sourced materials. The differentiating factor of this approach is that, instead of the abstract modeling artifice of a switchover cost, it looks at the detailed constraints specific sourcing contracts impose on the exercise of supplier switching strategies. We consider specific types of supply contracts that can be developed with the potential supplier. A "time flexible contract" allows the firm to specify the purchase amount over a given period of time without specifying the exact time of purchase. Besides time flexibility, the suppliers may offer "quantity flexibility" to the firm as well, i.e., purchase quantities could be within a prespecified quantity window (minimum and maximum order quantity contracts). Finally, "risk-sharing" features can be incorporated in the contract in terms of the purchase price that the firm eventually pays to a supplier. Within a prespecified price window the firm pays the realized price, but outside of it the firm shares, in an agreed way, added costs or benefits. Given the structure of a supply contract and for a two potential supplier sourcing environment, we study the firm's decision of when to purchase and how many units in each purchase from each supplier such that the expected net present value of the purchase cost plus inventory holding cost is minimized.

The structure of our paper is as follows. In Section 2 we present a numerical illustration to demonstrate how a flexible sourcing strategy can complement various "financial" instruments available to hedge against exchange rate uncertainty, and can actually be the most effective in the long run. In Section 3 we present a general framework for evaluating the multisupplier sourcing arrangements and formalize the global sourcing problem. In Section 4 we study the structure of dynamically optimal sourcing strategies when a firm is allowed to choose from two sources in different foreign countries. We also present numerical results that address the issues of vendor selection and capacity allocation between the two foreign suppliers. In Section 5 we develop a new modeling approach in addressing issues of supplier selection and switching of supplier strategies for specific forms of supply contracts (in particular time flexible and quantity flexible contracting arrangements with risk-sharing features). Finally, in Section 6 we synthesize our research results into useful managerial insights for global operations and logistic managers.

20.2 EFFECTIVENESS OF VARIOUS HEDGING STRATEGIES IN GLOBAL SOURCING: AN ILLUSTRATION

Carter and Vickery (1989) provide an excellent overview of various currency risk hedging mechanisms in global sourcing. According to the authors, the currency management strategies can be divided into two groups: macro level strategies and micro level strategies. Macro level strategies affect the sourcing decision itself through supplier selection or the structure of the sourcing contract (e.g., contractual sourcing flexibility exploited in the volume-timing of purchases), while micro level strategies are employed after the source has been selected and are typically short term financial strategies used to protect the buyer from the risk of exchange rate fluctuations. We first summarize some of the effective micro-level hedging strategies used in practice, and then proceed to demonstrate the effectiveness of the macro-level strategy that exploits the sourcing flexibilities in a portfolio of global suppliers.

In the case of a U.S. buyer, the simplest way to avoid the risk of volatile exchange rates is to pay in U.S. dollars. This strategy transfers all the risk of an adverse currency fluctuation from the buyer to the supplier, if the supplier is willing to accept it. The drawback here is that the purchasing firm is not able to take advantage of a favorable movement of the exchange rate.

An alternative strategy for a firm buying globally is to buy "forward" the needed foreign currency. A forward contract is an agreement to buy or sell foreign currency at a future time at a certain fixed exchange rate. The contract is between the firm sourcing from a foreign source and a financial institution. This practice removes the risk of an unfavorable currency fluctuation by providing the buyer with a firm future price. The forward contract is bought by paying a commission to the bank handling the transaction, and is set by the difference between the bank's buy and sell price. To engage in forward buying, the firm must have a line of credit with the bank. Because the expense of individual contracting is significant, the forward market is limited to large customers dealing in foreign trade.

Another hedging mechanism is to buy a "future," that is, a contract to purchase the needed amount of foreign currency from the currency commodity trade. The difference between a forward contract and a future contract is that the forward contract avoids currency risk by locking a fixed exchange rate, while the futures contract minimizes the currency risk by exploiting the movement of the exchange rate. If the buyer's selected currency of payment goes up in value in the interim, the buyer sells the contract for a profit and uses it to make up the difference when he makes sourcing payments. If, on the other hand, the buyer's selected currency falls in value, the buyer loses value when he sells the futures contract but makes up most of the differential from the sourcing contract. Apart from the above three methods of reducing exchange rate risks, involved parties also use various forms of risk-sharing contracts. One common type of risk-sharing contract stipulates that exchange rate losses, or gains, are to be shared equally (or by a prespecified percentage) by both parties.

We will now proceed to demonstrate the effectiveness of the various hedging strategies in an illustrative global sourcing example. The main objective of this illustration is to indicate the benefits of operationally hedging through a global supplier portfolio in sourcing environments fraught with currency risks. While it is essential for the firm to hedge financially in the short run to avoid catastrophic shocks of exchange rates, in the long run these strategies can be advantageously complemented with flexible global sourcing strategies.

Consider a U.S. firm that can source from two foreign suppliers, one in Canada and the other in Japan. The current date is January 1, and the goods are needed on January 1, April 1, July 1, October 1 of the current year and January 1 and April 1 of the following year. The current price quoted by both suppliers for each shipment is U.S. $100,000. Both suppliers have agreed to freeze their price at the current rate (in U.S. dollars) for the next six quarters. Assume that payment is made on the date of purchase.

The spot, forward and futures rates for the two currencies are given in Table 1a. The spot rate is the actual realized exchange rate between the currencies of the two countries, the forward rate is the rate at which a financial institution will sell forward currency contract and the futures rate is the rate at which currency futures are traded in the currency futures market.

Now consider the effect of various hedging strategies on sourcing from the Canadian supplier. Referring to the situation described above, if the firm pays throughout the six quarters in U.S. dollars (called base case strategy henceforth), the net present value (NPV) of its payment stream for the next six quarters is $479,078.68 irrespective of the realized exchange rates throughout the time period. All results related to this example are tabulated in Table 1b. All the reported costs in this table are in U.S. dollars and the discount rate used for NPV calculations 0.1 per quarter. If the firm pays throughout in Canadian dollars without any hedging, the NPV of its costs will be $449,546.64, making a net gain of $29,532.04 compared to the previous case. If it were to get into a risk-sharing agreement with its suppliers (sharing 50% of loss or gain because of the exchange rate movement) while paying in Canadian dollars, its NPV will be $464,312.66. The gain is less than the previous case because of the sharing of gains with the supplier.

If the firm decides to pay the Canadian supplier in Canadian dollars and hedge its risk by buying forward contracts from a financial institution, it can buy contracts at the forward rates and use these to settle its accounts. The NPV of its payments in this case is $456,986.90. Suppose the firm decides to hedge its risk by buying currency futures instead of forward contract. To keep the exposition simple, assume that one futures contract for the Canadian dollar is for the delivery of U.S.$100,000 and the delivery dates are the same as the dates the firm needs to make the payments. The firm therefore buys long one futures contract at the beginning of each quarter and uses this to hedge against the exchange rate risk for its payment that is due at the end of the quarter. (Please refer to Hull 1993, Chapter 2, for a discussion on hedging using currency futures). In this case, the NPV of the firm's payment is $459,354.38. Table 2 also reports the respective costs associated with sourcing from the Japanese supplier using the above hedging strategies. Using any of the above strategies, the firm could have kept its sourcing costs close to the NPV of a stable currency (i.e., $100,000 every quarter for the next six quarters), which we describe as the base case strategy. Now consider the situation where the firm switched between suppliers to take advantages of favorable exchange rates. Table 1c shows the optimal sourcing strategy for any given starting supplier. If the firm chooses to start with the Canadian firm, it will make four switches in six quarters. It would have to make five switches if it started sourcing from the Japanese supplier, The NPV of the payments is $441,465.37.

The numerical example, though simple, clearly illustrates that maintaining a portfolio of suppliers and switching among them in order to take advantage of the exchange rate fluctuations can be extremely effective in the long run. Because the currencies of the two countries move in opposite directions during the six quarters of the sourcing arrangement, the buyer can take advantage of the fluctuations of the exchange rates by always choosing the cheapest source. The above calculations do not take into account the transaction costs associated with implementing each of the above hedging strategies (i.e., costs of purchasing the futures or forward contracts, switchover costs between suppliers). It can be argued that the transaction costs for "financial" instruments may be substantially lower than the costs of switching between suppliers, but given the big difference in the NPV of the switching option and the other hedging mechanisms, the use of this strategy should not be ruled out. In particular, if the operational hedging is combined with appropriate financial hedging to avoid short term risks of macroeconomic shocks, the firm can obtain substantial long term profitability benefits. In our example, the combination of a "futures" contract and a supplier switching strategy leads to the lowest overall sourcing cost of $435,358.77 for the Japanese supplier.

20.3 GENERAL FRAMEWORK FOR EVALUATION OF GLOBAL SOURCING STRATEGIES IN A MULTI-SUPPLIER PORTFOLIO

We now present a methodology for (a) calculating the cost of a specific global sourcing arrangement in the presence of exchange rate uncertainty, and (b) finding the dynamically optimal sourcing solution for the global sourcing problem where the firm has the choice to source from n global suppliers. Let the exchange rate between the host country (U.S.) and the foreign country i be given by

$$e_i = \frac{U.S.\$}{n.c._i}, \ i \in \{1,2,...n\},$$ where $n.c._i$ is the currency of country i. Note than an increase in the exchange rate by this definition implies a depreciation of the U.S. dollar. In our further discussion we use, without loss of generality, e_i to denote both the sourcing cost from the i-th supplier and the exchange rate of that country. Let $e = (e_1, e_2, ..., e_n)$ be a vector of exchange rates. Also, let q_i be the percentage of demand sourced from the supplier in country i. We assume that all demand has to be met. Therefore, $\sum_1^n q_i = 1$. Let $q = (q_1, q_2, ...q_n)$ be a vector of quantities (expressed as percentages of demand) sourced from all suppliers. Let Q_i be the

capacity of the supplier in country i and $Q=(Q_1,Q_2,...Q_n)$ be a vector of capacity constraints of all the suppliers. Similar to quantities sourced, capacities are expressed as demand percentages. We assume that there is enough capacity in the supplier network to meet all demand, i.e., $\sum_i Q_i \geq 1$. Let $\pi(Q,q,e)$ be the cash outflow rate and $F(Q,q,e)$ be the infinite horizon cost of the optimal sourcing strategy when the vectors of supplier capacity, quantity sourced and exchange rates are Q, q and e respectively.

The firm can adjust its quantities sourced vector q, but incurs an adjustment cost (also referred to as switchover cost) each time it does so. Adjustment in quantity sourced may require setting-up new contracts or renegotiating existing ones, facing legal transaction costs due to contract termination or renegotiation, building up relationships with new suppliers or abandoning existing ones, having to establish local offices in new countries to handle sourcing details and to find and subcontract local freight forwarders. The adjustment cost captures such expenses, as well as internal processing costs (or benefits, if it so occurs) due to loss (gain) of economies of scale by decreasing (increasing) sourced volumes from specific regional ports, and other costs independent of purchase volume but specific to the contractual agreement (e.g., need to provide necessary tooling or training to inexperienced suppliers). Formally, the firm can change the quantity vector from q to q' instantaneously, but incurs a switchover cost of $k_{qq'}$ whenever it makes the switch. This switchover cost can have different functional forms depending on the specific situation. There are two general cases of the switchover cost modeling. In the first case, the switchover cost is independent of the adjustments in the sourced quantities. This case has two subcases. If, for instance, the switchover cost is incurred in home country currency and involves the payment of a fixed prenegotiated fee, it can be modeled as a constant independent of the fluctuations of the exchange rate. On the other hand, if the switchover cost entails transaction costs in the involved supplier's currencies, then it can be modeled as a linear combination of the exchange rates of their respective countries. In the second case, the switchover costs are linear functions of the adjustments in the quantity sourced from each supplier.

The exchange rate between the host country (h) and the supplier's country i, i.e., e_i, can be modeled by the following Ito process:

$$de_i = \mu_i(e_i,t)dt + \sigma_i(e_i,t)dz_i \tag{1}$$

where $\mu_i(e_i,t)=$ drift parameter, $\sigma_i(e_i,t)=$ variance parameter, and $dz_i=$ increments of a Wiener process. Let $\rho_{ij}(0\leq\rho_{ij}\leq 1)$ be the correlation coefficient between the exchange rates of countries i and j. Equation (1) implies that the current value of the exchange rate is known, but future values are always uncertain.

We now setup the needed assumptions for an application of contingent claims analysis to the problem. The securities market is assumed to be perfect, with trading in risky securities which are sufficient for dynamic spanning of the sourcing costs from the foreign suppliers. There is also trading in a riskless asset that has a rate of return equal to r. Spanning implies the existence of security portfolios which replicate the rates of change for the sourcing costs of the product. These portfolios have cost processes which follow the Ito process:

$$dm_i = \alpha_i\left(m_i,t\right)dt + \sigma_{m_i}\left(m_i,t\right)dz_i, \tag{2}$$

where m_i is the cost of the spanning portfolio replicating the sourcing cost of supplier i. The capital asset pricing model (CAPM) says that

$$\alpha_i\left(m_i,t\right) = \phi\rho_{e_i m}\sigma_i\left(e_i,t\right),$$

where ϕ is the market price of risk and $\rho_{e_i m_i}$ is the correlation coefficient between e_i and the entire market portfolio m. Let $\delta_i = \alpha_i\left(m_i,t\right) - \mu_i\left(e_i,t\right)$. Each of the terms δ_i represents the difference between the rate of change on the portfolio which replicates the cost of supplier i and the expected rate of change on the actual cost of that supplier. These terms, following the standard terminology in the finance literature, are called "equilibrium rates of return shortfalls" or "convenience yield" on the underlying variables. The "convenience yield" δ_i represents the benefits the sourcing firm derives from keeping supplier i in its portfolio. For example, if the currency of supplier i depreciates with respect to the U.S. dollar, the firm can reduce its total cost by switching to this supplier. Similarly, if there are significant disruptions in the home country supply network due to labor strikes or raw material shortages, or the firm faces pressures to increase prices by colluding home country suppliers, it can benefit from its option to source from the already established foreign supplier i.

The above assumptions allow the construction of a trading strategy which uses investments in the riskless asset and the security portfolios to replicate the cash outflows of the sourcing strategy. We can now use the standard arguments of contingent claims analysis to obtain a partial differential equation for the cost of the optimal global sourcing strategy. Let the portfolio consist of the following: 1 unit call option to source from a global portfolio of suppliers, which costs $F\left(Q,q,e\right)$ and l_i units long in the spanning portfolio m_i for all i, $i=1,2,\ldots,n$ which costs $\Sigma_i l_i e_i$. For simplicity we represent $F\left(Q,q,e\right)$ by F and $\sigma_i\left(e_i,t\right)$ by σ_i. In a short interval of time dt the expected growth of the portfolio, after application of Ito's Lemma, equals $d\left(F - \Sigma_i l_i e_i\right) = \left(\frac{1}{2}\Sigma_i \sigma_i^2 F_{ii} + \Sigma_{i\neq j} \rho_{ij}\sigma_i\sigma_j F_{ij}\right)dt$ $+ \Sigma_i\left(F_i - l_i\right)de_i$ where F_i represents the derivative of F with respect to e_i and F_{ij} denotes the partial derivative of F with respect to e_i and e_j. If we assume $l_i = F_i$, then the investment portfolio becomes risk-free since de_i is the only stochastic element in the growth term $d\left(F - \Sigma_i l_i e_i\right)$ of the portfolio. Convenience yield for holding the portfolio equals $\left(\Sigma_i l_i \delta_i e_i\right)dt$, which the firm saves because it holds the long position. The risk-free cost of the portfolio equals $r\left(F - \Sigma_i l_i e_i\right)dt$ because of the assumed trading in riskless assets.

During time dt, expected cash outflow (ECF) to the suppliers equals $\pi\left(Q,q,e\right)dt$. The expected cost growth (ECG) on the portfolio is the difference of the expected growth of the portfolio and the gain from holding the long position, i.e., its convenience yield. Therefore, in time dt, $d\left(F - \Sigma_i l_i e_i\right) - \Sigma_i F_i\delta_i e_i dt =$ $\left(\frac{1}{2}\Sigma_i \sigma_i^2 F_{ii} + \Sigma_{i\neq j} \rho_{ij}\sigma_i\sigma_j F_{ij} - \Sigma_i F_i\delta_i e_i\right)dt$. The sourcing firm's expected total cost (ETC) of sourcing from a global supplier network equals the sum of expected cash outflow (ECF), the expected cost growth (ECG), and any adjustment costs that may be incurred. The optimal sourcing policy is the one that minimizes this expected total cost. Also, ETC per unit time should be equivalent to the riskless cost in a risk-neutral economy. Therefore, the value of an optimal sourcing strategy should satisfy the following equation:

$$r\left(F - \Sigma_i\, F_i e_i\right) = \min_{q'}\left\{\pi\left(Q, q', e\right) + k_{qq'} + \tfrac{1}{2}\Sigma_i\, \sigma_i^2 F_{ii}\right.$$
$$\left. + \Sigma_{i\ne j}\, \rho_{ij}\sigma_i\sigma_j F_{ij} - \Sigma_i\, F_i\delta_i e_i\right\} \tag{3a}$$

$$s.t. \quad q'^T \le Q^T, \tag{3b}$$

where q'^T and Q^T are the transpose of vectors q' and Q, respectively. As a reminder, q is the vector of currently sourced quantities from all suppliers.

Equations (3) are the general formulation of the global sourcing problem with n suppliers. The constraint $q'^T \le Q^T$ ensures that the capacity constraint at each supplier is met. In the general case, Equation (3) will result in partial differential equations (PDEs), which have to be solved subject to value matching and smooth pasting conditions (to be explained later) at critical threshold levels of the exchange rates to obtain the optimal strategies (the actual quantity sourced from each supplier) and calculate the cost of such strategies. In the above expression $k_{qq'}$ can be either a constant k, i.e., $k_{qq'} = k$, a function of the exchange rate vector, i.e., $k_{qq'} = k_{qq'}(e)$, or a function of the adjustments in the quantity sourced. The optimal sourcing strategy and its cost depends on $q, Q, e, \pi(Q, q, e), k_{qq'}$ and various parameters that define the exchange rate. It is extremely difficult to obtain a closed form solution of the resulting PDE in the general n supplier case. However, some special cases of the problem yield mathematically tractable solutions, which allow us to build intuition on the form of global sourcing strategies. The next sections describe these cases.

20.4 SOURCING STRATEGIES WITH TWO SUPPLIERS AND CONSTANT SWITCHOVER COSTS

20.4.1 One Home and One Foreign Supplier

In this sub-section the firm has the choice to source from two suppliers, one in the home country (h) and the other in a foreign country (1). The capacity of the two suppliers are Q_h and Q_1, respectively. The firm incurs a constant cost of Kq_h dollars per unit of time if it sources q_h percent of its requirements from the home supplier and a stochastic cost of eq_1 dollars per unit time if it sources q_1 percent of its demand from the foreign supplier, where e is given by the following geometric Brownian motion:

$$de = \mu e\, dt + \sigma e\, dz. \tag{4}$$

The results for the above case can be easily inferred from similar results in the economics and international operations management literature (see Dixit (1989a,b),

Kouvelis and Sinha (1995), Kogut and Kulatilaka (1994) and Dasu and Li (1993)). According to these results, if the cost of adjusting the sourced quantity vector q is constant, linear or convex in adjustments in sourced quantities, then it is optimal for the firm to follow a "bang-bang" sourcing policy, that is, if at any point in time it is cheaper to source from one supplier, then the firm would source the maximum possible from this source. For example, if it is cheaper to source from the home supplier, then the firm would buy Q_h% from the home supplier and the rest $\left((1-Q_h)\%\right)$ from the supplier in country 1. If the exchange rate changes so that it becomes cheaper to source from supplier 1, then the firm will adjust q so that it sources Q_1% from supplier 1 and $(1-Q_1)$% from supplier h. We call the above adjustment a "switch" from supplier h to supplier 1. For notational simplicity, when adjustments in sourced quantity are easily predetermined by the nature of the optimal policy, we modify our switchover cost notation to k_{ij}, indicating switching from supplier i to j. Thus, switching from the home country supplier to foreign supplier incurs switchover cost k_{1h}.

Switchover costs and exchange rate uncertainty induce a band of "inaction," where firms wait for the exchange rate to increase beyond (or decrease below) critical values before switching. It has been shown that the hysteresis band increases with increase in the switchover costs (k_{1h}) and the volatility (σ) of the exchange rate process between the two countries, while an increase in the risk-free rate (r) or a decrease in the "rate of return shortfall" (δ) will cause the hysteresis band to shrink. The previously mentioned literature has discussed various methods of finding the critical switchover points and calculating the values of various policies. We refer the interested reader to the above articles for a complete discussion of this problem.

20.4.2 Suppliers in Two Foreign Countries

In this section we consider the case where the firm can choose from two suppliers located in different foreign countries. We call them supplier (in country) 1 and 2 respectively, and let Q_1 and Q_2 be their capacities. The exchange rate between the home country currency (in which the sales of the final product occur) and country i currency $(i \in \{1,2\})$ is given by:

$$de_i = \mu_i e_i dt + \sigma_i e_i dz. \qquad (5)$$

Following the notational conventions of Section 20.4.1, the switchover costs $k_{12}(k_{21})$ are applicable when a switch from supplier 1 to 2 (2 to 1) occurs, since from (3) we can easily see that in these cases quantities sourced are such that the capacity constraints $Q_2(Q_1)$ are binding, respectively. Also, the cost outflow rates when "primarily sourcing" from supplier 1 (i.e., the preferable supplier in terms of unit cost is 1) or supplier 2 are $\pi_1(Q, q_1, e) = e_1 Q_1 + e_2(1-Q_1)$ and $\pi_2(Q, q_2, e) = e_2 Q_2 + e_1(1-Q_2)$ respectively. We will assume below, if not otherwise indicated, that the switchover cost k_{ij} is constant, i.e., independent of the exchange rate process between the home and foreign country. As stated in Section 20.4.1, when the cost to adjust the sourced quantity vector q is linear or convex in the adjusted quantities, the capacity constraint is binding in at least one supplier, (i.e., when the preferred supplier is 1, the sourced quantity vector is $q_1 = (Q_1, 1-Q_1)$, and when the preferred supplier is 2, $q_2 = (1-Q_2, Q_2)$). We will not discuss the above case in detail in this paper, but our results can be easily extended to cover it.

We suppose the dependence of F on e_1 and e_2 for notational convenience in our further discussions, and let F^i represent the value of the optimal sourcing strategy when the current preferred supplier is in country i, $i \in \{1,2\}$. From (3) we can easily conclude that for a firm currently sourcing from supplier 1, and for exchange rates for which it is optimal to continue sourcing from the same supplier, it holds that

$$\tfrac{1}{2}\sigma_1^2 e_1^2 F_{11}^1 + \rho\sigma_1\sigma_2 e_1 e_2 F_{12}^1 + \tfrac{1}{2}\sigma_2^2 e_2^2 F_{22}^1 + (r-\delta_1)e_1 F_1^1 + (r-\delta_2)e_2 F_2^1$$
$$- rF^1 + e_1 Q_1 + e_2(1-Q_1) = 0. \tag{6}$$

Similarly, a firm that is sourcing from the supplier in country 2, and for exchange rates for which it is optimal to continue sourcing from the same supplier, it holds that

$$\tfrac{1}{2}\sigma_1^2 e_1^2 F_{11}^2 + \rho\sigma_1\sigma_2 e_1 e_2 F_{12}^2 + \tfrac{1}{2}\sigma_2^2 e_2^2 F_{22}^2 + (r-\delta_1)e_1 F_1^2 + (r-\delta_2)e_2 F_2^2$$
$$- rF^2 + e_1(1-Q_2) + e_2 Q_2 = 0. \tag{7}$$

It can be shown that (see Appendix 1):

$$F^1 = \frac{e_1 Q_1}{\delta_1} + \frac{e_2(1-Q_1)}{\delta_2} - A_1 e_1^{\beta_1} e_2^{1-\beta_1} - A_2 e_1^{\beta_2} e_2^{1-\beta_2} \tag{8}$$

where $\beta_1 \leq 0$ and $\beta_2 \geq 1$ are given by the roots of the following equation:

$$\sigma^2 \beta(1-\beta) + (\delta_1 - \delta_2)\beta + \delta_2 = 0 \tag{9}$$

and $\sigma^2 = \tfrac{1}{2}\sigma_1^2 - \rho\sigma_1\sigma_2 + \tfrac{1}{2}\sigma_2^2$. The term $A_1 e_1^{\beta_1} e_2^{1-\beta_1}$ in (8) is the value of the option to switch from supplier 1 to supplier 2 if e_1 increases. As $e_1 \to 0$, this option should have zero value because the firm will never switch to supplier 2, which implies $A_1 = 0$. Letting $A_2 = A$,

$$F^1 = \frac{e_1 Q_1}{\delta_1} + \frac{e_2(1-Q_1)}{\delta_2} - A e_1^{\beta_2} e_2^{1-\beta_2} \tag{10}$$

Using similar reasoning, it can now be shown that

$$F^2 = \frac{e_1(1-Q_2)}{\delta_1} + \frac{e_2 Q_2}{\delta_2} - B e_1^{\beta_1} e_2^{1-\beta_1}. \tag{11}$$

We use the following notation for the critical exchange rate levels at which supplier switching occurs according to an optimal sourcing strategy. If supplier 1 is the current supplier, then the firm switches over to supplier 2 if the exchange rate hits the barrier $e_1^{**}(e_2)$ (i.e., the switchover point e_1^{**} is a function of the current value of e_2). Similarly, if the current supplier is 2, then the firm switches over to supplier 1 if the exchange rate drops below the barrier $e_1^*(e_2)$. The presence of the "hysteresis band" will be proved below, after introducing some needed results.

Equations (12) and (13) below are the value matching and smooth pasting conditions, respectively, at the switchover points, and serve as the required boundary conditions in order to obtain a specific solution of (10) and (11):

$$F^1\left(e_1^{**}(e_2),e_2\right)=F^2\left(e_1^{**}(e_2),e_2\right)+k_{12}$$
$$F^2\left(e_1^*(e_2),e_2\right)=F^1\left(e_1^*(e_2),e_2\right)+k_{21}$$

(value matching) (12)

and

$$F_1^1\left(e_1^{**}(e_2),e_2\right)=F_1^2\left(e_1^{**}(e_2),e_2\right)$$
$$F_1^2\left(e_1^*(e_2),e_2\right)=F_1^1\left(e_1^*(e_2),e_2\right)$$

(smooth pasting) (13)

Substituting (10) and (11) in the value matching and smooth pasting conditions, we get a system of non-linear equations in terms of $e_1^*(e_2), e_1^{**}(e_2)$, A and B. Solving for these four unknowns, we can find the switchover thresholds and thus F^1 and F^2. This system of equations, after simple algebraic manipulation, is:

$$Ae_1^{**}(e_2)^{\beta_2}e_2^{1-\beta_2}-\frac{e_1^{**}(e_2)(Q_1+Q_2-1)}{\delta_1}+\frac{e_2(Q_1+Q_2-1)}{\delta_2}=Be_1^{**}(e_2)^{\beta_1}e_2^{1-\beta_1}-k_{12}$$

$$Be_1^*(e_2)^{\beta_1}e_2^{1-\beta_1}+\frac{e_1^*(e_2)(Q_1+Q_2-1)}{\delta_1}-\frac{e_2(Q_1+Q_2-1)}{\delta_2}=Ae_1^*(e_2)^{\beta_2}e_2^{1-\beta_2}-k_{21}$$

(14)

$$\beta_2Ae_1^{**}(e_2)^{\beta_2-1}e_2^{1-\beta_2}-\frac{Q_1+Q_2-1}{\delta_1}=\beta_1Be_1^{**}(e_2)^{\beta_1-1}e_2^{1-\beta_1}$$

$$\beta_2Ae_1^*(e_2)^{\beta_2-1}e_2^{1-\beta_2}-\frac{Q_1+Q_2-1}{\delta_1}=\beta_1Be_1^*(e_2)^{\beta_1-1}e_2^{1-\beta_1}$$

The above equations for defining the thresholds $e_1^*(e_2)$ and $e_1^{**}(e_2)$ are highly non-linear and do not have closed-form solutions, but the existence, and some general properties, of the "hysteresis band" can be proved.

Proposition 1 *Switchover costs induce a band of "inaction" in the exchange rate region, $e \in \left\{e_1^*(e_2), e_1^{**}(e_2)\right\}$, i.e., for exchange rate values between the two critical number firms wait for the exchange rates to increase beyond (or decrease below) the critical values $e_1^{**}(e_2)$ (or $e_1^*(e_2)$) before switching from supplier 1 to 2 (or from supplier 2 to 1).*

In the absence of switchover costs, and for a one period planning horizon, a firm that is currently sourcing the maximum possible from supplier 1 (i.e., $q=(Q_1,1-Q_1)$ will switch to supplier 2 if $e_1 \geq e_2$. Similarly, a firm that is sourcing the maximum possible from a supplier in country 2 (i.e., $q=(Q_2,1-Q_2)$ will switch to supplier 1 when $e_1 \leq e_2$. Thus, in this case the switchover points are $e_1^{**}(e_2)=e_1^*(e_2)=e_2$. According to Proposition 1 a firm that is sourcing from supplier 2 will wait until a critical exchange rate $e_1^*(e_2) < e_2$ before switching to supplier 1, while a firm sourcing from supplier 1 will wait until $e_1^{**}(e_2) > e_2$ before switching to supplier 2. The range $\left(e_1^*(e_2), e_1^{**}(e_2)\right)$ is the hysteresis band, or the band of "inaction" where the optimal policy is to continue sourcing from the present supplier. The optimal switchover levels $e_1^*(e_2)$ and $e_1^{**}(e_2)$ as functions of e_2 are as shown in Figure 2. In general, they depend on the parameters of the exchange rates between the two countries, and do not necessarily pass through the origin of the $(e_1 - e_2)$ plane. We note, parenthetically, that if the switchover cost is modeled as a linear combination of the exchange rates of the two countries where the suppliers are based (i.e., the cost of switching from supplier 1 to supplier 2, $k_{12} = c_1 e_1 + c_2 e_2$ and $k_{21} = c_3 e_1 + c_4 e_2$ and c_i's are known positive constants), then $e_1^*(e_2)$ and $e_1^{**}(e_2)$ are linear function of e_2. A rigorous argument proving the result is provided in the Appendix.

Using comparative statics, we can show the effect of changes in the switchover costs k_{12} and k_{21} on the hysteresis band.

Proposition 2 *An increase in k_{21} or k_{12} causes an increase in $e_1^{**}(e_2)$ and a decrease in $e_1^*(e_2)$, thereby widening the hysteresis band.*

As one would expect, the threshold levels $e_1^*(e_2)$ and $e_1^{**}(e_2)$ depend on the switching costs k_{21} and k_{12}. An increase in the switchover cost results in an increasing hysteresis band, which signifies the increased inertia of the firm in keeping the current supplier.

The hysteresis band is affected by both the variance parameters and the correlation coefficient of the exchange rates. These effects are best captured by studying the relevant variance term σ, to which we refer as "system variance," and is given by $\sigma^2 = \frac{1}{2}\sigma_1^2 - \rho\sigma_1\sigma_2 + \frac{1}{2}\sigma_2^2$ (as defined in (9)). If the change in σ_1, for example, causes σ to increase, then we observe an increase in the hysteresis band. If, on the other hand, a change in σ_1, causes σ to decrease, then the hysteresis band decreases. We were able to numerically illustrate the effect of the variance parameter (σ_1) and the correlation coefficient (ρ) on the hysteresis band. The results are graphically illustrated for a base case in Figure 3 (Base case: $\rho = 0$, $\sigma_1 = \sigma_2 = .1$, $\delta_1 = \delta_2 = .3$, $k_{12} = 1$, $k_{21} = .2$, $e_2 = 1$). Figure (3a) shows the effect of changes in the variance (σ_1) on the hysteresis band. When the variance parameter $\sigma_1 \leq 0.1$ the hysteresis band decreases with increasing σ_1. However, when $\sigma_1 \geq 0.1$, then the hysteresis band increases with increasing σ_1. Figure (3b) shows that the hysteresis band decreases with increasing correlation coefficient (ρ) between the two exchange rates. This result is clear from the expression for σ, where increasing ρ results in decreased variance of the system. In Figure (3c) we plot the upper and lower critical thresholds as a function of σ, the "system variance." As expected, the hysteresis band increases with increase in the "system variance."

In Figure (4) we plot F^1 vs. ρ to illustrate the effects of the correlation coefficient on the cost of the global sourcing strategy. For large values of the exchange rate e_1 (when it is closer to the point where the firm should be switching to supplier 2), the cost of sourcing from supplier 1 is the minimum when the correlation coefficient is 1. For $\rho=1$ we have a smaller hysteresis band, which will allow easy switching if the future evolution of exchange rates necessitates it. When the exchange rate is at a point where it is likely to stay favorable to continue sourcing from the current supplier, the cost of the policy is minimum when the correlation coefficient is -1. Figure 5 is a plot of F^1 vs. ρ for different values of the switchover cost. This plot illustrates the effect of the hysteresis band on the policy values. For relatively small switchover costs (i.e., narrow hysteresis band) the optimal sourcing strategy cost achieves its minimum at $\rho=1$ When the switchover costs are large, and subsequently the hysteresis band has increased, the cost is minimum at $\rho=-1$. The seemingly counterintuitive behavior of the policy for narrow hysteresis band can be explained by accounting for the fact that in this case the firms can switch between suppliers more easily in an adverse realization of current supplier exchange rate, and the use of perfectly correlated supplier exchange rate process makes it even easier and thus allows to better exploit the exchange rate movement through the flexibility of the sourcing arrangement.

In Figure 6 we plot the effects of variance on the sourcing costs. In Figure 6a, the cost of sourcing from either supplier increases initially with increasing variance (σ_1) of the exchange rate of supplier 1, and then starts to decrease. This effect can again be explained by the affect of variance σ_1 on the "system variance" or σ. For $\sigma_1 \leq 0.1$ the "system variance" (σ) decreases with increasing σ_1, thereby decreasing the value of the option to switch between suppliers. However, when $\sigma_1 \geq 0.1$, the "system variance" increases with increasing σ_1, which increases the value of the option to switch between suppliers. This result is better captured in Figure 6b, which shows that the sourcing cost decreases with increasing variance σ.

20.4.2.1 Capacity Allocation Issues

In this sub-section we investigate the effects of capacity constraints on the global sourcing problem. Specifically, we look at the effect of allocating a fixed capacity between suppliers in countries with different exchange rate characteristics. We also study the effect of increasing the total capacity of the supplier network. These results are valid for the case of constant switchover costs.

Proposition 3 *For a fixed total capacity of the supplier network (i.e., $Q_1+Q_2 =$ constant,*
(i) the critical switchover thresholds $e_1^*(e_2)$ and $e_1^{**}(e_2)$ are independent of the capacity allocation (Q_1 and Q_2) among the individual suppliers. Furthermore,
(ii) if $\frac{e_1}{e_2} \geq \frac{\delta_1}{\delta_2}$, the cost of the global sourcing strategy starting from supplier 1,

F^1, increases with increasing Q_1 and if $\frac{e_1}{e_2} \leq \frac{\delta_1}{\delta_2}$, F^1 decreases with increasing Q_1. Exactly the reverse holds for F^2 i.e.,

(iii) if $\frac{e_1}{e_2} \geq \frac{\delta_1}{\delta_2}$, F^2 decreases with increasing Q_1 and if $\frac{e_1}{e_2} \leq \frac{\delta_1}{\delta_2}$, F^2 increases with increasing Q_1.

Proposition 3 (i) shows that, no matter what the allocation of fixed capacity between the suppliers is, the firm's optimal sourcing strategy, as reflected in its supplier switching decision, remains the same. However, as Proposition 3 (ii) and 3 (iii) show the cost of the optimal sourcing strategy is affected and the actual realization of the exchange rates plays an important role in allocating capacity between the suppliers in a way that it minimizes the cost of the sourcing strategy. The cost of the global sourcing strategy (and subsequently the optimal capacity allocation) depends on the ratios $\frac{e_1}{e_2}$ and $\frac{\delta_1}{\delta_2}$. For low $\frac{e_1}{e_2}$ ratio (less than $\frac{\delta_1}{\delta_2}$), the firm is better off allocating higher capacity to supplier 1, while for a higher $\frac{e_1}{e_2}$ ratio (greater than $\frac{\delta_1}{\delta_2}$), the firm should allocate more capacity to the supplier in country 2, assuming it currently sources from supplier 1. The result reverses if the firm currently sources from supplier 2.

20.5 TWO-SUPPLIER SOURCING CONTRACTS WITH TIME AND QUANTITY FLEXIBILITY

We now consider the problem where two suppliers are available and their unit prices follow two correlated geometric Brownian motions. This situation arises when, for example, two different materials can be used to satisfy the demand. That is, each of these materials can be used to substitute the other, but the prices of these materials follow different stochastic processes. This extended model also applies to the case where two suppliers are located in two different foreign countries with different currencies, while the suppliers set the prices of their products based on their own currencies. We assume that a deterministic demand D at time T is given. Thus, the firm needs to obtain D units of the material from its suppliers at or before T in order to satisfy the demands. The firm can select the more favorable price by observing the fluctuation of the exchange rates of the two potential suppliers. We assume the unit prices of the two suppliers, P' and P'', follow the geometric Brownian motions

$$dP' = \mu'P'dt + \sigma'P'dz',$$

and

$$dP'' = \mu''P''dt + \sigma''P''dz'',$$

where dz' and dz'' are the increments of the Weiner processes, and we let ρ be the correlation coefficient of the two processes. Let P_t' and P_t'' denote the unit prices of the first and second suppliers, respectively, at time t. At time 0, the firm signs a contract with one or both suppliers by informing them how many units will be purchased during the time interval $[0,T]$. The firm will pay supplier 1 a purchase cost of $g'(P_t') > 0$ dollars per unit if the unit is purchased from supplier 1 at time t, and it will pay supplier 2 a purchase cost of $g''(P_t') > 0$ dollars per unit if the unit is purchased from supplier 2 at time t, where g', g'' are nondecreasing functions. The use of the purchase cost function g' and g'' allows us to incorporate risk sharing considerations in the sourcing contract, as we explain below.

The firm will pay a supplier $g(P_t) > 0$ dollars per unit when a unit is purchased at time t, where g is a nondecreasing function. In a simple supply contract, the cost function g can simply be the identity function $g(P) = P$, for all $P > 0$. In a more complicated pricing scheme, the function can be any nondecreasing function. For example, a *risk-sharing contract* can be defined as a contract such that the function g satisfies

$$g(P) = \begin{cases} P - \lambda(P - \overline{P}), & \text{if } P > \overline{P}; \\ P, & \text{if } \underline{P} \le P \le \overline{P}; \\ P + \lambda(\underline{P} - P), & \text{if } P < \underline{P}; \end{cases} \tag{15}$$

where $\overline{P}, \underline{P}$ are constants such that $\overline{P} > \underline{P}$ and $\lambda \in [0,1]$ represents how much the supplier is going to share the risk with the firm when the unit price rises above $\overline{P}$ and how much the firm is going to share the risk with the supplier when the unit price drops below $\underline{P}$. The contract becomes a non-risk-sharing contract when $\lambda = 0$. Throughout this paper, P is called the *unit price* and $g(P)$ is called the *unit purchase cost*. In a recent survey of hedging strategies for exchange rate risks in sourcing arrangements, Carter and Vickery(1988) reported that more than 50% of the surveyed firms used some form of risk-sharing agreements with their suppliers. The above-mentioned risk-sharing contract is identified in the article as a popular option referred to as an "exchange rate window." Our solution methods developed in this paper are not restricted to this purchase cost function, but they apply to any nondecreasing cost function g.

We consider the following types of supply contracts. A *time-inflexible contract* requires the firm to specify at time 0 how many units it intends to purchase and when those units will be purchased in the future. For example, if there is only one supplier and the firm does not want to hold any inventory at all, then it can specify that all D units will be purchased at time T. However, if the firm expects a significant price increase, then it can specify an earlier time that those units will be purchased. On the other hand, a *time-flexible contract* requires the firm to specify at time 0 how many units it intends to purchase from the supplier, but it does not require the firm to specify when those units will be purchased. After signing the time-flexible contract, the firm can observe the price movement and decide dynamically when to trigger a buy. Besides time flexibility, we consider supply contracts with *quantity flexibility* as well. A supply contract with quantity flexibility can be defined as follows: Suppose the firm signs a contract of Q units with a supplier and the contract has an $\alpha \times 100\%$ quantity flexibility, where $0 \le \alpha \le 1$. Then the supplier does not require the firm to purchase all Q units from it later. The firm can purchase a total of x units from this supplier, where $(1-\alpha)Q \le x < Q$. If the supply contracts are quantity-flexible, then the firm may order more than D units from the supplier at time 0, but eventually it will only purchase a total of D units from those suppliers. When $\alpha = 0$, the contract becomes quantity-inflexible and the firm should only order the amount that it will purchase later.

Suppose the firm signs a time-flexible contract of Q' units with supplier 1 and the contract has an $\alpha' \times 100\%$ quantity flexibility. Then the firm can purchase a total of x' units from this supplier in the future, where $(1-\alpha')Q' \leq x' \leq Q'$. Similarly, if the firm signs the contract of Q'' units with supplier 2 and the contract has an $\alpha'' \times 100\%$ quantity flexibility, then the firm can purchase a total of x'' units from supplier 2, where $(1-\alpha'')Q'' \leq x'' \leq Q''$. In Section 20.5.2 we develop a solution method for the case where $\alpha' = \alpha'' = 1$. In Section 20.5.3 the solution method is extended to solve the general two-supplier problem with $0 \leq \alpha'$, $\alpha'' \leq 1$. The two-supplier time-inflexible contract is discussed in Li and Kouvelis (1997). But before we get into the specifics of the analysis we elaborate in Section 20.5.1 on the cost objective the firm uses and the details of our cost modeling.

20.5.1 Cost Modeling Issues

Given a supply contract, the firm's decision is to decide when to purchase and how many units in each purchase such that the expected net present value (NPV) of the purchase cost plus inventory holding cost is minimized. We will refer the NPV of this total cost as the "discounted total cost." We assume that the purchase cost of the material and the inventory holding cost are discounted at a constant annual interest rate $r > 0$. This discount rate represents the firm's opportunity cost of capital. We also assume that the inventory holding cost is proportional to the purchase cost of the material and is continuously compounded at an annual rate $h(h > r)$. If the firm purchases one unit at time t and uses it to satisfy its demand at time T, then the purchase cost of this unit is $g(P_t)$ and the holding cost for this unit during the small time interval $[\tau, \tau + d\tau]$ is $g(P_t)e^{h(\tau - t)} \cdot h d\tau$, for $\tau \geq t$. Thus, the NPV of the purchase cost of such a unit is $g(P_t)e^{-rt}$, and the NPV of the holding cost for this unit is

$$\int_t^T g(P_t)e^{h(\tau - t)} \cdot h e^{-r\tau} d\tau = \frac{h}{h - r} \cdot g(P_t)e^{-rt}\left[e^{(h-r)(T-t)} - 1\right],$$

and hence, the discounted total cost per unit is

$$g(P_t)e^{-rt} + \frac{h}{h-r} \cdot g(P_t)e^{-rt}\left[e^{(h-r)(T-t)} - 1\right] = \frac{1}{h-r} \cdot g(P_t)e^{-rt}\left[he^{(h-r)(T-t)} - r\right].$$

(16)

Note that if $h \to r$, then the discounted total cost per unit approaches $\left[g(P_t) + rg(P_t)(T-t)\right]e^{-rt}$, which can be interpreted as follows: The quantity $rg(P_t)(T-t)$ represents the holding cost per unit of the material for a time period of length T-t, discounted to time t. Therefore, $\left[g(P_t) + rg(P_t)(T-t)\right]e^{-rt}$ represents the total cost per unit discounted to time 0. Our problem is to determine a purchasing strategy for the firm such that its expected discounted total cost is minimized. Note that our optimization criterion is the expected value of the discounted total cost, which does not take into account the risk resulted from the uncertainty of the unit price of the material. Thus, our model is appropriate when the amount of money involved in each supply contract is not very large and the firm, which has an objective of minimizing its long-term operating costs, would behave risk-neutrally when valuing each supply contract.

We proceed by providing some background on how a firm can value a simple supplier sourcing contract and optimize the timing of purchase decision in such a case. The discussion is based on the results in Li and Kouvelis (1997).

To obtain the optimal purchasing decision for the time-flexible contract with a general cost function g and no quantity flexibility, we can use the binomial lattice (or binomial tree) approach described as follows. The binomial lattice approach was introduced by Cox, Ross and Rubinstein (1979) and is a widely used numerical procedure for valuing financial options (see Hull 1997). It assumes that the price movements are composed of a large number of small binomial movements. We start by dividing the time interval $[0,T]$ into a large number of small time intervals of length Δt. Let N be the number of small time intervals. Thus, $\Delta t = T/N$. The time points $0, \Delta t, 2\Delta t, ..., N\Delta t$ are referred to as periods $0,1,2,...,N$, respectively. We assume that in each time interval the unit price moves from its initial value P to one of two new values, Pu and Pd, where $u = e^{\sigma\sqrt{\Delta t}}$ and $d = e^{-\sigma\sqrt{\Delta t}}$. The movement from P to Pu is called an "up" movement and the movement from P to Pd is called a "down" movement. The probability of an up movement is denoted by p, and the probability of a down movement is $q=1-p$. We set

$$p = \frac{1}{2}\left(1 + \frac{\tilde{\mu}}{\sigma}\sqrt{\Delta t}\right),$$

where $\tilde{\mu} = \mu - \frac{1}{2}\sigma^2$. It can be shown that the mean and the variance of the binomial price movement will coincide with that of the geometric Brownian motion as $N \to \infty$ (see Cox, Ross and Rubinstein 1979). Furthermore, the probabilities p and q must be nonnegative as $N \to \infty$.

We first calculate the unit prices P and the unit purchase costs $g(P)$ on the lattice. For example, if the function g is given by equation (1), and $\lambda = 1$, $D=1000$, $T=1$, $h=12\%$, $r=10\%$, $\mu = 0.25$, $\sigma = 0.25$, $P_0 = 1.0$, $\overline{P} = 1.6$, $\underline{P} = 0.75$, $N = 4$, then

$\Delta t = 0.25$, $u = e^{0.125} = 1.1331$, $d = e^{-0.125} = 0.8825$, $p = 0.71875$, $q = 0.28125$, and the unit purchase costs are given in the lattice depicted in Figure 7. Let $\tilde{P}_i(k)$ denote the unit price at the kth node in period i, for $i=0,1,...,N$ and $k=1,2,...,i+1$. Let $C_i(k)$ denote the minimum possible expected discounted cost per unit (discounted to time 0) if the unit price in period i is $\tilde{P}_i(k)$ and the firm has not made the purchase. The minimum expected discounted unit costs at the final nodes, $C_N(k)$, are equal to their unit purchase costs $g\left(\tilde{P}_N(k)\right)$ discounted by a factor of e^{-rT}. Thus, in this example, $C_4(1) = 1.3573$, $C_4(2) = 1.1618$, $C_4(3) = 0.9048$, $C_4(4) = 0.7047$, and $C_4(5) = 0.6786$. For $i=N-1,N-2,...,1,0$, the values of $C_i(k)$ are given as

$$C_i(k) = \min\left\{\frac{1}{h-r}\cdot g\big(\tilde{P}_i(k)\big)e^{-ri\Delta t}\Big[he^{(h-r)(N-i)\Delta t}-r\Big],\, pC_{i+1}(k)+qC_{i+1}(k+1)\right\}.$$

If $\frac{1}{h-r}g\big(\tilde{P}_i(k)\big)e^{-ri\Delta t}\Big[he^{(h-r)(N-i)\Delta t}-r\Big] \leq pC_{i+1}(k)+qC_{i+1}(k+1)$, then the optimal decision at this node is to make the purchase immediately, otherwise the optimal decision is to wait for another period. The minimum expected discounted unit costs of the nodes in this example, as well as their optimal decisions, are shown in Figure 7(b). In this figure, a "buy" indicates a decision of making the purchase immediately if a purchase has not been made, while a "wait" indicates a decision of not making the purchase immediately. The optimal expected discounted total cost of this time-flexible contract is $D\cdot C_0(1) = (1000)(1.1088) = 1108.8$.

Given an initial price P_0, if the purchase of D units is made t time units from now, then by (16), the expected discounted total cost is given as

$$\frac{D}{h-r}\cdot E\big[g(P_t)|P_0\big]\cdot e^{-rt}\Big[he^{h-r(T-t)}-r\Big].$$

Thus, the optimal purchasing decision for the time-inflexible contract with a general cost function and no quantity flexibility can be obtained by minimizing this cost function over all possible $t\in[0,T]$. This requires the evaluation of the expectation $E\big[g(P_t)|P_0\big]$ for every $t\in[0,T]$, which can be accomplished by using the binomial lattice as well. The probabilities of occurrence of each node are calculated using p and q, and are shown in Figure 7(c). Let $\tilde{p}_i(k)$ denote the probability of occurrence of the kth node in period i. Then, the expected value $E\big[g(P_{i\Delta t}|P_0)\big]$ is given by $\sum_{k=1}^{i+1}\tilde{p}_i(k)g\big(\tilde{P}_i(k)\big)$. In this example, we have $E\big[g(P_0)|P_0=1\big]=1.0000$, $E\big[g(P_{\Delta t})|P_0=1\big]=1.0627$, $E\big[g(P_{2\Delta t})|P_0=1\big]=1.1292$, $E\big[g(P_{3\Delta t})|P_0=1\big]=1.2014$, and $E\big[g(P_{4\Delta t})|P_0=1\big]=1.2364$. This implies that

$$\frac{1}{h-r}\cdot E\big[g(P_0)|P_0=1\big]\cdot e^0\Big[he^{(h-r)4\Delta t}-r\Big]=1.1212,$$

$$\frac{1}{h-r}\cdot E\big[g(P_{\Delta t})|P_0=1\big]\cdot e^{-r\Delta t}\Big[he^{(h-r)3\Delta t}-r\Big]=1.1304,$$

$$\frac{1}{h-r}\cdot E\big[g(P_{2\Delta t})|P_0=1\big]\cdot e^{-2r\Delta t}\Big[he^{(h-r)2\Delta t}-r\Big]=1.1389,$$

$$\frac{1}{h-r}\cdot E\big[g(P_{3\Delta t})|P_0=1\big]\cdot e^{-3r\Delta t}\Big[he^{(h-r)\Delta t}-r\Big]=1.1481,$$

$$\frac{1}{h-r}\cdot E\big[g(P_{4\Delta t})|P_0=1\big]\cdot e^{-4r\Delta t}\Big[he^0-r\Big]=1.1187.$$

Thus, the optimal decision is to purchase at time $4\Delta t=T=1$, and the minimum expected discounted total cost is $(1000)(1.1187)=1118.7$, which is 0.89% higher than that of the time-flexible contract.

Note that the above analysis is based on the setting of $N=4$, and the above result may not be accurate. A more accurate result can be obtained by using a larger value for N. In all computational studies in this paper we select $N=400$, which, from our computational experience, is the level at which the total costs of the contracts converge.

20.5.2 The Time-Flexible Contract with 100% Quantity Flexibility

In this subsection we consider the situation where two suppliers are offering time-flexible contracts with 100% quantity flexibility. In this case the optimal ordering strategy at time 0 is to order D units from each supplier. After placing these orders, the firm has the flexibility of purchasing between 0 and D units from each supplier. The optimal purchasing strategy can be found by using the 2-dimensional binomial lattice approach (see Boyle, Ervine and Gibbs 1989). We divide the time interval $[0,T]$ into N small time intervals of length $\Delta t = T/N$. Again, the time points $0, \Delta t, 2\Delta t, \ldots, N\Delta t$ are referred to as periods $0, 1, 2, \ldots, N$, respectively. For example, a 2-dimensional lattice with $N = 3$ is depicted in Figure 3, where the arrows between period 1 and period 3 are omitted. For $i = 0, 1, \ldots, N$, the nodes in period i are labeled (k, l), where $k, l = 1, 2, \ldots, i+1$. We assume that in each time interval the unit price of supplier 1 moves from its initial value P' to one of two new values $P'e^{\sigma'\sqrt{\Delta t}}$ and $P'e^{-\sigma'\sqrt{\Delta t}}$ and the unit price of supplier 2 moves from its initial value P'' to one of two new values $P''e^{\sigma''\sqrt{\Delta t}}$ and $P''e^{-\sigma''\sqrt{\Delta t}}$. Let

p_1 = probability that the prices of (supplier 1, supplier 2) will go (up, up),

p_2 = probability that the prices of (supplier 1, supplier 2) will go (up, down),

p_3 = probability that the prices of (supplier 1, supplier 2) will go (down, down),

p_4 = probability that the prices of (supplier 1, supplier 2) will go (down, up),

as shown in Figure 8. We set

$$p_1 = \frac{1}{4}\left[1+\left(\frac{\tilde{\mu}'}{\sigma'}+\frac{\tilde{\mu}''}{\sigma''}\right)\sqrt{\Delta t}+\rho\right],$$

$$p_2 = \frac{1}{4}\left[1+\left(\frac{\tilde{\mu}'}{\sigma'}-\frac{\tilde{\mu}''}{\sigma''}\right)\sqrt{\Delta t}-\rho\right],$$

$$p_3 = \frac{1}{4}\left[1+\left(-\frac{\tilde{\mu}'}{\sigma'}-\frac{\tilde{\mu}''}{\sigma''}\right)\sqrt{\Delta t}+\rho\right],$$

$$p_4 = \frac{1}{4}\left[1+\left(-\frac{\tilde{\mu}'}{\sigma'}+\frac{\tilde{\mu}''}{\sigma''}\right)\sqrt{\Delta t}-\rho\right],$$

where $\tilde{\mu}' = \mu' - \frac{1}{2}\sigma'^2$ and $\tilde{\mu}'' = \mu'' - \frac{1}{2}\sigma''^2$. It can be shown that the mean and variance of the binomial price movement will coincide with that of the two correlated geometric Brownian motions as $N \to \infty$ (see Boyle, Ervine and Gibbs (1989). Furthermore, the probabilities p_1, p_2, p_3, p_4 must be nonnegative as $N \to \infty$.

The following procedure can be used for the evaluation of these contracts. We first calculate the unit prices (P', P'') of the nodes in the lattice. Since the contracts are 100% quantity-flexible, the firm will select the supplier with a lower unit purchase cost whenever a purchase is made. Thus, if a node in period i has unit prices $\left(P', P''\right)$, then the unit purchase cost of that node is $\min\left\{g'\left(P'_{i\Delta t}\right), g''\left(P''_{i\Delta t}\right)\right\}$. Let $g_i(k,l)$ be the unit purchase cost of node (k,l) in period i. Let $C_i(k,l)$ be the minimum possible expected discounted cost per unit if the current state is at node (k,l) of period i. The minimum expected discounted unit costs at the final nodes, $C_N(k,l)$ are the same as their unit purchase costs discounted by a factor of e^{-rT}. For $i = N-1, N-2, \ldots, 1, 0$, the values of $C_i(k.l)$ are given as

$$C_i(k,l) = \min\left\{\frac{1}{h-r} \cdot g_i(k,l)e^{-ri\Delta t}\left[he^{(h-r)(N-i)\Delta t} - r\right],\right.$$

$$\left. p_1 C_{i+1}(k,l) + p_2 C_{i+1}(k,l+1) + p_3 C_{i+1}(k+1,l+1) + p_4 C_{i+1}(k+1,l)\right\}.$$

If the first term on the right hand side of this equation attains the minimum, then the optimal decision at this node is to make a purchase immediately, otherwise, the optimal decision is to wait for another period. The optimal expected total cost of the contract is $D \cdot C_0(1,1)$.

We now consider an example with $D = 1000$, $T = 1$, $h = 12\%$, $r = 10\%$, $\mu' = \mu'' = 0.25$, $\sigma' = \sigma'' = 0.25$, $\rho = 0.5$, $P'_0 = P''_0 = 1.0$, $N = 4$, where both suppliers have the same risk-sharing function g given by equation (1) with $\overline{P} = 1.5$, $\underline{P} = 0.75$, and $\lambda = 1$. Then, we have $\Delta t = 0.25$, $p_1 = 0.59357$, $p_2 = 0.125$, $p_3 = 0.15625$, $p_4 = 0.125$, and the unit purchase costs are given in the lattice depicted in Figure 9(a). The minimum expected discounted unit costs of the nodes, as well as their optimal decisions, are shown in Figure 9(b). The overall optimal strategy is to order 1000 units from each of the two suppliers at time 0, follow the "wait" and "buy" decisions shown in Figure 9(b), and purchase the 1000 units from the supplier with a lower unit price as soon as a "buy" decision is made. The optimal expected discounted total cost of this time-flexible contract is $D \cdot C_0(1,1) = (1000)(1.0346) = 1034.6$. Again, this result may not be accurate. A more accurate result can be obtained by using a larger value of N. For example, when $N = 400$, the optimal expected discounted total cost per unit is 1.0258, which implies that the optimal expected discounted total cost of the contract is 1025.8.

20.5.3 The General Time-Flexible and Quantity Flexible Contract

To solve the general two-supplier problem with purchase time flexibility, we first solve the case with $D = 1$ and $\alpha = 1$ (i.e., 100% quantity flexibility), and let $\hat{C}$ be the optimal solution value. Then, we solve the problem with $D = 1$ assuming that supplier 1 is the only supplier available, and let C' be the optimal expected discount total cost of this single-supplier problem. We also solve the problem with $D = 1$ assuming that supplier 2 is the only supplier available, and let C'' be the optimal expected discounted total cost. Let Q', Q'' be the purchase quantities that the firm informs suppliers 1 and 2, respectively (Q' and Q'' are decision variables). Thus, the firm has to purchase at least $(1-\alpha')Q'$ units from supplier 1 and at least

$(1-\alpha'')Q''$ units from supplier 2 during the time interval $[0,T]$. Also, the firm has to purchase at least $D-Q''$ units from supplier 1 and at least $D-Q'$ units from supplier 2. Let $x' = \max\{(1-\alpha')Q', D-Q''\}$ and $x'' = \max\{(1-\alpha'')Q'', D-Q'\}$. Then, x' units must be purchased from supplier 1 (with a minimum cost of C' per unit), x'' units must be purchased from supplier 2 (with a minimum cost of C'' per unit, and $D-x'-x''$ units can be purchased from either supplier (with a minimum cost of $\hat{C}$ per unit). Hence, the overall problem can be formulated as the following linear program:

$$\text{minimize } C'x' + C''x'' + \hat{C}(D-x'-x'')$$
$$\text{subject to } x' \geq (1-\alpha')Q'$$
$$x' \geq D-Q''$$
$$x'' \geq (1-\alpha'')Q''$$
$$x'' \geq D-Q'$$
$$x', x'', Q', Q'' \geq 0$$

This linear program can be rewritten as:

$$\text{minimize } (C'-\hat{C})x' + (C''-\hat{C})x''$$
$$\text{subject to } D-x'' \leq Q' \leq \frac{x'}{1-\alpha'}$$
$$D-x' \leq Q'' \leq \frac{x''}{1-\alpha''}$$
$$x', x'', Q', Q'' \geq 0$$

where the constant term $\hat{C}D$ has been dropped from the objective function. Obviously, it is optimal to set $Q' = \frac{x'}{1-\alpha'}$ and $Q'' = \frac{x''}{1-\alpha''}$. Then, we can eliminate Q' and Q'', and the first set of inequalities can be written as $D-x'' \leq \frac{x'}{1-\alpha'}$, or $x'+(1-\alpha')x'' \geq (1-\alpha')D$, while the second set of inequalities can be written as $D-x' \leq \frac{x''}{1-\alpha''}$, or $(1-\alpha'')x'+x'' \geq (1-\alpha'')D$. Thus, the linear program becomes:

$$\text{minimize } (C'-\hat{C})x' + (C''-\hat{C})x''$$
$$\text{subject to } x'+(1-\alpha')x'' \geq (1-\alpha')D$$
$$(1-\alpha'')x'+x'' \geq (1-\alpha'')D$$
$$x', x'' \geq 0$$

It is easy to check that the optimal solution to this two-variable linear program is given as

$$(x',x'') = \begin{cases} (0,D), & \text{if } \frac{1}{1-\alpha'}\left(C''-\hat{C}\right) \leq C'-\hat{C}; \\ (D,0) & \text{if } \left(C'-\hat{C}\right) \leq (1-\alpha'')\left(C''-\hat{C}\right); \\ \left(\dfrac{(1-\alpha')\alpha''D}{\alpha'+\alpha''-\alpha'\alpha''}, \dfrac{(1-\alpha'')\alpha'D}{\alpha'+\alpha''-\alpha'\alpha''}\right), \\ \qquad \text{if } (1-\alpha'')\left(C''-\hat{C}\right) < C'-\hat{C} < \dfrac{1}{1-\alpha'}\left(C''-\hat{C}\right); \end{cases}$$

and the corresponding optimal order quantities (Q', Q'') are given as $\left(0, \dfrac{D}{1-\alpha''}\right)$,

$\left(\dfrac{D}{1-\alpha'}, 0\right)$ and $\left(\dfrac{\alpha''D}{\alpha'+\alpha''-\alpha'\alpha''}, \dfrac{\alpha'D}{\alpha'+\alpha''-\alpha'\alpha''}\right)$. Among these $Q'+Q''$ units

ordered, $(1-\alpha')Q'$ units must be purchased from supplier 1 and $(1-\alpha'')Q''$ units must be purchased from supplier 2. The optimal purchase times for these $(1-\alpha')Q'+(1-\alpha'')Q''$ "inflexible" units can be obtained by solving two separate single-supplier problems. Out of the remaining $\alpha'Q'+\alpha''Q''$ units ordered, $D-(1-\alpha')Q'-(1-\alpha'')Q''$ units can be purchased from either supplier, where the optimal purchase times are obtained by solving a two-supplier problem with 100% quantity flexibility, while the other $Q'+Q''-D$ units will not be purchased at all.

Note that the above result can be interpreted as follows: All D units will be purchased from supplier 2 if $\dfrac{1}{1-\alpha'}\left(C''-\hat{C}\right) \leq C'-\hat{C}$. Similarly, all D units will be purchased from supplier 1 if $\left(C'-\hat{C}\right) \leq (1-\alpha'')\left(C''-\hat{C}\right)$. The quantity flexibility will benefit the firm only if $(1-\alpha'')\left(C''-\hat{C}\right) < C'-\hat{C} < \dfrac{1}{1-\alpha'}\left(C''-\hat{C}\right)$, or equivalently, $1-\alpha'' < \left(C'-\hat{C}\right)/\left(C''-\hat{C}\right) < \dfrac{1}{1-\alpha'}$ (provided that $\hat{C} \neq C''$), which happens when $C'-\hat{C}$ is close to $C''-\hat{C}$. This condition is more likely to be satisfied for supply environments with similar expected sourcing costs per unit (i.e., similar in C' and C''), but the price processes are not perfectly correlated with each other and the optimal purchase time is nonzero so that $\hat{C} < C'$ and $\hat{C} < C''$.

We now modify the example in Section 20.5.2 by changing the quantity flexibility of the two contracts to $\alpha' = 0.5$ and $\alpha'' = 0.25$. In this example $\hat{C} = 1.0258$ (from Section 20.5.2) and $C' = C'' = 1.1018$ (from Section 20.5.1). Thus, the optimal order quantities are given as $(Q', Q'') = \left(\dfrac{\alpha''D}{\alpha'+\alpha''-\alpha'\alpha''}, \dfrac{\alpha'D}{\alpha'+\alpha''-\alpha'\alpha''}\right) = (400, 800)$. Hence, there are 200 units which will be purchased from supplier 1 with the optimal purchase time obtained by using the single-supplier model. There are 600 units which will be purchased from supplier 2 with the optimal purchase time obtained by using the supplier model. The remaining 200 units will be purchased from either supplier with the optimal purchase time determined by the 2-dimensional lattice approach given in Section 20.5.2. The minimum expected discounted total cost of the supply contract is $200C' + 600C'' + 200\hat{C} = 1086.6$. For example, if we consider the case where $N = 4$ and suppose the unit prices of both suppliers drop from \$1 to \$0.8825 in period 1 (see Figure 7(a)), then 200 and 600 units should be purchased from suppliers 1 and 2, respectively, in period 1 (see Figure 7(b)). The current state is at node (2,2), and the remaining 200 units should not be purchased in period 1 (see Figure 9(b)). In period 2, suppose the unit price of supplier 1 increases to \$1 and the unit price of supplier 2 decreases to \$0.7788, then the current state is at node (2,3), and the remaining 200 units should be purchased from the more inexpensive supplier (i.e., supplier 2) in this period (see Figure 9(b)). In this example, the firm makes a single purchase from supplier 1 but two separate purchases from supplier 2.

Consider the example given in Section 20.5.2, i.e., $D = 1000$, $T = 1$, $h = 12\%$, $r = 10\%$, $\mu' = \mu'' = 0.25$, $\sigma' = \alpha'' = 0.25$, $\rho = 0.5$, $P_0' = P_0'' = 1.0$, $\overline{P} = 1.5$, $\underline{P} = 0.75$, $\lambda = 1$, $\alpha' = 0.5$, $\alpha'' = 0.25$. We now change the value of parameter ρ, and solve the problem for the time-flexible contract. We set $N = 400$, and we only consider the case where both suppliers have the same risk-sharing cost function g and the same parameters for their price processes. Figure 10 depicts the results of changing the value of ρ when the values of all other parameters are fixed. The corresponding minimum expected discounted total costs of the single-supplier contracts are also shown in the figure. When $\rho = 1$, the two-supplier time-flexible problem is equivalent to the single-supplier time-flexible problem. We notice that as ρ decreases, the minimum expected discounted total costs of the two-supplier time-flexible contract decreases. This is because if the correlation of the two price processes decreases, then the firm can have a bigger advantage by observing the price movement and always select the supplier with a lower price when quantity flexibility is available. We also observe that the minimum expected discounted total cost of the two-supplier problem is more sensitive to a change in ρ when ρ is large.

CONCLUSIONS

The impact of exchange rate uncertainty on the global sourcing problem can be significant. In this paper we develop a general framework for evaluating the total expected cost when sourcing from a global network of suppliers when the purchasing firm faces uncertain exchange rates. The key insight from our research is that firms are willing to continue to source from suppliers that are more expensive than others because of switchover costs and because of the option to switch between suppliers. The switchover cost induces a zone of inaction, also called the "hysteresis band," where the optimal decision for firms sourcing from one supplier is to continue sourcing from the same supplier. The extent of this "hysteresis band" depends on various parameters of the exchange rate processes and other factors such as the switchover cost. An increase in the switchover costs causes the hysteresis band to expand.

The behavior of the hysteresis band with respect to the variance parameters σ_i and correlation coefficient (ρ) has important implications on the design of a supplier portfolio that requires flexibility (Figure 2a). The effect of exchange rate parameter on the total system variance $\left(\frac{1}{2}\sigma_1^2 - \rho\sigma_1\sigma_2 + \frac{1}{2}\sigma_2^2 \right)$ is the key to understanding the above finding. Adding a supplier that reduces the total system variance is very important in designing a supplier network. In deciding capacity allocation between suppliers, the actual realization of the exchange rate plays a more important role than other parameters like the variance or the correlation coefficient.

In this paper we also illustrated valuation methodologies for risk-sharing sourcing contracts with time, quantity, and supplier flexibility under purchase price uncertainties. Our analysis illustrates that contractual flexibility in sourcing arrangements, when carefully exercised, can effectively reduce the sourcing cost in environments of price uncertainty.

Multi-supplier sourcing arrangements can help the firm lower its sourcing costs by exploiting available quantity flexibility in sourcing contracts. From our analysis of the two-supplier case, the firm benefits the most from the simultaneous use of two suppliers when there are substantial differences in the sourcing costs per unit of the two supply environments. In choosing a portfolio of suppliers for multi-supplier sourcing arrangements, the correlation of the price processes is the most important criterion. In the two-supplier case, as the correlation coefficient ρ of the two price processes decreases, the expected sourcing cost decreases at a decreasing rate. Finally, the firm benefits the most from exercising its quantity and supplier flexibility when the risk-sharing windows are reasonably large.

This research brings out some other important issues for future research. Within the present framework of the global sourcing problem, the issue of an optimal portfolio that hedges all risk can be investigated. Also strategies that consider both financial as well as operational hedging strategies should be explored. Our work also presented a first step in the detailed study of multi-supplier sourcing arrangements with specific types of supply contracts in uncertain price but deterministic demand environments. We focused on contracts for supplying a single demand point. The next challenge is to value and analyze supply contracts for multi-demand instances over a planning horizon. Our preliminary research indicates that the problem is not optimally decomposable to supply contracts with single demand points, and thus new valuation approaches have to be used for it.

APPENDIX 1

Derivation of Equation (8):

We will use the Method of Undetermined Coefficients to derive the solution of Equation (6). Let the solution to Equation (6) be of the form $F^1 = -Ae_1^{\beta}e_2^{1-\beta}$, where A and β are constants to be determined. Let F_i^1 denote the partial derivative of F^1 with respect to e_i and F_{ij}^1 be the partial derivative of F^1 with respect to e_i and $e_j, i, j \in \{1,2\}$. Then $F_1^1 = -\beta Ae_1^{\beta-1}e_2^{1-\beta}$, $F_2^1 = (\beta-1)Ae_1^{\beta}e_2^{-\beta}$, $F_{11}^1 = \beta(1-\beta)Ae_1^{\beta-2}e_2^{1-\beta}$, $F_{22}^1 = \beta(1-\beta)Ae_1^{\beta}e_2^{-1-\beta}$ and $F_{12}^1 = \beta(\beta-1)Ae_1^{\beta-1}e_2^{-\beta}$. Substituting for F_1^1, F_2^1, F_{11}^1, F_{22}^1 and F_{12}^1 in (6) and simplifying, we get

$$Ae_1^\beta e_2^{1-\beta}\left[\sigma^2\beta(1-\beta)+(\delta_1-\delta_2)\beta+\delta_2\right]=-e_1Q_1-e_2(1-Q_1)$$

where $\sigma^2=\tfrac{1}{2}\sigma_1^2-\rho\sigma_1\sigma_2+\tfrac{1}{2}\sigma_2^2$.

When β_1 and β_2 are given by Equation (9), then the left hand side (LHS) of the above equation is zero. When $A=\dfrac{Q_1}{\delta_1}$ and $\beta=1$, then the LHS of the above equation equals $-e_1Q_1$. When $A=-\dfrac{1-Q_1}{\delta_2}$ and $\beta=0$, then the LHS of the above equation equals $-e_2(1-Q_1)$. This means that the solution to Equation (6) is

$$F^1=\frac{e_1Q_1}{\delta_1}+\frac{e_2(1-Q_1)}{\delta_2}-A_1e_1^{\beta_1}e_2^{1-\beta_1}-A_2e_1^{\beta_2}e_2^{1-\beta_2}.$$

Proof of Proposition 1:

Define function $G(e_1,e_2)=F^1(e_1,e_2)-F^2(e_1,e_2)$. From now on we suppress the dependence of $G(.)$ and $F(.)$ on e_1 and e_2 whenever possible for notational convenience and denote the partial derivatives of G with respect to e_1 and e_2 by subscripts (i.e., G_1 is the first order derivative with respect to e_1, G_{11} is the second order derivative with respect to e_1, and so on). It holds that $G=Be_1^{\beta_1}e_2^{1-\beta_1}-Ae_1^{\beta_2}e_2^{1-\beta_2}+\dfrac{e_1(Q_1+Q_2-1)}{\delta_1}-\dfrac{e_2(Q_1+Q_2-1)}{\delta_2}$. For small values of e_1, the dominant term in G is $Be_1^{\beta_1}$; it is decreasing and convex in e_1. For large values of e_1, the dominant term in G is $Ae_1^{\beta_2}$, which is negative, decreasing and concave in e_1. Applying the boundary conditions at $e_1^*(e_2)$ and $e_1^{**}(e_2)$, we get $G\!\left(e_1^*(e_2),e_2\right)=-k_{21}$, $G\!\left(e_1^{**}(e_2),e_2\right)=k_{12}$, $G_1\!\left(e_1^*(e_2),e_2\right)=0$, and $G_1\!\left(e_1^{**}(e_2),e_2\right)=0$. This implies that the graph of G as a function of e_1 for a fixed e_2 should have an S shape in the range $\left[e_1^*(e_2),e_1^{**}(e_2)\right]$, tangential to the horizontal line at height k_{12} at $e_1^{**}(e_2)$, and tangential to the horizontal line at height $-k_{21}$ at $e_1^*(e_2)$ (see Figure 1).

Subtracting (7) from (6) we get

$$\tfrac{1}{2}\sigma^2e_1^2G_{11}+(r-\delta_1)G_1-rG+e_1(Q_1+Q_2-1)-e_2(Q_1+Q_2-1)=0.$$

At $e_1^{**}(e_2)$, substituting for G_1 and rearranging the terms in the above equation and then by recognizing that $G_{11}\le 0$ (since G is concave in e_1 in the neighborhood of $e_1^{**}(e_2)$), we get

$$e_1^{**}(e_2)>e_2+\frac{rk_{12}}{Q_1+Q_2-1}.$$

Therefore, switching costs induce an "inaction" on the part of the firm when currently sourcing from supplier 2 and exchange rates fall in the region $\left\{e_2, e_1^{**}(e_2)\right\}$. Similarly, for $e_1^{*}(e_2)$, we have that

$$e_1^{*}(e_2) < e_2 - \frac{rk_{21}}{Q_1 + Q_2 - 1}$$

and a similar inaction is observed in the region $\left\{e_1^{*}(e_2), e_2\right\}$ when the firm is currently sourcing from supplier 2. Thus, in the region $\left\{e_1^{*}(e_2), e_1^{**}(e_2)\right\}$ (observe that $e_1^{**}(e_2) > e_1^{**}(e_2)$) the firm continues pursuing its current sourcing policy, and the theorem is proved. $\qquad\square$

Proof that $e_1^{*}(e_2)$ and $e_1^{**}(e_2)$ are linear in e_2 when the switchover costs are linear combinations of exchange rates (see page 15):

The Partial Differential Equations are (6) and (7). The boundary conditions can now be written as:

$$F^1\!\left(e_1^{**}, e_2\right) = F^2\!\left(e_1^{**}, e_2\right) + c_1 e_1 + c_2 e_2$$
$$F^2\!\left(e_1^{*}, e_2\right) = F^1\!\left(e_1^{*}, e_2\right) + c_3 e_1 + c_4 e_2$$

(value matching)

and

$$F_1^1\!\left(e_1^{**}, e_2\right) = F_1^2\!\left(e_1^{**}, e_2\right) + c_1$$
$$F_2^1\!\left(e_1^{**}, e_2\right) = F_2^2\!\left(e_1^{**}, e_2\right) + c_2$$
$$F_1^2\!\left(e_1^{**}, e_2\right) = F_1^1\!\left(e_1^{**}, e_2\right) + c_3$$
$$F_2^2\!\left(e_1^{**}, e_2\right) = F_2^1\!\left(e_1^{**}, e_2\right) + c_4$$

(smooth pasting)

Let $x = \dfrac{e_1}{e_2}$. Then, using Equations (10-11), $e_2 V^1(x) = F^1(e_1, e_2)$ and $e_2 V^2(x) = F^2(e_1, e_2)$ where

$$V^1(x) = \frac{xQ_1}{\delta_1} + \frac{1 - Q_1}{\delta_2} - Ax^{\beta_2}$$

and

$$V^2(x) = \frac{x(1 - Q_2)}{\delta_1} + \frac{Q_2}{\delta_2} - Bx^{\beta_1}.$$

Using the notation of Proposition 1 for partial derivatives, we represent $V_x^i(x)$ as the derivative of $V^i(x)$ with respect to x, and $V_{xx}^i(x)$ as the second derivative of $V^i(x)$ with respect to x, $i \in \{1,2\}$. Then $F_1^i(e_1, e_2) = V_x^i(x)$, $F_2^i(e_1, e_2) = V^i(x)$

$$-xV_x^i(x), \quad F_{11}^i(e_1,e_2)=\frac{V_{xx}^i(x)}{e_2}, \quad F_{12}^i(e_1,e_2)=-\frac{xV_{xx}^i(x)}{e_2}, \quad \text{and} \quad F_{22}^i(e_1,e_2)=\frac{x^2V_{xx}^i(x)}{e_2}.$$

Substituting the above quantities in the PDEs and the boundary conditions, and simplifying, we get the following PDEs:

$$\frac{1}{2}\sigma^2x^2V_{xx}^1+(\delta_2-\delta_1)xV_x^1-(\delta_2+2r)V^1+xQ_1+(1-Q_1)=0$$

and

$$\frac{1}{2}\sigma^2x^2V_{xx}^2+(\delta_2-\delta_1)xF_x^2-(\delta_2+2r)V^2+x(1-Q_2)+Q_2=0$$

where σ^2 is as defined in the derivation of Equation (8). The boundary conditions

at the critical values $x^{**}(e_2)=\dfrac{e_1^{**}(e_2)}{e_2}$ and $x^{**}(e_2)=\dfrac{e_1^{**}(e_2)}{e_2}$

$$V^1(x^{**})=V^2(x^{**})+xc_1+c_2(i)$$
$$V^2(x^*)=V^1(x^*)+xc_3+c_4(iv) \qquad \textit{(value matching)}$$

and

$$V_x^1(x^{**}) \qquad\qquad =V_x^2(x^{**})+c_1(ii)$$
$$V_x^1(x^{**})-xV_x^1(x^{**})=V_x^2(x^{**})xV_x^2(x^{**})+c_2(iii)$$
$$V_x^2(x^*) \qquad\qquad =V_x^1(x^*)+c_3(v) \qquad \textit{(smooth pasting)}$$
$$V_x^2(x^*)-xV_x^2(v^*)=V_x^1(x^*)-xV_x^2(x^*)+c_4,(vi)$$

Equation (iii) can be derived from Equations (i) and (ii). Similarly, Equation (vi) can be derived from Equations (iv) and (v). Thus, Equations (iii) and (vi) can be ignored. We are now left with PDEs and boundary conditions that are functions of one variable x. The same PDEs and boundary conditions have been studied in Dixit (1989a). According to the results therein, there exist constants x_0^* and x_0^{**} such that $e_1^*(e_2)=x_0^*e_2$, and $e_1^{**}(e_2)=x_0^{**}e_2$. Thus $e_1^*(e_2)$ and $e_1^{**}(e_2)$ are linear functions of e_2. $\square$

Proof of Proposition 2:

Define the function $G(e_1,e_2)$ as in the proof of Proposition 1. We also make the same notational conventions as in the above proposition. It holds that

$$G = Be_1^{\beta_1}e_2^{1-\beta_1} - Ae_1^{\beta_2}e_2^{1-\beta_2} + \frac{e_1(Q_1+Q_2-1)}{\delta_1} - \frac{e_2(Q_1+Q_2-1)}{\delta_2}. \qquad \text{Also}$$

$$G(e_1^*(e_2),e_2)=k_{21}, \quad G(e_1^{**}(e_2),e_2)=k_{12}, \quad G_1(e_1^*(e_2),e_2)=0, \quad G_1(e_1^{**}(e_2),e_2)=0,$$

$G_{11}(e_1^{**}(e_2),e_2)\le0$ and $G_{11}(e_1^*(e_2),e_2)\ge0$. Now suppose that k_{21} changes by dk_{21}. We want to determine how the four endogenous variables $A, B, e_1^*(e_2)$ and

$e_1^{**}(e_2)$ respond. We differentiate the value matching and smooth pasting conditions totally. We get

$$G_A\Big(e_1^{**}(e_2),e_2\Big)dA + G_B\Big(e_1^{**}(e_2),e_2\Big)dB - dk_{21} = 0$$

$$G_A\Big(e_1^{*}(e_2),e_2\Big)dA + G_B\Big(e_1^{*}(e_2),e_2\Big)dB = 0$$

$$G_{11}\Big(e_1^{**}(e_2),e_2\Big)de_1^{**} + G_{1A}\Big(e_1^{**}(e_2),e_2\Big)dA + G_{1B}\Big(e_1^{**}(e_2),e_2\Big)dB = 0$$

$$G_{11}\Big(e_1^{*}(e_2),e_2\Big)de_1^{*} + G_{1A}\Big(e_1^{*}(e_2),e_2\Big)dA + G_{1B}\Big(e_1^{*}(e_2),e_2\Big)dB = 0$$

Solving for $de_1^{*}(e_2)$ and $de_1^{**}(e_2)$ in terms of dk_{21}, we obtain

$$\frac{de_1^{**}(e_2)}{dk_{21}} = \frac{1}{G_{11}\Big(e_1^{**}(e_2),e_2\Big)} \frac{\beta_2 e_1^{**}(e_2)^{\beta_2-1} e_1^{*}(e_2)^{\beta_1} - \beta_1 e_1^{**}(e_2)^{\beta_1-1} e_1^{*}(e_2)^{\beta_2}}{e_1^{**}(e_2)^{\beta_1} e_1^{*}(e_2)^{\beta_2} - e_1^{**}(e_2)^{\beta_2} e_1^{*}(e_2)^{\beta_1}}$$

$$\frac{de_1^{*}(e_2)}{dk_{21}} = \frac{1}{G_{11}\Big(e_1^{*}(e_2),e_2\Big)} \frac{(\beta_2-\beta_1) e_1^{*}(e_2)^{\beta_1+\beta_1}}{e_1^{**}(e_2)^{\beta_1} e_1^{*}(e_2)^{\beta_2} - e_1^{**}(e_2)^{\beta_2} e_1^{*}(e_2)^{\beta_1}}$$

Using the facts that $\beta_1 \le 0$ and $\beta_2 \ge 1$, $e_1^{*}(e_2) < e_1^{**}(e_2)$, $G_{11}\Big(e_1^{**}(e_2)\Big) < 0$, $G_{11}\Big(e_1^{*}(e_2)\Big) > 0$ and after simple, but tedious, algebraic manipulations, we can establish that $\dfrac{de_1^{**}(e_2)}{dk_{21}} > 0$ and $\dfrac{de_1^{*}(e_2)}{dk_{21}} < 0$. For k_{12}, following similar logic we obtain $\dfrac{de_1^{**}(e_2)}{dk_{12}} > 0$ and $\dfrac{de_1^{*}(e_2)}{dk_{12}} < 0$. This concludes the proof of proposition 2. $\square$

Proof of Proposition 3:

(i) From (14) we can see that if the sum $Q_1 + Q_2$ is constant, the specific values of Q_1 and Q_2 do not affect the solution of equations. Therefore, the critical values do not change.

(ii) From Equation (10),

$$F^1(e_1,e_2) = \frac{e_2}{\delta_2} + \left(\frac{e_1}{\delta_1} - \frac{e_2}{\delta_2}\right)Q_1 - Ae_1^{\beta_2}e_2^{1-\beta_2}$$

If $\dfrac{e_1}{e_2} \ge \dfrac{\delta_2}{\delta_1}$, then F^1 increases with increasing Q_1 and if $\dfrac{e_1}{e_2} \le \dfrac{\delta_1}{\delta_2}$, then F^1 decreases with increasing Q_1. Part (iii) can be proved in a similar manner. $\square$

REFERENCES

Baldwin, R. and P. Krugman (1989), "Persistent Trade Effects of Large Exchange Rate Shocks," *Quarterly Journal of Economics* 104, 635-654.

Boyle, P.P., J. Ervine and S. Gibbs (1989), "Numerical Evaluation of Multivariate Contingent Claims," *The Review of Financial Studies*, 2, 241-250.

Brennan, M.J. and E.S. Schwartz (1985), "Evaluating Natural Resource Investments," *Journal of Business* 58, 135-137.

Carter, J.R. and S.K. Vickery (1988), "Managing Volatile Exchange Rates in International Purchasing," *Journal of Purchasing and Materials Management*, Winter, pp. 13-20.

Carter, J.R. and S.K. Vickery (1989), "Currency Exchange Rates: Their Impact on Global Sourcing," *Journal of Purchasing and Materials Management*, Fall, pp. 19-25.

Cox, J.C., S.A. Ross and M. Rubinstein (1979), "Option Pricing: A Simplified Approach," *Journal of Financial Economics*, 7, 229-263.

Dasu, S. and L. Li (1997), "Optimal Operating Policies in the Presence of Exchange Rate Variability," *Management Science*, 43, 5, 705-722.

Dixit, A.K. (1989a), "Entry and Exit Decisions under Uncertainty," *Journal of Political Economy*, 97, 3, 620-638.

Dixit, A.K. (1989b), "Hysteresis, Import Penetration and Exchange Rate Pass Through," *Quarterly Journal of Economics*, 104, May, 205-228.

Dixit, A.K. and R. S. Pindyck (1994), *Investment Under Uncertainty*, Princeton University Press, NJ.

Gutierrez, G.J. and P. Kouvelis (1995), "A Robustness Approach to International Sourcing," *Annals of Operations Research*, 59, 165-193.

Huchzermeier, A. and M.A. Cohen (1993), "Valuing Operational Flexibility Under Exchange Rate Risk," *Operations Research*, 44, 1, 100-113.

Hull, J.C. (1993), *Options, Futures, and Other Derivative Securities*, Prentice Hall, NJ.

Kogut, B. and N. Kulatilaka (1994), "Operating Flexibility, Global Manufacturing, and the Option Value of a Multinational Network," *Management Science*, 40, 1, 123-139.

Kouvelis, P., K. Axarloglou, and V. Sinha (1997), "Exchange Rates and the Choice of Production Strategies in Supplying Foreign Markets," Working Paper, Olin School of Business, Washington University in St. Louis.

Li, C-L and P. Kouvelis (1997), "Flexible and Risk-Sharing Supply Contracts Under Price Uncertainty," Working Paper, Olin School of Business, Washington University in St. Louis.

Lu, L. (1995), "A One-Vendor Multi-Buyer Integrated Inventory Model," *European Journal of Operational Research*, vol. 81, 312-323.

McDonald, R. and D. Siegel (1986), "The Value of Waiting to Invest," *Quarterly Journal of Economics* 101, 707-728.

McDonald, R. and D. Siegel (1985), "Investment and the Valuation of Firms when There is an Option to Shutdown," *International Economic Review*, 26, 331-349.

Rosenthal, E.C., J.L. Zydiak and S.S. Chaudhry (1995), "Vendor Selection with Bundling," *Decision Sciences*, Vol. 26, No. 1, 35-48.

Scheller-Wolf, A. and S. Tayur (1997), "Reducing International Risk Through Quantity Contracts," Working Paper, GSIA, Carnegie Mellon University.

Shimko, D.C. (1992) *Finance in Continuous Time: A Primer*, Kolb Publishing Co., Miami, Florida.

Trigeorgis, L. (1993), "Real Options and Interactions with Financial Flexibility," *Financial Management Autumn, 202-224.*

Date	Spot	Canada Forward	Futures	Spot	Japan Forward	Futures
1-Jan	1.168	1.1803	1.181	158.856	158.813	158.721
1-Apr	1.2837	1.2991	1.297	150.2026	151.5765	150.998
1-Jul	1.29745	1.16	1.163	177.986	166.243	166.565
1-Oct	1.2941	1.311	1.3015	148.8967	147.966	147.7866
1-Jan	1.·299	1.29	1.2991	171.7867	171.9675	171.5675
1-Apr	1.3426	1.313	1.3152	146.9768	146.7867	146.8966

Table 1a: Assumed information on Spot, Forward and Futures rates for Canada and Japan.

	Canada	Japan
U.S Dollars	479078.7	479078.7
Foreign Currency	449546.6	480336.1
Forr. Curr. - Risk Sharing	464312.7	479707.4
Forward Contracts	456986.9	480026.5
Futures	459354.4	474229.5
Switching to Cheapest Supplier	441465.4	441465.4
Combination of Futures and Supplier Switching	451273.1	435358.8

Table 1b: Total Sourcing costs on various Sourcing Strategies.

Date	Sourcing Strategy	
1-Jan	Start: Canada	Start: Japan
1-Apr	Canada	Canada
1-Jul	Canada	Japan
1-Oct	Japan	Japan
1-Jan	Canada	Canada
1-Apr	Japan	Japan

Table 1c: Optimal Strategy.

Table 1: Illustrative Example on Effectiveness of Various Strategies in Global Sourcing.

660

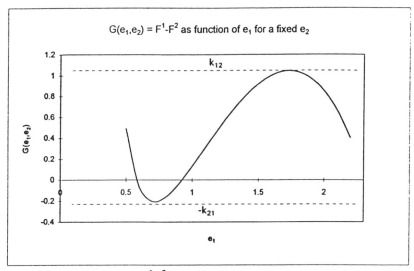

Figure 1: Shape of $G(e_1,e_2) = F^1 - F^2$ as function of e_1 for fixed e_2.

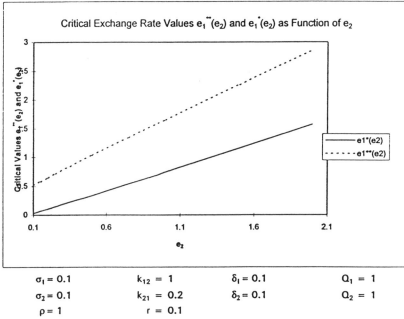

$\sigma_1 = 0.1$	$k_{12} = 1$	$\delta_1 = 0.1$	$Q_1 = 1$
$\sigma_2 = 0.1$	$k_{21} = 0.2$	$\delta_2 = 0.1$	$Q_2 = 1$
$\rho = 1$	$r = 0.1$		

Figure 2: Hysteresis Band for the two foreign supplier case with constant switchover costs.

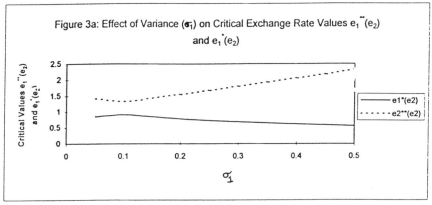

Figure 3a: Effect of Variance (σ_1) on Critical Exchange Rate Values $e_1{}''(e_2)$ and $e_1{}^*(e_2)$

$\sigma_2 = 0.1$	$k_{12} = 1$	$\delta_1 = 0.3$	$Q_1 = 1$
$e_2 = 1$	$k_{21} = 0.2$	$\delta_2 = 0.3$	$Q_2 = 1$
$\rho = 1$	$r = 0.1$		

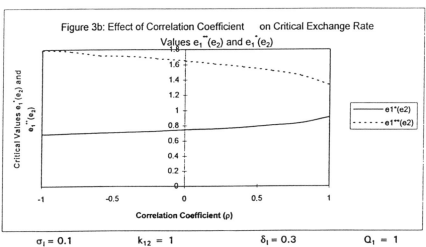

Figure 3b: Effect of Correlation Coefficient on Critical Exchange Rate Values $e_1{}''(e_2)$ and $e_1{}^*(e_2)$

$\sigma_1 = 0.1$	$k_{12} = 1$	$\delta_1 = 0.3$	$Q_1 = 1$
$\sigma_2 = 0.1$	$k_{21} = 0.2$	$\delta_2 = 0.3$	$Q_2 = 1$
$e_2 = 1$	$r = 0.1$		

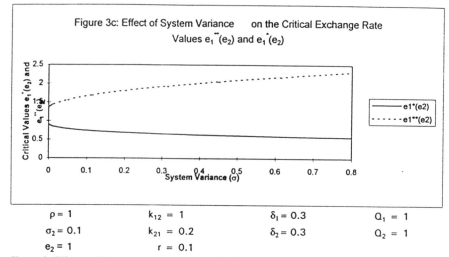

Figure 3c: Effect of System Variance on the Critical Exchange Rate Values e1**(e2) and e1*(e2)
Critical Values e1*(e2) and e1**(e2)
2.5
2
1.5
1
0.5
0
0
0.1
0.2
0.3
0.4
0.5
0.6
0.7
0.8
System Variance (σ)
e1*(e2)
e1**(e2)

$\rho = 1$ $\quad k_{12} = 1$ $\quad \delta_1 = 0.3$ $\quad Q_1 = 1$

$\sigma_2 = 0.1$ $\quad k_{21} = 0.2$ $\quad \delta_2 = 0.3$ $\quad Q_2 = 1$

$e_2 = 1$ $\quad r = 0.1$

Figure 3: Effects of Variance and Correlation Cofficient of Exchange Rates on the Hysteresis Band for the two supplier case.

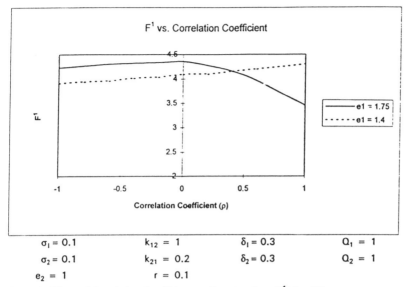

$\sigma_1 = 0.1$ $k_{12} = 1$ $\delta_1 = 0.3$ $Q_1 = 1$

$\sigma_2 = 0.1$ $k_{21} = 0.2$ $\delta_2 = 0.3$ $Q_2 = 1$

$e_2 = 1$ $r = 0.1$

Figure 4: Effect of Correlation Coefficient on Sourcing Cost (F^1) for different e_1 and a fixed e_2 (two foreign suppliers).

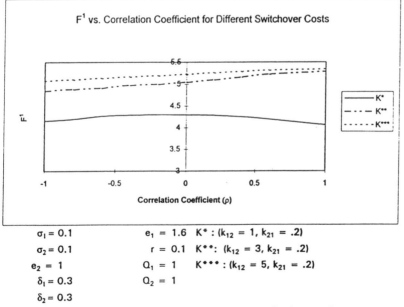

$\sigma_1 = 0.1$ $e_1 = 1.6$ K^* : ($k_{12} = 1$, $k_{21} = .2$)

$\sigma_2 = 0.1$ $r = 0.1$ K^{**}: ($k_{12} = 3$, $k_{21} = .2$)

$e_2 = 1$ $Q_1 = 1$ K^{***} : ($k_{12} = 5$, $k_{21} = .2$)

$\delta_1 = 0.3$ $Q_2 = 1$

$\delta_2 = 0.3$

Figure 5: Effect of Switchover Cost on Sourcing Costs (two foreign suppliers).

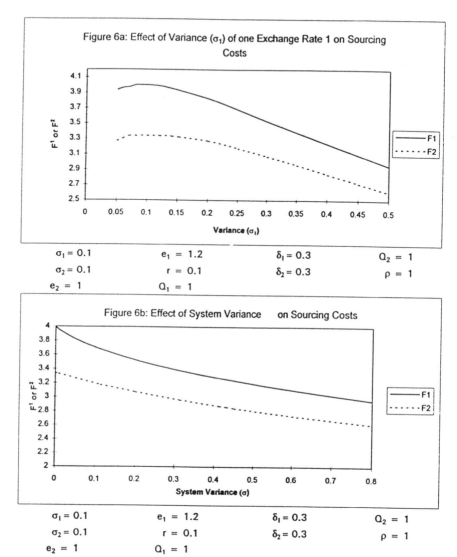

Figure 6: Effect of Exchange Rate Variance of supplier 1 and System Variance on optimal sourcing cost.

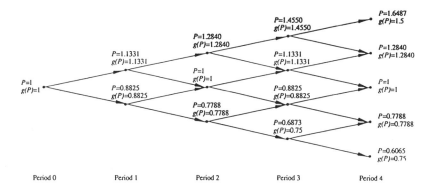

(a) unit prices and unit purchase costs

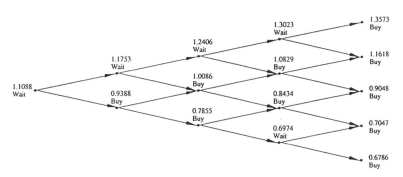

(b) minimum expected discounted unit costs and optimal decisions

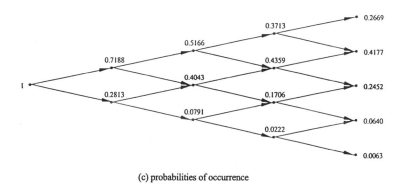

(c) probabilities of occurrence

Figure 7. Single-supplier example with risk-sharing and no quantity flexibility.

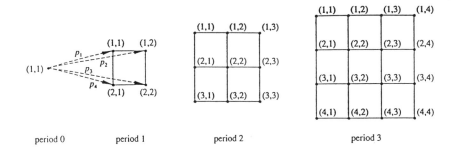

Figure 8. A two-dimensional lattice with $N = 3$.

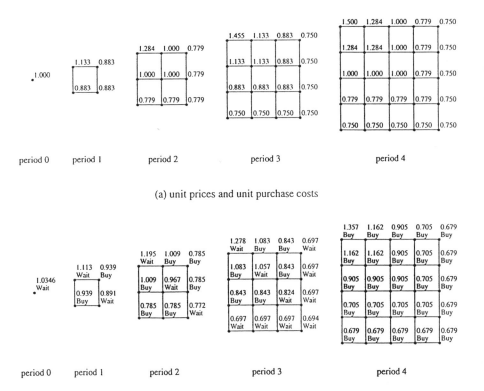

(a) unit prices and unit purchase costs

(b) minimum expected discounted unit costs and optimal decisions

Figure 9. Two-supplier example with 100% quantity flexibility.

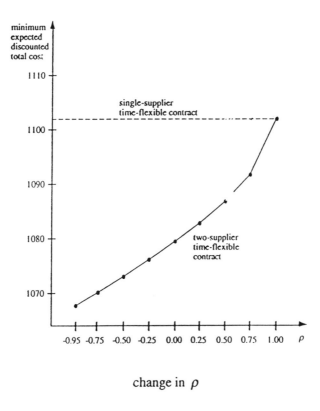

Figure 10. Computational results for the two-supplier problem with quantity flexibility.

21 GLOBAL SUPPLY CHAIN MANAGEMENT: A SURVEY OF RESEARCH AND APPLICATIONS

Morris A. Cohen* and Arnd Huchzermeier**

* University of Pennsylvania
The Wharton School
Operations and Information Management Department
1300 Steinberg Hall-Dietrich Hall
Philadelphia, PA 19104

** WHU Koblenz
Otto Beisheim Graduate School of Management
Production Management
Burgplatz 2
56179 Vallendar, Germany

21.1. Introduction

Today's global economic environment is in a state of transition. The principal changes include: 1) worldwide reduction of trade barriers and development of regional, multi-country economic zones, 2) converging consumer expectations for increased product value, variety and availability in all markets, 3) financial obligations to meet new standards for product safety, environmental protection and product recycling that are being adopted internationally, and 4) increased volatility in financial/currency markets. These developments coincide with the adoption of a new competitive strategy by leading multinational companies. This strategy is one of global supply chain management based on enhanced integration of suppliers and customers as well as increased coordination across multiple value-adding processes within the firm (MacCormack et al., 1994; Ohmae, 1995). Such strategies require firms to maintain core competency on a global scale for fundamental processes such as order fulfillment, supply management and new product development, (Majchrzak and Wang, 1995; Womack and Jones, 1996). The global supply chain strategy also requires that the flow of information, cash and material be managed on an international basis, (Porter, 1996; Preiss et al., 1996). As a consequence, the global supply chain strategy involves both operational and financial decisions. Its successful implementation can lead to more effective risk management and the leveraging of both firm-specific and location-specific advantages to yield lower costs and higher revenues.

In general, the cost of adopting a global supply chain strategy and the risks of its failure can be high. The potential payoffs, however, can be significant. Thus, in order to be successful, firms must develop a coherent and long-term implementation plan. These tradeoffs can be seen upon examination of a number of recent company examples.

The Ford Motor Company's 2000 program is an example of an ambitious global supply chain operating strategy. The company has restructured automobile manufacturing from a regional to a global organization, i.e., the North American, European, and, eventually, Latin American product development, purchasing, manufacturing, and sales functions in each region are centralized. Moreover, product teams now design "world cars". Production and distribution of individual products continue to take place at regional facilities, but the local business functions work under a single, global, product group's oversight and responsibility. The predicted results for Ford included a savings of three billion dollars, reduction of product development lead times from 37 to 24 months and reduction in order-to-delivery cycle times to less than 15 days by the year 2000 (*Business Week*, 1995). In fiscal year 1997, the strategy seems to be paying off, i.e., Ford shows record profits and its strongest balance sheet ever (*Wall Street Journal Europe*, 1998). Moreover, Ford's strategy for globalization under separate product groups with worldwide responsibility is now considered irreversible (*The Economist*, 1996).

Whirlpool Corporation, in its movement towards globalization, in the late 1980s constructed a variant of the globally integrated supply chain. It also developed unified product designs, but allowed for local differentiation. Again the products are produced and distributed by regional supply chains. The "world washer", for example, comes in three forms which are manufactured and marketed in three different geographic markets. At the same time, purchasing, production and distribution are managed regionally, in a highly coordinated fashion, in both North America and Europe. In spite of recent mixed earnings results, Whirlpool is maintaining its regionalized version of the global supply chain strategy in order to generate the cost savings and margins required to compete in the diversified European market (*Forbes*, 1996). The $3bn investment made by Whirlpool to acquire a foothold in the stagnant European market is viewed as an extremely risky gamble that has not paid off yet (*Financial Times*, 1997). Recently, some financial analysts have pronounced Whirlpool's globalization strategy to be a failure. Top management, however, insists that the problems of implementation are temporary (*Financial Times*, 1998).

Our final example of global supply chain management in action is Toyota's policy for production switching (*Wall Street Journal*, 1998). Due to the current economic crisis in Asia, market demand for locally produced cars has not materialized in countries such as Thailand, where the company has invested heavily in the construction of manufacturing facilities. Toyota is leveraging its assembly plant capacity in Thailand to become an exporter of parts for New Zealand, Australia, South America and even Africa. It is interesting to note that it was necessary for Toyota to shift both production and know-how quickly, (from assembly for local consumption to parts fabrication for export). Consequently, Toyota managers trained workers for different jobs and sent them to work in showrooms to observe the sales process. This practice of re-tooling workers is common in Japan but not in Thailand. According to the article, Toyota was able to crank up its Thai export capacity on schedule.

The principal thesis of this paper is that global supply chain strategies should be viewed as real compound options. Operational flexibility through global coordination of value-adding activities and the timing of operations investment decisions can enhance the firm's shareholder value significantly. Our focus is on dynamic investment strategies that include the option to wait/to defer, to abandon/to exit, to expand, to contract or to switch (Copeland et al., 1995; Trigeorgis, 1996). In addition, one can consider the option to improve (Huchzermeier and Loch, 1998). To date, the literature on real options has been limited to demonstrations that the option value of operational flexibility is greater than zero (Dixit and Pindyck, 1994; Sick, 1995). Furthermore, most research in this area has been restricted to evaluations of, entry or exit, options for simple supply chain network structures. Recently, Kogut and Kulatilaka (1994b) evaluated a compound production switching problem between two locations, (countries), assuming a simple production function. An integrated supply chain network, however, poses a more complex and constrained optimization problem. These networks, moreover, allow for a wide range of real options that can be exercised in response to the demand, price and exchange rate contingencies faced by firms in a global supply chain context. Such real, compound options include delaying final commitment of capacity and process technology investments, switching production among international locations contingent on real price/exchange rate scenarios, postponement of assembly and distribution logistics decisions in order to adjust service levels contingent on demand scenarios (Ghemawat, 1991; Huchzermeier, 1991; Huchzermeier and Cohen, 1996).

In this paper, we review the state-of-the-art of modeling approaches to support the design and management of global supply chain networks under multiple sources of uncertainty and different types of risk. In particular, we focus on the integration of supply chain network optimization with real options pricing methods. This unified approach demonstrates that international manufacturing and distribution systems can be viewed as compound real options that act to mitigate the impact of uncertainty due to fluctuations in factors such as market demand, market prices and foreign exchange rates. The paper introduces a framework that incorporates two types of real options for multi-period, multi-country settings: the option to postpone the deployment of resources and the option to switch production or the mode of operation. The paper reviews relevant literatures in operations management and finance, in the context of this integrated framework, and makes the following observations. First, a firm's downside risk can be reduced through proactive use of operational flexibility, contingent on demand and (real) price scenarios. Second, a firm's total supply chain cost and service performance can be improved significantly through functional coordination, on an international scale, in areas such as product design, manufacturing technology investment and distribution logistics strategy. Third, operational risk management strategies can be more effective than pure financial risk management strategies in the context of the global supply chain. Moreover, the combined deployment of operational and financial policies can enhance the firm's shareholder value even further. The paper concludes with a discussion of the gap between theory and practice and provide directions for future research.

Our review of the literatures associated with this new and challenging class of real options suggests that the benefit of using them in global supply chains can be clearly demonstrated. The barriers to successful implementation, however, continue to be large. Accordingly research should be directed towards providing insights and tools for realization of the value of the global supply chain option. Overall, research on global supply chain strategy problems is important and not well understood. At least three major sources of difficulty that we noted are: a) the complexity afforded by multi-location, multi-product network flows, b) the multiple sources of uncertainty and different types of risks that must be considered; including product prices, product demand, foreign exchange rates, technology developments and competitor actions and c) the need for consistency between operational and financial hedging policies, i.e., the proposed evaluation methods assume strong market efficiency and thus explicitly prohibit arbitrage opportunities. This paper reviews the emerging literatures in the area of network modeling and real options as it relates to global supply chain management. We conclude with a discussion of how to integrate these two aspects of the problem.

The paper is organized as follows. In section 21.2, we introduce a unified framework for the analysis of global supply chain strategies as a real compound option, which is consistent with shareholder value maximization. In section 21.3, we review the relevant literature. In section 21.4, we provide an illustration of the option value of stochastic recourse and production switching. The final section provides a general discussion of emerging research directions.

21.2. Applying options thinking to global supply chain networks

In this section, we provide a general discussion on how one applies options thinking to the analysis of global supply chain networks under risk. In general, there are six generic types of real options: 1) to wait/to postpone, 2) to expand, 3) to contract, 4) to exit, 5) to switch, and 6) to improve. In all cases, the expected increase in economic value is created primarily through dynamic investment policies that respond to a single underlying stochastic variable, e.g., a market price process. A firm's ability to anticipate and to respond flexibly to such changes reduces its expected downside risk and thus lowers costs, and/or or raises its revenues; which in turn enhances the firm's shareholder value. In what follows, we consider two types of real options that exploit such operational flexibility. We emphasize, in particular, the (compound) option value created through i) stochastic recourse, e.g., to postpone production and/or distribution logistics decisions until after the uncertainty of the stochastic variables is resolved, and ii) the ability to make multiple resource deployments, e.g., expand/contract production capacity within a global manufacturing and distribution network. A general approach to the evaluation of such types of options is presented below.

Financial Option	Real Option
Strike Price	Fixed Investment and Switching Costs
Spot Price	Expected Cash Flow From Undertaking the Project
Time to Expiration	Planning Horizon
Volatility of the Underlying Asset	Volatility of the Underlying Stochastic Processes
Interest Rate	Interest Rate

Table 1: Comparison of Financial and Real Options

Note that determination of optimal (capacity) investment decisions, under risk, is relevant in other management contexts, e.g., in finance to determine optimal portfolios or in marketing to determine optimal store assortments. Table (1) illustrates the similarities between financial and real options. In this paper, we demonstrate that there are important differences between these two kinds of options and that the general results of financial options pricing theory, e.g., an increase in the variability of the underlying stochastic process(es) leads to an increase in option value, is not always true for real options.

In this section, we demonstrate how a firm can jointly increase its shareholder value and the level of customer satisfaction through exploitation of real options within a global manufacturing and distribution network. In this context, customer satisfaction is measured by the effective level of service provided given the firm's production capacity investments and distribution logistics capabilities. Shareholder value explicitly takes into account the firm's global cash flows – measured by global after-tax profits/losses in a numeraire currency – subject to a discount rate adjusted for market risk. Overall, we study the impact of global operations and risk management policies on the firm's market value where financial markets are assumed to be perfectly efficient, i.e., arbitrage opportunities do not exist. Deviations from purchasing power parity are known to be persistent (Dornbusch, 1987), and hence the value of real options can even be larger.

21.2.1. The real option evaluation model

There are six key issues to be considered in the evaluation of global supply chain networks under risk. First, management needs to consider how operational flexibility within a global manufacturing and distribution network can contribute to the firm's competitive advantage. In this paper, we restrict our attention to the deployment of excess capacity as the source of operational flexibility. Note in Figure 1 how the set of operating strategies, e.g., investments in excess capacity, interact with network, stochastic, and dynamic elements to generate improvements related to returns, risk and customer satisfaction. These improvements, in turn, increase the firm's shareholder value.

Second, in each time period, the firm can use stochastic recourse to exercise the option to redeploy its value-adding activities and resources among the capacities available in the current supply chain configuration. For example, newly available market information, such as real prices or actual customer demand, can be utilized to rebalance the firm's production schedule at plants and its product flows to market regions. This type of option has not been considered in the literature to date. Such stochastic recourse can depend on either exchange rate/price scenarios, demand scenarios, or both. The problem becomes inherently complex due to the presence of fixed switching costs and fixed costs that lead to so-called "hysteresis bands" which delay switching from one global supply chain strategy to another.[1]

Third, within the multi-period planning horizon, the firm can dynamically exercise compound options, i.e., repeatedly switch sourcing, production and/or distribution in the network. The interplay, over time among the capacity decisions and stochastic elements, is captured in the dynamic programming model indicated in Figure 1.

---

[1] Economic hysteresis refers to the failure of investment decisions to reverse themselves when the underlying causes are fully reversed (Dixit and Pindyck, 1994).

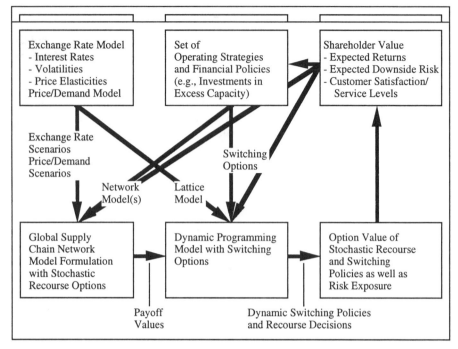

Figure 1: Options Thinking Applied to Global Supply Chain Networks

Fourth, the evaluation method must be consistent with standard financial option evaluation methods (e.g. Black-Scholes based). This is important since many firms will use a combination of real and financial options. In particular, the method for real option evaluation introduced here must resolve issues such as the use of discrete versus continuous time, and selection of an appropriate discount rate within a risk-neutral pricing framework.

Fifth, the modeling framework must be generalizable to multiple sources of uncertainty and different types of risk. Our review of previous research in this area (to be discussed in a subsequent section) reveals that few evaluation models consider more than one source of uncertainty. Relevant uncertainties include: exchange rates, market demand, market product prices, supplier capacity/capability, input prices, production yield, transportation delay, technology developments (for both products and processes), competitor actions and political changes that can affect the cost and feasibility of global supply chain solutions.

Sixth, a unified strategy for risk management involves use of both financial and operational, (real) options. Risk management ultimately requires a firm to position itself on its risk-return efficient frontier. The modeling framework shown in Figure (1), can be applied to evaluate a number of selected candidate supply chain policies. This allows managers to determine the shape of their efficient frontier and assists them in making supply chain strategy decisions.

In what follows, we describe the stochastic dynamic model formulation for the evaluation of real options and how it addresses the six issues raised here.

21.2.2. *Excess capacity as a source of operational flexibility*

Network, resource location and activity allocation models can capture the impact of operational flexibility within a firm's global manufacturing and distribution supply chain. Generally, nodes refer to facilities, (at a given location and with a fixed capacity), e.g., the component, semi-finished, and finished product plants, distribution and warehousing facilities and customer demand points. The connecting arcs represent permissible demand-supply linkages between active nodes. Material flows from node to node along the arcs. Such material flows generate cash flows and are supported by information flows.

Consider the formulation in (1). It presents a basic, single period, deterministic version of an international facility location and distribution logistics planning problem and accounts for exchange rates e, export/import tariffs Dut, and country tax rates Tax. The firm's objective is to maximize its global after-tax profit subject to capacity constraints in each plant (CAP) and demand/market requirements (DEM) in each market. Decision variables y refer to open/close decisions of manufacturing facilities and decision variables x refer to material flow, distribution logistics decisions. Prices p are quoted in the currency of the markets, whereas variable costs c, the markup/transfer price $Mark$ and the distribution logistics costs t are quoted in the currency of the plants. The firm incurs fixed operating costs F for opening manufacturing facilities only.

Define i, $i = 1,...,m$, as the index of plants and j, $j = 1,...,n$, as the index of market regions. Without loss of generality, we assume that $m = n$. The two-stage, single-product supply chain network problem (**SCNP**) defined in (1) maximizes the firm's global after-tax profits expressed in a numeraire currency.

$$\underset{X,Y}{Max} \sum_{i=1}^{m}(1 - Tax_i)[\sum_{j=1}^{n}\{e_j p_j - e_i(1 + Dut_{ij})c_i - e_i Mark_i - e_i t_{ij}\}x_{ij} - e_i F_i y_i]$$

$$s.t. \tag{1}$$

$$\sum_{j=1}^{n} x_{ij} \le CAP_i y_i, \quad i = 1...m$$

$$\sum_{i=1}^{m} x_{ij} = DEM_j, \quad j = 1...n$$

$$x_{ij} \ge 0, \quad i = 1...m, j = 1...n$$

$$y_i \in \{0,1\}, \quad i = 1...m$$

The first constraint ensures that outbound shipments do not exceed available capacity. The second constraint requires that market demand in each country be met. Under risk, this constraint may be relaxed to avoid delivery to unattractive markets or to enable generation of feasible schedules when demand exceeds supply (see section 4). The last two constraints are non-negativity constraints for shipments and an integrality constraint for the binary or plant open/close decisions. In general, (1) is a mixed integer-linear optimization problem whose size depends on the complexity of the supply chain network.

Note that the potential sources of uncertainty in this deterministic model are demand *DEM* and foreign exchange rates e. Without loss of generality, we can assume that the efficient forecast of future exchange rates is the current spot rate for each exchange rate, i.e., $E(e_{it}) = e_{i,t+1} = e_{it}$, where E is defined as the expectation operator (Meese and Rogoff, 1983). Thus, a typical approach to solving the SCNP for a particular future period is to replace future exchange rates with the current spot rates and to specify a future demand scenario.

Formulation (1) captures operational flexibility through the alternate supply chain material flow linkages and the potential to deploy capacity in excess of aggregate forecast market demand. The key tradeoff here is between the benefit of this operational flexibility and the fixed operating costs incurred to provide the capacity. More comprehensive model formulations have been developed, see, for example, Huchzermeier (1991) and Huchzermeier and Cohen (1996). The model formulation presented in (1) can easily be extended to account for multi-echelon and multi-product factors.

Proposition 1 follows immediately from the SCNP model formulation. It considers the case where exchange rates are treated as random variables. Assume that the optimal choice of facility locations in (1) is y_i^*, i.e., for all open facilities y_i^* is set to 1 and for all closed facilities y_i^* is set to 0.

Proposition 1: Assume that exchange rate changes are Normally distributed random variables and that the vector of spot exchange rates is a vector of ones. Assume that fixed costs are zero. Let γ_{ij} equal the sum of distribution logistics, import duty and variable costs (for using plant i to supply market j), and β_{ij} be the minimum of the reduced costs of the coefficients associated with decision variables x_{ij}, where $j=1,...,n$ and facility i is not utilized in the solution to (1), i.e., $y_i^* = 0$.

Then:
a) There exist realizations of the exchange rate e_i such that the optimal value of (1), (i.e. global after-tax profits), can be increased by utilizing plant i.
b) The minimum threshold value of the exchange rate such that plant i is utilized in an optimal solution to (1) is defined by $e_i' = -\beta_{ij}/\gamma_{ij} + 1$.
c) The probability that location i is open in an optimal solution to (1) is defined by $N^{-1}((e_i - e_i')/\sigma_i)$, where N^{-1} is the inverse of the cumulative standard Normal distribution function.

Proof: Part a) of this proposition follows immediately from the fact that the decision variable x_{ij} will become non-zero if and only if the objective function increases. Part b) results from solving the following equation: $(e_j p_j - e_i' \gamma_{ij}) - (e_j p_j -$

$e_i \gamma_{ij}) = \beta_{ij}$. Rearranging terms will lead to the stated result. This is also due to the fact that we are solving a linear program rather than a mixed-integer program as stated in (1). Part c) follows simply from the fact that the log of exchange rate changes are Normally distributed (see section 2.4). Moreover, since (1) is a maximization problem, the cost coefficients must be lowered (rather than increased).

Proposition 1 tells us that if a facility is closed in the solution to (1), for a specific set of forecast exchange rate values (e.g., the mean values equal to the spot rate), then profits (all else being equal) will increase by opening the closed facility, for some other realizations of the exchange rate e_i. These realizations, moreover, will occur with a non-zero probability. Thus, under exchange rate uncertainty it may be optimal to open "excess" capacity since with some probability, this excess will lead to a higher profit. In the presence of (finite) fixed costs, some of the expected gains will be offset and thus the threshold level will change accordingly.

We next consider the situation where demand is a random variable.

Proposition 2: When market demand is a random variable, investment in excess capacity can lead to higher expected sales and a higher level of customer service, which could lead to an increase in the firm's global after-tax profit.

Proof: Assume that the supply chain network is represented by one aggregate node. Furthermore, sales is defined by the minimum of demand and available capacity and service is defined by the probability that demand does not exceed capacity. Then it follows that an incremental investment in capacity will non-decrease revenue and service for all realizations of the random variable demand. The impact of this non-decrease on profit will depend, of course, on the fixed cost of the extra capacity and the expected increase in net revenue. This revenue benefit is determined by the variable costs, prices and the probability distribution of demand.

Overall, proposition (2) is a general statement that investment in excess capacity can increase sales and thus may increase expected profit. Firms will, in particular, keep additional facilities open when the expected increase in revenue plus the expected savings in transportation, import and variable costs outweigh the fixed operating costs of keeping the facility open. Optimal threshold levels or barrier policies for simple production switching policies in a two-country setting have been developed, see for example, by Kogut and Kulatilaka (1994b) and Dixit and Pindyck (1994). Note, however, that these models do not capture the option value of keeping multiple plants open.

The above propositions indicate that stochastic elements have to be considered explicitly and that excess capacity acts as an option in an uncertain world. Hence let us now consider a simple extension of the SCNP, with uncertainty. In what follows, we assume that both exchange rate and demand risk can be modeled by discrete (joint) scenarios with a probability distribution selected so as to match the first and second moments of the underlying distributions. In those scenarios where market demand exceeds available capacity, the level of customer service will be less than 100%. Moreover, in such cases, it may be optimal to abandon unattractive markets by not shipping their demand. As a consequence we must change the constraint for market requirements to be a less-or-equal sign, in order to permit solutions with a demand fill rate of less than 100%, i.e., where some markets are abandoned. We will discuss this issue in more detail in the example case study in section 21.4. In what follows, we treat the order fill rate as an output of the optimization model, rather than as a constraint that must be enforced under all scenarios.

Proposition 3: In an n-country world, incremental investments in capacity have diminishing returns, i.e., the tradeoff between expected sales/profits and capacity exhibits decreasing returns. Furthermore, expected sales/profit is higher when the variance is reduced, e.g., if demands are negatively correlated.

Proof: A formal proof is omitted. In general, the variance of aggregate demand is the sum of all elements of the variance-covariance matrix. Thus negative correlation, which reduces variance, leads to higher expected sales/profit for all values of capacity.

Proposition 3 states that the option value of operational flexibility depends critically on the correlation of market demand.

Proposition 4: In an n-country world, an incremental investment in aggregate capacity can lead to different option values, i.e., the option value of a fixed level of investment is not unique.

Proof: A formal proof is omitted. Basically, in an n-country world, the optimal location of the excess capacity will vary and thus placing it an arbitrary location will lead to a possible decrease in expected profit. Thus the value is "not unique".

This proposition says that real options differ fundamentally from financial options, e.g., the location of the investment does matter. In an efficient market setting, the price of a financial option must be unique to prohibit financial arbitrage. As shown here, this fundamental condition is not true for the supply chain type of real option. The general idea is that the global supply chain network differentiates capacity by location. The decision is more complicated than how much to invest overall in excess capacity. Rather, the firm must consider where to put that capacity and hence different decisions with the same overall investment requirement will lead to different option values. Thus, exploitation of the firm's supply chain network structure is essential.

21.2.3. The option value of stochastic recourse

Consider an arbitrary period in the firm's planning horizon. We assume that the beginning-of-period, or spot exchange rates e_t are known. The following three cases concerning the realization of uncertainty within the period must be considered: i) the end-of-period or future exchange rates e_{t+1} are random, ii) the end-of-period demand realizations DEM_{t+1} are random and iii) that both end-of-period exchange rates and demand realizations are random. In what follows, we assume case iii) and further that exchange rate and demand realizations can be approximated by discrete scenarios. These scenarios are defined so that the first and second moments of the underlying probability distributions (for demand and exchange rates) are matched. For example, demand can be approximated by a low, medium and high scenario and the future exchange rates can be approximated by a up and down movement from the spot exchange rate. The marginal probabilities of exchange rate realizations are denoted by $Prob(e)$ and the probabilities of the demand realizations are denoted by $Prob(DEM)$. Observe that dependencies between these two random variables can be captured through the correlation structure embodied in the joint distribution. Each scenario s thus is defined by a combination of end-of-period demand and/or exchange rate realizations. The probability of any demand-exchange rate scenario s then occurs with $Prob(s)$.

The stochastic supply chain network problem (**SSCNP**), (1'), is a reformulation of our basic model with uncertainty in market demand and exchange rates captured through a discrete set of scenarios.

$$\underset{X,Y}{Max} \sum_{s \in S} \Pr ob(s)[\sum_{i=1}^{m} (1 - Tax_i)[\sum_{j=1}^{n} \{e_{js} p_j - e_{is}(1 + Dut_{ij})c_i - e_{is} Mark_i - e_{is}t_{ij}\} x_{ijs} - e_{is} F_i y_i]]$$

s.t.

$$\sum_{j=1}^{n} x_{ijs} \le CAP_i y_i, \quad i = 1...m, s = 1...S \qquad (1')$$

$$\sum_{i=1}^{m} x_{ijs} = DEM_{js}, \quad j = 1...n, s = 1...S$$

$$x_{ijs} \ge 0, \quad i = 1...m, j = 1...n, s = 1...S$$

$$y_i \in \{0,1\}, \quad i = 1...m$$

$$\sum_{s \in S} \Pr ob(s) = 1$$

In this formulation, the order of events is as follows:
1) capacity decisions y_i are made
2) random scenario s is realized
3) activity allocation decisions x_{ijs} are made contingent on scenario realizations.

Note that the firm enhances its global after-tax profit by postponing resource allocation decisions x_{ijs} contingent on end-of-period realizations of demand and/or exchange rates. If resource location decisions were made after the realization of the scenarios, then the overall supply chain planning problem becomes separable by scenario and the problem reduces to the deterministic formulation (1) with the realized values for the coefficients in the objective function and right hand sides for the constraints.

We can reduce the complexity of the stochastic programming formulation, (1'), considerably by linking price risk and exchange rate risk as follows. Price changes in the local currency caused by currency fluctuations can be approximated by an exchange rate elasticity for prices. Conceptually, we permit deviations from the law of one price or from purchasing power parity, e.g., see Huchzermeier (1991). However, if changes in the exchange rates are assumed to follow a random walk, this violation does not lead to systematic product arbitrage opportunities (since deviations can go either way).

Also, observe that the optimal decision undertaken at the beginning of the period may not be optimal under any future scenario. Thus the firm will incur adjustment costs, even when the unfavorable state is realized. It is important to note that model formulation (1') will be embedded into a multi-period decision framework, i.e., the stochastic dynamic program (to be considered in section 21.2.4). Furthermore, the level of service/customer satisfaction is linked to the available capacity and the demand that can be met under each scenario. The weighted level of service across all scenarios then determines the firm's expected level of customer satisfaction.

In multi-echelon supply chain network configurations with more than two stages, stochastic recourse may be exercised in the following way. Only a subset of the continuous decision variables are subject to recourse, i.e., the ex-ante decision variables x allow for forward positioning of inventory up to a certain stage(s) of the firm's supply chain network and the ex-post decision variables x_s allow for delayed manufacturing and/or distribution logistics activities within the firm's global supply chain network. Strategies such as postponement and positioning of inventory decoupling points for delayed customization become relevant in this case. The firm thus can choose the products and locations in its global supply chain that will be allowed to react, via recourse adjustments, to realizations of the underlying random variables. Thus the firm can design its product, process and supply chain to figure out what should be treated as the second stage decision variable.

Proposition 5: When both x and y decisions are contingent on s, then the option value of stochastic recourse is zero, however, the firm's shareholder value is at a maximum.

Proof: A formal proof is omitted. It is clear that the problem becomes separable, by scenario, and that the decision problem under risk can be translated into s deterministic supply chain network optimization problems for which optimal solutions exist. Moreover, the firm does not incur costs for inventory disposal and/or product redistribution costs.

Proposition 5 states that a solution to (1') with no uncertainty (i.e., we know which s will be realized – which is problem (1)), provides an upper bound on profit.

Any solution to (1') with recourse, will have profit less than this upper bound. The solution to (1'), however will yield the y decisions which maximize the expected profit of the firm given that recourse decisions can be made. Moreover, ignoring the option value of stochastic recourse may present an important opportunity cost to a firm. It follows then that the task of capturing the full option value of the firm's supply chain cannot be left to financial risk managers alone.

Proposition 6: Increases in the volatility of the exchange rates only lead to a larger option value of stochastic recourse.

Proof: In general, the firm can capture more upside potential and thus the value of the stochastic recourse option increases with volatility.

This is a key point. The mechanism of stochastic recourse allows the firm to capture the upside potential in full, i.e., under the good outcome scenario. On the downside, one assumes that the firm is penalized equally. Consequently, the expected net gain should be zero. However, this ignores the fact that the firm can exploit its network structure and thus has the capability to rebalance and to mitigate the negative impact. The option value is increased due to the fact that under the high variance/bad outcome scenario, alternative replenishment strategies can be adopted (relative to the low variance/bad outcome scenario) and thus the option value is at least non-zero. The firm will exercise its real supply chain option, through recourse, only when the option is "in the money".

Proposition 7: Increases in the volatility of demand uncertainty lead to a reduced option value of stochastic recourse.

Proof: A formal proof is omitted. Without loss of generality, it can be stated that when capacity is binding, then expected sales decreases due to the fact that the downside is increased whereas the upside is truncated.

Proposition 7 alludes to the fact that global supply chain networks may act very differently than financial options, i.e., the option value may not increase when the volatility of the underlying risk factor increases. We argue that hedge (production) capacity limits the firm's ability to capture upturns of market demand. Thus the firm is left with a truncated demand distribution and with more downside risk, e.g., a higher probability of having to fulfill lower levels of demand with lower profit.

21.2.4. The compound option value of production switching

In general, it cannot be assumed that switching production is instant and/or costless. Therefore, a multi-period modeling framework needs to be adopted. In what follows, we assume that changes in a particular exchange rate does follow a random walk with drift, i.e.,

$$\ln(e_{t+dt}/e_t) = \mu\,dt + \sigma\,dz \tag{2}$$

where μ is the drift rate, σ is the volatility of exchange rate changes, dz is a standard Wiener disturbance term and e is the spot exchange rate, end-of-period exchange rate respectively. In section 2.5, we will discuss the risk-neutral pricing framework for real options and show that the drift rate should be replaced by the prevailing interest rate differential. The discrete time version of (2) can then be utilized to determine a set of equations using the mean and the standard deviation of the underlying stochastic process(es). For example, in the single-currency case, we have to solve the following set of equations.

$$\begin{aligned}
&\pi u + (1-\pi)d = \mu\Delta t \\
&\pi u^2 + (1-\pi)d^2 - (\mu\Delta t)^2 = \sigma^2 \Delta t \\
&u = \frac{1}{d} \\
&\pi \in (0,1)
\end{aligned} \tag{3}$$

Solving this set of equations leads to the size of the up- and down movements in each period as well as the corresponding transition probabilities. Solutions to equations (3) are as follows (Hull, 1997).

$$u = \frac{((\mu\Delta t)^2 + \sigma^2 \Delta t + 1) + \sqrt{((\mu\Delta t)^2 + \sigma^2 \Delta t + 1)^2 - 4(\mu\Delta t)^2}}{2\mu\Delta t} \tag{4}$$

$$\pi = \frac{\mu\Delta t - d}{u - d}$$

In case of a single exchange rate process, a binomial approximation model or lattice programming approach of exchange rate changes can be deployed (Hull, 1997; Ingersoll, 1987). The compound option evaluation model then utilizes a stochastic dynamic programming formulation, i.e., using backward recursion, to determine the option value for select operating strategies. A typical multi-period model formulation for a multi-period production switching or capacity investment planning problem is then defined as follows:

$$V(e_t, y) = \max_{y' \in \Omega}[-\delta_{y,y'} + SSCNP(e_t, y') + \alpha EV(e_{t+1}, y')] \tag{5}$$

where $V(e_t, y)$ is the maximum expected return of utilizing a particular supply chain network operating policy y in period t, when the beginning of period exchange rate is e_t. Typically y, y' respectively, refers to all structural and policy decisions within a firm's supply chain network including transfer pricing and cost allocation policies. For practical purposes, the choices are limited to the set of admissible policies Ω in each period. SSCNP(.) is the production/profit function for a single-period supply chain network configuration y', e.g., this may be the solution to a comprehensive supply chain network model formulation similar to (1'). $\delta_{y,y'}$ refers to the switching costs between policy y and y', and α is the one-period (risk-free) discount factor. Obviously, increases in the switching costs lead to less switching of operating policies, since the band of inaction widens, i.e., the hysteresis effect becomes more pronounced. In addition, the terminal value conditions need to be specified as well. A natural extension of this problem includes cases where a firm has the option to allocate production volumes among plants or where a firm has the option to switch production mode (say, among the export, joint venture, or wholly-owned subsidiary models) in order to serve a foreign market. Overall, this modeling approach is tractable, whereas most other approaches that have been presented in the literature are not. Furthermore, this approach can be extended to multiple sources of uncertainty and different types of risk (see section 21.2.7).

21.2.5. Risk neutral pricing and discounting

In what follows, we demonstrate that our proposed evaluation method for (any type) of real option is consistent with contingent claims analysis. We assume that the value of a derivative V depends only on the exchange rate e and time t, i.e., $V(e,t)$. The exchange rate process e is assumed to follow a geometric Brownian motion as specified above. From Ito's Lemma, it follows that the change in the value of the portfolio is then

$$dV = (V_t + \mu e\, V_e + \frac{\sigma^2}{2} e^2\, V_{ee})dt + \sigma e\, V_e dz \tag{6}$$

where $V_t = \partial V/\partial t$, $V_e = \partial V/\partial e$ and $V_{ee} = \partial^2 V/\partial^2 e$. Furthermore, we can construct the following portfolio Π, where we buy one security V, invest x in a foreign country (converted at the spot exchange rate e) and borrow $V+x$ in domestic currency, i.e., obtain a riskless loan. The value of the portfolio at time t is then

$$\Pi = -V - x + (V + x) \tag{7}$$

Define r and r_f as the domestic and foreign rate of interest. The change in the value of the portfolio over time dt is then

$$d\Pi = (V_t + \mu e\, V_e + \frac{\sigma^2}{2} e^2\, V_{ee})dt + \sigma e\, V_e dz +$$
$$[\frac{x}{e}(1 + r_f dt)(e + de) - x] - r(V + x)dt \tag{8}$$

Since the portfolio is financed by a riskfree loan, the change in the portfolio must be independent of risk, i.e., the Wiener disturbance term dz must drop out of the equation. This is a standard no-arbitrage argument, where one can not create a money pump through riskless financing. The portfolio becomes instantaneously riskless when x is defined by

$$x = -\frac{e\,V_e}{1 + r_f\,dt} \qquad (9)$$

Consequently equation (8) simplifies using (9). Furthermore, let dt go to zero, then we obtain the following partial differential equation for the price V of a derivative.

$$V_t + e\,V_e(r - r_f) + \frac{\sigma^2}{2}e^2\,V_{ee} = r\,V \qquad (10)$$

Observe that the drift rate μ is replaced by the interest rate differential $(r-r_f)$ and that the discount rate (under risk neutral pricing) is the riskfree rate of interest r. In addition, free and terminal boundary conditions need to be specified to obtain a closed form solution for this evaluation problem. Solutions to (10) have only been developed for the single-exchange rate case (Kogut and Kulatilaka, 1994b). We do not solve equation (10) directly. For practical application cases, we suggest to solve the stochastic dynamic programming formulation as presented in equation (5), which explicitly considers boundary conditions introduced by switching costs and the length of the planning horizon. Observe that the hysteresis effect is also captured in this formulation.

Thus, using standard options thinking, we can transform the general stochastic dynamic programming formulation for a global supply chain network into a tractable problem whose solution depends only on currently observable interest rates and the covariance matrix of exchange rates. Moreover, the approach can easily be applied to the more general case of n exchange rate processes (see the following section).

21.2.6. Extension to multiple sources of uncertainty and different types of risk

The proposed modeling framework can easily be extended to account for multiple sources of uncertainty, e.g., multiple exchange rate processes which are correlated, and different types or risk, e.g., demand risk, price risk and exchange rate risk. Similar to equation (3), we can determine sets of equations for the means and variances of the numeraire currency with all other currencies to determine the system state space. Furthermore, we assume that there is no triangular arbitrage and that financial markets are perfectly efficient.

Proposition 8: *(n-1)* equations for the mean and *(n-1)+(n-2)* equations for the variance and co-variance terms of the numeraire exchange rate processes are sufficient for a multi-nomial approximation model of exchange rate changes. Moreover, the solution to the set of equations determines values for all co-moments of the numeraire variance-covariance matrix.

Proof: See Huchzermeier and Cohen (1996).

The advantage of the multi-nomial approximation method is that the two-dimensional system state space only grows linearly in the number of exchange rate processes considered. In addition, the next proposition is essential for the numerical analysis of real options.

Proposition 9: The solution to the set of equations as defined in proposition 8 leads to non-negative transition probabilities.

Proof: See Huchzermeier and Cohen (1996).

Note that this property has often been violated in standard options pricing approaches, e.g., see Cox, Ross and Rubinstein (1979).

Additional types of risk, e.g., lead time risk, can easily be accounted for by modifying the single-period payoff function SSCNP(.) in equation (5) through a more complex stochastic programming formulation for the single-period supply chain optimization problem, as defined in section 21.2.3.

21.2.7. Operational and financial risk management

In this section, we address the issue of managing the firm's risk exposure and the integration of financial and operational risk management tools.

Proposition 10: In an *(n+1)*-country world with higher aggregate exchange rate risk, i.e., $\Sigma_{i=1...m} \Sigma_{j=1...n} Cov_{(i,j)} < \Sigma_{i=1...m+1} \Sigma_{j=1...n+1} Cov_{(i,j)}$, there can be increasing returns to incremental investments in capacity.

Proof: A formal proof is omitted. It is clear that the upside potential is increased when country *n+1* is added. Thus the firm's shareholder value can be increased further for finite fixed investment costs and distribution logistics costs (see proposition (1)).

Proposition (10) is an argument for globalization of the firm. In this context, globalization of operations refers to an increase of the firm's exposure to risk which is beneficial, i.e., shareholder value enhancing. Proposition (10) states that investments in riskier environments can be shareholder enhancing when the options are "properly" exercised, i.e., managed as a real option. In a single-currency world, there are diminishing returns to investing in excess capacity (Jordan and Graves, 1995). In a global and dynamic world, however, there can be increasing returns to investing in excess capacity (Huchzermeier, 1994). In general, investments in increasingly turbulent environments can expose the firm to more upside potential and an increased level of risk on the downside. As stated before, the firm's global manufacturing and distribution logistics network reduces/truncates the firm's risk effectively on the downside. Increasing returns can be achieved through investments in riskier currencies with higher volatility assuming that fixed investment and operating costs are small relative to the expected gains. Observe that capacity investments are not traded and thus are not value-priced. Of course, it is true that additional investments in the new country of operation will exhibit diminishing returns.

Proposition 11: If there exist states where costs are sufficiently high enough to lead to bankruptcy, then financial hedging can limit the system state space in such a way that insolvency (costs) can be avoided.

Proof: A formal proof is omitted. Basically, in case of a down state where the firm would incur insolvency, it pays to avoid those states through hedging exchange risk utilizing forward or futures contracts. Thus, bankruptcy is avoided and shareholder value is strictly increased. For a similar line of reasoning, we refer to Mello et al. (1995).

This last proposition states that in case of significant bankruptcy risk, financial hedging can lead to increases in the firm's shareholder value. Even in a financial market that is assumed to be perfectly efficient, i.e., there is no mispricing of financial risk management tools, the net effect of financial hedging on the firm's value can be strictly positive. It follows that the joint coordination of financial and operational risk management strategies can raise the firm's shareholder value even further.

To summarize, coordination of an integrated supply chain network provides a firm with an *option* to respond to uncertain events, such as exchange rate fluctuations, demand realizations and so forth. By exercising its option, the firm truncates its downside risk and captures upside potential. For example, by holding excess plant capacity in different countries, a firm can switch production among these plants, and thereby provide itself with both arbitrage and a leverage opportunities. This is an alternative to the purchase of long-term futures contracts and other financial instruments as a means of reducing risk due to exchange rate fluctuations. Accordingly, the approach is referred to as *operational hedging* in the literature (Huchzermeier, 1991; Huchzermeier and Cohen, 1996). Even when markets are assumed to be efficient, the benefit of operational hedging can be different from zero due to the fact that investments in capacity are non-traded risk management tools. As demonstrated above, various types of risk can have a very different impact on the firm's shareholder value. For example, increased demand risk may lead to lower sales. Thus the firm has to take its *operational hedge capacity* into account to manage its global risk exposure and to maximize its shareholder value. If, capacity is constrained, the increase in market volatility can have a negative impact on the shareholder value as well. This argument is counter to common wisdom that the value of an option always increases with risk.

21.3. Literature Review

The previous section presented a series of structural results concerned with applying real options thinking to global supply chain networks under risk. In this section, we review selections from current relevant literature on global supply chain management. We restrict our discussion to those articles which illustrate one or more of the properties of "options thinking" that was described in the previous section. We note that a number of more comprehensive reviews of the global supply chain management literature have been published elsewhere.

The current state-of-the-art of global supply chain network strategy planning models can be characterized by two fundamental approaches: (stochastic) network optimization models and options pricing methods. First, integrated supply chain network models are developed to capture the complexities of a multi-product, multi-echelon, multi-country, multi-period planning problem for the optimal choice of facility locations, capacity and technology used as well as sourcing, production and distribution decisions contingent on future states-of-nature. Also, the firm can leverage its integrated supply chain network through a revision of its decisions over time, contingent of the realization of uncertain outcomes, i.e., apply stochastic recourse to capture the upside potential of serving attractive markets. Financing and risk management strategies need to be considered as well. In a global environment, the focus of such models is on risk reduction due to portfolio effects, avoidance of tax payments through exploitation of global, after-tax cost tradeoffs through sourcing and of material and production, dynamic transfer pricing policies among entities in different countries, and/or exploitation of scale and scope economies associated with the production technology in place. The set of admissible resource location and allocation policies are numerous which are executed rather infrequently. Even small problem sizes can lead to complex optimization problems which can only be efficiently evaluated using Monte Carlo simulation methods (Eppen et al., 1989).

Second, options pricing approaches focus primarily on the evaluation of operational flexibility to enhance the firm's shareholder value. In general, global supply chain investment decisions are limited, e.g., production switching, but can be exercised frequently, to reduce the firm's downside risk. Few approaches consider multiple sources of uncertainty, e.g., price/exchange rate uncertainty or different types of risk, e.g., demand risk. In general, the system state space grows exponential in the number of scenarios considered, e.g., a Markov chain approximation with k states and l investment policies considered leads to a $(kl)^t$ paths to be considered. Consequently, there persists a significant gap in the literature on integrating the two modeling approaches. This polarization in research is due to the analytical complexities, i.e., availability of efficient solution methods for solving (stochastic) mixed-integer programming formulations in the first case and solving multi-dimensional stochastic dynamic programming problems or large sets of partial differential equations in the second case. In what follows, we limit our review of the literature to modeling approaches that account for both the complexities of international supply chain networks and the impact of real exchange rate risk. For detailed surveys of the recent literature, see also Ferdows (1989), Huchzermeier (1991), Verter and Dincer (1992), Cohen and Kleindorfer (1993), Flaherty (1996), Verter and Dincer (1996), Vidal and Goetschalckx (1997), and Cohen and Mallik (1998).

Huchzermeier (1991) develops a stochastic dynamic programming formulation to determine the option value of operational flexibility within a multi-country plant network of Apple computer. A tri-nomial approximation is developed to capture the variance-covariance matrix of the underlying stochastic processes. It is assumed that exchange rate changes do follow diffusion processes and that financial markets are perfectly efficient. The profit function as described above is derived from solving a multi-product, multi-stage supply chain network model. Moreover, there were four capacity expansion alternatives that needed to be simultaneously evaluated. Overall, it was shown that despite the complexity of a two-dimensional system state space, the proposed evaluation model using the multi-nomial approximation method was tractable even for the compound real option where multiple switching over time was permitted. The main results from the analysis are as follows. First, a global manufacturing and distribution logistics network provided a multinational firm with a natural and robust hedge against exchange rate uncertainty and demand risk. Second, operational flexibility can effectively reduce the firm's downside risk and enhance its shareholder value. Third, stochastic recourse with respect to assembly and distribution logistics postponement, i.e., forward positioning of inventory along the firm's global supply chain, allows the firm to mitigate against market risks such as demand risk and real price uncertainty.

Kogut and Kulatilaka (1994b) were probably the first to develop a stochastic dynamic programming model that explicitly treats supply chain as the equivalent of purchasing an option whose value is dependent upon the spot exchange rate. They consider a two-country production switching model with a simple production function only. Contrary to Kulatilaka (1988), the single exchange rate process is assumed to be mean-reverting which is modeled as a discrete time Markov chain. In this context, they analyze the hysteresis effect in the presence of switching costs and determine explicitly the hysteresis band. In general, the value of the real exchange rate which will induce the firm to switch from domestic to foreign country production is strictly less than the rate necessary to induce the switch to the domestic plant. Moreover, this hysteresis band widens with the degree of uncertainty and the value of the switching costs. This model does not consider detailed operational characteristics (e.g., multiple products or supply chain stages), and becomes intractable for more than one exchange rate process, i.e., for more than two countries.

Kouvelis and Sinha (1995) present a model that allows for switching of production modes (exporting from the home country, establishing a joint venture (JV) with local partners, or establishing a wholly owned subsidiary (WOS) for a firm in a foreign country. They formulated a profit maximizing, multi-period, stochastic dynamic program that allows the firm to switch among the modes of production within a planning horizon. They also identified a hysteresis phenomenon that characterizes switching behavior between the production strategies in the presence of a switching cost. The magnitude of the hysteresis band is a measure of inertia associated with keeping current strategy. It is affected by exchange rate variability and the market power of the entering firm. They concluded that a strongly depreciated home currency favors an export policy, while a strongly appreciated home currency favors a JV or WOS. The choice between JV or WOS depends on the transaction costs (including production, distribution and logistics costs) per unit demand in each of the modes of production as well as the switching costs from these modes to the export mode.

Huchzermeier and Cohen (1996) reported on a modeling framework that integrates both the supply chain network optimization and real option evaluation approaches to a global manufacturing strategy planning problem. This model maximizes discounted, expected, global after-tax profit for a multinational firm in terms of a numeraire currency. They propose a hierarchical approach for solving the problem. First, a multinomial approximation model of correlated exchange rate processes is used to specify a lattice model of n country exchange rate processes and the corresponding transition probabilities which are consistent with risk neutral pricing. It is important to note that the multi-nomial approximation utilizes n variance and $n-1$ covariance terms of the exchange rate processes only, but can easily be fitted to the entire variance-covariance matrix. In addition, the derived transition probabilities can only take on positive values between zero and one. This property is often violated by alternative approximation methods (Hull, 1997). Second, a set of global manufacturing strategy options defines available decision alternatives (in terms of local or global sourcing, production and distribution logistics) for the design of the firm's global supply chain network. Third, the value of a number of manufacturing options is determined by solving a mixed-integer programming model for each exchange rate scenario within each period. Finally, a stochastic dynamic program uses these values as inputs to determine the firm's discounted, expected, global, after-tax profit for each available policy option, over a multi-period planning horizon. It is important to note that solving this problem determines the (option) value of operational flexibility which is derived from switching among alternative strategies in response to exchange rate changes. Counter to financial options, the same (capacity) investment exhibits different option values for different multinational firms.

Dasu and Li (1997) studied optimal policies of a firm operating plants in different countries in the presence of exchange rate variability. They develop a two-country, single market, stochastic dynamic programming model where the combined capacity of the plants exceed the single product, deterministic demand. Thus, the firm can allocate production among the plants, depending on the exchange rate. Here, the state space of the dynamic program is the production quantity to be switched among the plants. When the switching costs are linear or step functions, they concluded that, regardless of whether the variable production cost function is concave or piece-wise linear convex, the optimal policy is always of the two level barrier type (i.e., each plant operates at either a minimum or a maximum output level). This result is analogous to the hysteresis result of Kogut and Kulatilaka (1994b), since it suggests that when production does shift between countries, the volume shifted will be greater or equal to some minimum batch size. While generalizations for multi-product and multi-market cases are proposed, we note that this model is limited to cost minimization rather than profit maximization, and that the effect of corporate tax rates is ignored.

21.4. Applichem revisited

In this chapter, we illustrate the global supply chain network dynamics using data from a Harvard case study Applichem (Flaherty, 1996). In the case, management must determine how much production capacity is needed and which plant(s) should be closed in their global supply chain. The main difficulty is that the network structure is complex, i.e., six plants serving six markets, and both real exchange rate risk and capacity constraints need to be considered. Detailed cost data is provided in the case and it is possible to conduct a detailed productivity analysis of each site. Based on this analysis, two strategies for reengineering the firm's global plant network emerge: i) close the plant in Japan due to its high operating costs or ii) close the plants in Japan, Canada and Venezuela and then restrict sales to the (unprofitable) Venezuelan market. The supply chain network analysis can be supported by a deterministic optimization model of the firm's distribution logistics network. Students analyzing the case, usually suggest local improvements to reduce costs at each plant site (based on a benchmark study). They then run the deterministic model to reoptimize the supply chain network. However, the recommended network strategy does not change. The main message of the case is that less capacity enhances the firm's global profitability. Contrary to this wisdom, we demonstrate that this analysis is misleading and that the recommended strategy will, in fact reduce the firm's shareholder value! We demonstrate, moreover, that the firm can derive option value from the operational flexibility of excess capacity within its global manufacturing and distribution logistics network through the use of stochastic recourse and production switching.

In what follows, we briefly summarize the structure of the decision problem and the data set provided in the case. Tables 2 and 3 show the volatility and the correlation of the log of exchange rate changes (IMF, 1998). We have updated the data to reflect more recent exchange rate movements, i.e., the volatilities of exchange rate changes are based on the years 1990 through 1998. The original case provides time series of exchange rates and wage rates. However, there is no explicit exchange rate model introduced in the case. It is left to the student to consider such uncertainty and to devise a way to include it in the global supply chain network analysis. We assume here that exchange rate changes are governed by a random walk model and that financial markets are perfectly efficient. i.e., there is no possibility for financial arbitrage. This implies that we do consider only exchange rates with the numeraire currency, e.g., U.S. dollar.

Can$/US$	DM/US$	Yen/US$	Pesos/US$	Bolivar/US$
0.0399	0.0809	0.0937	0.1996	0.1756

Table 2: Standard Deviation of Log-Exchange Rate Changes 1990-1998

	Can$/US$	DM/US$	Yen/US$	Pesos/US$	Bolivar/US$
Can$/US$	1.000				
DM/US$	0.328	1.000			
Yen/US$	-0.164	0.558	1.000		
Pesos/US$	-0.096	-0.716	-0.301	1.000	
Bolivar/US$	-0.251	-0.353	0.018	0.600	1.000

Table 3: Correlation Matrix of Log-Exchange Rate Changes 1990-1998

Table 4 summarizes the firm's prices in each market and cost data including duties. We used a matrix form for market prices to make it immediately apparent that the Japanese plant operates at a significant cost disadvantage.

PRICES						
FROM/TO	MEXICO	CANADA	VENEZUELA	GERMANY	US	JAPAN
MEXICO	$65.65	$65.65	$65.65	$65.65	$65.65	$65.65
CANADA	$62.62	$62.62	$62.62	$62.62	$62.62	$62.62
VENEZUELA	$50.50	$50.50	$50.50	$50.50	$50.50	$50.50
GERMANY	$50.50	$50.50	$50.50	$50.50	$50.50	$50.50
US	$66.66	$66.66	$66.66	$66.66	$66.66	$66.66
JAPAN	$63.13	$63.13	$63.13	$63.13	$63.13	$63.13

COSTS						
FROM/TO	MEXICO	CANADA	VENEZUELA	GERMANY	US	JAPAN
MEXICO	$59.49	$66.90	$93.78	$72.29	$69.31	$72.16
CANADA	$97.16	$56.46	$90.28	$68.96	$62.72	$67.91
VENEZUELA	$91.95	$60.28	$55.28	$67.03	$62.97	$65.75
GERMANY	$63.67	$42.42	$61,26	$36.67	$43.92	$45.52
US	$98.26	$61.25	$93.19	$69.33	$57.29	$68.98
JAPAN	$158.00	$101.41	$147.73	$111.02	$105.60	$93.28

Table 4: Market Prices and Cost Data

Table 5 shows the capacity limits and the demand requirements for each plant/market region. In addition, the optimal flow of product from each plant (country) to each market (country) is shown. These results were determined by solving the deterministic supply chain network problem, i.e., this configuration is the base case analysis. Observe that the Japanese plant is not utilized at all, and thus it should be shut down. Table 6 shows the cash flows and the global after-tax profit, accounting for differential taxes to be paid in each country of operation. The global after-tax profit of $410.39 (in millions) ignores the impact of risk, however, and does not capture the benefits of stochastic recourse or production switching.

CAPACITY	MEXICO	CANADA	VENEZUELA	GERMANY	US	JAPAN	TO/FROM
6.2	3.0	0.0	0.0	0.0	3.2	0.0	MEXICO
3.7	0.0	2.6	0.0	0.0	1.1	0.0	CANADA
4.5	0.0	0.0	4.5	0.0	0.0	0.0	VENEZUELA
47.0	0.0	0.0	11.5	20.0	3.6	11.9	GERMANY
18.5	0.0	0.0	0.0	0.0	18.5	0.0	US
0.0	0.0	0.0	0.0	0.0	0.0	0.0	JAPAN
SUPPLY	3.0	2.6	16.0	20.0	26.4	11.9	
DEMAND	3.0	2.6	16.0	20.0	26.4	11.9	

Table 5: Capacity Limits, Demand Requirements and Optimal Distribution Logistics Decisions for the Base Case

CASH FLOW	MEXICO	CANADA	VENEZUELA	GERMANY	US	JAPAN
MEXICO	$18.49	$0.00	$0.00	$0.00	-$11.73	$0.00
CANADA	$0.00	$16.01	$0.00	$0.00	-$0.11	$0.00
VENEZUELA	$0.00	$0.00	-$21.51	$0.00	$0.00	$0.00
GERMANY	$0.00	$0.00	-$123.68	$276.60	$23.69	$59.26
US	$0.00	$0.00	$0.00	$0.00	$173.38	$0.00
JAPAN	$0.00	$0.00	$0.00	$0.00	$0.00	$0.00
	$18.49	$16.01	-$145.19	$276.60	$185.23	$59.26

NET PROFIT
$410.39

Table 6: Optimized Cash Flows for the Base Case

Relaxing the constraint that all markets need to be served leads to a closure of the Canadian, Japanese and Venezuelan plants. Moreover, the Venezuelan market will no longer be serviced. The impact on the firm's global after-tax profit is significant, i.e., it can be increased by almost 50% to $600.72 (million). The dominant solution is to close plants and to vacate non-profitable markets. This strategy may seem radical, but it is in fact consistent with observed industry practice, e.g., compare against DEC's reengineering strategy in the 90s (see Arntzen et al., 1995).

The impact of risk on this supply chain network can be illustrated through simulation. Figures 2 and 3 show the impact of exchange risk on the firm's profitability assuming that the firm does not change its distribution logistics flows contingent on exchange rate realizations, i.e., it sticks to a fixed production mix and distribution logistics schedule. Figure 2 is based on the deterministic solution of the original problem where the Japanese plant is closed and all markets are being served. Figure 3 utilizes the schedule for the three plants (Germany, Mexico, US) serving five markets only (except Venezuela).

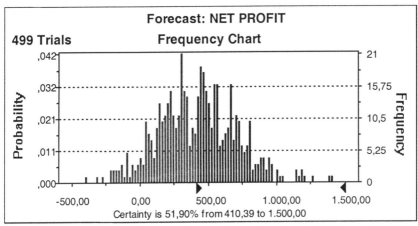

Figure 2: Keep all plants open and serve all markets ($\mu = 427.73$; $\sigma = 288.03$)

It is interesting to note that reducing the number of plants raises the firm's shareholder value, i.e., the expected returns are increased and the standard deviation of profits is also reduced. This is due to the fact that the firm no longer serves unprofitable markets.

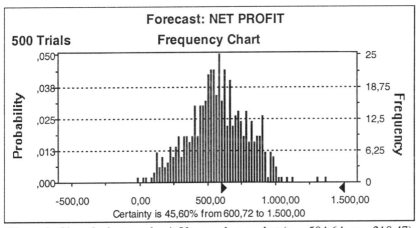

Figure 3: Close 3 plants and exit Venezuelan market ($\mu = 584.64$; $\sigma = 218.47$)

To illustrate the option value of stochastic recourse and production switching, we conducted the following supply chain network analysis under risk, where the setup of the single-period, single-product supply chain network management problem is as follows. First, the firm decides on which plants to open, close respectively. Second, uncertainty in spot exchange rates is resolved. Third, the firm optimizes its distribution logistics system contingent on the prevailing exchange rates to maximize its shareholder value, see Figure 4 below. This corresponds to solving repeatedly problem (1').

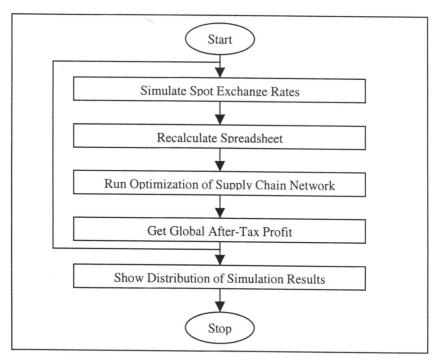

<u>Figure 4</u>: Supply Chain Network Analysis using Monte Carlo Simulation

In Figures 5a-5c, we consider the option value of stochastic recourse, i.e., the firm optimizes its distribution logistics flows contingent on exchange rate realizations. In addition, we require that world market demand is being met. For a given strategy, where a certain number of plants are being closed (ex-ante decision), the supply chain network is optimized (ex-post decision). It is our objective to demonstrate that the option value increases with the level of capacity deployed despite the fact that fixed costs prevail.

Figures 6a-6c also demonstrate the stochastic recourse option value of operational flexibility through market supply and distribution logistics decisions contingent on exchange rate realizations. In both Figures 5 and 6, we are solving model formulation (1'), i.e., use stochastic recourse to rebalance the product mix and material flows contingent on the realization of exchange rate/demand scenarios. The main difference between Figures 5 and 6 is that in Figure 6, the demand constraint as an equality is relaxed (100% fill).

Figures 5a-5c demonstrate that the option value of stochastic recourse increases with the number of open plants. Figures 6a-6c illustrate that stochastic recourse combined with the flexibility to choose to serve or not to serve a market, can enhance the firm's shareholder value even further. Relaxing the demand constraint can be viewed as introducing an option to "switch the mode of operation", i.e., to serve or not to serve a market.

The graphs in this section were obtained by integrating the spreadsheet add-in software *Crystal Ball Pro* (DECISIONEERING, 1998) with the optimization software *What's Best!* (LINDO SYSTEMS, 1998). It only took a few minutes to obtain the results of the simulation runs. Note that in this example, we considered only a single-period supply chain optimization problem under risk. Moreover, we ignored the impact of fixed operating costs. Our supply chain network analysis can easily be modified, however, to account for fixed operating costs by subtracting the fixed costs associated with a particular operating strategy from the expected returns. Compound option evaluation methods, however, need to be analyzed in a multi-period context with the dynamic programming methods as described in sections 21.2 or 21.3.

21.5. Outlook and conclusions

This paper presents a survey of the literature pertaining to analytic approaches for global supply chain strategy analysis and planning. The literature is quite recent, and we note that the research has not yet evolved in a coherent manner. A general criticism of the majority of reported models is that they lack practicality and would be difficult to implement. Few of the models, for example, incorporate the underlying complexity of a global supply chain network or account for international market dynamics, i.e., real price/exchange rate risk, demand risk, lead time risk and service requirements. Furthermore, the option value of stochastic recourse and production switching has not been exploited in most modeling approaches. In section 21.2, we propose a general modeling framework for evaluation of compound real options that is tractable and consistent with financial modeling approaches. We conclude the paper in this section by noting a number of specific areas for future research.

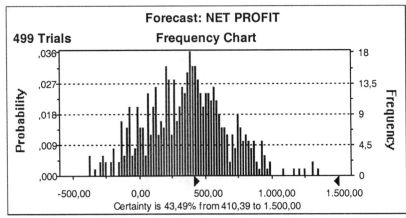

Figure 5a: 100% service with 3 plants ($\mu = 367.22$; $\sigma = 305.89$)

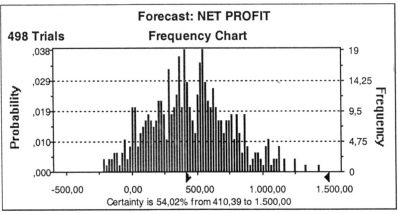

Figure 5b: 100% service with 5 plants ($\mu = 456.52$; $\sigma = 304.39$)

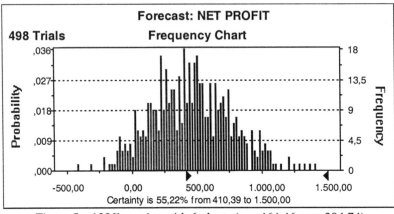

Figure 5c: 100% service with 6 plants ($\mu = 464.46$; $\sigma = 304.74$)

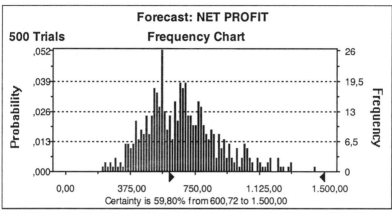

Figure 6a: Total flexibility with 3 plants ($\mu = 677.31$; $\sigma = 212.90$)

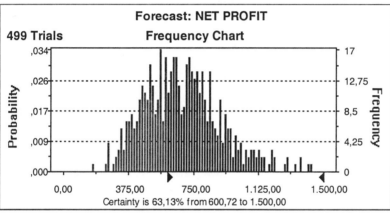

Figure 6b: Total flexibility with 5 plants ($\mu = 691.01$; $\sigma = 221.33$)

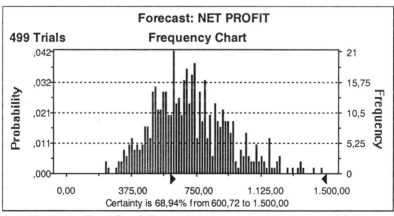

Figure 6c: Total flexibility with 6 plants ($\mu = 721.82$; $\sigma = 220.97$)

Studying operational flexibility versus economies of scale. The literature suggests that the firm can capitalize on production switching. However, there are few reported cases where firms do exercise this type of option. Further research is required on how the presence of economies of scale and the benefits of integrated supply chain management impact the firm's shareholder value in the long term.

Incentive issues. Many organizations adopt a decentralized or regional approach to the management of their operations. In general, a multinational firm faces the task of designing appropriate incentive systems to allow and encourage a global coordination of its operations. Moreover, one needs to study the barriers to production switching and the global transfer of product and process know-how.

Integrated financial and operational risk management. In many instances, we observe that firms are willing to buy insurance against financial risk. Rarely do they invest in slack capacity or excess inventory to mitigate against operational risk. Analysis of the costs and benefits of such risk reduction strategies, through the joint coordination of operational and financial risks, is an important area for future research.

Managing complexity versus product variety. Rather than leveraging the firm's product design and process knowledge on a global scale, management may decide to concentrate on new product development. Exploring the tradeoff between managing product variety and the cost of complexity requires both analytical and empirical research.

Impact of strategic alliances and inter-firm coordination. Few studies exist that deal with the impact on supply chain performance of multiple firms through the use of strategic alliances. Development of formal models and in-depth case studies are required.

Competitive facility location under risk. In many industries, excess capacity persists on a global scale, e.g., the automotive industry. Despite this fact, the first mover advantage of a firm entering in a newly opened markets is quickly eroded by the action of its major competitors that do follow. The inclusion of this issue poses new modeling and data analysis challenges in the context of realistic supply chain models.

Real options thinking in the context of global manufacturing. It is important to develop new business cases to apply and to quantify the results stated in section 21.2. Overall, we see it as an important challenge to managers to overcome the lack of understanding of the issue on i) how one integrates operational and financial risk management, and ii) how one develops tractable and consistent modeling approaches.

References

Arntzen, B.C., G.G. Brown, T.P. Harrison and L.L.Trafton (1995). "Global Supply Chain Management at Digital Equipment Corporation." *Interfaces*, January-February, 69-93.

Business Week (1995). "Ford: Alex Trotman's Daring Global Strategy," April 3.

Cohen, M.A. and P.R. Kleindorfer (1993). "Creating Value Through Operations." *Perspectives in Operations Management*, Boston: Kluwer Academic Press, 3-21.

Cohen, M.A. and S. Mallik (1997). "Global Supply Chains: Research and Applications." *Production and Operations Management* 6(2), 193-210.

Copeland, T., T. Koller and J. Murrin (1995). "Using Option Pricing Methods to Evaluate Flexibility." *Valuation*, New York: John Wiley, 446-475.

Cox, J.C., S.A. Ross and M. Rubinstein (1979). "Option Pricing: A Simplified Approach." *Journal of Financial Economics* 7, 229-263.

Dasu, S. and L. Li (1997). "Optimal Operating Policies in the Presence of Exchange Rate Variability." *Management Science*, 705-724.

Dixit, A.K. and R.S. Pindyck (1994). *Investment under Uncertainty*, Princeton: Princeton University Press.

Dornbusch, R. (1987). "Exchange Rates and Prices." *American Economic Review*, March, 93-106.

The Economist (1996). "The World that Changed the Machine," March 30.

Eppen, G.D., R.K. Martin and L. Schrage (1989). "A Scenario Approach to Capacity Planning." *Operations Research*, July-August, 517-527.

Ferdows, K. (1989). *Managing International Manufacturing*. Amsterdam: North Holland.

Financial Times (1997). "Rough and Tumble Industry," July 2.

Financial Times (1998). "Whirlpool Sticks to Its Global Guns," February 2.

Flaherty, T.M. (1996). Applichem (A). *Global Operations Management*, New York: McGraw Hill, 119-131.

Forbes (1996). "Whirlpool's Bloody Nose," March 11.

Ghemawat, P. (1991). "Flexibility: The Value of Recourse." *Commitment: The Dynamic of Strategy*. New York: The Free Press, 109-134.

Huchzermeier, A. (1991). Global Manufacturing Strategy Planning Under Exchange Rate Uncertainty, Unpublished Ph.D. Thesis, University of Pennsylvania, The Wharton School, Philadelphia, Pennsylvania.

Huchzermeier, A. (1994). Global Supply Chain Network Management under Risk, Working Paper, University of Chicago, Graduate School of Business, Chicago, Illinois.

Huchzermeier, A. and M.A. Cohen (1996). "Valuing Operational Flexibility under Exchange Rate Uncertainty." *Operations Research* 44(1), 100-113.

Huchzermeier, A. and C. Loch (1998). Evaluating R&D Projects as Real Options: Why More Variability is Not Always Better, Working Paper, WHU Koblenz, Otto-Beisheim Graduate School of Management, Vallendar, Germany.

Hull, J.C. (1997). *Options, Futures, and Other Derivatives*. Upper Saddle River: Prentice Hall.

Ingersoll, J.E. (1987). *Theory of Financial Decision Making*. Rowman and Littlefield.

International Monetary Fund (1998). "International Financial Statistics (IFS)," March.

Jordan, W.C. and S.C. Graves (1995). "Principles on the Benefits of Manufacturing Process Flexibility." *Management Science*, April, 577-594.

Kogut, B. and N. Kulatilaka, N. (1994a). "Option Thinking and Platform Investments: Investing in Opportunity." *California Management Review*, Winter, 52-71.

Kogut, B. and N. Kulatilaka (1994b). "Operating Flexibility, Global Manufacturing, and the Option Value of a Multinational Network." *Management Science*, 123-139.

Kouvelis, P. and V. Sinha (1995). "Exchange Rates and the Choice of Production Strategies." *Supplying Foreign Markets*, Duke University.

Kulatilaka, N. (1988). "aluing the Flexibility of Flexible Manufacturing Systems." *IEEE Transactions on Engineering Management.* November, 250-257.

MacCormack, A., L.J. Newman III and D.B. Rosenfield (1994). "The New Dynamics of Global Manufacturing Site Location." *Sloan Management Review*, Summer, 69-80.

Majchrzak, A. and Q. Wang (1996). "Breaking the Functional Mind-Set in Process Organizations." *Harvard Business Review*, September-October, 93-99.

Meese, R.A. and K. Rogoff. (1983). "Empirical Exchange Rate Models of the Seventies: Do They Fit Out of Sample?." *The Journal of International Economics*, 3-24.

Mello, A.S., J.E. Parsons and A.J. Triantis (1995). "An Integrated Model of Multinational Flexibility and Financial Hedging." *Journal of International Economics*, 27-51.

Nahmias, S. (1997). *Production and Operations Analysis*, Irwin.

Ohmae, K. (1995). "Putting Global Logic First." *Harvard Business Review*, January-February, 3-7.

Porter, M.E. (1996). "What is Strategy?." *Harvard Business Review*, November-December, 61-78.

Preiss, K., S.L. Goldman and R.N. Nagel, R.N. (1996). *Cooperate to Compete: Building Agile Business Relationships*. New York: Van Nostrand Reinhold.

Sick, G. (1995). "Real Options." Jarrow, R.A., V. Maksimovic and W.T. Ziemba, W.T. (Eds.), *Handbook in Finance*, Amsterdam: North-Holland, 631-691.

Trigeorgis, L. (1995). *Real Options in Capital Investment: Models, Strategies, and Applications*. Westport: Praeger.

Trigeorgis, L. (1996). *Real Options: Managerial Flexibility and Strategy in Resource Allocation*. Cambridge: MIT Press.

Verter, V. and M.C. Dincer, M.C. (1992). "An Integrated Evaluation of Facility Location, Capacity Expansion, and Technology Selection for Designing Global Manufacturing Strategies." *European Journal of Operational Research* 60, 1-18.

Verter,V. and M.C. Dincer (1996). "Global Manufacturing Strategy." *Facility Location: A Survey of Applications and Methods*. New York: Springer, 263-282.

Vidal, C.J. and M. Goetschalckx (1997). "Strategic Production-Distribution Models: A Critical Review with Emphasis on Global Supply Chain Models." *European Journal of Operational Research* 98, 1-18.

Wall Street Journal Europe (1998). "Ford Net Jumps 49% to Record in 4th Quarter," January 28.

Wall Street Journal (1998). "Japanese Auto Makers Shift Strategies to Keep Plants Viable in Southeast Asia."

Womack, J.P. and D.T. Jones (1996). "Beyond Toyota: How to Root Out Waste and Pursue Perfection." *Harvard Business Review*, September-October, 4-16.

Software:

Crystal Ball Pro, Version 4.0, DECISIONEERING, Denver, CO (1998).
What's Best!, Version 3.1., LINDO SYSTEMS INC., Chicago, IL (1998).

22 MANAGING SUPPLY CHAINS IN EMERGING MARKETS

Alan Scheller-Wolf and Sridhar Tayur

Graduate School of Industrial Administration
Carnegie Mellon University
Pittsburgh, PA 15213

22.1 INTRODUCTION

Manufacturing and retail companies, particularly those in North America, Europe and Japan, are moving towards global sourcing of their raw materials, components, and finished goods. Automobile firms were some of the initiators of this trend, followed by tool, steel and machinery companies such as Caterpillar, which currently sources products throughout Europe and North America. Still more recently apparel retailers have followed suit. Nike and Hush Puppies source products from Asia; in 1998 Laura Ashley announced plans to sell its European plants and do the same.

In many of these cases companies are looking to the emerging markets, such as those in India, the far east, and Mexico, to secure manufacturing capacity. A primary motivating force behind this move is the desire to take advantage of lower wage costs, but this shift is also encouraged by reduced tariff rates, improved international communication and transportation, and labor relations issues.

Such a shift can greatly complicate the management of a company's supply chain; often it becomes inaccurate to assume a static relationship will exist between the firm and a dedicated supplier providing goods at fixed rates. Rather, the inclusion of features such as alternate sourcing, limited and unreliable capacity, and fluctuating costs and demand intensities (possibly due to currency fluctuations) becomes central to any analysis of such a supply chain.

22.1.1 Motivating Examples

We present some motivating examples to illustrate a few of the unique situations which may arise in international operations:

1. A Canadian distributor supplying the North American Market buys products from a company in China. The price, quantity, and delivery date is set, including a known lead time. However, at some point before the delivery date the Chinese government may prevent its domestic company from shipping the goods; their foreign exchange reserve level is now high enough that the rate previously negotiated between the Chinese and Canadian companies is no longer acceptable. Thus the shipment is blocked by the Chinese government, and sits in port in China.

   As a contingency, the Canadian company also contracts with an Indian supplier, who will ship a quantity of the product if the Chinese firm reneges. This contingency shipment will be at a higher price, as labor costs in India are not as low as those in China. The Indian supplier is willing to act as a backup as he is attempting to enter the market; while he cannot compete with the Chinese company on price, he can do so through reliability.

   From the Canadian firm's viewpoint, the problem consists of determining how much to order from each of the two sources. It may also consider at what price, or alternately at what degree of unreliability, it is no longer optimal to source from China, instead ordering directly from the Indian company. Further, the Canadian company may wish to explore the possibility of entering into an agreement, possibly a quantity contract (guaranteeing that it will purchase no less than a certain amount of goods over a fixed period of time) which might either increase the reliability of the Chinese supplier, or motivate the Indian supplier to reduce his price to a competitive level.

2. An multinational heavy equipment firm is expanding their operations to include a new product line. This product will use materials sourced from the United States, Mexico, and Europe, will be manufactured and assembled at sites in the UK and United States, and will sell throughout North America and Europe. As they are introducing a new product, the company wishes to ensure that retailers carry their equipment so they can capture a portion of the highly competitive market. To this end, the firm must decide whether it wishes to construct a new multinational supply chain, attempt to adapt their current distribution network to include the new product, or somehow augment their existing system to handle the anticipated increase in volume.

   The decision to source overseas coupled with the considerable weight of the equipment creates the potential for both lead times and shipping costs to be large. In addition, the complex nature of the product dictates that some final manufacturing be performed within the supply chain. This constrains the company to a certain specified set of centers through which much of the material must pass. Complicating matters further is the use of facilities unique to the international import/export trade, such as bonded warehouses, which permit the importer to forgo paying duties while storing their wares. These issues make the construction of an efficient supply chain an extremely important, but not straightforward, undertaking.

   The company therefore needs an integrated model which it can use to explore its options. It must maximize the new product's potential to capture a significant share in the market while at the same time ensuring profitability. Finally, assuming that the products penetrate the market, the supply chain must be able to accommodate an anticipated increase in volume in the years to come.

3. An American clothing retailer sources its products from India. The Indian suppliers are fragmented (due to both governmental regulation and cultural factors) and individually have limited capacity, making it necessary to contract a number in parallel to obtain the total desired order. Since they are small companies and their credit histories are suspect, the suppliers either are unable to secure funding from lenders, or are able to secure capital only at very high interest rates. Their inventory holding costs are thus very high, leading them to carry little or no inventory. They

obtain raw materials only after an order has been received, increasing lead times to their American customer.

An additional difficulty facing the American firm is that it may enjoy very little loyalty from its suppliers. If the suppliers are presented with a better offer, they may transfer their services to the higher bidder, leaving their American customer without sufficient capacity to meet its needs. (This new offer may materialize due to an updated multi-fiber quota level, which governs the total amount of goods which may be imported annually from India to the United States. As the remaining quota drops, companies are more desperate to secure goods, for fear of being shut out of the market for the remainder of the quota's period of enforcement.)

To counter the long lead times and the unreliability of capacity, the American company may use an intermediary, in Hong Kong or Bombay for example, who can serve in one or both of the following roles:

- The intermediary may act as a lender, providing working capital to the Indian suppliers at a reasonable rate, thus permitting them to hold inventories of raw materials and reduce their lead times.

- The intermediary may also serve as a guaranteer of supplier capacity. This is possible as the intermediary is not only geographically closer to the suppliers, but also may have certain cultural levers at its disposal.

There are of course charges for these services. In order to make an informed decision whether to employ the intermediary, the American company must determine the value added to their operations through the intermediary's actions.

The American company may choose to bypass the intermediary by extending a letter of credit directly to the suppliers, committing the American firm to purchase a fixed number of goods, possibly over an extended time horizon. This will allow the Indian suppliers to secure raw materials, and also guarantees that they use the raw material to fulfill the American firm's order, as violation of the terms of a letter of credit is treated very seriously by the Indian government. One drawback of this strategy is that the exchange rate may be very volatile between India and the United States. If the Indian rupee falls in value as compared to the dollar, the American company may find itself locked into a contract specifying that they pay a price far higher for the capacity than they could find on the open market.

Thus the American company must evaluate their options and decide what length of lead time and level of capacity uncertainty they are willing to tolerate. If they decide to take action to reduce these factors they have to then decide whether they wish to hire an intermediary, which tasks they will expect the intermediary to accomplish, and at what price.

4. An American telecommunications firm is entering into an agreement to construct a network in an overseas country. As this is a large project with an expected duration of many years, the company is concerned with both escalations of their costs and depreciations of their returns over the project horizon. To try to alleviate these risks they are considering adding provisions to their contract which will adjust their costs and re-imbursements should the circumstances surrounding the project change appreciably.

After study of the issue, it becomes apparent that it is not enough for the American firm to only consider changes in the exchange rate between their currency and that of the customer. Rather, as significant costs are incurred in the patron country, the American company must also take into account that country's potential rate of inflation.

There are many cases to consider, depending on how the firm's payables and receivables are divided between dollars and the foreign currency. In certain cases, for example where costs are in the foreign currency and re-ceipts in dollars, different actions may be prescribed based on the financial policy of the project country. For countries with low inflation, moderate growth and floating currencies, such as those of Western Europe, some foreign exchange hedging may be called for. For high growth countries with sticky currencies against the dollar, such as Chile, inflation in labor costs are a major concern. Here price escalation clauses to keep pace with rising wage costs may be used. Finally, for high inflation countries with essentially floating currencies, such as Turkey, wage costs are expected to be offset by currency depreciations, and one recommended course of action is to *do nothing*.

It is thus apparent that this is a highly complex issue. Further compli-cations arise when one considers that many of the above strategies must be negotiated with, and accepted by, the customer.

22.1.2 The Questions

As the above examples above show, operations are entering an exciting phase, one in which new dynamics are being formed and new questions are being asked. These include:

- What factors must companies consider when deciding whether to source materials and products from overseas markets?

- What special opportunities are unique to overseas operations? Are they changing? How can these be taken advantage of?

- What risk factors must companies consider when doing business with overseas or emerging markets? Which of these are standard to all sourcing relations, and which are unique to international operations? When might these risks outweigh the opportunities?

■ What strategies are available to reduce these risks? In which situations are these strategies most appropriate, and how much should a company be willing to pay to engage in them?

To help answer these questions, a Markovian, periodic review production inventory model of a company's sourcing and sales operations may be used. This model can include alternate sourcing partners, each of which may have distinct static lead times which differ by one time unit, as well as differing state-dependent production capacities and minimum order requirements. Taken together, we refer to the minimum order quantities and finite capacities as "order bands".

Each supplier also may have unique state-dependent unit ordering, holding and backorder costs. (We assume full back-ordering.) In addition, the end item, or consumer demand which the company sees for its products may be stochastic, having a distribution which is likewise state-dependent. This can capture the phenomenon of *pass-through*, explained below.

For finite horizon problems, as well as discounted and average cost infinite horizon problems, the optimal policy for the company is a *modified base-stock policy*: In each period there exist state-dependent optimal inventory levels; the firm orders an amount from each supplier such that after ordering the inventory position is as close as possible to the prescribed levels. Note that the purchaser may not be able to reach the desired levels, or conversely may be forced to order an amount which carries their inventory above the optimal level, due to the order bands.

Establishing the structure of the optimal policy solves only half the problem; one must also be able to determine the appropriate order up-to levels. To this end, simulation based optimization, in particular Infinitesimal Perturbation Analysis, or IPA, may be used. This is true not only for lead times differing by one unit, but for the case of general lead times as well. (For the case of general lead times it is known that the optimal policy may depend on the amount of outstanding orders, and hence be arbitrarily complex. In this situation IPA can be used to find the optimal parameters within the class of order up-to policies.)

22.1.3 Exchange Rate Issues

It should be apparent that exchange rates are a major driver in international operations. Therefore we will discuss their behavior and effects in greater detail.

There is some disagreement as to the proper way to model the evolution of exchange rates. Within our framework we will consider three different models of their behavior: The random walk, mean reverting behavior, and a momentum model. Essentially, these models assume that rates evolve either randomly, or with a central or extreme tendency, respectively. The reader is referred to [35] for further explanation.

Exchange rates also may affect demand. Depending upon the nature of the market and a corporation's position in it, there may be different degrees of this effect, or *pass through*. If the corporation is a monopoly providing an essential good or service, a change in exchange rates causing the price of the product to rise will have little to no effect on demand, as the consumers have no choice but to purchase the product. We term this situation where demand is independent of state as the case of zero pass through. A more common situation is that in which a change in exchange rates affecting the price of the product does effect demand. Depending upon such factors as product differentiation, the concentration of foreign firms versus domestic firms, non-tariff barriers, and social dynamics (fashion and "mass psychology") the impact of exchange rates on demand can vary. We model two such levels of influence, which we call partial and full pass through, respectively. These reflect the situations where demand does vary with the exchange rate, but has a lesser or greater degree of stickiness (possibly due to the fact that price does). For a fuller discussion of the causes and effects of pass through, please see [12], [20], and [41].

Due to their influence on costs *and* demand, exchange rate fluctuations can have a significant influence on cash flow. Realizing this, companies have devised an assortment of techniques to try to dampen these effects. These include forward buying and hedging in the financial markets, management of currency choice and timing of transactions, and using parallel sources of cash flow. Additionally, using operational strategies ("real options") to counteract exchange rate variability has become more common; see [9], [21], [23].

Unfortunately, in many cases these strategies provide less than perfect solutions. For example, it may either be cost prohibitive, or simply impossible, to secure forward contracts for more than a few months in the desired currency. (This is particularly true of the most volatile ones.) Furthermore, many smaller companies may have few options as far as transaction timing and alternate cash streams. This can be particularly true when dealing with state-owned corporations.

22.1.4 Literature Review

This chapter summarizes material previously published, or submitted for publication, by various authors.

Studies of international operations issues are becoming more common. Two examples (of many) are Raman [32, 33] who has considered specific issues dealing with the apparel industry, and de Sousa [10] who looked at contractual provisions for risk minimization.

Works such as these have energized the search for models that capture the dynamics of international operations. To this end, the incorporation of dual sourcing, lead times, and state dependent order bands was accomplished in Scheller-Wolf and Tayur [37], as a generalization of earlier research of those same authors [36]; the interested reader is referred to these two papers for proofs of many of the results presented in this chapter. The models of [36, 37] were themselves extensions of those presented in Kapuściński and Tayur [22], Aviv and Federgruen [4], Federgruen and Zipkin [14, 15], Gavirneni, Kapuściński and Tayur [17] and Sethi and Cheng [38]. Similar work, but in a multi-stage setting, appears in Chen and Song [7].

The topic of supply modes with different lead times, or "emergency orders" has a history dating back almost forty years in the literature. In the single location setting, Barankin [5] and Daniel [8] considered this problem, with the first formal proof of the optimality of the order up-to structure for sources with differing lead times appearing in Fukuda [16]. More recently, Wittmore and Saunders [40], Moinzadeh and Nahmias [26], and Moinzadeh and Schmidt [27] have continued this direction of research. For multi-echelon systems, notable contributions have been made by Muckstadt and Thomas [28], Aggarwal and Moinzadeh [1], Pyke and Cohen [29] and Moinzadeh and Aggarwal [25].

Minimum order quantity contracts were first analyzed by Anupindi and Bassok [2] and Anupindi and Akella [3]; related models in a finite horizon, stationary setting are discussed therein. For a reference on Infinitesimal Perturbation Analysis, or IPA, the reader is referred to Glasserman and Tayur [18, 19].

22.2 DEFINITION OF THE MODEL

The general framework of the model is that of a company procuring materials from up to two suppliers having static leadtimes which differ by one unit, along with disparate unit ordering costs and order bands. The company also pays unit holding costs and penalty costs for not meeting *their* customers' (consumer) demand. All such demand is fully backordered. Time is taken to be discrete; at each period the state of the world is determined according to an underlying Markov chain, which may affect the consumer demand (seen by the purchaser), costs, and order bands in the period.

From this point forward we will examine the model where there are two suppliers having positive lead times. It should be understood that a single supplier is a restriction of this model, and all results hold for that case as well. We will call the orders *expedited* or *regular*, according to whether they have the shorter or longer of the two lead times, respectively.

In each period the sequence of events will be as follows: Period type is revealed, the expedited order arrives (if placed an appropriate number of period in the past), (consumer) demand is realized, costs are incurred, the regular order arrives (if placed an appropriate number of periods in the past), and then future orders are placed. Note that both orders which may arrive in any given period must have been placed in the same period; there is no order crossing in this model.

22.2.1 Motivating Examples Revisited

To help make the model concrete in the reader's mind, we revisit the motivating examples of Section 22.1.1, drawing correspondences between their characteristics and those of the model.

1. For the Canadian distributor, the state of the Markov chain would be the level of Chinese foreign exchange reserves. The regular and expedited prices and lead times would be those of the Chinese and Indian suppliers,

respectively. Order bands for one or both of the suppliers could also be included, if appropriate.

2. In the case of the heavy equipment manufacturer, the dealers may receive products along two different routes in the supply chain (having capacities). These routes may have different lead times, and hence one will be expedited and the other regular. Similarly, the cost of shipment along the different routes may differ.

   The state of the Markov chain is related to the number of unsatisfied orders in the last period. This does not affect costs or order bands, but can influence demand, as leaving customers unsatisfied may damage the reputation of the new product and impact sales in the current period.

3. In the case of the clothing retailer sourcing from India, the remaining multi-fiber quota would be the state of the chain, and the decision is between different order bands. In addition, if the retailer wishes to reduce leadtimes, then the expedited deliveries would be those under which the suppliers secured raw materials early, either via a letter of credit or through the actions of an intermediary.

4. The telecommunications firm must estimate the evolution of their costs and receivables over the horizon of the project. The state of the system would include both the exchange rate and the rate of inflation in the patron company. Capacities and minimum order quantities are not involved in this model; instead the company needs a forecast of their potential profits under different projected scenarios in order to evaluate the benefits of various contractual provisions.

22.2.2 Definitions

x_n: Inventory position at the start of period n. This is the inventory on hand, minus the amount on back-order (only one of these terms can be non-zero) plus the amount in transit.

y_n, w_n: Inventory position after placing period n orders of the expedited and regular type, respectively.

i_n: The state of the system in period n, $i \in S$. We assume that the states of i_n form a finite (K) state, irreducible Markov chain.

Q_i, C_i: Non-negative state-dependent minimum and maximum order sizes, respectively; $0 \leq Q_i \leq C_i$. These quantities pertain to the *sum* of both order types; this can easily be relaxed, as outlined in the note below.

l_0, l_1: The static lead-times for orders; $0 \leq l_0 = l_1 - 1 < \infty$. Let $l \stackrel{\text{def}}{=} l_0$.

c_i^0, c_i^1, p_i, h_i: Finite, state-dependent ordering, penalty, and holding costs, respectively, per unit of time. We assume these are not all identically zero (this would make the problem trivial). Holding costs are not paid on goods in transit, i.e. during the leadtimes. If this is not the case the expected pipeline inventory costs could be included in the unit ordering costs.

Note that the quoted unit price c_i^j for merchandise is lead time dependent. We assume $c_i^0 \geq c_i^1$ for all i; the customer pays a premium for expedited delivery.

D_i: A non-negative, state-dependent, time stationary discrete demand distribution. We denote as d_i a generic value sampled from the distribution D_i, and assume $\mathrm{E}[d_i^4] < \infty$ for all i.

D_i^l: The *random* l-fold convolution of demand distributions, starting with that of period $i + 1$. It is random as it is not known what the distributions convolved will be, but an expectation, based on the current state and the transition probabilities of the Markov chain is available. Again we denote by d_i^l a value sampled from this distribution.

$\Delta_i \stackrel{\text{def}}{=} C_i - Q_i$: The ordering flexibility, or width of the *ordering band* in a period. Define $\Delta \stackrel{\text{def}}{=} \max_{\{i \in S\}}\{\Delta_i\}$.

Note: If Q_i^e and Q_i^r are the minimum quantities for expedited and regular orders in period i (with the capacities defined similarly), then set $Q_i = Q_i^e + Q_i^r$ and $C_i = C_i^e + C_i^r$. Use these values to determine the particulars of the total order, and once this is set, constrain the expedited order in the obvious way.

22.2.3 Assumptions

Three primary assumptions are made in this model:

1. If we denote the stationary probabilities of the embedded Markov chain and the means of the demand distributions by π_i and μ_i respectively, for $1 \leq i \leq K$,

$$\sum_{i=1}^{K} \pi_i Q_i < \sum_{i=1}^{K} \pi_i \mu_i < \sum_{i=1}^{K} \pi_i C_i. \tag{22.1}$$

   This ensures that our process is stable.

2. For any $r \in \mathbf{R}$ and all $i \in S$, there exists paths with positive probability of arbitrary lengths N_r^1 and N_r^2 such that

$$\sum_{n=1}^{N_r^1} (d_{i_n} - C_{i_n}) \geq r. \tag{22.2}$$

$$\sum_{n=1}^{N_r^2} (Q_{i_n} - d_{i_n}) \geq r. \tag{22.3}$$

Without these assumptions there would be an upper or lower bound on the cumulative inventory level, under certain policies.

3. All demand distributions are what we call strongly (C-Q)-complete: If D_i has cumulative distribution function F_i, there exists a $\Gamma > 0$ such that for all i in S and all $\epsilon > 0$:

$$\inf_{\{x \le Q_i, x \in \mathcal{N}\}} \{F_i(x + \Delta_i) - F_i(x - \epsilon)\} \ge \Gamma.$$

Note if $C_i = Q_i$ for some i, this requires a point mass at all non-negative integers up to and including Q_i. This is a regularity condition which aids in our recurrence proofs.

22.3 FINITE HORIZON

We begin by considering finite horizon results.

22.3.1 Preliminaries

For this section define:

N: The length of the horizon.

β: The discount factor.

$v_n(x, i)$: The minimum total expected discounted cost from period $n + l$ onward, when starting with an inventory position of x and residing in state i. We seek a policy which returns a cost of $v_0(x, i)$ for all x and i.

Note that in period n the costs from period n to $n + l - 1$ can be viewed as fixed costs; the decision maker can take no action to influence them.

The expected discounted cost follows the recursion:

$$v_n(x, i) = \inf_{(y, w) \in A_i(x)} \{c_i^0(y - x) + c_i^1(w - y) + L^l(y, i) \tag{22.4}$$
$$+ \beta E[v_{n+1}(w - d_n, j)]\}, \quad x \in \mathbf{R}, \; i \in S, \; 0 \le n < N,$$
$$v_N(x, i) = 0 \; \forall \; x, \; i,$$

where $A_i(x)$ is the state-dependent (convex) set of feasible post-ordering inventory levels. In a similar manner, define $W_i(x) = [x + Q_i, x + C_i]$.

For all real y the expected holding plus penalty cost l periods in the future takes the form:

$$L^l(y, i) = \begin{cases} E[h_{i_{n+l}}(y - d_n^l)^+ + p_{i_{n+l}}(y - d_n^l)^-] & 0 \le n < N - l, \\ 0 & n \ge N - l. \end{cases} \tag{22.5}$$

Recall that d_n^l is the l-fold convolution of the demand values from periods $n+1$ to $n + l$.

Another representation of v_n, (22.10), following Fukuda [16], is derived below. This can be shown (inductively) to be equivalent to (22.4).

Define $c_i \overset{\text{def}}{=} c_i^0 - c_i^1$, and let $\hat{y}_i$ be the minimal point which minimizes $g_i(y) = c_i(y - x) + L^l(y, i)$. As the sum of convex functions in y, $g_i(y)$ has a unique minimizer which may or may not be finite, depending upon the parameters of the problem.

Define:

$$\tilde{L}^l(y, i) = \begin{cases} c_i(\hat{y}_i - y) + L^l(\hat{y}_i, i) & y \le \hat{y}_i, \\ L^l(y, i) & y > \hat{y}_i. \end{cases} \tag{22.6}$$

$$\Lambda(w, i) = \begin{cases} c_i(w - \hat{y}_i) + L^l(w, i) - L^l(\hat{y}_i, i) & w \le \hat{y}_i, \\ 0 & w > \hat{y}_i. \end{cases} \tag{22.7}$$

$$J_n(w, i) = c_i^1 w + \Lambda(w, i) + \beta E[\hat{v}_{n+1}(w - d_n, j)], \ y \in \mathbf{R}, \ i \in S, \ 0 \le n < N \tag{22.8}$$

$$I_n(x) = \min_{\{w \in W_i(x)\}} \{J_n(w)\}, \tag{22.9}$$

and finally

$$\hat{v}_n(x, i) = -c_i^1 x + \tilde{L}^l(x, i) + I_n(x, i). \tag{22.10}$$

Set all values equal to zero in period N.

22.3.2 Results

Our first two lemmata are instrumental in the proof of later results.

Lemma 22.3.1 *For all $0 \le n \le N$:*

(i) $L^l(y, i)$ is convex in y for all i.

(ii) $\tilde{L}^l(y, i)$ and $\Lambda(w, i)$ are convex in y and w respectively for all i.

(iii) J_n, $\hat{v}_n$, and I_n are convex in their first arguments, respectively, for all i.

(iv) $\hat{v}_n(x, i) \ge 0$ for all x and i.

(v) For all $i \in S$ and $0 \le n < N - l$,

(a) $\hat{v}_n(x, i) \to \infty$ when $|x| \to \infty$,

(b) $J_n(y, i) \to \infty$ when $y \to \infty$.

The structure of the optimal policy relies on the next lemma.

Lemma 22.3.2 *Given any (x, i) pair, and a solution which minimizes $v_n(x, i)$ (assuming for now that one exists), there will be an optimal solution ordering an expedited amount as close to $\hat{y}_i$ as permissible.*

As the first two terms of (22.10) are independent of the policy, one seeks to minimize $J_n(w, i)$ as defined in (22.8). This, along with Lemma 22.3.1, leads to the optimal finite-horizon policy.

Theorem 22.3.1 *For a given $i \in S$ and $0 \leq n < N - l$, let $w^*_{n,i}$ be the smallest value which minimizes $J_n(w, i)$. Then an optimal feasible policy for the __total order__ in period n is to order an amount u_n given by:*

$$w^*_n = \begin{cases} Q_{i_n} & \text{if} & w^*_{n,i} \leq x + Q_{i_n} \\ w^*_{n,i} - x & \text{if} & x + Q_{i_n} < w^*_{n,i} \leq x + C_{i_n} \\ C_{i_n} & \text{if} & x + C_{i_n} < w^*_{n,i}. \end{cases}$$

Furthermore, $w^*_{n,i} < \infty$ for all n and i (but $w^*_{n,i}$ may be $-\infty$).

Proof : From (22.10) it is clear that to minimize $v_n(x, i)$ it suffices to minimize $J_n(w, i)$. Further, from Lemma 22.3.1 we know that for each n and i, $J_n(w, i)$ is a convex function with minimum value less than $+\infty$. If the minimum is attainable, i.e. $w^*_{n,i} \in W(x)$ it is optimal to order up to $w^*_{n,i}$.

If the minimum is not attainable for a given x, because of the convex nature of $J_n(w, i)$ it is optimal to order a quantity as close to $w^*_{n,i}$ as possible. This will be either Q_{i_n} or C_{i_n}, depending upon whether x falls above or below the interval $[w^*_{n,i} - C_{i_n}, w^*_{n,i} - Q_{i_n}]$. ∎

Theorem 22.3.2 *For a given $i \in S$ and $0 \leq n < N - l$, an optimal feasible policy for expedited orders in period n is to order an amount u_n given by:*

$$u_n = \begin{cases} 0 & \text{if} & x \geq \hat{y}_i \\ \hat{y}_i - x & \text{if} & x < \hat{y}_i \leq w^*_n \\ w^*_n & \text{if} & w^*_n < \hat{y}_i, \end{cases}$$

where w^*_n is defined as in Theorem 22.3.1.

Proof : This follows immediately from Lemma 22.3.2 and Theorem 22.3.1. ∎

We next present three properties. The proof of the first two are straightforward, and are thus omitted.

Property 22.3.1 *For any $x \in \mathbf{R}$ and $i \in S$, $v_n(x, i)$ is non-increasing in n.*

Property 22.3.2 *For any $x \in \mathbf{R}$ and $i \in S$, $v_n(x, i)$ is nondecreasing in Q_i, for all i where $Q_i < C_i$.*

Property 22.3.3 *For any finite-horizon problem, consider a policy that follows an up-to z_n policy for total orders, and an up-to e_n policy for expedited orders in period n, $0 \leq n < N - l$. For all n, the cost of this policy is convex in z_n.*

Proof (Sketch): We examine the effect of increasing z_n by a small amount on a particular sample path. Assume we are in period n having order up-to level z_n. If $x_n + Q_n > z_n$ or $x_n + C_n \leq z_n$, a slight increase in z_n has no effect on our sample path, and hence no change in our cost function. Therefore we consider $x_n + Q_n \leq z_n < x_n + C_n$.

Let j be the largest $j < n$ such that $x_j + Q_j < z_j \leq x_j + C_j$ (implying, for all $j < k < n$, the policy on this sample path orders either C_k or Q_k). Let $j = 0$ if no such period exists. Note that j is defined in such a way as to make it the first period after n where a small increase in z_n does not influence w_j (and hence any of the periods afterwards).

It can be shown that z may be by perturbed by an appropriately small amount without changing the general structure of the optimal policy. In other words, there exists an $\epsilon > 0$ such that we can increase z_n to $z_n + \epsilon$ without:

(i) Changing the structure of the order policy for total orders along the sample path. (In periods where the minimum or maximum possible order was placed before the change, this will still be the case. The order in period n is increased by a certain amount, whereas the order in j is decreased by the same amount.)

(ii) Causing the policy to change order modes during the increase in any period. (The increase *may* cause a new mode to be used for the entire increase, if $z_n = \hat{y}_n$ for example.)

Therefore by increasing z_n by any amount $\nu < \epsilon$ we will raise the sample path by this amount in all periods up to period j without causing a switch in order modes during the increase. In period j and those after, no effect is seen due to the change.

By maintaining this condition, and using the linearity of our costs, we can show that the cumulative change in objective function is convex in ν along this particular sample path. Therefore along any particular sample path the objective function is convex in z_n. Taking expectation over all sample paths preserves this convexity. ∎

In a similar manner one can show:

Property 22.3.4 *For any finite-horizon problem, consider a policy that follows an up-to z_n policy for total orders, and an up-to e_n policy for expedited orders in period n, $0 \leq n < N - l$. For all n, the cost of this policy is convex in e_n.*

Property 22.3.5 *For any given vector of order-up-to values z_n, the solution value for this vector is convex in Q_i, for all i.*

22.4 INFINITE HORIZON

We treat the discounted cost case first.

22.4.1 Minimum Discounted Cost Problems

We define $m \stackrel{\text{def}}{=} argmax_{\{i \in S\}}\{C_i - Q_i\}$ (breaking ties arbitrarily), and a new policy, Z, as follows. Under Z, order a total amount u_n given by:

$$
u_n = \begin{cases}
Q_{i_n} & \text{if} & x \geq -Q_m - Q_{i_n} \\
-x & \text{if} & -Q_{i_n} > x \geq -C_{i_n} \\
-Q_m - x & \text{if} & -Q_m - Q_{i_n} > x \geq -Q_m - C_{i_n} \\
C_{i_n} & \text{if} & -C_m - C_{i_n} > x.
\end{cases}
$$

Note that both $-x$ and $(Q_m - x)$ are in $[Q_{i_n}, C_{i_n}]$ when x falls within the prescribed intervals. Therefore this is a valid policy. (It is immaterial whether u_n is ordered via expedited or regular order. Without loss of generality we will assume that all orders are made via expedited mode.)
Let

$$
p_0 = \min_{\{i \in S, 1 \leq j \leq K\}} \{p_{im}^{(j)} | p_{im}^{(j)} > 0\},
$$

where $p_{ik}^{(j)}$ is the j-step transition probability from i to k. As the state is governed by an irreducible finite state Markov chain, p_0, the minimal probability we will reach state m within K steps, is non-zero.

Lemma 22.4.1 *Policy Z is positive recurrent for some state i at the level $x = 0$.*

Proof (Sketch): Denote the inventory position at time zero by x_0 and follow policy Z.

Using Assumption 1 it can be shown that the inventory position under policy Z behaves as a random walk with drift, driving it back to $[-C_m, -Q_m]$ whenever it leaves this interval. This implies that the inventory position will return to this interval infinitely often, and in particular due to the structure of policy Z, it will up-cross $-C_m$ infinitely often.

Every time policy Z up-crosses $-C_m$, post demand, there is at least a $p_0 \Gamma^K$ probability that it will reach state zero. This is because (disregarding the chance it may be possible to order up to zero), once this threshold is crossed, policy Z in effect follows an order up to $-Q_m$ policy. Due to strong (C-Q)-completeness, at each step there is at least an Γ probability of remaining in the interval $[-C_m, -Q_m]$.

Therefore, as all states communicate, with probability no less than $p_0 \Gamma^K$ the process remains in this interval until state m is reached, from which it is possible to order up to zero.

By similar reasoning, the second moment of times before visits to zero, which we will call V_Z is likewise finite.

As the underlying Markov chain is finite, at least one of the $K < \infty$ states must be positive recurrent at zero, with mean period less than or equal to $K \mu_Z < \infty$. ∎

From the positive recurrence of policy Z it follows that:

Lemma 22.4.2 *For the discounted infinite horizon problem $(0 < \beta < 1)$, policy Z has finite cost.*

This implies:

Lemma 22.4.3 $\limsup_{N \to \infty} w^*_{0,i} < \infty \ \forall \ i \in S$.

Which leads to an optimal policy:

Theorem 22.4.1 *Let $0 < \beta \leq 1$. An optimal policy for the infinite horizon discounted problem is an order up-to policy with levels $\{(e_1, z_1), \ldots, (e_K, z_K)\}$ for the expedited and total orders, respectively.*

Proof : The state space chosen is stationary. Based on the non-negativity and convexity of the finite-horizon cost functions there is pointwise convergence:

$$\lim_{N \to \infty} \hat{v}_0(x, i) = \lim_{N \to \infty} v_0(x, i) = v^*(x, i),$$

where values of v^* are in $\mathbf{R}^+ \cup \{\infty\}$. As limits of convex functions, each $v^*(x, i)$ is convex in x for $i \in S$. Additionally, the v^* satisfy:

$$v^*(x, i) = -c^l_i x + \tilde{L}(x, i) + \min_{\{w \in W_i(x)\}} \{c^l_i w + \Lambda(w, i) + \beta E[v^*(w - d_n, j)]\}.$$

If $v^*(x_0, i_0) = \infty$ for some x_0 and i_0, then it can be shown that $v^*(x, i) = \infty$ for all x and i. Therefore any policy is optimal, including the policy in the hypothesis. (Note that this is the case for $\beta = 1$.)

If $v^* < \infty$, then the convexity of $v^*(x, j)$ for all $j \in S$, along with Lemma 22.3.1 implies the convexity of $c^l_i w + \Lambda(w, i) + \beta E[v^*(w - d_n, j)]$ in w, which implies that this function and hence $v^*(x, i)$ has a finite minimum. This minimum occurs at the order up-to level z_i (which has not yet been shown to be finite).

Lemma 22.3.2 provides the existence of the expedited order up-to levels e_i, via the fact that the v^* additionally satisfy:

$$v^*(x, i) = \inf_{(y,w) \in A_i(x)} \{c^0_i(y - x) + c^l_i(w - y) + L^l(y, i) + \beta E[v^*(w - d_n, j)]\}.$$

■

From Lemma 22.4.2, the optimal discounted infinite horizon policy must have finite cost, i.e. $v^*(x, i) < \infty$. This implies:

Lemma 22.4.4 *The up-to levels, z_i, are less than $+\infty$ for all $i \in S$.*

Note that $z_i = -\infty$ is possible. This may at first seem odd, but nevertheless it can be shown that for any fixed $\beta < 1$ and fixed $p_i, 1 \leq i \leq K$, there exists a c_i large enough such that we will never order in period i.

22.4.2 Infinite Horizon Average Cost

Similar reasoning as used in Lemma 22.4.2 can be used to prove:

Lemma 22.4.5 *For the infinite horizon average cost problem, policy Z has finite cost.*

This, along with finite horizon convexity implies:

Lemma 22.4.6 *For the infinite horizon discounted cost and average cost criterion, the class of up-to level policies has finite cost which is convex in the e_i's and z_i's.*

The next lemmata concern the structure of the optimal policy:

Lemma 22.4.7 *For the infinite horizon average cost case, there is an optimal policy which will order the minimum amount allowed, Q_i, whenever the inventory is above some finite level, A_i, for each i.*

Lemma 22.4.8 *If some $p_i > 0$, then there is an optimal policy having $z_j > -\infty, \forall j \in S$.*

Lemma 22.4.8 may fail to be true if $C_j = \infty$ for some state j; in this case it may be optimal to have $z_i = -\infty$ for some i. For example for a periodic Markov chain where $C_j = \infty$, $(c_i^0 - c_j^0)$ and $(c_i^1 - c_j^1)$ are large as compared to the penalty costs in the intermediate states between i and j.

Recall that we defined policy Z to evolve with the goal of reaching state $(0, i)$, for any i. We now define policies Z_x, $x \in \mathbf{R}$ that are constructed by shifting policy Z up by x units (so Z_x evolves with the goal of reaching state (x, i) for any i).

Lemma 22.4.9 *State (x_0, i) is positive recurrent under policy Z_{x_0}, for some i. Furthermore, there exists a policy $Z_{x_0,i,j}$ under which (x_0, i) and (x_0, j) communicate.*

Proof (Sketch): The fact that (x_0, i) is positive recurrent follows from the same reasoning used in Lemma 22.4.1. For the second statement of the lemma, note first that if $i = j$ the result follows from the first statement of the lemma. So assume i is different from j.

Starting from (x_0, i), assume that there is no path under policy Z_{x_0} having positive probability of reaching (x_0, j). As the embedded Markov chain is positive recurrent, i and j communicate; assume there is a path with positive probability from i to j under Z_{x_0}, but that it leads to $(x_0 \pm \delta, j)$, $\delta > 0$. Using this path and Assumption 2, it is possible to modify policy Z_{x_0}, maybe repeatedly, to construct a path such that (x_0, i) and (x_0, j) communicate with strictly positive probability. ∎

The conditions of [13] can be used to verify the form of the optimal policy.

Theorem 22.4.2 *For the model described previously, assuming discrete demand[1], the up-to policy is optimal for average cost criterion.*

Proof (Sketch): The previously defined model comprises a Markov process with denumerable state space. Define a space of policies to be considered: First define $\mathcal{G}$, which consists of policies (possibly randomized) such that for all $x > A_i$, no more than Q_i is ordered. From Lemma 22.4.7, any policy is dominated by a policy in $\mathcal{G}$, and therefore finding an optimal policy is equivalent to finding an optimal policy in $\mathcal{G}$. Expand the space of policies to be considered by defining the countable set $\mathcal{K} \stackrel{\text{def}}{=} \{-\infty, \ldots, A^*\} \times \{1, \ldots K\}$, where $A^* \stackrel{\text{def}}{=} \max_{\{i=1\ldots K\}}\{A_i\}$. Then define the new space of policies $\mathcal{F}$ to be the union of $\mathcal{G}$ and the sets of policies $\{Z_x | x \in \mathcal{K}\}$ and $\{Z_{x_0,i,j} | x_0, i, j \in \mathcal{K}\}$.

We must verify that conditions 1 through 5 of Federgruen, Schweitzer, and Tijms [1983] hold for policies in $\mathcal{F}$ and the set $\mathcal{K}$.

Assumptions 1 and 4 are satisfied for all Markov chains. Assumption 3 is satisfied as the cost functions are bounded from below.

Assumption 2a requires that the expected time and cost from any state (x, i) until the first visit to $\mathcal{K}$ be finite under all policies in $\mathcal{F}$. Policies in $\mathcal{F}$ but not in $\mathcal{G}$, like policy Z, are positive recurrent and thus have finite expected cost. For policies in $\mathcal{G}$, by using the busy periods of appropriately defined $GI/GI/1$ queues, it can be shown that the expected time (and hence the expected cost) until until $w_n \leq A^*$ is finite. Therefore both criteria of 2a are satisfied.

For 2b, we show that all states in $\mathcal{K}$ communicate under some policy in $\mathcal{F}$, using policy Z_x and/or $Z_{x,i,j}$ for appropriately defined x, i, and j.

Having shown that assumptions 1-4 are satisfied, we know that the system of equations has a solution. We now can verify condition 5 in a manner similar to that of 2a, which ensures that the optimal solution is attained at the minimizing point of the right hand side of the optimality equation. ∎

22.5 OPTIMIZATION

As mentioned previously, determining the structure of the optimal policy is of limited use if one is unable to specify the optimal policy parameters. Despite the fact that Theorem 22.4.2 establishes the existence of an optimality equation for the system parameters, the solution of such an equation is non-trivial. Fortunately, through the use of the simulation technique of Infinitesimal Perturbation Analysis, or IPA, these values can be estimated. If it is known that the leadtimes differ by exactly one unit, the order-up-to policy with the IPA derived parameters will be optimal. If this is not the case, such a solution will be optimal within the class of order-up-to policies.

---

[1]The vanishing discount approach in [6] can potentially be used to prove this result for continuous demands.

Prior to its use, the validity of applying IPA to Markovian problems with general dual leadtimes must be established. Note that this also establishes the validity of this technique for the single supplier problem as a special case.

22.5.1 Derivative Recursions

As the basis for IPA we need recursions for both the inventory process and some key derivatives. We first establish the simulation recursions, and then move to the derivative recursions. From Lemma 22.3.1 we know that J_n, I_n, and v_n are convex; therefore they are also continuous. This implies that one-sided derivatives exist at all points. We will use right derivatives. Differentiating the simulation recursions yields the derivative recursions, the validity of which are shown in Theorem 22.5.1.

Let $e_1, e_2, \ldots e_k$ and $z_1, z_2, \ldots z_K$ be the expedited and total base stock levels, respectively. Each period, n say, has the following sequence: The current inventory position X_n is known; expedited order E_{n-l} arrives (if one was placed in period $n-l$); demand d_n occurs; costs C_n are incurred based on the on-hand inventory, H_n; a regular order of $R_{n-l} = U_{n-l} - E_{n-l}$ arrives (if one was placed in period $n-l$); production of E_n expedited, and R_n regular of U_n total units occurs following the base stock policy with minimum order Q_n and maximum order C_n; the next period's type Y_{n+1} is revealed.

We start the system with $H_0 = X_0 = z_1$, assuming we have arbitrarily numbered the periods so that we begin in period type 1. We have the following recursions for inventory position, on-hand inventory, and the amount of each type ordered at time n:

$$
X_{n+1} = \begin{cases}
X_n - d_n + Q_n & \text{if} \quad X_n - d_n + Q_n > z_n \\
X_n - d_n + C_n & \text{if} \quad X_n - d_n + C_n < z_n \\
z_n & \text{otherwise.}
\end{cases}
$$

$$
U_n = \begin{cases}
Q_n & \text{if} \quad X_n - d_n + Q_n > z_n \\
C_n & \text{if} \quad X_n - d_n + C_n < z_n \\
z_n - (X_n - d_n) & \text{otherwise.}
\end{cases}
$$

$$
E_n = \begin{cases}
0 & \text{if} \quad X_n - d_n > e_n \\
U_n & \text{if} \quad X_n - d_n + U_n < e_n \\
e_n - (X_n - d_n) & \text{otherwise.}
\end{cases}
$$

$$
R_n = U_n - E_n.
$$

$$
H_{n+1} = H_n - d_{n+1} + E_{n+1-l} + R_{n-l}.
$$

Differentiating with respect to any z_i^* or e_i^* yields the derivative recursions in the obvious way for X_n', H_n', U_n', E_n' and R_n'. It should be understood that when we differentiate an equation, all derivatives which appear are taken with respect to the same parameter (z_i^* or e_i^* respectively).

With respect to z_i^*:

$$X'_{n+1} = \begin{cases} X'_n & \text{if} \quad X_n - d_n + Q_n > z_n \\ X'_n & \text{if} \quad X_n - d_n + C_n < z_n \\ I\{i^* = Y_n\} & \text{otherwise.} \end{cases}$$

$$U'_n = \begin{cases} 0 & \text{if} \quad X_n - d_n + Q_n > z_n \\ 0 & \text{if} \quad X_n - d_n + C_n < z_n \\ I\{i^* = Y_n\} - X'_n & \text{otherwise.} \end{cases}$$

$$E'_n = \begin{cases} 0 & \text{if} \quad X_n - d_n > e_n \\ U'_n & \text{if} \quad X_n - d_n + R_n < e_n \\ -X'_n & \text{otherwise.} \end{cases}$$

$$R'_n = U'_n - E'_n.$$

$$H'_{n+1} = H'_n + E'_{n+1-l} + R'_{n-l}.$$

With respect to e_i^*:

$$X'_n = 0.$$

$$U'_n = 0.$$

$$E'_n = \begin{cases} 0 & \text{if} \quad X_n - d_n > e_n \\ 0 & \text{if} \quad X_n - d_n + R_n < e_n \\ I\{i^* = Y_n\} & \text{otherwise.} \end{cases}$$

$$R'_n = -E'_n.$$

$$H'_{n+1} = H'_n + E'_{n+1-l} + R'_{n-l}.$$

Note that due to the way the policy is defined, the total order amount U_n is independent of the expedited order up-to level, e_n. This is because U_n constrains E_n by definition.

The cost in period n is

$$C_n = h_n(H_n)^+ + p_n(H_n)^- + c_n^0 E_n + c_n^1 R_n.$$

The average cost over N periods is

$$C^N = \frac{1}{N} \sum_{n=1}^{N} C_n,$$

where the infinite horizon counterpart C^∞ is arrived at by letting N approach infinity in C^N.

Differentiating C_n with respect to some z_i^* (or e_i^*):

$$C_n' = \begin{cases} h_n H_n' + c_n^0 E_n' + c_n^1 R_n' & \text{if } H_n > 0 \\ -p_n H_n' + c_n^0 E_n' + c_n^1 R_n' & \text{if } H_n \le 0. \end{cases}$$

The justification of the derivatives is given by

Theorem 22.5.1 *For the above model, for any up-to policy with vectors* $\{e_1, \ldots, e_K\}$ *and* $\{z_1, \ldots, z_K\}$ *of up-to levels:*

1. **Finite Horizon**

    (a) *The right derivatives* X_n', H_n', U_n', E_n', R_n' *and* C_n' *exist with probability one.*

    (b) $E[X_n'] = E[X_n]'$, $E[H_n'] = E[H_n]'$, $E[U_n'] = E[U_n]'$, $E[E_n'] = E[E_n]'$, $E[R_n'] = E[R_n]'$, $E[C_n'] = E[C_n]'$.

    (c) $\frac{1}{N} \sum_{n=1}^{N} C_n'$ *is a valid estimate for* $E[C^N]'$.

2. **Discounted Infinite Horizon** *Results analogous to the above hold.*

3. **Infinite Horizon Average Cost**

    (a) $(X_n, H_n, U_n, E_n, R_n, i_n)$ *is a Markov chain which converges to a proper steady-state distribution.*

    (b) $(X_n, X_n', H_n, H_n', U_n, U_n', E_n, E_n', R_n, R_n', i_n)$ *converges to a proper steady-state distribution, with* $\{X_n'\}, \{H_n'\} \in \{0, 1\}$ *and* $\{U_n'\}, \{E_n'\},$ $\{R_n'\} \in \{-1, 0, 1\}$.

    (c) $\lim_{N \to \infty} \frac{1}{N} \sum_{n=1}^{N} C_n'$ *is a valid estimate for the derivative of* $\lim_{N \to \infty} \frac{1}{N} \sum_{i=1}^{N} C_n$.

Proof (Sketch):

1. (a) Follows from Lemma 22.3.1, and the fact that demands are continuous, as in [19].

    (b) We can rewrite the recursion for X_n as:

    $$X_{n+1} = X_n - d_n + Q_n + \min\{(z_n - X_n + d_n - Q_n)^+, C_n - Q_n\}.$$

    This makes it clear that X_n is a composition of Lipschitz functions, and therefore is itself Lipschitz. Similar expressions can be written for U_n and E_n, making them Lipschitz as well. Finally, as R_n, H_n, and C_n are combinations of these functions, they too are Lipschitz.

Arguments similar to those in [19] can be used to show that these functions have integrable moduli, and thus Lemma 3.2 of that same paper can be applied to yield the desired result.

(c) Follows from the fact that C_n is Lipschitz, the Dominated Convergence Theorem, and Lemma 3.2 of [19].

2. These results follow directly from the finite horizon results via Lemma 4.3 and Theorem 4.4 of [19].

3. (a) We first show the positive recurrence of the Markov chain $\{X_n, i_n\}$. Note that under all policies in $\mathcal{F}$, the inventory position returns to the interval (z_{min}, z_{max}) infinitely often. Therefore to prove positive recurrence of $\{X_n, i_n\}$ it is sufficient to show that from any state within this interval we return to (z_m, m) with probability one, within a finite amount of time, where $m \stackrel{\text{def}}{=} argmax_{\{i \in S\}}\{C_i - Q_i\}$. This is accomplished by finding paths having probabilities strictly bounded away from zero which map any (x_0, j) such that $x_0 \in (z_{min}, z_{max})$ to (z_m, m).

Once we establish the positive recurrence of $\{X_n, i_n\}$ we have established the existence of a proper steady-state distribution; therefore this process admits coupling. Given any two initial inventory positions and on hand inventory levels, if the inventory positions couple at time T, then the on hand inventories (and in fact the entire process $\{X_n, H_n, U_n, E_n, R_n, i_n\}$) will couple by time $T + l$, establishing the existence of a stationary distribution for this process. Thus, the overall process, and hence C_n, has a proper steady-state distribution.

(b) Follows proof of Theorem 4.7 in [19], in light of (b) and arguments similar to those used earlier in this proof. ∎

22.6 APPLICATIONS

In this section we describe the results of applying IPA to two different international operations problems:

- The derivation of optimal ordering levels for use in the supply chain of a multinational heavy equipment company.

- The solution to various experiments investigating the trade-off between mean income and deviation of cash flow for a small multinational corporation.

22.6.1 The Heavy Equipment Supply Chain

The genesis of this problem is described in brief below; for more information the reader is referred to articles in *The Financial Times* [24] and *Business Week* [39], where the strategic implications are described in greater detail.

A heavy equipment manufacturer has made the decision to introduce a new product line starting in 1999. This equipment consists of two different machines and some seventy work-tools and attachments which are sourced in various locations throughout North America and Europe. These products will be sold by a network of North American dealers in 58 districts. Generally speaking, the issue is to determine how to configure the distribution network so as to maximize expected profits while ensuring a high service level for the initial year, 1999, and five years later, in 2004.

For both years, the firm has demand forecasts for each of the items, as well as customer patience data gathered through surveys of their dealers. The manufacturer also is able to provide cost and logistical data which include shipping times and rates, source costs, source capacities, minimum order sizes, holding cost rates at the different locations, and handling costs.

To facilitate distribution of products to the dealers the company could use up to seven possible transshipment locations, in addition to direct shipment (DS) from sources. These transshipment locations are grouped into two disjoint sets: three *Tool Facilities* (TF's, which had not yet been constructed) and four *Parts Distribution Centers* (PDC's, which were already built and in use). Thus there are four options for the construction of the network:

1. Use of both TF's and PDC's.

2. Use of PDC's only.

3. Use of TF's only.

4. Use of neither TF's or PDC's; allowing only direct shipment (DS) of products.

A crucial element of this model is the fact that since the company is entering a new market, it is very important that they capture a very high percentage of customer demand for their product. The dealer surveys implied that the company's product was highly substitutable with a competitor's, thus it is imperative that dealers have stock available on hand to immediately satisfy demand. It therefore becomes important not simply to find the minimum cost channel for a product, but to provide an additional channel for expedited delivery to the dealer, likely at a higher cost, should their inventory drop below a certain threshold. Further, the manufacturer believes the percentage of customers left unsatisfied in a particular period influences the behavior of customers in the following period, adding a non-stationary (Markovian) element to the problem.

By applying the model presented above, and through the use of IPA, optimal solutions within the class of order-up-to policies were generated for each of the four scenarios. These solutions consisted of order-up-to levels for each of

Table 22.1 Percentage of Optimal Profit Across Different Scenarios.

Year || TF & PDC	PDC Only	TF Only	DS Only
2000 || 100.00	96.76	88.99	77.58
2004 || 100.00	97.98	89.38	81.23

the products, locations, and methods of shipment which ensured that there was a sufficiently high proportion of customers served. Not surprisingly, the optimal profit results are achieved by option 1, using the entire network. The profit results for the work-tools for each of the two study years across the four scenarios are presented in Table 22.1, as a percentage of the optimal. All of the optimal solutions captured virtually 100% of the simulated customer demand. Based on these results, the company decided to implement option 2, use of PDC's and direct shipment only. They felt that the marginal increase in profits did not justify the building and use of the Tool Facilities.

Implementation was set for May 15, 1998.

22.6.2 Mean-Variance Cash Flow Trade-Off

The problem under consideration in this section is that of a small manufacturer supplying a product to a large foreign multinational corporation (MNC), where transactions are undertaken in the manufacturer's domestic currency. The multinational follows a periodic review inventory policy. The state varies as the exchange does, evolving as a finite state Markov chain with a transition matrix taking the form of a random walk, mean reverting, or momentum process. The demand is subject to variation as the state changes, according to a level of zero, partial, or full pass through. The corporation may order at most C units form the supplier in any period, which arrive with zero lead time. In addition, there is a specified non-negative minimum order quantity, Q, which the corporation agrees to order every period. There is only one order type in this model. The exporter has dual objectives, keeping mean cash flow high while reducing its variance through contractually manipulating Q.

The stipulated order band influences the order-up-to levels, z_i, which in turn affect the average cost of the policy. Therefore if the manufacturer wishes to receive a contractual agreement to increase Q, which will increase the MNC's costs, she must drop the unit price. The exact amount of decrease is not immediately apparent though: Once the manufacturer decreases the unit price the customer's order up to levels change and so too does the average cost. Therefore an iterative procedure must be implemented to find the reduction in price which will ensure that a customer following an optimal policy for an increased Q will encounter the same average costs as if Q were zero.

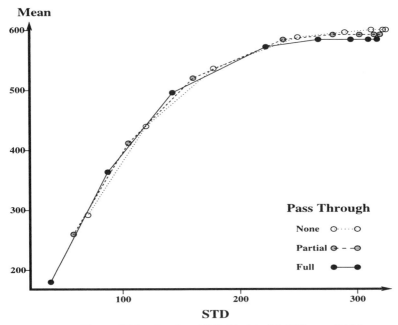

Figure 22.1   Random Walk Model with Different Q Values.

The experiments consider the effect of adopting different values of Q for each of the demand and pass through models, illustrating the effects with mean/standard deviation frontier plots. In addition, the effect of varying the capacity is also investigated.

22.6.2.1   Procedure.   A five state Markov chain was used for the exchange rate level, according to one of the three models (random walk, mean reverting, or momentum); $p = 20$ and $h = 5$ for all periods. The demand was chosen to have a Gamma(2,15) distribution for the base case (so expected base demand was 30). For partial pass through the pass through coefficient, β, was chosen to be 0.4; for full 0.8. In cases where pass through was present, in the middle state there was base demand, and in other states the mean demand increased or decreased by 0.2β per state as the process moved from center. The unit cost vector was $\{16, 18, 20, 22, 24\}$ for the $Q = 0$ case; it changed as solutions were found for higher Q values.

22.6.2.2   Results for Variable Minimum Order Quantities.   Figure 1.1 shows the mean/standard deviation frontier formed by increasing the minimum order quantity, Q, while keeping the customer's total cost constant, for the random walk model. (The plots of the other models show similar general characteristics; significant individual differences are mentioned below.) The value of Q increases as we move *left*; from a value of zero through the values of 5,

10, 15, 20, 25, 27, and 28. This causes the simulated estimates of both the expectation and standard deviation of income per period to decrease.

Looking to Figure 1.1, for zero pass through and $Q = 0$ the mean income is $599.36 with a standard deviation of $324.76, the point in the upper right-hand corner of the graph. As Q changes a concave mean/standard deviation frontier is formed, which indicates that the manufacturer can decrease his risk with virtually no impact on his expected income, for relatively small values of Q. For example, for this case with $Q = 15$, the mean income is reduced by only two dollars, to $597.41, while the deviation decreases by slightly more than ten percent, to $290.48. This holds true for all types of pass through and exchange rate models we considered.

Once Q increases to a level above twenty, the expected income begins to decrease rapidly. Note in particular the final (leftmost) three points on each of the plots, for the Q values of 25, 27 and 28. For every exchange rate model and pass through level, the decrease in the supplier's mean income is greater in this span of three units than in the previous span of 25. (For the momentum model the supplier's mean income is reduced to a level below the bottom of the graph.) This is caused by the proximity of the minimum order quantity and the expected demand, which equals thirty. Having these two parameters so close increases instability in the system, which drives up the customer's inventory/penalty costs. To compensate for this the producer must reduce his unit price considerably, which cuts into the profits.

Turning to individual exchange rate models, our experiments show a marked insensitivity to pass through in the mean-reverting model; virtually identical paths are traced by increasing Q for each of the three pass through scenarios. This is as one would expect, as this model spends the majority of its time in the mean or middle state, where pass through has no effect. This is in contrast to the momentum model, where the proportion of time the rates spend in the extremely high or low state is larger. In this case a definite pattern of domination by the zero pass through case can be seen: Both income and variance tend to be higher for equal values of Q when compared to the other two pass through models. This becomes acute as Q grows large, due to the inherent instability in the momentum model. This same effect is also apparent, to a lesser degree, in the random walk model of Figure 1.1.

It is hypothesized that this increased income for the zero pass through case follows from the insensitivity of end demand to exchange rate, which forces the MNC to maintain high stocks even when the price he must pay the producer is high. If he could expect a decrease in demand due to pass through he could buy less at times when the exchange rate didn't favor him, which would decrease his costs and the producer's income. This is seen as pass through increases. The difference in standard deviation may simply reflect the larger mean income.

Remarks.

(1) Experiments to see how Q affects the MNC's cost variance showed that the change in variance is not monotone: There is an interval where the

variance goes down; the increase in holding cost variance is more than offset by decrease in purchasing cost variance.

(2) To isolate the effects of exchange rate fluctuations on supplier's revenue variance, experiments were also performed for the special case when the end demand has no variance in any period, but depends on the exchange rate. Not surprisingly, the supplier's variance comes down. The benefit of the minimum order quantity at this lower variance remains, but is not as significant as when there is end demand variance.

22.6.2.3 Comparison of Exchange Rate Models. One may also compare the different exchange rate models for the same pass through parameter. In general, these comparisons indicate that less volatile exchange rate models, favor the producer: The mean/variance trade-off curve is higher for the mean reverting model than in the momentum model. The benefit of minimum order quantities, however, appears uniform.

One exception of note is for the zero pass through case, where the simulated income is often greatest for the momentum model. This again is due, it is believed, to the fact that in this case the customer is often forced to order when the unit price has been inflated by exchange rates. This drives his costs, and the producer's profits, up on average.

22.6.2.4 Variable Minimum Order Quantities and Capacities. Figures 1.2 and 1.3 show the effect varying both capacity and the minimum order quantity have on the momentum model with zero and full pass through, respectively. The unit costs were altered to keep the MNC's total average cost constant across all points in each figure. The different lines correspond to different capacities, evaluated for values of Q which increase as before, from right to left. From these graphs it is apparent that while increasing capacity does increase the expected income, it likewise has the effect of increasing the producer's risk.

Looking first to Figure 1.2, where the darker circle in the plot of $C = 40$ represents two data points, all three plots have the same general shape. Furthermore, while increasing the capacity from forty to fifty units can increase the expected income appreciably, from \$534.40 to \$605.62 for $Q = 0$, it likewise increases the standard deviation of income from \$235.95 to \$337.54. So as one would expect, the increase in expected profits does not come without a cost, that of augmented risk. This effect is also present when we increase from fifty to sixty units of capacity; the incremental increase in risk is smaller, but so too are the supplemental profits.

This exposes an often overlooked effect of increasing capacity: By giving the customer a greater choice in the amount ordered, the producer is exposed to greater fluctuation in cash flow, and thus greater risk. It must be remembered that this is in addition to any risk undertaken due to capital expenditure on the producer's part to increase capacity. So in effect, the uncertainty incurred

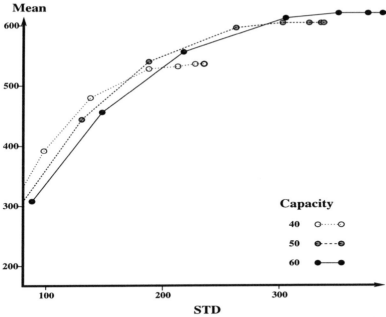

Figure 22.2 Momentum Model, No Pass Through, with Different Q and C Values.

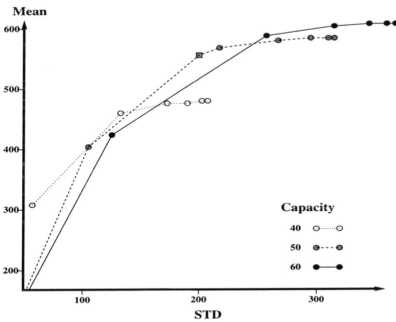

Figure 22.3 Momentum Model, Full Pass Through, with Different Q and C Values.

through the increase of capacity may be twofold; that brought on by any investment necessary to upgrade or expand production, coupled with the increase in flexibility afforded to the customer.

In such a situation it becomes vital for the producer to hedge against such cash flow fluctuations. Contractual minimum order quantities can play an important role in this effort. By simultaneously increasing the minimum order quantity with the capacity, the producer is able to expand production while moderating the cash flow variance. This allows her to provide the MNC with a greater sense of security through a greater ceiling on capacity, while also exacting a concession which will dampen the producer's own risk.

This point is reinforced by the data illustrated in Figure 1.3, where the benefits which can be reaped by securing an agreement for a minimum order quantity are clearly shown. If the producer increases capacity from forty with no minimum order quantity to fifty while also securing an agreement for a minimum order quantity of 21 units (shown as a square on the plot of $C = 50$ in Figure 1.3), she can increase the expected income from \$478.93 to \$556.37 while having the standard deviation decrease from \$207.63 to \$201.10. Thus *both* the expected cash flow and the risk position improve, without passing on any greater expected costs to the customer.

This type of a technique appears to be more of an option for the full pass through model, where there is a greater vertical separation between the three plots than in the zero pass through case. Indeed, the manufacturer could attain the same dual benefit by increasing capacity from fifty to sixty, and the minimum order quantity from zero to fifteen.

To test the theory that the pass through was a significant factor in determining the benefits of this strategy, experiments were performed with the most stable of the models, that of mean reverting exchange rates. Once again, the benefit of increasing the minimum order quantity along with capacity is pronounced for the high pass through case. In this instance the dominating solution appears at $Q = 22$. It seems that the variability of the full pass through model makes the increase in capacity quite valuable to the customer, who in exchange for this increased capacity might be willing to accept a larger increase in unit price. This in turn grants the producer a substantial increase in profits.

22.7 SUMMARY

As the magnitude and complexity of international operations grows, so too does the need for accurate, yet parsimonious models to aid in their analysis. In this chapter we present one such model, which captures the interaction between a company and its overseas suppliers. We establish the optimal policy for this model, and provide a method for calculating the optimal policy parameters.

Work in this field continues apace. For example, Raman, Scheller-Wolf and Tayur [30] use IPA to study the utility of different intermediary schemes, while Rao, Scheller-Wolf and Tayur[31] devise an international supply chain configuration for a heavy equipment manufacturer, slated for implementation in 1998. Finally, de Sousa, Scheller-Wolf and Tayur [11] examine the appropriate use of contingency clauses to minimize risks in overseas bids and contracts, taking into account both currency fluctuations and inflation.

References

[1] P. K. AGGARWAL AND K. MOINZADEH. Order Expedition in Multi-Echelon Production/Distribution Systems, *IIE Transactions*, Vol. 2, pp. 86-96, 1994.

[2] R. ANUPINDI AND Y. BASSOK. Analysis of Supply Contracts with Forecasts and Flexibility, *Working paper, Northwestern University*, (revised) April 1996.

[3] R. ANUPINDI AND R. AKELLA. An Inventory Model with Commitments, *Working paper, Northwestern University*, August 1993 (under revision).

[4] Y. AVIV AND A. FEDERGRUEN. Stochastic Inventory Models with Limited Production Capacities and Varying Parameters, *Probability in The Engineering and Informational Science*, Vol. 11, pp. 107-135, 1997.

[5] E.W. BARANKIN. A Delivery Lag Inventory Model with an Emergency Provision, *Naval Research and Logistics*, Vol. 38, pp. 64-69, 1961.

[6] BEYER, D. AND S. P. SETHI. Average Cost Optimality in Inventory Models with Markovian Demand, forthcoming in *Journal of Optimization Theory and Applications*, Vol. 92, No. 3, 1997.

[7] F. CHEN AND J.S. SONG. Optimal Policies for Multi-Echelon Inventory Problems with Nonstationary Demand, *Working paper, Graduate School of Business, Columbia University*, 1997.

[8] K. H. DANIEL. A Delivery Lag Inventory Model with Emergency Provision, *Multistage Inventory Models and Techniques*, Scarf, Gillford and Shelly (Eds.), Stanford University Press, Stanford, CA, 1962.

[9] S. DASU AND L. LI. Optimal Operating Policies in the presence of Exchange Rate Variability, *Working paper, Anderson Graduate School of Management*, 1993.

[10] F. DE SOUSA. Minimizing Risks in Bids and Contracts Abroad, *Operations Management Seminar, GSIA, Carnegie Mellon University*, 1997.

[11] F. DE SOUSA, A. SCHELLER-WOLF AND S. TAYUR. Minimizing Risks in Bids and Contracts Abroad, *Working Paper*, 1998.

[12] R. DORNBUSCH. Exchange Rates and Prices, *American Economic Review*, Vol. 77, no. 1, 1987.

[13] A. FEDERGRUEN, P. SCHWEITZER AND H. TIJMS. Denumerable Undiscounted Decision Processes with Unbounded Rewards, *Mathematics of Operations Research,* Vol. 8, No. 2, 1983, pp.298-314.

[14] A. FEDERGRUEN AND P. ZIPKIN. An Inventory Model with Limited Production Capacity and Uncertain Demands I: The Average-cost Criterion, *Mathematics of Operations Research,* Vol. 11, No. 2, 1986, pp. 193-207.

[15] A. FEDERGRUEN AND P. ZIPKIN. An Inventory Model with Limited Production Capacity and Uncertain Demands II: The Discounted-cost Criterion, *Mathematics of Operations Research,* Vol. 11, No. 2, 1986, pp. 208-215.

[16] Y. FUKUDA. Optimal Policy for the Inventory Problem with Negotiable Leadtime, *Management Science,* Vol. 12, pp. 690-708, 1964.

[17] S. GAVIRNENI, R. KAPUŚCIŃSKI AND S. TAYUR. Value of Information in Capacitated Supply Chains, *GSIA Working Paper,* October 1996.

[18] P. GLASSERMAN AND S. TAYUR. The Stability of Capacitated, Multi-echelon Production-inventory System under a Base-stock Policy, *Operations Research,* Vol. 42, No.5, 1994, pp 913-925.

[19] P. GLASSERMAN AND S. TAYUR. Sensitivity Analysis for Base-stock levels in Multi-echelon Production-inventory Systems, *Management Science,* Vol. 42, No. 5, 1995, pp. 263-281 .

[20] P. HOOPER AND C. L. MANN. Exchange Rate Pass Through in the 1980's: The Case of US Imports of Manufacturers, *Brookings Papers on Economic Activity,* 1, 1989, pp. 297-329.

[21] A. HUCHZERMEIER AND M. COHEN. Valuing Operational Flexibility Under Exchange Rate Risk, *Operations Research,* Vol. 44, no. 1, 1996, pp. 100-113.

[22] R. KAPUŚCIŃSKI AND S. TAYUR. A Capacitated Production-Inventory Model With Periodic Demand, *GSIA Working Paper,* January 1996; revised December 1996. .

[23] P. KOUVELIS AND V. SINHA. Exchange Rates and the Choice of Production Strategies in Supplying Foreign Markets, *Working paper,* 1996.

[24] P. MARSH. *"Caterpillar Digs Into Compact Market",* The Financial Times, US Edition, February 12, 1998.

[25] K. MOINZADEH AND P. K. AGGARWAL. An Information Based Multiechelon Inventory System with Emergency Orders, *Operations Research,* Vol. 45, No. 5, pp. 694-701, 1997.

[26] K. MOINZADEH AND S. NAHMIAS. A Continuous review Model for an Inventory System with Two Supply Modes, *Management Science,* Vol. 34, No. 6, pp. 761-773, 1988.

[27] K. MOINZADEH AND C. P. SCHMIDT. An (S-1,S) Inventory System with Emergency Orders, *Operations Research,* Vol. 39, pp. 308-321, 1991.

[28] J. A. MUCKSTADT AND L. J. THOMAS. Are Multi-Echelon Inventory Models Worth Implementingin Systems with Low Demand Rate Items?, *Management Science,* Vol. 26, pp. 483-494, 1981.

[29] D.F. PYKE AND M. A. COHEN. Multi-Product Inrtegrated Production Distribution Systems, Working Paper No. 254, The Amos Tuck School of

23 BOTTOM-UP VS. TOP-DOWN APPROACHES TO SUPPLY CHAIN MODELING
Jeremy F. Shapiro

Sloan School of Management
Massachusetts Institute of Technology
30 Wadsworth Street; E53-307
Cambridge, MA 02142

23.1. Introduction

The term **"supply chain management"** crystallizes concepts about **integrated planning** proposed by operations research practitioners, logistics experts, and strategists over the past 40 years (e.g., Hanssman (1959), LaLonde et al (1970), Porter (1985)). Integrated planning refers to **functional coordination** within the firm, between the firm and its suppliers, and between the firm and its customers. It also refers to **inter-temporal coordination** of supply chain decisions as they relate to the firm's operational, tactical and strategic plans.

Practical developments in supply chain management have grown in the 1990's due to advances in information technology (IT). Astonishing gains in PC computing speed, coupled with improvements in communications and the flexibility of data management software, have promoted a range of applications. **Enterprise Resource Planning** (ERP) systems, which have appeared within the past five years, offer the promise of homogeneous, transactional databases that will enhance integration of supply chain activities. In many companies, however, the scope and flexibility of installed ERP systems have been less than desired or expected (Bowersox et al (1998)).

Still, competitive advantage in supply chain management is gained not simply through faster and cheaper communication of data. As many managers have come to realize, ready access to transactional data does not automatically lead to better decision making. A guiding principle is:

*To effectively apply IT in managing its supply chain, a company must distinguish between the form and function of **Transactional IT** and **Analytical IT***

Transactional IT is concerned with acquiring, processing and communicating raw data about the company's past and current supply chain operations, and with the compilation and dissemination of reports summarizing these data. Typical examples are POS recording systems, general ledger systems, quarterly sales reports, and ERP systems.

By contrast, Analytical IT evaluates supply chain decisions based on models constructed from **supply chain decision databases**, which are largely, but not wholly, derived from the company's transactional databases. Analytical IT is comprised of these supply chain decision databases, plus modeling systems and communication networks linking corporate databases to the decision databases. It is concerned with analyzing decisions over short, medium and long term futures. Typical examples of this type of IT are modeling systems for scheduling weekly production, forecasting demand for next month and allocating it to manufacturing facilities, or locating a new distribution center. Differences between Transactional and Analytical IT are discussed further in Shapiro (1998).

As we argue in this paper, **mathematical programming models** are critical elements of Analytical IT. Such models can unravel the complex interactions and ripple effects that make supply chain management difficult and important. They are the only analytical tools capable of fully evaluating large numerical data bases to identify optimal, or demonstrably good, plans. They can also measure tradeoffs among cost or revenues and service, quality and time. Linear and mixed integer programming models are the types most commonly used for supply chain management problems. In practice, mixed integer programming models are best combined with approximation and heuristic methods, especially for scheduling applications.

Managers use the term **optimization** when pursuing IT and business process changes that will improve supply chain management practices. For most managers, however, the term has a vague connotation. They are often unaware of the rigor of mathematical programming methods and their potential to significantly improve the bottom line. Nevertheless, in promoting mathematical programming models, the operations research specialist is better served by referring to them as optimization models. For this reason we will henceforth refer to systems using mathematical programming models as **optimization modeling systems**

Inter-temporal coordination of supply chain decisions has received far less attention than functional coordination. Moreover, current efforts to improve supply chain management using IT and business process re-design have focused on two domains, operations and strategy, with radically different time frames, planning concerns and organizational needs. Companies are seeking to improve **supply chain operations** by acquiring off-the-shelf software systems for manufacturing and distribution planning that are add-ons to ERP. These systems focus largely on scheduling problems in providing myopic evaluations of portions of a company's supply chain. Their analytical capabilities usually do not include mathematical programming models and methods. These systems' capabilities are at best only partially customized to the peculiarities of the company's operational problems.

At the strategic level, an increasing number of companies are employing optimization modeling systems to study **resource acquisition decisions** that will strongly affect their long-term competitiveness. Such decisions include the location and mission of new facilities, perhaps to support new products, bid prices for mergers and acquisitions, and consolidation of existing facilities under conditions of excess capacity. These optimization modeling systems may be developed in-house, acquired from strategy consultants, or from third-party vendors. They are rarely regarded as permanent tools that the company can and should employ on a repetitive basis to review strategic decisions.

Thus far, little effort has been made to link analytic tools and databases at the two extreme levels of planning. The development of optimization modeling systems for **tactical supply chain management**, which would bridge the gap, has been largely ignored. The goal of a company employing such a system would be to adapt organizational processes to exploit global insights provided by it on a routine basis, say once a month. The potential rewards of doing so are enormous. Limited applications have shown that a company can expect to reduce total supply costs by at least 5% through the use of tactical optimization modeling systems.

The purpose of this chapter is to review conceptual and practical developments in the bottom-up and top-down approaches to integrated supply chain management. Particular attention will be given to the present and future roles of optimization modeling systems in achieving such integration. Related developments in IT and business process re-engineering will also be discussed.

23.2. Hierarchy of Supply Chain Optimization Modeling Systems

In the previous section, we emphasized the importance of inter-temporal integration of supply chain activities as well as their functional integration. Integration can be most fully achieved by the application of a suite of optimization modeling systems to the gamut of operational, tactical and strategic planning decision problems faced by the company. These Analytical IT systems are linked to overlapping supply chain decision databases created in large part from data provided by Transactional IT systems.

23.2.1 Components of the Supply Chain System Hierarchy

In Figure 1, we display the Supply Chain System Hierarchy comprised of optimization modeling systems and transactional systems responsible for inter-temporal and functional integration of supply chain activities in a manufacturing and distribution company with multiple plants and distribution centers. As shown in the Figure, the six types of optimization modeling systems are Analytical IT and the four other systems are Transactional IT. Strictly speaking, the Demand Forecasting and Order Management System is a hybrid with analytical capabilities for forecasting demand and transactional capabilities for handling customer orders.

Distinctions among the transactional and scheduling systems displayed in Figure 1 have become blurred. Software companies offering ERP systems have either acquired or entered into alliances with companies offering operational modeling systems. Similarly, some DRP systems include modules for vehicle scheduling and forecasting. For the purposes of discussion, we choose to maintain separation among the form and function of these various component systems.

The transactional and scheduling systems in the System Hierarchy represent the current bottom-up thrust in supply chain management. IT developments are the driving force for innovation, with business process re-design as a natural consequence. The area is red hot with annual sales of software in the hundreds of millions of dollars and growing rapidly (e.g., see Naj(1996)).

Strategic optimization modeling systems in the System Hierarchy reflect the top-down thrust in supply chain management. The driving force is senior management's need for strategic analysis in the face of globalization of the company's markets and supply chains, and competition in cost and service. An increasing number of studies, which employ strategic optimization modeling systems, are being performed by consultants to provide management with insights into the evolution and re-design of their supply chains, and answers to "what if" questions about the long-term future.

Long-term and short-term tactical supply chain planning, and systems in the System Hierarchy to support them, have thus far been mainly ignored. They are the most difficult areas in which to develop better planning methods, based in part on the use of optimization modeling systems, because they require radical business process re-design. We return to a discussion of this point in the next sub-section where we more explicitly address timing issues associated with exercising the systems in the System Hierarchy.

This Supply Chain System Hierarchy is hypothetical. To the best of our knowledge, no company has implemented and integrated all ten types of systems, although many companies have implemented several of them. As IT for supply chain management continues to improve and modeling applications expand, we expect that companies will in the near future implement versions of the entire System Hierarchy.

Starting from the bottom-up, the following are synopses of the capabilities of each system type.

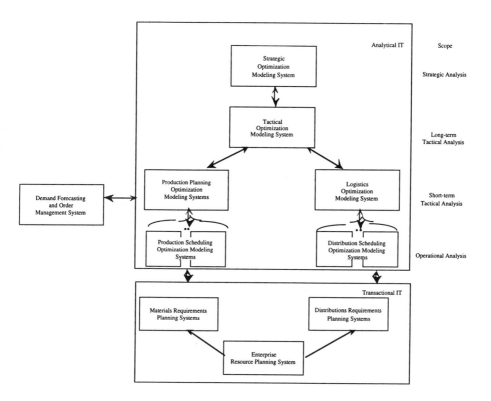

Supply Chain Modeling System Hierarchy
Figure 1

Enterprise Resource Planning (ERP) System: The ERP System manages the company's transactional data on a continuous, real-time basis. This System standardizes the company's data and information systems for order entry, financial accounting, purchasing, and many other functions, across multiple facilities and business units. Despite the claim implied by the term ERP, effective "resource planning" across the "enterprise" can be identified only by optimization models created by the modeling systems using data from the ERP System. See Bowersox et al (1998) for further discussion of ERP systems and their impact on supply chain management. ■

Materials Requirement Planning (MRP) System: Analysis with the MRP System begins with a **master production schedule** of finished products needed to meet demand in each period of a given planning horizon. Using these data, along with a **balance on hand** of inventories of raw materials, work-in-process and finished goods, and a **bill of materials** description of the company's product structures, the MRP System develops **net requirements** of raw materials and intermediate products to be manufactured or ordered from vendors to meet demand for finished products. Products at all stages of manufacturing are analyzed by the MRP System at the SKU level. See Baker (1993) and Sipper and Bulfin (1997) for more discussion about MRP systems and the role of optimization models in determining a master production schedule. ■

Distribution Requirements Planning (DRP) System: Analysis with a DRP System begins with **forecasts** of finished products to be transported, a **balance on hand** of inventories of these products at plants and distribution centers, and **inventory management data** such as safety stock requirements, replenishment quantities, and replenishment times. In conjunction with the Distribution Scheduling Optimization Model Systems, the DRP System then schedules in-bound, inter-facility, and out-bound shipments through the company's logistics network, taking into account a wide range of **transportation factors** such as vehicle loading and routing, consolidations, modal choice, channel selection, and carrier selection. Products throughout the logistics network are analyzed by the DRP System at the SKU level. See Stegner (1994) for more discussion about DRP systems. ■

Demand Forecasting and Order Management System: This System combines data about current orders with historical data to produce requirements for finished products to be met by the operational, tactical and strategic plans. For operational and short-term tactical planning, an important challenge is to manage the transition from forecasts, which have a significant degree of uncertainty, to orders, which have much less uncertainty. Longer-term planning requires linkages to data on industry and economic factors that have a high degree of uncertainty. See

Makridakas and Wheelwright (1989) or Rosenfield (1994) for further discussions of demand forecasting. ■

Production Scheduling Optimization Modeling Systems: These are modeling systems located at each plant in the company's supply chain that address operational decisions such as the sequencing of orders on a machine, the timing of major and minor changeovers, or the management of work-in-process inventories. The models must fit the environment, which may be discrete parts manufacturing, process manufacturing, job-shop scheduling, or some hybrid. A single facility may require different modeling systems at different stages of manufacturing; for example, fine paper production at a mill involves process manufacturing to produce mother rolls of paper followed by job-shop scheduling to produce the final products. See Shapiro (1993a) for a review of relevant mathematical programming models. Naj (1996) discusses business applications for such systems and the capitalization of firms offering them, including i2, Red Pepper, and Manugistics. ■

Distribution Scheduling Optimization Modeling Systems: A manufacturing and distribution company faces a variety of vehicle and other scheduling and operational planning problems. In addition to local delivery of products to customers, some companies must decide on a short-term basis which distribution center should serve each market based on inventory availability. As with production scheduling, distribution scheduling problems and models vary significantly across industries. See Golden and Assad (1988) for a broad treatment of vehicle routing algorithms and applications, and Hall and Partyka (1997) for a survey of off-the-shelf packages for vehicle routing. ■

Production Planning Optimization Modeling Systems: Each plant in the company's supply chain uses its version of this optimization modeling system to determine a master production plan for the next quarter for each stage of manufacturing, along with resource levels and resource allocations for each stage, that minimize avoidable manufacturing costs. As part of the optimization, the model also determines work-in-process inventories and major machine changeovers. The models used by this System will be multi-period as well as multi-stage. Therefore, for reasons of computational necessity, products are aggregated into product families. These aggregations are reversed when the System hands off the master schedule to the plant's Production Scheduling and MRP Systems. Although many papers have appeared in the academic literature discussing production planning models with this broad scope (e.g., see Thomas and McClain (1993)), few modeling systems based on them have yet been implemented. Exceptions are the systems developed at Harris Corporation (Leachman et al (1996)) and at Sadia (Taube-Netto (1996)). ■

Logistics Optimization Modeling System: This System determines a logistics master plan for the entire supply chain that analyzes how demand for all finished products in all markets will be met over the next quarter. Specifically, it focuses on the assignment of markets to distribution centers and other facilities responsible for sourcing them. Its goal is to minimize avoidable transportation, handling, warehousing and inventory costs across the entire logistics network of the company. Again, for reasons of computational necessity, finished products are aggregated into product families and markets are aggregated into market zones. These aggregations are reversed when the System hands off the master schedule to the plant's Distribution Scheduling and DRP Systems. This type of optimization modeling system has also not yet been widely implemented.

Tactical Optimization Modeling System: This System determines an integrated supply/manufacturing/distribution/inventory plan for the company's entire supply chain over the next 12 months. Its goal may to be minimize total supply chain cost of meeting fixed demand, or to maximize net revenues if product mix is allowed to vary. Raw materials, intermediate products and finished products are aggregated into product families. Similarly, markets are aggregated into market zones. This is another type of optimization modeling system that has not yet been widely implemented.

Strategic Optimization Modeling System: This System is used to analyze resource acquisition and other strategic decisions faced by the company such as the construction of a new manufacturing facility, the break-even price for an acquisition, or the design of a supply chain for a new product. Its goal may be to maximize net revenues or return on investment. A number of off-the-shelf packages, with varying degrees of modeling capabilities, are available for this type of application. See the software guide edited by Haverly and Whelan (1996). Examples of strategic studies include Shapiro (1992), Shapiro et al (1993b), Arntzen et al (1998), Barlund et al (1995).

In Figure 1, we have shown data being passed in both directions between each optimization modeling system and those that are immediately above and below it in the hierarchy. Clearly, the decision data bases of these systems overlap. Moreover, data directed downward is disaggregated whereas data directed upward is aggregated. See Shapiro (1998) for more details.

For production planning problems, communication among inter-temporal problems and models has been called hierarchical planning (Bitran and Tirupati (1993)). The approach is valid for more general supply chain problems, and to other areas of business planning. Hierarchical planning includes mathematical programming decomposition methods that formalize communication between levels; Graves (1982) provides a good example of this approach. Although such methods may be beyond the reach of most practical implementations, they provide a very useful way of thinking about supply chain problem linkages.

Finally, the System Hierarchy mechanizes the suggestion by Hammer and Champy (1993; p. 93) that IT advances permit "businesses (to) simultaneously reap the benefits of centralization and decentralization." This is a main theme underlying the widespread interest in integrated supply chain management. But, if it is to become more than a vague and ultimately unachieved goal, companies must actively re-engineer their business processes to promote analysis and communication as depicted in Figure 1.

23.2.2 Frequency of Analysis, Cycle Times and Run Times of Supply Chain Systems

In the previous sub-section, we did not address details about the intended mode for using components of the System Hierarchy. We remedy this deficiency by discussing several timing features of these systems:

- **frequency of analysis** – the number of times each year, quarter or month that managers and planners use the system
- **cycle time** – how long it takes to complete analysis with the system each time it is used
- **run time** – batch time required for each run of the system

These times are displayed in Table 1. They are representative, rather than definitive, and may vary significantly from company to company, although within limits. The frequency of analysis will be much longer than once a week for the Tactical Supply Chain Modeling System, and much shorter than once a quarter for the Production Scheduling Modeling System.

The planning horizons of the systems overlap, which facilitates coordination and communication among them. Note also that all optimization modeling systems are intended to be exercised on a **rolling planning horizon**; that is, the planning horizon of each model is longer than the frequency of analysis. As we descend in Table 1 from strategic to operational systems, the planning horizon becomes shorter while the description of time in the models, as measured by the number of periods, becomes more detailed.

In addition, the objective function shifts from net revenue maximization to avoidable cost minimization as we move from strategic to operational planning. In principle, net revenue maximization should be sought at all planning levels, but the company may have few options to affect revenue at the operational level. One can expect or hope that, in the coming years, net revenue maximization will work its way down the

	Planning Horizon	Model Structure	Objective Function	Frequency of Analysis	Cycle Time	Run Time
Strategic Optimization Modeling System	1 - 5 years	Yearly snapshots	maximize net revenues or returns on assets	Once a year	1 - 2 months	1 0 - 60 min
Tactical Optimization Modeling System	12 months	3 months, 3 quarters	minimize total cost of meeting forcasted demand or maximize net revenue by varying product mix	Once a month	1 week	60 - 120 min
Production Planning Optimization Modeling System	13 weeks	4 weeks, 2 months	minimize avoidable production and inventory costs	Once a week	1 day	10 - 30 min
Logistics Optimization Modeling System	13 weeks	4 weeks, 2 months	minimize avoidable logistics costs	Once a week	1 day	10 - 30 min
Production Scheduling Optimization Modeling Systems	7 days to 28 days	7 days to 28 days	minimize myopic production costs	Once a day	30 min	10 min
Distribution Scheduling Optimization Modeling Systems	7 days to 28 days	7 days to 28 days	minimize myopic distribution costs	Once a day	30 min	10 min
MRP Systems	7 days to 28 days	7 days to 28 days	not applicable	Once a week	1 - 3 hours	60 min
DRP System	7 days to 28 days	7 days to 28 days	not applicable	Once a week	1 - 3 hours	60 min
Forecasting and Order Management System	1 week - 5 years	Varied	not applicable	Varied	Varied	10 sec - 10 min
Enterprise Resource Plainng System	not applicable			Real-time or Continuous		not applicable

Features of Analytical and Transactional Systems

Table 1

hierarchy as companies improve their capabilities to exploit modeling systems. For example, the Production Planning Optimization Modeling System could be employed to maximize short-term net revenues by identifying which customized orders to accept or reject, or to price such orders, so as to guarantee healthy margins. Such a change in using this System would require changes in business processes to support both the requisite analysis and negotiations with customers.

After improvements in a company's supply chain have been realized by implementing operational systems at the bottom of the Hierarchy, the three types of tactical planning systems in the middle of the Hierarchy offer the greatest opportunities for true integrated supply chain management. But, as we mentioned above, their implementation requires radical business process re-design. In particular, the organizational issues to be resolved when using a tactical modeling system on a repetitive basis include:

• What are effective IT and business processes for acquiring the necessary data from across the supply chain on a repetitive basis in a timely and accurate manner?

• Who uses these data to perform tactical analyses?

• How can we successfully incorporate managerial judgment into plans suggested by the models?

• Once a tactical plan has been determined, how is it disseminated and implemented?

• How can managerial incentives be changed to reflect global efficiencies?

Space limitations prevent further discussion of these issues (see Shapiro (1998)).

23.3. Supply Chain Decision Databases

As we stated in the introduction, the application of any optimization modeling system in the System Hierarchy requires inputs from a supply chain decision database that is created by transforming transactional data found in the ERP, MRP and DRP systems. The decision database also contains data derived from the Forecasting and Order Management System. Furthermore, supply chain decision databases for the various optimization modeling systems overlap and should be created and managed in a holistic manner.

The importance of decision databases, their differences from transactional databases, and the optimization models they generate have been ignored by vendors of ERP systems. Either by omission or commission, they wrongly suggest that integrated supply chain management in the firm can be readily achieved once an ERP system has been successfully installed there. In the paragraphs that follow, we outline key principles for creating and exploiting decision databases. Space does not permit us to delve deeply into their form and content; see Shapiro (1993c, 1997, 1998) for more details.

23.3.1 Adapt Managerial Accounting Principles in Computing Costs

The information revolution has facilitated advances in **management accounting**, which is defined by Atkinson et al (1997; p. 3), as "the process of identifying, measuring, reporting, and analyzing information about the economic events of the (firm)" for the purposes of internal planning and control. Its focus is different than that of **financial accounting**, which is concerned with reporting historical results to external audiences. Financial accounting data is often inaccurate for the purposes of supply chain decision making because it employs allocations of indirect and fixed costs based on historical volumes that will change in the future.

The management accounting community has developed a new method called **activity-based costing**, which seeks to allocate indirect costs, otherwise called **activities,** to cost objects, such as product and customer costs. This allocation is based on **cost drivers** that determine how the activities contribute to the total cost of the cost objects; see Atkinson et al (1997) for more details and Elram et al (1994) for applications in logistics. Indirect cost drivers may be volumetric, such as a cost driver for the receiving department that equals the number of parts handled by the department during the planning period. They may also be non-volumetric, such as a cost driver for machine set-up costs that equals the number of times a machine is set-up during the planning period. The allocation of direct costs involves natural and obvious drivers, such as machine hours for machine cost, and is much more straightforward.

For the purposes of decision making, the managerial accounting or modeling practitioner must develop **cost relationships** of direct and indirect costs, rather than mere point estimates of them. These are functions that describe how costs will vary in the future as a function of the values of their cost drivers. Nonlinear and discontinuous functions of a single cost driver may be approximated by linear and mixed integer programming constructions. If a cost relationship involves multiple cost drivers with cross-product terms, nonlinear programming modeling techniques may be required.

The connection between activity-based costing and supply chain optimization models is important, but complicated. For our purposes here, the key construction needed to create costs for the supply chain decision database is a mapping of indirect costs in the company's general ledger into natural categories of supply chain costs with associated cost relationships, drivers and resources. A cost driver corresponds to a resource if its availability is limited. Otherwise, the cost driver is merely an accounting device for tracking costs.

A serious difficulty too often encountered by modeling practitioners when they wish to create an optimization model of a multiple facility supply chain is that, for historical reasons, the firm uses different accounting systems at each facility. The practitioner is then faced with the complex task of homogenizing and calibrating costs across facilities so that the optimization model will make even-handed comparisons when it computes an optimal plan. Fortunately, implementation of an ERP system in such a firm will eliminate these planning and modeling difficulties, along with many others.

Activity-based costing and optimization modeling play complementary roles in identifying costs and cost relationships for supply chain decision making. A supply chain optimization model provides a template for costs and cost relationships in terms of generic planning elements such as processes, resources, facility costs, and transformation recipes. Activity-based costing analysis of general ledger and other raw cost data determines the specific nature of these generic elements, and parameters defining their cost relationships.

23.3.2 Aggregate Products, Customer, Suppliers

For the purposes of strategic and tactical planning, the modeling of supply chain operations should incorporate suitable **aggregations** of products, customers, and sometimes suppliers. It is not necessary or desirable to describe operations at the individual SKU level for the purposes of strategic or tactical supply chain analysis. Thus, a supply chain model for a retailing company may be based on 100 product families that are derived by aggregating the 50,000 SKU's sold by the company. The importance of product aggregation to effective modeling was recognized by researchers long before supply chain management became important; for example, see Zipkin (1982) and Axsater and Jonsson (1984).

Aggregation of products into product families is the most difficult type of aggregation. It requires wide and deep knowledge of the company's product lines. Products selected for a single family should have similar characteristics with respect to supply chain costs, resource utilization, transformations, and so on. Fortunately, the practitioner need not force unnatural aggregations to ensure a manageable number of product families because the process is self correcting according to the 80/20 rule. Thus, a natural aggregation resulting in 250 product families may easily be reduced to 60 families by aggregating the 200 smallest families, which may constitute only 20% of total volume, into 10 miscellaneous families.

Customer aggregation may be done in a hybrid manner, depending on the company, the industry, and the scope of the modeling analysis.

Large customers should be retained as separate entities, whereas several small customers in a confined geographical area can be aggregated into a market zone. Customer aggregation can be reversed by a post-optimality analysis that creates detailed management reports of customer sourcing plans. Supplier aggregations may be made in a similar manner.

To accurately capture planning details, aggregations must become less crude as we descend in the System Hierarchy. Still, the numbers of families of products, customers and suppliers should remain manageable due to the decreasing scope of the models. Even for scheduling models, product aggregation may be necessary and desirable. For example, when scheduling production on a paper machine, the difficult decisions are when to make major changeovers from one type of paper to another and how much total volume of the new type to produce. By contrast, optimizing the sequence of grades to be produced within a given type is straightforward and does not require a model. Thus, in this case, the scheduling model can ignore details about product grades. Once an optimal plan has been computed, a simple post-optimality routine can be applied to determine the timing and size of production for the various grades.

23.3.3 Incorporate External Data About Suppliers, Markets, the Industry and Domestic and Foreign Economies

This requirement of the supply chain decision database reminds us of the obvious fact that transactional data about the company's operations is not sufficient in scope for supply chain analyses of a strategic and tactical nature. Depending on the analysis, an optimization model will require data about supplier costs and capacities, and market conditions for the company's products. For strategic planning purposes, economic data about long-term prospects for the company's industry and national economies in which the company operates its supply chain may also be required.

23.3.4 Develop Data Describing the Future

As we discussed in the introduction, transactional data in ERP, MRP and DRP systems portray the firm's past and present, whereas analytical data in the supply chain decision databases address the firm's future. These data describing future operations of the company must be extrapolated from historical, transactional data. The extent of this extrapolation is greatest for strategic planning, but some extrapolation may be needed even for scheduling purposes; for example, short-term forecasting of demand from large customers who have not yet placed firm orders.

Uncertainty about the future, especially as it relates to strategic planning, can be analyzed by optimizing the models under different scenarios of data. Scenario construction must be based on managerial judgment about the firm's major areas of strategic initiative. The emerging discipline of scenario planning proposes guidelines for these constructions; for example, see Georgantzas and Acar, 1995.

At virtually all levels of planning, the supply chain decision databases will contain data describing options not currently included in the company's operations. Strategic options might include potential acquisitions or mergers, the construction of new facilities, the development of new technologies, or supply chain planning for a new product. Tactical options might include supply contracts with new vendors, or manufacturing exchange agreements with other firms.

Preparing data to evaluate such options is not simply a matter of forecasting or extrapolation. For example, evaluation of a potential acquisition requires integrating its supply chain decision database with the decision database of the company to create combined inputs for an optimization model of the merged supply chains. Similarly, a model to evaluate the re-design of a distribution network needs data describing the locations of potential distribution centers, along with costs for constructing or renting facilities in these locations, and transportation rates from these locations to the company's markets. In summary, the decision database often requires data about supply chain options for which the firm has no historical data.

23.3.5 Incorporate Parameters Reflecting Management Policies

Supply chain analysis at all levels of planning requires data and structural inputs reflecting company policies and managerial judgments about risk. The decision database must be extended to include these data, and optimization models require decision variables and constraints that mechanize them. For a global corporation planning its strategy for next year, the CEO may wish to impose a constraint limiting the manufacture of any product family at any single facility to no more than 75% of total forecasted volume for the year.

For a distribution company developing its tactical plans for the coming quarter, the VP for Marketing and Sales may wish to limit the maximal distance between any distribution center and any market it serves to 300 miles, which roughly equals the limit of one day deliveries. For a manufacturing firm, the Manufacturing VP may wish to limit the dollar volume of out-sourced production for next year in a particular department to no more than 25% of the department's annual budget. For a company making local deliveries of products, the general manager may wish to limit the number of part-time drivers scheduled each week.

In performing strategic and tactical analyses, model structures reflecting management policy such as these may be soft implying the need to make several model runs to measure the tradeoffs of cost or revenue against the other criteria. Such multi-objective analysis requires the managers to specify a range of permissible values for the non-monetary criteria. On the other hand, for scheduling applications, the tradeoffs among criteria must be hard-wired to ensure rapid computation of a final plan for execution.

23.3.6 Integrate Model Outputs with Model Inputs

A supply chain decision database will include output from an optimization model as well as the inputs used in generating it. This data integration has a few subtleties although it is mainly straightforward. Only non-zero output data need be stored, such as the positive flows of products from the distribution centers to the markets. Moreover, technical output, such as values of zero-one variables, can be suppressed. The values of shadow prices on resources and reduced costs on activities are useful data. They need to be combined with the appropriate input data on resources and activities.

23.3.7 Provide Graphical Displays and On-line Analytical Processing of Model Inputs and Outputs

For strategic and tactical supply chain planning, it is important that an optimization model system provide graphical displays of data, including graphical mapping of inputs and outputs using a geographical information system (GIS). These displays do not add to the system's inherent analytical capabilities. Still, they are very useful for communicating data, problems and solutions, especially to managers who may be too busy to study detailed, tabular data. Camm et al (1997) report on a successful modeling study of Procter and Gamble's supply chain that relied heavily on graphical mapping displays to extract judgments from product strategy teams.

For a company doing business in the continental U.S., a GIS would be used to display the results of a logistics network design analysis. For example, the GIS could produce a map of the U.S. for each product showing

- the optimal location of distribution centers supplying the product
- color-coded data indicating which markets are served by each distribution center
- links of variable thickness indicating the relative flows of product between distribution centers and markets

For scheduling applications, graphical displays allow human verification and manipulation of plans computed by an optimization model. An open issue is the appropriate level of human interaction with the scheduling system. The present style of many systems relies too heavily on such interaction. As a result, scheduling plans are inferior to those that could be found by rigorous optimization models and solution methods. Moreover, the human scheduler may spend excessive time manipulating trial solutions to find one that is acceptable. A related difficulty is that some off-the-shelf scheduling systems rely too heavily on the human scheduler to customize solutions to the company's scheduling environment. In many cases, the company would benefit significantly by investing in a more expensive scheduling system with customized analytic methods.

On-line analytical processing (OLAP) refers to the application of data management and data mining tools to organize, analyze and display complex input and output databases. A key feature is Multidimensional Data Viewing (MDV), which allows an analyst to develop graphical displays of complex data in a relational database. This capability is also valuable for comparing and contrasting plans for multiple scenarios.

23.4. Conclusions

Applications of the optimization modeling systems for integrated supply chain management discussed in this paper are work-in-process. In many companies, ambitious tasks to implement ERP systems, and other transactional and analytical systems at the bottom of the System Hierarchy of Figure 1, have yet to be successfully completed. It is not surprising, therefore, that most companies have not seriously begun initiatives for moving up the System Hierarchy to develop and use optimization modeling systems for tactical planning.

Similarly, the number and scope of applications of systems at the top of the System Hierarchy continues to grow, but we have seen few initiatives to move down the Hierarchy to develop and use optimization modeling systems for the same or related tactical planning problems. For companies that have successfully used strategic optimization modeling systems to perform studies, the lack of interest in extending them to tactical applications is particularly frustrating for modeling practitioners. From a technical perspective, such extensions are easy to accomplish once the studies have been performed because they achieved validation of the model and creation of the supply chain decision database.

This reluctance is not surprising since repetitive use of a tactical optimization modeling system requires considerable business process re-design. Moreover, tactical applications require the development of permanent supply chain decision databases, which, as we already observed, are not yet well understood. Despite the difficulties, we can be optimistic about the ultimate breakthrough of tactical modeling applications because the potential rewards are so great. Such applications in a given company may stem either from extensions of bottom-up or top-down system evolutions.

An important issue that merits further investigation, which is closely related to those discussed above, is the extent and sustainability of competitive advantage accruing to a firm as the result of transactional and analytical IT used to integrate its supply chain activities. The academic community has suggested that IT innovations are necessary, but not sufficient, for achieving competitive advantage (Mata et al (1995), Powell and Dent-Metcallef (1997)). They must be combined with complementary organizational and business initiatives if competitive advantage is to be achieved. This observation is consistent with the approach taken by most software firms offering systems in the System Hierarchy. They seek to sell their systems as off-the-shelf commodities, thereby providing all companies with the same capabilities and little competitive advantage. They are reluctant to customize their systems to the peculiarities of supply chain management in a client company because multiple, customized systems are difficult to support. Some companies have come to realize that it pays to invest in customization of selected components of the System Hierarchy, as well as in training and business process re-design that allows IT to be more fully exploited.

References

Arntzen, B. C., D. W. Mulgrew and G. L. Sjolander, " Redesigning 3M's Worldwide Product Supply Chains," *Supply Chain Management Review*, 1 (1998), 4,16-27.

Atkinson, A. A., R. D. Banker, R. S. Kaplan, and S. M. Young, *Management Accounting*, Second Edition, Prentice-Hall, 1997.

Axsater, S. and H. Jonsson, "Aggregation and Disaggregation in Hierarchical Production," *European Journal of Operational Research*, 17 (1984), 338-350.

Baker, K. R., "Requirements Planning," Chapter 11 in *Handbooks in Operations Research and Management Science: Logistics of Production and Inventory*, edited by S. C. Graves, A. H. G. Rinnoy Kan, P. H. Zipkin, North-Holland, 1993.

Bitran, G. R. and D. Tirupati, "Hierarchical Production Planning," Chapter 10 in *Handbooks in Operations Research and Management Science: Logistics of Production and Inventory*, edited by S. C. Graves, A. H. G. Rinnoy Kan, P. H. Zipkin, North-Holland, 1993.

Barlund, R. G., J. Kerkhoff, R. T. Schmidt, and J. F. Shapiro, "An Optimization Modeling System for Integrated Sourcing, Manufacturing and Distribution Planning," presentation at 1995 Annual Conference of the Council of Logistics Management, San Diego.

Bowersox, D. J., D. J. Closs, and C. T. Hall, "Beyond ERP – The Storm Before the Calm," *Supply Chain Management Review*, 1 (1998), 4, 28-37.

Camm, J. D., T. E. Chorman, F. A. Dill, J. R. Evans, D. J. Sweeney, and G. W. Wegryn, "Blending OR/MS, Judgment, and GIS: Restructuring P&G's Supply Chain," *Interfaces*, 27 (1997), 128-142.

Ellram, L., M. J. Kwolek, B. J. La Londe, S. P. Siferd, T. L. Pohlen, D. G. Waller, and W. R. Wood, "Understanding the Implications of Activity-based Costing for Logistics Management," *Journal of Business Logistics*, (1994), 11-25.

Georgantzas, N. C. and W. Acar, *Scenario-driven Planning: Learning to Manage Strategic Uncertainty*, Quorum Books, 1995.

Golden, B. L. and A. A. Assad, editors, *Vehicle Routing: Methods and Studies*, North-Holland, 1988.

Graves, S. C., 'Using Lagrangean Techniques to Solve Hierarchical Production Planning Problems," *Management Science*, 28(1982), 260-275.

Hall, R. W. and J. G. Partyka, "On the Road to Efficiency," *OR/MS Today*, 24 (1997), 3, 38-47.

Haverly, R. C. and J. F. Whelan, editors, *Logistics Software, 1996 Edition*, Andersen Consulting.

Hammer, M. and J. Champy, *Reengineering the Corporation*, HarperBusiness, 1993.

Hanssman, F. "Optimal Inventory Location and Control in Production and Distribution Networks," *Operations Research*, 7 (1959), 483-498.

758

LaLonde, B. J., J. R. Grabner and J. F. Robeson, "Integrated Distribution Systems: A Management Perspective," *International Journal of Physical Distribution Management*, (1970), 40.

Leachman, R. C., R. F. Benson, C. Liu and D. J. Raar, " IMPReSS: An Automated Production-Planning and Delivery-Quotation System at Harris Corporation – Semiconductor Sector," *Interfaces*, 26 (1996), 1, 6-37.

Makridakis, S. and S. Wheelwright, *Forecasting Methods for Management, 5th Edition*, Wiley, 1989.

Mata, F. J., W. L. Fuerst and J. B. Barney "Information Technology and Sustained Competitive Advantage: A Resource-based Analysis," MIS Quarterly, 19 (1995), 487-505.

Naj, A. K., "Manufacturing Gets a New Craze from Software: Speed," *Wall Street Journal*, August 13, 1996.

Porter, M., *Competitive Advantage: Creating and Sustaining Superior Performance*, The Free Press, MacMillan, 1985.

Powell, T. C., and A. Dent-Metcallef., "Information Technology as Competitive Advantage: The Role of Human, Business and Technology Resources," *Strategic Management Journal*, 18 (1997), 375-405.

Rosenfield, D. B., "Demand Forecasting," Chapter 14 in *The Logistics Handbook*, edited by J. F. Robeson, W. C. Copacino, and R. E. Howe, The Free Press, 1994.

Shapiro, J. F., "Integrated Logistics Management, Total Cost Analysis, and Optimization Models," *Asia Pacific International Journal of Business Logistics*, (1992).

Shapiro, J. F., "Mathematical Programming Model and Methods for Production Planning and Scheduling," Chapter 8 in *Handbooks in Operations Research and Management Science: Logistics of Production and Inventory*, edited by S. C. Graves, A. H. G. Rinnoy Kan, P. H. Zipkin, North-Holland, 1993a.

Shapiro, J.F., V. M. Singhal, and S. N. Wagner, "Optimizing the Value Chain," *Interfaces*, 23 (1993b), 2, 102-117.

Shapiro, J. F., "The Decision Data Base," WP# 3570-93-MSA, Sloan School of Management, MIT, June, 1993c.

Shapiro, J. F., "On the Connections Among Activity-based Costing, Mathematical Programming Models for Analyzing Strategic Decisions, and the Resource-based View of the Firm," December, 1997, to appear in the *European Journal of Operational Research*.

Shapiro, J. F., *Optimization Modeling for Integrative Supply Chain Management*, manuscript in progress to be published by Duxbury Press, 1998.

Sipper, D. and R. L. Bulfin, Jr., *Production: Planning, Control and Integration*, Mc-Graw Hill, 1997.

Stenger, A. J., "Distribution Resource Planning," Chapter 17 in *The Logistics Handbook*, edited by J. F. Robeson, W. C. Copacino, and R. E. Howe, The Free Press, 1994.

Taube-Netto M., "Integrated Planning for Poultry Production at Sadia," *Interfaces*, 26 (1996), 38-53.

Thomas, L. J. and J. O. McClain, "An Overview of Production Planning," Chapter 7 in *Handbooks in Operations Research and Management Science: Logistics of Production and Inventory*, edited by S. C. Graves, A. H. G. Rinnoy Kan, P. H. Zipkin, North-Holland, 1993.

P. Zipkin, "Exact and Approximate Cost Functions for Product Aggregates," *Management Science*, 28 (1982), 1002-1012.

24 INVENTORY PLANNING IN LARGE ASSEMBLY SUPPLY CHAINS

Gerald E. Feigin

IBM T.J. Watson Research Center
Yorktown Heights, NY

24.1 INTRODUCTION

Large assembly supply chains, such as those found in the computer, consumer electronics, and automobile industries, usually support the production of multiple end-products where each end-product has a complex multi-level bill of material (BOM) consisting of hundreds, if not thousands, of components and subassemblies with widely varying lead times and costs. The end-products typically have many of these components and subassemblies in common. The supply chains are subject to demands for end-products which are highly volatile and notoriously difficult to forecast, yield and other quality problems, periodic engineering changes, frequent new product introductions, rapid obsolescence of end-products and components, and geographically dispersed production and vendor locations.

Because of the large size of such supply chains, firms typically have many millions of dollars of capital tied up in inventory throughout their networks. This inventory takes numerous forms and exists for a variety of reasons. Even the most efficient supply chain contains a large quantity of inventory, if only in the form of work in process (WIP). Beyond this minimal level which is required because of the time needed to produce and deliver products, the justification for holding inventory anywhere in a supply chain can only be based on two broad considerations: to enable economies of production and procurement and to hedge against various forms of risk including fluctuations in lead times, uncertain customer demand, variable product quality, and changes in raw material prices. Forward buying opportunities, volume discounts, transportation economies of scale (e.g., full truck load (FTL) unit rates vs. less than full truck load (LTL) unit rates), manufacturing unit cost reductions through batching and set-up reductions are examples of economies of production and procurement. Also included under the category of production economies are savings realized through resource contraint considerations. For example, consider the decision to keep a manufacturing line running even if supply temporarily outstrips demand because the alternative of keeping machines idle and perhaps laying workers off is uneconomical. As another example, consider the decision of a plant manager to build finished goods inventory of a product in advance of a high volume selling season because the alternative of temporarily expanding capacity during the high season is too costly. Uncertainties that plague supply chains which can be hedged against by holding inventory include unforeseen late shipments, manufacturing delays due to machine breakdowns, higher than forecasted demand for products, and higher than expected defect rates.

Holding inventory as a result of exploiting production and procurement economies or protecting against variability can only be justified if the resultant inventory holding costs are less than savings garnered from the instigating action. In the case of production and procurement economies, providing such a justification is usually quite straightforward, at least in principle. [1] However, in the case of variability, the justification may not be quite so obvious, particularly because the protection offered by safety stock may not lead to any directly measurable cost savings or increases in revenue. On the other hand, the introduction of safety stock can lead to a quantifiable improvement in customer service, measured in any number of ways, and frequently, customer service, though not reported on any financial statement, is of critical importance to the management of a corporation – not because of any altruistic urge on the part of corporate managers but because customer service acts as a surrogate measure for maintaining and increasing sales. Put another way, poor service levels in the form of late shipments, shipments of defective parts, or any other error resulting in a customer's dissatisfaction, may place a firm's revenue in jeopardy. Holding safety stock to protect against these unforeseen but inevitable errors may well be in the firm's best interest.

In the context of a complex supply chain, a perennial management challenge is knowing where and in what quantities to hold safety stock in a network to protect against variability and to ensure that target customer service levels are met. The difficulty of this problem arises from the fact that the quantity of inventory held at one stocking location in a supply chain and the policy governing replenishment of inventory at that location will have an impact on all supply chain participants directly or indirectly interacting with that stocking location, whether customers who obtain goods from that location or suppliers who produce parts that ultimately end up as products, parts, or sub-assemblies stored at that location. In their quest to improve the efficiencies of their supply chains, companies must come to appreciate the interdependency of all links in their supply chain and must take as broad a view as is reasonably possible in understanding how decisions on inventory management at one location in a supply chain have an impact on other participants in the supply chain.

In this chapter, we describe a model of supply chains designed to help quantify the tradeoff between service levels provided to customers and the investment in safety stock inventory required to support these service levels. Because our intent is to assess the quantity of inventory needed to hedge against uncertainy in supply chains, we deliberately exclude from our model the other major consideration that leads to holding inventory, namely economies of production and procurement. As a result, our model is based on an extremely simple but effective inventory control mechanism: the base-stock policy. The base-stock policy is essentially a continuous review, one-for-one replenishment mechanism. As such, it represents an ideal manufacturing and distribution environment where information flows instantaneously and lot sizes are exactly the size of customer orders.

---

[1] In practice, complications arise in quantifying the tradeoff between production economies and inventory holding costs. For example, it may not be a trivial matter to separate fixed costs from variable costs in this analysis.

Because of this ideal inventory control mechanism, a typical objection to our model might be: "Our supply chain does not operate in the way your model assumes. For example, our material managers don't follow a continuous review base-stock policy for placing replenishment orders for parts. They place orders on a weekly basis and they typically purchase larger quantities than what might strictly be required because of economies of scale they realize in the form of volume purchase discounts or reduced unit transportation costs. So how can we reasonably expect that our inventory levels match those that your model recommends?" The answer to this objection is that you cannot expect that your inventory levels will match those that the model recommends. However, the model still provides useful information because it tells the material managers exactly how much their volume purchase discounts are costing them in increased inventory investment. For example, suppose the model indicated that $10 million was the appropriate quantity of inventory to have on-hand for a particular part number at a particular location and the end of month inventory for that part number at that location was actually $20 million. If the material manager justified the additional $10 million on the basis of savings due to volume discounts, he had better be prepared to claim that the volume discount more than compensated for the holding costs of the additional $10 million in inventory. Broadly speaking, the results of the model are intended to show what level of inventory you ideally could have in your supply chain. If your actual inventory levels deviate significantly, then you should either take some corrective action or be prepared to justify these discrepancies in a convincing way.

Another way in which the results of our model can be of value, even if the modeling assumptions do not match the operational practices of a firm, is to make use of the derived service levels from the model as benchmarks for operational performance. For example, if the supply chain model indicates that a fill rate of 93% is the desired target for a particular inventory location in the supply chain, this target can be used to assess whether actual performance at that store should be modified.

Literature Review

The literature relating to the type of model we describe is extensive and we make no attempt here to provide a comprehensive review of related research. For excellent overviews of models and algorithms related to inventory management in multi-echelon supply chains, see Axsater, 1993, Muckstadt and Roundy, 1993, and Federgruen, 1993.

Much of the research on models for inventory management in supply chains has focused on distribution networks. Extending these models to treat assembly networks is difficult. Successful research efforts in this direction include Ernst and Pyke, 1992, Cohen and Lee, 1988, Glasserman and Wang, 1997, Lee and Billington, 1986, Ettl et al., 1997, Graves and Willems, 1998. The work in Ettl et al., 1997 describes in detail the model we present in this chapter. The models presented in Lee and Billington, 1986, and Graves and Willems, 1998 are similar in spirit to the model described here.

24.2 THE SUPPLY CHAIN MODEL

Overview

Stores. Our supply chain model consists of a collection S of stocking locations, called *stores*, each of which stocks a single stock keeping unit (sku). Each site in the supply chain in general has two types of stores: *input* stores and *output* stores. The input stores model the stocking of different types of components received from upstream suppliers, and the output stores model the stocking of finished-product inventory at the site. Every site consists of at least one output store and zero or more input stores.

Arcs. Stores in our model are connected by directed arcs which indicate the flow of goods through the supply chain. Arcs connect input stores to output stores and output stores to input stores. The arcs connecting input stores to output stores within a site indicate what input store material is used to produce the output store finished products. Associated with an arc connecting an input store i to an output store j is a *usage count*, u_{ij}, which indicates the number of skus in the input store required to produce a sku in the output store. Thus, the arcs connecting input stores to an output store together with their usage counts constitute the single-level BOM for that output store. The arc connecting an output store of one site to an input store of another site indicates that the output store provides replenishments to the specified downstream input store.

Replenishment Cycle Times. Associated with each store $i \in S$ is a replenishment cycle time, called the *nominal lead time*, L_i, which is the time required to obtain a replenishment for a sku in that store, assuming that the required upstream store material is in stock. In the case of output stores, the nominal lead time models the production time required to produce a sku in the output store from the corresponding raw material input stores. In the case of input stores, the nominal lead time models the transportation or transit time required to move product from one location to another. We assume that the nominal lead time of a store is *not* a function of the number of orders outstanding. As a result, all nominal lead times in our model are represented as independent, identically distributed (i.i.d.) random variables with non-negative support and finite moments. The distributions and associated parameters defining these random variables are specified as part of the input data to the model.

It is worth mentioning that while our model can be used to represent manufacturing plants, it can also be used to model distribution centers and warehouses. For distribution centers and warehouses, both the input stores and output stores represent the material stored at the site (i.e., no physical transformation of material occurs) but the nominal lead time would be used to model the time required to retrieve parts from bins and prepare orders for shipment.

Product Demand. Our supply chain model, just like real supply chains, is driven by customer demands for products. We represent demand as independent streams of orders for skus stored at any of the output stores in the model. Each stream models the demands of a customer or group of customers over time. Multiple demand streams can exist for the same store, representing the fact that multiple customers may order items from the same location in a supply chain.

Let $\mathcal{M}$ denote the set of all customer demand streams in our model. Associated with each demand stream are several pieces of input data: the output store from which this demand stream obtains goods, str(m); a forecast or statement of demand in the form of a sequence of independent random variables $\{D(m, 1), D(m, 2), ...\}$, each random variable representing the aggregate demand for the sku at store str(m) by customer stream m in a given period [2]; a transit or delivery time, T_m, which models the time to deliver the product from the output store to the customer; and finally an optional statement of the target or desired service level for that customer stream. To this end, let W_m denote the waiting time to receive a typical order from customer stream m; then the service-level requirement for class stream m is given by $P[W_m \leq \beta_m] \geq \alpha_m$, where α_m and β_m are input parameters: β_m is a possibly random due date for class m orders, while α_m is the fraction of class m orders that are filled before the due date.

We assume independence among the demand streams, and mutual independence among the demand streams, the transit times and the nominal lead times.

Source Stores and End Stores. There are two special types of stores in our model: source stores denoted by $\mathcal{S}_s$ and end stores denoted $\mathcal{S}_0$. Source stores are output stores that are connected to no upstream input stores. They represent the upstream boundaries, or raw material sources, of our supply chain model. End stores are output stores that have at least one customer demand stream associated with them. They may or may not be connected to downstream input stores. In particular, an output store can be both a supplier to downstream input stores in the model and a supplier to external customer demand streams.

Inventory Control. As stated earlier, the inventory at each store in our model is controlled by a base-stock policy. To explain how a base-stock policy works, we first define the inventory position at a store. At any store, whether an input store or output store, there is, at any time, some quantity of skus which are currently in inventory. We refer to this quantity as on-hand inventory. There is also a certain number of outstanding orders which we define to be the total quantity of replenishment stock which has been ordered from upstream suppliers but which has not been physically delivered to the store. We refer to

---

[2]The length of a period in our model is arbitrary. It could be a day, week, month, or any other duration, depending on the modeller's preference.

this as the on-order quantity. Finally, there is a certain quantity of skus from this store that have been ordered by downstream customers but not filled. We refer to this quantity as the back-order quantity. The inventory position at a store is then defined as the total of on-hand inventory plus on-order minus backorders.

The base-stock mechanism that triggers replenishment orders at each store works as follows. On a continuous basis, the inventory position is compared against a specified base-stock level. If the inventory position falls below the base-stock level, a replenishment order is placed for a quantity that will bring the inventory position just back to the base-stock level. In this way, the inventory position is maintained over time at the base-stock level. A replenishment order may be triggered by any event that causes the inventory position to be less than the base-stock level, such as the receipt of an order or a change in the base-stock level. It is through the specification of the base-stock level that the inventory level at a store is controlled.

Replenishment Execution. When a store receives an order, it is filled immediately if there is enough on-hand inventory at the store to fill the order. If not, it is placed in a first-come-first-served back-order queue. As on-order inventory arrives to the store and becomes on-hand inventory, the back-order queue is checked and as many of the orders as possible are filled. For output stores, an order that is filled is immediately shipped to the downstream customer or store that placed the order. For input stores at assembly sites, the situation is somewhat more complex. If the order was triggered by an output store which has a BOM consisting of more than one type of sku, the assembly operation will only proceed when sufficient quantities of all input skus are available. Thus, assembly does not begin until sufficient raw material is present at all input stores.

Performance Analysis

The primary purpose of the supply chain model is to help in quantifying the tradeoff between inventory investment and customer service levels. The first issue we address is, given our supply chain model, how to estimate the inventory levels in the network and the corresponding service levels. More precisely, we want to answer the following question: Given the input data for the model and given that we've specified the base-stock level at every store in the network, what is the estimated inventory in the network and what is the resultant service level provided to customers?

Many of the equations and formulas that we use in this section have been derived and fully explained in Ettl et al., 1997. Our intent here is to summarize those results, not rederive them.

Determining Demand at all Stores. As we previously described, we represent customer demand for products in our model as a set of independent demand streams for skus at given stores. Our base-stock replenishment model requires translating these customer demand streams into effective demand streams at each store i, denoted $\{D_i(1), D_i(2), D_i(3), ...\}$. To this end, first define the *off-set* associated with customer class m as $o_m := \mathsf{E}[T_m]$. Next, for each end store $i \in S_0$, define the effective demand stream as:

$$D_i(t) := \sum_{m:\mathrm{str}(m)=i} D(m, o_m + t) \quad t = 1, 2, 3, ...$$

Now we can proceed recursively, moving upstream in the BOM, for all other stores in the network. Specifically, define the offset of store i to be $o_i := \mathsf{E}[L_i]$, for all $i \in S$. Then, for all stores $i \notin S_0$, define the effective demand streams as:

$$D_i(t) := \sum_{j \in S_{i>}} u_{ij} D_j(o_j + t) \quad t = 1, 2, 3, ...$$

where $S_{i>}$ is the set of all stores which are immediately downstream of store i. If i is an input store, then $S_{i>}$ consists of all the output stores that call for i in their BOMs. If i is an output store, then $S_{i>}$ consists of all input stores which receive replenishments from store i. The effective demand streams, $\{D_i(t); i \in S, t = 1, 2, ...\}$, are the starting point for analyzing each of the stores in our supply chain model.

Analyzing Performance of a Source Store. Consider a source store $i \in S_s$ operating under a base-stock control mechanism with aggregate demands $\{D_i(1), D_i(2), D_i(3), ...\}$. The source store, by definition, has no upstream stores; its nominal lead time is the time to replenish an item to the store. Suppose that the base-stock level for this store is R_i. We would like to estimate: the fill rate of orders arriving to this store, f_i; the stockout probability, p_i; the average inventory level at the store, $\mathsf{E}[I_i]$; the average on-order level, $\mathsf{E}[N_i]$; and the average back-order level, $\mathsf{E}[B_i]$.

To proceed, we need to make some operational assumption about how the aggregate demand stream is realized in the form of an order arrival process. To keep things analytically tractable, we assume that the order process for each time period follows a compound Poisson process and require that the mean and variance of the aggregate demand in period t be matched by the corresponding compound Poisson process. To this end, we require that

$$\mathsf{E}(D_i) = \lambda_i \mathsf{E}(X_i), \quad \text{and} \quad \mathsf{Var}(D_i) = \lambda_i \mathsf{E}(X_i^2) \tag{24.1}$$

where λ_i corresponds to the Poisson arrival rate for the period, $\mathsf{E}(X_i)$ and $\mathsf{E}(X_i^2)$ are the first two moments of the order (or batch) size. Because there are three degrees of freedom in specifying the compound Poisson process and only two constraint equations (24.1), we have some freedom in specifying the parameters of the process so long as these constraints are satisfied.

Let us assume for the moment that the demand to the store is stationary with demand per unit time denoted by D_i. Later, we will modify our estimates to account for non-stationarity. Then a store in our model behaves like an

$M^X/G/\infty$ queue. Let N_i be the total number of jobs in the queue $M^X/G/\infty$ (in equilibrium). In our base-stock model, we interpret N_i as the on-order level at store i. Following the results in Liu et al., 1990, the mean and the variance of N_i are:

$$\mu_i := \mathsf{E}(N_i) \;=\; \lambda_i \mathsf{E}(X_i)\mathsf{E}(L_i) = \mathsf{E}(D_i)\mathsf{E}(L_i), \tag{24.2}$$

$$\sigma_i^2 := \mathsf{Var}(N_i) \;=\; \mathsf{E}(N_i) + \lambda_i[\mathsf{E}(X_i^2) - \mathsf{E}(X_i)]\int_0^\infty \bar{F}_{L_i}^2(y)dy$$

$$=\; \mu_i + [\mathsf{Var}(D_i) - \mathsf{E}(D_i)]\int_0^\infty \bar{F}_{L_i}^2(y)dy, \tag{24.3}$$

where $F_{L_i}(y) = 1 - \bar{F}_{L_i}(y)$ denotes the distribution function of the source store's nominal lead time, L_i.

Although the generating function of N_i is available in Liu et al., 1990, from which one can in principle derive the distribution of N_i, here we choose to simply approximate the distribution as normal with mean μ_i and variance σ_i^2 as derived above. This way, we can write

$$N_i = \mu_i + \sigma_i Z, \tag{24.4}$$

where Z denotes the standard normal variate.

Estimating On-Hand Inventory and Back-Orders. Let I_i and B_i be, respectively, the level of on-hand inventory and the number of backorders at store i. These relate to N_i and R_i through the following relations:

$$I_i = [R_i - N_i]^+, \qquad B_i = [N_i - R_i]^+, \tag{24.5}$$

where $[x]^+ := \max\{x, 0\}$. In view of (24.4), we can write

$$R_i = \mu_i + k_i\sigma_i, \tag{24.6}$$

where k_i is the *safety factor*. Note that since k_i and R_i relate to each other as in (24.6), we can use either value to specify the base-stock level. Then,

$$\mathsf{E}(B_i) = \mathsf{E}[N_i - R_i]^+ = \sigma_i\mathsf{E}[Z - k_i]^+ = \sigma_i\int_{k_i}^\infty (z - k_i)\phi(z)dz,$$

where $\phi(z) = \exp(-z^2/2)/\sqrt{(2\pi)}$ is the density function of Z. Denoting

$$G(k_i) := \int_{k_i}^\infty (z - k_i)\phi(z)dz, \tag{24.7}$$

we have

$$\mathsf{E}(B_i) = \sigma_i G(k_i).$$

Similarly, letting

$$H(k_i) := k_i + G(k_i) = \int_{-\infty}^{k_i} (k_i - z)\phi(z)dz, \tag{24.8}$$

we have

$$\mathsf{E}(I_i) = \sigma_i H(k_i).$$

Estimating the Stockout Probability and Fill Rate of a Store. The stockout probability at store i, denoted p_i, is simply

$$p_i := P[I_i = 0] = P[N_i \geq R_i] = P[Z \geq k_i] = \bar{\Phi}(k_i). \tag{24.9}$$

That is, p_i is the fraction of time that the on-hand inventory at store i is zero.

The fill rate at store i, denoted f_i, is the fraction of customer orders that is filled by on-hand inventory. Let $\bar{f}_i := 1 - f_i$. Observe that when $N_i < R_i$, the difference, $R_i - N_i$, is the on-hand inventory; and when $N_i > R_i$, there is a backorder queue of $N_i - R_i$ units. In equilibrium, we can equate the rate of orders entering the backorder queue and the rate of orders leaving the backorder queue, to arrive at the following relation:

$$\lambda_i E(X_i)\bar{f}_i = [E(\tilde{L}_i)]^{-1} E[N_i | N_i > R_i] P[N_i > R_i]. \tag{24.10}$$

(When there are $N_i = n$ orders in processing, the rate of output is $n/[E(\tilde{L}_i)]$; conditioning on $N_i > R_i$ yields the right hand side above.)

After some simplification (see Ettl et al., 1997 for details), we have

$$\bar{f}_i = \sigma_i \phi(k_i)/\mu_i + \bar{\Phi}(k_i). \tag{24.11}$$

Modifications to Handle Non-Stationary Demand. Our performance estimates above relied on the assumption that demand at a store is stationary. We need to modify our estimates to handle non-stationary demand, but would like to do so in as simple a manner as possible. Our approach is to modify our characterization of μ_i and σ_i. Denote

$$D_i^c(\ell) := D_i(1) + \cdots + D_i(\ell),$$

where the superscript "c" stands for "cumulative."

From (24.2), it is clear that μ_i is simply the mean of the cumulative demand over the lead time: $\mu_i = E[D_i^c(L_i)]$.

For σ_i^2, we use the following approximation based on an upper bound:

$$
\begin{aligned}
\sigma_i^2 &= E\{ \text{Var}[D_i^c(L_i)|L_i] \} \\
&= E\{ \sum_{m=1}^{L} \text{Var}[D_i(m)] \} \\
&= \int_\ell \text{Var}[D_i^c(\ell)] dF_L(\ell). \tag{24.12}
\end{aligned}
$$

Analyzing Performance of a Non-Source, Non-Assembly Store. Non-source stores have the added complexity that their performance depends on the reliability of their upstream suppliers. It is precisely this dependence that makes the analysis of supply chains complex. To analyze the performance of stores that have a single upstream supplier, we need to modify our analysis to account for the possibility that the upstream supplier may not be 100% reliable. In other words, the actual lead time of the store may differ from the nominal lead time because of stockouts at the upstream store. Thus, to modify our estimates to account for this possibility, we use the same basic model as above but rather than using the nominal lead time as the service time for our $M^X/G/\infty$ model we use the actual lead time, which is defined as the nominal lead time plus possibly some additional time that corresponds to the delay experienced by an order when the upstream store has a stockout. Specifically, let i correspond to the store we are analyzing and let j be the single upstream store of store i. Then we define the actual lead time of store i as:

$$\tilde{L}_i \equiv \begin{cases} L_i & \text{w.p.} & \bar{f}_j \\ L_i + \tau_j & \text{w.p.} & f_j \end{cases} \qquad (24.13)$$

In words, the actual lead time of a store is the nominal lead time if the order is filled right away by the upstream store; otherwise it is the nominal lead time augmented by an additional delay term, τ_j, which is the "typical" delay experienced by a back-order at store j. To characterize this additional delay term, we resort to another approximation, based on the simpler $M/M/\infty$ queue (see Ettl et al., 1997 for details):

$$\tau_j = \tilde{L}_j r_j, \quad \text{where} \quad r_j := \frac{\mathsf{E}(B_j)}{p_j(R_j + 1)}. \qquad (24.14)$$

Note that our definition of the actual lead time is recursive as it depends on the actual lead time, stockout probability, and expected back-orders of upstream stores. Thus, in evaluating the performance of a network, we start with source stores and work our way forward in the network one store at a time until all stores have been analyzed.

Analyzing Performance of a Non-Source Assembly Store. To analyze output stores which require parts from more than one input store (as in the case of an assembly operation, for example), we proceed as in the case of a single upstream supplier but obtain a more complex expression for the actual lead time, reflecting the fact that the assembly can only proceed when all parts in the upstream stores are available. Let i denote the store that we are analyzing and let $\mathcal{S}_{>i}$ denote the set of upstream (input) stores that are used in the assembly of output store i. Our approximate expression for the actual lead time becomes:

$$\tilde{L}_i = \begin{cases} L_i & \text{w.p.} & \pi_{0i} \\ L_i + \tau_j & \text{w.p.} & \pi_{ji}, \ j \in \mathcal{S}_{>i} \end{cases}, \qquad (24.15)$$

where

$$\pi_{0i} := \left(1 + \sum_{h \in \mathcal{S}_{>i}} \bar{f}_h/f_h\right)^{-1}, \quad \pi_{ji} := (\bar{f}_j/f_j)\left(1 + \sum_{h \in \mathcal{S}_{>i}} \bar{f}_h/f_h\right)^{-1}. \qquad (24.16)$$

Relating Fill Rates to Service Levels. If a customer stream m has a target or desired service level specified in the form $P[W_m \le \beta_m] \ge \alpha_m$, one question we would like our performance analysis module to answer is whether the supply chain will be able to deliver this level of service to this customer stream, once a set of reorder points has been specified. To answer this question, we recognize first that when the orders from customer stream m are filled from on-hand inventory at store $i \equiv str(m)$, the delay is simply the transit time T_m to go from the store to the customer. When there is no on-hand inventory (i.e. a stockout situation exists), an additional delay is incurred of τ_i. Hence,

$$P[W_m \le \beta_m] = f_i P[T_m \le \beta_m] + (1 - f_i) P[T_m + \tau_i \le \beta_m]. \qquad (24.17)$$

Thus, if the right hand side of (24.17) has a value that exceeds the requirement α_m, then we can state that the required service level will be met.

Estimating the Value of Supply Chain Inventory. Our model estimates on-hand inventory at store i as $E[I_i] = c_i \sigma_i H(k_i)$. If we assume that c_i is input as the cost per sku at store i, our estimate of the dollar value of on-hand inventory at store i is $c_i \sigma_i H(k_i)$. We can also estimate WIP immediately upstream of store i as being approximately equal to the estimated number of skus on order at store i, $E[N_i]$, which is equal to the mean demand over the lead time for store i [3]. To value the WIP, we can either specify as input data the dollar value of a unit of WIP upstream of store i (for all stores), or, to lessen the burden of data input, we can apply an average cost based on the sum of the costs of the input stores and the cost of a sku in the output store:

$$\hat{c}_i := \frac{1}{2}(c_i + \sum_{j \in \mathcal{S}_{>i}} c_j u_{ji}).$$

Then the dollar value of WIP immediately upstream of store i is estimated to be $\hat{c}_i E[N_i] = \hat{c}_i \mu_i$. For source stores, which have no upstream stores, we can either define $\hat{c}_i \equiv c_i$ or $\hat{c}_i \equiv 0$. The appropriate choice depends on the specific application. Also, for stores that model external suppliers, the value of the inventory at those stores may or may not be included in the valuation of inventory in the network. The choice again depends on the specific application.

The total dollar value of inventory in the supply chain, C, is just the sum of the WIP and finished goods inventory at each store:

$$C = \sum_{i \in \mathcal{S}} [\hat{c}_i \mu_i + c_i \sigma_i H(k_i)]. \qquad (24.18)$$

Supply Chain Optimization

Within the context of our supply chain model, we can formulate two kinds of optimization problems. The first is to find the minimal dollar value of inventory

---

[3] Whether we use the mean demand over the nominal lead time or the actual lead time in this estimate is subject to debate. Currently, we use the mean demand over the actual lead time.

(finished goods and WIP) required to support a stated set of customer service levels, one for each customer stream defined. The second is, roughly speaking, to maximize the service levels provided to customers subject to a budgetary constraint on overall inventory investment.

In the first case, we proceed by first finding the minimal fill rate needed to satisfy all customer streams at each end store. (Recall that an end store is by definition a store that has at least one customer stream associated with it.) To do this, we first calculate for each customer stream the required fill rate, f_m, which we define via equation (24.17):

$$f_m \equiv \frac{\alpha_m - \mathsf{P}[T_m + \tau_i \leq \beta_m]}{\mathsf{P}[T_m \leq \beta_m] - \mathsf{P}[T_m + \tau_i \leq \beta_m]}. \tag{24.19}$$

Then, we define the minimal or required fill rates at end store i, f_i^r, to be

$$f_i^r \equiv \max_{m \in \mathcal{M}:\mathrm{str}(m)=i} f_m$$

Letting $\mathbf{k}$ denote the vector of safety factors for our supply chain model, the formulation of our optimization problem is:

$$\max_{\mathbf{k}} C(\mathbf{k})$$

subject to:

$$f_i^r \geq f_i \quad \forall i \in \mathcal{S}_0$$

We have written the objective function C explicitly as a function of the decision variables, $\mathbf{k}$. This is a constrained non-linear optimization problem and in Ettl et al., 1997, we describe a method for solving this problem by first obtaining analytical estimates of the partial derivative of the objective function with respect to each decision variable $k \in \mathbf{k}$ and then using the conjugate gradient method to find an optimal solution.

To proceed with the second formulation, we propose a minimax formulation which attempts to minimize the maximum deviation of the fill rate at an end store from its target fill rate. Specifically, our formulation becomes:

$$\min \max_{i \in \mathcal{S}_0} |f_i^r - f_i|$$

subject to:

$$C(\mathbf{k}) < B,$$

where B is the stated budgetary limit on inventory. Many variants on this formulation are possible. For example, one might want to give preference to achieving target fill rates for certain stores or certain customer streams. This would entail modifying the objective function to reflect this type of preference, possibly by introducing weighting factors.

24.3 CASE STUDY

Introduction

JCN Computer Systems Inc. (a fictitious name and company), a producer of personal computers for homes and businesses, currently markets six models of desktop personal computers, each model based on an Intel^{TM} processor of a given clock speed and generation. For each model, there is a base configuration which potential buyers can modify based on their preferences. Computers are then built to order based on these preferences.

In the North American market, which is the focus of this case study, JCN sells primarily through direct channels: customers place orders either by calling a toll-free number or through an internet web site. The computers are then shipped directly from a consolidation center to the customers by a third party shipping agent. Once an order has been packaged, shipment to customers takes two days.

JCN owns three manufacturing facilities in the US: one for final assembly of PCs, one for manufacturing mother boards which are used in final assembly, and one for manufacturing double-inline-memory-modules (DIMMs), also used in final assembly. In addition, JCN owns a consolidation center and warehouse adjacent to the final assembly site in which orders are configured and packaged for final shipment to the customer. Consolidation includes merging of monitors and printers with the final order if these items have been included in the order.

The final assembly site has a capacity for producing up to 25000 personal computers per month. Both of the other manufacturing sites have ample capacity for providing the subassemblies to meet the production needs at the assembly site. All of the remaining components that are needed in the assembly process are sourced externally, some from US manufacturers and others from foreign suppliers, mostly located in the far east.

JCN's management team responsible for operations has an on-going focus on the dual objectives of improving customer service and increasing inventory turnover. JCN's customer service target, articulated by the marketing department, is to consistently deliver customer orders within 5 business days of order receipt. The operations team is interested in knowing what inventory levels are consistent with meeting this target customer service level. To address this question, we shall use the supply chain model to study the tradeoff between customer service levels and inventory investment for JCN's North American supply chain.

Constructing the Model

Our model of JCN's supply chain is depicted schematically in Figure 24.1. In constructing our model, we have elected to focus on a subset of JCN's parts and manufacturing operations. This subset comprises 90% of JCN's total cost of goods sold. The remaining 10% consists largely of numerous low cost parts, which for the sake of simplicity, we have decided to exclude from our analysis.

The model consists of a set of four manufacturing sites: mother board assembly, DIMM assembly, PC assembly, and order consolidation.

The mother board assembly site is modelled as a simple assembly of two items: a panel and a chip set. JCN manufactures many types of mother boards but only three principal ones are used in the assembly of their desktop models. Each mother board requires a specific set of chips, which are connected to the panel using both pin-through-hole technology and surface-mount technology. Both the panels and the chip sets used in the assembly of mother boards are obtained from external suppliers.

The DIMM assembly process is similar to the mother board assembly process. It entails mounting synchronous dynamic random access memory (SDRAM) chips on small electronic cards (DIMMs). Each PC can be ordered with 1, 2, or 4 DIMMs, corresponding to 32, 64, or 128 MB of memory. The SDRAM modules that are used in the DIMM assembly process are obtained from an external supplier. Because the electronic cards on which the chips are mounted are inexpensive relative to the chips, we choose to model the DIMM assembly operation as a processing operation involving the chips only.

The PC assembly process is modelled as a manufacturing site that produces 6 models of finished PCs: Box 1 - Box 6. Each box is assembled from 11 components, though the precise makeup of a PC is not known until an order is placed, because of the ability of customers to customize their orders. As a result, the BOM for each PC consists of fractional usages, also called *feature ratios*. The feature ratios indicate the proportion of orders for a particular product that request a specific component. The BOM and feature ratios for each model are given in Table 24.1. For example, for all PC models, the feature ratios specify that of the three types of chasses available, 50% of orders request the mini-tower, 25% request the desktop and 25% request the mid-tower. These feature ratios are obtained from the forecasting group (a part of the marketing organization), which estimates these numbers based largely on statistical analysis of past sales.

The order consolidation process is modeled as a site in which keyboards, monitors, and printers are merged with the appropriate boxes for shipment to the end customer. The feature ratios listed in Table 24.1 list the proportion of orders that request each type of keyboard, monitor, and printer. The final products at order consolidation – identified as PC 1 – PC 6 – represent the packaged products which are ready for shipment to the customer.

To model the manufacturing lead times for each of JCN's assembly processes, we obtain estimates from each of the manufacturing sites on what the minimum, most likely, and maximum times are for assembly. These times, in units of days, are displayed in Table 24.2. Based on these three values, we fit the manufacturing lead times to a normal distribution with a mean equal to the most likely value and a standard deviation equal to (maximum − minimum)/6. Also displayed in Table 24.2 are the unit costs for each product that JCN produces.

The supplier production lead times, in-bound transit times, and unit costs for all externally sourced parts are listed in Table 24.3. As with manufacturing lead times, we obtain estimates from the procurement organization on the minimum, most likely, and maximum times for both in-bound transit times and supplier lead times. We then fit these values to normal distributions.

To characterize the customer demand for products, we obtain from the forecasting group a three month outlook for each model, displayed in Table 24.4. The outlook consists of estimates of the minimum, most likely, and maximum sales volume for each period and each model. The forecasting group derives these estimates by analyzing past sales patterns and by obtaining input from marketing and sales personnel. As with lead times, we fit the forecast for each period to a normal distribution.

The transit time for moving assembed PCs from the PC assembly site to the order consolidation site is modeled as taking a half day. Moving mother boards from MB assembly to PC assembly requires one day as does moving DIMMs from DIMM assembly to PC assembly.

Analysis Using Supply Chain Model

To carry out our analysis, we use a software application that implements the supply chain model described in section §24.2. To use the software, we translate the data for JCN's supply chain summarized in Tables 1–4 into the appropriate input format for the software. This translation is straightforward except for a few items that need further elaboration. For example, the service requirement to consistently deliver customer orders within 5 business days of order receipt is translated into the requirement that 95% of all orders be delivered within 5 days. In doing this translation, we have interpreted "consistent" as meaning "95% of the time" and we have ignored any additional time involved in processing an order, such as the time required for order entry.

We ran the model to minimize the total investment in inventory subject to the above service level constraint. Overall, the solution indicates that for JCN's supply chain the maximum possible inventory turnover that can be achieved is approximately 18, based on an annualized cost of goods sold of $500 million. This includes inventory held by external suppliers at the supplier locations. If we exclude the inventory held by external suppliers, the number of inventory turns rises to 44. Further, if we suppose that the external suppliers provide line-side stocking of their components at JCN's assembly sites and that JCN does not therefore own any of the on-hand component inventory at its assembly sites, the number of inventory turns soars to 108.

The detailed output of the model is shown in Tables 24.5 and 24.6. Table 24.5 shows the target fill rates, estimated on-hand inventory, and WIP for each end-store in our supply chain model – each end-store corresponding to one of the 6 completed PCs at order consolidation. Table 24.6 shows the same quantities for all other stores in the supply chain. The on-hand inventory and WIP quantities are all expressed in terms of days of supply (DOS). As a general observation, the contribution of on-hand inventory to overall inventory investment is much smaller than the contribution of WIP. Table 24.6 reveals that the quantity of on-hand inventory as a percentage of WIP contributed by that store averages only 5%. Thus, a relatively small quantity of inventory is needed in this example to hedge against uncertainty in order to provide the desired end-customer service level. In keeping with this observation is the fact that the the fill rates for all stores average a relatively low 50%.

Examining the output more closely reveals some results which make intuitive sense and others which are not so easy to explain. Table 24.5 reveals that zero DOS should be held as final PCs and that the target fill rate for these stores should be 0. This result makes sense because only three days are required to pack and ship orders to customers – 1 day for packing the order, and 2 days for shipping – and this is well below the 5 day service target.

The target fill rates and DOS at other locations displayed in Table 24.6 are, at first glance, somewhat puzzling. For example, 13 inch, 15 inch and 17 inch monitors at the order consolidation site have target fill rates of 0.3, 0.6, and 0.4 respectively. The corresponding on-hand DOS are 0.1, 0.2, and 0.1. The target fill rates appear to be fairly low as do the corresponding DOS. On the other hand, the mean transit lead time for monitors is seven days. Thus, one might conclude that frequent stockouts of monitors at order consolidation would result in long delays in packing orders. The explanation for this apparent anomaly is that the lead time for monitors given that a stockout exists is significantly less than seven days because of the WIP that is in transit from the supplier site. As a result, given the base-stock mechanism that underlies our supply chain model, it is reasonable for the fill rates to be set as low as they are.

Another observation is that in this example the costs associated with supplier-held inventory (at the supplier location) and JCN-owned inventory (at the assembly sites) are assumed to be the same. For example, the microprocessor inventory stored at the PC assembly site is valued at the same unit cost as the microprocessors held at the supplier location. Thus, in minimizing total inventory in the supply chain, the optimal solution does not necessarily favor keeping inventory further upstream at the supplier locations. If we modify the optimization so that it excludes from the objective function the inventory held by suppliers, the optimal solution pushes more inventory to the suppliers, resulting in higher fill rates at the supplier locations. This result makes sense because if JCN is only liable for inventory held at its assembly sites, JCN has every incentive to have its suppliers provide as high a service level as possible.

The results also reveal that the target fill rate for Box 4 at the order consolidation center should be 0.7. This suggests building at least some boxes at PC assembly and shipping them to the order consolidation center prior to receipt of firm orders. From a manufacturing operations perspective, however, this action may not be desirable because of the high risk and expense associated with building boxes in advance of receiving actual orders. In fact, this recommendation flies in the face of the operations strategy of JCM which is to assemble all boxes to order. As a result, we might want to modify the optimization in this case to force the reorder points for Box 1–6 at PC assembly and at order consolidation to be zero. Such a modification would then result in a solution which minimizes total inventory but which respects the build-to-order assembly strategy.

24.4 CONCLUSION

We have described a model of supply chains intended to analyze tradeoffs between inventory investment and service levels in complex supply chains. A perennial concern that arises in constructing a mathematical model is whether the model is credible. Underlying this concern are two questions. First, are the approximations that we have used to derive our performance estimates robust? And second, even if no approximations were used, does the model represent supply chains in a reasonable fashion and can the results of the model be reliably used to drive changes in supply chains?

Regarding the first question, we first note that, indeed, we did use a number of approximations in section §24.2 to derive performance estimates. For example, consider the approximations used to estimate the actual lead time of stores, as in equations (24.13) – (24.16) and the approximations used to handle non-stationary demand. One way to test the validity of these approximations is to compare results with those obtained via simulation. While comparisons based on a simple supply chain presented in Ettl et al., 1997 indicate that the model agrees quite well with simulation results, further experiments on more complex supply chains are necessary. If the approximations do not in fact hold up well in more complex scenarios, an alternative is to implement the model as a simulation, rather than use the analytical performance estimates. Simulation, however, presents its own challenges, particularly with regard to solving the optimization problems described in section §24.2.

The second and more fundamental question about the validity of the model is much more difficult to address. Even if we put aside the objections raised in the introduction concerning operational issues that the model purposely ignores such as lot-sizing practices and capacity constraints, we still confront the issues of whether the model captures in a credible fashion the sources of uncertainty in a supply chain, the way these sources of uncertainty affect supply chain performance, and the mechanism by which safety stock inventory serves to mitigate the effects of variability. In constructing our model, we have been torn by the need to faithfully capture the effects of uncertainty on supply chain performance and by the conflicting need to keep the model simple and the data input requirements modest. The resulting model is a compromise between these needs. One manifestation of this compromise is the way in which we capture uncertainty in lead times by requiring as input minimum, most likely, and maximum values and then fitting these values to a normal distribution. No doubt there are better ways to characterize uncertainty but implementing these methods would certainly entail greater data collection and analysis. Also worth mentioning is the fact that in our model formulation, we have not incorporated a source of uncertainty mentioned in the introduction that is important in many manufacturing environments, namely yield variability. Further research could perhaps shed light on improved methods to model yield variability and other forms of uncertainty in supply chains.

Acknowledgments

The author wishes to thank Drs. M. Ettl, G. Lin, and D. D. Yao, all of whom contributed significantly to the development of the model described in this chapter as well as to the software implementation of the model, which was used to carry out the analysis presented in the case study.

References

Axsater, S. (1993). Continuous review policies. In Graves, S., Rinooy Kan, A., and Zipkin, P., editors, *Logistics of Production and Inventory*, chapter 4. North Holland.

Cohen, M. and Lee, H. (1988). Strategic analysis of integrated production-distribution systems: Models and methods. *Operations Research*, 36:216 – 228.

Ernst, R. and Pyke, D. (1992). Component part stocking policies. *Naval Research Logistics*, 39:509 – 529.

Ettl, M., Feigin, G., Lin, G., and Yao, D. (1997). A supply network model with base-stock control and service requirements. RC 20473, IBM Corp. Submitted to *Operations Research*.

Federgruen, A. (1993). Centralized planning models for multi-echelon inventory systems under uncertainty. In Graves, S., Rinooy Kan, A., and Zipkin, P., editors, *Logistics of Production and Inventory*, chapter 3. North Holland.

Glasserman, P. and Wang, Y. (1997). Inventory-leadtime trade-offs in assemble-to-order systems. *Operations Research*. To appear.

Graves, S. and Willems, S. (1998). Optimizing strategic safety stock placement in supply chains. Technical report, MIT.

Lee, H. and Billington, C. (1986). Material management in decentralized supply chains. *Operations Research*, 41:835 – 847.

Liu, L., Kashyap, B., and Templeton, J. (1990). On the $GI^X/G/\infty$ system. *J. Appl. Prob.*, 27:671 – 683.

Muckstadt, J. and Roundy, R. (1993). Analysis of multistage production systems. In Graves, S., Rinooy Kan, A., and Zipkin, P., editors, *Logistics of Production and Inventory*, chapter 2. North Holland.

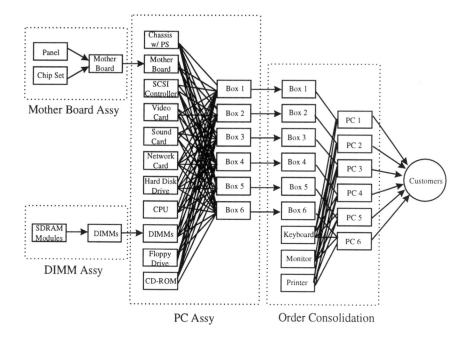

Figure 24.1 Model of JCN's supply chain.

Model	Chassis			CPU					
	Mini-tower	Desktop	Mid-tower	200Mhz	200MMX	233MMX	233II	266II	300II
PC 1	0.5	0.25	0.25	1	0	0	0	0	0
PC 2	0.5	0.25	0.25	0	1	0	0	0	0
PC 3	0.5	0.25	0.25	0	0	1	0	0	0
PC 4	0.5	0.25	0.25	0	0	0	1	0	0
PC 5	0.5	0.25	0.25	0	0	0	0	1	0
PC 6	0.5	0.25	0.25	0	0	0	0	0	1

Model	Video Card		Sound Card		Network Card		Hard Drive		
	VC 1	VC 2	Sound 1	Sound 2	Network 1	Network 2	2.1 GB	3.2 GB	4.3 GB
PC 1	0.75	0.25	0.75	0.25	0.8	0.2	0.33	0.33	0.33
PC 2	0.75	0.25	0.75	0.25	0.8	0.2	0.2	0.4	0.4
PC 3	0.75	0.25	0.75	0.25	0.8	0.2	0.1	0.4	0.5
PC 4	0.45	0.65	0.5	0.5	0.8	0.2	0.2	0.5	0.3
PC 5	0.25	0.75	0.5	0.5	0.8	0.2	0.1	0.3	0.6
PC 6	0.25	0.75	0.5	0.5	0.8	0.2	0.1	0.2	0.7

Model	Memory			Mother board			SCSI Cont. Card	Floppy Drive
	32 MB	64 MB	128 MB	MB 1	MB 2	MB 3		
PC 1	0.75	0.15	0.1	1	0	0	1	1
PC 2	0.75	0.15	0.1	0	1	0	1	1
PC 3	0.4	0.4	0.2	0	1	0	1	1
PC 4	0.4	0.4	0.2	0	0	1	1	1
PC 5	0.4	0.4	0.2	0	0	1	1	1
PC 6	0.4	0.4	0.2	0	0	1	1	1

Model	ROM		Monitor			Printer		Keyboard	
	CD	DVD	15"	17"	19"	Ink Jet	Laser	KB 1	KB 2
PC 1	0.9	0.1	0.4	0.5	0.1	0.2	0.2	0.95	0.05
PC 2	0.9	0.1	0.4	0.5	0.1	0.2	0.2	0.95	0.05
PC 3	0.9	0.1	0.4	0.5	0.1	0.2	0.2	0.95	0.05
PC 4	0.9	0.1	0.4	0.5	0.1	0.2	0.2	0.95	0.05
PC 5	0.9	0.1	0.2	0.4	0.4	0.2	0.2	0.95	0.05
PC 6	0.9	0.1	0.2	0.4	0.4	0.2	0.2	0.95	0.05

Table 24.1 Bill of material and feature ratios for JCN's personal computers. .

| Product | Location | Manufacturing Lead Time | | | Unit Cost |
		Min	Most Likely	Max	
Box 1	PC Assy	0.5	1	1.5	$ 1,312
Box 2	PC Assy	0.5	1	1.5	$ 1,402
Box 3	PC Assy	0.5	1	1.5	$ 1,555
Box 4	PC Assy	0.5	1	1.5	$ 1,783
Box 5	PC Assy	0.5	1	1.5	$ 1,945
Box 6	PC Assy	0.5	1	1.5	$ 2,260
PC 1	Ordr Cons	1	1	1	$ 1,854
PC 2	Ordr Cons	1	1	1	$ 1,948
PC 3	Ordr Cons	1	1	1	$ 2,108
PC 4	Ordr Cons	1	1	1	$ 2,348
PC 5	Ordr Cons	1	1	1	$ 2,627
PC 6	Ordr Cons	1	1	1	$ 2,959
MB 1	MB Assy	2	3	4	$ 100
MB 2	MB Assy	2	3	4	$ 130
MB 3	MB Assy	2	3	4	$ 190
DIMM	DIMM Assy	3	5	7	$ 80

Table 24.2 Planned manufacturing lead times and weighted avg. unit costs for JCN's products.

Component	Category	In-Bound Transit Time			Supplier Lead Time			Unit Cost
		Min	Most Likely	Max	Min	Most Likely	Max	
Mini-tower	Chassis	2	3	4	20	25	30	$ 22
Desktop	Chassis	2	3	4	20	25	30	$ 25
Mid-tower	Chassis	2	3	4	20	25	30	$ 35
200mhz	CPU	1	2	3	18	20	22	$ 180
200M	CPU	1	2	3	18	20	22	$ 215
233M	CPU	1	2	3	18	20	22	$ 290
233II	CPU	1	2	3	22	25	28	$ 390
266II	CPU	1	2	3	22	25	28	$ 510
300II	CPU	1	2	3	22	25	28	$ 769
VC 1	Elec. Card	6	8	10	3	5	7	$ 80
VC 2	Elec. Card	6	8	10	3	5	7	$ 150
SCSI Cont.	Elec. Card	6	8	10	3	5	7	$ 175
Sound 1	Elec. Card	6	8	10	3	5	7	$ 55
Sound 2	Elec. Card	6	8	10	3	5	7	$ 79
Network 1	Elec. Card	6	8	10	3	5	7	$ 69
Network 2	Elec. Card	6	8	10	3	5	7	$ 79
2.1 GB	Hard Drive	4	5	6	20	25	30	$ 140
3.2 GB	Hard Drive	4	5	6	20	25	30	$ 180
4.3 GB	Hard Drive	4	5	6	20	25	30	$ 220
Modules	SDRAM	6	8	10	18	20	22	$ 75
FDD	Floppy Dr.	6	8	10	6	8	10	$ 15
CD	ROM	6	8	10	6	8	10	$ 70
DVD	ROM	6	8	10	6	8	10	$ 110
15 inch	Monitor	5	7	9	10	13	16	$ 205
17 inch	Monitor	5	7	9	10	13	16	$ 405
19 inch	Monitor	5	7	9	10	13	16	$ 620
Ink Jet	Printer	5	7	9	9	11	13	$ 150
Laser	Printer	5	7	9	9	11	13	$ 330
KB 1	Keyboard	3	4	5	4	5	6	$ 10
KB 2	Keyboard	3	4	5	4	5	6	$ 25
Panel 1	Panel	1	2	3	6	8	10	$ 20
Panel 2	Panel	1	2	3	6	8	10	$ 30
Panel 3	Panel	1	2	3	6	8	10	$ 40
Chip Set 1	Chip Set	6	8	10	20	25	30	$ 50
Chip Set 2	Chip Set	6	8	10	20	25	30	$ 60
Chip Set 3	Chip Set	6	8	10	20	25	30	$ 80

Table 24.3 Transit times, supplier lead times, and unit costs for externally sourced components.

Model	June			July			August		
	Min	Most Likely	Max	Min	Most Likely	Max	Min	Most Likely	Max
PC 1	2800	3100	3400	2900	3500	4100	2700	3700	4700
PC 2	1700	2100	2500	1800	2300	2800	1800	2500	3200
PC 3	1500	1900	2300	1400	2000	2600	1300	2100	2900
PC 4	1800	2000	2200	1700	2200	2700	1900	2500	3100
PC 5	3800	4500	5200	3600	4700	5800	3500	5100	6700
PC 6	2400	2800	3200	2600	3100	3600	2600	3300	4000
Totals:	14000	16400	18800	14000	17800	21600	13800	19200	24600

Table 24.4 Monthly demand forecast for JCN's personal computers.

Product	Location	Target Fill Rate	On-Hand (DOS)	WIP (DOS)
PC 1	Ordr Cons	0.0	0.0	1.3
PC 2	Ordr Cons	0.0	0.0	1.3
PC 3	Ordr Cons	0.0	0.0	1.3
PC 4	Ordr Cons	0.0	0.0	1.3
PC 5	Ordr Cons	0.0	0.0	1.3
PC 6	Ordr Cons	0.0	0.0	1.3

Table 24.5 Model results for JCN's final products.

Component	Location	Target Fill Rate	On-Hand (DOS)	WIP (DOS)	Supplier Location	Target Fill Rate	On-Hand (DOS)	WIP (DOS)
CD-ROM	PC Assy	0.5	0.1	8.2	CD Supplier	0.4	0.1	8.0
DVD-ROM	PC Assy	0.4	0.1	8.8	CD Supplier	0.0	0.0	8.0
Desktop	PC Assy	0.1	0.0	3.0	Chassis Supplier	0.9	0.8	25.0
Mid-tower	PC Assy	0.7	0.2	3.6	Chassis Supplier	0.2	0.1	25.0
Mini-tower	PC Assy	0.7	0.2	4.0	Chassis Supplier	0.1	0.0	25.0
Chip Set 1	MB Assy	0.2	0.1	11.1	Chipset Supplier	0.0	0.0	25.0
Chip Set 2	MB Assy	1.0	1.5	8.0	Chipset Supplier	1.0	2.7	25.0
Chip Set 3	MB Assy	0.6	0.3	8.7	Chipset Supplier	0.3	0.2	25.0
DIMM	PC Assy	0.7	0.1	1.2	DIMM Assy	0.2	0.0	5.1
Floppy Drive	PC Assy	0.7	0.3	8.4	FD Supplier	0.1	0.0	8.0
2.1 GB	PC Assy	0.8	0.2	5.1	HD Supplier	0.7	0.4	25.0
3.2 GB	PC Assy	0.2	0.0	5.1	HD Supplier	0.8	0.7	25.0
4.3 GB	PC Assy	0.7	0.2	5.1	HD Supplier	0.7	0.6	25.0
200M	PC Assy	0.5	0.4	5.2	Intel	0.0	0.0	20.0
200mhz	PC Assy	0.9	0.5	3.4	Intel	0.1	0.0	20.0
233II	PC Assy	0.5	0.1	2.8	Intel	0.3	0.2	25.0
233M	PC Assy	0.9	0.8	2.9	Intel	0.4	0.5	20.0
266II	PC Assy	0.8	0.4	2.3	Intel	0.7	1.0	25.0
300II	PC Assy	0.1	0.0	2.5	Intel	0.5	0.5	25.0
KB 1	Ordr Cons	0.2	0.0	4.0	KB Supplier	0.7	0.2	5.0
KB 2	Ordr Cons	0.5	0.1	4.1	KB Supplier	0.4	0.1	5.0
MB 1	PC Assy	0.2	0.0	1.2	MB Assy	0.3	0.1	3.5
MB 2	PC Assy	0.8	0.3	1.1	MB Assy	0.8	0.4	3.2
MB 3	PC Assy	0.6	0.1	1.1	MB Assy	0.5	0.1	3.1
15 inch	Ordr Cons	0.3	0.1	7.1	Mtr Supplier	0.7	0.3	13.0
17 inch	Ordr Cons	0.6	0.2	7.2	Mtr Supplier	0.5	0.2	13.0
19 inch	Ordr Cons	0.4	0.1	7.2	Mtr Supplier	0.5	0.3	13.0
Network 1	PC Assy	0.2	0.0	8.1	NC Supplier	0.5	0.1	5.0
Network 2	PC Assy	0.7	0.3	8.1	NC Supplier	0.4	0.1	5.0
Panel 1	MB Assy	0.1	0.0	2.1	Panel Supplier	0.6	0.3	8.0
Panel 2	MB Assy	0.5	0.2	2.1	Panel Supplier	0.8	0.7	8.0
Panel 3	MB Assy	0.5	0.1	2.2	Panel Supplier	0.4	0.2	8.0
Box 1	Ordr Cons	0.0	0.0	0.7	PC Assy	0.2	0.0	1.3
Box 2	Ordr Cons	0.3	0.1	0.6	PC Assy	0.7	0.4	1.3
Box 3	Ordr Cons	0.3	0.1	0.8	PC Assy	0.3	0.2	1.3
Box 4	Ordr Cons	0.7	0.1	0.7	PC Assy	0.2	0.0	1.3
Box 5	Ordr Cons	0.7	0.2	0.7	PC Assy	0.3	0.1	1.3
Box 6	Ordr Cons	0.1	0.0	0.5	PC Assy	0.9	0.5	1.5
Ink Jet	Ordr Cons	0.5	0.1	7.1	Prtr Supplier	0.7	0.3	11.0
Laser	Ordr Cons	0.2	0.0	7.1	Prtr Supplier	0.7	0.3	11.0
Sound 1	PC Assy	0.6	0.2	8.3	SC Supplier	0.1	0.0	5.0
Sound 2	PC Assy	0.4	0.1	8.3	SC Supplier	0.2	0.0	5.0
SCSI Controller	PC Assy	0.9	0.4	8.1	SCSI Supplier	0.6	0.2	5.0
Modules	DIMM Assy	0.4	0.1	8.0	SDRAM Supplier	0.8	0.4	20.0
VC 1	PC Assy	0.2	0.1	8.1	VC Supplier	0.4	0.1	5.0
VC 2	PC Assy	0.3	0.1	8.1	VC Supplier	0.6	0.2	5.0

Table 24.6 Model results for components at internal stocking locations and at supplier locations.

25 MANAGING INVENTORY FOR FASHION PRODUCTS

Ananth Raman

Graduate School of Business Administration
Harvard University
Boston, MA 02163

25.1 Introduction

With lifecycles becoming shorter and demand more uncertain in growing numbers of product categories, more and more companies, in industries as diverse as personal computers, toys, and even agricultural chemicals, are being forced to deal with stockouts and markdowns, problems usually associated with businesses like apparel and footwear. These changes have been attended by increased attention by managers in these industries to fill rates, inventory turns, and product obsolescence, topics studied extensively in management science models that traditionally have focused on fashion products.[1]

Changes in information technology and the structure of the retailing industry have occasioned circumstances conducive to the adoption of such model-based approaches. Technological changes have facilitated the collection, propagation, and analysis of large volumes of sales data that constitute a vital input to management science models. Concurrently, with retail consolidation giving rise to large chains of stores that rely on centralized inventory management, it has become virtually impossible for a human being to interpret the streams of data emanating from point-of-sale scanners as a basis for determining optimal inventory without support from appropriate management science models.

The growing importance of inventory-related costs in the fashion industry ought to contribute to the value-added by these management science models. Department store markdowns, for example, averaged 31.5% of sales in 1995, a result of steady growth during the last two decades, compared to 8% of sales in 1970. Stockouts have become increasingly common as well; some retailers, in private interviews, estimate that approximately one in three customers looking for a particular product does not find it in stock.

Although we believe that management science can contribute significantly to improving managerial decisions in these industries, adoption of these management science models is limited to a handful of applications and demonstrated success is rare. Our field research on retail merchandising practices for fashion products[2] revealed that few retail merchants have adopted or are even aware of the management science models currently available. We attribute the lack of implementation of these models to shortcomings in the current research.

Fashion products have been studied by many authors in management science. We organize their papers into, and describe the issues addressed in each of, several different categories. We consider first papers that deal with production and inventory planning for fashion products. Papers that focus on the impact of pricing and operational changes are examined next because we believe that the failure to adopt management science approaches might stem in part from inadequate attention to decisions that relate closely to production and inventory planning. Similarly, because implementing production and inventory planning approaches recommended by management scientists is complicated by the difficulty of estimating parameters of the demand distribution that are vital inputs to the proposed management science models, we review papers that address the estimation of these parameters.

25.2 Production and inventory planning

Planning and inventory production models for fashion products date to the 1960s. The models reviewed by us share many common features. First, all model demand uncertainty explicitly and are, hence, necessarily stochastic. Second, they consider a finite selling period at the end of which unsold inventory is marked-down in price and sold at a loss. In this sense, these models are similar to the classic newsboy model. Third, they model multiple production commitments such that sales information is obtained and used to update demand forecasts between planning periods. Production commitments in each of these periods is usually constrained by production capacity and minimum order quantities. The last two characteristics – finite selling periods and multiple production commitments – differentiate style goods inventory models from other stochastic inventory models.

We review papers that deal with production and inventory planning for fashion products in this section. We first review models that capture the inventory decision of a style goods manufacturer. The later part of this section focuses on the inventory decision for a style goods retailer.

25.2.1 Manufacturer inventory management

To understand the different aspects of inventory management for fashion products, consider the following model introduced in <u>Fisher and Raman</u> (1996). A fashion manufacturer that designs and manufactures n products in a given season makes two production commitments each year for product i. The first, denoted as x_{i0}, is made before any demand information is obtained, hence, forecasts are poor. The second production commitment, $x_i - x_o$ (x_i denotes total season production), is made after some demand information has been obtained, hence, the season demand forecast is much more accurate. We use x_0 to denote the vector of x_{i0} and x to denote the vector of x_i. Let D_{i0} represent the demand for product i observed before the second production commitment and D_i represent total units demanded for product i during periods 1 and 2 together (hence, $D_i - D_{i0}$ represents demand for product i during the first period alone). Understandably, the firm would like to postpone production commitments until later in the season when better demand forecasts can be obtained; however, its ability to do so is constrained by its second period production capacity K. This situation is represented graphically, in Figure 1:

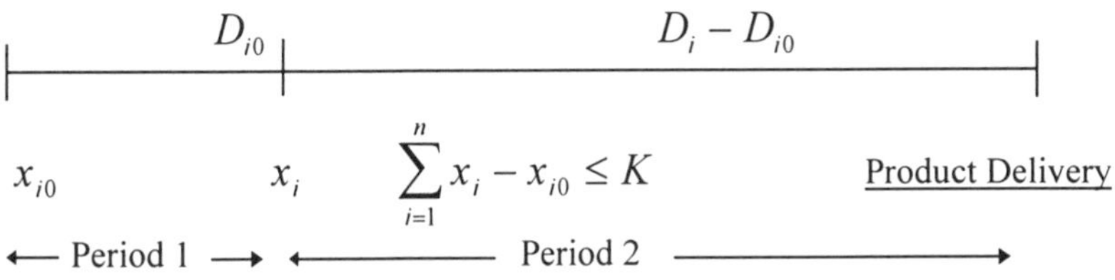

Figure 1: Manufacturer's Production Commitment

If O_i is the unit overproduction cost for product i and U_i the unit underproduction cost for product i, the firm's decision problem (**P**) at the beginning of period 1 can be written as:

$$\underset{x_0 \geq 0}{Min}\, Z(x_0) = E_{D_0}\, \underset{x \geq x_0}{Min}\, E_{D|D_0} \sum_{i=1}^{n}\left[O_i(x_i - D_i)^+ + U_i(D_i - x_i)^+\right]$$

$$\sum_{i=1}^{n} x_i \leq K + \sum_{i=1}^{n} x_{i0}$$

Notice that in model (**P**) the manufacturer makes the second production commitment (i.e., $x_i - x_0$) after observing demand during the first production period D_0. The authors present another model in the paper that incorporates production minimums. Their solution procedure consists of two steps. First they derive a revised problem in which the second period capacity constraint is replaced with a minimum production quantity constraint in the first period. It is easy to solve the revised problem using standard convex programming techniques since the marginal cost of producing an additional unit of a product can be computed numerically. The authors also estimate parameters for the demand distribution and demonstrate that their approach improved profits by 60% at a skiwear manufacturer.

Other authors have studied similar decision problems. In this section, we review papers by Murray and Silver (1966), Hausman and Peterson (1972), Bitran et al. (1986), and Matsuo (1990).

Murray and Silver, like Fisher and Raman, model the decision problem faced by a vendor planning to stock or produce an item with a finite selling season and uncertain demand. It is possible to acquire the item at various times (denoted by T_j) during the season and the cost of acquisition at each of these times is known in advance. In time period j (i.e., between T_{j-1} and T_j), the vendor observes sales, thus obtaining information about the market. The authors model demand uncertainty and forecast improvement as follows: in any period j, N_j customers purchase one of the items stocked by the vendor. The authors assume N_j to be known with certainty. Randomness in demand occurs due to p, the fraction of customers that will buy a particular item. The vendor does not know, for sure, what p is. As the season progresses and sales figures become known, the vendor obtains more nearly perfect knowledge about the numerical value of p. The authors formulate the decision problem as a dynamic program, which they suggest can be solved by mapping a range of demand onto a single quantity. They do not consider production capacity constraints. The authors' solution procedure would probably be difficult to implement with large problems.

Hausman and Peterson formulate a general mathematical model for the problem of scheduling, in a common production facility with limited production capacity, production quantities for a set of fashion items with seasonal, stochastic demand. Although their paper includes a multi-product formulation, the single-product version also presented in their paper is more easily understood, hence, we summarize the latter here.

The single-product formulation considers a production season that comprises n planning periods, each denoted by the subscript j. X_j denotes the forecast for total season demand, with the forecast made at the beginning of period j; $Z_j = \dfrac{X_{j+1}}{X_j}$ follows the log-Normal distribution. If Y_j is the initial inventory at the beginning of period j and K the production capacity for each period, the optimization problem at the beginning of period j can be written as:

$$g_j(X_j, Y_j) = Min_{Y_j < Y_{j+1} < Y_j + K} \left\{ \int_0^\infty g_{j+1}(X_j Z_j, Y_{j+1}) f_{LN}(Z_j | \mu_j, \sigma_j) dZ_j \right\} \text{ and}$$

$$g_{N+1}(X_{N+1}, Y_{N+1}) \quad = C_0(Y_{N+1} - X_{N+1}), \text{ if } Y_{N+1} > X_{N+1}$$
$$= C_u(X_{N+1} - Y_{N+1}), \text{ if } Y_{N+1} \le X_{N+1}.$$

The multi-product formulation is a generalized version of the above problem. The authors suggest three heuristic procedures, termed H-1, H-2, and H-3, for deriving the production quantity for each period. None accounts for forecast revisions. H-1 treats the remaining capacity as only that available in the current period; in other words, a target order-up-to level (i.e., it represents the solution to a capacity constrained newsboy problem with capacity equal to the capacity available in the particular period). H-2 obtains target order-up-to quantities for the various products by solving a capacity constrained newsboy problem with capacity equal to the total available capacity in the remaining periods. Actual production quantities for the particular period are determined from the target order-up-to quantities by assuming that the same fraction of the products is produced in each of the subsequent periods. H-3 is identical to H-2 except that the current period production is allocated among the various products in proportion to their currently perceived need.

The heuristics, although simple to implement, will probably not perform well under most circumstances because they fail to account for forecasts revisions. They also fail to account for the fact that forecast accuracy might be expected to improve more significantly for one product than for another after demand is observed, in which case it would be valuable to delay production of the product for which the forecast is likely to improve significantly upon observing early demand information. Moreover, the heuristics do not generalize easily to accommodate setup costs and minimum order quantities, common constraints in the fashion products manufacturing industry.

Bitran et al. take setup costs into account in an elegant manner. They use a two-level hierarchical structure to study the problem characterized by families and items. Production changeover costs from one family to another are higher than other costs, whereas setup costs between items in the same family are negligible. In deriving a solution method, the authors assume inter-family setup costs to be so high as to constrain families from being produced more than once during a given season. Hence, the authors restrict the production of each style to a single period during the production season. Having imposed this restriction, the solution procedure works at two levels. The first level determines the period in which each product family should be produced, the second determines the production quantities for each item of the various families that are to be produced in a given period. The authors demonstrate the efficacy of their approach using disguised data from a consumer electronics company in which their approach performed within 7% of the optimum. The approach will work reasonably well when setup costs are high enough to restrict production of a particular style to a single period, but will not generalize to situations in which this is not the case.

Matsuo addresses the same problem addressed by Bitran et al., but unlike Bitran et al. and other authors who have studied fashion production planning, his formulation does not require discrete time periods. He employs the same hierarchical structure and capacity restrictions seen in Bitran et al. and, like them, proposes a two stage production sequencing algorithm (in the first stage, a production sequence of families and the production start time of each family are determined at the start of the season while the production quantity for each item is determined in the second stage of the algorithm). Matsuo proposes a simple heuristic sequencing rule followed by a procedure for determining production quantities for the families. The sequencing rule is based on the ratio of inventory holding cost plus expected value of additional information per unit of time to expected resource consumption of a family. Given the production start time for each family, the quantity to be produced of each style is determined dynamically at the scheduled time using a variant of the newsboy formula.

All of the foregoing papers treat fashion inventory from the manufacturer's perspective. In these problems, the manufacturer is not expected to have the product when a retailer order is received, but needs to be able to deliver the product before the advent of the retailer's selling season. Thus, all these papers deal with the "terminal delivery" problem wherein the manufacturer incurs neither a penalty nor backorder cost if the product stocks out before the delivery date. Fashion retailers (both catalog and store) face a somewhat different problem in that they need to have inventory available when demand occurs, failing which they lose sales or incur a backorder cost.

25.2.2 Retailer inventory management

We review papers by Rajaram (1998), Bradford and Sugrue (1990), and Eppen and Iyer (1997a) in this section.

Rajaram studies the fashion retailer's problem in the context of a catalog retailer's operations. Like Fisher and Raman, he considers a two-stage planning problem. The retailer makes an initial production commitment at the beginning of period 1, when forecasts are poor, and a second production commitment after some sales data have been observed. But Rajaram's model, unlike the models considered so far in this chapter, assigns to the retailer a backorder or stockout cost if demand exceeds supply even before the end of the season. Figure 2 represents the problem graphically.

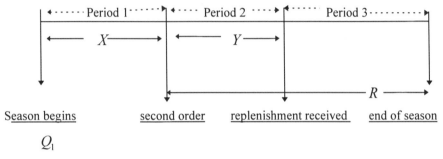

Figure 2: Retailer Production Commitments

If I represents the firm's inventory position at the beginning of period 2 immediately after a replenishment order has been placed, C_b denotes the cost of backordering demand during the season, C_u the cost of lost sales, and C_o the cost of overstocking inventory at end of season, the retailer's decision at the beginning of the first period can be written as:

$$\underset{Q_1 \geq 0}{Min}\, C(Q_1) = E_X\big[C_1(Q_1, X) + C_2(Q_1, X)\big],$$

where

$$C_1(Q_1, X) = C_b(X - Q_1)^+, \text{ and}$$

$$C_2(Q_1, X) = \underset{I \geq (Q_1 - X)^+}{Min}\left\{ \begin{bmatrix} C_b \min\{(Y - (Q_1 - X)^+)^+, (I - (Q_1 - X)^+)\} \\ + E_{R/X}\big[C_u(R - I)^+ + C_o(I - R)^+\big] \end{bmatrix} \right\}$$

The expression for the retailer's objective function has two components. The first component, $C_1(Q_1, X)$, represents the firm's backorder cost during the first period. The second cost component, $C_2(Q_1, X)$, comprises two parts: the first (i.e., $C_b \min\{(Y - (Q_1 - X)^+)^+, (I - (Q_1 - X)^+)\}$) refers to the expected backorder cost while the firm is waiting for replenishment from its supplier (i.e., in period 2), the second (i.e., $E_{R/X}\left[C_u(R-I)^+ + C_o(I-R)^+\right]$) to the expected understocking and overstocking cost incurred during period 3. The backorder costs incurred in period 2 reflect the firm's backordering of demand that it cannot meet from inventory, but can meet from replenishment units that will be received at the beginning of period 3. If $(Q_1 - X)^+ < Y < I$, the first $(Q_1 - X)^+$ units are satisfied from inventory and the retailer backorders the remaining $Y - (Q_1 - X)^+$ units of second-period demand. If $Y > I$, the first $(Q_1 - X)^+$ units will be satisfied from inventory, $I - (Q_1 - X)^+$ units will be backordered, and $(Y - I)$ units will be turned away. In other words, the retailer stops backordering demand if second-period demand (i.e., Y) exceeds I, since this demand cannot be satisfied even after replenishment is received at the beginning of period 3. Rajaram suggests a heuristic procedure to solve this problem and compares its performance to historical performance at the catalog retailer.

Bradford and Sugrue consider a two-period stocking policy model for a retailer. The retailer places an order at the beginning of period 1 and an opportunity for replenishment occurs at the beginning of period 2. Replenishment lead time is instantaneous and no capacity constraints are placed on order size. As in models that consider the manufacturer's inventory problem, the demand distribution at the beginning of period 1 is updated after period 1 demand is observed. The solution method involves iterating over values of first and second period inventory. The authors' enumerative procedure works well for small problems, but would be time-consuming for real world applications.

Eppen and Iyer consider the problem of buying fashion goods for a catalog retailer's ``big book.'' The retailer also issues "sale-catalogs" and owns outlet stores and thus has the opportunity to divert inventory originally purchased for the big book to other sources of distribution. The retailer gathers demand information as the season progresses and thus faces a lower level of demand uncertainty than at the beginning of the season. The authors construct a dynamic programming model that permits the decision maker to both "buy" and "dump" items at the beginning of each period. The sequence of decisions is formulated as a dynamic program. The solution method proposed by the authors applies to the problem in which no "buy" decisions are permitted after the first period. However, the decision maker may dump items at the beginning of any period. The heuristic proposed by the authors is intuitive, easy to implement, and performs well in computational comparisons.

25.2.3 Implementation in industry

It is clear from the foregoing papers that management scientists have made significant progress in modeling, and deriving heuristics to solve, the production and inventory planning problem for style goods. Their progress has been significantly aided by advances in computer technology that have enabled the application of computer-intensive algorithms to this problem. It is impressive, moreover, that management scientists have incorporated features such as setup times and production minimums in their models.

Why then are these models rarely applied to planning fashion inventory? In our descriptive study of retail merchandising practices over the last two years, we found that few fashion companies are even aware of, let alone use, these mathematical models for fashion inventory planning. Even the papers surveyed in this section mostly fail to document impact on management decisions. In other words, there are few demonstrated applications of these models and, at best, limited evidence of their ability to influence managerial decisions.

We believe this disconnect between good theoretical papers and practice on this topic stems from two sources. First, the surveyed papers focus on production and inventory planning to the exclusion of decisions (e.g., pricing and operational changes) that are closely related in practice. Inventory and pricing decisions, particularly decisions related to timing of markdowns, are linked closely to one another; the analysis in the literature needs to acknowledge this linkage. Second, the success of these models depends to a large extent on accurate estimation of various cost and demand distribution parameters. For example, most of these models require as inputs the costs associated with inventory holding and product stockout and the mean and variance of the underlying demand distribution. Yet few of the papers cited thus far consider how these parameters might be estimated, a vital ingredient for the successful implementation of management science models in fashion production and inventory planning.

25.3 Related decisions and parameter estimation

In this section, we survey first papers that deal with pricing and operational changes for fashion products and then papers that identify estimation techniques for various demand parameters. We do not survey papers that identify techniques for estimating stockout, obsolescence and inventory holding cost because these approaches, though important, have not, to our knowledge, attracted significant attention from management scientists.

25.3.1 Pricing

Merchants who determine the stocking quantity of individual stock-keeping units (SKUs) also determine how to price these SKUs at various stages of the product lifecycle. Understandably, pricing is affected by inventory on hand, expected demand, competitors' actions, and, especially when markdowns are offered, the price-elasticity of demand. Gallego and Van Ryzin (1994), Bitran and Mondschein (1997), and Subrahmanyan and Shoemaker (1996) have dealt with pricing and markdowns for fashion products.

Gallego and Van Ryzin model pricing over time for a fashion retailer that has a known quantity of short lifecycle product in stock. Demand (λ_s) for the product at time s is unknown, but influenced by price (p_s); that is, λ_s is a function of p_s. The authors investigate the problem of dynamically adjusting price to maximize sales revenue (which, because all costs are sunk in their model, is equivalent to maximizing profits) when demand is price-sensitive and stochastic. They assume that no replenishment of product is possible. Mathematically, the problem is formulated as finding the pricing policy, u, that maximizes $J_u(n,t) = E_u\left[\int_0^t p_s\, dN_s\right]$, where

N_s denotes the number of items sold up to time s. Using this formulation, the authors derive an upper bound on expected revenue based on a deterministic version of the problem. They use this upper bound to evaluate the performance of some simple, fixed-price policies, which they show to be asymptotically optimal as the volume of expected sales tends towards infinity. Surprisingly, in what we consider the central contribution of this paper, these simple fixed-price policies perform very well in numerical tests even when expected demand is not very high.

Bitran and Mondschein examine the performance of a continuous review model (similar to that of Gallego and Van Ryzin) and compare it to that of a periodic pricing model in which prices are allowed to change at periodic intervals (i.e., the retailer has only a limited number of opportunities to alter selling price). The authors study the impact of varying demand uncertainty, time horizon, and initial inventory on a firm's pricing policies. They note that greater demand uncertainty leads to higher initial prices, more substantial discounts, and greater volume of unsold inventory at the end of the product lifecycle. Not surprisingly, they also find that higher initial inventory and shorter planning horizons lead to lower prices.

Subrahmanyan and Shoemaker formulate a stochastic, dynamic program in which a retailer jointly determines both pricing and inventory policies. Although their model is fairly general, their proposed solution procedure is computationally intensive and would not translate to larger problems. Their formulation of the retailer's decision problem (with both inventory and pricing determined concurrently) captures an important aspect of retail inventory management.

The literature on pricing for style goods can be criticized on two counts. First, as none of the models has been applied to real companies, it remains to be seen how the proposed methods would work in practice. A role model for a proposed application would be the approach described in Smith et al. (1992), which describes American Airlines' Yield Management system. Second, the papers have not adequately connected pricing decisions with stocking and replenishment decisions. Forecast updating, a theme that takes center stage in the style goods production planning literature surveyed earlier, is almost completely ignored by the fashion pricing literature.

25.3.2 Operational changes

Operational changes such as reducing lead time also affect production and inventory planning for fashion products. The degree to which a fashion company can make production and inventory commitments after observing sales information affects its production plans and stockout and obsolescence costs. That early sales data are a good predictor of later demand has been noted by, among others, Fisher et al. (1994). Not surprisingly, a company that can make production and inventory commitments after early sales data have been obtained is likely to achieve superior inventory performance.

Fisher et al. (1997) identify numerous ways to increase a firm's reactive capacity. For example, firms can reduce transportation and factory throughput times, approaches emphasized in the Quick Response movement in the apparel industry (see Hammond (1992) for an overview). Other approaches to augmenting reactive capacity, such as increasing production capacity in the factory and holding raw material inventory, have not received equivalent attention. Given the difficulty of understanding and quantifying the bottom-line impact of such operational changes, management science models that incorporate that impact would greatly benefit managers in the fashion industry.

Eppen and Iyer (1997b) study the impact of a very interesting arrangement, termed a "backup agreement," between a catalog retailer and its suppliers. Backup agreements permit retailers to reorder, after observing initial demand, up to a certain pre-specified "backup quantity" from a supplier that holds it as inventory. The authors find that both retailer and supplier performance improves for a range of parameters with the use of backup agreements. Intuitively, backup agreements might increase retailer's profits given that they permit retailers to make inventory commitments after receiving market signals and, hence, are similar to reactive capacity as described in Fisher et al. The authors explain the counter-intuitive notion that such agreements might increase supplier profits by noting that suppliers' cost of manufacturing (and hence, overstocking cost) is much lower than retailers' purchase cost. Thus, in their model, backup agreements foster better contracts between suppliers and retailers giving rise to better channel performance.

Raman and Kim (1997), using data from a school-uniform manufacturer, document the impact of working capital cost on production and inventory planning for fashion products. Expensive working capital incurs high inventory carrying cost which induces a firm to make smaller production commitments early in the season (since early production incurs higher inventory carrying cost) at the expense of higher stockouts later in the season. This insight is particularly valuable to small firms and firms in developing countries for which working capital costs tend to be high (see for example, Raman (1996)).

25.3.3 Demand parameter estimation

All management science models that deal with fashion inventory require demand density as an input. This is understandable given that demand for fashion products tends to be extremely unpredictable. Few of the papers that develop these models, however, offer methods for estimating demand distribution; most assume the distribution to be known. This, in our opinion, is the Achilles' heel of fashion production and inventory planning approaches. Much work is needed in this area, and the applicability of management science models is constrained by the dearth of techniques for estimating parameters that are important to them.

The traditional approach to estimating demand distributions consists of using historical demand data for the *same* product. This approach has limited value for fashion products, which, being by definition short lifecycle products, lack historical demand data. Although historical demand data for *similar* products have been, and continue to be, used to estimate some parameters in the demand distribution, other techniques (e.g., expert judgment) must be used to estimate other parameters. To be useful for estimating demand distributions for fashion products, historical data must be analyzed creatively. Fisher and Raman (1996), for example, use historical data to estimate the correlation between early and season demand and the fraction of lifecycle demand observed up to a certain time, but use expert forecasts to estimate the mean and standard deviation of demand for individual products.

Three major demand parameter estimation problems have considerable relevance for fashion production and inventory planning:

1. estimating standard deviation of demand,
2. estimating mean demand when sales are observable but demand is not, and
3. updating forecasts after early sales have been observed.

Estimating standard deviation is difficult in the absence of historical data. Techniques that rely on experts to estimate the standard deviation or, equivalently, the probability distribution of demand run into difficulty because experts are poorly calibrated. Alpert and Raiffa (1982) have shown that even experts who have been exposed to one or more courses in statistics grossly underestimate standard deviation. What is more troubling is that the extent to which experts underestimate standard deviation varies from situation to situation and, hence, their estimates cannot be scaled up in a simple way. The problem is compounded in the fashion industry, most fashion experts never having been exposed to formal training in statistics and being unfamiliar with terms such as probability and standard deviation.

A few authors have demonstrated success in estimating standard deviation from expert forecasts. Raman and Fisher (1998), who estimate standard deviation of demand from the variance of expert forecasts, find that the standard deviation of a random variable can be estimated using the standard deviation among expert forecasts. If θ_{ij} is expert $j's$ forecast for random variable X_i, then they show that

$Var(X_i)$ can be estimated by $\sum_{i=1}^{n} k(\theta_{ij} - \overline{\theta}_i)^2$, where $\overline{\theta}_i$ is the average of n different θ_{ij}, and k is a constant estimated using historical data. Shleifer (1993) relies on experts to classify products as either "new" or "never-outs" and uses historical demand data to estimate standard deviation estimates for each category.

Estimating mean demand retrospectively, though trivial at first glance, is complex because retailers (excepting catalog retailers), being unable to observe demand, must infer demand levels from sales data. In the newsboy problem, for example, a newsboy who sells all 100 newspapers stocked on a particular day knows sales (=100), but cannot estimate demand precisely (knowing only that it exceeded 100 units). If D_i is the demand on day t and I_t the inventory available on day t, then S_t the sales on day is equal to $\min(D_t, I_t)$. Most retailers lack formal mechanisms for extracting demand parameters from sales data.

Among management science papers that have examined the problem of estimating demand from sales and inventory data, Wecker (1976) was probably the first to point out that using sales data to estimate demand could produce erroneous results. More recently, Nahmias (1994) and Agrawal and Smith (1996) have studied this problem when demand follows either the Normal or the Negative-Binomial distribution. Both papers provide techniques for estimating demand parameters under the restrictive assumption that starting inventory level on each day is constant. This limits the applicability of this idea. To be relevant, we need to develop approaches that make fewer restrictive assumptions and are easy to implement in real-world situations.

Fashion inventory models exploit the fact that forecasts become much more accurate after some sales data has been observed. Hence, forecast updating techniques play an important role in the application of fashion inventory models. Murray and Silver (1966), Chang and Fyffe (1971), and Guerrero and Elizondo (1997) are among a number of authors who have proposed forecast updating procedures. Guerrero and Elizondo describe and then, using data from multiple sources, compare the performance of a number of approaches for updating forecasts. They categorize forecast updating approaches into "multiplicative" (e.g., the naïve multiplicative model scales up early demand by a constant factor to derive the forecast for total demand), "additive" (e.g. the naïve additive model adds a constant to the early demand observation to derive the forecast for total demand), and approaches that combine features of both the foregoing approaches (e.g., if D_t denotes the total accumulated demand value at time t, $D_{t,k}$ denotes the total unaccumulated value observed k time steps before the end of period t, and

$$D_t^k = \sum_{i=0}^{L-k} D_{t,L-i},$$

then a general forecasting formula is given by $D_t = \beta_{k0} + \beta_{k1}D_t^k$). The authors suggest variations on these approaches for variables with "non-constant levels" (i.e., data that demonstrate trends over time). For example, the impact of trend can be captured by modifying the above formula to obtain $D_t = \beta_{k0} + \beta_{k1}D_t^k + \beta_{k2}t$. Although the paper deserves to be commended for comparing the performance of a number of different approaches using real data, it is unclear under what circumstances particular approaches might perform better than others.

25.4 Conclusion

The problem of managing fashion inventory has generalized to many industries, including, as we noted earlier, personal computers, toys, and agricultural chemicals. Moreover, the costs associated with fashion inventory management – stockouts, inventory holding, and markdowns or obsolescence—have become significant for these businesses.

Were management scientists to develop suitable decision support tools to aid inventory management, the considerable complexity of these decisions might be mitigated and decision making improved substantially. A review of the literature reveals that management scientists have made significant progress in addressing a number of issues that affect fashion inventory management. Further progress will likely rely on techniques such as stochastic optimization, dynamic programming, and statistical estimation, in which management scientists have considerable expertise.

We attribute the lack of implementation and adoption of the models that have been developed to date at least partially to the failure of management scientists to integrate pricing, operational changes, scheduling, and inventory levels in their decision models and to identify techniques for estimating parameters in the demand distribution. To have an impact on this problem, more of us in the management science community must seek out real-world instances of it, and implement models that integrate its various facets. Such efforts are needed to bridge the gap between theoretical models and real-world practice in fashion inventory management.

Endnotes

[1] A distinction is made between products such as apparel and personal computers that have short lifecycles and products such as tomatoes that have a long lifecycle, but, as individual units of inventory, short lives. Some papers have used the term "style goods" to refer to short lifecycle or "fashion" products.

[2] Details on the "Harvard-Wharton Retail Merchandising Effectiveness Project," can be found in "In Search of Rocket-Science Retailing: Proposal to the Sloan Foundation," by Marshall Fisher and Ananth Raman, *Harvard Business School Working Paper*.

References

[1] Agrawal, Narendra and Stephen Smith. "Estimating Negative Binomial Demand for Retail Inventory Management with Unobservable Lost Sales." *Naval Research Logistics* 43, 1996.

[2] Alpert, W. and Howard Raiffa. "A Progress Report on the Training of Probability Assessors." in Daniel Kahneman and Amos Tversky eds. *Judgment Under Uncertainty: Heuristics and Biases*, Cambridge University Press, 1982.

[3] Bitran, Gabriel R., Elizabeth Haas and Hirofumi Matsuo. "Production Planning of Style Goods with High Setup Costs and Forecast Revisions." *Operations Research* 34(2), March-April 1986.

[4] Bitran, Gabriel R., and Susana V. Mondschein. "Periodic Pricing of Seasonal Products in Retailing." *Management Science* 43(1), January 1997.

[5] Bradford, John W. and Paul K. Sugrue. "A Bayesian Approach to the Two-period Style-goods Inventory Problem with Single Replenishment and Heterogeneous Poisson Demands." *Journal of Operations Research Society* 41(3), 1990.

[6] Eppen, Gary D. and Ananth V. Iyer. "Improved Fashion Buying using Bayesian Updates." *Operations Research* 45(6), November-December 1997a.

[7] Eppen, Gary D. and Ananth V. Iyer. "Backup Agreements in Fashion Buying -- The Value of Upstream Flexibility." *Management Science* 43(11), November 1997b.

[8] Fisher, Marshall, Janice Hammond, Walter Obermeyer, and Ananth Raman. "Configuring a Supply Chain to Reduce the Cost of Demand Uncertainty." *Journal for Production and Operations Management Society*, Special Issue on Supply Chain Coordination and Integration 6(3), Fall 1997.

[9] Fisher, Marshall and Ananth Raman. "Reducing the Cost of Demand Uncertainty through Accurate Response to Early Sales." *Operations Research* 44(4), January-February 1996.

[10] Fisher, Marshall, Janice Hammond, Walter Obermeyer, and Ananth Raman. "Making Supply Meet Demand in an Uncertain World." *Harvard Business Review* 72(3), May/June 1994.

[11] Gallego, Guillermo and Garrett Van Ryzin. "Optimal Dynamic Pricing of Inventories with Stochastic Demand over Finite Horizons." *Management Science* 40(8), August 1994.

[12] Guerrero, Victor M. and J. Alan Elizondo. "Forecasting a Cumulative Variable Using Its Partially Accumulated Data." *Management Science* 43(6), June 1997.

[13] Hammond, Janice H. "Quick Response in the Apparel Industry." *Harvard Business School Note* No. 690-038, 1992.

[14] Hausman, Warren H. and Rein Peterson. "Multiproduct Production Scheduling for Style Goods with Limited Capacity, Forecast Revisions and Terminal Delivery." *Management Science* 18(7), March 1972

[15] Matsuo, Hirofumi. "A Stochastic Sequencing Problem for Style Goods with Forecast Revisions and Hierarchical Structure." *Management Science* 36(3), 1990.

[16] Murray, George R. Jr. and Edward A. Silver. "A Bayesian Analysis of the Style Goods Inventory Problem," *Management Science* No. 11, July 1966.

[17] Nahmias, Stephen. "Demand Estimation in Lost Sales Inventory Systems." *Naval Research Logistics* 41, 1994.

[18] Rajaram, Kumar. "Merchandising Planning in Fashion Retailing: Models, Analysis and Applications." *Unpublished Ph.D Dissertation at the Wharton School*, submitted April 1998.

[19] Raman, Ananth and Marshall Fisher. "Estimating Uncertainty in Judgmental Forecasts." *Harvard Business School Working Paper*, 1998.

[20] Raman, Ananth and Bowon Kim. "Impact of Inventory Holding Cost on Production and Inventory Planning for Fashion Products." *Harvard Business School Working Paper*, 1998.

[21] Raman, Ananth. "Apparel Exports and the Indian Economy." *Harvard Business School Case No.* 696-065, 1996.

[22] Schleifer, Arthur. "L.L. Bean." *Harvard Business School Case* No. 893-003, 1993.

[23] Smith, Barry C., John F. Leimkuhler and Ross M. Darrow. "Yield Management at American Airlines." *Interfaces* 22(1), January-February 1992.

[24] Subrahmanyan, Saroja and Robert Shoemaker. "Developing Optimal Pricing and Inventory Policies for Retailers Who Face Uncertain Demand." *Journal of Retailing* 72(1), 1996.

26 INVENTORY CONTROL FOR JOINT MANUFACTURING AND REMANUFACTURING

E.A. van der Laan[1], M. Fleischmann[2], R. Dekker[3], and L.N. Van Wassenhove[1]

[1] Technology Management Area
INSEAD
77305 Fontainebleau Cedex
France

[2] Faculty of Business Administration
Erasmus University Rotterdam
PO Box 1738, 3000 DR Rotterdam
The Netherlands

[3] Faculty of Economics
Erasmus University Rotterdam
PO Box 1738, 3000 DR Rotterdam
The Netherlands

26.1   INTRODUCTION

The growing environmental burden of a 'throw-away-society' has made apparent the need for alternatives to landfilling and incineration of waste. Opportunities have been sought to reintegrate used products and materials into industrial production processes. Recycling of waste paper and scrap metal have been around for a long time. Collection and reuse of packages and recovery of electronic equipment are more recent examples. Efforts to efficiently reuse products and/or materials have introduced a wide range of novel and complex issues that affect the complete supply chain of recoverable products. Supply chain management in the light of product reuse is what we refer to as *'Reverse Logistics Management'*.

What makes efficient reuse even more complicated is the fact that the bulk of products that appear on the market are not designed for reuse. Product design is crucial since it determines to a large extent whether products and components can be easily disassembled, cleaned, tested and repaired if necessary. Nevertheless, product recovery has been successfully implemented for a wide variety of products. This indicates that even under imperfect conditions product recovery can be beneficial from an ecological point of view as well as from an economical point of view.

A specific type of product recovery is *product remanufacturing*. Product remanufacturing is the process that restores used products or product parts to an 'as good as new' condition, after which they can be resold on the market of new products. Examples of products that are actually being remanufactured in practice are automobile parts, commercial and military aircraft, diesel, gasoline and turbine engines, electronic equipment, machine tools, medical equipment, and railroad locomotives. The industrial operations involved with remanufacturing, like disassembly, testing, cleaning, repair, overhaul and refurbishing, are of a very stochastic nature due to the uncertainty in timing, quantity and quality of returned products (see [8, 28]). This results in a large uncertainty regarding the availability of inputs, and highly variable processing times.

Since remanufacturing restores a used product to an 'as good as new' condition it can serve as an alternative input resource in the fabrication of new products, but also *vice versa*. The former situation applies for instance to the electronics industry, where returned modules can be reused in new products. The latter may apply to the automobile industry (see [29]), where spare parts are made out of used parts and manufacturing of new parts is only used at times that the supply of remanufacturable parts is too low. A general framework for the situation in which demand can be supplied by both manufacturing and remanufacturing is given in Figure 26.1.

In this chapter we focus our attention on quantitative models for inventory control and production planning that apply to the situation depicted in Figure 26.1. Note that we are only concerned with simultaneously controlling the remanufacturable inventory, the serviceable inventory, the manufacturing source and the remanufacturing source, to satisfy *end item* demand. We do not address the issues related to shop floor scheduling and capacity planning

Figure 26.1 A hybrid system with manufacturing and remanufacturing operations, and stocking points for remanufacturables and serviceables.

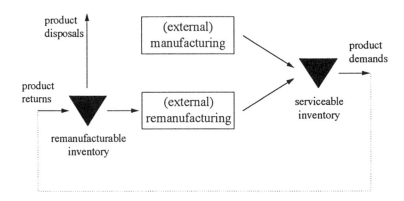

for (re)manufacturing operations (see [3, 7, 8, 9, 10, 11, 15, 16, 27]). This means, for example, that we will not enter the discussion whether and how to use MRP (Material Requirements Planning) logic in a remanufacturing environment. For one view on this discussion we refer to [8]. For a general overview of quantitative models for reverse logistics we refer to [4].

We feel that any quantitative model that addresses our situation should at least contain the following characteristics:

I End item demand can be satisfied from two sources, i.e., the manufacturing source and the remanufacturing source, both having quite different characteristics (see II and III).

II In contrast with the manufacturing facility, the input of the remanufacturing facility is characterized by limited and random availability. This is due to the uncertainty in the *timing* of the returns, the *quantity* of the returns, and the *quality* of the returns.

III Due to the uncertainty in the *quality* of subassemblies and components, the operations with respect to disassembly, cleaning and repair are highly variable. Therefore, the remanufacturing facility is typically characterized by substantial and highly variable processing times.

The above indicates that we cannot just aggregate the manufacturing source together with the remanufacturing source, since both have quite different characteristics. The remanufacturing source may be seen as more unreliable than the manufacturing source, due to its limited availability of inputs and its variable holding costs. Since remanufacturing lead-times may be substantial, they need to be modeled explicitly. Assuming zero lead-times would be very inappropriate.

The complexities and the various process interactions that are prominent in a joint manufacturing/remanufacturing facility make efficient control a difficult task. Since little is known about the structure of the optimal policy, the main objective of this chapter is to obtain more insights in this issue.

The remainder of this chapter is organized as follows. In Section 26.2 we investigate if there are quantitative models in the literature that, although not originally developed for product remanufacturing, capture the above characteristics of a joint manufacturing/remanufacturing facility. Since the answer to this question appears to be negative, we present a quantitative framework that does capture the above characteristics (Section 26.3). The framework enables us to investigate the influence of system parameters, such as lead-times, the holding costs structure, and the return rate, on system performance under heuristic control policies. This may give us some clues about the structure of the optimal control policy and which information it should make use of (Section 26.4). We end this chapter with a summary and concluding remarks in Section 26.5.

26.2 RELATED MODELS THAT DO NOT APPLY TO REMANUFACTURING

A considerable number of articles in the production planning and inventory control literature have explicitly modeled both a return process and a demand process. Some of these articles deal with situations that appear to have a clear analogy with product remanufacturing. This literature comprises articles on repair and spare parts models, articles on traditional two-source models, and articles, originating from finance, on the so-called cash-balancing models. This tempts some people to believe that remanufacturing issues have already been studied extensively and therefore any recent literature that specifically deals with remanufacturing is redundant and obsolete. We show that this belief is false by reviewing the models in question, indicating why they do not capture the unique characteristics of a situation with joint manufacturing/remanufacturing. Recall that these characteristics are (i) two exogeneous supply sources, that is manufacturing and remanufacturing, (ii) time-dependent availability of the remanufacturing source, and (iii) non-zero stochastic lead-times for remanufacturing.

26.2.1 Repair and spare parts models

The literature dealing with spare parts management and repair is quite extensive. The models considered all apply to a closed system consisting of a fixed number of parts or products that are subject to failure. If a part or product fails it needs to be replaced by a spare part. If possible, the failed part is then repaired in order to satisfy future demand. Otherwise parts are procured from outside. The objective is to determine the number of spare parts in the system for which total system costs are minimized and/or for which prespecified service levels are guaranteed. An example is the well-known family of METRIC models introduced by [25].

Unfortunately, the above models do not apply to remanufacturing. A common assumption in these models is that demands for new products are generated by product failures only, i.e., product demands and product returns are *perfectly* correlated. This is not the case for remanufacturing, where product returns may or may not be caused by a product replacement. Even in case of a product replacement it may not be clear when the used product will be available for remanufacturing. The product need not be returned immediately after use, or the collection process may delay and randomize its time of arrival to the remanufacturing company. In general it may even be more appropriate to assume complete *independence* between the demand and return flows, than to assume correlation.

Another difference between models for spare parts management and models for remanufacturing lies in the objective: with spare parts management the objective is to determine the *fixed* number of spare parts in the system, such that the associated long-run average costs are minimized. With remanufacturing the objective is to develop a policy on when and how much to remanufacture, dispose, and produce, such that some cost function is minimized. Essential in remanufacturing is that the number of products in the system may vary over time.

Reviews on spare parts and repair management can be found in [23, 20, 2, 17].

26.2.2   Cash-balancing models

Another type of models that show some resemblance with remanufacturing stem from finance: the cash-balancing models. These models usually consider local cash of a bank with incoming money flows (customer deposits), and outgoing money flows (customer withdrawals). Local cash can be increased by ordering money from central cash, or decreased by transferring money to central cash. The objective is to determine the time and quantity of the cash transactions, such that the sum of fixed and variable transaction costs, backlogging costs, and interest costs related to the local cash are minimized. There exist continuous review and periodic review cash-balancing models.

The reason why we feel these models are not suitable for remanufacturing is that their analysis does not handle storage of returns (customer deposits) and lead-times for (re)manufacturing operations. Although storage of returns makes no sense from the perspective of cash-balancing, it is typical for remanufacturing systems. Additionally, the relation between the manufacturing and remanufacturing lead-times appears to be crucial. Therefore we believe that one should favor an approach that can handle a more detailed modeling of the remanufacturing source.

An extensive overview of cash-balancing models is given by [13].

26.2.3   Classical two-source models

Inventory models in which outside procurement orders may be placed at two different suppliers, so called two-source models (see e.g. [1, 18, 34]), show some analogy with hybrid remanufacturing systems in which product demand may be served by both manufacturing (supplier one) and remanufacturing (supplier two). Usually these models deal with the situation in which there is the regular supplier and an alternative supplier that offers smaller lead-times against a higher price.

The reason why these models are not suitable for remanufacturing lies in the fact that both suppliers are assumed to be continuously available. In remanufacturing though, the availability of the remanufacturing source varies over time, since it depends on the uncertain flow of product returns. As a consequence, the two types of models make different trade-offs in optimizing their associated costs. In classical two-source models the faster source is chosen if inventories are running dangerously low. In remanufacturing models the manufacturing source is chosen at times that a replenishment should occur but the remanufacturing source is not available. If disposal of excess remanufacturable products is allowed, one also has to make a trade-off between manufacturing and remanufacturing costs. However, instead of bringing the models closer together, the option of disposal introduces even more complexity in terms of control policy structure and system analysis.

A small subgroup of the multiple-source models considers random availability of suppliers ([12, 21]). However, these models contain simplifying assumptions such as deterministic demand, zero lead-times, and equal characteristics for all suppliers, so they do not match with our criteria.

26.3   AN ANALYTICAL FRAMEWORK FOR JOINT MANUFACTURING AND REMANUFACTURING

In this section we follow the analytical framework of [29], who developed an *exact* procedure to study the system in Figure 26.1. This framework satisfies our three criteria and it allows to study various control policies under different conditions, such as stochastic lead-times, correlation between the return and demand flows, and Coxian-2 distributed return and demand flows. Formally, the characteristics of the system studied are as follows:

The *demand and return processes* are stochastic and may be modeled by any Markovian arrival process, although it is common practice to assume (compound) Poisson arrivals. In a straightforward manner we may use Coxian-2 arrival processes. These processes enable us to do a three moment fit of an arbitrary arrival process. If disposal of remanufacturable products is not allowed, we assume that the return intensity λ_R (the average number of returns per unit of time) is smaller than the demand intensity λ_D.

Testing process. It is assumed that every returned item is already tested and satisfies the quality requirements for remanufacturing.

The *remanufacturing process* has unlimited capacity. The remanufacturing lead-time L_r, which is the time that passes between the time at which a remanufacturing order is released to the remanufacturing facility and the time of actual delivery, is a random variable with mean μ_{L_r} and variance $\sigma^2_{L_r}$. The fixed remanufacturing costs are c_r^f per batch, and the variable remanufacturing costs are c_r^v per product. A remanufacturing release moves a batch of remanufacturables from the remanufacturable inventory to the remanufacturing work-in-process (WIP). After remanufacturing all products immediately enter the serviceable inventory.

The *manufacturing process* has unlimited capacity. The manufacturing costs consist of a fixed component c_m^f per order, and a variable component of c_m^v per product. The manufacturing lead-time L_m, which is the time that passes between the time at which a manufacturing order is placed and the time of actual delivery, is a random variable with mean μ_{L_m} and variance $\sigma^2_{L_m}$. Manufactured products enter the serviceable inventory.

The *inventory process* consists of two stocking points: one to keep remanufacturable inventory and one to keep serviceable inventory. The holding costs in remanufacturable inventory are c_r^h per product per time-unit, and the holding costs in serviceable inventory are c_s^h per product per time-unit. Although there is quite some controversy with respect to the valuation of these holding cost parameters, it is reasonable to assume that $c_r^h \leq c_s^h$. If disposal is not allowed, both stocking points have unlimited capacity.

If disposals are allowed, the *disposal process* depends on the actual control policy employed. Variable disposal costs are c_d^v per disposed product and fixed disposal costs are c_d^f.

Customer service: demands that cannot be fulfilled immediately are backordered against backorder costs c_b per product per unit of time.

Under some predefined control policy $\mathcal{P}$ with decision vector V, the long-run average system costs per unit of time, denoted $\overline{C}_\mathcal{P}(V)$, are the summation of the following components:

c_s^h	$\times$	average serviceable inventory per time unit
c_r^h	$\times$	average remanufacturable inventory per time unit
c_r^v	$\times$	average number of remanufactured products per time unit
c_r^f	$\times$	average number of remanufacturing batches per time unit
c_m^v	$\times$	average number of manufactured products per time unit
c_m^f	$\times$	average number of manufacturing batches per time unit
c_b	$\times$	average backordering position per time unit
c_d^v	$\times$	average number of disposed products per time unit
c_d^f	$\times$	average number of disposal batches per time unit

The objective is to choose the parameters of policy $\mathcal{P}$ such that total long run average costs per unit of time are minimized. However, it is still unclear what $\mathcal{P}$ should be, since the structure of the optimal policy is unknown. To start our investigations we therefore have to rely on heuristic control policies.

26.3.1 Definition of control policies

The two heuristic control policies that we have chosen to implement are all natural extensions of the classical (s, Q) policy. While manufacturing orders are controlled by an (s, Q) policy, remanufacturing batches are either being pushed through the remanufacturing facility, or they are being pulled when they are really needed. Although we suspect that the optimal policy will be a mixture of push and pull we look at two extremes. The next two paragraphs formally define our policies.

Figure 26.2 A schematic representation of the PUSH policy.

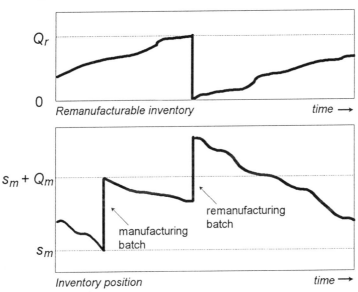

The (s_m, Q_m, Q_r) **PUSH policy.** Remanufacturing starts whenever a batch of returned products of size Q_r is available at the stocking point for remanufacturables. The remanufacturing order will arrive at the on-hand serviceable inventory after L_r time units. Manufacturing of Q_m products starts whenever the serviceable inventory position drops to the level s_m. At that time, the inventory position is increased to $s_m + Q_m$. The manufacturing order will arrive at the on-hand serviceable inventory after L_m time units (see Figure 26.2).

This policy is named *PUSH* policy, since remanufacturable inventory *pushed* through the remanufacturing process as soon as possible.

The (s_m, Q_m, s_r, S_r) PULL policy. Remanufacturing starts whenever t serviceable inventory position is at or below s_r, and sufficient remanufactura inventory exists to increase the serviceable inventory position to S_r. The manufacturing batch is exactly the amount that is necessary to increase the ventory position to S_r. If we denote this amount by Q then a remanufactur order decreases the remanufacturable inventory with Q products, and increa the remanufacturing WIP and the inventory position with Q products. T remanufacturing order will arrive at the on-hand serviceable inventory af L_r time units. Manufacturing of Q_m products starts whenever the servicea inventory position drops to the level s_m. At that time, the inventory posit is increased to $s_m + Q_m$. The manufacturing order will arrive at the on-ha serviceable inventory after L_m time units (see Figure 26.3).

This policy is named *PULL* policy, since remanufacturable inventory is *pu* into the remanufacturing process only when needed to fulfill customer dema for serviceables. Note that s_m should always be smaller than s_r, since oth wise the remanufacturing option would never be chosen, and remanufactura inventory would accumulate infinitely.

Figure 26.3 A schematic representation of the PULL policy.

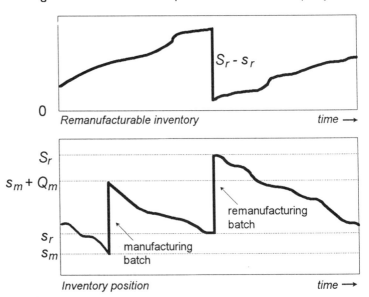

26.3.2 Mathematical analysis

Next we outline a procedure to calculate the long run average costs $\overline{C}_P(V)$. The notation used in this outline is specified in Table 26.1.

Table 26.1 Notation used for the analysis.

$I_s^{net}(t)$	$=$	The net serviceable inventory at time t, defined as the number of products in on-hand serviceable inventory minus the number of products in backorder at time t
$I_s(t)$	$=$	The serviceable inventory position at time t, defined as the net serviceable inventory plus the number of products in manufacturing work-in-process plus the number of products in remanufacturing work-in-process
$I_r^{OH}(t)$	$=$	The number of products in remanufacturable on-hand inventory at time t
$W_m(t)$	$=$	The number of products in manufacturing work-in-process at time t
$W_r(t)$	$=$	The number of products in remanufacturing work-in-process at time t
$D(t_0, t_1)$	$=$	The demands in time interval $(t_0, t_1]$
$Z(t - t_0, t - t_1)$	$=$	The number of products ordered to be (re)manufactured in the interval $(t - t_0, t - t_1]$ that enter serviceable inventory at or before time t minus the demands in the interval $(t - t_0, t - t_1]$
$\overline{B}$	$=$	The long-run average backordering position
$\overline{I}_s^{OH}$	$=$	The long-run average on-hand serviceable inventory
$\overline{I}_r^{OH}$	$=$	The long-run average on-hand remanufacturable inventory
$\overline{O}_r$	$=$	The long-run average number of remanufacturing orders
L^{min}	$=$	The minimum of all possible realisations of the manufacturing and remanufacturing lead-time
$L^{max}(< \infty)$	$=$	The maximum of all possible realisations of the manufacturing and remanufacturing lead-time

Under the assumption that the control policy only depends on the inventory position and/or the remanufacturable inventory, the state transitions of the joint manufacturing/remanufacturing system at hand can be formulated as a continuous time Markov chain. This Markov chain, say $\mathcal{M}$, has a two-dimensional state variable $X(t) = \left(I_s(t), I_r^{OH}(t)\right)$ with discrete state space $S = \{I_s(t)\} \times \{I_r^{OH}(t)\}$. By definition, $X(t) = (i_s, i_r^{OH}) \in S$ whenever $I_s(t) = i_s$ and $I_r^{OH}(t) = i_r^{OH}$.

The limiting joint probability distribution $\pi(i_s, i_r^{OH})$, defined as

$$\pi(i_s, i_r^{OH}) = \lim_{t \to \infty} \Pr\{I_s(t) = i_s, I_r^{OH}(t) = i_r^{OH}\} \tag{26.1}$$

is obtained by solving the balance equations (inflow equals outflow) that are associated with $\mathcal{M}$. Although these balance equations are usually easy to write down, it is generally quite difficult to find a closed form expression for (26.1). Therefore, we have to rely on numerical procedures instead. More complicated is the calculation of the average on-hand serviceable inventory and the average backorder position, i.e.,

$$\overline{I}_s^{OH} = \sum_{i_s^{net} > 0} i_s^{net} \lim_{t \to \infty} \Pr\{I_s^{net}(t) = i_s^{net}\}, \tag{26.2}$$

and

$$\overline{B} = - \sum_{i_s^{net} < 0} i_s^{net} \lim_{t \to \infty} \Pr\{I_s^{net}(t) = i_s^{net}\}. \tag{26.3}$$

These cannot be calculated directly using a Markov chain formulation, since the transitions of I_s^{net} are not Markovian. However, we can derive an expression for $I_s^{net}(t)$ from which we can calculate its long run distribution. By definition, we have

$$I_s^{net}(t) = I_s(t) - W_m(t) - W_r(t). \tag{26.4}$$

Additionally, we have the following relation for the net inventory process:

$$
\begin{aligned}
I_s^{net}(t) = \ & I_s^{net}(t - L^{max}) \\
& + \text{every ordered product that is delivered in } (t - L^{max}, t] \\
& - \text{every demand that arrives in } (t - L^{max}, t]
\end{aligned}
\tag{26.5}
$$

Here we mean by 'every ordered product' all products that were ordered for manufacturing *and* remanufacturing.

The number of ordered products that arrive in $(t - L^{max}, t]$ can be split into two groups: The products that are in (re)manufacturing WIP at time $t - L^{max}$ (these will all arrive before or at time t, since lead-times are never larger than L^{max}), and the number of products that were ordered during the interval $(t - L^{max}, t]$ *and also* arrived before time t. So, (26.5) becomes

$$I_s^{net}(t) = I_s^{net}(t - L^{max})$$

$\qquad +$ every product in (re)manufacturing WIP at time $t - L^{max}$

$\qquad +$ every product that is both ordered *and* delivered

$\qquad\qquad$ in $(t - L^{max}, t]$

$\qquad -$ every demand that arrives in $(t - L^{max}, t]$

$$= I_s^{net}(t - L^{max}) + W_m(t - L^{max}) + W_r(t - L^{max})$$

$\qquad +$ every product that is both ordered *and* delivered

$\qquad\qquad$ in $(t - L^{max}, t]$

$\qquad -$ every demand that arrives in $(t - L^{max}, t]$

$$(26.6)$$

Using (26.4) we simplify (26.6) as

$$I_s^{net}(t) = I_s(t - L^{max})$$

$\qquad +$ every product that is both ordered *and* delivered

$\qquad\qquad$ in $(t - L^{max}, t]$

$\qquad -$ every demand that arrives in $(t - L^{max}, t]$

Note that all products that are ordered after time $t - L^{min}$ arrive *after* time t. So finally we have

$$I_s^{net}(t) = I_s(t - L^{max})$$

$\qquad +$ every product ordered in $(t - L^{max}, t - L^{min}]$ *and* delivered

$\qquad\qquad$ before time t

$\qquad -$ every demand that arrives in $(t - L^{max}, t - L^{min}]$

$\qquad -$ every demand that arrives in $(t - L^{min}, t]$

$$= I_s(t - L^{max}) + Z(t - L^{max}, t - L^{min}) - D(t - L^{min}, t) \quad (26.7)$$

We can use relation (26.7) to derive the long run distribution of $I_s^{net}(t)$ if we take into account the following stochastic (in)dependencies between $I_s(t - L_{max})$, $I_r^{OH}(t - L_{max})$, $Z(t - L_{max}, t - L_{min})$, and $D(t - L^{min}, t)$:

- $Z(t - L_{max}, t - L_{min})$ is correlated with $I_s(t - L_{max})$ and $I_r^{OH}(t - L_{max})$,

- $I_s(t - L_{max})$ and $I_r^{OH}(t - L_{max})$ are correlated,

- $D(t - L^{min}, t)$ is uncorrelated with $I_s(t - L_{max})$, $I_r^{OH}(t - L_{max})$, and $Z(t - L_{max}, t - L_{min})$.

Substituting $u = t - L^{max}$ and $\Delta L = L^{max} - L^{min}$, the long-run distribution of the net inventory is derived from (26.7) as

$$\lim_{t \to \infty} \Pr\{I_s^{net}(t) = i_s^{net}\}$$

$$= \sum_\Omega \lim_{t \to \infty} \Pr\{D(t - L^{min}, t) = d\}$$

$$\times \lim_{u \to \infty} \Pr\{I_s(u) = i_s, I_r^{OH}(u) = i_r^{OH}, Z(u, u + \Delta L) = z\}$$

$$= \sum_\Omega \exp^{-\lambda_D L^{min}} \frac{(\lambda_D L^{min})^d}{d!} \times \pi(i_s, i_r^{OH}) \times h_{z|(i_s, i_r^{OH})}(\Delta L), \qquad (26.8)$$

where

$$\Omega = \{(i_s, i_r^{OH}, z, d) | i_s + z - d = i_s^{net}\}, \qquad (26.9)$$

and

$$h_{z|(i_s, i_r^{OH})}(\Delta L) = \lim_{u \to \infty} \Pr\{Z(u, u + \Delta L) = z | I_s(u) = i_s, I_r^{OH}(u) = i_r^{OH}\}. \qquad (26.10)$$

The conditional probability $h_{z|(i_s, i_r^{OH})}(\Delta L)$ is calculated as

$$h_{z|(i_s, i_r^{OH})}(\Delta L) = \sum_{(k, \ell) \in \mathcal{S}} q_{(k, \ell, z)|(i_s, i_r^{OH}, 0)}(\Delta L) \qquad (26.11)$$

where $q_{(k, \ell, z)|(i_s, i_r^{OH}, 0)}(\Delta L)$ is the conditional probability that during the interval $(t - L^{max}, t - L^{min}]$ the initial system state changes from state

$$\{I_s(t - L^{max}) = i_s, I_r^{OH}(t - L^{max}) = i_r^{OH}, Z(t - L^{max}, t - L^{max}) = 0\}$$

into state

$$\{I_s(t - L^{min}) = k, I_r^{OH}(t - L^{min}) = \ell, Z(t - L^{max}, t - L^{min}) = z\}.$$

This conditional probability can be calculated with the transient analysis of an appropriate Markov chain, using a discretization technique. Stochastic lead-times complicate this technique, since the transition rates of the under-lying Markov chain are not time-independent. However, for discretely distributed lead-times one can identify a series of time-intervals for which a time-independent Markov chain exists. Details of this approach can be found in [29].

Relations (26.8)–(26.11) now enable to calculate the average net inventory (26.2) and the average backorder position (26.3). All other cost function components are derived using (26.1) and the PASTA (Poisson Arrivals See Time Averages) property (see [35]).

Although the above analysis is *exact*, we have to evaluate the cost function *numerically*. The main problem here involves the truncation of infinite sums. While in some situations there may exist general rules for truncation, in other situations we have to commit ourselves to heuristic bounds and stopping criteria.

Note that the analysis does not depend on any specific assumptions regarding the demand and return processes, and the control policy involved, as long as they are Markovian. Main complicating factors are the dimensions of the Markov chain involved and the truncation of infinite state spaces. The reader should keep in mind however that this framework is meant as a means to compute expected costs for a general remanufacturing setting. It is not meant as an efficient numerical recipe. For the latter we refer to [19] and [31]. Details on how to model Coxian-2 arrival processes and how to incorporate correlation between the demand and the return processes can be found in [29].

Remark All the numerical examples presented in the next section were calculated using the analysis of this section and the following parameter settings: $\lambda_D = 1.0$, $\lambda_R = 0.7$, $L_m \equiv 2.0$, $L_r \equiv 2.0$, $c_s^h = 1.0$, $c_r^h = 0.5$, $c_b = 50$ unless specified otherwise. We initially choose $c_m^v = c_r^v = c_d^v = 0$, since these values are only relevant if product disposals are allowed (section 4.3). The influence of fixed costs are not considered in this paper, so we set $c_m^f = c_r^f = c_d^f = 0$.

26.4   ON THE STRUCTURE OF OPTIMAL POLICIES

Now that we have a framework for a joint manufacturing/remanufacturing system we may ask what kind of control policies are reasonable, or even optimal, for such a system. Since this is not an easy question to answer, we start with studying this system under some simplifying assumptions. These assumptions are (i) stocking of remanufacturables is not allowed, (ii) disposal of remanufacturables is not allowed, and (iii) the remanufacturing lead-time is zero. This system applies to situations where products are returned because of delivery errors or special return agreements between supplier and customer. Except perhaps for repackaging, no remanufacturing operations need to take place and the products can enter the serviceable inventory immediately. In this light, disposal of returned products does not seem to make sense as long as handling costs are reasonably low. The above system is studied by [6]. We discuss the main outcomes below.

Since returned items immediately enter serviceable inventory the model considered comes down to a variant of a standard stochastic inventory model where demand may be both positive or negative. First, assume that demands and returns are generated by independent Poisson processes. Since disposal is not considered in this model the return rate is restricted to be smaller than the demand rate. Let $\gamma = \lambda_R/\lambda_D < 1$ denote the return ratio. Finally, assume the manufacturing lead-time to be constant.

Under these assumptions [6] show average–cost optimality of an (s, Q) control policy for manufacturing. Analogous to standard inventory control models with backordering it is easy to see that it suffices to consider control policies depending only on the inventory position $I_s(t)$. Note that in terms of the framework depicted in Figure 26.1 this is due to the fact that there is only one process to be controlled, namely manufacturing.

Let $G(l)$ denote the decision relevant expected variable costs, namely the expected holding and backorder costs at time $t + \mu_{L_m}$ when $I_s(t)$ equals l. Conditioning the cost function on $I_s(t)$ yields

$$G(l) = (c_s^h + c_b) \sum_{j=-\infty}^{l-1} D_j + c_b[(\lambda_D - \lambda_R)\mu_{L_m} - l],$$

where D_j is the probability that the net demand during a lead-time period is at most j.

It should be noted that in the above model any stationary policy based on the inventory position is an (s, Q) policy upto a transient start–up phase. (Take s as the largest value of $I_s(t)$ for which a manufacturing order is placed.) Therefore, the optimality proof boils down to showing that there exists an average cost optimal stationary policy in the above model. This can be achieved by applying results from general Markov decision theory. Using the results by [24], two conditions need to be verified: (i) the inventory levels for which it is optimal not to place an order under an α–discounted cost criterion are bounded below uniformly in α, and (ii) there exists a stationary policy inducing an irreducible, ergodic Markov chain yielding finite average costs in steady state. In the above model (i) can be shown by using convexity of $G(.)$ and the fact that $G(l) \to \infty$ for $l \to -\infty$ whereas the costs incurred during a first passage from level l to $l-1$ are bounded in α. (ii) is proven by showing that an arbitrary (s, Q) policy verifies this condition. Since the analysis provides some interesting insights we discuss this last step in somewhat more detail.

Due to [19] the process $I_s(t)$ in the above model under a given (s, Q) policy is known to be ergodic and to have the following stationary distribution

$$\lim_{t \to \infty} P\{I_s(t) = s + k\} = \begin{cases} \dfrac{1 - \gamma^k}{Q}, & \text{for } 1 \le k \le Q \\[2mm] \dfrac{(\gamma^{-Q} - 1)\,\gamma^k}{Q}, & \text{for } Q < k. \end{cases}$$

Note that in contrast with standard inventory models the state space is unbounded above. The long–run average costs for a given (s, Q) policy in this model, denoted by $C(s, Q)$, can now be written as

$$C(s,Q) = \frac{1}{Q}[c_m^f(\lambda_D - \lambda_R) + \sum_{l=1}^{Q}(1-\gamma^l)G(s+l) + (\gamma^{-Q}-1)\sum_{l=Q+1}^{\infty}\gamma^l G(s+l)].$$

The analysis of this expression is complicated due to the infinite sum whose coefficients depend on both control parameters s and Q. This difficulty can be overcome by introducing

$$H(k) := (1-\gamma)\sum_{l=0}^{\infty}\gamma^l G(k+l).$$

It is easy to verify that

$$C(s,Q) = [c_m^f(\lambda_D - \lambda_R) + \sum_{k=s+1}^{s+Q} H(k)] / Q.$$

This expression is remarkable since it can be interpreted as the average cost of a standard (s, Q) inventory model with expected backorder and holding cost function $H(.)$ (compare [36]). This shows that the above return flow inventory model can be transformed into an equivalent standard (s, Q) model. One consequence is that standard methods can be applied to compute optimal values of the control parameters.

The above approach can be extended in several directions. For example, considering compound Poisson demand and return processes leads to an (s, S) model in an analogous way. This case is discussed in detail in [5]. Furthermore, a remanufacturing lead–time that is shorter than the manufacturing lead–time can be incorporated, using a well-known approach from standard inventory models. Letting $\tilde{D}$ denote demand in time interval $(t, t + L_m]$ minus returns in time interval $(t, t + L_m - L_r]$ we have $I_s^{net}(t + L_m) = I_s(t) - \tilde{D}$ where the last two terms are independent. Hence, the decision relevant costs are $\tilde{G}(y) := E[G(I_s^{net}(t + L_m))|I_s(t) = y] = E[G(y - \tilde{D})]$ which are again convex in y. This puts the extended model in the form discussed above. See [5] for a more detailed discussion of this approach.

The above analysis shows that introducing return flows alone does not entail major changes in the mathematical structure of standard inventory models. In the next sections we see that adding a controllable remanufacturing process makes the model considerably more complex. The question arises how the structure of the optimal policy changes if we explicitly model the remanufacturing process by introducing remanufacturing lead-times (Section 26.4.1), if we introduce different holding costs for remanufacturables, work-in-process, and serviceables (Section 26.4.2), or if we allow disposal of returned products (Section 26.4.3).

26.4.1 The influence of lead-times on the structure of the optimal policy

To address the influence of lead-times on the structure of the optimal policy, we consider a counter-intuitive result that is reported in [33]. The authors show that an increase in the remanufacturing lead-time may result in lower system costs under a PUSH policy.

Figure 26.4 shows a graphical representation of this effect for the PUSH policy. Initially, as the remanufacturing lead-time L_r increases from 0, optimized costs $C^*_{PUSH}(V; L_r)$, where $V = (s_m, Q_m, Q_r)$, monotonously decrease until L_r reaches a certain level L^* from which costs start to increase monotonously.

Figure 26.4 The counter-intuitive effect of decreasing costs with increasing remanufacturing lead-time.

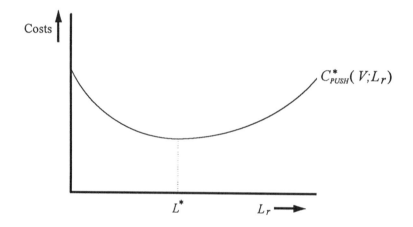

To understand this counter-intuitive effect it is important to note that at the time a remanufacturing batch is started, the inventory position is larger than the reorder point for manufacturing, s_m. Also note that in principle s_m is used as safety stock to protect against the demands during the *manufacturing* lead-time. Suppose for the sake of argument that a remanufacturing batch is started at the time the inventory position lies somewhere between s_m and $s_m + Q_m$. At least for moderate values of the return rate this is very probable. If the remanufacturing lead-time is very small compared to the manufacturing lead-time this batch will arrive in the serviceable inventory when the on-hand inventory is well above zero (see Figure 26.5), since the safety stock s_m is meant to protect against the longer manufacturing lead-time. In other words, each time a remanufacturing batch comes in there is too much safety stock, and therefore there are excessive holding costs (assuming that $c^h_r < c^h_s$).

On the other hand, if the remanufacturing lead-time is very large compared to the manufacturing lead-time this batch will arrive in the serviceable inventory when the net inventory is well below zero (see Figure 26.6), since the safety

Figure 26.5 A remanufacturing order coming in too early.

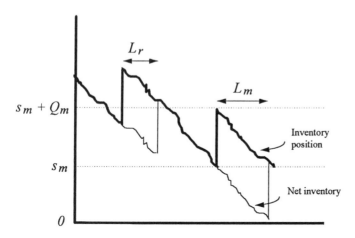

stock s_m is meant to protect against the much shorter manufacturing lead-time. In other words, each time a remanufacturing batch comes in, there may not have been enough safety stocks to protect against shortage, and therefore we may have excessive shortage costs.

Figure 26.6 A remanufacturing order coming in too late.

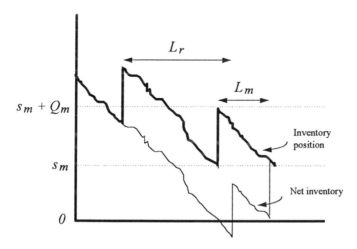

Of course, one could argue that the optimal value of s_m will be adjusted to meet the above issues, but then we will have similar problems with the manufacturing batches. Whatever the value of s_m, it seems that we can never time the manufacturing batches *and* the remanufacturing batches efficiently if L_r differs a lot from L^*. Setting the value of s_m always results in some kind of compromise, making the sum of total holding costs and backorder costs larger than seems necessary. Note that the lead-time 'imbalance' effect will only occur for moderate values of the return rate. If the return rate is very small, the small number of remanufacturing batches will only have a limited effect on system performance. Similarly, if the return rate is close to the demand rate, the influence of the manufacturing batches will be relatively small.

From the above heuristic argument we may conclude that we can improve the PUSH policy by changing the *timing* of the incoming remanufacturing batches.

An alternative policy. In order to improve the PUSH policy we will first consider the situation in which the remanufacturable holding costs are zero and we do not have holding costs for remanufacturables in WIP either. That is, one may keep any number of products in remanufacturable inventory or WIP against zero cost.

Note that if L^* is the 'optimal' lead-time for remanufacturing, we may improve the PUSH policy by altering the remanufacturing lead-time, i.e., by altering the time-interval between the time that an order is put into inventory position and the time that the order arrives in the serviceable inventory. Note that for the PUSH policy this time-interval is always equal to the processing-time L_r.

If L_r is smaller than L^* we would like to increase the remanufacturing lead-time. We can do this by increasing the time that an order spends in WIP (this means increasing the processing time) against WIP holding cost c_w^h. But this does not make sense if $c_r^h \leq c_w^h \leq c_s^h$, which is a reasonable assumption. Instead, we wait a fixed time $L^* - L_r$ before the order is released to the remanufacturing facility.

If L_r is larger than L^* we would like to decrease the remanufacturing lead-time. Assuming that we cannot decrease the remanufacturing processing time we do the following. As soon as a remanufacturing batch is available it is released to the remanufacturing facility, but we wait a fixed time $L_r - L^*$ before the order is put into the inventory position.

We now have the following control policy as an alternative to the PUSH policy:

Case A $(L_r < L^*)$ The (s_m, Q_m, Q_r) PUSH policy is employed with the following alteration: As soon as Q_r remanufacturables become available they will enter the inventory position, but remanufacturing will only start after time $L^* - L_r$. In this way the remanufacturing processing time is still L_r, but the remanufacturing lead-time has changed into L^*. The system dynamics and costs under this policy are exactly the same as those under a PUSH policy with remanufacturing lead-time L^*. Therefore, $C_{Alt}^*(V, L^*; L_r) \equiv C_{PUSH}^*(V; L^*) \leq C_{PUSH}^*(V; L_r)$. So, the alternative policy dominates the PUSH policy with respect to system costs as long as $L_r < L^*$.

Case B ($L_r \geq L^*$) The (s_m, Q_m, Q_r) PUSH policy is employed with the following alteration: As soon as Q_r remanufacturables become available, they will be remanufactured, but they will only enter the inventory position after time $L_r - L^*$. In this way the remanufacturing processing time is still L_r, but the remanufacturing lead-time has changed into L^*. The system dynamics and costs $C^*_{Alt}(V, L^*; L_r)$ that result from this alternative policy are exactly the same as those under the PUSH policy with remanufacturing lead-time L^*. Therefore, $C^*_{Alt}(V, L^*; L_r) \equiv C^*_{PUSH}(V; L^*) \leq C^*_{PUSH}(V; L_r)$, so the alternative policy dominates the PUSH policy with respect to system costs as long as $L_r \geq L^*$.

This control policy dominates the (s_m, Q_m, Q_r) PUSH policy with respect to costs for all values of L_r. Moreover, the associated costs do not depend on the remanufacturing lead-time (see Figure 26.7).

Figure 26.7 The alternative policy compared to the (s_m, Q_m, Q_r) PUSH policy in absence of holding costs for remanufacturables and WIP.

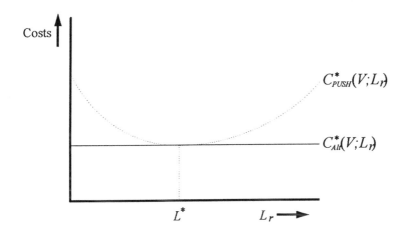

If we let go of the assumption of zero holding costs for remanufacturable inventory and we introduce holding costs for WIP, things are slightly more complicated. The cost function $C^*_{PUSH}(V; L_r)$ now contains the additional cost component $c^h_w \lambda_R L_r$, and $C^*_{Alt}(V, L; L_r)$ the cost component $c^h_w \lambda_R L_r + c^h_r 1_{\{L_r < L\}} \lambda_R (L - L_r)$ with L the chosen remanufacturing lead-time. Here, $1_{\{L_r < L\}}$ is an indicator function, which is assigned the value one if $L_r < L$ and the value zero otherwise. Consequently, we have

$$C^*_{Alt}(V, L; L_r) = C^*_{PUSH}(V; L) + c^h_w \lambda_R (L_r - L) + c^h_r 1_{\{L_r < L\}} \lambda_R (L - L_r) \quad (26.12)$$

828

Minimizing $C^*_{Alt}(V, L; L_r)$ for L we find that the local optima satisfy the equation

$$\frac{dC^*_{PUSH}(V; L)}{dL} = (c^h_w - c^h_r 1_{\{L_r < L\}})\lambda_R. \tag{26.13}$$

Using (26.13) it is not hard to see that the global minimum of (26.12) is reached for $L_1 > L_r$, where L_1 is that value of L for which the tangent with slope $(c^h_w - c^h_r)\lambda_R$ to $C^*_{PUSH}(V; L)$ as a function of L, intersects with $C^*_{PUSH}(V; L)$, or for $L_2 < L_r$, where L_2 is that value of L for which the tangent with slope $c^h_w \lambda_R$ to $C^*_{PUSH}(V; L)$ as a function of L, intersects with $C^*_{PUSH}(V; L)$.

We can now generalize the alternative policy for non-zero holding costs for remanufacturables and WIP:

- For $L_r < L_1$ we adopt the alternative policy as in Case A

- For $L_1 \leq L_r \leq L_2$ we adopt the (s_m, Q_m, Q_r) PUSH policy

- For $L_r > L_2$ we adopt the alternative policy as in Case B.

The optimal costs resulting from this policy are as in Figure 26.8.

Figure 26.8 The alternative policy compared to the (s_m, Q_m, Q_r) PUSH policy in presence of holding costs for remanufacturables and WIP.

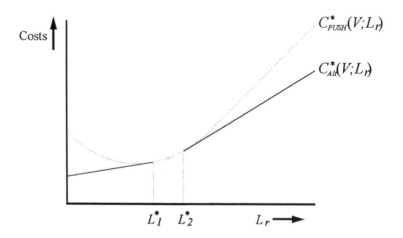

Note that an increase in the remanufacturing lead-time results in an increase in costs for all L_r as long as $c^h_r < c^h_w$, i.e., the slope $(c^h_w - c^h_r)L_r$ is positive.

Implications. In this section we have established two things: *(i)* The effect of decreasing costs with increasing remanufacturing lead-time is due to the non-optimality of the (s_m, Q_m, Q_r) PUSH policy, possibly together with an unrealistic setting of c_w^h, and *(ii)* the PUSH policy is a non-optimal policy in general, because it is fully dominated by the alternative policy.

A derivative but more important result of the above is that *any* policy for which the ordering of a remanufacturing batch coincides with updating the inventory position, is a non-optimal policy for large remanufacturing lead-times. This type of policy will result in increasing backorder costs plus serviceable holding costs as the remanufacturing lead-time is increased. This is not the case for the alternative PUSH-policy, since its costs grow linearly in the remanufacturing lead-time. Hence, the alternative policy dominates all 'standard' policies for large remanufacturing lead-times.

26.4.2 *The influence of holding costs on the structure of the optimal policy*

In the previous section we found that holding costs may have an influence on the choice of the optimal policy parameters. But do they also influence the *structure* of the optimal policy? Up to now we have only considered push type policies, i.e., (batches of) incoming returns are remanufactured as soon as possible. However, in [29] it is shown that such a policy is not optimal if holding costs for remanufacturables are valued sufficiently lower than holding costs for serviceables. In that case a pull type policy may be considered instead.

A numerical example that compares the behaviour of the optimized costs under the PULL policy, $C_{PULL}^*(W)$, where $W = (s_m, Q_m, s_r, S_r)$, to that of the PUSH policy under different holding cost assumptions is given in Figure 26.9. As expected, the PULL policy performs better than the PUSH policy as long as holding costs for remanufacturables are valued sufficiently lower than holding costs for serviceable products.

Figure 26.9   The effect of the remanufacturable holding cost on system costs under the (s_m, Q_m, Q_r) PUSH policy and the (s_m, Q_m, s_r, S_r) PULL-policy.

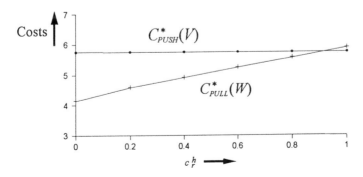

830

If we recall the conclusions from the previous section, we may expect that the PULL policy performs worse than the *alternative* PUSH policy if the remanufacturing lead-time differs considerably from the manufacturing lead-time. However, we can improve the PULL policy in a similar way as we did for the PUSH policy. In case of small remanufacturing lead-times we may delay remanufacturing for a fixed time interval; in case of large remanufacturing lead-times we can introduce a mixture of a push and pull policy: initially push the remanufacturables through a number of (re)manufacturing operations, store them in a work-in-process buffer and eventually pull them through the remaining operations when they are really needed. This is particularly useful with respect to disassembly and repair times, which are often very variable. Pushing the remanufacturables through these variable processes, and then pulling them further through the less variable assembly operations does not only improve system costs but also other performance measures, such as flow-time and lateness ([9]).

26.4.3 The influence of the return rate on the structure of the optimal policy

In the introduction of Section 26.4 we saw that one of the conditions for optimality of the classical (s, Q) policy is that disposal of incoming returns is not allowed. A typical picture of the effect of the return rate on costs behaviour under the (s_m, Q_m, Q_r) PUSH policy is given in Figure 26.10 (see e.g. [4, 32]).

Figure 26.10 The effect of the return rate on system costs under the (s_m, Q_m, Q_r) PUSH policy; $c_m^v = 10$, $c_r^v = 5.0$, $c_d^v = 0$.

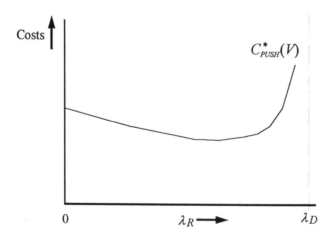

Initially costs decrease as the return rate increases (assuming that the marginal cost of remanufacturing is lower than that of manufacturing), but after some point costs start to increase. As λ_R approaches λ_D the 'load' of the system grows bigger and bigger, leading to extremely high holding costs. A disposal policy can overcome this problem. For instance, one could extend the (s_m, Q_m, Q_r) PUSH policy with an extra parameter s_d, which specifies the inventory position at which returns should be disposed of upon arrival. Even this simple disposal policy may result in a dramatic improvement of system performance. This can be seen from the numerical example in Figure 26.11. Allthough in this example variable disposal costs are put to zero, one should realise that for any finite value of c_d^v average costs will tend to a constant if γ tends to λ. As a contrast, if disposals are not taken into consideration, costs will tend to infinity as γ tends to λ. Hence, even if disposal is costly, a disposal policy may reduce average costs considerably. It should be noted that $c_d^v = 0$ is not an extreme case, since returned items can have a positive scrap value (i.e. $c_d^v < 0$), and neither is $\gamma \approx \lambda$, since towards the end if a product's life cycle the number of product returns may exceed product demand.

Figure 26.11 The (s_m, Q_m, Q_r) PUSH policy compared to the (s_m, Q_m, Q_r, s_d) PUSH-disposal policy; $c_m^v = 10$, $c_r^v = 5.0$, $c_d^v = 0$.

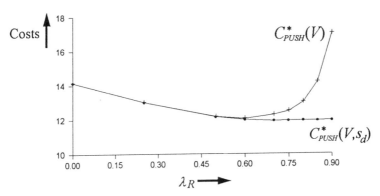

Furthermore, in [32] it is shown that a disposal policy reduces the variability in the inventory processes by taking care of excess returns. Clearly, the above strongly suggests that the structure of the optimal policy should include the option of disposal. For a more detailed account of the effects of various disposal policies, we refer to [30] and [32].

26.4.4 Insights from periodic review models

In *periodic review* models the planning horizon is divided into discrete planning periods. At the beginning of each planning period n, decisions are taken according to the values of the following decision variables.

$Q_d^{(n)}$ =   the quantity (batch-size) of remanufacturable products that is disposed of in planning period n,

$Q_p^{(n)}$ =   the quantity (batch-size) of products that is procured outside or internally produced in planning period n,

$Q_r^{(n)}$ =   the quantity (batch-size) of products that is remanufactured in planning period n.

All decision variables are assumed to be integer. The objective in periodic review models is to determine values for the decision variables, such that the total expected costs over the entire planning horizon are minimized.

Within this category, [14] considers a model with assumptions and characteristics as listed in Table 26.2. One of the objectives is to determine the structure of the optimal policy. It is argued that a nice structure can only be obtained for some special cases. These special cases relate to the assumptions made about the stocking policy of the returned products, and to the values of the manufacturing lead-time L_m and the remanufacturing lead-time L_r.

Table 26.2 Assumptions and characteristics of the remanufacturing model by [14].

demands/returns	All returns and demands per period are continuous time-independent random variables. The inter-arrival distributions are arbitrary distribution functions, which may be stochastically dependent.
testing	No testing facility.
remanufacturing	The remanufacturing lead-time L_r is non-stochastic and equal to μ_{L_r}; the remanufacturing facility has infinite capacity and variable remanufacturing costs.
production	The procurement lead-time L_m is non-stochastic and equal to μ_{L_m}; there are variable production costs.
inventories	Type I and Type II inventory buffers have infinite capacity; Type I and Type II inventory have (different) variable inventory holding costs.
disposal	Variable disposal costs.
service	Modeled in terms of backorder costs.
control policy	At the beginning of each period n decisions are taken on $Q_d^{(n)}$, $Q_p^{(n)}$, and $Q_r^{(n)}$, such that the total expected costs over the planning horizon are minimized.

For instance, for the case that returned items are not allowed to be stocked the following results hold:

- If $L_m = L_r$ the structure of the optimal policy can be formulated as a so-called (L, U) policy:

$$Q_p^{(n)} = L^{(n)} - x, \quad Q_r^{(n)} = x_r, \quad Q_d^{(n)} = 0, \quad \text{for} \quad x < L^{(n)},$$
$$Q_p^{(n)} = 0, \quad Q_r^{(n)} = x_r, \quad Q_d^{(n)} = 0, \quad L^{(n)} \le x \le U^{(n)},$$
$$Q_p^{(n)} = 0, \quad Q_r^{(n)} = U^{(n)} - x_s, \quad Q_d^{(n)} = x - U^{(n)}, \quad x > U^{(n)}.$$

Here, x_r is the remanufacturable inventory, x_s is the inventory position of serviceables, and $x = x_s + x_r$. In words this policy states that every return is remanufactured in order to increase the inventory position to $U^{(n)}$. If there are not enough returns available to increase the inventory position up to $L^{(n)}$ an additional manufacturing order is placed to do so. Disposal of returns occurs such that the inventory position plus the number of returns never exceeds $U^{(n)}$.

- If $L_m = L_r + 1$ the structure of the optimal policy can be formulated as a so-called $(L, U, \hat{U})$ policy. IF $\hat{U} > U$ this policy is equal to the L, U policy. Otherwise, if the inventory is smaller than $\hat{U}$, which does not depend on period n, then procure up to $L^{(n)}$. If the inventory exceeds $\hat{U}$ then dispose down to $\hat{U}$, remanufacture the remaining remanufacturable products and procure $L^{(n)} - \hat{U}$ new products.

- If $L_m < L_r$ or $L_m > L_r + 1$ the structure of the optimal policy is not expected to be of a simple form.

Note that if returned products are not allowed to be stocked, all returned products will be remanufactured or disposed of at the end of each period. If returned products are allowed to be stocked there can be a more subtle control of the production, the remanufacturing and the disposal process. For the latter case Inderfurth derives the following results with respect to the structure of the optimal policy.

- If $L_m = L_r$ the structure of the optimal policy can be formulated as a so-called (L, M, U) policy:

$$Q_p^{(n)} = L^{(n)} - x, \quad Q_r^{(n)} = x_r, \quad Q_d^{(n)} = 0, \quad \text{for} \quad x < L^{(n)},$$
$$Q_p^{(n)} = 0, \quad Q_r^{(n)} = x_r, \quad Q_d^{(n)} = 0, \quad L^{(n)} \le x < M^{(n)},$$
$$Q_p^{(n)} = 0, \quad Q_r^{(n)} = M^{(n)} - x_s, \quad Q_d^{(n)} = 0, \quad M^{(n)} \le x \le U^{(n)},$$
$$Q_p^{(n)} = 0, \quad Q_r^{(n)} = M^{(n)} - x_s, \quad Q_d^{(n)} = x - U^{(n)}, \quad x > U^{(n)}.$$

Note that if $M^{(n)} = U^{(n)}$ this policy is equal to the abovementioned (L, U) policy. If $M^{(n)} < U^{(n)}$ and $M^{(n)} \le x \le U^{(n)}$, a fraction of the remanufacturables $(x - M^{(n)})$ is kept in stock, while the rest is remanufactured. If $x > U^{(n)}$ a fraction is also disposed of. This generalizes the results of [26] who studied this system for zero (re)manufacturing lead-times.

- If $L_m \neq L_r$ the structure of the optimal policy is much more difficult to obtain and becomes very complex, even if the manufacturing lead-time and remanufacturing lead-time differ only one period.

The above results show that under general conditions the optimal policy will be very complex and difficult to identify. Only if the (re)manufacturing lead-times are equal, a simple policy that does not depend on the timing of individual (re)manufacturing orders is optimal. This coincides with our findings in section 4.1, where it was shown that traditional policies can be improved considerably by redefining the inventory position in such a way that differences in (re)manufacturing lead-times are accounted for.

26.5 SUMMARY AND CONCLUSIONS

In this paper we have addressed the issue of inventory control for joint manufacturing and remanufacturing. A short review of the existing literature on models that are related, but not specifically meant to apply to remanufacturing, showed that these models do not quite capture all of the characteristics that are typical for these types of systems. These characteristics are (i) two exogenous supply sources, that is manufacturing and remanufacturing, (ii) time-dependent availability of the remanufacturing source, and (iii) non-zero stochastic lead-times for remanufacturing. One framework that fits the above criteria was presented in Section 26.3. This framework enables the analysis of a broad variety of control policies under various system characteristics, like stochastic lead-times, Coxian-2 arrival processes, and correlation between the demand and return processes.

With respect to the structure of the optimal policy it was shown that the characterization of the (re)manufacturing lead-times plays an important role. Traditional policies can be improved considerably if additional lead-time information is used in the control policy and/or the definition of the inventory position. Other factors include the return and demand rate, which define the 'load' of the system. If this load is high it may be necessary to implement a disposal policy. The valuation of holding costs largely determines the choice between push and pull control systems. The analysis of periodic review models shows that, in general, the structure of the optimal policy will be very complex, even if the (re)manufacturing lead-times differ only slightly. The traditional (s, S) policy will only be optimal if stocking and disposal of remanufacturables is not allowed, and remanufacturing lead-times are sufficiently short.

More research is needed to identify the structure of the optimal policy for more general situations and to use the performance of the optimal policy as a benchmark for the performance of heuristic policies.

References

[1] E.W. Barankin (1961). A delivery-lag inventory model with an emergency provision. *Naval Research Logistics Quarterly*, 8:285–311.

[2] D.I. Cho and M. Parlar (1991). A survey of maintenance models for multi-unit systems. *European Journal of Operational Research*, 51:1–23.

[3] S.D.P. Flapper (1994). Matching material requirements and availabilities in the context of recycling: An MRP-I based heuristic. In *Proceedings of the Eighth International Working Seminar on Production Economics*, Vol. 3:511–519, Igls/Innsbruck, Austria.

[4] M. Fleischmann, J.M. Bloemhof-Ruwaard, R. Dekker, E. van der Laan, J.A.E.E. van Nunen and L.N. Van Wassenhove (1997). Quantitative models for reverse logistics: A review. *European Journal of Operational Research*, 103:1–17.

[5] M. Fleischmann and R. Kuik (1998). On optimal inventory control with stochastic item returns. Management Report Series 21–1998, Erasmus University Rotterdam, The Netherlands.

[6] M. Fleischmann, R. Kuik and R. Dekker (1997). Controlling inventories with stochastic item returns: A basic model. Management Report Series 43(13), Erasmus University Rotterdam, The Netherlands.

[7] V.D.R. Guide, Jr. (1996). Scheduling using drum-buffer-rope in a remanufacturing environment. *International Journal of Production Research*, 34(4):1081–1091.

[8] V.D.R. Guide, Jr., M.E. Kraus and R. Srivastava (1997). Scheduling policies for remanufacturing. *International Journal of Production Economics*, 48(2):187–204.

[9] V.D.R. Guide, Jr. and R. Srivastava. Inventory buffers in recoverable manufacturing. Forthcoming in *Journal of Operations Management*, December 1998.

[10] V.D.R. Guide, Jr., R. Srivastava and M.S. Spencer (1997). An evaluation of capacity planning techniques in a remanufacturing environment. *International Journal of Production Research*, 35(1):67–82.

[11] S.M. Gupta and K.N. Taleb (1994). Scheduling disassembly. *International Journal of Production Research*, 32(8):1857–1866.

[12] Ü. Gürler and M. Parlar (1997). An inventory problem with two randomly available suppliers. *Operations Research*, 45(6):904–918.

[13] K. Inderfurth (1982). Zum Stand der betriebswirtschaftlichen Kassenhaltungstheorie. *Zeitschrift für Betriebswirtschaft*, 3:295–320 (in German).

[14] K. Inderfurth (1997). Simple optimal replenishment and disposal policies for a product recovery system with leadtimes. *OR Spektrum*, 19:111–122.

[15] M.R. Johnson and M.H. Wang (1995). Planning product disassembly for material recovery opportunities. *International Journal of Production Research*, 33(11):3119–3142.

[16] H.R. Krikke, A. van Harten and P.C. Schuur (1996). On a medium term product recovery and disposal strategy for durable assembly products. To appear in *International Journal of Production Research*.

[17] M.C. Mabini and L.F. Gelders (1991). Repairable item inventory systems: A literature review. *Belgian Journal of Operations Research, Statistics and Computer Science*, 30(4):57–69.

[18] K. Moinzadeh and S. Nahmias (1988). A continuous review model for an inventory system with two supply modes. *Management Science*, 34(6):761–773.

[19] J.A. Muckstadt and M.H. Isaac (1981). An analysis of single item inventory systems with returns. *Naval Research Logistics Quarterly*, 28:237–254.

[20] S. Nahmias (1981). Managing repairable item inventory systems: A review. *TIMS Studies in the Management Sciences*, 16:253–277.

[21] M. Parlar and D. Perry (1996). Inventory models of future supply uncertainty with single and multiple sources. *Naval Research Logistics*, 43:191–210.

[22] K.D. Penev and A.J. de Ron (1996). Determination of a disassembly strategy. *International Journal of Production Research*, 34(2):495–506.

[23] W.P. Pierskalla and J.A. Voelker (1976). A survey of maintainance models: the control and surveillance of deteriorating systems. *Naval Research Logistics Quarterly*, 23:353–388.

[24] L.I. Sennot (1989). Average cost optimal stationary policies in infinite state markov decision processes with unbounded costs. Operations Research, 37(4):626–633.

[25] C.C. Sherbrooke (1968). METRIC: a multi-echelon technique for recoverable item control. *Operations Research*, 16:122–141.

[26] V.P. Simpson (1978). Optimum solution structure for a repairable inventory problem. *Operations Research*, 26:270–281.

[27] K.N. Taleb and S.M. Gupta (1996). Disassembly of multiple product structures. To appear in *Computers and Industrial Engineering*.

[28] M.C. Thierry (1997). *An Analysis of the Impact of Product Recovery Management on Manufacturing Companies*. PhD thesis, Erasmus University Rotterdam, The Netherlands.

[29] E.A. van der Laan (1997). *The Effects of Remanufacturing on Inventory Control*. PhD thesis, Erasmus University Rotterdam, The Netherlands.

[30] E.A. van der Laan, R. Dekker and M. Salomon (1996). Product remanufacturing and disposal: A numerical comparison of alternative strategies. *International Journal of Production Economics*, 45:489–498.

[31] E.A. van der Laan, R. Dekker, M. Salomon and A. Ridder (1996). An (s,Q) inventory model with remanufacturing and disposal. *International Journal of Production Economics*, 46–47:339–350.

[32] E.A. van der Laan and M. Salomon (1997). Production planning and inventory control with remanufacturing and disposal. *European Journal of Operational Research*, 102:264–278.

[33] E.A. van der Laan, M. Salomon and R. Dekker (1998). An investigation of lead-time effects in manufacturing/remanufacturing systems under simple PUSH and PULL control strategies. To appear in: *European Journal of Operational Research*.

[34] A.S. Whittmore and S. Saunders (1977). Optimal inventory under stochastic demand with two supply options. *SIAM Journal of Applied Mathematics*, 32:293–305.

[35] R.W. Wolff (1982). Poisson arrivals see time averages. *Operations Research*, 30:223–231.

[36] Y.-S. Zheng (1992). On properties of stochastic inventory systems. Management Science, 38(1):87–103.

27 A TAXONOMIC REVIEW OF SUPPLY CHAIN MANAGEMENT RESEARCH

Ram Ganeshan, Eric Jack,
Michael J. Magazine, and Paul Stephens

QAOM Department
The University of Cincinnati
Cincinnati, OH 45221-0130

27.1 Introduction

In the previous chapters, we focused largely on quantitative approaches to solving Supply Chain Management (SCM) problems including such issues as: inventory management, supply contracts, information flow, product variety, and international operations. In this chapter, we will broaden our focus to include other approaches to SCM problems, by presenting a broad taxonomy for understanding SCM research.

27.1.1. Why present a Taxonomy on SCM Research?

Efforts to describe and explain supply chain management (SCM) have recently led to a plethora of research and writing in this field. At the same time, the level of attention SCM now receives in business practices also heavily influences the growing interest in SCM research. SCM is now seen as a governing element in strategy (Fuller, O'Conner and Rawlinson, 1993) and as an effective way of creating value for customers. However, despite the growing interest in supply chain management, there is still a lack of cohesive information that explains the SCM concept and emphasizes the variety of research work being accomplished in this area.

Other researchers have also provided taxonomies and frameworks that help both practitioners and academics understand how to manage supply chains. For example, Bowersox (1969), in addition to reviewing relevant streams of thought in physical distribution, suggested that the distribution function can provide a competitive advantage through channel-wide integration beyond the firm. Shapiro (1984) provided a prescriptive framework that can enable a company to gain leverage by ensuring a good fit between its logistics system and competitive strategy. Houlihan (1985) made a strong case for viewing the supply chain as a single entity by incorporating a logistics focus into the strategic decisions of the firm. Langley (1992) cast the evolution of logistics into three specific contexts: past (1950-1964), present (1965 - present) , and future. Since that time, a variety of authors [Stevens, 1989, Masters and Pohlen, 1994, Mourits and Evers, 1995, and Thomas and Griffin, 1996] have added to this body of literature by providing integrative frameworks to help design and manage supply chains.

Despite the efforts of previous authors, we believe that the growing literature in SCM warrants a close re-examination of published SCM works to date. The purpose of such an update is to carefully chart the historical development of SCM and to synthesize future directions of research. Simply starting off with a literature search into this SCM field quickly becomes overwhelming due to the amount of work being done that seemingly falls into the subject of supply chain management. Therefore, it would be beneficial, to both the novice and even those familiar with the subject, for us to discuss, describe, and categorize the work being done in supply chain management.

We have intentionally limited our focus to articles that have already been published at the time of this writing as an attempt to understand the state of research at a fixed point in time. Naturally, in an emerging field like SCM, there is much research in the pipeline and a new "chapter" is needed periodically. We hope this is useful for today's researchers.

27.1.2. How this chapter is organized

In section 2, we explore the basics of SCM from a conceptual perspective by tracing the roots of the definition and the origins of the concept from a broad stream of literature. Recognizing that there is not a clear consensus on the definition of supply chain management, we look at the various approaches to defining SCM from the 1980's to the present, and provide our own interpretation based on the literature.

In section 3, we show that the paths leading to the current state of SCM has evolved over the past four decades. In particular, we characterize SCM as evolving over the years from materials management, physical distribution, and integrated logistics. We also show that SCM benefits from a variety of concepts that were developed in several different disciplines including marketing, economics, operations research, management science, operations management, and logistics.

In section 4, we summarize the volume of SCM research into three broad categories -- competitive strategy, firm focused tactics, and operational efficiencies -- based on the level and detail of the SCM problems being addressed.

In section 5, we focus on the research methodologies and solution approaches that have been used to address SCM problems. In particular, we categorize these research methodologies into four broad categories: concepts, case-oriented, frameworks, and models.

Finally, in the appendix, we provide a database of selected papers to help summarize the volume of research that has been done in the SCM arena.

27.2   The Concept of Supply Chain Management

27.2.1. Supply Chain "Defined"

The SCM literature offers many variations on the same theme when defining a supply chain. The most common definition [see for example, Houlihan (1985), Jones and Riley (1984), Stevens (1989), Scott and Westbrook (1991), Lee and Billington (1993), and Lamming (1996)] is a system of suppliers, manufacturers, distributors, retailers, and customers where materials flow downstream from suppliers to customers and information flows in both directions.

Our working definition of a supply chain is from Stevens (1989) who defines it as:

". . . a connected series of activities which is concerned with planning, coordinating and controlling materials, parts, and finished goods from supplier to

customer. It is concerned with two distinct flows (material and information) through the organization."

Several authors include strategic decision making as a differentiating virtue of a supply chain. For example, Oliver and Webber (1992) state that a supply chain should be viewed as a single entity that is "guided by strategic decision-making." They emphasize that systems integration, not just interface, is the key to success in SCM. Some researchers also include the carriers in the supply chain [Gentry (1996)]. Still others [O'Brien and Head (1995)] include governments as part of the chain since, as a global concept, managing the supply chain would also include all of the issues associated with government regulations and customs.

27.2.2. The "Management" in SCM - Scope of Responsibility

Trying to assess the actual scope of supply chain management is much more difficult than simply defining a supply chain. Towill (1997) argues that the definition needs to be flexible because it "applies right across the business spectrum ranging from international supply chains down to a number of related sequential activities undertaken under one roof but covering a number of independent cost centers." In our view, this is essentially how it has been applied throughout the literature. Houlihan (1985) is credited with first coining the term "supply chain," but it seems that researchers have varying interpretations of exactly what managing a supply chain means.

A firm, be it manufacturing or service, belongs to at least one supply chain. How widely or narrowly the chain is managed is an indicator of the extent to which supply chain management is being practiced. On one hand, supply chains can be managed as a single entity through a dominant member (referred to as the "predator" by Towill, 1997), and on the other, through a system of partnerships requiring well-developed cooperation and coordination. Cooper, Ellram, Gardner, and Hanks (1997) suggest that the span of management control should be determined by the added value of any relationship to the firm. Additionally, Forrester (1961) suggests that the five flows of any economic activity -- money, orders, materials, personnel, and equipment -- are interrelated by an information network, which gives the "system," what has now come to be called a supply chain, its own character. Therefore, the scope of responsibility for managing a supply chain seems highly specific to the firm and its myriad of relationships with its suppliers, vendors and customers.

27.2.3. Who Manages the Supply Chain?

We can break down SCM into various its various elements but one must not lose sight of the fact that SCM is rooted in senior-level decision making. Otherwise, SCM may well be reduced to its component functions of purchasing, distribution, materials management or even integrated logistics. Of course, SCM includes implementation and operational aspects in which day to day operations are managed below the senior management level.

One can argue that a proliferation of interpretations of what SCM means has led to some confusion among researchers and practitioners. There should be some characteristics unique to supply chain management that differentiates it from past research that previously fell under the aegis of integrated logistics. Houlihan (1985) makes it clear that the differentiating factor is the strategic decision making aspect of supply chain management. SCM reaches out beyond the boundaries of cost containment and links operating decisions to strategic considerations within and beyond the company. In the past, these issues were primarily in the domain of middle management. But, in channel-wide supply chain management, these issues are now part of the responsibilities of upper management.

27.3   The Evolution of Supply Chain Management

27.3.1. The Evolutionary Paths Leading to SCM

The literature suggests that SCM has its roots in the evolutionary path followed through materials management and physical distribution after WWII, functional logistics (different managers for all functions), and integrated logistics (one manager for all the functions).

Forrester (1958) justifies the first step beyond functional logistics by using a systems analysis approach to describe the forces that determine growth, fluctuation, and decline. He develops a complete company model that described the flows of information, materials, manpower, capital, equipment, and money. Bowersox (1969) discusses the evolution of integrated logistics and touches upon what will become known as the supply chain. He states that related responsibilities seldom terminate when product ownership transfer occurs and that firms are linked together in cooperative vertical marketing systems providing total channel-wide performance. Langley (1992) suggests four stages of development: (i) cost control, (ii) profit-center orientation recognizing the positive impact on sales, (iii) view of logistics as key to product differentiation, and (iv) as a principal strategic advantage. Masters and Pohlen (1994) describe the evolution of logistics into three phases: (1) functional management (1960 - 1970) - functions such as purchasing, shipping, and distribution are each managed separately, (2) internal integration (1980s) - the management of supply chain functions of a single facility are unified and become the responsibility of a single individual, and (3) external integration (1990s) - the management of supply chain functions throughout the chain are unified requiring cooperation and coordination between links in the chain. La Londe (1994) describes the evolution of integrated logistics in three phases: (1) physical distribution - the distribution of goods is all that needs to be managed by a logistics manager, (2) internal linkages - it is important for the logistics manager to control both internal supply functions and physical distribution, and (3) external linkages - logistics management requires cooperation in management with upstream and downstream entities in order to maximize the benefits of the total logistic system.

27.3.2. Contributions from Various Disciplines

Many different disciplines, including marketing, economics/systems dynamics, operations research / management science, and operations management, have contributed concepts that originated outside the original SCM theory but are used throughout the SCM literature.

Marketing has spawned such ideas as EDLP (every day low pricing) [Blattberg, Eppen and Lieberman (1981)] and postponement [Alderson (1957)]. The SCM literature has largely developed postponement in conjunction with inventory management and control [Jones and Riley (1984), Lee and Billington (1995), Zinn and Levy (1988), and Zinn and Bowersox (1988)].

Economics and systems dynamics has contributed with Forrester's (1958, 61) work describing growing variation upstream in a supply chain and is now popularized as the "Bullwhip Effect" or the "Forrester Effect" [Lee, Padmanabhan and Whang (1997), Berry and Naim (1996)].

OR/MS and Operations Management, the primary focus of this book, are used in several areas including: (i) multi-echelon inventory models [Clark and Scarf (1960), Clark (1972)], (ii) plant and distribution center location models [Geoffrion and Graves (1974), Cohen and Lee (1988), Revelle and Laporte (1996), Camm et al. (1996)], (iii) order allocation schemes [Anupindi and Akella (1993)], (iv) lean manufacturing [Lamming (1996), Levy (1997)], (v) quick response (QR) [Fisher (1997], (vi) vendor managed inventories (VMI) [Cachon and Fisher (1997)], (vii) enterprise & distribution resource planning [Hammel and Kopczak (1993), Verwijmeren, Viist, and Karel (1996)], and (viii) JIT supply [Leenders, Nollet, and Ellram (1994), O'Brien and Head (1995)].

Logistics pioneered the concept of: (i) integrated logistics that eventually came to be called SCM [Bowersox (1969), Slater (1976)], and (ii) partnership building and management [Slater (1976), Gentry (1996), and Walton (1996)].

27.3.3. Key Factors that Influence SCM Research

In analyzing these papers from a historical perspective, we found an interesting stream of factors that the authors listed as being influential to supply chain management problems that they addressed in their papers.

In the early years, the importance of cost control and internal efficiencies were influenced by the economic climate of the late 1950s. Bowersox (1969), posits that the cost control concerns that emerged in the 50's acted as a catalyst to the renewed interest in logistics management. Additionally, the introduction of computers and the adoption of many mathematical models and other optimization tools had a great impact upon the development of SCM. This was predicted by Forrester who wrote that electronic data processing, decision making, simulation, feedback control, and systems analysis were "the tools of process" that will influence the future direction of the management of information through the five flows of economic activity alluded to earlier.

In the 60's, Bowersox (1969) notes that computers emerged from their infancy and found fertile applications in physical distribution. At this time, Bowersox also addresses the concept of integrated physical distribution and argues that physical distribution has the potential for system integration beyond the firm into the total cooperative channels of distribution. Slater (1976) states that management's emphasis on liquidity, cost reduction and the impact on change has led to the recognition of logistics as an important method of improving the bottom line. Slater argues that a total systems approach to the logistics channel will reduce total cost and considerably improve the overall quality of the operations.

Gradually, the growing importance of logistics is noted as potential competitive advantage in response to increasing competition and growing customer requirements. Fuller (1993) sums it up nicely by stating:

". . . logistics has the potential to become the next governing element of strategy as an inventive way of creating value for customers, an immediate source of savings, a discipline on marketing, and a critical extension of production flexibility."

In the last two decades, as logistics slowly evolved into SCM, several researchers (see for example Houlihan (1988) Copacino and Rosenfield (1992), Lee and Billington (1993), Fuller (1993), Thomas and Griffin (1996)) have tried to account for the increasing awareness and implementation of supply chain management. We note seven factors here that have consistently been mentioned in the literature as influencing supply chain/strategic logistics planning: (1) new customer service requirements, (2) competitive pressure, (3) changing costs (for example, the cost of logistics estimated at 30% of cost of goods sold) (4) pressure for improved financial performance, (4) need to redesign and improve logistics systems, (5) regulatory changes, (6) improved communications, and (7) information technology.

In recent times, changing environmental awareness have also influenced supply chain management research. For example, Bloemhof-Ruwaard, Van Beek, Hordijk, and Van Wassenhove, (1995) cite factors that might influence future direction of SCM: (1) legal requirements or consumer pressure to reduce waste, (2) green supply chain management to include waste treatment, (3) reuse of materials and packaging, (4) recovery of products, (5) adaptation of new materials, (6) product redesign, and (7) process changes.

27.3.4   The SCM Time-line

The historical development of SCM can be illustrated through the use of a time line
(See Figure 1) that highlights a few of the many significant research papers that have
led to the development of SCM as a new concept. The vertical axis of the time line
represents, as the ensuing discussion will show, our method of categorization of
supply chain research. The dots show the distribution of this research from our
sample of articles.

Figure 1. SCM's Evolutionary Timeline

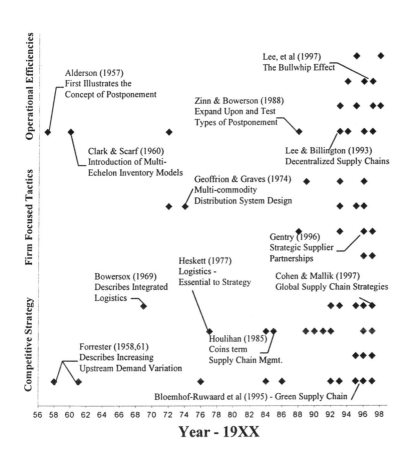

27.4   Classifying Research on Supply Chain Management

In this section, we classify the supply chain management research into three broad perspectives: competitive strategy, firm focused tactics, and operational efficiencies.

Competitive strategy issues have a long range impact on the firm, firm focused tactical issues operate in a shorter time frame and operational efficiencies involve day to day decisions that can be altered quickly. Stevens (1989) offers us some simple guidelines to follow in categorizing the perspective (competitive strategy, firm focused tactics and operational efficiencies) in which each research effort fits. These are fundamental descriptions of each perspective found in many textbooks but not necessarily universally accepted. We believe that the basic structure is simple, well known and lends itself nicely to taxonomic categorization.

Readers should refer to appendix A for a listing of the articles that were reviewed for classification. Appendix A provides information beyond the author and year of publication such as summaries of: (1) key factors that influenced the research, (2) the main problems or concerns addressed by the research, (3) the solution methodology used and (4) the authors' conclusions.

27.4.1. Research on Competitive Strategy in SCM

Competitive decisions are made within multiple planning horizons, usually monthly, annually, or over multi-year planning horizons to achieve an enterprise wide, or supply chain wide, optimal solution which reflects global objectives considering the tradeoffs among functional or organizational demand management structure requirements. Examples of competitive strategy decisions include, but are not limited to: site selection, new product introductions, go/no go decisions on new production/distribution resources, or new markets, as well as overall guidelines for firm wide objectives.

Research which addresses company strategy: (1) should develop objectives and policies for the entire supply chain AND clearly analyze how these support the needs of the firm, (2) should determine the shape of the supply chain in terms of design, and (3) should discuss how supply chain management can enhance the competitiveness of the firm.

We classify each research paper dealing with competitive strategy issues by considering the competitive advantage sub-categories we have outlined in Table 1.

27.4.2. Research on Firm-Focused Tactics

Tactical planning reflects decisions for the coming days, weeks, or months. Higher level decisions and cross-functional objectives/metrics have already been determined at the strategic level; but, actual demand may deviate from plan. Local opportunities require more detailed planning solutions within the local problem domain; this is needed to realign the availability of people, materials and other resources to meet actual demand and bring the operation back within business objectives.

Research which addresses company tactics: (1) should focus on the implementation of strategic decisions, (2) are functional in nature, and may deal with only a few players in the overall chain and (3) may involve systems (MRP, DRP, JIT, etc.) necessary to manage the supply chain.

	Strategic Subcategory	Definition	List of Papers on Strategic SCM Issues
A	Objectives	Understanding the dynamics of the supply chain and the development of objectives for the entire supply chain that includes analysis of how such goals support the needs of the firm. Includes contextual evaluation of supply chain alternatives.	Beamon (1996); Bloemhof-Ruwaard, et al. (1995); Davis (1993); Ernst and Kamrad (1996); Fisher (1997); Forrester (1958); Forrester (1961); Fuller, et al. (1993); Gopal (1992); Oliver and Webber (1992); Shapiro (1984); Slater (1976)
B	Design	Should determine the shape of the supply chain. Includes the design of supply chains or location decisions. Needs to focus on the objectives of the design and not just the development of a tool used in decision making.	Arntzen et al (1995); Berry and Naim (1996); Camm et al (1997); Mourits and Evers (1995); Revelle and Laporte (1996); Towill et al (1992)
C	Competitive Advantage	How supply chain management can enhance the competitiveness of the firm. Includes strategic planning tools.	Cohen and Mallik (1997); Copacino & Rosenfield (1992); Heskett (1977); Houlihan (1985); Jones and Riley (1984); McMullan (1996); Roberts (1990); Scott and Westbrook (1991); Stevens (1989); Towill (1997)
D	Historical Perspectives	Evolutionary or historical perspectives which give us insight to the strategic nature of supply chain management.	Bowersox (1969); Carter & Narasimhan (1996); LaLonde (1993); Langley (1992); Lee and Billington (1995); Masters and Pohlen (1994); Thomas and Griffin (1996)

Table 1: Summary of Research on Competitive Strategy

We subcategorized the tactical decisions into the four main areas that determine how supply chains achieve higher strategic goals and objectives. The categories are:

- Relationship Development
- Integrated Operations
- Transportation and Distribution
- Systems

Therefore, we classified each of the papers that deal with the tactical issues by considering the major issue the paper focuses on. For example, we categorized Choi and Hartley's (1996) paper, which compares supplier-selection practices based on a survey of companies at different levels in the auto industry, as tactical (subcategory, 21). Similarly, we categorized Geoffrion and Graves' (1974) paper on *Multicommodity Distribution Systems* as tactical (subcategory: transportation and distribution) because it focuses on distribution system design. Table 2 gives a summary of the SCM research in the tactical area listed by subcategory.

	Firm Focused Tactics Subcategory	Definition	List of Papers on Tactical SCM Issues
A	Relationship Development	Developing upstream and downstream relations, third-party issues	Choi and Hartley (1996); Gentry (1996); Henig et al (1997); Holmlund and Kock (1996); Prida and Gutierrez (1996); Tagaras and Lee (1996); Walton (1996)
B	Integrated Operations	Managing firm operations as an integrated unit while achieving efficiencies in operations management, including engineering, manufacturing, purchasing & may include immediate up & downstream links	Cohen and Lee (1988); Lamming (1996); Leenders et al. (1994); Roy and Potter (1996); Viswanathan and Mathur (1997)
C	Transportation and Distribution	Achieving efficiencies in managing transportation and physical distribution as an integrated system.	Anupindi & Bassok (1996); Bowersox (1972); Caputo and Mininno (1996); Geoffrion and Graves (1974); Geoffrion and Powers (1995); Min (1996); Robinson et al (1993); Satterfield and Robinson (1996);
D	Systems	Development of operations and information systems or the use of information to aid the achievement of strategic objectives.	Bhaskaran (1996); Bowersox & Morash (1989); Hammel & Kopczak (1993); Verwijmeren et al (1996)

Table 2: Summary of Research on Firm-Focused Tactics

27.4.3. Research on Operational Efficiency:

Typically, operational decisions reflect day to day operations up to two weeks ahead. This area is concerned with the daily operation of a facility such as a plant or distribution center to ensure that the most profitable way to fulfill actual order requirements is considered and executed.

Research that addresses the operational perspective: (1) is concerned with the efficient operation of the company within the supply chain and (2) focuses on controls and performance measures (inventory investment, service level, throughput efficiency, supplier performance and cost).

We sub-categorized these operational problem areas in the following manner:

- Inventory Management and Control
- Production, Planning and Scheduling
- Information Sharing, Coordination, and Monitoring
- Operational Tools

As an example, we categorized Clark and Scarf's (1960) paper, in which they examine policies for a multi-echelon inventory problem, as operational (sub-category, inventory management and control). Table 3 gives a summary of the SCM research in the operational area listed by subcategory.

	Operational Subcategory	Definition	List of Papers on Operational SCM Issues
A	Inventory Management and Control	In terms of the operating efficiency of the supply chain, determining & measuring the performance of inventory. Also includes inventory investment, service levels, allocation schemes & multi-echelon inventory theory	Alderson (1957); Anupindi and Akella (1993); Cachon and Fisher (1997); Clark (1972); Clark and Scarf (1960); Garg and Tang (1997); Glasserman and Tayur (1995); Lee and Billington (1993); Stenger (1994); Stenger (1996); Zinn and Levy (1988); Zinn and Bowersox (1988)
B	Production, Planning and Scheduling	Determining & measuring the performance of production, planning and scheduling to aid the efficient operation of the supply chain.	Graves et al. (1998); Kruger (1997); Lederer and Li (1997); Levy (1997); O'Brien and Head (1995)
C	Information Sharing, Coordination, and Monitoring	Specifies schemes for coordination and control in the sharing of information needed in the efficient operation of the supply chain.	Fisher and Raman (1996); Gavirneni et al (1998); Lee et al (1997); Moinzadeh and Aggarwal (1997); Srinivasan et al (1994)
D	Operational Tools	Development of tools which aid in the efficient operation of the supply chain	Bagahana and Cohen (1998); Slats et al. (1995)

Table 3: Summary of Research on Operational Efficiency

Solution Methodology	List of SCM Papers Sorted by Solution Methodology
Concepts and Non-Quantitative Models	Alderson (1957); Beamon (1996); Bowersox (1969); Forrester (1958, 1961); Gopal (1992); Heskett (1977); Houlihan (1985); Lamming (1996); Lee et al (1997); Lenders et al (1993); Min (1996); Scott and Westbrook (1991); Slats et al (1995); Towill (1997); Verwijmeren et al (1996);
Case Oriented and Empirical study	Bagahana and Cohen (1998); Cachon and Fisher (1997); Caputo and Mininno (1996); Carter and Narasimhan (1996); Choi and Hartley (1996); Davis (1993); Fuller et al (1993); Hammel and Kopczak (1993); Holmlund and Kock (1996); Jones and Riley (1984); Lee and Billington (1995); Levy (1997); McMullan (1996); O'Brien and Head (1995); Oliver and Webber (1992); Revelle and Laporte (1996); Roberts (1990); Roy and Potter (1996); Srinivasan et al (1994); Stenger (1996); Walton (1996);
Frameworks, Taxonomies, and Literature Reviews	Bloemhof-Ruwaard et al (1995); Clark (1972); Cohen and Mallik (1997); Copacino and Rosenfield (1992); Ernst and Kamrad (1996); Fisher (1997); Geoffrion and Powers (1995); La Londe (1994); Langley (1992); Masters and Pohlen (1994); Mourits and Evers (1995); Prida and Gutierrez (1996); Shapiro (1984); Slater (1976); Stenger (1994); Stevens (1989); Thomas and Griffin (1996); Zinn and Levy (1988)
Quantitative Models	Anupindi and Akella (1993); Anupindi and Bassok (1996); Arntzen et al (1995); Berry and Naim (1996); Bhaskaran (1996); Bowersox (1972); Bowersox et al (1989); Camm et al (1997); Clark and Scarf (1960); Cohen and Lee (1988); Fisher and Raman (1996);Garg and Tang (1997); Gavirneni et al (1998); Gentry (1996); Geoffrion and Graves (1974); Glasserman and Tayur (1995); Graves et al (1998); Henig et al (1997); Kruger (1997); Lederer and Li (1997); Lee and Billington (1993); Moinzadeh and Aggarwal (1997); Robinson et al (1993); Satterfield and Robinson (1996); Tagaras and Lee (1996); Towill et al (1992); Viswanathan and Mathur (1997); Zinn and Bowersox (1988)

Table 4: Summary of Solution Methodologies

Overall, we recognize that some of these papers can also be placed in multiple categories. However, we elected to place each paper in one specific category based on the overriding focus of the work presented in the paper. We recognize that cross-referencing these papers into several categories could be beneficial. Therefore, we view this as an opportunity for further clarification.

27.5 Solution Methodology

27.5.1 Categorizing Research Methodologies

For each research paper, we categorized the solution methodology used by the author to address the specific SCM problem areas. We divided the solution methodologies into the following areas:

1. Concepts and Non-Quantitative Models - research that analyzes the supply chain in an attempt to define, describe, and develop methods for the management of the supply chain without using quantitative models.
2. Case Oriented and Empirical study - research that works with specific firms or industries and uses data collected by the researcher or another qualified source to aid in the management of the supply chain.
3. Frameworks, Taxonomies, and Literature Reviews - research that categorizes or explains concepts in SCM as an effort in the understanding of the breadth and depth of the concept.
4. Quantitative Models (to include optimization, simulation, stochastic models, and heuristics) - research that attempts to develop methods for the management of the supply chain using quantifiable models.

For example, Steven's (1989) paper on Integrating the Supply Chain, we categorize as concepts and non-quantitative models because it presents a three-phase analytical but non-quantitative process that can be used to develop an integrated supply chain strategy. Similarly, we categorized the solution methodology presented in Camm et al's (1997) paper on restructuring Procter and Gamble's supply chain as a quantitative model (LP-based) solution methodology. Table 4 represents a categorization of the solution methodologies presented in this research.

In addition, Figure 2 illustrates the comparison between the research methodologies with the type of SCM problem addressed in our classification scheme. As one might expect, most of the strategic work is of a conceptual nature while quantitative models are mostly found dealing with operational and tactical issues. The numbers correspond with the first column of the table in Appendix A so that the reader can quickly find where a specific article lies in the matrix.

854

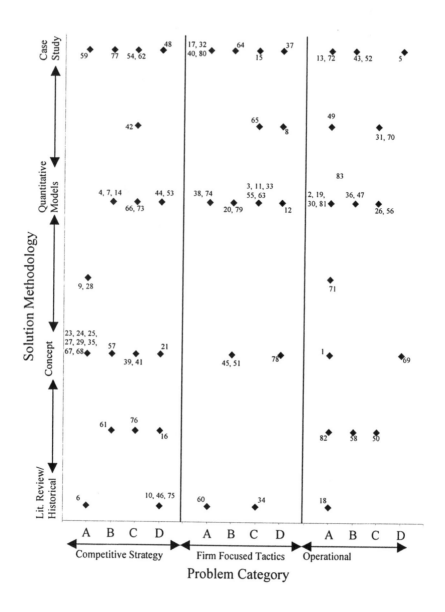

Figure 2: Problem Category vs. Solution Methodology

27.6   Conclusions

This paper recognizes that many other frameworks have been developed and presented in SCM literature. However, the value of this framework is as a tool to help researchers synthesize the volume and breadth of being accomplished on SCM.

Defining supply chain management requires that we answer three questions; (1) **what is a supply chain?** - There appears to be some convergence in the literature as to what a supply chain is and can be generally summed up as a series of interconnected activities which are concerned with planning, coordinating and controlling materials, parts, and finished goods from supplier to customer, (2) **what managing the supply chain means** - Supply chains are managed as a single entity either through a dominant member or through a system of partnerships requiring well developed cooperation and coordination. The scope of responsibility for managing a supply chain is specific to the firm and its myriad of relationships with its suppliers, vendors and customers, and (3) **who manages the supply chain** - We can break down SCM into various parts but one must not lose sight of the fact that SCM is rooted in senior level decision making. Otherwise, what is to distinguish SCM from purchasing, distribution, materials management or even integrated logistics? Of course, SCM includes implementation and operational aspects in which day to day operations are managed below the senior management level.

From an analysis of the key factors that have influenced SCM research in the past, we conclude that external market forces have largely driven SCM. As many of the papers suggest, the keys to success in SCM require heavy emphasis on integration of activities, cooperation, coordination, and information sharing throughout the entire supply chain. We do not expect this to change. The realities of implementing a system that requires integration, coordination and cooperation will, however, call for a major change in business culture. Bowersox (1997) argues that the main challenge for SCM will be the management of intense relationships across enterprises that involve such issues as collaboration, information sharing, partitioning, diverse corporate cultures, shared risks, cost sharing, integrity, and trust. Bowersox calls for comprehensive performance metrics to manage, measure and reward performance on an integrated basis. He also suggests that these measures should link the supply chain's performance directly to the relevant stakeholders since logistics cost represents 10 percent of the national gross domestic product.

To have the most benefit, the supply chain must be managed as a single entity. Firms must avoid sub-optimization through self-interest at any link in the chain by managing the entire chain as a single entity while simultaneously dealing with the power relationships that are inherent in the chain. On the other hand, we note that many firms belong to multiple supply chains and little has been written to address the issues associated with multiple relationships.

It appears that the thrust of the SCM research has been focused in logistics and operations management areas. However, we have found that many other fields have contributed to this research and this trend should continue in the future. The synthesis of research from as many fields as possible helps us to establish a better definition of the concept of SCM and to show how it can potentially be managed more effectively.

Researchers recognize how well developed the field has become in the last decade and they point to some of the accomplishments of the recent past and to the needs of the future. For example, Baganha and Cohen (1998) point out that amplification of the variability of demand up the supply chain has been recognized and described. Bhaskaran (1996) notes that manufacturers have recognized the need to optimize the performance of the supply chain connecting raw material to finished product. Ernst and Kamrad (1996) discuss that the concepts of postponement and modularization have been well researched and they add to this by discussing the combined potential of these two approaches. Geoffrion and Powers (1995) point out that using optimization to design distribution systems became technically feasible a little more than two decades ago, and developments have occurred at a rapid rate ever since. They conclude that creative logistics analysts and planners have found ingenious and unanticipated ways to use and embellish classic models. They believe that this trend should continue, as these tools become more widely used and accessible through personal computers. Gavirneni, Kapuscinski, and Tayur (1998) note that the focus of managing the supply chain has led to radical changes in thinking about supplier/customer relations.

Cohen and Mallik (1997) review the state of knowledge and practice of SCM and find that in terms of practice, what's been written about supply chain management is conceptual and somewhat impractical, inspirational but sometimes vague, or too company specific and therefore too hard to apply to other situations. They illustrate that many attempts at modeling supply chains are overwhelmed by oversimplified underlying assumptions. Researchers point out that firms are able to share information because IT costs have been reduced dramatically and the advantages of cheap information are widely accepted but their value has yet to be quantitatively explored. Fisher (1997) finds that "the performance of supply chains have never been worse despite implementation of new concepts (quick response, mass customization, lean manufacturing, agile manufacturing), and the application of new technology." Also, as globalization increases among firms, the need will be even more pressing in the future for usable supply chain management tools. Fortunately, interest in supply chain management remains high and research continues to develop at a rapid pace. The state of knowledge, practice and the tools that are developed will continue to improve, for these are the daunting challenges that spur on good research.

Appendix A   SCM Quick Reference

LN	Year	Author	Title	Problems Addressed	Conclusions
1	57	Alderson, W.	Marketing Behavior and Executive Action: A Functionalist Approach to Marketing Theory	Marketing efficiency within a complete system of distribution can be promoted through application of the principal of postponement.	Postponement can reduce the cost of carrying inventory and reduce marketing risk. It is a major analytical tool that can be derived from the view that inventory management is an essential marketing function.
2	93	Anupindi, R. & Akella, R.	Diversification Under Supply Uncertainty	Dual (or more) sourcing strategies provide a hedge against supplier quality and delivery uncertainty. The operational issue of quantity allocation between two suppliers needs to be explored due to impacts on inventory policy and costs.	Optimal inventory policies can be determined using these models but application to real world situations is questioned and limitations are discussed.
3	96	Anupindi, R. & Bassok, Y.	Distribution Channels, Information Systems and Virtual Centralization	What are some of the challenges that present themselves when retailers decide to explore this horizontal link? The transhipment of goods and distribution needs to be reexamined under this new condition.	The system of decentralized retailers with info. sharing (virtual centralization) gives more revenues to the manufacturer. All retailers will gain from such a system but not all equally.
4	95	Arntzen, B.C., Brown, G.G., Harrison, T.P. and Trafton, L.	Global SCM at Digital Equipment Corporation	shows how DEC evaluates global SC alternatives and determines worldwide manufacturing and distribution strategy using the global SC model	GSCM is a very general approach to modeling SCs applicable to to any firm involved in multistage, multiproduct manufacturing (using the global bill of materials)
5	98	Baganha, M.P. & Cohen, M.A.	The Stabilizing Effect of Inventory in Supply Chains	Develop a model to explain the observations of the bullwhip effect and indicate mechanisms which can promote stabilization at various points in the chain.	Wholesalers can introduce a degree of stabilization into the supply chain by transmitting an order process to manufacturers with variability lower than the variability inherent in the retailer replenishment order process.

LN	Year	Author	Title	Problems Addressed	Conclusions
6	96	Beamon, B. M.	Performance Measures in Supply Chain Management	Evaluate existing supply chain performance measures and establish a general framework for their development and application	Effective performance measures are often neglected in the design and analysis of supply chains. This paper providesa general framework for their effective development and appliation
7	96	Berry, D. and Naim, M.	Quantifying the relative improvements of redesign strategies in a PC supply chain	outlines the dev. of simulation models that describe implications of SC redesign	dynamic performance improvements can be achieved at each successive stage of the redesign process
8	96	Bhaskaran, S.	Simulation Analysis of a Manufacturing Supply Chain	Need a tool to manage the transmission of schedule instability and the resulting inventory fluctuation. Controlling fluctuations in forecasted demand can have a great impact on inventory levels throughout the SC.	Operations in series in the pipeline need to be coordinated. Kanban can use forecasts to help reduce demand errors. The model can help to 1)simulate differences between MRP & kanban systems 2)simulate the effects of mfg. smoothing at the top of the pipe.
9	95	Bloemhof-Ruwaard, J.M., Van Beek, P., Hordijk, L., and Van Wassenhove, L.N.	Interactions between operational research and environmental management	How to incorporate environmental issues when analyzing supply chains	OR may be a suitable science to cope with the cradle-to-grave approaches in SCM and with the global problems in the environmental chain. More complicated models are needed to cope with recovery mgt and regional problems in env. chain approach.
10	69	Bowersox, D.J.	Physical Distribution Development, Current Status, and Potential	A flurry of attention has focused upon the concept of integrated physical distribution since the mid-1950's and a synthesis of contemporary physcial distribution thought is needed.	Physical distribution has evolved into an issue of competitive advantage and includes not only single firm issues of total cost and system integration but also integration beyond the firm into the total cooperative channels of distribution.

LN	Year	Author	Title	Problems Addressed	Conclusions
11	72	Bowersox, D.J.	Planning Physical Distribution Operations with Dynamic Simulation	This article reviews a dynamic simulation model, LREPS, which is capable of simulating the physical distribution system of a manufacturing firm engaged in national distribution of packaged goods.	The LREPS model has proven to be of assistance in the planning of physical distribution systems. Using this model, a synthesis of results into some general priciples to guide integrated physical distribution system design will emerge.
12	89	Bowersox, D. and Morash, E.	The Integration of Marketing flows in Channels of Distribution	Integrating channelwide marketing strategies can provide enhanced potential for strategic leveraging of channel efficiency and effectiveness	Answers to questions include:amount of slack time (postponement potential), cost ramifications of network modifications, advisability of "crash programs" (flow acceleration) on the critical path
13	97	Cachon, G. & Fisher, M.	Campbell Soup's Continuous Replenishment Program: Evaluation and Enhanced Inventory Decision Rules	Describe how a continuous replenishment program works in practice. Could more sophisticated inventory rules improve performance.	Under the revised system it was found that retailer inventories were reduced 66% without reducing service level. Cost of goods sold fell by 12%. These savings could have been achieved without VMI.
14	97	Camm, J., Chorman, T., Dill, F., Evans, J., Sweeney, D., and Wegryn, G.	Restructuring P&G's Supply chain	choose the best location and scale of operation for making each product: provide modeling support tool to ensure best possible solution across SC	Two models were tied using aggregation and parameterization to avoid significant suboptimization. Synergy of OR/MS and GIS led to high level of acceptance
15	96	Caputo, M. and Mininno, V.	Internal, vertical and horizontal logistics integration in Italian grocery distribution	focus on branded products industry and large-scale retail trade business to increase operating efficiency and improve interfunctional and interorganizational coordination in logistics channels in Italian grocery distribution	Successful strategy depends on clear delineation of responsibility at each echelon and the quickness of processing and exchange of information (EDI); vertical integration (systematic coord of physical and informational flows)

LN	Year	Author	Title	Problems Addressed	Conclusions
16	96	Carter and Narasimhan	North American vs European Future Purchasing Trends	Document the changes and emerging trends in sourcing and supply mgt. and their impacts on supply mgt.	Most significant trends: greater strategic importance; reliance on Info Tech.; EDI; Env sensitive purchasing; sourcing teams
17	96	Choi, T. and Hartley, J.	An exploration of supplier selection practices across the SC	Compare supplier-selection practices based on a survey of companies at different levels in the auto industry	No difference found (among auto assemblers, direct and indirect suppliers) for the importance placed on consistency (quality and delivery), reliability, relationship, flexibility, price, and service. Price is one of the least important selection items.
18	72	Clark, A.J.	An Informal Survey of Multi-Echelon Inventory Theory	To fully appreciate the directions of multi-echelon research, it is useful to have a basic familiarity with analytical approaches to the problems.	It is probable that research in multi-activity inventory therory has reached a point where marginal returns from further work are likely to diminish. The principle work to be done lie in refinements and extensions of previous results.
19	60	Clark, A.J. & Scarf, H.	Optimal Policies for a Multi-Echelon Inventory Problem	There is a need to consider the problem of determining optimal purchasing quantities in a multi-installation model.	Optimal solutions for multi echelon inventory models are possible.
20	88	Cohen, M. and Lee, H.	Strategic Analyis of Integrated Production-Distribution Systems	measuring cost/service/ flexibility tradeoffs in production/ distribution systems	methodology considers relationships between production and distribution control policies that affect inventory control, plant product mix, and production sched. Other manuf. strategy decisions (e.g. facility location, capacity planning) are assumed fixed

LN	Year	Author	Title	Problems Addressed	Conclusions
21	97	Cohen, M.A. & Mallik, S.	Global Supply Chains: Research and Application	The objective of the paper is to review the state of knowledge & practice of SCM. A global supply chain will realize competitive advantage only if management is coordinated & value added activities are flexible in response to changes in market conditions.	Majority of models lack practiaclity & would be difficult to implement. Globalization has increased considerably among big US firms. Evidence of the growth of SC coordination is less clear. Paper concludes with several specific areas for future work.
22	92	Copacino, W. and Rosenfield, D.	Analytical tools for Strategic Planning	Focus on analytical tools for strategic logistics planning	None Given
23	93	Davis, T.	Effective Supply Chain Management	Hewlett-Packard has developed a framework for addressing the uncertainty that plagues the performance of suppliers, the reliability of manufacturing and transportation processes, and the changing desires of customers.	The author describes several cases in which entire product families have been reevaluated in a supply chain context. The methodology employed should help others manage their supply chain.
24	96	Ernst, R. & Kamrad B.	A Conceptual Framework for Analyzing Supply Chain Stuctures	Evaluate different supply chain structures in the context of modularization and postponement.	Specific examples of companies that fall into each category are cited. Verticle integration along the supply chain is not desirable. The framework provides a way to start the decision process by helping to contrast the chain design options.
25	97	Fisher, M.	What is the right supply chain for your product? Effective Supply Chains	Managers lack the framework for deciding which ideas and technologies best fit their particular company's situation	To take the right approach, comapnies should first determine whether their products are functional (predictable demand) or innovative. Then select either an efficient (cost) SC or a responsive (time) SC strategy

LN	Year	Author	Title	Problems Addressed	Conclusions
26	96	Fisher, M. and Raman, A.	Reducing the Cost of Demand Uncertainty through accurate response to early sales	How to avoid stockouts and inventory obsolescence because long lead times coupled with a concentrated selling season force all or at least most production to be committed before demand information is available.	Relative to the cost that would have been incurred if no response were used, optimized response reduces cost by enough to quadruple profits.
27	61	Forrester, J. W.	Industrial Dynamics	Industrial Dynamics studies the behavior of industrial systems to show how policies, decisions, structures, and delays are interrelated to influence growth and stability.	The five flows are interrelated by an info network; I.D. recognizes the critical importance of this info network in giving the system its own dynamic character. Uses models to show how info. and policy create the character of the organization.
28	58	Forrester, J.W.	Industrial Dynamics	Predict the specific kinds of progress which will be achieved and describe the concepts which will make it posssible	Companies will come to be recognized not as a collection of separate functions but as a system in which the flows of information, materials, manpower, capital equip., and money set up forces than determine growth, fluctuation and decline
29	93	Fuller, J., O'Conner, J., & Rawlinson, R.	Tailored Logistics: The Next Advantage	The goal of logistics strategy is to organize companies to compete across the span of their markets without overcharging customers or underserving others	Logistics have become central to product strategy because it is increasingly clear, products are not just things-with-features. they are TWF bundeled with services
30	97	Garg, A. & Tang C.S.	On Postponement Strategies for Product Families with Multiple Points of Differentiation	There is a need to develop research which studies products with more than one point of differentiation. Discovering the conditions when one type of postponement is the most beneficial. Extenstion of Eppen & Schrage (1981)	Centralized model - variability & correlation of demands play an important role in which postponement strategy to use. Decentralized model - magnitudes of lead times have a strong impact on inventory. A simple way to ID the optimal point to be postponed.

LN	Year	Author	Title	Problems Addressed	Conclusions
31	98	Gavirneni, S., Kapuscinski, R. & Tayur, S.	Value of Information in Capacitated Supply Chains	The degree of cooperation varies significantly from one supply chain to another. How much cooperation and coordination is needed for firms to derive the benefits of improved relationships?	Info. is most beneficial @ moderate variances. The benefit of info. flow is higher at higher capacities. If the variance of the demand seen by the customer is small, we can expect the benefit of info. flow to increase with increase in penalty cost.
32	96	Gentry, J.	The Role of the Buyer-Supplier Strategic Partnerships: A SCM Approach	Reports on the nature and degree of carrier involvement within buyer-supplier strategic partnerships	Study indicates that three-way relationships are typical. Therfore, buyer-supplier partnerships can establish a solid foundation for the formation of successful SCs over time. Sholld integrat carrier operations into overall buyer-supplier planning/comm.
33	74	Geoffrion, A. and Graves, G. W.	Mulitcommodity Distribution System Design by Benders Decomposition	distribution system design: optimal location of intermediate distribution facilities between plants and customers	Remarkable effectiveness of Benders Decomposition as a computational strategy for static multicommodity intermediate location problems
34	95	Geoffrion, A.M & Powers R.F.	Twenty Years of Strategic Distribution System Design: An Evolutionary Perspective	An overview of the state of design in distribution systems would be helpful in understanding why the tools we use today exist in the form that they do.	Creative logistics analysts and planners have found ingenious & unanticipated ways to use and embellish classic models. This should continue as these tools become more widely used and accessible thru personal computers.
35	92	Gopal, C.	Manufacturing systems for a competitive global strategy	Explores aspects of an integrated system and the logistics chain, from purchasing to distribution, necessitated by a global policy	Benefits of a global system include: better coordination of customer service in multiple markets, fast and accurate info transfer, less difficulty complying with local laws and regs, conformance of both central and local planning to overall objectives

LN	Year	Author	Title	Problems Addressed	Conclusions
36	98	Graves, S., Kletter, D. & Hetzel, W.	A Dynamic Model for Requirements Planning with Application to Supply Chain Optimization	Develops a new model for requirements planning multi-stage production inventory systems.	Provides some evidence of the value of taking a corporate wide view by optimizing the supply chain rather than sub-optimizing each of the pieces.
37	93	Hammel, T. and Kopczak	Tightening the Supply Chain	describes how HP successfully used its "Frontier Program" through product redesign and DRP to improve its series 700 terminals operations	product redesign was essential (including CIM); enhanced product availability through demand pull DRP; warranty cost in check with US manufacture
38	97	Henig, M., Gerchak, Y., Ernst, R., and Pyke, D.	An Inventory Model Embedded in Designing Supply a Contract	Explore the joint optimization of contract parameter and inventory control policy in an environment characterized by demand uncertainty in order to specify the frequency of, and volume for, future deliveries	Show that the difference in costs can be significant when comparing the costs of suboptimal policies, in conjunction with the best contract volume, to those of the optimal inventory policy and associated contract volume.
39	77	Heskett, J.L.	Logistics - Essential to Strategy	Many firms develop competitive logistics strategies based in part on concepts like postponement, standardization, consolidation, & differentiation. Firms redesign systems to provide more effective support for corporate strategy.	Finds that logistics managment must participate in strategic decision making. Argues that globalization will have a huge impact on the importance of good logistics design and development within corporate strategy.
40	96	Holmlund, M. & Kock, S.	Buyer Dominated Relationships in a Supply Chain - A Case Study of Four Small Sized Suppliers	There is a need to analyze the relationships, and bonds between a dominating buyer and small sized suppliers in a supply chain.	Although the relationships between the buyer and the small vendors was quite old, few strong bonds were found. Contracts remained relatively short term. Buyer has no incentive to help vendor max. profits and vendor was forced to make unprofitable parts.

LN	Year	Author	Title	Problems Addressed	Conclusions
41	85	Houlihan, J. B.	International Supply-Chain Management	Describes the concepts underlying the new approaches to managing change in international chains, the barriers to be overcome and the lessons learned	A holistic approach to Int SCM requires the incorporation of a logistics focus into the strategic decisions of the firm;
42	84	Jones, T. and Riley, D.	Using Inventory for competitive andantage through SCM	Focus on the myths of the past	Barriers to SCM are traditional, organizational, legal, and non-integrated mgt systems--mgt objectives and measures are in fundamental conflict. Key to success: mutually advantegous relationships that make the chain work more smoothly at lower costs.
43	97	Kruger, G. A.	The Supply Chain Approach to Planning and Procurement Management	The supply chain approach models stochastic events influencing a manufacturing organization's shipment and inventory performance in the same way that an engineer models tolerance buildup in a new product design.	The actual performance the factory experiences will depend upon whether the supply chain performs according to the inputs provided the statistical model.
44	94	La Londe, B. J.	Evolution of the Integrated Logistics Concept	Describes the evolution of Integrated Logistics: (1) Physical distbn; (2) Internal linkages; (3) external linkages;	None Given
45	96	Lamming, R.	Squaring lean supply with SCM	To understand lean production better, we must investigate SCM.	Challenge: redesign the way in which responsibility for value mgt is shared. Precept of vantage point and customer superiority that are central to SCM are directly contrary to those of lean supply; need shared beliefs; customer infallibility-problematic

LN	Year	Author	Title	Problems Addressed	Conclusions
46	92	Langley, C.J.	The Evolution of the Logistics Concept	Casts the evolution of logistics into three specific contexts: past (1950-1964); present (1965-); future	Four stages in development of Logistics function: (1) cost control; (2) profit-center orientation recognizing positive impact on sales; (3) view logistics as key to product differentiation; (4) principal strategic adv. revolves around logistics.
47	97	Lederer, P. and Li, L.	Pricing, production, scheduling, and delivery-time competition	studies competition between firms that produce goods or services for customers sensitive to delay time. Firms compete by setting prices and production rates for each type of customer and by choosing scheduling policies.	A faster, lower variability and lower cost firm always has a larger market share, higher capacity utilization, and higher profits. Also, customers with higher waiting costs pay higher full prices, and that each firm charges a higher price and delivers faster to more impatient customers.
48	95	Lee, H. and Billington, C.	The Evolution of SCM models and practice at HP.	Show how HP successfuly used an integrated team approach to implement SCM and improve customer satisfaction	SCM is a business fundamental
49	93	Lee, H.. & Billington, C.	Material Management in Decentralized Supply Chains	Need a decentralized model that allows for 1) generalized network structure 2)uncertainties (demand, process & supply) 3)simplicity in computation 4)capacitated production systems.	A simple model based DSS can be used by practioners to evaluate alternative supply chain designs and determine the most practical inventory control under uncertain conditions in a system of decentralized control.
50	97	Lee, H., Padmanaabhan, V. & Whang S.	Information Distortion in a Supply Chain: The Bullwhip Effect	An analysis of the sources of the bullwhip effect and strategies that can be employed to lessen the negative impact of the effect are explored.	Demand distortion may arise as a result of optimizing behaviors of individuals in the supply chain. Cooperation and coordination among members of the chain is necessary to combat this problem. Net benefits of such actions can be shared by members.

LN	Year	Author	Title	Problems Addressed	Conclusions
51	94	Leenders, Nollet, and Ellram	Adapting Purchasing to SCM	purchasing function must become fully integrated into the customer-employer SC	suppliers and the way in which we relate to them must provide a strategic advantage. The PM should manage the SC by integrating the org's internal and external ops.
52	97	Levy, D.L.	Lean Production in an International Supply Chain	While the business press has championed both globalization and lean production as inevitable and valuable, there has been little investigation into the interaction of the two.	Lean production requires frequent, rapid flows of info. & goods along the value chain, which is costly when value chain activities are geographically dispersed. Two key elements of lean production; design for mfg. and low defect levels stabilize the SC.
53	94	Masters J. & Pohlen, T.	Evolution of the Logistics Profession	Describes the evolution of Logistics profession into 3 pahases: functional mgt (1960-70s); Interanl integration (1980s); External integration (1990s)	None Given
54	96	McMullan. A.	Supply chain management practices in Asia pacific today	Examines how managers in Asia Pacific are responding to pressures of competition and the strategies they are implementing to enhance SCM (mgt issues, roles and responsibilities, competitive strategies, performance mgt)	To successfully implement SCM, many firms will have to change their organizational structures, SCM relationships, and performance measures. Also requires implementation of new information technology.
55	96	Min, H.	Distribution channels in Japan	Effective way of enahacing ability of US co's to penetrate Japanese market is to study Japanese practices (e.g. an indigeneous channel dist. which disfavors foreign firms due to legal impeds & "locked-up relationships"	Strategies for successful penetration include: targeting specialized niche markets; selling under Japanese brand names; empahasize follow-up service; piggyback; sell in non-keiretsu system; direct marketing

LN	Year	Author	Title	Problems Addressed	Conclusions
56	97	Moinzadeh, K and Aggarwal, P.	An Information Based Multiechelon Inventory System with Emergency Orders	(S-1,S) multiechelon inventory system where all the stocking locations have the option to replinish their inventory through either a normal or a more expensive emergency resupply channel	Incorporating information on the remaining leadtimes of the outstanding orders, when selecting a resupply mode, can result is considerable cost savings when compared to policies which allow a single resupply mode.
57	95	Mourits, M. and Evers, J.	Distribution Network Design: An Integrated Planning Support Framework	Many logistics support systems have been developed to assist design but there are too many shortcomings--need coherent approach. Goal of paper is to make optimization easier to use & make solutions meaningful to customers.	IPSF provides a systematic approach to the challenges firms face in designing their supply chain.
58	95	O'Brien, C. & Head, M.	Developing a Full Business Environment to Support JIT Logistics	Establish a business cycle (the transactions needed to complete a purchase of goods or services) appropriate to JIT supply to the motor industry. The business cycle also includes the communication between all partners in the chain.	EDI was seen to be necessary to support a JIT full business cycle. There are significant savings to be gained from the use of the concept when the FBC is integrated throughout the SC. More work is needed to support the financial aspects of SCM.
59	92	Oliver, K. and Webber, M.	SCM: Logistics catches up with strategy	Booz-Allen study of a variety of industries worldwide found that traditional approaches of seeking tradeoffs among the various conflicting objectives of key functions along the SC no longer worked well	Unlike traditional approaches (manipulation of inventories and improved material mgt), SCM requires and underlying strategic focus along with the involvement and commitment of top mgt.
60	96	Prida, B. & Gutierrez, G.	Supply Management: From Purchasing to External Factory Management	The role of the purchasing function (buyer) has changed dramatically with the onset of SCM. The authors propose to discuss the challenges that face purchasing employees.	The evolution in supply chain management has occured in three stages:1)traditional purchasing role 2)SCM thru subcontracting 3)SCM through innovation.

LN	Year	Author	Title	Problems Addressed	Conclusions
61	96	Revelle, C. S. & Laporte, G.	The Plant Location Problem: New Models and research Prospects	Objectives needed: 1) Capacitated Plant, Fixed charge Transport 2) Max. ROI Location has been ignored in past 3)Multiobjective Problemss should be considered 4)Multiple product/Multiple Machine Problems 5) Probs with spatial interaction models	Plant location problems can be expanded and restrictions can be relaxed to bring more realism into models while providing new challenges for solving plant location problems.
62	90	Roberts, J. H.	Formulating and Implementing a Global Logistics Strategy	Provide the background scenario for which ILC is developing their corporate logistics strategy	None Given
63	93	Robinson, E.P., Gao, L., and Muggenborg, S.D.	Designing and Integrated Distribution System at DowBrands, Inc.	Developed an optimization-based decision support system for designing two-echelon, multi-product distribution systems and applied it to DowBrands, Inc.	Optimization procedure gave management the analytical support it needed to eliminate uncertainties and develop guidelines for change
64	96	Roy, R. & Potter, S.	Managing Engineering Design in Complex Supply Chains	Literature on "supplier partnerships" has largely overlooked the implications for managing design and development.	The case studies identify several factors that determine the extent to which it is appropriate to devolve design and development to suppliers. These factors include; the type of industry, firm and product and level of innovation.
65	96	Satterfield, R.K. & Robinson, E.P.	Designing Distribution Systems to Support Vendor Strategies in Supply Chain Management	The interactions between cost, distribution service, market share, and revenue are largely ignored by existing optimization based system design models.	Successful application of framework for designing a distribution system. Model draws upon research from varied disciplines and includes revenue considerations in addition to cost minimization. Integrated approaches for vendor dist. design are justified.

LN	Year	Author	Title	Problems Addressed	Conclusions
66	91	Scott, C. and Westbrook, R.	New Stategic Tools for SCM	How to overcome major barriers to SC Integration	Celebrated examples of SCM are more admired than emulated. Managers should ask three questions: (1) what is the current shape of the SC?; (2) How collaborative are the relationships?; (3) which combination of physical & info. processing to use?
67	84	Shapiro, R. D.	Get Leverage from Logistics	Managers should understand precisely what their companies are trying to do and bring their logistical capabilities in line with corporate purpose-- What must our logistics system do well?	A Company can gain leverage by ensuring a good fit between its logistics system and competitive strategy
68	76	Slater, A.	The Significance of Industrial Logistics	There is a need for a framework which fully describes the role and organization of industrial logistics. A definition, the structure and a discussion of it's significance to the bottom line are needed.	A total systems approach to the logistics channel will reduce total costs and considerably improve the overall quality of the operation. The isolation and individual aims of channel members should be replace by cooperation.
69	95	Slats, P.A., Bhola, B., Evers, J.J., and Dijkhuizen, G.	Logistic Chain Modelling	Analyze and evaluate the role OR can paly in logistic chain integration and BPR	OR models and techniques are well suited to analyze the local performance of logistic sub-chains and processes.
70	94	Srinivasan, K., Kekre, S. & Mukhopadhyay, T.	Impact of Electronic Data Interchange Technology on JIT Shipments	Does the use of EDI enhance the shipment performance of suppliers in a JIT environment?	EDI technology facilitiates accurate and timely information which aids in the coordination of material movement and leads to better shipment performance. Firms pursuing a single source strategy are prone to greater shipment errors.

LN	Year	Author	Title	Problems Addressed	Conclusions
71	94	Stenger, A.	Inventory Decision Framework	Provides a framework within which to view inventory decisions in logistics	Inventory decisions must be made within the context of the efficient functioning of the entire SC. Approach include: eliminate need for inventory; maintain min inventory required; use scientific approaches to manage inventory
72	96	Stenger, A.	Reducing Inventories in a Multi-Echelon Manufacturing Firm	Presents a method to help a firm identify the relative impact of various determinants of inventories as a means of setting priorities for inventory reduction.	rough-cut models seem to offer potential to better gauge the magnitude of the opportunities in a given situation
73	89	Stevens, G.	Integrating the Supply Chain	Develop an integrated supply chain strategy	companies that consider the supply chain during the strategic debate, will be more successful in terms of increased market share and lower asset base.
74	96	Tagaras, G. & Lee, H. L.	Economic Models for Vendor Evaluation with Quality Cost Analysis	Vendor selection criteria needs to go beyond costs and delivery performance and also address the quality of incoming materials.	Vendors must not be evaluated only on the basis of their own quality and prices. The attractiveness of a vendor also depends on the quality and cost characteristics of the purchaser.
75	96	Thomas, D. and Griffin, P.	Coordinated SCM	Review the literature addressing coordinated planning between two or more stages of the supply chain, placing emphasis on models that address a total supply chain model.	Strategic models based on case studies are popular; most of these models are based on complicated integer LPs with underlying network structure that can be exploited using decomposition; advancement in comm and info tech present many SCM opportunities

LN	Year	Author	Title	Problems Addressed	Conclusions
76	97	Towill, D.R.	The Seamless Supply Chain - The Predator's Strategic Advantage	The paper reviews the techniques available to "predators" seeking to gain competitive advantage for their supply chains.	Must go beyond the lean supply chain to the seamless supply chain. Partnerships between members of the supply chain take a attitudnal change as well as a change of ownership in the process.
77	92	Towill, Naim and Wikner	Industrial Dynamics Simulation Models in the Design of Supply Chains	Review the operation of supply chains, and provide simple conclusions about ways of reducing demand amplification	By using a simulation model of proposed supply chains, different strategies can be compared and costed.
78	96	Verwijmeren, M., van der Vlist, P., & van Donselaar, K.	Networked Inventory Management Information Systems: Materializing Supply Chain Management	Need to elevate the SCM research from global concepts to some tangible information systems for inventory managment in practice.	A group of networked systems can act as a single integrated decision system thus removing amplification in the supply chain.
79	97	Viswanathan, S. and Mathur, K.	Integrating Routing and Inventory Decisions in One-warehouse Mulitretailer Multiproduct Distribution Systems	Consider distributions systems with a central warehouse and many retailers that stock a number of different products. Objective is to determine replinishment policies that specify delivery quantities and vehicle routes to minimize long-run inv. and trans costs.	Proposed heuristic is an excellent planning tool to control inventory and transportation costs in a multiechelon inventory/distribution system.
80	96	Walton, L.W.	Partnership Satisfaction: Using the Underlying Dimensions of Supply Chain Partnership to Measure Current and Expected Levels of Satisfaction	The question of partnership satisfaction has generally gone unanswered. Need to assess current satisfaction and future expectation of partnerships as perceived by business executives.	Planning dimension found to be stat. sig., limited sharing of benefits between partners, managers are generally pleased with value of partnerships. Managers are not satisfied with their current level of communication and info. exchange with partners.

LN	Year	Author	Title	Problems Addressed	Conclusions
81	88	Zinn, W. & Bowersox, D.J.	Planning Physical Distribution with the Principle of Postponement	To provide effective support, managers need to serve an increased number of delivery destinations, while simultaneously supporting a growing number of SKU's. The principal of postponement can help in keeping costs down in this chaotic environment.	Under specific conditions, the principle of postponement offers an opportunity for management to improve the productivity of physical distribution systems by reducing cost associated with anticipatory distribution.
82	88	Zinn, W. & Levy M.	Speculative Inventory Management: A Total Channel Perspective	Determine the best place in the supply chain for speculative inventory.	CFS is found to be useful in explaining actual company behavior in cases in which mgmt.'s objective is to minimize costs. Postponement/Speculation is more suitable for deciding under what circumstances a speculative inventory should appear in the chain.
83	95	Glasserman, P. & Tayur S.	Sensitivity Analysis for Base-stock Levels in Multiechelon Production-inventory Systems.	Effective management of inventories in large scale production and distribution systems requires methods for brining model solutions closer to the complexities of real systems.	For various cost and performance measures in capacitated, multiechelon production inventory systems, derivatives with respect to base stock levels can be consistently estimated from simulation, or even real data.

References

Alderson, W. Marketing Behavior and Executive Action: A Functionalist Approach to Marketing Theory. Irwin:Homewood, Il. 1957.

Anupindi, R. and Akella, R. Diversification under supply uncertainty. *Management Science* 39:944-963, 1993.

Anupindi, R. and Bassok, Y. Distribution Channels, Information Systems and Virtual Centralization. Proceedings of MSOM Conference: 87-92, 1996.

Arntzen, B.C., Brown, G.G., Harrison, T.P., and Trafton, L. Global supply chain management at Digital Equipment Corporation. *Interfaces* 25:69-93, 1995.

Bagahana, M.P. and Cohen, M.A. The Stabilizing Effect of Inventory in Supply Chains. To appear in *Operations Research*. 1998.

Beamon, B.M. Performing Measures in Supply Chain Management. From Rensellaer Polytechnic University conference on agile manufacturing, Albany, New York, October 2-3, 1996.

Berry, D. and Naim, M.M. Quantifying the relative improvements of redesign strategies. *International Journal of Production Economics* 46-47, Dec: 181-196, 1996.

Bhaskaran, S. Simulation Analysis of Manufacturing Supply Chain. Presented at Supply Chain Linkages Symposium, Indiana University, 1996.

Blattberg, R.C., Eppen, G.D., and Lieberman, J. A Theoretical and Empirical Evaluation of Price Deals for Consumer Non-Durables. *Journal of Marketing* 45:116-129, 1981.

Bloemhof-Ruwaard, J.M., Van Beek, P., Hordijk, L., and Van Wassenhove, L.N. Interactions between operational research and environmental management. *European Journal of Operational Research* 85:229-243, 1995.

Bowersox, D.J. Readings in Physical Distribution Management: The Logistics of Marketing. Eds. Bowersox, D.J. , La Londe, B.J., and Smykay, E.W., New York: MacMillan, 1969.

Bowersox, D.J. Planning Physical Distribution Operations with Dynamic Simulation. *Journal of Marketing* 36:17-25, 1972.

Bowersox, D.J. and Morash, E.A. The Integration of Marketing Flows in Channels of Distribution. *European Journal of Marketing* 23:58-67, 1989.

Bowersox, D. J. Integrated Supply Chain Management: A Strategic Imperative, presented at the Council of Logistics Management 1997 Annual Conference, 5-8 Oct. Chicago, IL., 1997.

Cachon, G.P. and Fisher, M.L. Campbell Soup's Continuous Replenishment Program: Evaluation and Enhanced Inventory Decision Rules. *Production and Operations Management* 6:266-276, 1997.

875

Camm, J.D., Chorman, T., Dill, F., Evans, J., Sweeney, D., and Wegryn, G. Blending OR/MS, Judgement, and GIS: Restructuring P&G's Supply Chain. *Interfaces* 27:128-142, 1997.

Caputo, M. and Mininno, V. Internal, Vertical, and Horizontal Logistics Integration in Italian Grocery Distribution. *International Journal of Physical Distribution & Logistics Management* 26:64-89, 1996.

Carter, J.R. and Narasimhan, R. A comparison of North American and European future purchasing trends. *International Journal of Purchasing & Materials Management* 32:12-22, 1996.

Choi, T.Y. and Hartley, J.L. An exploration of supplier selection practices across the supply chain. *Journal of Operations Management* 14:333-343, 1996.

Clark A. and Scarf H. Optimal Policies for a Multi-Echelon Inventory Problem. *Management Science* 6:475-490, 1960.

Clark A. An Informal Survey of Multi-Echelon Inventory Theory. *Naval Res. Logistics Quarterly* 19:621-650, 1972.

Cohen, M.A. and Lee, H.L. Strategic Analysis of Integrated Production-Distribution Systems Models and Methods. *Operations Research* 36:216-228, 1988.

Cohen, M.A. and Mallik, S. Global Supply Chains: Research and Applications. *Production and Operations Management* 6:193-210, 1997.

Cooper, M.C. Logistics in the Decade of the 1990s. from The Logistics Handbook, Eds. Roberson, Capcino & Howe, Free Press:New York, 1994.

Cooper, M.C., Ellram, L.M., Gardner, J.T., and Hanks, A.M. Meshing Multiple Alliances. *Journal of Business Logistics*, 18:67-89, 1997.

Copacino, W.C. and Rosenfield, D.B. Analytical Tools for Strategic Planning. from Logistics: The Strategic Issues, edited by Christopher, M., 1992.

Davis, T. Effective Supply Chain Management. *Sloan Management Review* 34:35-46, 1993.

Ernst, R. and Kamrad, B. A Conceptual Framework for Analyzing Supply Chain Structures. Proceedings of 1996 MSOM Conference: 1 -7, 1996.

Fisher, M.L. and Raman, A. Reducing the Cost of Demand Uncertainty through Accurate Response to Early Sales. *Operations Research* 44:87-99, 1996.

Fisher, M.L. What is the right supply chain for your product? *Harvard Business Review* 75:105-116, 1997.

Forrester, J. Industrial Dynamics: A Major Breakthrough for Decision Makers. *Harvard Business Review* 36:37-66.

Forrester, J. Industrial Dynamics. MIT Press, 1961.

Fuller, J., O'Connor, J., and Rawlinson, R. Tailored Logistics: The Next Advantage. *Harvard Business* Review 71:87-93,1993.

Garg, A. and Tang, C.S. On Postponement Strategies for Product Families with Multiple Points of Differentiation. *IIE Transactions* 29:641-650, 1997.

Gavirneni, S., Kapuscinski, R., and Tayur, S. Value of Information in Capacitated Supply Chains. from Quantitative Models for Supply Chain Management, Eds. Magazine, M.J., Tayur, S. and Ganeshan, R., Kluwer: Cambridge, 1998.

Gentry, J.J. The role of carriers in buyer-supplier strategic partnerships : A supply chain management approach. *Journal of Business Logistics* 17:33-55, 1996.

Geoffrion, A.M. and Graves, G.W. Multi-Commodity Distribution Design by Benders Decomposition, *Management Science* 20:822-844, 1974.

Geoffrion, A.M. and Powers, R.F. Twenty years of strategic distribution system design : An evolutionary perspective. *Interfaces* 25:105-127, 1995.

Glasserman, P. and Tayur, S. Sensitivity Analysis for Base-Stock Levels in Multiechelon Production-inventory Systems. *Management Science* 41:263-279, 1995.

Gopal, C. Manufacturing Logistics Systems for a Competitive Global Strategy. from Logistics: TheStrategic Issues, edited by Christopher, M. 1992.

Graves, S.C, Kletter, D.B. and William, H.B. A Dynamic Model for Requirements Planning with Application to Supply Chain Optimization. To be published in *Operations Research*. 1998.

Hammel, T.R. and Kopczak, L.R. Tightening the supply chain. *Production & Inventory Management Journal* 34:63-70, 1993.

Henig, M., Gerchak, Y., Ernst, R., Pike, D., An Inventory Model Embedded in Designing a Supply Contract. *Management Science* 43:184-197, 1997.

Heskett, J.L. Logistics- Essential to Strategy. *Harvard Business Review* 55:85-96, 1977.

Holmlund, M. and Kock, S. Buyer dominated relationships in a supply chain--A case study of four small-sized suppliers. *International Small Business Journal* 15:26-40, 1996.

Houlihan, J.B. International Supply Chain Management. *International Journal of Physical Distribution & Materials Management* 15:22-38, 1985.

Houlihan, J.B. International Supply Chain Management: A New Approach. *Management Decision* 26:13-19, 1988.

Iyer, A.V and Bergen, M.E. Quick Response in Manufacturer-Retailer Channels. *Management Science* 43:559-570, 1997.

Jones, T.C. and Riley, D.W. Using Inventory for Competitive Advantage through Supply Chain Management. *International Journal of Physical Distribution and Materials Management* 15:16-26, 1984.

Kruger, G.A., The Supply Chain Approach to Planning and Procurement Management. *Hewlett-Packard Journal* February:28-38, 1997.

La Londe, B.J. Evolution of the Integrated Logistics Concept. from The Logistics Handbook, Eds. Roberson, Capcino & Howe, Free Press:New York, 1994.

La Londe, B.J. and Masters, J.M. Emerging logistics strategies : Blueprints for the next century. *International Journal of Physical Distribution & Logistics Management* 24:35-47, 1994.

Lamming, R. Squaring Lean Supply with Supply Chain Management. *International Journal of Operations & Production Management* 16(2):183-196, 1996.

Langley, C.J. The Evolution of the Logistics Concept, from Logistics: The Strategic Issues, edited by Christopher, M. 1992.

Lederer, P.J. and Li, L. Pricing, Production, Scheduling, and Delivery-Time Competition. *Operations Research* 45:407-420, 1997.

Lee, H.L. and Billington, C. Material management in decentralized supply chains. *Operations Research* 41:835-847, 1993.

Lee, H.L. and Billington, C. The evolution of supply-chain-management models and practice at Hewlett-Packard. *Interfaces* 25:42-63, 1995.

Lee, H.L., Padmanabhan, V., and Whang, S. Information Distortion in a Supply Chain: The Bullwhip Effect. *Management Science* 43:546-558, 1997.

Leenders, M.R., Nollet, J., and Ellram, L.M. Adapting purchasing to supply chain management. *International Journal of Physical Distribution & Logistics Management* 24:40-42, 1994.

Levy, D. Lean Production in an International Supply Chain. *Sloan Management Review* 38:94-102, 1997.

Masters, J.M. and Pohlen, T.L. Evolution of the Logistics Profession. from The Logistics Handbook, Eds. Roberson, Capcino & Howe, Free Press:New York, 1994.

McMullan, A. Supply Chain Management Practices in Asia Pacific Today. *International Journal of Physical Distribution & Logistics Management* 26:79-95, 1996.

Min, H. Distribution Channels in Japan (Challenges and Opportunities for the Japanese Market Entry). *International Journal of Physical Distribution & Logistics Management* 26:22-35, 1996.

Moinzadeh, K. and Aggarwal, P.K. An Information Based Multi-Echelon Inventory System with Emergency Orders. *Operations Research* 45:694-701, 1997.

Mourits, M. and Evers, J.J. Distribution network design. *International Journal of Physical Distribution & Logistics Management* 25:43-57, 1995.

O'Brien, C. and Head, M. Developing a full business environment to support just-in-time logistics. *International Journal of Production Economics* 42:41-50, 1995.

878

O'Laughlin, K.A. and Copacino, W.C. Logistics Strategy. from The Logistics Handbook, Eds. Roberson, Capcino & Howe, Free Press:New York, 1994.

Oliver, K.R. and Webber, M.D. Supply-Chain Management: Logistics Catches Up With Strategy. from Logistics: The Strategic Issues, edited by Christopher, M. 1992.

Prida, B. and Gutierrez, G. Supply management : From purchasing to external factory management. *Production & Inventory Management Journal* 37:38-43, 1996.

Revelle, C.S. and Laporte, G. The plant location problem : New models and research prospects. *Operations Research* 44:864-874, 1996.

Roberts, J.C. Formulating and Implementing a Global Logistics Strategy. *International Journal of Logistics Management* 1:53-58, 1990.

Robinson, E.P., Gao, L., and Muggenborg, S.D. Designing an integrated distribution system at DowBrands, Inc. *Interfaces* 23:107-117, 1993.

Roy, R. and Pofter, S. Managing engineering design in complex supply chains. *International Journal of Technology Management* 12:403-420, 1996.

Satterfield, R. and Robinson, E.P. Designing Distribution Systems to Support Vendor Strategies in Supply Chain Management. Presented at Supply Chain Linkages Symposium, Indiana University, 1996.

Scott, C. and Westbrook, R. New Strategic Tools for Supply Chain Management. *International Journal of Physical Distribution & Logistics Management* 21:22-33, 1991.

Shapiro, R.D. Get leverage from logistics. *Harvard Business Review* 62:119-126, 1984.

Slater, A. The Significance of Industrial Logistics. *International Journal of Physical Distribution* 7:70-112, 1976.

Slats, P.A., Bhola, B., Evers, J.J., and Dijkhuizen, G. Logistic Chain Modelling. *European Journal of Operational Research* 87:1-20, 1995.

Srinivasan, K., Kekre, S., and Mukhopadhyay, T. Impact of electronic data interchange technology on JIT shipments. *Management Science* 40:1291-1304, 1994.

Stenger, A.J. Reducing Inventories in a Multi-Echelon Manufacturing Firm: A Case Study. *International Journal of Production Economics* 45:239-249, 1996.

Stenger, A.J. Distribution Resource Planning. from The Logistics Handbook, Eds. Roberson, Capcino & Howe, Free Press:New York, 1994.

Stevens, G.C. Integrating the Supply Chain. *International Journal of Physical Distribution & Materials Management* 19:3-8, 1989.

Tagaras, G. and Lee, H.L. Economic models for vendor evaluation with quality cost analysis. *Management Science* 42:1531-1543, 1996.

Thomas, D.J. and Griffin, P.M. Coordinated supply chain management. *European Journal of Operational Research* 94:1-15, 1996.

Towill, D.R. The seamless supply chain - The predators strategic advantage. *International Journal of Technology Management* 13:37-56, 1997.

Towill, D.R., Naim, N.M., and Wikner, J. Industrial Dynamics Simulation Models in the Design of Supply Chains. *International Journal of Physical Distribution and Logistics Management* 22:3-13, 1992.

Verwijmeren, M., Viist, P.v., and Karel, D.v. Networked Inventory Management InformationSystems: Materializing Supply Chain Management. *International Journal of Physical Distribution & Logistics Management* 26:16-31, 1996.

Viswanathan, S. and Mathur, K. Integrating Routing and Inventory Decisions in One-Warehouse Muti-Retailer, Multi-Product Distribution Systems. *Management Science* 43:294-312, 1997.

Walton, L.W. Partnership Satisfaction: Using the underlying dimensions of supply chain partnership to measure current and expected levels of satisfaction. *Journal of Business Logistics* 17:57-75, 1996.

Zinn, W. and Bowersox, D.J. Planning Physical Distribution with the Principle of Postponement. *Journal of Business Logistics* 9:117-136, 1988.

Zinn, W. and Levy, M. Speculative Inventory Management: A Total Channel Perspective. *International Journal of Physical Distribution & Materials Management* 18:34-39, 1988.

INDEX

accounting inventory 130
activity-based costing 750-751
address models 504,507
advance Information 219
allocation rules 306,307,322,323
analytical information technology
   739,740
approximation algorithm 150
assemble-to-order 603
assemble-to-order system 62
assembly system 33, 75
assortment problem 483
asymmetric information 271,277-
   280,285,288
asymptotic approximation algorithm
   151
auto-regressive time series model
   574,575

backorders 770
backup agreement 342,800
base order quantities 86
base stock policy
   10,44,120,472,481,519,
   567,568,578,764
basic category 522-524
batch ordering 421
Beer Game 385,412
Bin Packing Problem (BPP) 152
binary probit 504
binomial lattice approach 645,647
bin-packing constant 152,162,168,180
bonded warehouse 706
Brownian Motion (Wiener Process)
   387,385,413
bullwhip effect 385,413,419,420
bundled discount scheme 227
business cycles 383,384,385,412,413
buy-back contracts 116,247-255,260
buyback 306,307,321,322

(C-Q) completeness 714
capacitated 444
Capacitated Vehicle Routing Problem
   (CVRP) 152
capacity investments 674
capacity limits 564,576
capacity pooling 575

capital equipment suppliers
   383,384,386,387,393,396,397,400,
   403-408,411
capital lifetime 386-387,391,394,396-
   398,404-405,407
cash balancing models 812
category management 493
central control 304,308,323
centralized control 471,475,476
centralized information 431
channel coordination
   305,312,318,322,325
Classifying Research on Supply Chain
   Management 848
competition penalty 127
competitive facility location 700
component commonality 588
compound options 674
conditional distribution 340
consumer surplus 535
contingent claims 628,634
contract, linear 275,278,279-
281,285,288
contract, nonlinear
   275,278,287,285,288
contract, one-part 275,277,279-
   280,285,288
contract, two-part 275,278-
   281,285,288
Contributions from Various Disciplines
   845
control theory 385,387,409,412
coordinating contracts 246,249-
   251,258-260
coordination 199,226
correlation 801
correlation parameter 422,426
coupled systems 170
coupling 19
cumulant generating function (cgf) 50
customer aggregation 751,752
customer demand 767
customer satisfaction 675
customer service 764
cyclical demand 16

dam model 13
decentralized control 304,308,
   471,475,477
decentralized information 432

decision rights
301,306,307,308,310,443,470,595
delayed product differentiation
555,560,561,562,567,570,571,575,
578
delivery commitment 201
demand correlation 602
demand estimation 800
demand signal processing 385-386
demand uncertainty
679,792,793,797,799
derivative recursions 25, 722
design for customization 558
design for localization 558,559
design for postponement 558,569
direct shipping 174
discount rate 686
disposal 813,821,831
distribution requirements planning
744,746,752
divergent flow network 471
dollar volume commitment 204,211
double marginalization
114,245,250,305,308,319,321
downside risk 673
dynamic programming 343

early postponement 475
Echelon reorder points 85
Echelon-stock policies 75,78
Effective Standard Deviation (ESD)
570-574,577
efficient forecasts of exchange rates
678
elasticity 240-241,244,258
elasticity of prices 682
Electronic data interchange (EDI)
92,361,443
electronic invoicing 370
electronic order 370
emerging markets 705
employee productivity
384,388,395,396-404,410,411
employee training 384-385,389,395-
396,397,398,399-403,404,405,407
enterprise resource planning
739,740,744,750,752
Evolution of SCM 844
exchange rate
628,631,636,637,641,642,651,709
exchange rate risk 684
expedited orders 711
Exponential family 59

exponential smoothing forecast
420,425,429
Exponential tails 50-51
externalities 114
Extremal Partitioning Algorithm (EPA)
160

Factors Influencing SCM Research 845
family differentiation point 474
fashion category 522-524
features 595
fill rate 45,61,62,771
final assembly 587
financial risk management 687
finite selling periods 792
fixed partition policies 180
flexibility 201
forecasting 385,387,408,412,419,420,
427,801
franchising 245-246
frontier plots 731

game theory 113
general framework 200
generalized failure rate 240-243,258
Geographical Information System 754
geometric Brownian motion 636,647
global sourcing 627,629,631,635,641,
651,652,705
global supply chain 630
global supply chain network 680

Harris recurrence 31
heavy traffic approximations 58-60
hierarchical planning 390-391,412
holding costs 810,829
hysteresis 629,637,639,640,641,
651.675

IIA Property 507
incentive-compatibility 276
Incentives 199
incomplete information 272
Increasing failure rate (IFR)
240,244,445
individual rationality 276
Infinitesimal Perturbation Analysis
(IPA) 20,446,721
information 199,340
information asymmetry 271,277-280
information sharing 201,443
information technology 92
information, value of 280-281

installation-stock policies 75
intermediary 707
inter-temporal correlation 574
inventory control 443,809,811
inventory planning 763
inventory position 12,712
inventory turnover 372
inventory-leadtime tradeoffs 60
investment accelerator 386,405
Ito Process 634

late customization 470
late postponement 475
lead times
    302,306,307,322326,421,428,435,
    767,810,824
learning effect 556,564,569,572,575
lifecycles 791,803
linear contracts 127
linear trend demand model 430
Lipschitz function 26
lost sales 795
Luce model 507

machine tools 384-
    385,386,388,400,406,411
majorization ordering 522-523
make-to-order 472-474
make-to-stock 472,474,603
manufacturer substitution 497
manufacturing 810,813
marginal cost pricing 115
marginal cost pricing 271,272,280,284-
    285
markdowns 791,798
market efficiency 686
Markov Decision Process (MDP) 566
Markov-modulated 16
material flow complexity 365
materials requirements planning
    744,745,752
maximal vanilla box 600
maximum likelihood estimation
    542,545
Mean Absolute Percent Error (MAPE)
    392-393
mean reversion 690
mean-variance cash flow trade-off 730
minimum purchase commitments
    306,307,313,314
modified base-stock policy 709
Modified Circular Partitioning (MCRP)
    157

modular design 557
monopoly, bilateral 274
Monte carlo simulation 696
moving average forecast 420,423,429
multi-depot CVRP 165
multi-fiber quota 707
multi-nomial approximation 687
multinomial logit 505-507,515-
    517,534,542,543
multiple retailers 426
multistage systems 53-58
$M^X/G/\alpha$ queue 770

Nash equilibrium 125
newsboy heuristics 535,537
Next-Fit Heuristic (NF) 163,164
non-linear optimization 774
non-stationary demand 9,16
non-stationary demand 444,771

obsolescence 791,798,800,803
obsolete inventory 372
one-warehouse multi-retailer 75
on-hand inventory 770
operational flexibility 675
operational hedge capacity 688
operational hedging 627,629
operational risk management 687
optimal 443
optimal assortment 513,519
optimal policies 821
optimality proof 102
optimization 81
option evaluation 674
option thinking 674
options 205,221,595
order band 709
order complexity 369
order process variability 218
order up-to policy 420,423,443,562

parameter estimation 798,800
parametric monotonicity 531
pareto improving 341
part commonality 470
pass through 709
penalty methods 261-262
perfect packing 152,163,187
performance evaluation 81
periodic review 444
periodical commitment 204,215
periodicity of ordering 200
postponement 470,682

postponement of operations 557
postponement strategies 555,556,558
power-of-two policies 137,176,186,189
preference relation 499
price variations 422
price, wholesale 271,284-285
price-only contracts 237-245,263
price-sensitive demand 275
pricing 301,306,307,310-
    313,315,316,318,329,798
probabilistic substitution 511
process sequencing 470,477,479,484
product aggregation 751
product differentiation point 474
product line design 495-496
product recovery 809
product variety 587,700
production line structuring 470,486
production rule 481
production scheduling 745
production switching 683
pseudo component 614
PULL policy 816
purchasing power parity 675
PUSH policy 815,824

quality 304,307,308,312,326-328
quantity commitment 200
quantity contract 706
quantity discounts 118
quantity fixing 308-311
quantity flexibility (QF)
    306,307,314,315,319,630,643,648
quantity flexibility contracts 140,255-
    260
quantity forcing 246
queues 45,46
quick response 470,486,555,800

(R,nQ) policies 75
random utility 504,514,529,543
random walk 51-53,387-388,412,683
reactive capacity 800
recursive procedure 84
reengineering 694
regeneration 31
region partitioning
    152,154,156,170,177
remanufacturing 809,813
repair models 811
replenishment lead time 472
resale price maintenance (RPM) 307-
    309,321

resequencing of operations 557
response time 473
responsibility tokens 134
return rate 811,830
returns policies 247-255,263-
    264,306,307,321
revelation principle 277
reverse logistics 809,810
risk neutral pricing 686
risk pooling 564
risk sharing 251,630,643,651
risk-return efficient frontier 676
rolling Horizon Flexibility 204,216

(s,S) policy 443,570,575
safety stock inventory 764
sample path gradient 533
scale economies 522
scenario planning 753
scheduling 794,803
SCM Time-line 847
scope of responsibility 843
sensitivity analysis 77,83
serial system 21,74
serial supply chain 119
service commitment 341
shareholder value 675
shelf space allocation 498
shipment discrepancy 364
shortage game 138
shortfall 13
shortfall formulation 44
signal analysis 410,413
simple discount scheme 227
simulation recursion 722
simulation-based optimization 20
single-location systems 76
social welfare 535
split/unsplit demands 152,169
spreadsheet optimization 697
spreadsheet simulation 697
standard setting 262-263
standardization
    470,471,479,480,556,563,575,576
static choice assumptions 509
stationary nested joint replenishment
    policies 175
statistical economies of scale 564,572
stochastic dynamic programming 686
stochastic recourse 681
stockout policy 769
stockouts 791,800,802
strategic partnerships 435

style goods 792,797,799
sub-gradient optimization 591
submodular functions 189
substitution demand 526
substitution matrix 512
supply chain 339,443
supply chain "defined" 842
supply chain decision database 739,749-755
supply chain design 405-408,412
supply chain inventory game 119
supply chain network 676
supply chain optimizations 773
supply chain system hierarchy 741,742,747,749,752,755,756
supply chain total cost 384,400,401-403,405-406
supply contract 271,443,630,643,644
systems 75,78

task assembly sequence 606
time flexibility 630,643,648
total minimum quantity commitment 204,206
transactional information technology 739,741
Traveling Salesman Problem (TSP) 151
two stage stochastic program 587
two-echelon systems 185
two-source models 813

utility function 500-502,504,515,529,543

value of centralized demand information 94
value of information 433
vanilla box problem 483,595
variability 764
variance estimation 801
variant component 614
Vehicle Routing Problem with Deliveries and Backhauls (VRPDB) 166
vehicle scheduling 745
vendor managed inventory 141
vendor selection 628
Vertical restraints 245
Virtual Vending Machine 545

Who manages the supply chain? 843
Wiener Filters (Predictors) 393-394,396,408
working capital 800
zero inventory ordering 181